Special Issue of the Manufacturing Engineering Society 2019 (SIMES-2019)

Special Issue of the Manufacturing Engineering Society 2019 (SIMES-2019)

Special Issue Editors

Eva M. Rubio
Ana M. Camacho

MDPI • Basel • Beijing • Wuhan • Barcelona • Belgrade • Manchester • Tokyo • Cluj • Tianjin

Special Issue Editors
Eva M. Rubio
Universidad Nacional de Educación a Distancia (UNED)
Spain

Ana M. Camacho
Universidad Nacional de Educación a Distancia (UNED)
Spain

Editorial Office
MDPI
St. Alban-Anlage 66
4052 Basel, Switzerland

This is a reprint of articles from the Special Issue published online in the open access journal *Materials* (ISSN 1996-1944) (available at: https://www.mdpi.com/journal/materials/special_issues/SIMES_2019).

For citation purposes, cite each article independently as indicated on the article page online and as indicated below:

LastName, A.A.; LastName, B.B.; LastName, C.C. Article Title. *Journal Name* **Year**, *Article Number*, Page Range.

ISBN 978-3-03936-360-5 (Pbk)
ISBN 978-3-03936-361-2 (PDF)

© 2020 by the authors. Articles in this book are Open Access and distributed under the Creative Commons Attribution (CC BY) license, which allows users to download, copy and build upon published articles, as long as the author and publisher are properly credited, which ensures maximum dissemination and a wider impact of our publications.

The book as a whole is distributed by MDPI under the terms and conditions of the Creative Commons license CC BY-NC-ND.

Contents

About the Special Issue Editors ix

Eva María Rubio and Ana María Camacho
Special Issue of the Manufacturing Engineering Society 2019 (SIMES-2019)
Reprinted from: *Materials* **2020**, *13*, 2133, doi:10.3390/ma13092133 1

**Elena Verdejo de Toro, Juana Coello Sobrino, Alberto Martínez Martínez,
Valentín Miguel Eguía and Jorge Ayllón Pérez**
Investigation of a Short Carbon Fibre-Reinforced Polyamide and Comparison of Two Manufacturing Processes: Fused Deposition Modelling (FDM) and Polymer Injection Moulding (PIM)
Reprinted from: *Materials* **2020**, *13*, 672, doi:10.3390/ma13030672 5

**Gustavo Medina-Sanchez, Rubén Dorado-Vicente, Eloísa Torres-Jiménez and
Rafael López-García**
Build Time Estimation for Fused Filament Fabrication via Average Printing Speed
Reprinted from: *Materials* **2019**, *12*, 3982, doi:10.3390/ma12233982 19

**Jorge Barrios-Muriel, Francisco Romero-Sánchez, Francisco Javier Alonso-Sánchez and
David Rodríguez Salgado**
Advances in Orthotic and Prosthetic Manufacturing: A Technology Review
Reprinted from: *Materials* **2020**, *13*, 295, doi:10.3390/ma13020295 35

Amabel García-Domínguez, Juan Claver, Ana María Camacho and Miguel A. Sebastián
Considerations on the Applicability of Test Methods for Mechanical Characterization of Materials Manufactured by FDM
Reprinted from: *Materials* **2020**, *13*, 28, doi:10.3390/ma13010028 51

Irene Buj-Corral, Alejandro Domínguez-Fernández and Ramón Durán-Llucià
Influence of Print Orientation on Surface Roughness in Fused Deposition Modeling (FDM) Processes
Reprinted from: *Materials* **2019**, *12*, 3834, doi:10.3390/ma12233834 69

**Ma-Magdalena Pastor-Artigues, Francesc Roure-Fernández, Xavier Ayneto-Gubert,
Jordi Bonada-Bo, Elsa Pérez-Guindal and Irene Buj-Corral**
Elastic Asymmetry of PLA Material in FDM-Printed Parts: Considerations Concerning Experimental Characterisation for Use in Numerical Simulations
Reprinted from: *Materials* **2020**, *13*, 15, doi:10.3390/ma13010015 85

Natalia Beltrán, David Blanco, Braulio José Álvarez, Álvaro Noriega and Pedro Fernández
Dimensional and Geometrical Quality Enhancement in Additively Manufactured Parts: Systematic Framework and A Case Study
Reprinted from: *Materials* **2019**, *12*, 3937, doi:10.3390/ma12233937 109

**J. Antonio Travieso-Rodriguez, Ramon Jerez-Mesa, Jordi Llumà, Oriol Traver-Ramos,
Giovanni Gomez-Gras and Joan Josep Roa Rovira**
Mechanical Properties of 3D-Printing Polylactic Acid Parts subjected to Bending Stress and Fatigue Testing
Reprinted from: *Materials* **2019**, *12*, 3859, doi:10.3390/ma12233859 133

Ana María Camacho, Álvaro Rodríguez-Prieto, José Manuel Herrero, Ana María Aragón, Claudio Bernal, Cinta Lorenzo-Martin, Ángel Yanguas-Gil and Paulo A.F. Martins
An Experimental and Numerical Analysis of the Compression of Bimetallic Cylinders
Reprinted from: Materials 2019, 12, 4094, doi:10.3390/ma12244094 153

Juan A. García-Manrique, Bernabé Marí, Amparo Ribes-Greus, Ll úcia Monreal, Roberto Teruel, Llanos Gascón, Juan A. Sans and Julia Marí-Guaita
Study of the Degree of Cure through Thermal Analysis and Raman Spectroscopy in Composite-Forming Processes
Reprinted from: Materials 2019, 12, 3991, doi:10.3390/ma12233991 173

Luis M. Alves, Rafael M. Afonso, Frederico L.R. Silva and Paulo A.F. Martins
Deformation-Assisted Joining of Sheets to Tubes by Annular Sheet Squeezing
Reprinted from: Materials 2019, 12, 3909, doi:10.3390/ma12233909 185

Leire Godino, Jorge Alvarez, Arkaitz Muñoz and Iñigo Pombo
On The Influence of Rotary Dresser Geometry on Wear Evolution and Grinding Process
Reprinted from: Materials 2019, 12, 3855, doi:10.3390/ma12233855 199

David Blanco, Eva María Rubio, Marta María Marín and Joao Paulo Davim
Repairing Hybrid Mg–Al–Mg Components Using Sustainable Cooling Systems
Reprinted from: Materials 2020, 13, 393, doi:10.3390/ma13020393 211

César Ayabaca and Carlos Vila
An Approach to Sustainable Metrics Definition and Evaluation for Green Manufacturing in Material Removal Processes
Reprinted from: Materials 2020, 13, 373, doi:10.3390/ma13020373 233

Enrique García-Martínez, Valentín Miguel, Alberto Martínez-Martínez, María Carmen Manjabacas and Juana Coello
Sustainable Lubrication Methods forthe Machining of Titanium Alloys: An Overview
Reprinted from: Materials 2019, 12, 3852, doi:10.3390/ma12233852 255

Rosario Domingo, Beatriz De Agustina and Marta María Marín
Study of Drilling Process by Cooling Compressed Air in Reinforced Polyether-Ether-Ketone
Reprinted from: Materials 2020, 13, 1965, doi:10.3390/ma13081965 277

Santiago Yagüe García and Cristina González Gaya
Reusing Discarded Ballast Waste in Ecological Cements
Reprinted from: Materials 2019, , 3887, doi:10.3390/ma12233887 293

Alejandro Sambruno, Fermin Bañon, Jorge Salguero, Bartolome Simonet and Moises Batista
Kerf Taper Defect Minimization Based on Abrasive Waterjet Machining of Low Thickness Thermoplastic Carbon Fiber Composites C/TPU
Reprinted from: Materials 2019, 12, 4192, doi:10.3390/ma12244192 303

Pedro F. Mayuet Ares, Franck Girot Mata, Moisés Batista Ponce and Jorge Salguero Gómez
Defect Analysis and Detection of Cutting Regions in CFRP Machining Using AWJM
Reprinted from: Materials 2019, 12, 4055, doi:10.3390/ma12244055 321

David Repeto, Severo Raul Fernández-Vidal, Pedro F. Mayuet, Jorge Salguero and Moisés Batista
On the Machinability of an Al-63%SiC Metal Matrix Composite
Reprinted from: Materials 2020, 13, 1186, doi:10.3390/ma13051186 337

Sergio Aguado, Pablo Pérez, José Antonio Albajez, Jorge Santolaria and Jesús Velázquez
Configuration Optimisation of Laser Tracker Location on Verification Process
Reprinted from: *Materials* **2020**, *13*, 331, doi:10.3390/ma13020331 361

Jorge Moreno, Pascual Simon, Eduardo Faleiro, Gabriel Asensio and Jose Antonio Fernandez
Estimation of an Upper Bound to the Value of the Step Potentials in Two-Layered Soils from
Grounding Resistance Measurements
Reprinted from: *Materials* **2020**, *13*, 290, doi:10.3390/ma13020290 375

Óscar de Francisco Ortiz, Manuel Estrems Amestoy, Horacio T. Sánchez Reinoso
and Julio Carrero-Blanco Martínez-Hombre
Enhanced Positioning Algorithm Using a Single Image in anLCD-Camera System by Mesh
Elements' Recalculation and Angle Error Orientation
Reprinted from: *Materials* **2019**, *12*, 4216, doi:10.3390/ma12244216 387

Alberto Mínguez Martínez and Jesús de Vicente y Oliva
Industrial Calibration Procedure for Confocal Microscopes
Reprinted from: *Materials* **2019**, *12*, 4137, doi:10.3390/ma12244137 401

Francisco Javier Brosed, Raquel Acero Cacho, Sergio Aguado, Marta Herrer,
Juan José Aguilar and Jorge Santolaria Mazo
Development and Validation of a Calibration Gauge for Length Measurement Systems
Reprinted from: *Materials* **2019**, *12*, 3960, doi:10.3390/ma12233960 429

Carolina Bermudo Gamboa, Sergio Martín-Béjar, F. Javier Trujillo Vilches, G. Castillo López
and Lorenzo Sevilla Hurtado
2D–3D Digital Image Correlation Comparative Analysis for Indentation Process
Reprinted from: *Materials* **2019**, *12*, 4156, doi:10.3390/ma12244156 441

Jose Calaf-Chica, Marta María Marín, Eva María Rubio, Roberto Teti and Tiziana Segreto
Parametric Analysis of the Mandrel Geometrical Data in a Cold Expansion Process of Small
Holes Drilled in Thick Plates
Reprinted from: *Materials* **2019**, *12*, 4105, doi:10.3390/ma12244105 455

Magdalena Ramirez-Peña, Francisco J. Abad Fraga, Alejandro J. Sánchez Sotano
and Moises Batista
Shipbuilding 4.0 Index Approaching Supply Chain
Reprinted from: *Materials* **2019**, *12*, 4129, doi:10.3390/ma12244129 473

Martin Folch-Calvo, Francisco Brocal and Miguel A. Sebastián
New Risk Methodology Based on Control Charts to Assess Occupational Risks in
Manufacturing Processes
Reprinted from: *Materials* **2019**, *12*, 3722, doi:10.3390/ma12223722 491

About the Special Issue Editors

Eva M. Rubio is a Professor at the Department of Manufacturing Engineering of the National Distance Education University (UNED, Spain). She received her MSc in Aeronautical Engineer (1997) from the Polytechnic University of Madrid and PhD in Industrial Engineering (2002) from the National University of Distance Education (UNED). Her research focuses on the field of manufacturing engineering, specifically the analysis of machining processes, the analysis of metal forming processes, industrial metrology, and teaching and innovation in engineering.

Ana M. Camacho is a Professor at the Department of Manufacturing Engineering of the National Distance Education University (UNED, Spain). She received her MSc in Industrial Engineering from the University of Castilla-La Mancha (UCLM) in 2001 and the PhD in Industrial Engineering from the UNED in 2005. Her main research interests are innovation in manufacturing engineering and materials technology, especially focused on the analysis of metal forming and additive manufacturing techniques through computer-aided engineering tools and experimental testing, and the development of methodologies for materials selection in demanding applications.

Editorial

Special Issue of the Manufacturing Engineering Society 2019 (SIMES-2019)

Eva María Rubio * and Ana María Camacho

Department of Manufacturing Engineering, Industrial Engineering School, Universidad Nacional de Educación a Distancia (UNED), St/Juan del Rosal 12, E28040 Madrid, Spain; amcamacho@ind.uned.es
* Correspondence: erubio@ind.uned.es; Tel.: +34-913-988-226

Received: 23 April 2020; Accepted: 24 April 2020; Published: 5 May 2020

Abstract: The Special Issue of the Manufacturing Engineering Society 2019 (SIMES-2019) has been launched as a joint issue of the journals "Materials" and "Applied Sciences". The 29 contributions published in this Special Issue of Materials present cutting-edge advances in the field of manufacturing engineering focusing on additive manufacturing and 3D printing, advances and innovations in manufacturing processes, sustainable and green manufacturing, manufacturing of new materials, metrology and quality in manufacturing, industry 4.0, design, modeling, and simulation in manufacturing engineering and manufacturing engineering and society. Among them, these contributions highlight that the topic "additive manufacturing and 3D printing" has collected a large number of contributions in this journal because its huge potential has attracted the attention of numerous researchers over the last years.

Keywords: additive manufacturing; 3D printing; forming; machining; metrology; industry 4.0; green manufacturing; modeling and simulation; quality in manufacturing; technological and industrial heritage

---

After the complete success of the first edition [1] with 48 contributions on emerging methods and technologies, the Special Issue of the Manufacturing Engineering Society 2019 (SIMES-2019) [2] was launched as a joint issue of the journals "Materials" and "Applied Sciences".

Once again, this Special Issue was promoted by the Manufacturing Engineering Society (MES) [3] of Spain, with the aim of covering the wide range of research lines developed by the members and collaborators of the MES and other researchers within the field of manufacturing engineering.

In this Special Issue of the journal Materials, 29 contributions to cutting-edge advances in different fields of the manufacturing engineering have been collected. In particular, in additive manufacturing and 3D printing [4–11]; sustainable and green manufacturing [12–16]; metrology and quality in manufacturing [17–21]; advances and innovations in manufacturing processes [22–25]; manufacturing of new materials [26–28]; design, modeling, and simulation in manufacturing engineering [29,30]; industry 4.0 [31] and manufacturing engineering and society [32].

Among all of them, the topic additive manufacturing and 3D printing stands out for the number of contributions it has had in this Special Issue showing the interest that this topic arouses, currently, among researchers, the industry and the public in general. The works focus on the manufacturing processes [4,5], the characterization of materials and parts [6–8], the estimation of times [9,10] and the dimensional and geometrical quality of the obtained pieces [11]. The next topics by the number of contributions are sustainable and green manufacturing and metrology and quality in manufacturing with five contributions each. The first one gathers four works about new sustainable lubrication/cooling techniques used in removal processes [12–15] and the other about reusing waste in ecological cement [16]. The second one collects research about the optimization of laser tracker location on verification process [17], estimation of an upper bound to the value of the step potentials from grounding resistance measurements [18], enhanced positioning algorithm by mesh elements, recalculation and angle error orientation [19], industrial calibration procedure for confocal microscopes [20] and, finally, the development and validation of a calibration gauge for length measurement systems [21]. The Special Issue also collects four pieces of work in the topic advances and innovations in manufacturing processes. In particular, an experimental and numerical analysis of the compression of bimetallic cylinders [22], a study about thermal analysis and Raman spectroscopy in composite-forming processes [23], a new deformation-assisted joining of sheets to tubes by annular sheet squeezing [24] and an analysis of the influence of the rotary dresser geometry on wear evolution and on the grinding process [25]. In addition, the Special Issue shows three papers in the topic manufacturing of new materials. Concretely, two about thermoplastic carbon fiber composites C/TPU [26,27] and, other, about an Al-SiC metal matrix composite [28]. It counts also with two works in design, modeling, and simulation in manufacturing engineering about the indentation process [29] and the cold expansion process [30]; one, in the topic industry 4.0 about the application of the supply chain to shipbuilding [31] and the other in the topic manufacturing engineering and society about a new risk methodology based on control charts to assess occupational risks in manufacturing processes [32].

Finally, it remains to highlight that in just four months since the publication of the first work [18], all the papers present prominent activity in their "article metrics"; being remarkable that some of the papers, belonging to this Special Issue, have already had more than five hundred abstract and full-text views, which is clear evidence of the interest readers have for all these topics.

Funding: This research received no external funding.

Conflicts of Interest: The authors declare no conflict of interest.

References

1. Rubio, E.M.; Camacho, A.M. Special Issue of the Manufacturing Engineering Society (MES). *Materials (Basel)* **2018**, *11*, 2149. [CrossRef] [PubMed]
2. Special Issue of the Manufacturing Engineering Society 2019 (SIMES-2019). Available online: https://www.mdpi.com/journal/applsci/special_issues/society_2019 (accessed on 21 February 2020).
3. Sociedad de Ingeniería de Fabricación. Available online: http://www.sif-mes.org/ (accessed on 20 February 2020).
4. Verdejo de Toro, E.; Coello Sobrino, J.; Martínez Martínez, A.; Miguel Eguía, V.; Ayllón Pérez, J. Investigation of a Short Carbon Fibre-Reinforced Polyamide and Comparison of Two Manufacturing Processes: Fused Deposition Modelling (FDM) and Polymer Injection Moulding (PIM). *Materials (Basel)* **2020**. [CrossRef] [PubMed]
5. Medina-Sanchez, G.; Dorado-Vicente, R.; Torres-Jiménez, E.; López-García, R. Build time estimation for fused filament fabrication via average printing speed. *Materials (Basel)* **2019**, *12*, 1–16. [CrossRef] [PubMed]
6. Barrios-Muriel, J.; Romero-Sánchez, F.; Alonso-Sánchez, F.J.; Rodríguez Salgado, D. Advances in Orthotic and Prosthetic Manufacturing: A Technology Review. *Materials (Basel)* **2020**, *13*, 295. [CrossRef] [PubMed]
7. García-Domínguez, A.; Claver, J.; Camacho, A.M.; Sebastián, M.A. Considerations on the Applicability of Test Methods for Mechanical Characterization of Materials Manufactured by FDM. *Materials (Basel)* **2019**, *13*, 28. [CrossRef]
8. Buj-Corral, I.; Domínguez-Fernández, A.; Durán-Llucià, R. Influence of print orientation on surface roughness in fused deposition modeling (FDM) processes. *Materials (Basel)* **2019**, *12*. [CrossRef]
9. Pastor-Artigues, M.-M.; Roure-Fernández, F.; Ayneto-Gubert, X.; Bonada-Bo, J.; Pérez-Guindal, E.; Buj-Corral, I. Elastic Asymmetry of PLA Material in FDM-Printed Parts: Considerations Concerning Experimental Characterisation for Use in Numerical Simulations. *Materials (Basel)* **2019**, *13*, 15. [CrossRef]
10. Beltrán, N.; Blanco, D.; Álvarez, B.J.; Noriega, Á.; Fernández, P. Dimensional and geometrical quality enhancement in additively manufactured parts: Systematic framework and a case study. *Materials (Basel)* **2019**, *12*. [CrossRef]

11. Travieso-Rodriguez, J.A.; Jerez-Mesa, R.; Llumà, J.; Traver-Ramos, O.; Gomez-Gras, G.; Rovira, J.J.R. Mechanical properties of 3D-printing polylactic acid parts subjected to bending stress and fatigue testing. *Materials (Basel)* **2019**, *12*. [CrossRef]
12. Camacho, A.M.; Rodríguez-Prieto, Á.; Herrero, J.M.; Aragón, A.M.; Bernal, C.; Lorenzo-Martin, C.; Yanguas-Gil, Á.; Martins, P.A.F. An Experimental and Numerical Analysis of the Compression of Bimetallic Cylinders. *Materials (Basel)* **2019**, *12*, 4094. [CrossRef]
13. García-Manrique, J.A.; Marí, B.; Ribes-Greus, A.; Monreal, L.; Teruel, R.; Gascón, L.; Sans, J.A.; Marí-Guaita, J. Study of the degree of cure through thermal analysis and Raman spectroscopy in composite-forming processes. *Materials (Basel)* **2019**, *12*. [CrossRef] [PubMed]
14. Alves, L.M.; Afonso, R.M.; Silva, F.L.R.; Martins, P.A.F. Deformation-assisted joining of sheets to tubes by annular sheet squeezing. *Materials (Basel)* **2019**, *12*. [CrossRef] [PubMed]
15. Godino, L.; Alvarez, J.; Muñoz, A.; Pombo, I. On the influence of rotary dresser geometry on wear evolution and grinding process. *Materials (Basel)* **2019**, *12*. [CrossRef] [PubMed]
16. Blanco, D.; Rubio, E.M.; Marín, M.M.; Davim, J.P. Repairing Hybrid Mg–Al–Mg Components Using Sustainable Cooling Systems. *Materials (Basel)* **2020**, *13*, 393. [CrossRef] [PubMed]
17. Ayabaca, C.; Vila, C. An Approach to Sustainable Metrics Definition and Evaluation for Green Manufacturing in Material Removal Processes. *Materials (Basel)* **2020**, *13*, 373. [CrossRef] [PubMed]
18. García-Martínez, E.; Miguel, V.; Martínez-Martínez, A.; Manjabacas, M.C.; Coello, J. Sustainable Lubrication Methods for the Machining of Titanium Alloys: An Overview. *Materials (Basel)* **2019**, *12*, 3852. [CrossRef]
19. Domingo, R.; de Agustina, B.; Marín, M.M. Study of Drilling Process by Cooling Compressed Air in Reinforced Polyether-Ether-Ketone. *Materials (Basel)* **2020**, *13*, 1965. [CrossRef]
20. García, S.Y.; Gaya, C.G. Reusing discarded ballast waste in ecological cements. *Materials (Basel)*. **2019**, *12*. [CrossRef]
21. Sambruno, A.; Bañon, F.; Salguero, J.; Simonet, B.; Batista, M. Kerf Taper Defect Minimization Based on Abrasive Waterjet Machining of Low Thickness Thermoplastic Carbon Fiber Composites C/TPU. *Materials (Basel)* **2019**, *12*, 4192. [CrossRef]
22. Mayuet Ares, P.F.; Girot Mata, F.; Batista Ponce, M.; Salguero Gómez, J. Defect Analysis and Detection of Cutting Regions in CFRP Machining Using AWJM. *Materials (Basel)* **2019**, *12*, 4055. [CrossRef]
23. Repeto, D. On the Machinability of Al-SiC Metal Matrix Composite. *Materials (Basel)* **2020**, *13*, 1186. [CrossRef] [PubMed]
24. Aguado, S.; Pérez, P.; Albajez, J.A.; Santolaria, J.; Velázquez, J. Configuration Optimisation of Laser Tracker Location on Verification Process. *Materials (Basel)* **2020**, *13*, 331. [CrossRef] [PubMed]
25. Moreno, J.; Simon, P.; Faleiro, E.; Asensio, G.; Fernandez, J.A. Estimation of an Upper Bound to the Value of the Step Potentials in Two-Layered Soils from Grounding Resistance Measurements. *Materials (Basel)* **2020**, *13*, 290. [CrossRef] [PubMed]
26. de Francisco Ortiz, Ó.; Estrems Amestoy, M.; Sánchez Reinoso, H.T.; Carrero-Blanco Martínez-Hombre, J. Enhanced Positioning Algorithm Using a Single Image in an LCD-Camera System by Mesh Elements' Recalculation and Angle Error Orientation. *Materials (Basel)* **2019**, *12*, 4216. [CrossRef] [PubMed]
27. Mínguez Martínez, A.; de Vicente y Oliva, J. Industrial Calibration Procedure for Confocal Microscopes. *Materials (Basel)*. **2019**, *12*, 4137. [CrossRef] [PubMed]
28. Brosed, F.J.; Cacho, R.A.; Aguado, S.; Herrer, M.; Aguilar, J.J.; Mazo, J.S. Development and validation of a calibration gauge for length measurement systems. *Materials (Basel)* **2019**, *12*. [CrossRef]
29. Bermudo Gamboa, C.; Martín-Béjar, S.; Trujillo Vilches, F.J.; Castillo López, G.; Sevilla Hurtado, L. 2D–3D Digital Image Correlation Comparative Analysis for Indentation Process. *Materials (Basel)* **2019**, *12*, 4156. [CrossRef]
30. Calaf-Chica, J.; Marín, M.M.; Rubio, E.M.; Teti, R.; Segreto, T. Parametric Analysis of the Mandrel Geometrical Data in a Cold Expansion Process of Small Holes Drilled in Thick Plates. *Materials (Basel)* **2019**, *12*, 4105. [CrossRef]

31. Ramirez-Peña, M.; Abad Fraga, F.J.; Sánchez Sotano, A.J.; Batista, M. Shipbuilding 4.0 Index Approaching Supply Chain. *Materials (Basel)* **2019**, *12*, 4129. [CrossRef]
32. Folch-Calvo, M.; Brocal, F.; Sebastián, M.A. New risk methodology based on control charts to assess occupational risks in manufacturing processes. *Materials (Basel)* **2019**, *12*. [CrossRef]

© 2020 by the authors. Licensee MDPI, Basel, Switzerland. This article is an open access article distributed under the terms and conditions of the Creative Commons Attribution (CC BY) license (http://creativecommons.org/licenses/by/4.0/).

Article

Investigation of a Short Carbon Fibre-Reinforced Polyamide and Comparison of Two Manufacturing Processes: Fused Deposition Modelling (FDM) and Polymer Injection Moulding (PIM)

Elena Verdejo de Toro [1], Juana Coello Sobrino [1,2,*], Alberto Martínez Martínez [2], Valentín Miguel Eguía [1,2] and Jorge Ayllón Pérez [2]

[1] Faculty of Industrial Engineering, 02071 Albacete, Spain; elena.verdejo@alu.uclm.es (E.V.d.T.); valentin.miguel@uclm.es (V.M.E.)
[2] Materials Science and Engineering, Instituto de Desarrollo Regional, 02071 Albacete, Spain; alberto.martinez@uclm.es (A.M.M.); jorge.ayllon@uclm.es (J.A.P.)
* Correspondence: juana.coello@uclm.es; Tel.: +34-967-599-200

Received: 31 October 2019; Accepted: 31 January 2020; Published: 3 February 2020

Abstract: New technologies are offering progressively more effective alternatives to traditional ones. Additive Manufacturing (AM) is gaining importance in fields related to design, manufacturing, engineering and medicine, especially in applications which require complex geometries. Fused Deposition Modelling (FDM) is framed within AM as a technology in which, due to their layer-by-layer deposition, thermoplastic polymers are used for manufacturing parts with a high degree of accuracy and minimum material waste during the process. The traditional technology corresponding to FDM is Polymer Injection Moulding, in which polymeric pellets are injected by pressure into a mould using the required geometry. The increasing use of PA6 in Additive Manufacturing makes it necessary to study the possibility of replacing certain parts manufactured by injection moulding with those created using FDM. In this work, PA6 was selected due to its higher mechanical properties in comparison with PA12. Moreover, its higher melting point has been a limitation for 3D printing technology, and a further study of composites made of PA6 using 3D printing processes is needed. Nevertheless, analysis of the mechanical response of standardised samples and the influence of the manufacturing process on the polyamide's mechanical properties needs to be carried out. In this work, a comparative study between the two processes was conducted, and conclusions were drawn from an engineering perspective.

Keywords: additive manufacturing; fused deposition modelling; composites; 3D printing; polymer injection moulding; polyamide; CFRP

1. Introduction

Fused Deposition Modelling (FDM) is a promising technology framed within Additive Manufacturing (AM) and has been implemented in the manufacturing processes of functional structures with complex geometries, showing interesting responses in different fields [1]. Medicine, biotechnology, automotive industry or aerospace are domains that use such technologies to obtain parts made of polymers or polymer–matrix composites with high stiffness and low weight values, also reducing material waste during the process [2]. Among the different technologies that are part of AM, those expected to be most suited to the manufacturing industry are Selective Laser Sintering (SLS), Selective Laser Melting (SLM) and FDM [3]. The last of these technologies is considered to be more developed than the other two [4]. FDM technology consists of the layer-by-layer deposition of a thermoplastic polymer previously driven into a semi-liquid state through the nozzle that allows the deposition of the material. The most widely processed polymers are acrylonitrile butadiene styrene (ABS), polylactic acid (PLA), polyvinyl alcohol (PVA), polyamides (PA) or polyether ether ketone (PEEK) [5]. Three-dimensional printing is also considered a sustainable production technology, which means that material waste is reduced in the manufacturing process, or even avoided [6]. In addition, the high geometrical accuracy in the dimensions of the printed parts is also a determining factor [7]. Design flexibility is considered an advantage of the FDM process as it takes into account not only the idea of modifying the prototype as much as desired, but also considers the wide variety of parameters that this technology allows to be used [8]. Infill density, infill pattern, layer height or printing velocity are the parameters traditionally considered [9]. Nevertheless, other aspects that are sometimes neglected might have an influence on the part being manufactured [10]. For example, Chacón et al. demonstrated the influence of the raster angle and the placement of the part and established the optimum location to obtain the best properties, studying the behaviour of parts under tensile and bending loads [11]. It is necessary to study the influence of these parameters on the mechanical response of printed parts to create a summary that finds the best combination for each application, depending on different stress types [12]. Some authors have also studied the anisotropy of AM parts, determining the significant effect of the placement of the part on the build plate [13].

The main disadvantages of printed parts that need to be addressed include the poor mechanical properties limited by the polymeric matrix and the poor adhesion between consecutive layers during deposition [14]. Some works have studied the influence of parameters on polymers, such as PA12 (ultimate tensile strength (UTS) = 33 MPa) or reinforced PA12 [15]. However, the mechanical properties of PA12 are lower than PA6 (UTS higher than 50 MPa), which has been considered a more suitable choice for replacing metal parts [16]. Higher stiffness and strength values are obtained in PA6 than in PA12, which is more widely used in FDM research, considering the limitations on the melting temperature according to the thermal resistance of printed parts [17]. Other materials, such as ABS and PLA, have been compared to PA, but have yielded worse results [18]. Nylon 6 showed a tensile strength of ~80 MPa vs. ~50 MPa for PLA and ~65 MPa for ABS considering injected parts. In the case of AM manufactured parts, lower results were obtained: a tensile strength of ~50 MPa for nylon 6, ~45 MPa for PLA and ~25 MPa for ABS [18].

Alone, matrices do not reach the expected requirements in mechanical behaviour. To solve the low strength and stiffness problem, the use of reinforcement has been considered the best alternative [19]. Carbon and glass fibres (short or continuous) and nanoparticles, such as nanospheres, are used to reinforce the polymeric matrix [20]. For reinforced polyamide, Chabaud et al. obtained mechanical properties similar to aluminium alloys [21]. Carbon fibres have been chosen as a favourite reinforcement due to their high stiffness and strength, and low weight [9,21].

Despite having shown higher mechanical properties as a matrix in composites, printed parts made of short carbon fibre-reinforced PA6 have not been fully studied. Furthermore, previous research has focused on evaluating the tensile or dynamic properties but has not considered the compressive properties. Moreover, knowledge is needed about which manufacturing technique fits better with the requirements of parts. In addition, some new technologies, such as FDM, must be compared with traditional ones, in this case polymer injection moulding, to determine whether the latter might offer a more effective solution. Taking into account all the above-mentioned considerations, this experimental work focuses on evaluating and comparing the response of printed and injected samples, while also considering the possibility of dealing with compressive loads. The combination, or even the replacement of traditional technologies with new ones, will be essential for the manufacturing process of different functional parts in forthcoming years. This work presents this comparison and provides results for different mechanical tests summarised from an engineering perspective.

2. Materials and Methods

2.1. Raw Material and Equipment

The short carbon-fibre-reinforced material used as feedstock in the printing process was CarbonXTM CRF-Nylon (3DX Tech, Grand Rapids, MI, USA), a filament of PA6 with a diameter of 2.85 mm. Fibre content was 20 weight percent (wt %). Furthermore, an Olsson Ruby nozzle of 0.4 mm replaced the printer's default brass nozzle as it has previously been proven that fibre damages brass after a few printed samples. The printer was an Ultimaker 2 Extended + (Ultimaker, Utrecht, Netherlands). Samples were designed using Autodesk Inventor (2016 version, Autodesk, St. Raphael CA, USA), and the slicing program was Cura 3.5.1 (Ultimaker, Utrecht, The Netherlands). In the case of the injection moulding, the same material was used but in granulated form. Although the granulate could have been obtained directly from the filament, we decided to use the granulated composite from the same supplier in order to work with a homogeneous size of grain.

2.2. Parameters for Manufacturing

As our aim was to study the differences in the behaviour of samples manufactured by injection moulding and 3D-printing technologies, we needed to distinguish and set different parameter types corresponding to each manufacturing technique.

Three-dimensional-printing parameters were set following the recommendations of the feedstock manufacturer. The nozzle temperature was 260 °C, the build plate temperature was 80 °C, the printing speed was 50 mm/s and the layer height and nozzle diameter were 0.1 mm and 0.4 mm, respectively. In order to compare the samples manufactured by injection moulding, we manufactured samples with 100% infill density using the longitudinal pattern. The 60% infill density parts were manufactured by employing different patterns (triangles, lines ±45° and longitudinal; see Figure 1) to compare them in bending, compression and tensile tests.

Figure 1. Stereomicroscope images (×1.25) of the appearance of the injected and different patterned printed samples.

The mould temperature was set at 80 °C, whereas the polymer temperature during injection was 260 °C. The pressure in the process was 8.5 bar. Injection moulding parameters were chosen to be as close as possible to those used in the printing process.

2.3. Measurement Techniques

The influence of both processes on fibre length was first investigated. The specimens manufactured by both 3D printing and injection moulding were extracted and burnt in Thermogravimetry-Differential Thermal Analysis (TG-DTA) equipment (EXSTAR 6200, Seiko Instruments, Chiba, Japan). The process started at 23 °C and went to 900 °C in an inert nitrogen atmosphere to avoid fibre degradation. The heating rate was 20 °C/min at an inert gas flow of 200 mL/min. Having obtained fibres, measurements were taken with a microscope. Three specimens of each manufactured sample were measured, obtaining the results shown in Figure 2.

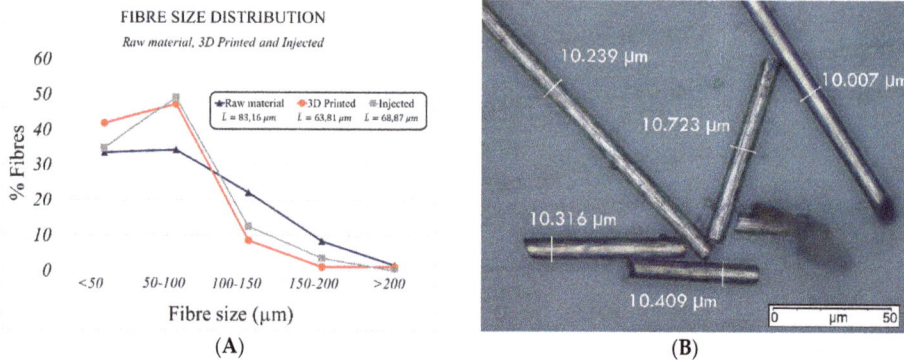

Figure 2. Results of the fibre length distribution in the raw material, injected and printed samples (**A**) and measurement of diameters in fibres using 400× with a microscope (**B**).

In many experimental studies, the printer's thermal environment affects the degree of crystallinity of the composite's polymeric matrix, which strongly influences its mechanical properties. In general, the higher the degree of crystallinity, the greater is the strength and the less is the deformation generated in mechanical tests. To obtain reliable results in mechanical tests, a previous analysis of the thermal environment in the printing process was carried out. To this end, three build plate temperatures were chosen in order to determine whether the thermal environment had an impact on the mechanical response of printed parts: 110 °C, 60 °C and 25 °C, which was considered representative of the room condition. DSC tests of samples after printing were conducted and bending tests were carried out to study whether crystallinity appears in the material and the response of the printed parts to mechanical stresses. Although in all three cases, the build plate temperature was lower than the initial characteristic crystallisation temperature determined for the material, the gradient of temperature from the printing temperature (260 °C) to the temperature set for the build plate was found to promote crystallinity.

All mechanical tests were conducted under room conditions (23 °C, HR 50%). The tensile properties of the samples were obtained according to UNE EN ISO 527-2, with a Zwick Z010 TN2S using an extensometer also produced by Zwick with a calibrated length (L0) of 25 mm. Data were collected and processed with the testXpert machine software (8.1 version, ZwickRoell, Ulm, Germany). The distance between grips was 115 mm, and test velocity was set at 1 mm/min.

A compression test was conducted as standard, but recommendations were taken from UNE EN ISO 14126:2001. The cylindrical samples ($\phi s = 12$ mm, Ls = 18 mm) were manufactured by both injection moulding and 3D-printing. To do so, a new steel mould with the sample's chosen dimensions had to be manufactured according to the requirements of the above-mentioned injection equipment. Compression velocity was also set at 1 mm/min. Two flat carbon steel discs ($\phi = 16$ mm) were used to compress the sample. The same Servosis machine as in the bending test was used to carry out the test, following UNE EN ISO 178:2011. Molybdenum (IV) sulphide was placed on the surface of the flat discs to reduce the friction between samples and discs, and to prevent the sample from becoming barrel-shaped during the test.

The moulds for injection were designed with only one 6 mm gate; no runners were used in order to obtain a direct material flow into the cavity. The gate was placed at one end of the part to be obtained, which in the worst case was 160 mm (tensile specimens). These designs were elaborated to make the flow of the material easier in spite of the maximum available pressure of the equipment.

3. Results and Discussion

3.1. Fibre Length Analysis

After the calcination process in an inert atmosphere, a hundred fibres of each specimen were obtained and measured. The results of the histograms of the lengths of the initial filament (raw material), the 3D printed and the injected material, are shown in Figure 2A. Both processes had a negative influence on fibre length. The average initial feedstock filament length was $\overline{L}$ = 83.16 µm. However, after printing and injection moulding, fibre length decreased by up to 23.27%.

This may have a number of consequences on the behaviour of the manufactured samples if the critical length (L_c) is greater than the fibre length of the processed parts, since the reinforcing effect is considered to be less effective. In order to obtain the critical fibre length, it was necessary to determine the tensile strength of the fibre (σ_f), its diameter (ϕ_f) and the interfacial shear strength (IFSS) fibre-matrix. This last property can only be found by performing a pull-out test of the composite material. Once the necessary parameters were identified, the critical length was obtained by applying Equation (1) [16].

$$L_c = \frac{\sigma_f \cdot \phi_f}{2 \cdot IFSS} \tag{1}$$

The value of σ_f = 2.2 GN·m^{-2} was taken from the literature [16]. The experimental measurement of the diameter of the fibre was found to be ϕ_f = 10.12 ± 0.94 µm (Figure 2B). The shear stress of the interfacial bonding between the carbon fibre and the matrix (IFSS = 44 MPa) was taken from experimental results reported in different research works [1,22,23] because it was not possible to run the pull-out test in this experimental study. Carbon fibres were considered to be unsized.

The critical length obtained by Equation (1) was L_c = 253 µm. Therefore, as the length of the fibres inside the matrix was shorter than the critical theoretical value, the reinforcing effect would be lower than for a fibre length longer than the critical one, especially in the tensile test. Nevertheless, fibres with a longer length than the critical one were observed after both manufacturing processes; that is, 3D-printing and injection moulding. As a result, the reinforcing effect would take place, but to a lower extent and only as a result of the longest fibres. Moreover, some researchers [9] have found that the 3D-printing process reveals a greater influence on the mechanical behaviour of the part than the reinforcement effect itself. In addition, some studies report that the highest reinforcing volumes can only be reached by using short fibres and taking into account that in AM, the mixing–printing process itself tends to result in a length shorter than the critical one [17].

3.2. Study of Variation in Crystallinity in the Printed Samples

The DSC tests were carried out with the three printed samples of each type (at different build plate temperatures). The results showed no variation in crystallisation peak, starting point or ending point. The samples behaved the same in the thermal analysis and the results obtained were satisfactory, as shown in Figure 3B,C.

DSC analysis of the raw material, i.e., the CarbonXTM CRF-Nylon wire, showed a crystallisation peak (Figure 3A), whereas this was not seen in the DSC analysis carried out for 3D printed parts (Figure 3B,C) independently. For this reason, it can be concluded that the AM manufacturing technique promotes the crystallinity of the PA6 matrix.

There was no variation between the top and bottom parts of the same sample, which were those farthest from and closest to the build plate, respectively (Figure 4). This result is also supported by Figure 3 showing that the DSC curves corresponding to bottom and top surfaces are identical; the initial parts of the curves are typically different according to the initial humidity conditions of the samples and/or the stabilisation time of the furnace chamber. In the bending test, the top and bottom parts in the same samples were tested with bending stresses as a fracture appeared in the stretched part. Both parts were tested to check the influence of the thermal gradient during the printing process.

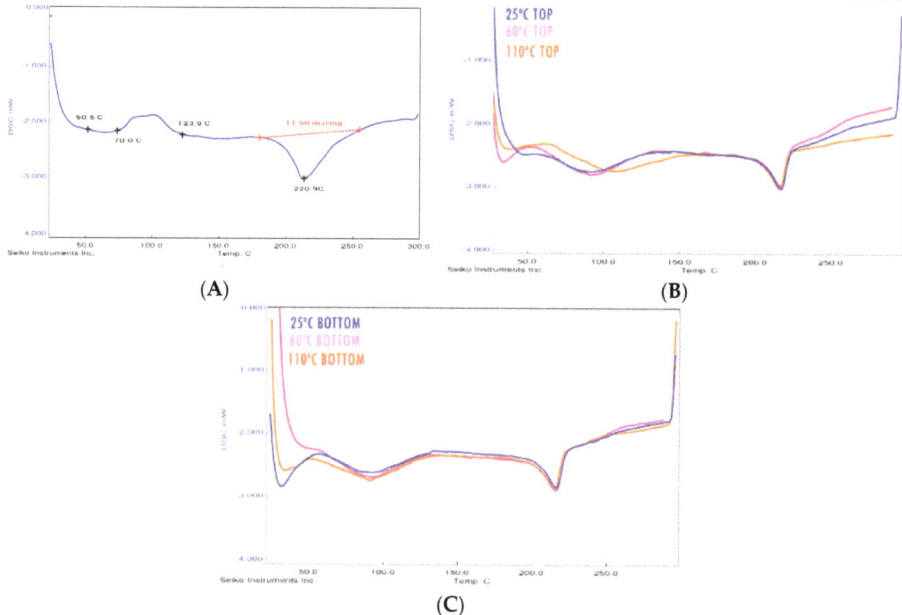

Figure 3. The DSC analysis of the samples printed at different built plate temperatures to analyse the influence of the thermal environment on the degree of crystallinity. The DSC analysis of raw material is shown in (**A**), the top parts are the analyses shown in (**B**) and the bottom parts are the analyses shown in (**C**).

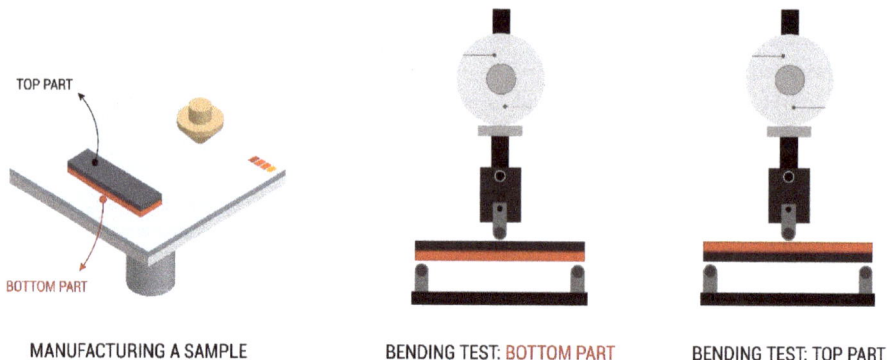

Figure 4. Positioning samples in the bending test, labelled (top and bottom) according to printing placement.

The results obtained for the maximum flexural strength (σ_{fM}), Young's Modulus (E_f) and yield strength (Y_f) showed a negligible variation between tested specimens (Table 1) and, consequently a negligible influence of the build plate temperature on the flexural behaviour of samples (see Figure 5). The degree of crystallinity was not affected by the thermal gradient, which appeared in the manufacturing process under the set conditions.

Table 1. Mechanical properties obtained from the bending tests.

Build Plate Temperature	110 °C		60 °C		25 °C	
Mechanical Properties	Top	Bottom	Top	Bottom	Top	Bottom
E_f (MPa)	2078.45	2147.71	2147.71	1873.63	2437.25	2404.65
Y_f (MPa)	47.39	47.61	46.10	43.29	52.77	50.79
σ_{fM} (MPa)	68.95	71.04	67.99	62.81	73.63	71.60

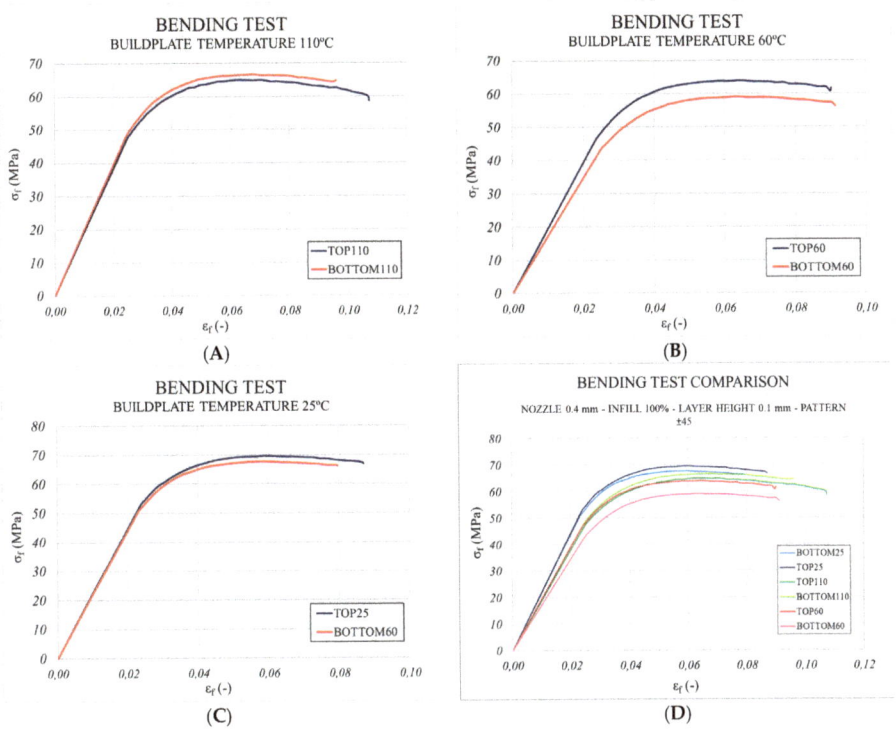

Figure 5. Analysis of the mechanical response of the samples manufactured by 3D printing at different build plate temperatures to study if anisotropy was caused by the thermal environment. (**A**) Build plate temperature: 110 °C, (**B**) build plate temperature: 60 °C, (**C**) build plate temperature: 25 °C and (**D**) bending test comparison of (**A**–**C**).

Homogeneous behaviour due to the uniformity of each sample regarding its crystallinity was expected. Therefore, the mechanical properties of parts would be independent of the thermal gradient appearing in the 3D printing process and the build plate temperature would not be considered a variable affecting the part.

3.3. Determination of Properties in the Tensile Test

The tensile test results showed marked differences between the 60% filled parts and the 100% filled ones. As expected, infill density had a significant impact on the three mechanical parameters under study (see Figure 6). Ultimate tensile strength (UTS) was higher when the completely filled unidirectional pattern was used, which allowed 63% more stress than the 60% filled one. The results obtained for Young's Modulus showed improvements of up to 62% for the completely filled samples. For yield stress, the same occurred as in tensile strength.

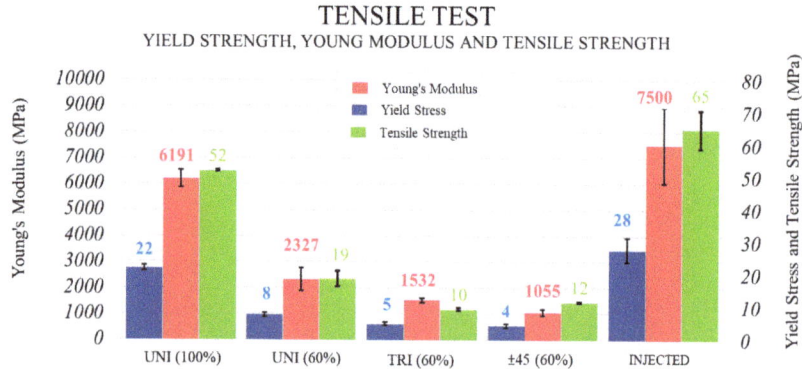

Figure 6. Tensile test. Results obtained for Young's Modulus, yield strength and tensile strength of the injected and 3D printed samples.

When studying the influence of pattern, the unidirectional pattern behaved better than the triangular one (improvements of 47%) and the linear ±45° (up to 37%). For yield stress and Young's Modulus, the same occurred between hollow patterns. Due to the non-deformability of the triangles in the triangular pattern, the samples built using that distribution of filaments displayed greater levels of stiffness, which might be due to a higher Young's Modulus compared to the alternative linear ±45° pattern (30% higher in the former). Nevertheless, the best result was obtained with the unidirectional pattern (improvements up to 55%) due to filament orientation according to the preferable distribution of stress in the axial direction in the tensile test. However, the ±45° pattern is used in applications where the direction of the stresses is unknown or known but non-axial, whereas the unidirectional pattern might have a worse response to the mechanical stresses loading the part.

Consequently, not only infill density but also the infill pattern was a determining factor in the results obtained.

When comparing the printed and injected parts (Figure 7), it can immediately be seen that the injected samples behaved better under axial stresses. However, differences were not as large as expected. Yield stress was 21% higher, whereas Young's Modulus was 17% greater and tensile strength at the break point was only 20% higher in the injected samples. Comparing FDM and IM, Lay et al. obtained similar results [18], while Blok. et al. tested a printed short carbon fibre-reinforced nylon (6 wt % of fibre weight content) and reported a UTS = 33 MPa and E = 1900 MPa [9]. In the case of the current study, for the unidirectional pattern, UTS = 52 MPa and E = 6191 MPa, but the percentage of fibre inside the polymeric matrix is also higher (20 wt %). Furthermore, greater deviations from the average were obtained with the injected samples, in which more homogeneous behaviour was expected than for the 3D printed samples. It can be concluded that the printed parts would be suitable to replace the injected ones when working under stretching loads.

From the fractographies in Figure 8, a sound structure for the parts obtained by injection moulding (Figure 8A) can be observed. Properly, the fractography is dominated by the matrix and fibres fracture, and to some extent, by the fracture of the matrix–fibre interface. This reveals that the grade of polymerisation obtained in the injection moulding process is reasonable despite the low-pressure injection condition (8.5 bar). The fractures of the samples obtained by 3D printing show that pores and fibres have been separated from the matrix interface in a greater proportion (Figure 8C,D). This is a consequence of the short length fibre used leading to an insufficient interface area with the PA6 matrix. From this viewpoint, the interface is more suited to the injection moulding samples.

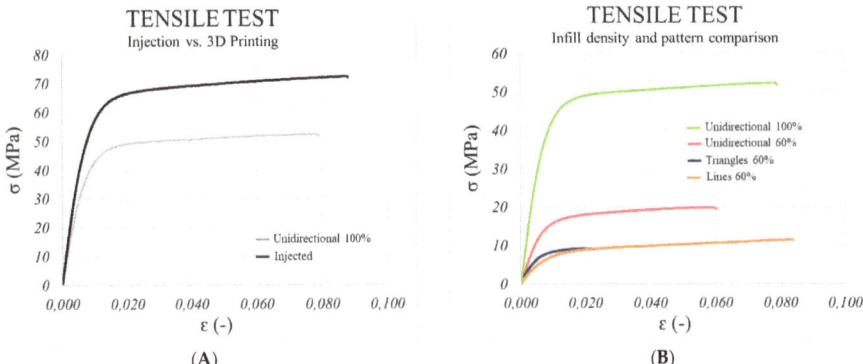

Figure 7. Comparison of the behaviour of the injected and printed samples with stretching stresses in the tensile test (**A**), and the influence of infill density and printing pattern on the mechanical behaviour of the printed parts (**B**).

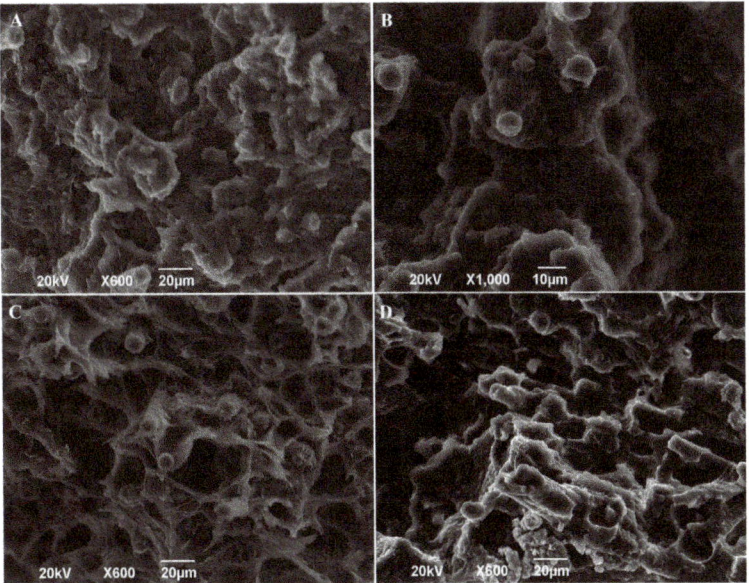

Figure 8. Fractographies of tested tensile samples: (**A**) Injection moulding 600×; (**B**) injection moulding 1000×, (**C**) 3D printed unidirectional 0° 600×; (**D**) 3D printed ±45° 600×.

3.4. Determination of Properties in the Compression Test

The compression test results are shown in Figure 9 and indicate improvements of only 4% in the injected samples in comparison with the 3D printed ones (100% infill). It is worth noting that Young's Modulus is higher in the printed parts (50% improvement with respect to the injected samples). This suggests that when compressive loads are applied, the 3D printing process leads the parts to show higher stiffness than in the IM process. In compression tests, this is due to the direction of the force being opposite to that in the tensile test and the separation of consecutive layers of the printed parts is more difficult. Thus, using high infill density values avoids an early breakage between consecutive layers. Qualitatively, it can be observed that the behaviour of the 3D printed part decreases as of a determined strain. Furthermore, the decline appears just when the compressive stresses of both samples are equal. This decrease is a consequence of the separation taking place between consecutive layers. This conclusion is supported by the enormous separation observed in Figure 10B corresponding to 60%-filled parts, while this occurred incrementally in completely filled samples. In the 60%-filled non-unidirectional samples (Figure 10C,D), polymeric hardening took place under compression stresses (Figure 11). This can be an advantage for parts that work under compressive stresses with no deformability constraints. Inversely, the unidirectional pattern broke down due to the presence of gaps between consecutive layers. Improvements up to 73% in yield strength and 33% in Young's Modulus were reached when 100% instead of 60% infill density was used, as seen in Figure 9.

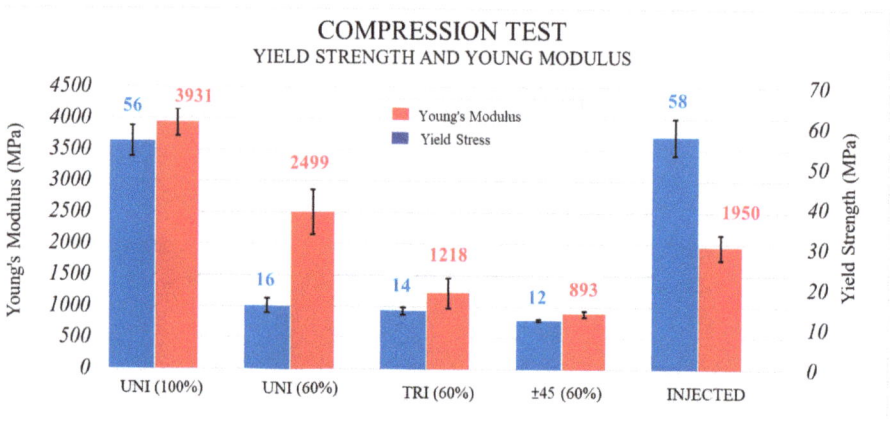

Figure 9. Compression test. Results obtained for yield strength, tensile strength and Young's Modulus of the injected and 3D printed samples.

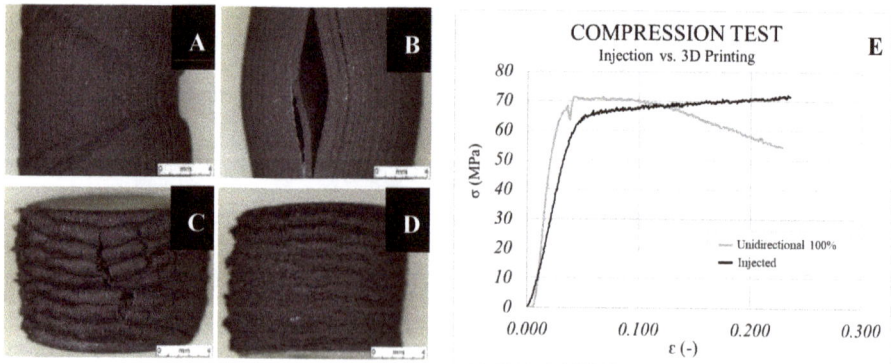

Figure 10. Fracture images of samples under compressive stress. (**A**) Unidirectional 100%, (**B**) unidirectional 60%, (**C**) triangles 60% and (**D**) linear ±45. (**E**) Comparison between the injected and printed samples in the compression test.

It was possible to see a slight influence of infill pattern for yield strength (unidirectional had a higher yield strength than triangles, and these were higher than the ±45° alternate linear one). These smaller differences were due mainly to the stress direction in the test, which tends to compress samples. In this case, the bonding between different layers was not as important as in the tensile test, where the commitment of transmitting load between consecutive layers came into play, and its low failure strength was a determining factor for the properties. However, Young's Modulus was strongly affected by pattern. Filament orientation was due to pattern choice, which acted as a determining factor to consider the preferential direction of the stresses in each test and, consequently, in each functional application. Both the tensile and compression tests justify this conclusion.

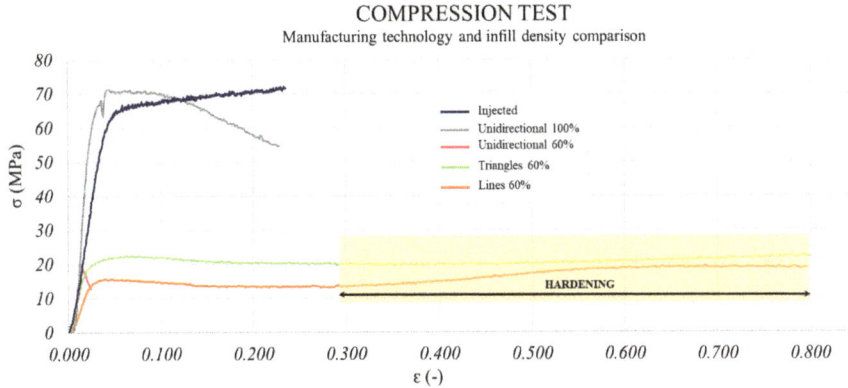

Figure 11. Compression test: influence of pattern on the behaviour of sample under compressive stresses.

4. Conclusions

This work describes a comparative analysis of 3D printing and injection moulding technologies to establish the influence of the process on the behaviour of parts made of short carbon-fibre-reinforced PA6. Both the 3D printing and IM processes led to breakage of fibres, making them shorter by up to 24%. The thermal environment has been shown to have no influence on the behaviour of 3D printed parts.

Tensile tests of printed parts showed worse results than for the IM parts, but differences did not exceed 21% for either yield strength, tensile strength or Young's Modulus. These results are consistent with those obtained by other authors, which validates the IM process used as a reference. A non-linear relationship between infill density and mechanical properties is supported. The unidirectional pattern is the best choice when lower densities are used.

Compared to tensile tests, compression tests revealed a more similar behaviour of 3D printed parts and IM parts (only 4% improvement). Printed samples had higher stiffness values than the IM parts. This phenomenon has not been reported by other researchers. The selection of pattern is a determinant in this case since a hardening effect with the strain appeared for those manufactured using a non-unidirectional pattern, but without reaching the best results.

The comparison between the tensile and compression tests revealed that this reinforced polyamide did not behave the same under compressive and stretching loads, regardless of the manufacturing process. Consequently, as previously reported in the literature, it is crucial to analyse the application for which parts are designed. These outcomes are a novel contribution to the existing literature.

Author Contributions: All authors are active members of the research group Science and Engineering of Materials at the Castilla-La Mancha University in Spain and they have actively contributed to the development of the paper. Conceptualization, E.V.d.T., V.M.E., J.C.S. and A.M.M.; data curation, E.V.d.T. and J.C.S.; formal analysis, E.V.d.T., V.M.E., J.C.S., A.M.M. and J.A.P.; funding acquisition, V.M.E., J.C.S. and A.M.M.; investigation, E.V.d.T., V.M.E., A.M.M. and J.A.P.; methodology, E.V.d.T., A.M.M. and J.A.P.; project administration, V.M.E.; resources, V.M.E., J.C.S. and A.M.M.; software, A.M.M. and J.A.P.; supervision, J.C.S., V.M.E.; validation, V.M.E. and J.C.S.; visualization, A.M.M.; writing—original draft, E.V.d.T. and V.M.E.; writing—review and editing, V.M.E., E.V.d.T. and J.C.S. All authors have read and agreed to the published version of the manuscript.

Funding: This research received no external funding.

Conflicts of Interest: The authors declare no conflicts of interest.

References

1. Ma, Y.; Yan, C.; Xu, H.; Liu, D.; Shi, P.; Zhu, Y.; Liu, J. Enhanced interfacial properties of carbon fiber reinforced polyamide 6 composites by grafting graphene oxide onto fiber surface. *Appl. Surf. Sci.* **2018**, *452*, 286–298. [CrossRef]
2. Van de Werken, N.; Reese, M.S.; Taha, M.R.; Tehrani, M. Investigating the effects of fiber surface treatment and alignment on mechanical properties of recycled carbon fiber composites. *Compos. Part A Appl. Sci. Manuf.* **2019**, *119*, 38–47. [CrossRef]
3. Dizon, J.R.C.; Espera Jr, A.H.; Chen, Q.; Advincula, R.C. Mechanical characterization of 3D-printed polymers. *Addit. Manuf.* **2018**, *20*, 44–67. [CrossRef]
4. Puerto Pérez-Pérez, M.; Gómez, E.; Sebastián, M.A. Delphi prospection on additive manufacturing in 2030: Implications for Education and Employment in Spain. *Med. Math.* **2018**, *11*, 1500. [CrossRef] [PubMed]
5. Parandoush, P.; Lin, D. A review on additive manufacturing of polymer-fiber composites. *Compos. Struct.* **2017**, *182*, 36–53. [CrossRef]
6. Frank, A.G.; Dalenogare, L.S.; Ayala, N.F. Industry 4.0 technologies: implementation patterns in manufacturing companies. *Int. J. Prod. Econ.* **2019**, *210*, 15–26. [CrossRef]
7. Long, T.E.; Williams, C.B.; Bortner, M.J. Introduction for polymer special issue: Advanced polymers for 3D Printing/additive manufactuing. *Polymer (Guildf)*. **2018**, *152*, 2–3. [CrossRef]
8. Gao, W.; Zhang, Y.; Ramanujan, D.; Ramani, K.; Chen, Y.; Williams, C.B.; Wang, C.C.L.; Shin Yung, C.; Zhang, S. The status, challenges and future of additive manufacturing in engineering. *Comp. Aided-Des.* **2015**, *69*, 65–89. [CrossRef]
9. Blok, L.G.; Longana, M.L.; Yu, H.; Woods, B.K.S. An investigation into 3D printing of fibre reinforced thermoplastic composites. *Addit. Manuf.* **2018**, *22*, 176–186. [CrossRef]
10. Sagias, V.D.; Giannakopoulos, K.I.; Stergiou, C. Mechanical properties of 3D printed polymer specimens. *Procedia Struct. Integr.* **2018**, *10*, 85–90. [CrossRef]
11. Chacón, J.M.; Caminero, M.A.; García-Plaza, E.; Núñez, P.J. Additive manufacturing of PLA structures using fused deposition modelling: Effect of process parameters on the mechanical properties and their optimal selection. *Mater. Des.* **2017**, *24*, 143–157. [CrossRef]
12. Rankouhi, B.; Javadpour, S.; Delfanian, F.; Letcher, T. Failure Analysis and Mechanical Characterization of 3D Printed ABS with Respect to Layer Thickness and Orientation. *J. Fail. Anal. Prev.* **2016**, *16*, 467–481. [CrossRef]
13. Dawoud, M.; Taha, I.; Ebeid, S. Mechanical behaviour of ABS: An experimental study using FDM and injection moulding techniques. *J. Manuf. Process.* **2016**, *21*, 39–45. [CrossRef]
14. Ngo, T.D.; Kashani, A.; Imbalzano, G.; Nguyen, K.T.Q.; Hui, D. Additive manufacturing (3D printing): A review of materials, methods, applications and challenges. *Compos. Part B Eng.* **2018**, *143*, 172–196. [CrossRef]
15. Türk, D.A.; Brenni, F.; Zogg, M.; Meboldt, M. Mechanical characterization of 3D printed polymers for fiber reinforced polymers processing. *Mater. Des.* **2017**, *118*, 256–265. [CrossRef]
16. Hull, D. *Materiales Compuestos*, 1st ed.; Reverté: Barcelona, Spain, 1987.
17. Van der Werken, N.; Tekinalp, H.; Khanbolouki, P.; Ozcan, S.; Williams, A.; Tehrani, M. Additively manufactured carbon fiber-reinforced composites: state of the art and perspective. *Addit. Manuf.* **2019**. [CrossRef]
18. Lay, M.; Thajudin, N.L.N.; Hamid, Z.A.A.; Rusli, A.; Abdullah, M.K.; Shuib, R.K. Comparison of physical and mechanical properties of PLA, ABS and nylon 6 fabricated using fused deposition modeling and injection molding. *Comp. Part B.* **2019**, *176*, 107341. [CrossRef]
19. Chapiro, M. Current achievements and future outlook for composites in 3D printing. *Reinf. Plast.* **2016**, *60*, 372–375. [CrossRef]
20. Fu, S.-Y.; Lauke, B.; Mäder, E.; Yue, C.-Y.; Hu, X. Tensile properties of short-glass-fiber- and short-carbon-fiber-reinforced polypropylene composites. *Comp. Part A.* **2000**, *31*, 1117–1125. [CrossRef]
21. Chabaud, G.; Castro, M.; Denoual, G.; Le Duigou, A. Hygromechanical properties of 3D printed continuous carbon and glas fibre reinforced polyamide composite for outdoor structural applications. *Addit. Manuf.* **2019**, *26*, 94–115. [CrossRef]

22. Na, W.; Lee, G.; Sung, M.; Han, N.; Yu, W.R. Prediction of the tensile strength of unidirectional carbon fiber composites considering the interfacial shear strength. *Compos. Struct.* **2017**, *168*, 92–103. [CrossRef]
23. Na, W.; Yin, X.; Zhang, J.; Zhao, W. Compressive Properties of 3D Printed Polylactic Acid Matrix Composites Reinforced by Short Fibers and SiC Nanowires. *Adv. Eng. Mater.* **2018**. [CrossRef]

© 2020 by the authors. Licensee MDPI, Basel, Switzerland. This article is an open access article distributed under the terms and conditions of the Creative Commons Attribution (CC BY) license (http://creativecommons.org/licenses/by/4.0/).

Article

Build Time Estimation for Fused Filament Fabrication via Average Printing Speed

Gustavo Medina-Sanchez, Rubén Dorado-Vicente *, Eloísa Torres-Jiménez and Rafael López-García

Department of Mechanical and Mining Engineering, University of Jaén, EPS de Jaén, Campus Las Lagunillas s/n, 23071 Jaén, Spain; gmedina@ujaen.es (G.M.-S.); etorres@ujaen.es (E.T.-J.); rlgarcia@ujaen.es (R.L.-G.)
* Correspondence: rdorado@ujaen.es; Tel.: +34-953-212-439

Received: 25 October 2019; Accepted: 29 November 2019; Published: 1 December 2019

Abstract: Build time is a key issue in additive manufacturing, but even nowadays, its accurate estimation is challenging. This work proposes a build time estimation method for fused filament fabrication (FFF) based on an average printing speed model. It captures the printer kinematics by fitting printing speed measurements for different interpolation segment lengths and changes of direction along the printing path. Unlike analytical approaches, printer users do not need to know the printer kinematics parameters such as maximum speed and acceleration or how the printer movement is programmed to obtain an accurate estimation. To build the proposed model, few measurements are needed. Two approaches are proposed: a fitting procedure via linear and power approximations, and a Coons patch. The procedure was applied to three desktop FFF printers, and different infill patterns and part shapes were tested. The proposed method provides a robust and accurate estimation with a maximum relative error below 8.5%.

Keywords: 3D printing; rapid prototyping; efficiency; printing time; experimental model

1. Introduction

1.1. About Additive Manufacturing

Since the first 3D printer was developed in the early 80s, the number of additive manufacturing (AM) solutions, often called 3D printing methods in a non-technical context, and their applications do not stop increasing. It is noteworthy the potential of AM in different applications such as bio-printing [1,2], replicating broken objects or custom parts [3], experimental and educational demonstrators [4], rapid tooling [5] and so on.

According to the standard ISO/ASTM 52900-15 [6], AM solutions produce objects by joining materials, usually layer by layer, from 3D models. This standard classifies the existing solutions in seven types of processes considering how materials are deposited and bonded: material extrusion, binder jetting, material jetting, directed energy deposition, vat photopolymerization, powder bed fusion, and sheet lamination. The main AM advantages over traditional methods are low product development time, material savings, and capability to produce objects with complex shapes, enhanced density, and interior structures [7].

Fused filament fabrication (FFF) based on material extrusion processes is the most widespread AM technique [8], and low price models dominate the shipment numbers [9]. FFF consists of heating a thermoplastic filament, extruding the resulting melt and filling layer by layer a part following 2D paths while the plastic solidifies. Although less accurate than other AM technologies, FFF printers are broadly used because of their price [10], the wide range of plastics that can be printed, and the strength of the obtained parts [1,11].

1.2. Build Time

AM processes have several issues that limit their potential applications. AM issues in the spotlight are: development of new compatible materials [12], dimensional accuracy and surface roughness [13–15], mechanical properties of printed parts, discretization of CAD model and printed object, printer capabilities, maintenance, optimization of shape and part orientation, and time to manufacture a part (build time) [16,17].

Regarding build time, an accurate estimation can help firms to enhance their processes planning, and to compare AM solutions [18]. Moreover, these estimations can lead to more meaningful results in works dealing with the optimization of process parameters with the target to reduce the operation time [19], as well as those studies that consider the influence of printing speed and, therefore, time in the printed part appearance [20] or strength [21]. Finally, the costs of AM machine-hours depends on build time [22], and it stands to reason that, its contribution to the overall costs will increase the price of the used AM technology. Despite the previous arguments, build time has received less attention than other issues, such as dimensional accuracy and mechanical properties [23].

Time estimation is not a simple task because it depends on the printer and its control characteristics, as well as the printing parameters and the machine path planning. Moreover, time prediction is, in general, not accurate [24,25]. The simplest build time estimation is calculated as the total motion path length divided by the programmed printing speed and, in some cases, can differ more than 30% from the actual build time.

During the last decade, several researchers are concerned about build time estimation in additive manufacturing. According to the detailed work of Zhang et al. [26], there are three main strategies to determine the build time:

- Analytical-approaches: Define complex analytical models that describe in detail the printer kinematics and, therefore, allow build time estimation. These are the most accurate solutions, but the construction is complex (it depends on many data and on knowing the printing path in advance), it is applicable to a specific system and the prediction depends on the nominal values of the machine parameters and its control, which could differ from their actual values, increasing the estimation error.
- Parameter-approaches: Determine simple analytical relations between time and a selected set of factors that depends mainly on part geometry such as height, surface, and volume. Although its implementation is simple, the accuracy is low and, again, depends on each system.
- Experimental-approaches: Fit the real system response for different values of a set of parameters, usually shape factors such as in parameter-approaches.

Experimental solutions are more accurate than parameter ones and simpler than analytical ones, but there are no rules for data selection neither for the fitting strategy and, therefore, the repeatability is low. A change in the printing parameters forces to construct a new response function, therefore, experimental methods are not flexible.

The method developed by Zhang et al. [26], which is based on Grey theory, is an interesting improvement of the experimental approaches. The authors claim that their estimation error has an average value of 10% and it is better than other existing approaches. On the other hand, it is not well established how to select the input factors and many shape parameters are needed (part volume, support volume, part surface, part height, support height, and part projected area).

Later works to those reported in Zhang's paper determine the build time according to the aforementioned strategies. Zhu et al. [27] developed both, an experimental and a parametric model. The last one is based on a reduced number of printing parameters (volume, height, and density) selected depending on their influence on the build time.

Different strategies are used to build experimental models. For example, the experimental solution explained by Zhu et al. [27] proposes a multi-factor regression, and Mohamed et al. [28] used a Q-optimal response surface methodology.

Examples of analytical methods are the estimations proposed by Habib and Khoda [29] or Komineas et al. [30]. The model developed by Komineas et al. [30] for material extrusion processes is based on a trapezoidal speed profile. It considers that tangential acceleration and deceleration are equal and it does not take into account the influence of normal acceleration limit (direction changes) in the printer speed. On the other hand, Habib and Khoda [29] also propose a simple trapezoidal speed profile model to estimate build time, nevertheless, the goal is not to make an accurate estimation of build time, but to use it to optimize the deposition direction. Moreover, some current computer applications, such as Pronterface [31], provide an analytical time estimation for FFF machines based on the printer characteristics, a trapezoidal speed profile and a cornering algorithm.

1.3. A Blended Solution

This work explains a new build time estimation model for FFF machines, which combines the analytical and experimental approaches. The proposed model takes into account the kinematics of the problem (as the analytical strategies) and defines it by fitting a low number of printing speed experimental observations. The solution is simpler than that of the analytical models and does not need to know the machine nominal parameters and how the printer is controlled. In contrast to parameter and experimental-approaches, the proposed solution requires only two parameters and few experimental tests to provide an accurate time prediction (estimation error below 8.5% in the printed examples). The experimental procedure requires a low-cost setup and can be easily accomplished with the explanation included in this paper. Moreover, unlike the experimental solutions, once the printing speed is approximated, the proposed method can be applied regardless of the chosen printing parameters.

Because of the printing path in FFF and tool path in a milling process are similar, our build time approach is obtained by modifying and extending the mechanistic model of Coelho et al. [32]. The proposed solution is based on a printing speed model, which not only considers the interpolation segment length, as the mechanistic model does, but also the path shape via the changes of direction. Path planning is key in the resulting build time of material extrusion processes [33]. Direction changes and small interpolation segments have a noteworthy impact on the actual time. Based on experimental observations, we show two different fitting strategies to determine the aforementioned printing speed model.

Some assumptions are considered. The method provides the time to print a part, without including the setup and heating times. Because of the fact that FFF motions are generally based on linear interpolation, the study is limited to this kind of interpolation scheme.

The remaining paper is organized as follows. Section 2 discusses the estimation procedure and Section 3 describes the experimental methods. Section 4 shows the approximation speed surfaces and validation examples. Finally, the main conclusions are drawn in Section 5.

2. Build Time Estimation Model

In this work, the way proposed to obtain the build time of actual parts is via the estimation of the actual printing speed. Through time measurements of known paths, we obtain information about how path definition (interpolation segment lengths and direction changes), machine characteristics, and its control influence the actual speed between consecutive interpolation points of the printing path. Known an approximation of the actual printing speed, time estimation is simple: reading the CNC printing code and using the speed approximation to estimate the real time of each path segment.

2.1. Average Printing Speed

We assume hereafter that the main factors that influence the printing speed f are the interpolation segment length s and the direction change α. According to Figure 1a, s is the Euclidean distance between the two consecutive interpolation points. The direction change α is the angle between the direction of three consecutive interpolation points. To understand how s and α influence the printing speed, we measure the average speed along circular arcs (Figure 1b), which are built via repeating interpolation segments with the same length and direction change (a line with $\alpha = 0°$ corresponds to a circle of infinite radius). The speed measurement procedure is explained in Section 3.

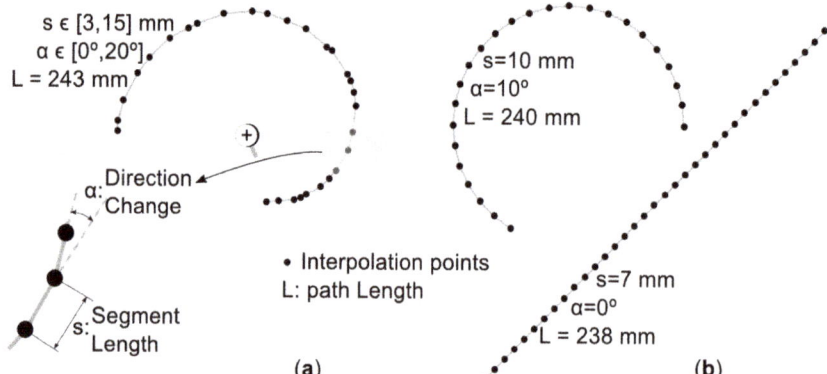

Figure 1. Examples: (**a**) Printing path with random segment lengths and direction changes, (**b**) printing paths for experimental speed estimation.

The speed-segment length relation evolves from linear to a power law (Figure 2a). The linear relation becomes smaller with increasing α, whereas the slope does not change. On the other hand, we have different power laws for each α; printing speed asymptotically decreases as α increases (Figure 2b).

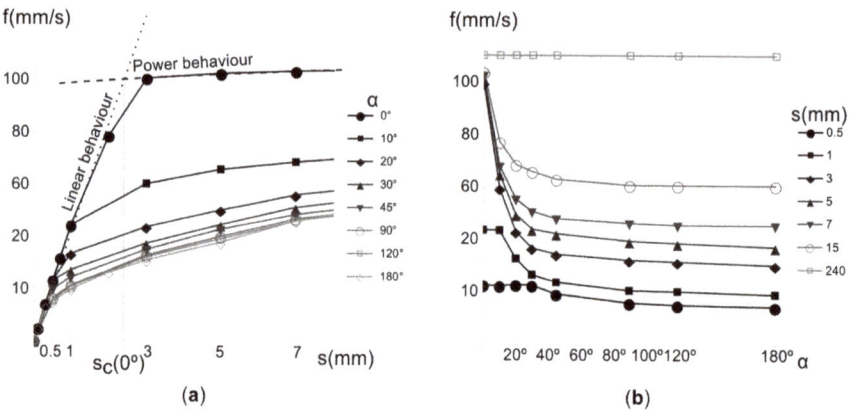

Figure 2. Example of experimental measurements of printing speed f. (**a**) Speed vs. s curves for different direction changes. (**b**) Speed vs. α for different segment lengths.

Note that, we measure f from $s = 0.1$ mm to a maximum value s_{Max}. We choose s_{Max} so that the circular path defined by interpolation segments with s_{Max} and $\alpha = 10°$ is the maximum circle within the printer bed.

Above s_{Max} it is more difficult to measure the printing speed. For this reason and based on the previous discussion, if $s > s_{Max}$ we assume a power law. This power model is computed by interpolation of $f(\alpha, s_{Max})$ and $f(0, b)$, where b is the printer bed diagonal length.

2.2. Printing Speed Surface

This section is devoted to determining an approximation surface that provides the printing speed for given values of s and α. There are different possibilities to define the printing speed surface $f(s, \alpha)$. The interpolation of the measurement points by means of a degree 1x1 polynomial spline patch provides a straightforward solution, but this approach requires many measurements to accurately predict time.

In order to reduce the number of needed measurements, we propose two alternatives:

- A linear-power (LP) surface that approximates the linear and power relation of f with respect to s as a function of the direction of change α.
- A spline of Coons patches (CP). Each Coons patch is defined by linear interpolation of four boundary curves.

To determine the isocurves α = constant on the LP surface, we approximate the segment length $s_c(\alpha)$ where linear and power approximations intersect (Figure 2). The procedure consists of the next steps:

Step 1. Approximate the speed measurements at $\alpha = 0$ and $s < s_c(\alpha = 0) = s_{c,0}$ by a line $f(0, s) = L(s) = m \cdot s + n$. The speed profiles, such as those portrayed in Figure 2a, show that if $s < s_c(\alpha)$, then the linear relation $f(0, s)$ does not change with α.

Step 2. Fit the speed measurements, within the interval $s_{c,0} < s \leq s_{Max}$, at k angles $0° \leq \alpha_i \leq 180°$, $i = [0, 1, \ldots, k-1]$ by power curves $P_i(s) = a_i s^{b_i}$. Although s_c varies with α, the relation $s_{c,0} > s_c(\alpha_i)$ is satisfied, so that we always are in the region ruled by the power law, and fewer measurements are needed. Note that, by increasing k the approximation improves at the expense of accomplishing more measurements.

Step 3. Compute s at the intersection of $L(s)$ and the power curves $P_i(s)$. Fitting the resulting data, for example by means of a degree-2 polynomial spline, we obtain the curve $s_c(\alpha)$.

Step 4. Build the curves $f(\alpha, s_c(\alpha))$ and $f(\alpha, s_{Max})$ and interpolate them using a power function $P(\alpha, s) = a(\alpha, s) s^{b(\alpha, s)}$.

Step 5. Define the speed surface as a piecewise function:

$$f(\alpha, s) \equiv \{L(s),\ s \leq s_c(\alpha);\ P(\alpha, s),\ s_c(\alpha) < s \leq s_{Max}\} \tag{1}$$

Regarding the CP, the idea is to define $f(\alpha, s)$ as a spline of two Coons patches. A Coons patch is a surface determined by its boundary curves, i.e., it is a way of filling the space between the curves.

The easiest Coons construction is a bilinear blend of two ruled surfaces and a bilinear interpolation surface [34]. Let $f(\alpha_0, s), f(\alpha_1, s)$ and $f(\alpha, s_0), f(\alpha, s_1)$ be the four parametric boundary curves, then it is easy to build a linear interpolation surface with each couple of curves:

$$\begin{aligned} r_1(\alpha, s) &= \left(1 - \frac{\alpha - \alpha_0}{\alpha_1 - \alpha_0}\right) f(\alpha_0, s) + \frac{\alpha - \alpha_0}{\alpha_1 - \alpha_0} f(\alpha_1, s) \\ r_2(\alpha, s) &= \left(1 - \frac{s - s_0}{s_1 - s_0}\right) f(\alpha, s_0) + \frac{s - s_0}{s_1 - s_0} f(\alpha, s_1) \end{aligned} \tag{2}$$

On the other hand, we can compute the bilinear interpolation of the four patch corners:

$$r_{1,2}(\alpha, s) = \left[1 - \frac{\alpha - \alpha_0}{\alpha_1 - \alpha_0},\ \frac{\alpha - \alpha_0}{\alpha_1 - \alpha_0}\right] \begin{bmatrix} f(\alpha_0, s_0) & f(\alpha_0, s_1) \\ f(\alpha_1, s_0) & f(\alpha_1, s_1) \end{bmatrix} \left[\left(1 - \frac{s - s_0}{s_1 - s_0}\right),\ \frac{s - s_0}{s_1 - s_0}\right]^t. \tag{3}$$

Finally, the Coons surface is:

$$CP(\alpha, s) = r_1(\alpha, s) + r_2(\alpha, s) - r_{1,2}(\alpha, s). \tag{4}$$

A unique Coons patch with $f(0, s), f(180, s)$ and $f(\alpha, 0), f(\alpha, s_{Max})$ is unable to adequately reproduce the actual printing speed surface. To overcome this drawback, we define the printing speed $f(\alpha, s)$ as a spline of two Coons joined at the value of $s_{c,0}$ defined in the same way as in the LP construction.

The following steps summarize the CP procedure:

Step 1. Measure the average speed at $\alpha = 0°$ and $\alpha = 180°$ for different segment lengths s, and at $s = s_{c,0}$ and $s = s_{Max}$ for several α.
Step 2. Fit the experimental data to obtain the boundary curves: $f(0°, s), f(180°, s), f(\alpha, s_{c,0}), f(\alpha, s_{Max})$. We approximate the experimental data by B-splines curves.
Step 3. Build two Coons patches: CP_A with $f(0°, s), f(180°, s), f(\alpha, 0), f(\alpha, s_{c,0})$, and CP_B with $f(0°, s)$, $f(180°, s), f(\alpha, s_{c,0}), f(\alpha, s_{Max})$.
Step 4. Compute the speed surface by the following spline function:

$$f(\alpha, s) \equiv \{CP_A(\alpha, s), s \leq s_{c,0}; CP_B(\alpha, s), s_{c,0} \leq s \leq s_{Max}\} \quad (5)$$

Note that, the approximation improves by increasing the number of Coons, but it requires increasing the number of measures.

Either for LP surface or the CP approximation, for $s > s_{Max}$, the idea is to interpolate the curves $f(\alpha, s_{Max})$ and $f(\alpha, b)$ (that we assume equal to $f(0°, b)$) using a power function.

Section 4 portraits the resulting LP and CP approximation surfaces and shows the measurements used in both surfaces: 22 measures for LP and 25 for CP.

2.3. Build Time Estimation from G-Code

Once we have the printing speed surface, it is possible to determine the build time from the path G-code. Observe that, the time required for heating the filament and the bed, and the time needed by the hot-end to go home (setup time) are not considered.

The computation process consists of:

- Read the ISO code and obtain the printing path in each layer.
- For each interpolation segment j, determine the programmed printing speed f_{pj}, its length s_j, and the direction change α_j of the segment respect to the previous one.
- Choose the printing speed surface according to the machine and estimate the actual printing speed $f(\alpha_j, s_j)$.

We take the programmed speed for z movements, as well as for the hot-end reposition when motors in x-y-z axis work at the same time because only the printing speed of x-y motors is measured.
- Finally, if the path has l segments, the estimated build time t is:

$$t = \sum_{j=1}^{l} \frac{s_j}{f_a}, \quad f_a = \{f(\alpha_j, s_j), \text{ if } f(\alpha_j, s_j) < f_{pj}; f_{pj}, \text{Otherwise}\}. \quad (6)$$

In order to accomplish the above procedure, we write a Mathematica® function. This function reads a G-code file, distinguishes each layer and detects the programmed printing speed along the layer paths. After that, it obtains the coordinates of the interpolation points and computes segments lengths s and, using the dot product, angles α. Finally, the developed function uses Equation (6) to estimate the actual build time.

3. Materials and Methods

We use the previously described estimation method (Section 2) to obtain the build time in three low-cost FFF machines: BQ Hephestos®, Witbox®, and Airwolf 3D HD® printers. 2D dimensional paths with random lengths and random direction changes, hereafter referred to as "random paths," and actual 3D printed parts are designed to validate our time estimation procedure.

3.1. Printers

Table 1 shows the main characteristics of the tested 3D printers. Extrusion and movements along x, y, and z axes are powered by standard stepper motors. We use the same travel speed for all printers: 120 mm/s.

3.2. Speed Measurement Procedure

In order to measure the speed at a specific s and α, we drive the printer hot-end through interpolation segments of length s and direction change α (circular paths with total length $L \approx 240$ mm, Figure 1b), measure the build time t, and finally compute the average speed as the ratio of L to t.

Table 1. Main technical characteristics of tested 3D printers.

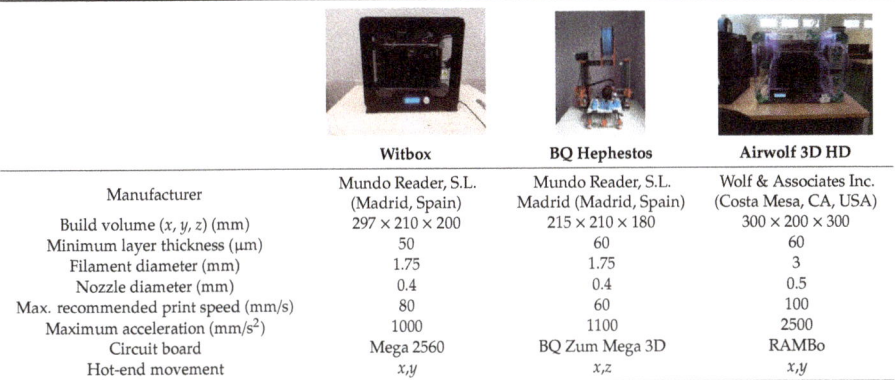

	Witbox	BQ Hephestos	Airwolf 3D HD
Manufacturer	Mundo Reader, S.L. (Madrid, Spain)	Mundo Reader, S.L. Madrid (Madrid, Spain)	Wolf & Associates Inc. (Costa Mesa, CA, USA)
Build volume (x, y, z) (mm)	297 × 210 × 200	215 × 210 × 180	300 × 200 × 300
Minimum layer thickness (μm)	50	60	60
Filament diameter (mm)	1.75	1.75	3
Nozzle diameter (mm)	0.4	0.4	0.5
Max. recommended print speed (mm/s)	80	60	100
Maximum acceleration (mm/s^2)	1000	1100	2500
Circuit board	Mega 2560	BQ Zum Mega 3D	RAMBo
Hot-end movement	x,y	x,z	x,y

A chronometer can lead to inaccurate time measurements, and even more for short paths. Thus, we decided to implement a measurement procedure based on producing a sound at the ends of the printing path, which leads to a clear identification of the build time. For printers with a Marlin firmware, such as those studied in this work, this means to add the following line before the first printing path position and after the last one:

"M300 P200 S440; play a 440 Hz tone during 200 ms."

A speaker to run the previous instruction is required. For a printer without a speaker, it is easy to connect a buzzer to an empty port of its electronic card and use the previous command. Witbox machine has a speaker, meanwhile, the Airwolf and the Hephestos machines need a buzzer.

Sound is recorded by a microphone connected to a PC, which allows distinguishing the time between the start and end. Each test was run three times. To determine the speed measurement uncertainty $u(f)$, we apply the combined standard uncertainty [35] to the equation $f = L/t$. The maximum $u(f)$ obtained was lower than 0.2 mm/s for the three studied machines.

3.3. Experimental Tests

In addition to the speed observations for building the approximation surfaces discussed in Section 3.2, we design two types of validation tests: random paths and printing examples.

All tests were conducted in the Pronterface application. Pronterface, similar to other 3D printing applications, provides an analytical print time estimation based on the planner functions used by the printer firmware to define the printer kinematics. These functions are a model of the speed, which by default is a trapezoidal profile, and a cornering algorithm that deals with the direction changes, and it is usually based on a limit jerk equation. The performance of analytical models depends on the nominal parameters of motors and control, whose values can differ from the real ones, and an adequate definition of factors such as the jerk limit.

It is interesting to compare the proposed model with a 3D printing software estimation, as many researchers [36–39] trust on those predictions to conduct costs and process optimization studies.

3.3.1. Random Path Tests

A set of random paths are tested to assess the performance of the proposed estimation procedure in specific s-α regions. Six paths, with a similar total length of 2400 mm, composed of segments with random lengths and directions (random paths) are conducted at travel speed without extrusion in each printer. A Mathematica function is implemented to provide the random paths and to write the needed G-code file. This function defines x-y positions and the reference printing speed (travel speed).

3.3.2. Printing Examples

We design three 12 mm high prisms with simple bases (triangle, pentagon, and star) at two scales (1:1, 2:1), and print them using two printer configurations. The well-known software Ultimaker CURA® (Free and open source LGPLv3 application developed and maintained by David Braam for Ultimaker, a 3D printer manufacturer based in Utrecht, Netherlands) provides the G-code. This software provides a similar time estimation to that of Pronterface, but it does not allow estimating times from G-codes obtained outside the software, such as for the random paths. For this reason, we compare our results to Pronterface predictions instead of Ultimaker CURA estimations.

Table 2 shows the factors and levels considered, and Table 3 summarizes the 12 tests performed corresponding to all possible combinations for the considered factors and levels, and the actual printing time measures. Note that, geometry, size, and printing parameters modify the s-α values and therefore, the resulting build time.

Table 2. Factors (parameters) and levels (factor values) to define the printing examples.

Factors	Levels		
	−1	0	1
Shape	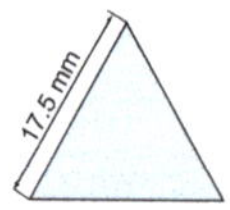 (triangle, 17.5 mm)	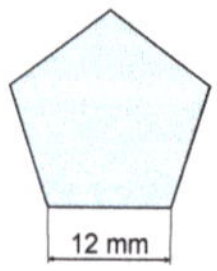 (pentagon, 12 mm)	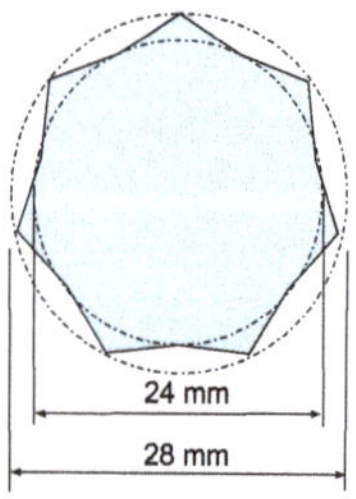 (star, 24 mm / 28 mm)
Scale	1:1	-	2:1
Printer configuration	Pattern: Zig-zag Top-Bottom thickness: 0 mm	-	Pattern: Concentric Top-Bottom thickness: 0.2 mm
Default configuration for the tested printers:			

- 0.2 mm Layer Height
- Material: PLA
- Brim: 2 mm
- Infill density 15%
- Wall thickness 0.5 mm
- No support

- Reference Speeds (mm/s)
  - Print: 40
  - Infill: 80
  - Wall: 20
  - Top-Bottom: 15
  - Travel: 120

Table 3. Actual printing time measured for the accomplished tests of Table 2.

Test	Factors			Real Printing Time (s)		
	Shape	Scale	Configuration	Air-wolf	Witbox	Hephestos
1	−1	−1	−1	324	344	353
2	0	−1	−1	339	367	386
3	1	−1	−1	469	509	551
4	−1	−1	1	346	364	372
5	0	−1	1	389	424	447
6	1	−1	1	608	702	769
7	−1	1	−1	932	1014	1094
8	0	1	−1	1318	1417	1524
9	1	1	−1	2351	2498	2671
10	−1	1	1	1035	1103	1155
11	0	1	1	1560	1709	1836
12	1	1	1	2936	3346	3693

4. Results and Discussion

The present section is organized as follows: Section 4.1 shows the experimental results required to build the two proposed approximation surfaces described in Section 2.2 (LP and CP surfaces), as well as the resulting surface models. Section 4.2 presents several tests for assessing the accuracy of the proposed models. This validation procedure consists of carrying out several paths where the printing time is recorded and compared to that provided by each approximation model. In Section 4.2.1, this procedure is applied to random paths (without extrusion of printing material), with the target of facilitating the variation in direction and segment length of a trajectory and analyzing their influence on printing time estimation. In Section 4.2.2., the validation procedure is also applied to several printed parts with different geometries in order to find out if the results provided by the approximation surfaces are also accurate for actual examples. A discussion regarding a comparison between the proposed methods, as well as between them and some usual methods for estimating the printing time, such as the Pronterface and theoretical estimations, is included at the end of the present section.

4.1. Printing Speed Measurements

Figures 3 and 4 show the printing speed measures and the proposed speed estimation models: LP and CP surfaces obtained. Black dots depict the measures used for the approximations, and grey dots represent additional measures used to verify the goodness of the approximation.

The mean absolute error (MAE) of all measurements and the determination coefficient R^2 of the approximations are included in Figures 3 and 4. The goodness of the approximations is high for all printers since R^2 is close to 1 and MAE value is low.

Note that, we choose a reduced number of measures to be fitted (similar for LP and CP approximations). The idea is to take measurements close to $s = 0$, s_c and s_{Max} at $\alpha = 0°$, 180°, and at $s_{c,0}$ and s_{Max} for different angles (black dots in Figures 3 and 4). The linear region, for the tested printers, is always obtained for segment lengths lower than 1 mm so that, to capture this behavior we take measurements every 0.1–0.2 mm from $s = 0$. On the other hand, the power region is wider than the linear one and, therefore, we use steps of 2–5 mm from s_c and s_{Max}. Regarding the data at s_c and s_{Max} for different angles, it is better to take more measurements between $\alpha = 0°$ and $\alpha = 60°$ because the greatest variations are registered within this range. Considering the previous suggestions, we obtain similar surfaces when fitting different experimental data.

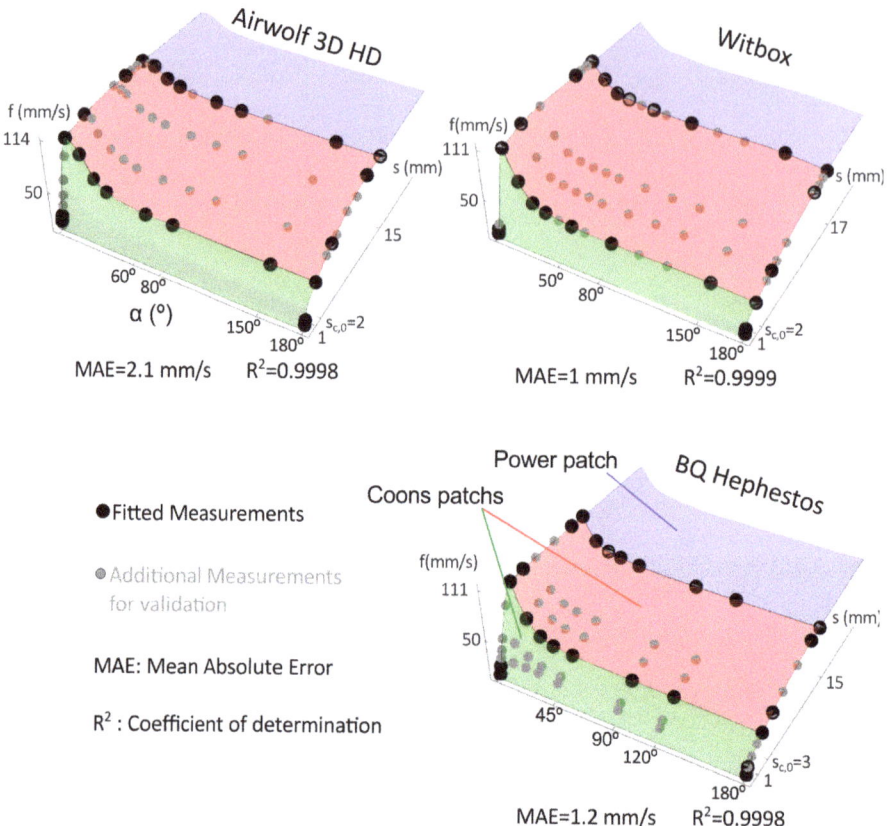

Figure 3. Coons patches (CP) printing speed surfaces for the tested printers (different colors are used to identify each patch of the CP spline surface). Dots represent the experimental measurements, CP surfaces fit black dots, and additional measurements (gray dots) are represented to visualize the goodness of the approximation.

LP and CP surfaces are similar and evolve as it is expected considering a trapezoidal speed profile and the machine acceleration limits. With respect to s, the surfaces evolve from linear to power, and that is explained because in each interpolation segment the printer accelerates to reference speed and decelerates up to the segment end (trapezoidal speed profile). On the other hand, the speed decreases with α (mainly between 0° and 60°). It stands to reason that the curvature and printing speed have a quadratic relation, which agrees with the experimental data and with the LP and CP surfaces obtained.

Machine characteristics influence the resulting speed surface. Although the surfaces have similar shapes, the Airwolf surfaces (LP and CP) provide the highest speed values and the Hephestos the lowest values in the considered s-α domain. In the Airwolf machine, the linear region grows steeper than in the other printers whereas the power region grows smoother. It stands to reason that the actual average speed depends on the printer acceleration limits (see Table 1) and this explains the above results.

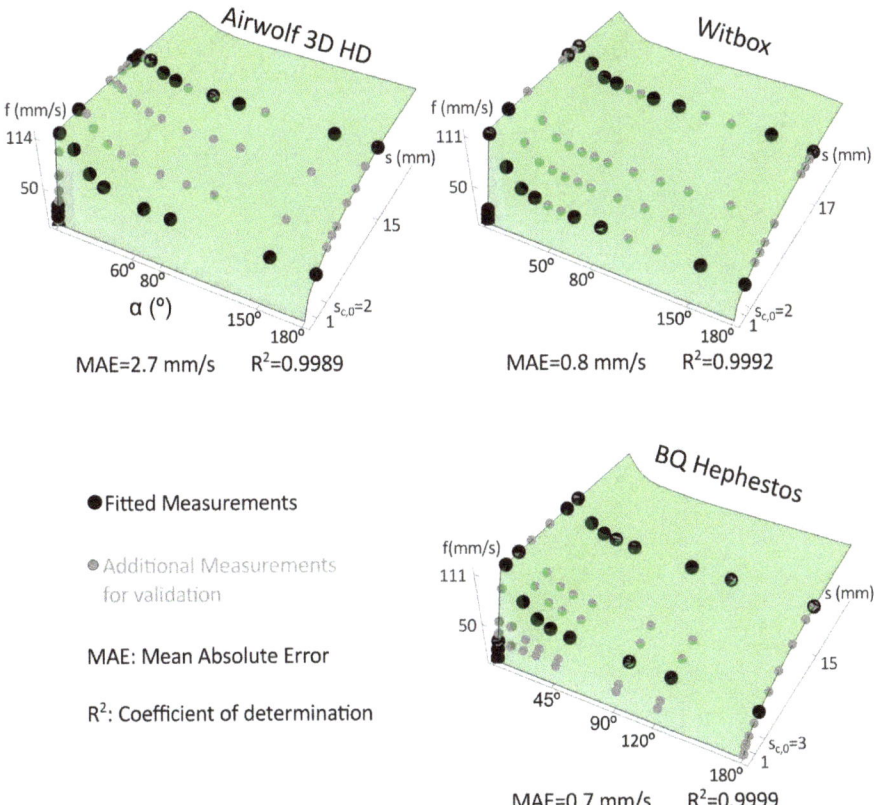

Figure 4. Linear-power (LP) printing speed surfaces for the tested printers. Dots represent the experimental measurements, LP surfaces fit black dots, and gray dots are additional measurements to visualize the goodness of the approximation.

4.2. Validation Tests

4.2.1. Random Paths

Six random paths (Section 3.3.1) were printed using the Pronterface application. Figure 5 shows the resulting actual printing time, the theoretical estimation (sum of ratios of segment length to programmed speed), the Pronterface prediction and the print time estimation provided by the LP and CP surfaces in the tested printers.

Comparing the six examples, the proposed approximations provide the most accurate estimations. In each printer, LP and CP average errors are similar and always lower than 5.5%. This value improves Pronterface and theoretical average errors, which are up to 48% and 59% respectively. The dispersion observed in the error values is a consequence of how close the approximations are to the actual printing speed surfaces at each region s-α.

Speed mainly changes within the linear region, in the transition from linear to power and because of direction changes up to 60° (see Figures 3 and 4). Thus, Pronterface, LP, and CP errors have maximum values at R1 and R2. Regarding the theoretical error, it does not consider s and α variations, which leads to a maximum error in R4. On the other hand, theoretical and Pronterface predictions are quite similar in regions R1 to R4, but Pronterface estimation improves when α increases (regions R5 to R6) because it considers the cornering algorithm used by the printer control.

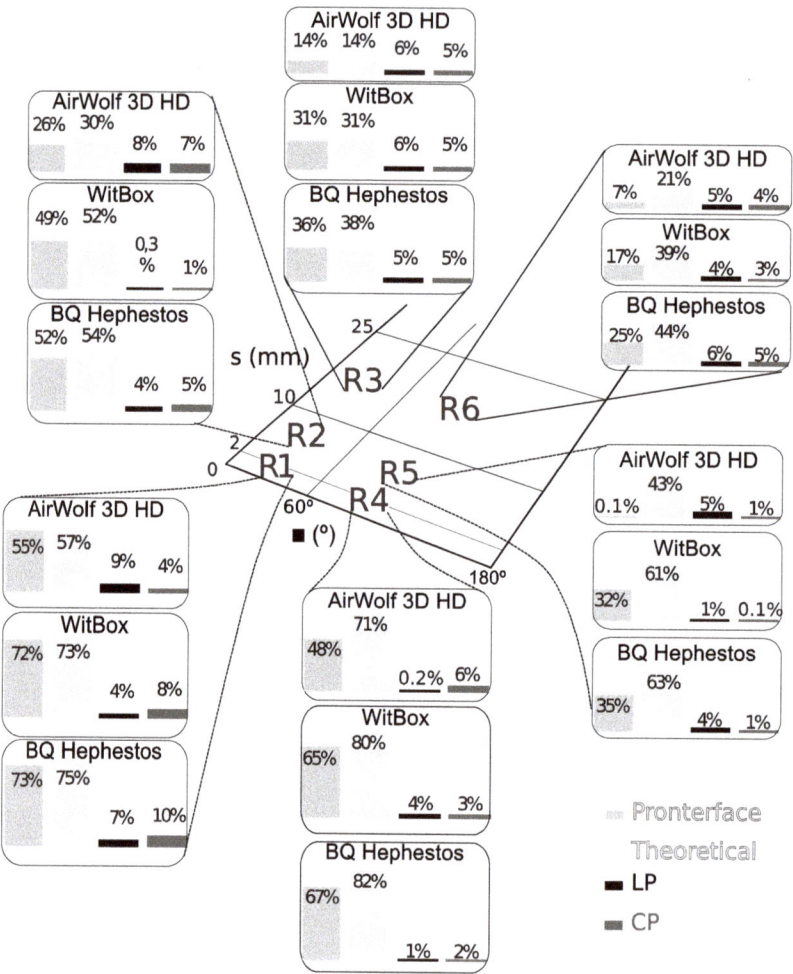

Figure 5. Estimation relative error for six random paths at different regions of the s-α domain.

Finally, comparing the LP and CP approximations in the printers, the Airwolf speed surface has a MAE greater than those obtained for the Witbox and Hephestos printers (see Figures 3 and 4), and this leads to the differences observed in the time estimation error values.

4.2.2. Printing Examples

To assess the performance of the proposed estimations in real parts, 12 prisms are printed (see Table 3). Examples in the previous Section 4.2.1 point out that estimation error depends on the s-α region, but for actual printed parts, s and α values are not concentrated in a specific region and that can reduce the prediction errors. Another fact that contributes to differentiate random paths and printing examples is the programmed printing speed. While for random paths, the programmed speed is 120 mm/s, which is always greater than the experimental maximum speed measured, in the printing examples the programmed speed changes along the printing path and can be beneath the actual maximum speed surface.

Figure 6 portrays the relative errors corresponding to the theoretical estimation (considering the programmed printing speeds), the Pronterface prediction and the estimations provided via our LP and CP surfaces.

According to Figure 6, it is easy to observe the improvement in printing time prediction provided by the proposed approaches in comparison with theoretical and Pronterface estimations. For the 12 printed examples, while theoretical and Pronterface approaches show dispersed error values with maximum values in samples 6 and 12, and minimum values in samples 1 and 4, LP and CP solutions show similar relative errors that never exceed 8.5%.

Comparing the printers, the Hephestos shows the worst results followed by the Witbox, and the Airwolf shows the minimum error. With respect to LP and CP estimations, this result differs from those obtained for the random paths, but it makes sense considering that, for the printing examples, the programmed speed can have a value below the maximum actual speed surface. In this case, the estimations consider a speed equal to the programmed speed (see Equation (6) for LP and CP solutions). Estimation errors decrease with the difference between programmed and average actual speed. Printer acceleration determines that difference, in a manner that, the fastest printers show the lowest estimation errors in the printing examples.

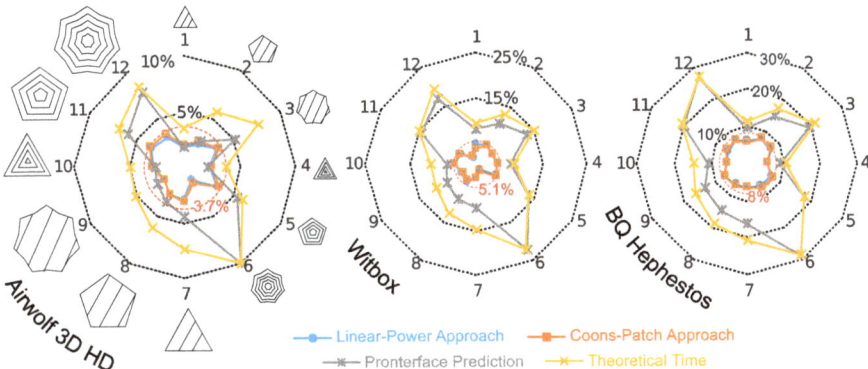

Figure 6. Relative error of the proposed method, theoretical estimation and Pronterface prediction for the twelve printed samples presented in Table 3.

5. Conclusions

The printing speed predictions showed in this paper lead to accurate build time estimations (maximum relative error of 8.5 % in the printed examples). The experimental methodology devised to build a printing speed surface can be straightforwardly applied to any FFF or similar machines, by measurement printing times for different segment lengths and direction changes along linear and circular paths. The proposed fitting procedures, LP and CP approaches, provide good mean printing speed approximations (mean absolute error lower than 2.7 mm/s) even for a reduced number of experimental data (22 measurements for LP and 25 measurements for CP).

The estimation procedure requires to read the G-code that defines the printing path. For each interpolation segment, the method compares and chooses the lowest speed between the programmed and predicted one, and computes the required time to travel the segment length at that speed.

It is noteworthy that the proposed method was successfully applied to three low-cost FFF printers. In the experimental tests accomplished (six random paths and twelve printed prisms), LP and CP estimations provide the minimum errors. In many cases, these errors are well below those provided by theoretical and Pronterface (analytical) estimations. Moreover, for all tested printers, while the theoretical and Pronterface estimation errors show high dispersion, CP and LP errors hardly change.

Hence, the infill pattern and the component shape and size do not modify the accuracy of the proposed approach.

LP and CP surfaces are defined for a specific maintenance state of the printers, and it stands to reason that wears or maintenance problems could increase time estimation error. Further effort is required to study this fact, which could help to find out when to start maintenance tasks in a FFF machine.

Author Contributions: Conceptualization, G.M.-S., R.D.-V. and E.T.-J; methodology, G.M.-S. and R.D.-V.; software, G.M.-S. and R.D.-V.; validation, G.M.-S. and R.D.-V.; formal analysis, G.M.-S. and R.D.-V.; investigation, G.M.-S., R.D.-V., and E.T.-J.; resources, G.M.-S., R.D.-V., E.T.-J., and R.L.-G.; data curation, G.M.-S. and R.D.-V.; writing—original draft preparation, G.M.-S., R.D.-V., E.T.-J., and R.L.-G.; writing—review and editing, G.M.-S., R.D.-V., and E.T.-J.; visualization, G.M.-S., R.D.-V., and E.T.-J.; supervision, R.D.-V. and R.L.-G.; project administration, R.D.-V. and R.L.-G; funding acquisition, G.M.-S., R.D.-V., E.T.-J., and R.L.-G.

Funding: This research received no external funding.

Acknowledgments: This work is supported by the Spanish Ministerio de Economía, Industria y Competitividad, under research grant DPI2015-65472-R.

Conflicts of Interest: The authors declare no conflict of interest.

References

1. Gibson, I.; Rosen, D.; Stucker, B. *Additive Manufacturing Technologies: 3D Printing, Rapid Prototyping, and Direct Digital Manufacturing*; Springer: New York, NY, USA, 2010.
2. Bechthold, L.; Fischer, V.; Hainzlmaier, A.; Hugenroth, D.; Ivanova, L.; Kroth, K.; Römer, B.; Sikorska, E.; Sitzmann, V. *3D Printing: A Qualitative Assessment of Applications, Recent Trends and the Technology's Future Potential*; Studien zum deutschen Innovationssystem: Belin, Germany, 2015.
3. Shewbridge, R.; Hurst, A.; Kane, S.K. Everyday Making: Identifying Future Uses for 3D Printing in the Home. In Proceedings of the 2014 Conference on Designing Interactive Systems, Vancuver, BC, Canada, June 21–25 2014; pp. 815–824.
4. Snyder, T.J.; Andrews, M.; Weislogel, M.; Moeck, P.; Stone-Sundberg, J.; Birkes, D.; Hoffert, M.P.; Lindeman, A.; Morrill, J.; Fercak, O. 3D Systems' Technology Overview and New Applications in Manufacturing, Engineering, Science, and Education. *3D Print. Addit. Manuf.* **2014**, *1*, 169–176. [CrossRef] [PubMed]
5. Boparai, K.S.; Singh, R.; Singh, H. Development of rapid tooling using fused deposition modeling: A review. *Rapid Prototyp. J.* **2016**, *22*, 281–299. [CrossRef]
6. ISO/ASTM52900-15. *Standard Terminology for Additive Manufacturing—General Principles—Terminology*; ASTM International: West Conshohocken, PA, USA, 2015.
7. Schniederjans, D.G. Adoption of 3D-printing technologies in manufacturing: A survey analysis. *Int. J. Prod. Econ.* **2017**, *183*, 287–298. [CrossRef]
8. Wohlers, T.; Campbell, R.; Caffrey, T. *Wohlers Report 2016: 3D Printing and Additive Manufacturing State of the Industry: Annual Worldwide Progress Report*; Wohlers Associates: Fort Collins, CO, USA, 2016.
9. Shanler, M.; Basiliere, P. *Hype Cycle for 3D Printing*; Gartner Research Group: Stamford, CT, USA, 2018.
10. Akande, S.O.; Dalgarno, K.W.; Munguia, J. Low-Cost QA Benchmark for Fused Filament Fabrication. *3D Print. Addit. Manuf.* **2015**, *2*, 78–84. [CrossRef]
11. Llewellyn-Jones, T.; Allen, R.; Trask, R. Curved Layer Fused Filament Fabrication Using Automated Toolpath Generation. *3D Print. Addit. Manuf.* **2016**, *3*, 236–243. [CrossRef]
12. Banerjee, S.S.; Burbine, S.; Kodihalli Shivaprakash, N.; Mead, J. 3D-Printable PP/SEBS Thermoplastic Elastomeric Blends: Preparation and Properties. *Polymers* **2019**, *11*, 347. [CrossRef]
13. Barrios, J.M.; Romero, P.E. Decision Tree Methods for Predicting Surface Roughness in Fused Deposition Modeling Parts. *Materials* **2019**, *12*, 2574. [CrossRef]
14. Pérez, M.; Medina-Sánchez, G.; García-Collado, A.; Gupta, M.; Carou, D. Surface quality enhancement of fused deposition modeling (FDM) printed samples based on the selection of critical printing parameters. *Materials* **2018**, *11*, 1382. [CrossRef]
15. García Plaza, E.; López, P.J.N.; Torija, M.Á.C.; Muñoz, J.M.C. Analysis of PLA geometric properties processed by FFF additive manufacturing: Effects of process parameters and plate-extruder precision motion. *Polymers* **2019**, *11*, 1581. [CrossRef]

16. Oropallo, W.; Piegl, L.A. Ten challenges in 3D printing. *Eng. Comput.* **2016**, *32*, 135–148. [CrossRef]
17. Thompson, M.K.; Moroni, G.; Vaneker, T.; Fadel, G.; Campbell, R.I.; Gibson, I.; Bernard, A.; Schulz, J.; Graf, P.; Ahuja, B. Design for Additive Manufacturing: Trends, opportunities, considerations, and constraints. *CIRP Ann.* **2016**, *65*, 737–760. [CrossRef]
18. Wang, Y.; Blache, R.; Xu, X. Selection of additive manufacturing processes. *Rapid Prototyp. J.* **2017**, *23*, 434–447. [CrossRef]
19. Dey, A.; Yodo, N. A Systematic Survey of FDM Process Parameter Optimization and Their Influence on Part Characteristics. *J. Manuf. Mater. Process.* **2019**, *3*, 64. [CrossRef]
20. Micó-Vicent, B.; Perales, E.; Huraibat, K.; Martínez-Verdú, F.M.; Viqueira, V. Maximization of FDM-3D-Objects Gonio-Appearance Effects Using PLA and ABS Filaments and Combining Several Printing Parameters: "A Case Study". *Materials* **2019**, *12*, 1423. [CrossRef]
21. Kuznetsov, V.E.; Tavitov, A.G.; Urzhumtsev, O.D.; Korotkov, A.A.; Solodov, S.V.; Solonin, A.N. Desktop Fabrication of Strong Poly (Lactic Acid) Parts: FFF Process Parameters Tuning. *Materials* **2019**, *12*, 2071. [CrossRef]
22. Minguella-Canela, J.; Morales Planas, S.; Gomà Ayats, J.; de los Santos López, M. Assessment of the potential economic impact of the use of AM technologies in the cost levels of manufacturing and stocking of spare part products. *Materials* **2018**, *11*, 1429. [CrossRef]
23. Bikas, H.; Stavropoulos, P.; Chryssolouris, G. Additive manufacturing methods and modelling approaches: A critical review. *Int. J. Adv. Manuf. Technol.* **2016**, *83*, 389–405. [CrossRef]
24. Baumann, F.W.; Kopp, O.; Roller, D. Abstract API for 3D printing hardware and software resources. *Int. J. Adv. Manuf. Technol.* **2017**, *92*, 1519–1535. [CrossRef]
25. Di Angelo, L.; Di Stefano, P. A neural network-based build time estimator for layer manufactured objects. *Int. J. Adv. Manuf. Technol.* **2011**, *57*, 215–224. [CrossRef]
26. Zhang, Y.; Bernard, A.; Valenzuela, J.M.; Karunakaran, K. Fast adaptive modeling method for build time estimation in Additive Manufacturing. *CIRP J. Manuf. Sci. Technol.* **2015**, *10*, 49–60. [CrossRef]
27. Zhu, Z.; Dhokia, V.; Newman, S.T. A new algorithm for build time estimation for fused filament fabrication technologies. *Proc. Inst. Mech. Eng. Part B J. Eng. Manuf.* **2016**, *230*, 2214–2228. [CrossRef]
28. Mohamed, O.A.; Masood, S.H.; Bhowmik, J.L. Mathematical modeling and FDM process parameters optimization using response surface methodology based on Q-optimal design. *Appl. Math. Model.* **2016**, *40*, 10052–10073. [CrossRef]
29. Habib, M.A.; Khoda, B. Attribute driven process architecture for additive manufacturing. *Robot. Comput. Integr. Manuf.* **2017**, *44*, 253–265. [CrossRef]
30. Komineas, G.; Foteinopoulos, P.; Papacharalampopoulos, A.; Stavropoulos, P. Build Time Estimation Models in Thermal Extrusion Additive Manufacturing Processes. *Procedia Manuf.* **2018**, *21*, 647–654. [CrossRef]
31. Pronterface. Printrun: Pure Python 3d Printing Host Software. Available online: http://www.pronterface.com (accessed on 1 April 2019).
32. Coelho, R.T.; de Souza, A.F.; Roger, A.R.; Rigatti, A.M.Y.; de Lima Ribeiro, A.A. Mechanistic approach to predict real machining time for milling free-form geometries applying high feed rate. *Int. J. Adv. Manuf. Technol.* **2010**, *46*, 1103–1111. [CrossRef]
33. Jin, Y.; He, Y.; Du, J. A novel path planning methodology for extrusion-based additive manufacturing of thin-walled parts. *Int. J. Comput. Integr. Manuf.* **2017**, *30*, 1301–1315. [CrossRef]
34. Farin, G.E. *Curves and Surfaces for CAGD. A Practical Guide*; Morgan Kaufmann: San Francisco, CA, USA, 2002.
35. JCGM. Evaluation of Measurement Data—Guide to the Expression of Uncertainty in Measurement. Available online: https://www.bipm.org/utils/common/documents/jcgm/JCGM_100_2008_E.pdf (accessed on 2 September 2019).
36. Galicia, J.A.G.; Benes, B. Improving printing orientation for Fused Deposition Modeling printers by analyzing connected components. *Addit. Manuf.* **2018**, *22*, 720–728. [CrossRef]
37. Piili, H.; Happonen, A.; Väistö, T.; Venkataramanan, V.; Partanen, J.; Salminen, A. Cost estimation of laser additive manufacturing of stainless steel. *Phys. Procedia* **2015**, *78*, 388–396. [CrossRef]

38. Brajlih, T.; Valentan, B.; Balic, J.; Drstvensek, I. Speed and accuracy evaluation of additive manufacturing machines. *Rapid Prototyp. J.* **2011**, *17*, 64–75. [CrossRef]
39. Henrique Pereira Mello, C.; Calandrin Martins, R.; Rosa Parra, B.; de Oliveira Pamplona, E.; Gomes Salgado, E.; Tavares Seguso, R. Systematic proposal to calculate price of prototypes manufactured through rapid prototyping an FDM 3D printer in a university lab. *Rapid Prototyp. J.* **2010**, *16*, 411–416. [CrossRef]

© 2019 by the authors. Licensee MDPI, Basel, Switzerland. This article is an open access article distributed under the terms and conditions of the Creative Commons Attribution (CC BY) license (http://creativecommons.org/licenses/by/4.0/).

Article

Advances in Orthotic and Prosthetic Manufacturing: A Technology Review

Jorge Barrios-Muriel [†], Francisco Romero-Sánchez [*,†], Francisco Javier Alonso-Sánchez [†] and David Rodríguez Salgado [†]

Department of Mechanical Engineering, Energy and Materials, University of Extremadura, 06006 Badajoz, Spain; jorgebarrios@unex.es (J.B.-M.); fjas@unex.es (F.J.A.-S.); drs@unex.es (D.R.S.)
* Correspondence: fromsan@unex.es; Tel.: +34-924-289-600
† Current address: Escuela de Ingenierías Industriales, Universidad de Extremadura, Avda. de Elvas s/n, 06006 Badajoz, Spain.

Received: 31 October 2019; Accepted: 31 December 2019; Published: 9 January 2020

Abstract: In this work, the recent advances for rapid prototyping in the orthoprosthetic industry are presented. Specifically, the manufacturing process of orthoprosthetic aids are analysed, as thier use is widely extended in orthopedic surgery. These devices are devoted to either correct posture or movement (orthosis) or to substitute a body segment (prosthesis) while maintaining functionality. The manufacturing process is traditionally mainly hand-crafted: The subject's morphology is taken by means of plaster molds, and the manufacture is performed individually, by adjusting the prototype over the subject. This industry has incorporated computer aided design (CAD), computed aided engineering (CAE) and computed aided manufacturing (CAM) tools; however, the true revolution is the result of the application of rapid prototyping technologies (RPT). Techniques such as fused deposition modelling (FDM), selective laser sintering (SLS), laminated object manufacturing (LOM), and 3D printing (3DP) are some examples of the available methodologies in the manufacturing industry that, step by step, are being included in the rehabilitation engineering market—an engineering field with growth and prospects in the coming years. In this work we analyse different methodologies for additive manufacturing along with the principal methods for collecting 3D body shapes and their application in the manufacturing of functional devices for rehabilitation purposes such as splints, ankle-foot orthoses, or arm prostheses.

Keywords: rapid prototyping; additive manufacturing; orthoses; prostheses; fused deposition modeling; laminated object manufacturing; selective laser sintering

1. Introduction

Assistive technologies, such as orthotic or prosthetic devices, have existed for many centuries. Orthotic devices have been widely used not only to provide immobilization, support, correction, or protection, but also to treat musculoskeletal injuries or dysfunctions [1]. In the 1970s, new techniques like plastic coating were developed, due to the demand of orthotic devices with a more attractive appearance, by applying a tinted rubber-based plastic film [2], allowing the improvement of orthoses appearance and comfort. In the early 1980s the rise of additive manufacturing technologies (AMT), popularly known as 3D printing technologies in a manufacturing environment, with the introduction of the stereolithography technique, based on the cure of photopolymer resin in thin layers with a UV laser allowed construction of 3D models. In the following years, other AMT were introduced, such as: fused deposition modeling (FDM), laminated object manufacturing (LOM), selective laser sintering (SLS), 3D printing, and variable rapid prototyping (Polyjet Technology), among others.

The AMTs are included in the field of rapid prototyping techniques (RPT), producing fully functional parts directly from a three-dimensional model without a machining process. A radical change in manufacturing of orthotic devices is already happening due to the exponential growth of RPT over recent decades [3,4]. A quick search in Scopus of 3D printing provides only 122 results before year 2000, 303 results between 2001 and 2005, 756 results from 2006 and 2010, 4521 results for the period 2011–2015, and 22,513 results in the last five years. In the biomedical engineering context, the developments have evolved quickly due to the need for individualized devices able to adapt properly to the patient's anatomical shapes [5–9]. For this reason, RPT may be helpful in the orthoprosthetic industry, as these devices must adapt perfectly to the body, not only to accomplish their rehabilitative function, but to avoid disuse, as many of these devices produce blistering, ulcers, or discomfort [10,11]. These techniques have been already applied to the manufacturing of spinal braces [11,12], exoskeleton parts [13,14], and passive orthoses [15–17], and the application in the medical and dental industry represents one of largest serving industries in the world [18]. Moreover, RPT offer advantages in the design of custom orthotic devices (Figure 1): The orthotic and prosthetic devices are highly customizable, as in Zuniga et al. [19], it is possible to fit the devices to complex geometrical features, with high accuracy, and these devices are manufactured efficiently in terms of cost, lead-time, and product quality [20].

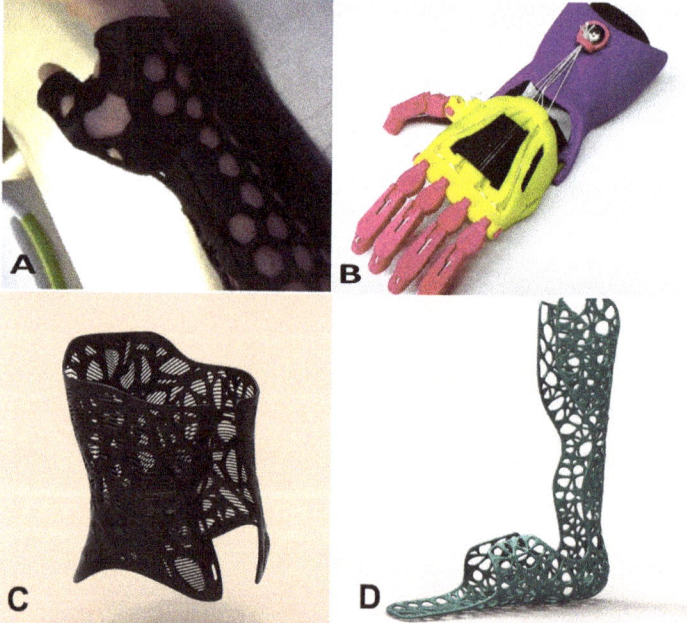

Figure 1. Examples of 3D printed orthotics (**a**) Forearm static fixation (courtesy of Fitzpatrick et al. [21]). (**b**) Cyborg beast hand prosthesis—a low-cost 3D-printed prosthetic hand for children licensed under the CC-BY-NC license (courtesy of Zuniga et al. [19]). (**c**) Spinal brace (courtesy of Andiamo company [22]). (**d**) Ankle-foot orthosis (courtesy of Andiamo company [22]).

Currently, most rehabilitation devices are designed and hand-crafted by orthopaedists. Therefore, the quality of the product depends on the specialists' skills and experience [23]. The manufacturing process requires time and depends on the expertise of the specialist to obtain products with functional features that match the unique gait dynamics of each subject. Thus, the need for custom-made products such as orthoses and assistive devices is an explicit need considering the evolution of the technology during the beginning of this century [6,24,25].

Regarding orthoprosthetic manufacturing, the first step is to acquire the morphology of the body segment. In the traditional manufacturing process, the subject's morphology is usually acquired by means of foam or plaster moulds. A prototype is obtained by using a computerized numerical control (CNC) or a milling machine in the thermosetting polyurethane model obtained by the mould. Lastly, the specialist performs several modifications in the device to adjust it to the subject (Figure 2a). However, CNC and milling present some limitations as they are not able to reproduce complex surface designs or to deal with different thickness and materials.

On the contrary, in the new RPT approach, the manufacturing process (Figure 2b) begins with the acquisition of subject's morphology by means of 3D scanning technologies. Then, computer aided design (CAD)-computed aided engineering (CAE) tools are applied to obtain subject specific designs; whereas, functionality is studied by testing different materials and structures. Lastly, the design is easily exported to an additive manufacturing machine where the prototype is obtained. Manufacturing time may vary between several weeks in the traditional process, to a couple of days in the RPT approach. Therefore, the use of RPT, together with the new 3D acquisition methodologies, represents an alternative in the orthoprosthetic industry.

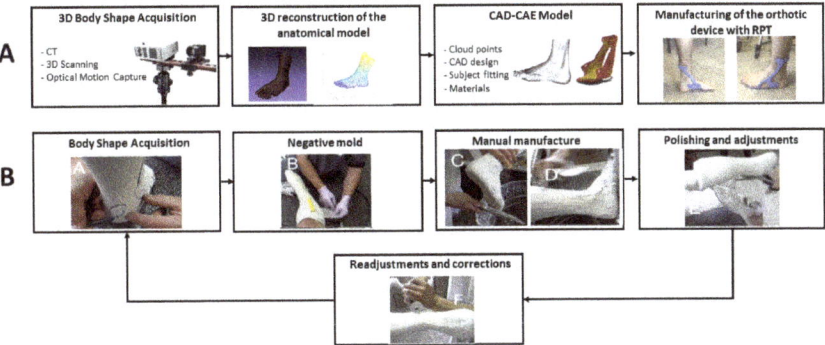

Figure 2. Phases of the manufacturing process of custom-fit orthotic devices. (**A**) Rapid prototyping techniques (RPT) methodology (courtesy of J. Barrios-Muriel). (**B**) Traditional methodology (courtesy of Mavroidis et al. [26] under CC-BY License.) Computer aided design (CAD)-computed aided engineering (CAE), computed tomography (CT).

Many other applications of additive manufacturing (AM) and RPT are in the field of manufacturing of medical instruments [27], drug delivery systems [28], engineered tissues [29,30], scaffolds for bone regeneration [31,32], dental implants [33,34], prosthetic sockets, [35] or surgery [36–38]. In this work, we present a review of the developments in the manufacturing of orthotic devices, especially those related to the use of RPT to improve the quality and manufacture times in the rehabilitation field, as in splints, ankle-foot orthoses, or arm prostheses. This review provides comprehensive coverage on the different methodologies ready to be used in the orthotic and prosthetic industry. New 3D data acquisition techniques and the use of different materials are also referred to. This work is intended to be a reference guide on the techniques in this field for practitioners, but also for experienced readers who are interested in pursuing further research.

2. 3D Anatomical Data Acquisition Technologies

Applications of RPT combined with different techniques for measuring and modelling the human body are useful to generate new criteria for the orthotic device design. Depending on the data acquisition method used, the data can be expressed as a point cloud, voxels (3D volumetric pixels), or three dimensional coordinates of different anatomical points. Up to date, there is no standardized morphology acquisition procedure, however, there are several acquisition methods to support fabrication using RPT within the field of orthotic devices modelling, including computer tomography, 3D scanning, and different optical motion capture systems.

2.1. Computed Tomography

Computed tomography (CT) is a powerful technique to facilitate diagnostics and for surgical planning. Traditionally, the recorded images were in the axial or transverse plane. Currently, modern scanners record images along different planes, enabling volumetric reconstructions for 3D representations. Several studies have applied CT for the manufacture of orthotic devices. For example, Tang et al. [39] recently proposed the use of CT combined with AM techniques to manufacture insoles for diabetes. In their work, they studied pressure and tissue strain along the plantar foot to correlate these variables with the therapeutic effect of footwear and custom-made orthotic inserts, being able to reduce peak plantar pressure by 33.67%. Artioli et al. [40] studied the use of different acquisition techniques to manufacture 3D printed silicone ear prostheses, concluding that the use of CT and AM (using polylactic acid or polylactide, PLA, resolution 100 µm) produce differences of 0.1% between the manufactured prosthesis and the objective model. Liacouras et al. [41,42] used CT to acquire the morphology of the patients stump and to develop the strategies for designing transtibial prosthetic sockets. Moreover, the data of CT allowed a finite element analysis of the prosthetic model socket to calculate the structural stresses and strain at the sockets, as well as the contact pressure at the fibula head. The high image resolution between tissues is one of the greatest advantages of CT, along with the capacity to improve contrast and reduce noise. However, several drawbacks are worthy of mentioning. Radiation is the main concern and the exposition is directly proportional to the duration of the scanning. Other drawbacks are the partial pixel effect, leading to a blurred boundary, as the different densities share common pixels [43].

2.2. 3D Scanning

To capture human topography or the external shape, 3D scanning arise as the most practical and comfortable solution. 3D scanning systems use light based techniques to determine the three-dimensional position in space of the different points that integrate the surface of an object. Computer software is then used to reconstruct surfaces from the point cloud and then, the CAD model is obtained.

Currently, 3D scanners for human measurement are available, including the use of single image for reconstruction, structured light technologies, lasers, and different algorithms for stereo reconstruction [44–46]. The most common technologies used to reconstruct human body shape are laser and structured light technologies [44]. The laser technique uses a projected laser dot or line from a hand-held device. A sensor measures the distance to the surface, typically a charge-coupled device or a position sensitive device. For static objects, data is collected in relation to an internal coordinate system and, for dynamic conditions, the position of the scanner must be determined to correctly define the point cloud [47]. Structured light methods use a projector-camera system with pre-defined light patterns projected on the moving object. However, a drawback of this technology is the inability to capture certain topography sections of human anatomy with intricate creases and folds, such as between fingers when the hand is in a neutral position, the back of the knee when flexed, or the armpits. The gathered information is more precise, however, and noise is reduced. Recent techniques explore the feasibility of 4D acquisition [48], but to the best knowledge of the authors, there is no report on its use for the design of orthotic and prosthetic devices yet.

Processing time is significantly reduced compared to magnetic resonance imaging (MRI) and CT, as well as the size of the data files [44]. The use of MRI and CT is mostly used to reconstruct of internal organs or tumours with high accuracy for surgical guidance [49]. Recording time and resolution may vary between different 3D scanners, ranging from 3–5 min and a tenth of millimetres for high accuracy systems to a couple of minutes [50] and millimetres for low cost systems [51]. Other advantages of 3D scanning methods are affordable hardware and software, minimal training requirement, availability, accessibility, and efficiency [44].

Several authors suggest the use of reverse engineering software to obtain a refined model by repairing the data. In the pioneering work of Chee Kai et al. [52], a 3D scanning method was selected over the traditional methods for prosthesis modeling, such as plaster-of-Paris impressions, MRI, and CT. Mavroidis et al. [26] used 3D laser scanning to create patient-specific foot orthoses. Surface data of the patient anatomy was manipulated to an optimal form using computer aided design (CAD) software and was fabricated using a rapid prototyping machine. The prototype properly fit the subject's anatomy compared to a commercial ankle foot orthosis. Paterson [53] investigated the 3D anatomical data acquisition methods to establish a clinically valid, standardized method. He concluded that laser scanning appears to be the most suitable method to reduce the acquisition of ambiguous data and with a high performance in terms of cost, resolution, speed, accuracy, patient safety, cost, and overall efficiency. More recent works, such as those of Mali and Vasistha [54] or Agudelo-Ardila et al. [55], present efficient solutions for the manufacturing of lower and upper limb orthoses, respectively, using reverse engineering.

2.3. Optical Motion Capture System

Recently, techniques to measure the topography of the human body in dynamic movements are receiving attention, as the design of orthotic devices should not be designed only for static conditions, as most of them will be used in dynamic conditions to increase rehabilitation. As previously mentioned, there are many commercial 3D systems that are able to measure 3D shapes with high accuracy; however, most of them cannot acquire human motion. An optical motion capture system is a popular technology to capture human movement. These optical systems use several cameras recording in 2D to reconstruct the 3D position of a set of reflective markers placed in anatomical landmarks. These markers should be seen by two or more cameras calibrated to provide overlapped projections. However, the optical motion capture system has a drawback due to the strong limitations related to the number and density of markers [56]. Although only three markers are needed to reconstruct each body segment as a rigid body (e.g., to perform a kinematic analysis), this number is increased if body shape must be also retrieved. The number usually depends on the camera resolution but it does not exceed 60–80 markers per body segment. The main application of the marker-based acquisition technology is in the assessment of the manufactured devices due to the standardized protocols to acquire kinematics [57].

Research has also been focused into structured light using this method [58]. Unkovskiy et al. [59] used a portable structured light scanner to retrieve the topology of the nose cavity and the face to design and manufacture a prosthetic nose. A portable projector was used to project in the region of interest an arbitrary light pattern with a colour code. The system performs the triangulation between the projection pattern and the camera image and to retrieve the correspondence between images. The advantage of this system is the noise reduction compared to the video capture image. The matching of the stereo projection pattern and the camera recorded image is less affected by noise than multiple stereo matching of camera images. Regretfully, synchronous measurement of the entire body shape using multiple projector-camera systems has not been reported yet using this technology. This is essential for capturing the 3D shape motion and to analyse, in the case of gait for example, foot width, length, circumferences, and arch changes during movement [60]. However, there is no commercial system including this technology and thus, more studies should be performed with different methods to compare accuracy and precision between these technologies. The cutting edge technologies in this field are the markerless systems. Chatzitofis et al. [61] proposed recently a low-cost robust and fast system to acquire body kinetics. Although this work still uses reflective straps it could be combined with 4D scanning systems, such as those proposed by Joo et al. [62], to obtain accurate topology and motion of the body segment to rehabilitate.

3. Rapid Prototyping Technologies for Orthotic Devices

The combination of CAD and computed aided manufacturing (CAM) is a well-known approach that is receiving increasing attention in the field of orthotic devices to replace the traditional craft practices. As stated by Ciobanu et al. [63], customized manufacturing through RPT requires: 3D scanning of the anatomic surface, 3D surface reconstruction, CAD modeling, conversion to stereolithography format (STL) and, lastly, machining using a special rapid prototyping machine (i.e., a 3D printer) controlled through computer. RPT offer advantages in manufacturing processes of custom-fit orthotic devices, in terms of greater design freedom, ability to create functional elements, superior accuracy and cost efficiency, shorter delivery time, and better user experience of the final product.

In an RPT manufacturing process, a representative virtual 3D CAD model is formed layer upon layer to form a physical object [9]. In RPT, a virtual model of the part is designed through CAD and is converted to a STL file format, which is the default standard file format for RP systems [64,65]. AMT can be categorized in different ways depending on the nature of the fabrication process, such as laser, printer technology, and extrusion technology [7]. There are many different AM processes. Kruth [66] proposed in the early 90's the use of different methodologies for additive manufacturing, classified according to the material used for the prototype (see Table 1). Nevertheless, Paterson et al. [67] demonstrated that only a few of them could be used to manufacture orthotic and prosthetic devices.

Liquid based process, such as stereolitography (SLA), solid ground curing (SGC), UV light curing (ULC), and ballistic particle manufacturing (BPM); or solid based process, such as laminated object manufacturing (LOM) are methodologies used in the manufacturing of orthoses and prostheses. Nevertheless, the most used methodologies to manufacture orthotic and prosthetic devices are fused deposition modelling (FDM), selective laser sintering (SLS), and powder bed and inkjet head 3D printing (3DP). These methodologies represent an optimized trade-off between cost, delivery time, accuracy, and comfort.

Table 1. Rapid prototyping techniques available for orthoprosthetics.

Material	Process
Liquid base	Stereolithography (SLA) Solid ground curing (SGC) UV light-curing (ULC) Ballistic particle manufacturing (BPM)
Solid base	Laminated object manufacturing (LOM) Fused deposition modeling (FDM)
Powder base	Selective laser sintering (SLS) 3D printing (Polymer injection)

3.1. Fused Deposition Modeling (FDM)

In the FDM process (see Figure 3a), a semi-molten material is extruded through an extrusion head that traverses in the X and Y axes to create each two dimensional layer of the piece to be manufactured. Two extrusion nozzles compose the movable extrusion head: one to deposit the build material and the other one that contains the support material [68].

In general, the perimeter of each layer is extruded first and then the delimited zone by the previous extrusion is filled by the extruder head by following a pre-defined pattern [68]. Once the layer is completed, the support platform lowers and another layer is extruded. The process continues layer by layer until the piece is complete.

The most common materials for FDM are polycarbonate (PC) and Acrylonitrile butadiene styrene (ABS) or a mixture of them. These materials have similar properties to thermoplastic material for injection moulding [69]. Other materials such as polymers or nylon-based materials may be used. The main advantage of FDM technology is in the use of low-cost materials. Tan et al. [70] were pioneers in the use of used FDM for tibial prosthesis manufacturing and concluded that the functional characteristics of prosthesis were valid for clinical purposes. On the contrary, manufacturing times are high. Since then, an increasing number of applications have arisen for FDM in the biomedical field for upper [17] and lower limb orthoses [50,71], hand prostheses [19], facial prosthesis [40,72], and drug delivery systems [73].

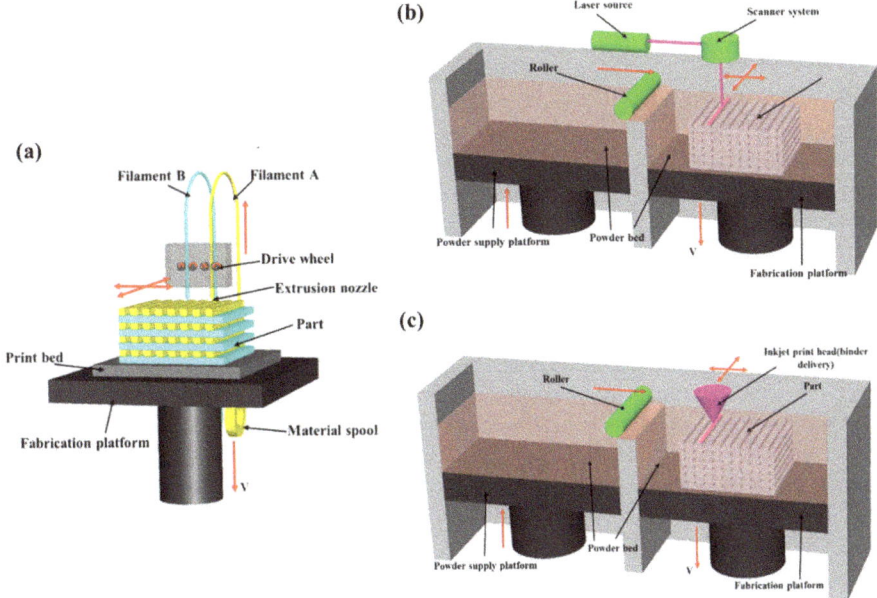

Figure 3. Comparison of the proposed schemes dor rapid prototyping. (**a**) Fused deposition modeling (FDM). (**b**) Selective laser sintering (SLS). (**c**) 3DP. Image adapted from Wang et al. [74] with permission of Elsevier Ltd.

3.2. Selective Laser Sintering (SLS)

The DTM Corporation (now a part of 3D Systems) introduced the first SLS system in the 90s [75]. The SLS technique creates three-dimensional solid objects or parts by selectively fusing powdered polymer-based materials such as nylon/polyamide with a CO_2 laser, turning powder material into solid objects (Figure 3b). A CO_2 laser selectively sinters defined regions by traversing across the powder bed in the X and Y-axes to form a 2D profile [76]. Once the 2D profile has been completed, the platform lowers, a new layer of powder is distributed and the sintering process is repeated. Subsequently, the process is termed a powder-based fusion process. In general, all materials used are thermoplastics, the most common being polyamide 12 (PA), acrylonitrile butadiene styrene (ABS), and polycarbonate (PC) [77]. These materials lead to a considerable weight reduction improving the usability of the rehabilitation devices.

As an example of the use of this technology in the manufacturing of custom orthoses, Schrank and Stanhope [10] evaluated the accuracy of SLS manufacturing process of ankle foot orthoses (AFO). In this work the discrepancy between the CAD model and final product manufactured with SLS was measured through the Faroarm 3D scanner (accuracy ±25 μm). The results showed values below 1.5 mm (SD = 0.39 mm). Deckers et al. [78] developed and tested an SLS-based AFO, highlighting the need to properly characterize the mechanical characteristics of the AFO such as strength, fatigue, and resistance to impacts. Vasiliauskaite et al. [79] tested a polyamide-based orthoses manufactured with SLS and concluded that the features were similar to a thermoformed polypropylene orthosis, the first one being stiffer than the second but enough for the purpose of rehabilitation.

On the other hand, [80] manufactured a splint for upper extremities using the SLS method. They also concluded that the results of this manufacturing technique were good but did not make any clinical validation of the device. Another similar application of splints using SLS is the design proposed by Evill [81]. In this work, several aspects such as ventilation, hygiene, and aesthetics were improved through CAD. Although there are not conclusive results that confirm these purposes, the new design parameters considerably improve comfort.

3.3. Powder Bed and Inkjet Head 3D Printing: 3DP

3DP refers to three dimensional printing, which is understood as the process in which the manufactured product is made by means of powder layers stuck with adhesive. In this process, first, a powder layer is spread on the build platform. Second, a liquid binder is deposited selectively through an inkjet printhead by following a patterned layer in the XY plane. Once the 2D pattern is formed, the platform lowers, the next powder layer is spread and so on. This process is sometimes referred as powder bed and inkjet head 3D printing (or 3DP). It should not be confused with the widespread definition of 3D printing, that involves all additive manufacturing process that result in the manufacturing of tree-dimensional objects. The 3DP process is somewhat similar to SLS (see Figure 3b,c): In 3DP, a printing head places liquid adhesive in the material; whereas, in SLS a CO_2 laser is used to fuse the layers. The accuracy of this process is lower than in SLS; nonetheless, this method is preferred due to its low cost and quickness. These qualities have lead 3DP to have a predominant role in the prototyping industry.

The employed materials (mainly thermoplastics as ABS) have the required properties to be used in orthotic and prosthetic applications. Herbert et al. [82] investigated whether this technology was suitable to produce functional prosthesis, and they suggested that, although the manufacturing levels were limited, patients felt more comfortable with prostheses made with 3DP machine (Corporation Z402) than the traditional handmade ones. Regretfully, the resistance was not studied in that work and therefore the durability of the product is unknown. Saijo et al. [83] used this technology to develop patient-specific maxilofacial implants reporting a reduction in operation times. As stated by Ventola [84], 3DP is particularly interesting in tissue engineering and regenerative medicine because of its digital precision, control, and versatility.

To summarize, Table 2 shows a comparison of the different RPT by using the commercial models used in the works described in the bibliography.

Table 2. Characteristics of the most used machines for AMT.

	FDM	SLS	3DP
Model	Dimension STT 768	spro 60 SD SLS	uPrint System
Production Time (h)	7	3	7
Active volume (mm)	203 × 203 × 305	381 × 330 × 457	203 × 152 × 152
Material	ABS P400	Duraform PA (Nylon 12)	ABS P430
Material consumption (g)	40	20.15	55
Cost ($/kg)	190	90	30

4. Variable Property Rapid Prototyping

Variable rapid prototyping differs from other AMT as it aims at producing objects of varied properties. In this sense, each material used to manufacture the object provides specific values of strength, strain, heat deflection temperature, etc. [85]. Object geometries recently introduced 3D printers that use polyjet Matrix^{TM} technology to allow the generation of composite material prototypes of varying stiffness and dual material prototypes. The Connex500^{TM} 3D printer operates by using inkjet heads with two or more photopolymer materials. The material is extruded in 16 μm thick layers. Each photopolymer layer is cured by UV light immediately after the extrusion [86].

A carpal skin was produced by Oxman [80], exploring the multiple material building capabilities available with this technology. Campbell et al. [87] explored the benefits of multiple materials integrated into a wrist splint compared to traditional custom-fitted wrist splints of qualified and experienced clinicians. The work focused on the attempt to place multiple materials to behave as hinges or cushioned features as opposed to traditional fabrication processes where a similar approach would be very difficult to replicate. A drawback for this technology is that the actual commercial CAD software is not efficient to apply the design potential and few computation tools manage the physical interaction between material properties. Nevertheless, recent advances in the use of additive manufacturing of hybrid composites, as recently presented for dental implants [88], may lead to a substantial revolution in the field of orthotics, were hybrid exoskeletons are currently changing from rigid structures to wearable garments.

5. Material Selection for Orthotic Devices

The choice of material when designing an orthotic device is vital to its success. Physical properties of the orthotic materials include their elasticity, hardness, density, response to temperature, durability, flexibility, compressibility, and resilience [89]. It should be mentioned that each physical property cannot be used alone as a single factor for assessing materials for orthotic devices. A hard material as well as an incorrect aspect in the design may result in an uncomfortable device or a biomechanically detrimental orthotic device. Thermoplastics, composites, and foams are the main materials used to manufacture orthotic devices through AMT [89].

The most known materials used in rapid prototyping manufacturing are ABS (Acrylonitrile butadiene styrene) and PLA (polylactic acid) [90]. ABS is a polymer commonly used to produce car bumpers due to its toughness and strength. PLA is a biodegradable thermoplastic that has been derived from renewable resources such as starch prepared from the grains of corn. These materials are used for the majority 3D printing machines. Rigid and semi-rigid structures can be manufactured with these materials. Depending on the designed thickness, these materials may have different properties.

Soft parts and some semi-rigid parts of orthotic devices are made with foamed materials, usually with open or closed cell structure. The first type allows the movement of gas between the cells whereas the second encloses the gas within the cell walls allowing for a water-tight material. This is desirable for orthotic manufacture as sweat will not be able to penetrate into the material to cause premature degradation. Paton et al. [91] investigated the physical properties of soft materials used to fabricate orthoses designed for the prevention of neuropathic diabetic foot ulcers. They concluded that the most clinically desirable dampening materials tested were Poron®96 and Poron®4000 (thickness of 6 mm) and the material with the best properties for motion control was ethylene vinyl acetate (EVA). With the evolution of AMT, insoles with variable porous structure and adjustable elastic modulus are being manufactured to adapt to the different needs of patients with diabetes [92].

In the SLS manufacture process, polymer powders and ceramics are mixed to form composites. Rilsan™ D80 DuraForm™ PA and DuraForm™ GF are examples of materials used in SLS techniques. Faustini et al. [93] explored the feasibility of using RPT for AFO manufacturing process. The study determined that the optimal SLS material for AFO to store and release elastic energy was Rilsan™ D80, considering minimizing energy dissipation through internal friction is a desired material characteristic. A relatively recent work from Walbran et al. [94] compares yield stress of custom made orthoses made by SLS of nylon with carbon fibres and FDM with PLA. Although PLA was chosen in terms of mechanical properties and cost, they concluded that the AMT have a strong potential not only to obtain consistent and repeatable models of the subjects affected limb independently of the operator's skill and restraint of the subject, but to further automate the AFO design process.

Materials used in FDM are those with similar properties to thermoplastic materials for injection moulding. In general polycarbonate (PC) and ABS or their combinations, as mentioned above (PC-ABS, PC-ISO or ABSi), nylon-based materials and other polymers can be used. The main advantage of these materials is the low cost. Finally, in the case of variable properties rapid prototyping base materials, digital materials or composite digital materials can be used. The material properties can be modified by combinations and distribution of different types.

6. Discussion

The traditional manufacturing processes for orthotic and prosthetic devices is still mostly hand-crafted and requires special abilities from the othopaedist to obtain a quality product. Nevertheless, this manufacturing process, in general, produces discomfort to the patient. The acquisition of the morphology of the subject is not a clean process as the use of plaster is required to obtain the mould. Additionally, the final product may produce blistering on the subject's skin, as the morphology is acquired in static conditions.

Alternatively, RPT and AMT have a strong potential to change not only the way in which orthotic and prosthetic products are designed, but also the manufacturing process and the specialist profile. The use of RPT in the orthoprosthetic industry may suppose a considerable change in the know-how; however, it also leads to important benefits. The goal is to accelerate the reconstruction process of 3D anatomical models and biomedical objects for the design and manufacturing of medical products and simulate 3D body shape to design the most suitable orthoses for the patient. Thus, the use of CAD and RPT facilitate the design and fabrication of custom-fit orthotic products with a number of advantages over traditional methods: the use of new materials, customized designs, virtual testing, etc.

The application of these technologies may lead to a significant improvement in the orthotic manufacturing process as production times are lower, morphology acquisition is faster and more pleasant for the patient, as plaster moulds are suppressed and manufacturing errors are minimized. Considerable effort has been applied in the application of AMT to the medical industry and specifically in the design and manufacturing of orthoses and prostheses for rehabilitation purposes [89], to mitigate the effects of aging [95], in the design of active wearable exoskeletons [96], and also to bring this technology closer to the general public [19,97].

The inclusion of these manufacturing methods requires a high investment in equipment, materials, and training that may cause hesitation from investors or orthotic and prosthetic technicians. Moreover, healthcare specialists show some reservations to change their own work routines [86]. Nevertheless, the eruption of 3D printers into the market and the continuous improvements made in this field will give an impulse to the implementation of this technology in the orthoprosthetic industry. In addition, the experience shown in other medical fields, such as in dental implant manufacturing will make possible the implementation of this technology to produce orthoses and prosthesis to reduce waiting lists.

The mass production of these aids must include a series of key points leading to high quality patterns in the manufacturing process. These key points are: acquisition of the subject's anthropometric data, product design, material selection, manufacturing process planning, and product service, among others. The correct application of these steps reduces the total production time and, therefore, the delivery times. Thus, the inclusion of RPT to produce high quality and short delivery time orthotic and prosthetic devices, that also satisfies functionality and patients comfort is now a reality. Finally, the inclusion of manufacturing technology in a traditional environment such as the orthoprosthetic industry will lead to better products to satisfy the specifications required in rehabilitation.

7. Conclusions

In this work, a review of the different RPT applied to the orthoprosthetic industry has been presented. Specifically, the manufacturing process to manufacture orthoses and prostheses have been analysed and the main works in this field have also been presented. These techniques have been shown to have an exponential growth in the following years in the biomedical field. The new advances in the subject's morphology acquisition as well as the use of RPT can improve the accuracy of the final device, leading to a better rehabilitation process. RPT will help us to optimize the manufacturing process and improve both the design and functionality of assistive devices. Thus, RPT combined with CAD-CAM tools provide a major control in the design and manufacture processes. Finally, the future lines of development in this field will be based on the design of new structures and materials to improve comfort, which will grant the success of the new orthoprosthetic aids.

Author Contributions: Conceptualization, J.B.-M. and F.R.-S.; Formal analysis, J.B.-M.; Funding acquisition, F.J.A.-S. and D.R.S.; Methodology, J.B.-M. and F.R.-S.; Project administration, F.J.A.-S. and D.R.S.; Supervision, F.R.-S., F.J.A.-S. and D.R.S.; Writing—original draft, J.B.-M.; Writing—review & editing, F.R.-S., F.J.A.-S. and D.R.S. All authors have read and agreed to the published version of the manuscript.

Funding: This research was funded by the Consejería de Economía e Infraestructuras de la Junta de Extremadura and the European Regional Development Fund "Una manera de hacer Europa" under project IB18103.

Conflicts of Interest: The authors declare no conflict of interest.

References

1. Webster, J. *Atlas of Orthoses and Assistive Devices*; Elsevier: Amsterdam, The Netherlands, 2019.
2. Meyer, P.R., Jr. Lower limb orthotics. *Clin. Orthop. Relat. Res.* **1974**, *102*, 58–71. [CrossRef] [PubMed]
3. Mikołajewska, E.; Macko, M.; Szczepański, Z.; Mikołajewski, D. Reverse Engineering in Rehabilitation. In *Encyclopedia of Information Science and Technology*, 4th ed.; IGI Global: Hershey PA, USA, 2018; pp. 521–528.
4. Jiang, R.; Kleer, R.; Piller, F.T. Predicting the future of additive manufacturing: A Delphi study on economic and societal implications of 3D printing for 2030. *Technol. Forecast. Soc. Chang.* **2017**, *117*, 84–97. [CrossRef]
5. Singh, S.; Ramakrishna, S. Biomedical applications of additive manufacturing: present and future. *Curr. Opin. Biomed. Eng.* **2017**, *2*, 105–115. [CrossRef]
6. Gibson, I.; Srinath, A. Simplifying medical additive manufacturing: Making the surgeon the designer. *Proced. Technol.* **2015**, *20*, 237–242. [CrossRef]
7. Thompson, M.K.; Moroni, G.; Vaneker, T.; Fadel, G.; Campbell, R.I.; Gibson, I.; Bernard, A.; Schulz, J.; Graf, P.; Ahuja, B.; et al. Design for Additive Manufacturing: Trends, opportunities, considerations, and constraints. *CIRP Ann.* **2016**, *65*, 737–760. [CrossRef]
8. Espalin, D.; Arcaute, K.; Rodriguez, D.; Medina, F.; Posner, M.; Wicker, R. Fused deposition modeling of patient-specific polymethylmethacrylate implants. *Rapid Prototyp. J.* **2010**, *16*, 164–173. [CrossRef]
9. Lantada, A.D.; Morgado, P.L. Rapid Prototyping for Biomedical Engineering: Current Capabilities and Challenges. *Ann. Rev. Biomed. Eng.* **2012**, *14*, 73–96. [CrossRef]
10. Schrank, E.S.; Stanhope, S.J. Dimensional accuracy of ankle-foot orthoses constructed by rapid customization and manufacturing framework. *J. Rehabil. Res. Dev.* **2011**, *48*, 31–42. [CrossRef]
11. Cobetto, N.; Aubin, C.E.; Clin, J.; Le May, S.; Desbiens-Blais, F.; Labelle, H.; Parent, S. Braces Optimized With Computer-Assisted Design and Simulations Are Lighter, More Comfortable, and More Efficient than Plaster-Cast Braces for the Treatment of Adolescent Idiopathic Scoliosis. *Spine Deform.* **2014**, *2*, 276–284. [CrossRef]
12. Redaelli, D.F.; Storm, F.A.; Biffi, E.; Reni, G.; Colombo, G. A Virtual Design Process to Produce Scoliosis Braces by Additive Manufacturing. In *International Conference on Design, Simulation, Manufacturing: The Innovation Exchange*; Springer: Berlin, Germany, 2019; pp. 860–870.
13. Bourell, D.; Stucker, B.; Cook, D.; Gervasi, V.; Rizza, R.; Kamara, S.; Liu, X. Additive fabrication of custom pedorthoses for clubfoot correction. *Rapid Prototyp. J.* **2010**, *16*, 189–193. [CrossRef]
14. Langlois, K.; Moltedo, M.; Bacek, T.; Rodriguez-Guerrero, C.; Vanderborght, B.; Lefeber, D. Design and development of customized physical interfaces to reduce relative motion between the user and a powered ankle foot exoskeleton. In Proceedings of the 2018 7th IEEE International Conference on Biomedical Robotics and Biomechatronics (Biorob), Enschede, The Netherlands, 26–29 August 2018; pp. 1083–1088.
15. Gibson, K.S.; Woodburn, J.; Porter, D., Telfer, S. Functionally Optimized Orthoses for Early Rheumatoid Arthritis Foot Disease: A Study of Mechanisms and Patient Experience. *Arthritis Care Res.* **2014**, *66*, 1456–1464. [CrossRef] [PubMed]
16. Pallari, J.H.P.; Dalgarno, K.W.; Woodburn, J. Mass customization of foot orthoses for rheumatoid arthritis using selective laser sintering. *IEEE Trans. Biomed. Eng.* **2010**, *57*, 1750–1756. [CrossRef] [PubMed]
17. Baronio, G.; Harran, S.; Signoroni, A. A critical analysis of a hand orthosis reverse engineering and 3D printing process. *Appl. Bionics Biomech.* **2016**, *2016*, 8347478. [CrossRef] [PubMed]
18. Campbell, I.; Diegel, O.; Kowen, J.; Wohlers, T. *Wohlers Report 2018: 3D Printing and Additive Manufacturing State of the Industry: Annual Worldwide Progress Report*; Wohlers Associates: Fort Collins, CO, USA, 2018.

19. Zuniga, J.; Katsavelis, D.; Peck, J.; Stollberg, J.; Petrykowski, M.; Carson, A.; Fernandez, C. Cyborg beast: A low-cost 3d-printed prosthetic hand for children with upper-limb differences. *BMC Res. Notes* **2015**, *8*, 10. [CrossRef]
20. De Carvalho Filho, I.F.P.; Medola, F.O.; Sandnes, F.E.; Paschoarelli, L.C. Manufacturing Technology in Rehabilitation Practice: Implications for Its Implementation in Assistive Technology Production. In *Proceedings of the International Conference on Applied Human Factors and Ergonomics*; Springer: Berlin, Germany, 2019; pp. 328–336.
21. Fitzpatrick, A.P.; Mohanned, M.I.; Collins, P.K.; Gibson, I. Design of a patient specific, 3D printed arm cast. In Proceedings of the International Conference on Design and Technology, Geelong, Australia, 5–8 December 2016; pp. 135–142.
22. Parvez, N.; Parvez, S.. Andiamo Project. Available online: http://andiamo.io/ (accessed on 30 October 2019).
23. Chevalier, T.; Chockalingam, N. Effects of foot orthoses: How important is the practitioner? *Gait Posture* **2012**, *35*, 383–388. [CrossRef]
24. Dalgarno, K.; Pallari, J.; Woodburn, J.; Xiao, K.; Wood, D.; Goodridge, R.; Ohtsuki, C. Mass customization of medical devices and implants: state of the art and future directions. *Virtual Phys. Prototyp.* **2006**, *1*, 137–145. [CrossRef]
25. Baumers, M.; Tuck, C.; Bourell, D.; Sreenivasan, R.; Hague, R. Sustainability of additive manufacturing: Measuring the energy consumption of the laser sintering process. *Proc. Inst. Mech. Eng. Part B J. Eng. Manuf.* **2011**, *225*, 2228–2239. [CrossRef]
26. Mavroidis, C.; Ranky, R.G.; Sivak, M.L.; Patritti, B.L.; DiPisa, J.; Caddle, A.; Gilhooly, K.; Govoni, L.; Sivak, S.; Lancia, M.; et al. Patient specific ankle-foot orthoses using rapid prototyping. *J. Neuroeng. Rehabil.* **2011**, *8*, 1. [CrossRef]
27. Culmone, C.; Smit, G.; Breedveld, P. Additive manufacturing of medical instruments: A state-of-the-art review. In *Additive Manufacturing*; Elsevier: Amsterdam, The Netherlands, 2019.
28. Palo, M.; Holländer, J.; Suominen, J.; Yliruusi, J.; Sandler, N. 3D printed drug delivery devices: perspectives and technical challenges. *Expert Rev. Med. Devices* **2017**, *14*, 685–696. [CrossRef]
29. Jeon, O.; Lee, Y.B.; Jeong, H.; Lee, S.J.; Wells, D.; Alsberg, E. Individual cell-only bioink and photocurable supporting medium for 3D printing and generation of engineered tissues with complex geometries. *Mater. Horiz.* **2019**, *9*, 1625–1631. [CrossRef]
30. Derakhshanfar, S.; Mbeleck, R.; Xu, K.; Zhang, X.; Zhong, W.; Xing, M. 3D bioprinting for biomedical devices and tissue engineering: A review of recent trends and advances. *Bioact. Mater.* **2018**, *3*, 144–156. [CrossRef] [PubMed]
31. Koffler, J.; Zhu, W.; Qu, X.; Platoshyn, O.; Dulin, J.N.; Brock, J.; Graham, L.; Lu, P.; Sakamoto, J.; Marsala, M.; et al. Biomimetic 3D-printed scaffolds for spinal cord injury repair. *Nat. Med.* **2019**, *25*, 263. [CrossRef] [PubMed]
32. Egan, P.F. Integrated design approaches for 3D printed tissue scaffolds: Review and outlook. *Materials* **2019**, *12*, 2355. [CrossRef] [PubMed]
33. Dawood, A.; Marti, B.M.; Sauret-Jackson, V.; Darwood, A. 3D printing in dentistry. *Brit. Dent. J.* **2015**, *219*, 521. [CrossRef]
34. Tahayeri, A.; Morgan, M.; Fugolin, A.P.; Bompolaki, D.; Athirasala, A.; Pfeifer, C.S.; Ferracane, J.L.; Bertassoni, L.E. 3D printed versus conventionally cured provisional crown and bridge dental materials. *Dent. Mater.* **2018**, *34*, 192–200. [CrossRef]
35. Nguyen, K.T.; Benabou, L.; Alfayad, S. Systematic Review of Prosthetic Socket Fabrication using 3D printing. In Proceedings of the 2018 ACM 4th International Conference on Mechatronics and Robotics Engineering, Valenciennes, France, 7–11 February 2018; pp. 137–141.
36. Chae, M.P.; Rozen, W.M.; McMenamin, P.G.; Findlay, M.W.; Spychal, R.T.; Hunter-Smith, D.J. Emerging applications of bedside 3D printing in plastic surgery. *Front. Surg.* **2015**, *2*, 25. [CrossRef]
37. Malik, H.H.; Darwood, A.R.; Shaunak, S.; Kulatilake, P.; Abdulrahman, A.; Mulki, O.; Baskaradas, A. Three-dimensional printing in surgery: A review of current surgical applications. *J. Surg. Res.* **2015**, *199*, 512–522. [CrossRef]
38. Tack, P.; Victor, J.; Gemmel, P.; Annemans, L. 3D-printing techniques in a medical setting: A systematic literature review. *Biomed. Eng. Online* **2016**, *15*, 115. [CrossRef]
39. Tang, L.; Wang, L.; Bao, W.; Zhu, S.; Li, D.; Zhao, N.; Liu, C. Functional gradient structural design of customized diabetic insoles. *J. Mech. Behav. Biomed. Mater.* **2019**, *94*, 279–287. [CrossRef]

40. Artioli, B.O.; Kunkel, M.E.; Mestanza, S.N. Feasibility Study of a Methodology Using Additive Manufacture to Produce Silicone Ear Prostheses. In *World Congress on Medical Physics and Biomedical Engineering 2018*; Springer: Berlin, Germany, 2019; pp. 211–215.
41. Liacouras, P.; Garnes, J.; Roman, N.; Petrich, A.; Grant, G.T. Designing and manufacturing an auricular prosthesis using computed tomography, 3-dimensional photographic imaging, and additive manufacturing: a clinical report. *J. Prosthet. Dent.* **2011**, *105*, 78–82. [CrossRef]
42. Liacouras, P.C.; Sahajwalla, D.; Beachler, M.D.; Sleeman, T.; Ho, V.B.; Lichtenberger, J.P. Using computed tomography and 3D printing to construct custom prosthetics attachments and devices. *3D Print. Med.* **2017**, *3*, 8. [CrossRef] [PubMed]
43. Diwakar, M.; Kumar, M. A review on CT image noise and its denoising. *Biomed. Signal Process. Control* **2018**, *42*, 73–88. [CrossRef]
44. Haleem, A.; Javaid, M. 3D scanning applications in medical field: A literature-based review. *Clin. Epidemiol. Global Health* **2019**, *7*, 199–210. [CrossRef]
45. Górski, F.; Zawadzki, P.; Wichniarek, R.; Kuczko, W.; Żukowska, M.; Wesołowska, I.; Wierzbicka, N. Automated Design of Customized 3D-Printed Wrist Orthoses on the Basis of 3D Scanning. In *International Conference on Computational & Experimental Engineering and Sciences*; Springer: Berlin, Germany, 2019; pp. 1133–1143.
46. Grazioso, S.; Caporaso, T.; Selvaggio, M.; Panariello, D.; Ruggiero, R.; Di Gironimo, G. Using photogrammetric 3D body reconstruction for the design of patient–tailored assistive devices. In Proceedings of the IEEE 2019 II Workshop on Metrology for Industry 4.0 and IoT (MetroInd4. 0&IoT), Naples, Italy, 4–6 June 2019; pp. 240–242.
47. Barrios-Muriel, J.; Romero Sánchez, F.; Alonso, F.; Salgado, D. Design of Semirigid Wearable Devices Based on Skin Strain Analysis. *J. Biomech. Eng.* **2019**, *141*, 021008.
48. Bogo, F.; Romero, J.; Pons-Moll, G.; Black, M.J. Dynamic FAUST: Registering human bodies in motion. In Proceedings of the IEEE Conference on Computer Vision and Pattern Recognition, Honolulu, HI, USA, 21–26 July 2017; pp. 6233–6242.
49. Thompson, A.; McNally, D.; Maskery, I.; Leach, R.K. X-ray computed tomography and additive manufacturing in medicine: A review. *Int. J. Metrol. Qual. Eng.* **2017**, *8*, 17. [CrossRef]
50. Dal Maso, A.; Cosmi, F. 3D-printed ankle-foot orthosis: a design method. *Mater. Today Proc.* **2019**, *12*, 252–261. [CrossRef]
51. Parrilla, E.; Ballester, A.; Solves-Camallonga, C.; Nácher, B.; Antonio Puigcerver, S.; Uriel, J.; Piérola, A.; González, J.C.; Alemany, S. Low-cost 3D foot scanner using a mobile app. *Footwear Sci.* **2015**, *7*, S26–S28. [CrossRef]
52. Chee Kai, C.; Siaw Meng, C.; Sin Ching, L.; Seng Teik, L.; Chit Aung, S. Facial prosthetic model fabrication using rapid prototyping tools. *Integr. Manuf. Syst.* **2000**, *11*, 42–53. [CrossRef]
53. Paterson, A. Digitisation of the Splinting Process: Exploration and Evaluation of a Computer Aided Design Approach to Support Additive Manufacture. Ph.D Thesis, Loughborough University, Loughborough, UK, 2013.
54. Mali, H.S.; Vasistha, S. Fabrication of Customized Ankle Foot Orthosis (AFO) by Reverse Engineering Using Fused Deposition Modelling. In *Advances in Additive Manufacturing and Joining*; Springer: Berlin, Germany, 2020; pp. 3–15.
55. Agudelo-Ardila, C.; Prada-Botía, G.; PH, R. Orthotic prototype for upper limb printed in 3D. A efficient solution. In *Journal of Physics: Conference Series*; IOP Publishing: Bristol, UK, 2019; Volume 1388, p. 012016.
56. Kimura, M.; Mochimaru, M.; Kanade, T. Measurement of 3D foot shape deformation in motion. In Proceedings of the ACM 5th ACM/IEEE International Workshop on Projector Camera Systems, Bali Way, CA, USA, 20 August 2008; p. 10. [CrossRef]
57. Mo, S.; Leung, S.H.; Chan, Z.Y.; Sze, L.K.; Mok, K.M.; Yung, P.S.; Ferber, R.; Cheung, R.T. The biomechanical difference between running with traditional and 3D printed orthoses. *J. Sports Sci.* **2019**, *37*, 2191–2197. [CrossRef]
58. Liberadzki, P.; Adamczyk, M.; Witkowski, M.; Sitnik, R. Structured-Light-Based System for Shape Measurement of the Human Body in Motion. *Sensors* **2018**, *18*, 2827. [CrossRef]
59. Unkovskiy, A.; Spintzyk, S.; Brom, J.; Huettig, F.; Keutel, C. Direct 3D printing of silicone facial prostheses: A preliminary experience in digital workflow. *Journal Prosthet. Dent.* **2018**, *120*, 303–308. [CrossRef] [PubMed]

60. Schmeltzpfenning, T.; Plank, C.; Krauss, I.; Aswendt, P.; Grau, S. Dynamic foot scanning. Prospects and limitations of using synchronized 3D scanners to capture complete human foot shape while walking. In *Advances in Applied Digital Human Modeling, Proceedings of the 3rd International Conference on Applied Human Factors and Ergonomics, Miami, FL, USA, 19–22 July, 2010*; Karwowski, W., Salvendy, G., Eds.; CRC Press: Boca Raton, FL, USA, 2010.
61. Chatzitofis, A.; Zarpalas, D.; Kollias, S.; Daras, P. DeepMoCap: Deep Optical Motion Capture Using Multiple Depth Sensors and Retro-Reflectors. *Sensors* **2019**, *19*, 282. [CrossRef] [PubMed]
62. Joo, H.; Simon, T.; Sheikh, Y. Total capture: A 3d deformation model for tracking faces, hands, and bodies. In Proceedings of the IEEE Conference on Computer Vision and Pattern Recognition, Salt Lake City, UT, USA, 18–22 June 2018; pp. 8320–8329.
63. Ciobanu, O.; Ciobanu, G.; Rotariu, M. Photogrammetric scanning technique and rapid prototyping used for prostheses and ortheses fabrication. *Appl. Mech. Mater.* **2013**, *371*, 230–234. [CrossRef]
64. Kudelski, R.; Dudek, P.; Kulpa, M.; Rumin, R. Using reverse engineering and rapid prototyping for patient specific orthoses. In Proceedings of the IEEE 2017 XIIIth International Conference on Perspective Technologies and Methods in MEMS Design (MEMSTECH), Lviv, Ukraine, 20–23 April 2017; pp. 88–90.
65. Vitali, A.; Regazzoni, D.; Rizzi, C.; Colombo, G. Design and additive manufacturing of lower limb prosthetic socket. In Proceedings of the ASME 2017 International Mechanical Engineering Congress and Exposition, Tampa, FL, USA, 10 January 2017.
66. Kruth, P.P. Material Incress Manufacturing by Rapid Prototyping Techniques. *CIRP Ann. Manuf. Technol.* **1991**, *40*, 603–614. [CrossRef]
67. Paterson, A.M.; Bibb, R.; Campbell, R.I.; Bingham, G. Comparing additive manufacturing technologies for customised wrist splints. *Rapid Prototyp. J.* **2015**, *21*, 230–243. [CrossRef]
68. Chua, C.K.; Leong, K.F. *3D Printing and Additive Manufacturing: Principles and Applications (with Companion Media Pack) of Rapid Prototyping*, 4th ed.; World Scientific Publishing Company: Singapore, 2014.
69. Petrovic, V.; Vicente Haro Gonzalez, J.; Jorda Ferrando, O.; Delgado Gordillo, J.; Ramon Blasco Puchades, J.; Portoles Grinan, L. Additive layered manufacturing: Sectors of industrial application shown through case studies. *Int. J. Prod. Res.* **2011**, *49*, 1061–1079. [CrossRef]
70. Tan, K.C.; Lee, P.V.S.; Tam, K.F.; Lye, S.L. Automation of prosthetic socket design and fabrication using computer aided design/computer aided engineering and rapid prototyping techniques. In Proceedings of the First National Symposium of Prosthetics and Orthotics, Singapore, Republic of Singapore, 28 October 1998; pp. 19–22.
71. Leite, M.; Soares, B.; Lopes, V.; Santos, S.; Silva, M.T. Design for personalized medicine in orthotics and prosthetics. *Proced. CIRP* **2019**, *84*, 457–461. [CrossRef]
72. Abdullah, A.M.; Mohamad, D.; Din, T.N.D.T.; Yahya, S.; Akil, H.M.; Rajion, Z.A. Fabrication of nasal prosthesis utilising an affordable 3D printer. *Int. J. Adv. Manuf. Technol.* **2019**, *100*, 1907–1912. [CrossRef]
73. Melocchi, A.; Parietti, F.; Loreti, G.; Maroni, A.; Gazzaniga, A.; Zema, L. 3D printing by fused deposition modeling (FDM) of a swellable/erodible capsular device for oral pulsatile release of drugs. *J. Drug Deliv. Sci. Technol.* **2015**, *30*, 360–367. [CrossRef]
74. Wang, X.; Jiang, M.; Zhou, Z.; Gou, J.; Hui, D. 3D printing of polymer matrix composites: A review and prospective. *Compos. Part B Eng.* **2017**, *110*, 442–458. [CrossRef]
75. Beaman, J.; Deckard, C. Selective laser sintering with assisted powder handling U.S. Patent No. 4,938,816, 3 July 1990.
76. Gibson, I.; Rosen, D.W.; Stucker, B. *Additive Manufacturing Technologies*; Springer: Berlin, Germany, 2015.
77. Kruth, J.P.; Mercelis, P.; Van Vaerenbergh, J.; Froyen, L.; Rombouts, M. Binding mechanisms in selective laser sintering and selective laser melting. *Rapid Prototyp. J.* **2005**, *11*, 26–36. [CrossRef]
78. Deckers, J.P.; Vermandel, M.; Geldhof, J.; Vasiliauskaite, E.; Forward, M.; Plasschaert, F. Development and clinical evaluation of laser-sintered ankle foot orthoses. *Plast. Rubber Compos.* **2018**, *47*, 42–46. [CrossRef]

79. Vasiliauskaite, E.; Ielapi, A.; Deckers, J.; Vermandel, M.; De Beule, M.; Van Paepegem, W.; Forward, M.; Plasschaert, F. Selective laser sintered ankle foot orthosis can support drop foot gait. In *16th National Day on Biomedical Engineering*; National Committee on Biomedical Engineering (NCBME): Brussels, Belgium, 1 December 2017; p. 23.
80. Oxman, N. Material-Based Design Computation. Ph.D Thesis, Massachusetts Institute of Technology, Cambridge, MA, USA, 2010.
81. Evill, J. Cortex. 2013. Available online: http://www.evilldesign.com/cortex (accessed on 29 October 2019).
82. Herbert, N.; Simpson, D.; Spence, W.D.; Ion, W. A preliminary investigation into the development of 3-D printing of prosthetic sockets. *J. Rehabil. Res. Dev.* **2005**, *42*, 141–146. [CrossRef] [PubMed]
83. Saijo, H.; Igawa, K.; Kanno, Y.; Mori, Y.; Kondo, K.; Shimizu, K.; Suzuki, S.; Chikazu, D.; Iino, M.; Anzai, M.; et al. Maxillofacial reconstruction using custom-made artificial bones fabricated by inkjet printing technology. *J. Artif. Org.* **2009**, *12*, 200–205. [CrossRef] [PubMed]
84. Ventola, C.L. Medical Applications for 3D Printing: Current and Projected Uses. *Pharm. Ther.* **2014**, *39*, 704–711.
85. Oxman, N. Variable property rapid prototyping. *Virtual Phys. Prototyp.* **2011**, *6*, 3–31.
86. Peng, H.; Mankoff, J.; Hudson, S.E.; McCann, J. A Layered Fabric 3D Printer for Soft Interactive Objects. In Proceedings of the ACM 33rd Annual ACM Conference on Human Factors in Computing Systems, Seoul, Korea, 18–23 April 2015; pp. 1789–1798. [CrossRef]
87. Campbell, I.; Bourell, D.; Gibson, I. Additive manufacturing: Rapid prototyping comes of age. *Rapid Prototyp. J.* **2012**, *18*, 255–258. [CrossRef]
88. Silva, M.; Felismina, R.; Mateus, A.; Parreira, P.; Malça, C. Application of a hybrid additive manufacturing methodology to produce a metal/polymer customized dental implant. *Proced. Manuf.* **2017**, *12*, 150–155. [CrossRef]
89. Vaish, A.; Vaish, R. 3D printing and its applications in Orthopedics. *J. Clin. Orthop. Trauma* **2018**, *9*, S74–S75. [CrossRef]
90. Ciobanu, O.; Soydan, Y.; Hizal, S. Customized foot orthosis manufactured with 3D printers. In Proceedings of the 8th International Symposium on Intelligent Manufacturing Systems, Antalya, Turkey, 27–28 September 2012; 7p.
91. Paton, J.; Jones, R.B.; Stenhouse, E.; Bruce, G. The physical characteristics of materials used in the manufacture of orthoses for patients with diabetes. *Foot Ankle Int.* **2007**, *28*, 1057–1063. [CrossRef]
92. Ma, Z.; Lin, J.; Xu, X.; Ma, Z.; Tang, L.; Sun, C.; Li, D.; Liu, C.; Zhong, Y.; Wang, L. Design and 3D printing of adjustable modulus porous structures for customized diabetic foot insoles. *Int. J. Lightweight Mater. Manuf.* **2019**, *2*, 57–63. [CrossRef]
93. Faustini, M.C.; Neptune, R.R.; Crawford, R.H.; Stanhope, S.J. Manufacture of passive dynamic ankle-foot orthoses using selective laser sintering. *IEEE Trans. Biomed. Eng.* **2008**, *55*, 784–790. [CrossRef] [PubMed]
94. Walbran, M.; Turner, K.; McDaid, A. Customized 3D printed ankle-foot orthosis with adaptable carbon fibre composite spring joint. *Cogent Eng.* **2016**, *3*, 1227022. [CrossRef]
95. Trauner, K.B. The emerging role of 3D printing in arthroplasty and orthopedics. *J. Arthroplast.* **2018**, *33*, 2352–2354. [CrossRef]
96. Tsai, W.; Yang, Y.; Chen, C.H. A wearable 3D printed elbow exoskeleton to improve upper limb rehabilitation in stroke patients. In *Smart Science, Design & Technology: Proceedings of the 5th International Conference on Applied System Innovation (ICASI 2019), Fukuoka, Japan, 12–18 April 2019*; CRC Press: Boca, Raton, FL, USA, 2019; p. 231.
97. Ang, B.W.; Yeow, C.H. Print-it-Yourself (PIY) glove: a fully 3D printed soft robotic hand rehabilitative and assistive exoskeleton for stroke patients. In Proceedings of the 2017 IEEE/RSJ International Conference on Intelligent Robots and Systems (IROS), Vancouver, BC, Canada, 24–28 September 2017; pp. 1219–1223.

 © 2020 by the authors. Licensee MDPI, Basel, Switzerland. This article is an open access article distributed under the terms and conditions of the Creative Commons Attribution (CC BY) license (http://creativecommons.org/licenses/by/4.0/).

Article

Considerations on the Applicability of Test Methods for Mechanical Characterization of Materials Manufactured by FDM

Amabel García-Domínguez, Juan Claver, Ana María Camacho * and Miguel A. Sebastián

Department of Manufacturing Engineering, Universidad Nacional de Educación a Distancia (UNED), 28040 Madrid, Spain; agarcia5250@alumno.uned.es (A.G.-D.); jclaver@ind.uned.es (J.C.); msebastian@ind.uned.es (M.A.S.)
* Correspondence: amcamacho@ind.uned.es; Tel.: +34-913-988-660

Received: 15 November 2019; Accepted: 17 December 2019; Published: 19 December 2019

Abstract: The lack of specific standards for characterization of materials manufactured by Fused Deposition Modelling (FDM) makes the assessment of the applicability of the test methods available and the analysis of their limitations necessary; depending on the definition of the most appropriate specimens on the kind of part we want to produce or the purpose of the data we want to obtain from the tests. In this work, the Spanish standard UNE 116005:2012 and international standard ASTM D638–14:2014 have been used to characterize mechanically FDM samples with solid infill considering two build orientations. Tests performed according to the specific standard for additive manufacturing UNE 116005:2012 present a much better repeatability than the ones according to the general test standard ASTM D638–14, which makes the standard UNE more appropriate for comparison of different materials. Orientation on-edge provides higher strength to the parts obtained by FDM, which is coherent with the arrangement of the filaments in each layer for each orientation. Comparison with non-solid specimens shows that the increase of strength due to the infill is not in the same proportion to the percentage of infill. The values of strain to break for the samples with solid infill presents a much higher deformation before fracture.

Keywords: additive manufacturing; FDM; ABS; anisotropy; infill density; layer orientation; ASTM D638–14:2014; ISO 527–2:2012

1. Introduction

Currently, the characterization of parts obtained by additive manufacturing (AM) technologies is a very prolific research field. This fact is a consequence, not only of the significant increase of the presence and the importance of additive technologies in a wide range of industries, but it is also due to the lack of specific standardization, as claimed in previous works such as the one by Rodríguez-Panes et al. [1] and in the review made by Popescu et al. [2], who specifically demanded that test standards for Fused Deposition Modelling (FDM) parts should be developed, including the definition of printing parameters such as layer thickness, perimeters, and raster dimensions. Thus, the most commonly applied international standards for tensile tests are ISO 527–2 [3] and ASTM D638–14 [4], but, in both cases, the guidelines provided respond to the characteristics of plastic injection parts, which represents a very different start context. Contrary to the continuity and isotropy of injected plastics, the layer by layer structure of the pieces obtained by additive manufacturing is discontinuous and anisotropic, so the polymeric nature of the material is the only common feature for both technologies. It is also worth mentioning that, while ISO 527–2 and ASTM D638–14 are identified as the consultation documents for tensile tests by the main standards focused on additive manufacturing technologies, as ISO 17296–3:2014 [5] and ISO/ASTM 52921:2013 [6] are, the Spanish standard UNE 116005:2012 [7] has also been considered; this standard is only focused on tensile tests and polymer materials, and it is based on ISO 527–2. Despite both standards being similar, UNE 116005:2012 is developed as a specific standard for additive manufacturing, and it is not a standard recovered from other productive sectors. The national standards must follow the guidelines of the international ones and act as complementary information sources to guide researchers and professionals in specific tasks of applicative character, as in this case, with performing mechanical tests for additive manufacturing purposes.

The experiences developed by researchers when applying these standards in additive contexts have a key role, since the obtained results help to clarify the applicability or not of these standards in different additive scenarios, as well as to provide action guidelines. A significant variety of approaches have been faced in order to determinate the influence of manufacturing parameters on the mechanical response of additive products; for example, ABS specimens for unmanned aerial systems produced by FDM were characterized under tension [8] and under compression [9]; Banjanin et al. [10] evaluated the mechanical response of specimens of polylactide (PLA) and acrylonitrile styrene butadiene (ABS), concluding that the results in ABS showed a higher repeatability under tension than under compression, and vice versa. Chacón et al. [11] assessed the effect of build orientation, layer thickness, and feed rate on the mechanical performance of PLA specimens, stating that, for optimal mechanical performance, low layer thickness and high feed rate values are recommended. In the work of Tanikella et al. [12], a set of seven materials has been tested for determination of tensile strength of FDM specimens fabricated with an open-source 3D printer, concluding that this property highly depends on the mass of the specimen. The influence of the type of infill pattern and density, printing temperature, among other parameters on selected mechanical properties of PLA and ABS samples, were evaluated by Ćwikła et al. [13], showing that, if the maximum strength is the priority, shell thickness should be increased. In the work of Kuznetsov et al. [14], the influence of geometric process parameters on the strength of the sample was also evaluated proposing a new methodology, and concluding that the best combinations of printing parameters allows for obtaining interlayer cohesion close to the one of the feedstock material. Rajpurohit and Dave found a close relationship between the raster angle and failure mode [15]. In addition, Zaldivar et al. [16] investigated the effect of the build orientation on the mechanical properties of a polyetherimide thermoplastic blend as material, concluding that FDM materials behave more like composite structures than isotropic cast resins and that designs should define the build configurations allowable.

However, sometimes, a complementary and previous question is needed because the kind of part we want to produce or the purpose of the data we want to obtain from the tests can guide the definition of the most appropriate specimens. Figure 1 tries to express this idea graphically. A basic geometry usually obtained as a continuous solid by plastic injection (Figure 1a) can be materialized in very different ways when considering additive manufacturing. This includes:

- a layer by layer solid configuration (Figure 1(2)), which means an infill percentage of 100%,
- configurations with different infill percentages (Figure 1(3)),
- and cellular structures (Figure 1(4)).

On the other hand, for more complex geometries that are not possible to manufacture with plastic injection processes but are possible in additive scenarios (Figure 1b), these different approaches could be chosen as alternatives for the materialization of the parts which traditionally would be understood as solid. The differences between the internal material continuity that the selection of one or another of these approaches introduces are significant. Thus, this selection implies different ways of understanding the material since, although in every case mentioned before the polymer used can be the same, the behavior of the piece is highly conditioned by internal morphology of those parts traditionally understood as solid.

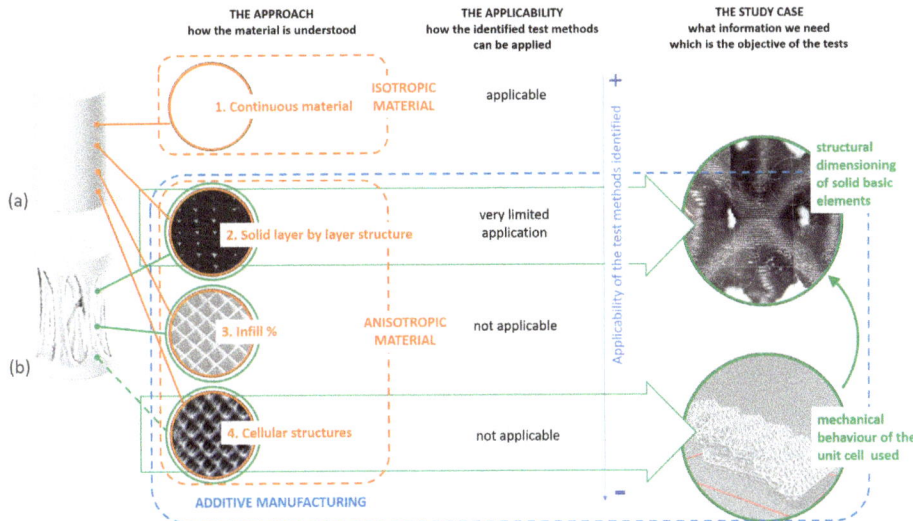

Figure 1. Applicability of conventional test methods for material characterization of parts obtained by additive manufacturing techniques considering their internal morphology: (**a**) basic geometry suitable both for traditional and additive manufacturing technologies. (**b**) complex geometry suitable for additive manufacturing technologies.

As exposed before, from the point of view of the applicability of the mentioned standards, only a continuous material could be characterized. However, it is interesting to reflect the different levels of applicability that the selection of each approach introduces in the context of additive processes. In that sense, the applicable criterion is the continuity or not of the internal morphology of the piece. Thus, a solid layer by layer structure would represent the closest conditions to the ones considered in these standards. In the case of predefined infill patterns, these standards are not really applicable, but, in practice, many studies use this kind of specimens and analyze their behavior by applying them. Finally, cellular structures are similar to these infill patterns in terms of discontinuity, but this group includes lattice structures and approaches in which the sizing of some parts or elements can change along the piece and those fluctuations provide graded densities.

This way, the design of each case study makes each approach more or less appropriate, and it is essential to take it into account. In that sense, the methodology optimization developed by the authors in a previous work [17] and its data needs can be a good example. The mentioned methodology uses optimized lattice structures for the infill areas where each bar is sized according to the stress it can withstand [18]. In order to calculate the section needed in each case, values of the material strength are requested by the structure of the methodology. Since these bars have diameters between 1 and 2.5 mm, infill patterns or cellular structures cannot be implemented into them, so the layer by layer solid configuration is the only option. In addition, as each bar has a different section as a consequence of the different stress values calculated, the lattice structure cannot be considered as uniform along the piece. Thus, those approaches that estimate the mechanical behavior of cellular structures through the testing of specimens out of the standards and that contain a certain number of unit cells would not be of application in this particular case, as in the works of Chen et al. [19], who developed a finite element mesh based method for optimization of lattice structures for AM; Hussein [20], who studied the development of lightweight cellular structures for additive manufacturing with metal; Mahmoud and Elbestawi [21], where lattice structures were fabricated by additive manufacturing in orthopedic implants; Maliaris and Sarafis [22], where lattice structures were modelled using a generative algorithm; Panda [23] and Weeger et al [24], who worked in the design and development of cellular structures for

AM; or Vannutelli [25], where the mechanical behavior of lattice structures was analyzed. As Figure 1 exposes, a solid layer by layer configuration represents the most consistent alternative with the nature of this particular case study and the data that are needed. This way, the identified standards could be applied, always considering their limitations in additive contexts.

Accordingly, the aim of this work is to analyze the mechanical behavior of solid structures manufactured by FDM (as a typology of interest in different design contexts supported by additive technologies, as the ones described above) compared to non-solid ones, and to determine which standard provides better results for the mechanical characterization of materials manufactured by FDM, giving some guidelines in the application of them due to the lack of specific standards for characterization of materials manufactured by FDM. To achieve this goal, a series of tensile tests are carried out using solid specimens obtained by FDM, and according to the specifications of UNE 116005:2012 (based in ISO 527–2) and ASTM D638–14. The obtained results are interpreted together with the layer structures observed in the specimens after fracture.

2. Materials and Methods

2.1. Work Methodology

According to the aim of this work, and in order to have a clear view of the research methodology followed in this work, Figure 2 shows the approach definition based in the different situations presented in Figure 1, and a summary of the main steps of the methodology. Four main steps are established, and then basic information related to the particular context of this work is indicated for each one of these stages. As shown in the work diagram, at the end of the process, the obtained results and the conclusions derived from them must be analyzed in order to assess whether the results fit the objectives of the study. The main aspects of the work developed in these stages are exposed in the following sections. Afterwards, the results are compared with the ones obtained by the authors in a previous work in which the specimens used were fabricated using infill patterns [26].

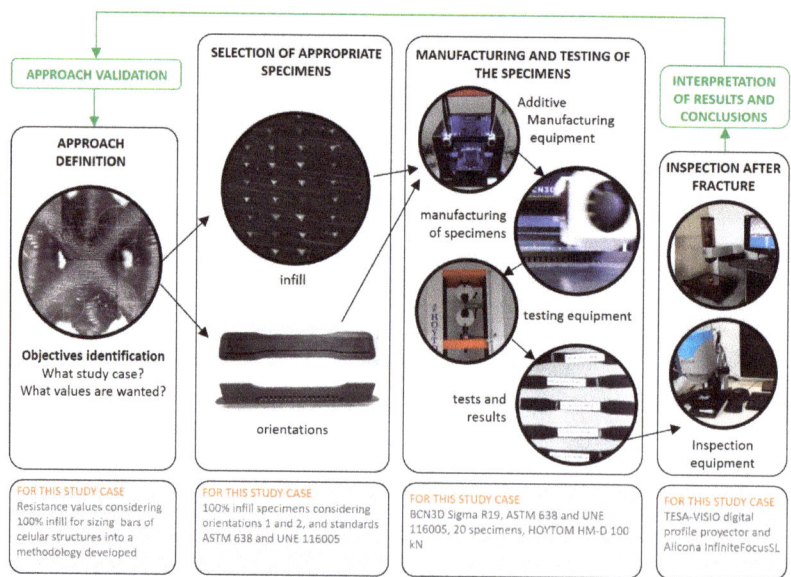

Figure 2. Approach definition based in Figure 1 and sketch of the work methodology.

2.2. Materials and Equipment

The material used in this work is ABS Pro of the commercial house BCN3D (Castelldefels, Spain), whose main physical and mechanical properties are presented in Table 1. The FDM printer to be used is a dual extruder BCN3D R19 printer (BCN3D, Castelldefels, Spain), presented in Figure 3a, whose main technical characteristics are shown in Table 2. The software Cura (Ultimaker, Geldermalsen, The Netherlands) is typically used to export the three-dimensional models of the specimens to G-code. In this paper, the version BCN3D Cura 2.1.4 adapted by BCN3D (Castelldefels, Spain) has been used. Figure 3b shows the equipment for the mechanical testing, a universal testing machine (model Hoytom HM–D 100 kN, Hoytom, S.L., Leioa, Spain).

Table 1. Main mechanical and physical properties of the filaments according to the manufacturer.

Material (Manufacturer)	Color	Tensile Strength (MPa)	Hardness, HB (MPa)	Modulus of Elasticity (GPa)	Density (kg/m^3)
ABS (BCN3D)	Black	45	97	2.3	1040

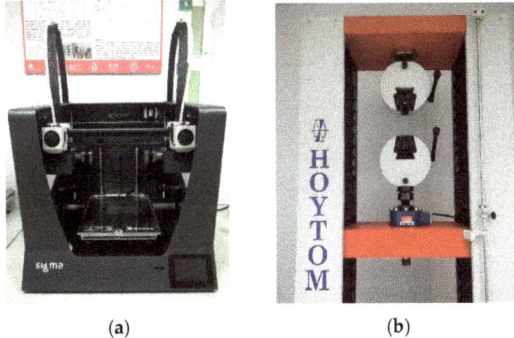

(a) (b)

Figure 3. Equipment for additive manufacturing and testing of specimens: (**a**) a dual extruder BCN3D R19 printer; (**b**) detail of the working area in Hoytom HM-D 100kN Universal testing machine.

Table 2. Main technical characteristics of the Fused Deposition Modelling (FDM) printer BCN3D R19.

Characteristic	Value
Maximum printing volume (mm^3)	210(x) × 297(y) × 210(z)
Firmware	BCN3D Sigma – Marlin
Nozzle diameter (mm)	0.3 mm/0.4 mm (Standard)/0.5 mm (Special)/0.6 mm/0.8 mm/1.0 mm
Resolution (mm)	0.05–0.5
Maximum printing velocity (mm/s)	50
Maximum displacement velocity (mm/s)	200

For analysis of the surface, a digital profile projector TESA-VISIO (TESA SA, Renens, Switzerland) and an optical micro-coordinate measurement machine IF-SL ALICONA (Bruker Alicona, Graz, Austria) are used (see Figure 4).

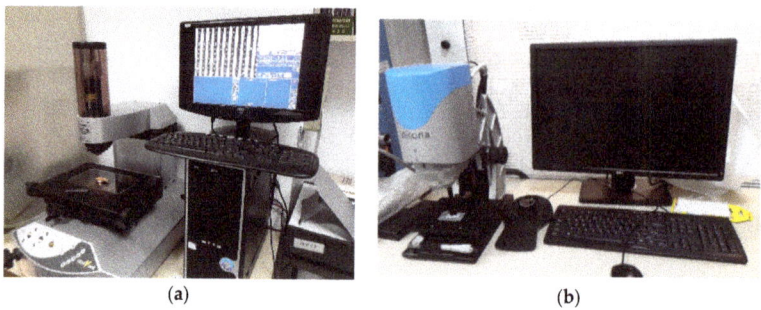

Figure 4. Equipment for surface analysis: (**a**); digital profile projector TESA-VISIO; (**b**) optical micro-coordinate measurement machine IF-SL ALICONA.

2.3. Printing Parameters and Definition of a Work Plan

The general manufacturing parameters for the FDM printing of the specimens are presented in Table 3.

Table 3. Manufacturing parameters with FDM printer BCN3D R19.

Material	Extruder Temp. (°C)	Bed Temp. (°C)	Wall/Infill Speed (mm/s)	Layer Height (mm)	Adhesion Platform	Retraction Dist. (mm)/Speed (mm/s)	Wall/Infill Pattern
ABS (BCN3D)	260	90	50	0.1	YES (8 mm)	4/40	Solid

Due to the lack of specific standards for material characterization of FDM parts, as explained before, two different standards are going to be used for mechanical behavior testing: UNE 116005:2012 (based in ISO 527–2) [7] and ASTM D638-14:2014 [4]; the geometry of the specimens according to both standards is shown in Figure 5, where Type 1A and Type 1 are the geometries chosen for both standards, respectively.

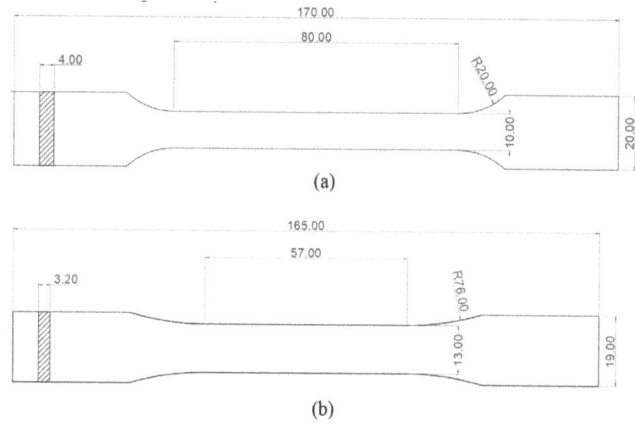

Figure 5. Geometry of the specimens according to different international standards: (**a**) Type 1A: UNE 116005 [7]; (**b**) Type I: ASTM D638–14 [4].

As explained by Rodríguez-Panes et al. [26], the build orientation of the test specimen is one of the most influencing parameters on the mechanical properties of FDM parts. Figure 6a presents the three possible orientations, orientation 1 (flat) and 2 (on-edge) being the ones chosen in this work since orientation 3 is not appropriate for tensile testing due to the arrangement of the filaments perpendicular to the direction of the load. Figure 6b and c shows the two orientation of the samples used in this work.

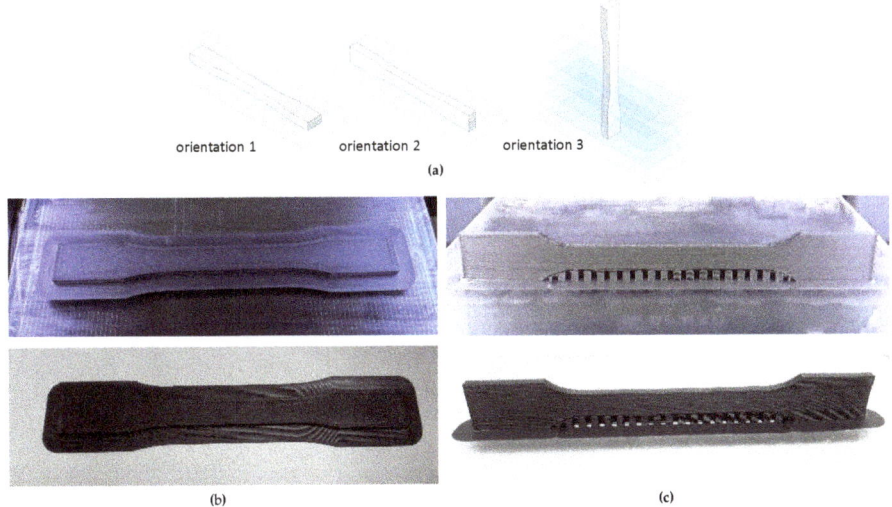

Figure 6. (**a**) typical build orientation of the samples; (**b**) specimen printed with orientation 1; (**c**) specimen printed with orientation 2.

Five specimens for each orientation (1 and 2) are going to be tested according to either standard UNE 116005:2012 and ASTM D638–14:2014. A summary of the experimental work plan and nomenclature used in experiments is presented in Table 4.

Table 4. The experimental work plan and nomenclature of the specimens.

Standard	Orientation 1	Orientation 2
UNE 116005:2012	1 PROB ABS BK 1 1 PROB ABS BK 2 1 PROB ABS BK 3 1 PROB ABS BK 4 1 PROB ABS BK 5	3 PROB ABS BK 1 3 PROB ABS BK 2 3 PROB ABS BK 3 3 PROB ABS BK 4 3 PROB ABS BK 5
ASTM D638–14:2014	ASTM_D638_PROB1_1 ASTM_D638_PROB1_2 ASTM_D638_PROB1_3 ASTM_D638_PROB1_4 ASTM_D638_PROB1_5	ASTM_D638_PROB2_1 ASTM_D638_PROB2_2 ASTM_D638_PROB2_3 ASTM_D638_PROB2_4 ASTM_D638_PROB2_5

Results from the experimental testing of specimens in Table 4 are going to be compared to those ones obtained ii [26] in specimens with non-solid infill; the material of the experiments was ABS (PrintedDreams) in blue and the test procedure according to standard ASTM D638–14:2014 with type I general usage test specimen. The aim is to observe the influence of the infill in the characterization of materials obtained by FDM and for validation purposes. A summary of the cases for comparison is gathered in Table 5.

Table 5. Nomenclature of the specimens and printing parameters by Rodriguez-Panes et al. [26].

Nomenclature	Layer Height (mm)	Orientation	Infill (%)
S1	0.1	1	20
S2	0.1	1	50
S3	0.1	2	20

2.4. Experimental Procedure for Tensile Testing

The specimens in Table 4 are printed according to printing parameters in Table 3. Figure 7a shows the group of samples according to ASTM D638–14:2014. Each specimen is fixed by the grips as it is seen in Figure 7b, showing a specimen just before tensile testing.

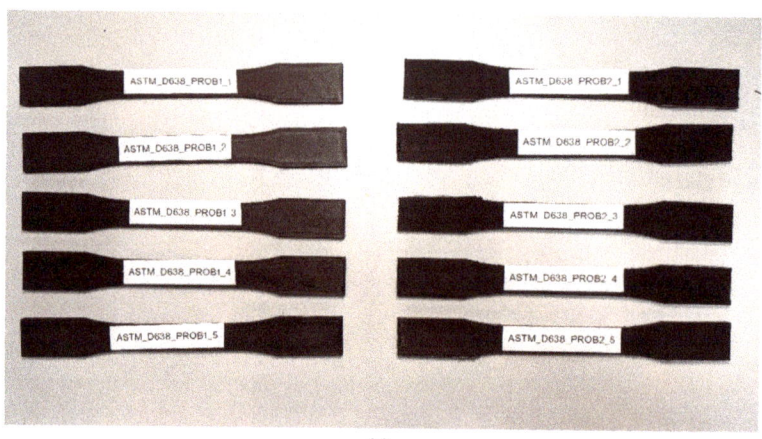

(a)

(b)

Figure 7. (a) the test specimens manufactured according to ASTM D638–14; (b) specimen placement before tensile testing.

The tests are performed at a velocity of 5 mm/min until breakage. The mechanical properties to be obtained are: tensile strength at yield (σ_y), tensile strength at break (σ_U) and nominal strain at break (ε_t). Equations (1) to (3) show how these parameters can be obtained, as defined in the standard ASTM D638-14 [4]:

$$\sigma_y = \frac{F_{max}}{A_0} \quad (1)$$

$$\sigma_U = \frac{F_{break}}{A_0} \quad (2)$$

$$\varepsilon_t(\%) = \frac{\Delta L}{L_0} \times 100 \tag{3}$$

being:

F_{max}: the maximum force sustained by the specimen,
F_{break}: the force sustained by the specimen at breakage,
L_0: original grip separation,
ΔL: extension (change in grip separation).

3. Results and Discussion

3.1. Mechanical Behavior of Solid Specimens

The specimens after tensile testing are presented in Figure 8. Most of them break in close to the radius of fillet close to the grips, which is common for FDM specimens [27], and it is due to stress concentrations at fillet areas [10], being the kind of fracture brittle, with no plastic deformation observed. As Wu et al. reported [28], craze is the main plastic deformation mechanism of ABS, and a great number of crazes are generated perpendicular to the load direction. This brittle behavior is typical in FDM parts, and it has been explained by other authors due to the presence of voids that help the crack initiation and propagation resulting in abrupt rupture [15]; the presence of voids will be analyzed further in Section 3.2.

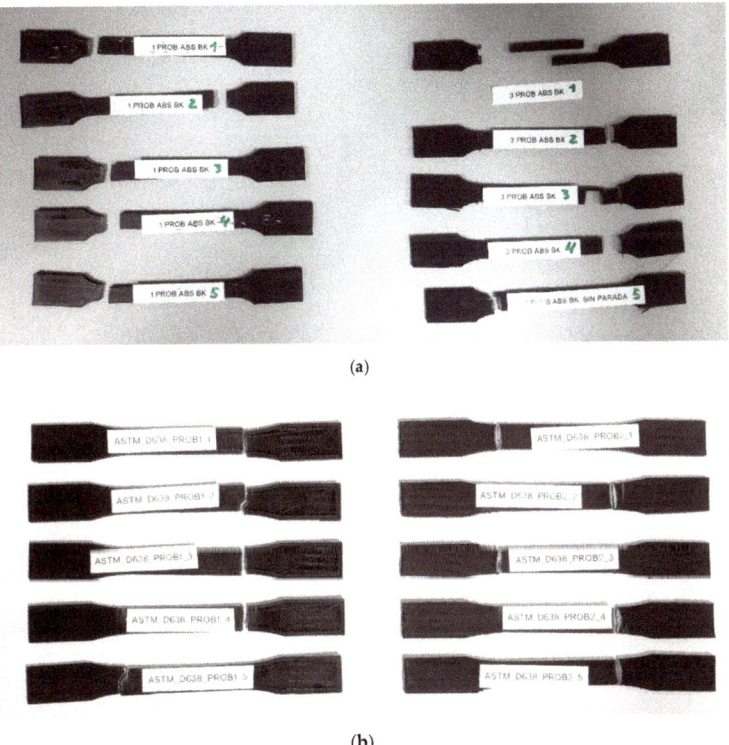

Figure 8. Specimens after tensile testing: (**a**) according to UNE 116005:2012; (**b**) according to ASTM D638–14:2014.

Nominal stresses and strains are obtained for all the tests performed and presented in Figure 9 as stress–strain curves. Figure 9a presents the results according to the standard UNE 116005:2012 for printing orientations 1 and 2; and Figure 9b according to the standard ASTM D638–14, also for both orientations. A comparison of results with the intermediate values of each series of tests is also shown in Figure 9c.

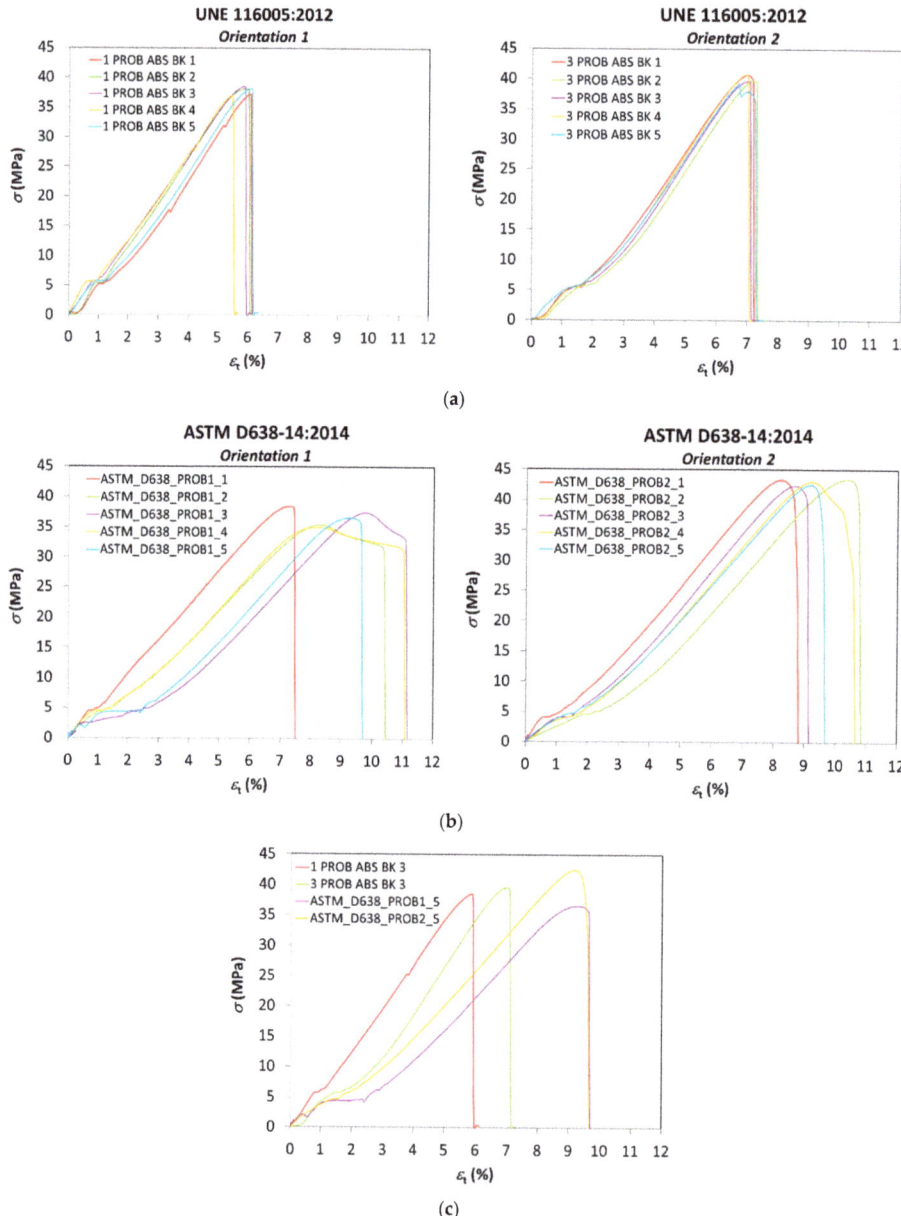

Figure 9. Stress–strain curves. (**a**) according to UNE 116005:2012; (**b**) according to ASTM D638-14; (**c**) comparison of results with the intermediate values of each series of tests.

The mechanical properties (nominal strain at break, tensile strength at yield, tensile strength at break,) obtained from the tensile tests are presented in Table 6.

Table 6. Mechanical properties and nomenclature of the group of tests.

Group	Nomenclature	ε_t (%) Mean (Standard Deviation)	σ_y (MPa) Mean (Standard Deviation)	σ_U (MPa) Mean (Standard Deviation)
UNE 116005:2012 (Orientation 1)	UNE-O1	5.89 (0.26)	37.82 (0.66)	37.82 (0.66)
UNE 116005:2012 (Orientation 2)	UNE-O2	7.05 (0.11)	39.65 (1.42)	39.64 (1.42)
ASTM D638-14:2014 (Orientation 1)	ASTM-O1	9.87 (1.57)	36.61 (1.42)	33.38 (3.03)
ASTM D638-14:2014 (Orientation 2)	ASTM-O2	9.61 (0.86)	43.00 (0.49)	40.09 (2.38)

Tests performed according to standard UNE 116005:2012 present a much better repeatability than the ones according to ASTM D638–14. This is a very important finding since the UNE standard is specifically designed for specimens fabricated by additive manufacturing and proves that the geometry used is more appropriate for characterizing materials obtained by FDM than the one used from international standards such as ASTM D638–14, very often used in the scientific literature for these purposes. Except specimen 3 PROB ABS BK 5, all the specimens experience a brittle breakage without plastic deformation. On the contrary, results for ASTM D638–14 show a higher variability, especially for orientation 1, where some specimens exhibit some plastic deformation before fracture (ASTM_D638_PROB1_2, ASTM_D638_PROB1_3 y ASTM_D638_PROB1_4); elongations are particularly higher than the ones obtained by the standard UNE 116005:2012.

As a general trend, orientation 2 provides higher strength to the parts obtained by FDM according to results for both standards, but the influence of the orientation is more important in the case of the standard ASTM. This behavior is coherent with the orientation of the filaments in each layer for orientations 1 and 2, as shown in Figure 10. In this figure, the fracture surface of two of the specimens tested for each orientation (see designation in Figure 8a) is observed, showing the differences in the inner distribution of the filaments as well. These observations are in consonance with the work of Aliheidari et al. [29], who claimed that the nature of the layered structure, and, particularly, the adhesion between the layers, have a direct impact on the mechanical properties.

Observing the images in detail and placing the specimens in the direction of vertical growth of layer (Figures 11 and 12), it is possible to distinguish in both cases two lateral areas (shell) where the beads deposited during the printing have the same arrangement in all the layers, being parallel to the longitudinal axis of the specimen and a central area, in the middle of them, where the arrangement of the beads is different in alternate layers, being in one layer the longitudinal direction of the specimen, and in the above and below layers, the transverse direction. The layers in the longitudinal direction benefit from the alignment of polymer molecules along the stress axis [30].

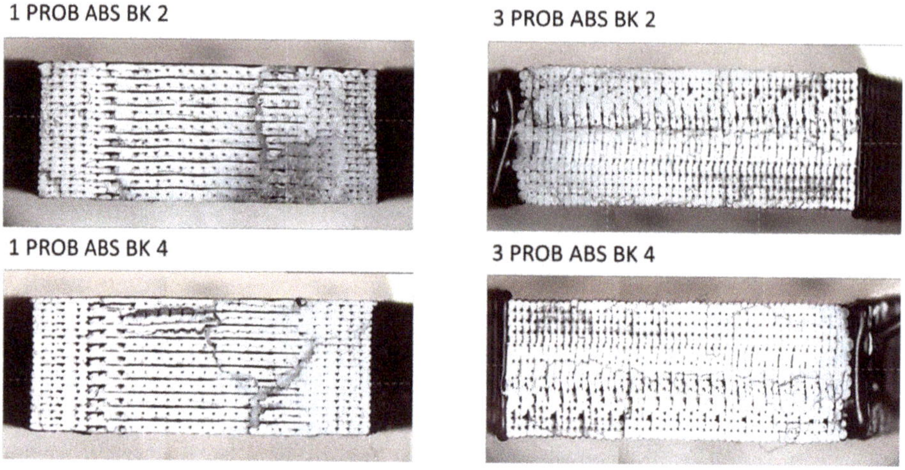

Figure 10. View of the fracture surface of two specimens tested with orientation 1 (**left**) and orientation 2 (**right**) with digital profile projector TESA-VISIO.

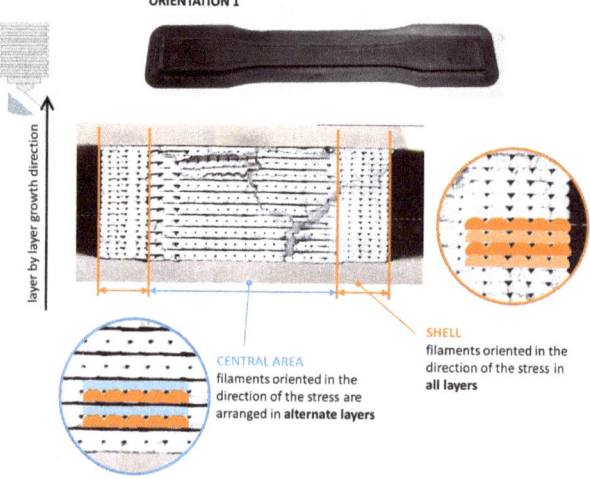

Figure 11. Layout of the filaments related to the longitudinal stress direction for specimens with orientation 1. Identification of central area and shell.

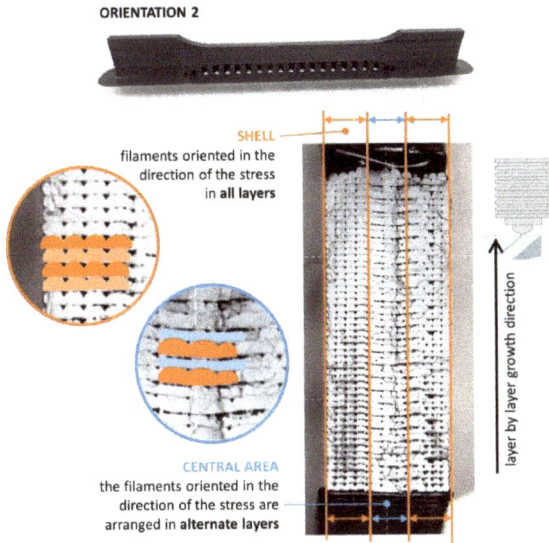

Figure 12. Layout of the filaments related to the longitudinal stress direction for specimens with orientation 2. Identification of central area and shell.

The areas where the filaments are oriented in the direction of the load (shell) are expected to have a higher strength because the filaments contribute together to the strength of the specimen. On the contrary, the central areas present a different behavior: on one hand, there are layers where the filaments are in the same direction than the load, but there are also layers where the filaments are arranged perpendicular to the stress direction, so the cohesion between filaments is crucial to keeping the integrity of the specimen. This situation is analogous to a composite reinforced by continuous or discontinuous fibres, respectively [31]. In this sense, the percentage of section with higher strength provided by the lateral areas (shell), where the filaments are arranged in the same direction than the load, is clearly higher in specimens with orientation 2; this fact justifies their better performance under tension as presented in Figure 9 and Table 6.

In addition, Figures 11 and 12 allow for appreciating a pattern in the fracture of the specimens with both orientations. Figure 13 shows this situation with more detail. This fibre discontinuity has been reported by other authors [15] as the reason for premature failure of the part, resulting in brittle fracture.

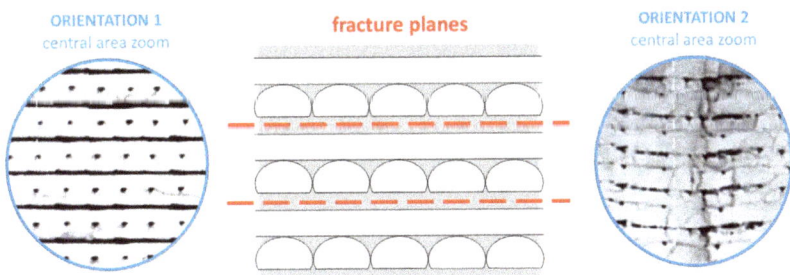

Figure 13. Fracture planes in central areas in specimens with both orientations.

Throughout the entire central area of both types of specimens, the loss of cohesion between the layers occurs in the same planes. Thus, as shown in Figure 13, the plane between the bottom of a layer (in which the filaments are in the same direction of the load) and the top of the layer below it (in which the filaments are arranged perpendicular to the load) results in being a weak point of these structures. As it is possible to appreciate in Figure 13, and also in Figures 11 and 12, the cohesion between layers along these planes fails in all cases. In non-solid specimens, the brittle interface fracture leading to delamination between layers in FDM specimens is explained by the weak interlayer bonding or to interlayer porosity, as explained in the work by Ziemian et al. [30], who emphasized that the tensile strength is greatly affected by the fiber to fiber fusion and any air gap between the fibers. Figure 14 tries to illustrate the reason of this recurring behavior in these planes. As shown in Figure 14b, when the deformation of the specimen starts the filaments parallel to the load are the ones that are really affected by the load and so they are the first to be deformed. On the other hand, the bonding between the filaments arranged perpendicular to the load is not strong enough, and it fails easily.

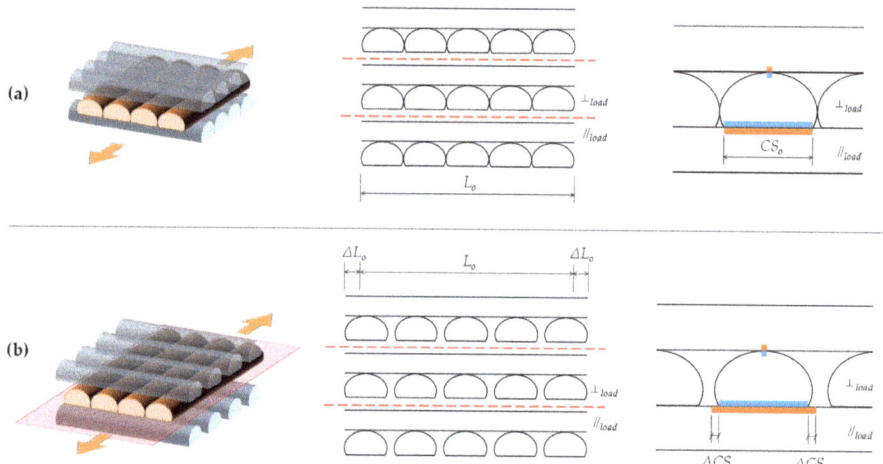

Figure 14. Elongations and displacements between layers. (a) state before deformation; (b) state after deformation.

If the elongation of the filaments oriented in the direction of the load is related to the contact surface between the filaments of adjacent layers, very different situations can be observed. As shown in Figure 14a, before the deformation of the specimen, the contact surface (CS) between two filaments located on adjacent layers is clearly defined. However, when deformation starts (Figure 14b), the filaments in the same direction of the load will suffer certain elongation while the ones arranged perpendicular to the load not. Thus, the dimensions of that initial contact surface become different because the load supported by the filaments of each layer depends on their orientation and the cohesion between filaments of different layers is not strong enough to cause equivalent deformation in alternant layers.

Considering as the initial reference a layer with the filaments oriented in the same direction of the load and given that they are the ones which mainly support the load applied and which are deformed, Figure 14b also shows as this phenomenon is very different if the upper or lower layer is considered. In the first case, the contact surface is significantly smaller, and once the union between filaments of layers oriented perpendicular to the load has failed, these filaments can move away maintaining their punctual unions with the filaments parallel to the load of the adjacent layers, which are being deformed. Thereby, as shown in Figure 14b, these relative displacements between the contact surfaces of filaments located on adjacent layers become relevant when the lower layer is considered, due to the larger initial contact surface and its different evolution in each layer. For this reason, the fracture planes shown in Figure 13 appear. Recent studies combined computational and experimental techniques [32] to evaluate the cohesive strengths between filaments and seems to be a promising field of research in fracture mechanisms of FDM parts.

3.2. Comparative Analysis with Conventional Samples with Pattern Infill

Before the comparison is done, it is important to clarify that, although the specimens in this work have been printed as solid parts, the additive manufacturing technique leaves some gaps between layers as presented in Figure 15, so the infill is supposed to be close to 100% infill, but not completely. These triangular air voids are responsible for the decrease of the tensile strength because of a decrease in the cross-sectional area of the specimen [30] compared to bulk materials. Moreover, the lower strength of FDM samples compared with injection-molded samples have been attributed to the gaps between filaments and inner pores within them [28].

Figure 15. Visualization of the gaps between layers in a specimen performed by an optical micro-coordinate measurement machine IF-SL ALICONA.

A comparison of the results to those ones obtained by Rodríguez-Panes et al. [26], where a pattern infill (non-solid) was used in the FDM specimens, is presented in this section according to Table 5. Comparison is realized with results from standard ASTM D638–14:2014 (Figure 16), as it was the same standard used in the previous work.

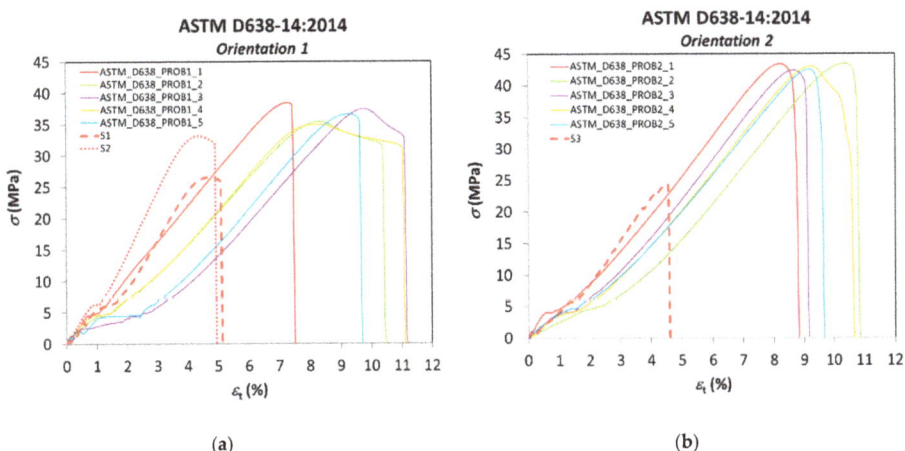

Figure 16. Comparison of results with non-solid specimens (Table 5) tested by Rodríguez-Panes et al. [26]: (a) specimens with orientation 1 and 20 and 50% infill; (b) specimen with orientation 2 and 20% infill.

Table 7 presents the data of mechanical properties obtained from graphs of Figure 16.

Table 7. Comparison of mechanical properties in samples with solid (ASTM–O1, ASTM–O2) and non-solid (S1, S2, S3) infill.

Group	Nomenclature	ε_t (%)	σ_y (MPa)	σ_U (MPa)
ASTM D638–14:2014 (Orientation 1)	ASTM–O1	9.87	36.61	33.38
	S1	5.09	26.56	25.83
	S2	4.90	33.21	32.06
ASTM D638–14:2014 (Orientation 2)	ASTM–O2	9.61	43.00	40.09
	S3	4.54	24.38	24.38

Tensile strengths of ABS parts obtained by FDM have been reported to be in the range of 11–40 MPa [10]; the explanation to this wide range is associated to their anisotropic behavior. Results are in good agreement with previous works such as the one by Tymrak et al. [33], with tensile strengths around 30 MPa, or the work by Banjanin et al. [10], with an average value of tensile stresses for ABS samples of 31 MPa, considering that they used non-solid specimens, so they are expected to have lower values than the ones obtained in our work with solid ones; in fact, these reference values are in very good agreement with the ones obtained in our previous work with non-solid samples [26], used for comparison in this section.

As expected, for the specimens with an infill percentage of 20% (S1 and S3), the mechanical properties are the poorest ones, the effect of the infill being more significant in the case of orientation 2, where the differences in tensile strength at yield and break are particularly high (almost double). On the other hand, the sample with an infill percentage of 50% (S2) shows a higher strength than samples S1 and S3, but lower than the strength provided by the solid infill of samples ASTM-O1. Nevertheless, the increase of strength due to the infill does not seem to be in the same proportion to the percentage of infill; that is, for an increase of the infill percentage of almost 400% (from 20% to almost 100% infill for solid specimens), the increase of tensile strength is only 37.83%, and, for an increase of infill percentage of 150%, the increase of tensile strength is 25.04%. In the case of the strain to break, the values for the samples with solid infill presents a much higher deformation before fracture.

4. Conclusions

In general, tests performed according to the specific Spanish standard for additive manufacturing UNE 116005:2012 present a much better repeatability than the ones according to the general test standard ASTM D638–14, which proves that the geometry of sample and procedure used in the standard UNE is more appropriate for characterizing materials obtained by FDM and for comparison between different materials. All the specimens according to the standard UNE experience a brittle breakage without plastic deformation. Results for ASTM D638–14 show a higher variability, especially for orientation 1, where some specimens exhibit some plastic deformation before fracture and elongations are particularly higher than the ones obtained by the standard UNE.

As a general trend, orientation 2 provides higher strength to the parts obtained by FDM, but the influence of the orientation is more significant in the case of the standard ASTM. This behavior is coherent with the arrangement of the filaments in each layer for each orientation, the percentage of the section with higher strength provided by the lateral areas (shell) being clearly higher in specimens with orientation 2, which justifies their better performance under tension.

Comparison with non-solid specimens (and different percentage of infill) show that, for samples with an infill percentage of 20% (S1 and S3), the mechanical properties are the poorest ones, the effect of the infill being more significant in the case of orientation 2. The sample with an infill percentage of 50% (S2) shows a higher strength than samples S1 and S3, but it is lower than the strength provided by the solid infill of samples, although the increase of strength due to the infill does not seem to be in the same proportion to the percentage of infill. In the case of the strain to break, the values for the samples with solid infill presents a much higher deformation before fracture.

As a general remark and given that national standards follow the guidelines of the international ones (acting as complementary information sources), positive experiences with national standards as the one presented in this work could be considered in future updates of international standards.

Author Contributions: Conceptualization, A.G.-D., J.C., A.M.C., and M.A.S.; methodology, A.G.-D. and J.C.; formal analysis, A.G.-D., J.C., and A.M.C.; investigation, A.G.-D., J.C., and A.M.C.; resources, J.C. and A.M.C.; writing—original draft preparation, A.G.-D., J.C., and A.M.C.; writing—review and editing, A.G.-D., J.C., A.M.C., and M.A.S.; supervision, M.A.S.; project administration, J.C. and A.M.C.; funding acquisition, J.C. and A.M.C. All authors have read and agreed to the published version of the manuscript.

Funding: This research was funded by the Annual Grants Call of the E.T.S.I. Industriales of UNED through the projects of references [2019–ICF04] and [2019–ICF09].

Acknowledgments: This work has been developed within the framework of the "Doctorate Program in Industrial Technologies" of the UNED and in the context of the project DPI2016–81943–REDT of the Ministry of Economy, Industry, and Competitiveness. We would like to extend our acknowledgement to the Research Group of the UNED "Industrial Production and Manufacturing Engineering (IPME)" and the technical staff of the Department of Manufacturing Engineering for the given support during the development of this work.

Conflicts of Interest: The authors declare no conflict of interest.

References

1. Rodríguez-Panes, A.; Claver, J.; Camacho, A.M.; Sebastián, M.A. Análisis normativo y evaluación geométrica de probetas para la caracterización mecánica de piezas obtenidas por fabricación aditiva mediante FDM. In *Actas del XXII Congreso Nacional de Ingeniería Mecánica*; Pedrero, J.I., Ed.; UNED: Madrid, Spain, 2018.
2. Popescu, D.; Zapciu, A.; Amza, C.; Baciu, F.; Marinescu, R. FDM process parameters influence over the mechanical properties of polymer specimens: A review. *Polym. Test.* **2018**, *69*, 157–166. [CrossRef]
3. ISO 527-2:2012. *Plastics-Determination of Tensile Properties-Part 2: Test Conditions for Moulding and Extrusion Plastics*; International Organization for Standardization: Geneva, Switzerland, 2012.
4. ASTM D638-14:2014. *Standard Test Method for Tensile Properties of Plastics*; ASTM International: West Conshohocken, PA, USA, 2014.
5. ISO 17296-3:2014. *Additive manufacturing-General Principles. Part 3: Main Characteristics and Corresponding Test Methods*; International Organization for Standardization: Geneva, Switzerland, 2014.
6. ISO/ASTM 52921:2013. *Standard Terminology for Additive Manufacturing-Coordinate Systems and Test Methodologies*; International Organization for Standardization: Geneva, Switzerland, 2013.
7. AENOR/UNE 116005. *Fabricación por adición de capas en materiales plásticos. Fabricación aditiva. Preparación de probetas*; AENOR: Madrid, Spain, 2012.
8. Ferro, C.G.; Brischetto, S.; Torre, R.; Maggiore, P. Characterization of ABS specimens produced via the 3D printing technology for drone structural components. *Curved Layer. Struct.* **2016**, *3*, 172–188. [CrossRef]
9. Brischetto, S.; Ferro, C.G.; Maggiore, P.; Torre, R. Compression Tests of ABS Specimens for UAV Components Produced via the FDM Technique. *Technologies* **2017**, *5*, 20. [CrossRef]
10. Banjanin, B.; Vladić, G.; Pál, M.; Baloš, S.; Dramićanin, M.; Rackov, M.; Kneţević, I. Consistency analysis of mechanical properties of elements produced by FDM additive manufacturing technology. *Rev. Mater.* **2018**, *23*, 4. [CrossRef]
11. Chacón, J.M.; Caminero, M.A.; García-Plaza, E.; Núñez, P.J. Additive manufacturing of PLA structures using fused deposition modelling: Effect of process parameters on mechanical properties and their optimal selection. *Mater. Des.* **2017**, *124*, 143–157. [CrossRef]
12. Tanikella, N.G.; Wittbrodt, B.; Pearce, J.M. Tensile strength of commercial polymer materials for fused filament fabrication 3D printing. *Addit. Manuf.* **2017**, *15*, 40–47. [CrossRef]
13. Ćwikła, G.; Grabowik, C.; Kalinowski, K.; Paprocka, I.; Ociepka, P. The influence of printing parameters on selected mechanical properties of FDM/FFF 3D-printed parts. *IOP Conf. Ser. Mater. Sci. Eng.* **2017**, *227*, 1–11. [CrossRef]
14. Kuznetsov, V.; Solonin, A.; Urzhumtsev, O.; Schilling, R.; Tavitov, A. Strength of PLA components fabricated with fused deposition technology using a desktop 3D printer as a function of geometrical parameters of the process. *Polymers* **2018**, *10*, 313. [CrossRef]

15. Rajpurohit, S.R.; Dave, H.K. Analysis of tensile strength of a fused filament fabricated PLA part using an open-source 3D printer. *Int. J. Adv. Manuf. Technol.* **2019**, *101*, 1525–1536. [CrossRef]
16. Zaldivar, R.J.; Witkin, D.B.; McLouth, T.; Patel, D.N.; Schmitt, K.; Nokes, J.P. Influence of processing and orientation print effects on the mechanical and thermal behavior of 3D-Printed ULTEM 9085 Material. *Addit. Manuf.* **2017**, *13*, 71–80. [CrossRef]
17. García-Domínguez, A.; Claver-Gil, J.; Sebastián, M.A. Proposals for the optimization of pieces produced by additive manufacturing. *Dyna* **2018**, *94*, 293–300. [CrossRef]
18. García-Domínguez, A. Metodología para la optimización de piezas producidas por fabricación aditiva en estrategias de mass customization (unpublished Doctoral Dissertation). Ph.D. Thesis, Universidad Nacional de Educación a Distancia (UNED), Madrid, Spain, 2019.
19. Chen, W.; Zheng, X.; Liu, S. Finite-element-mesh based method for modeling and optimization of lattice structures for additive manufacturing. *Materials* **2018**, *11*, 2073. [CrossRef] [PubMed]
20. Hussein, A.Y. The Development of Lightweight Cellular Structures for Metal Additive Manufacturing. Ph.D. Thesis, University of Exeter, Devon, UK, 2013.
21. Mahmoud, D.; Elbestawi, M. Lattice Structures and Functionally Graded Materials Applications in Additive Manufacturing of Orthopedic Implants: A Review. *J. Manuf. Mater. Process.* **2017**, *1*, 13. [CrossRef]
22. Maliaris, G.; Sarafis, E. Mechanical behavior of 3D printed stochastic lattice structures. In Proceedings of the 8th International Conference on Materials Structure and Micromechanics of Fracture, Brno, Czech Republic, 27–29 June 2016.
23. Panda, B.N. Design and development of cellular structure for Additive Manufacturing. Master's Thesis, Instituto Superior Técnico, Universidade de Lisboa, Lisbon, Portugal, 2015.
24. Weeger, O.; Boddeti, N.; Yeung, S.; Kaijima, S.; Dunn, M.L. Digital Design and Manufacture of Soft Lattice Structures. *Addit. Manuf.* **2019**, *25*, 39–49. [CrossRef]
25. Vannutelli, R. Mechanical Behavior of 3D Printed Lattice-Structured Materials. Master's Thesis, Youngstown State University, Youngstown, OH, USA, 2017.
26. Rodríguez-Panes, A.; Claver, J.; Camacho, A. The Influence of Manufacturing Parameters on the Mechanical Behaviour of PLA and ABS Pieces Manufactured by FDM: A Comparative Analysis. *Materials* **2018**, *11*, 1333. [CrossRef] [PubMed]
27. Masood, S.H.; Mau, K.; Song, W. Tensile properties of processed fdm polycarbonate material. *Mater. Sci. Forum* **2010**, *654–656*, 2556–2559. [CrossRef]
28. Wu, W.; Geng, P.; Li, G.; Zhao, D.; Zhang, H.; Zhao, J. Influence of Layer Thickness and Raster Angle on the Mechanical Properties of 3D-Printed PEEK and a Comparative Mechanical Study between PEEK and ABS. *Materials* **2015**, *8*, 5834–5846. [CrossRef]
29. Aliheidari, N.; Tripuraneni, R.; Ameli, A.; Nadimpalli, S. Fracture resistance measurement of fused deposition modeling 3D printed polymers. *Polym. Test.* **2017**, *60*, 94–101. [CrossRef]
30. Ziemian, C.; Sharma, M.; Ziemian, S. Anisotropic mechanical properties of ABS parts fabricated by fused deposition modelling. In *Mechanical Engineering*; Murat, G., Ed.; IntechOpen: Rijeka, Croatia, 2012; pp. 159–180. ISBN 978-953-51-0505-3.
31. Chacón, J.M.; Caminero, M.A.; Núñez, P.J.; García-Plaza, E.; García-Moreno, I.; Reverte, J.M. Additive manufacturing of continuous fibre reinforced thermoplastic composites using fused deposition modelling: Effect of process parameters on mechanical properties. *Compos. Sci. Technol.* **2019**, *181*, 107688. [CrossRef]
32. Li, J.; Yang, S.; Li, D.; Chalivendra, V. Numerical and experimental studies of additively manufactured polymers for enhanced fracture properties. *Eng. Fract. Mech.* **2018**, *204*, 557–569. [CrossRef]
33. Tymrak, B.M.; Kreiger, M.; Pearce, J.M. Mechanical properties of components fabricated with open-source 3-D printers under realistic environmental conditions. *Mater. Des.* **2014**, *58*, 242–246. [CrossRef]

© 2019 by the authors. Licensee MDPI, Basel, Switzerland. This article is an open access article distributed under the terms and conditions of the Creative Commons Attribution (CC BY) license (http://creativecommons.org/licenses/by/4.0/).

Article

Influence of Print Orientation on Surface Roughness in Fused Deposition Modeling (FDM) Processes

Irene Buj-Corral *, Alejandro Domínguez-Fernández and Ramón Durán-Llucià

School of Engineering of Barcelona (ETSEIB), Department of Mechanical Engineering, Universitat Politècnica de Catalunya (UPC), Avinguda Diagonal, 647, 08028 Barcelona, Spain; alejandro.dominguez-fernandez@upc.edu (A.D.-F.); ramon.duran@estudiant.upc.edu (R.D.-L.)
* Correspondence: Irene.buj@upc.edu; Tel.: +34-934054015

Received: 22 October 2019; Accepted: 19 November 2019; Published: 21 November 2019

Abstract: In the present paper, we address the influence of print orientation angle on surface roughness obtained in lateral walls in fused deposition modelling (FDM) processes. A geometrical model is defined that considers the shape of the filaments after deposition, in order to define a theoretical roughness profile, for a certain print orientation angle. Different angles were considered between 5° and 85°. Simulated arithmetical mean height of the roughness profile, Ra values, were calculated from the simulated profiles. The Ra simulated results were compared to the experimental results, which were carried out with cylindrical PLA (polylactic acid) samples. The simulated Ra values were similar to the experimental values, except for high angles above 80°, where experimental roughness decreased while simulated roughness was still high. Low print orientation angles show regular profiles with rounded peaks and sharp values. At a print orientation angle of 85°, the shape of the profile changes with respect to lower angles, showing a gap between adjacent peaks. At 90°, both simulated and experimental roughness values would be close to zero, because the measurement direction is parallel to the layer orientation. Other roughness parameters were also measured: maximum height of profile, Rz, kurtosis, Rku, skewness, Rsk, and mean width of the profile elements, Rsm. At high print orientation angles, Rz decreases, Rku shifts to positive, Rsk slightly increases, and Rsk decreases, showing the change in the shape of the roughness profiles.

Keywords: Fused Deposition Modeling; roughness; Polylactic Acid; print orientation angle; build angle

1. Introduction

In the fused deposition modelling (FDM) process, a filament is heated and then the material is deposited by a nozzle onto a printing bed. FDM printed parts are used in different applications, for example medical, electrical, aerospace, etc. For example, it allows printing patterns for investment casting of biomedical implants [1]. In addition, highly metallic-filled conductive composites can be prepared by FDM to be used in electromagnetic shielding, sensors, and circuit printing [2]. As for aerospace, carbon fiber reinforced PLA printed composites can be used [3].

FDM allows a wide range of materials, and the printed parts have effective mechanical properties. However, printing speed is low and the layer-by-layer building of parts leads to poor surface roughness due to the stair stepping effect [4,5]. When the lateral walls of a certain workpiece are inclined, the use of printing supports is required. In addition, the inclination of the lateral walls will have an effect on surface roughness, since the wall will not be perpendicular to the layer plane.

Different authors have studied the effects of printing parameters on surface roughness. For example, Pérez et al. [6] considered layer height, printing speed, temperature, printing path, and wall thickness. They found that layer height and wall thickness had the greatest influence on arithmetical mean height, Ra. Reddy et al. [7] used layer thickness, material infill, and printing quality as factors. They also considered build inclination. Both layer thickness and build inclination turned out to be the most influential factors on roughness. Peng and Yan [8] optimized roughness and energy consumption. They employed layer thickness, printing speed, and infill ratio as factors, with layer height being the most important parameter influencing roughness. Kovan et al. [9] studied the effect of layer height and printing temperature on surface roughness. You [10] studied infill ratio, printing temperature, and printing speed. They found that roughness increases with printing speed and decreases with infill ratio. Altan et al. [11] studied the effect of printing processes on surface roughness and tensile strength, with layer thickness and deposition head velocity being the most influential parameters on roughness. Mohamed et al. [12] investigated the effect of printing parameters on the dynamic mechanical properties of polycarbonate–acrylonitrile butadiene styrene (PC-ABS) printed parts. The main factors were layer height, air gap, and the number of contours. Luis studied Ra and Rq values obtained through experimental tests in FDM processes [13].

Regarding previous geometrical models for roughness in FDM processes, Pandey et al. obtained a semiempirical model for roughness, in which they took into account both layer thickness and build orientation [14]. Ahn et al. considered the filaments to have the shape of elliptical curves which overlap in the vertical direction [15]. Boschetto et al. approximated the roughness profiles of printed parts as a sequence of circumference arcs [16]. Ding et al. obtained roughness profiles from the overlapping of different surfaces representing beads [17]. Kaji and Barari obtained roughness profiles from the cusp geometry of the lateral walls of parts, taking into account both straight lines and degree two polynomial curves [18]. On the other hand, the Slic3r manual considers the shape of the cross section of the deposited filaments to be a rectangle with round ends, in which the initial area of the filament is equal to its final area [19]. A similar approach was employed by Jin et al. However, the length of the rectangle in their cross-section model is calculated based on the volume conservation, taking into account the plastic flow-rate and speed-rate [20]. From the assumptions made in [19], Buj et al. calculated pore size from the nozzle diameter, infill, and layer height of printed samples [21]. Other authors take into account the overlapping among filaments, due to diffusion when printing high melting temperature thermoplastic polymers such as polyether ether ketone (PEEK) [22]. However, this effect is not so important with low melting temperature polymers like polylactic acid (PLA) and acrylonitrile butadiene styrene (ABS).

Regarding print orientation, Bottini and Boschetto investigated the effect of deposition angle and interference grade on the assembly and disassembly forces in the interference fit of FDM printed parts [23]. They found that assembly forces depend on both parameters, while disassembly forces do not depend on deposition angle, as surface morphology is modified as a result of assembly. In addition, different authors have studied the influence of print orientation on the mechanical strength of parts [24]. Domingo-Espín et al. studied six different orientations and determined stiffness and tensile strength of polycarbonate (PC) samples [25]. They recommended that, when the yield strength of a material is exceeded, the parts should be oriented in a way that the greater tensile stresses are aligned with the direction of the longest contours, in order increase their tensile strength. Knoop et al. studied the effect of building orientation on the tensile, flexural, and compressive strength of polyamide (PA) parts [26]. As a general trend, they found higher tensile strength for build orientation X of the tensile test specimens (on its edge), than for build orientation Y (flat lying), or Z (upright). Uddin et al. studied the effect of print orientation on the tensile and compressive strength of ABS parts [27]. They obtained the highest stiffness and failure strength for layer thickness 0.09 mm, printing plane YZ and horizontal print orientation. Chacón et al. [28] studied the influence of print orientation on the tensile and flexural strength of PLA parts. They observed that low layer thickness and high feed rate values improved mechanical performance. Sood et al. [29] investigated the effect of layer thickness, build orientation, raster angle, raster width, and air gap on the compressive strength of parts. They found that an artificial neural network (ANN) model was better for modeling compressive strength than a regression model. The optimal value for layer orientation, giving higher compressive strength, was 0.036°. McLouth et al. [30] analyzed the influence of print orientation and raster pattern on the fracture toughness of ABS parts. They concluded that samples with layers that are parallel to the crack plane turned out to have lower fracture toughness than samples with other print orientations. As for the influence of print orientation on roughness, Chaudhari et al. studied the surface finish of ABS parts printed with different layer thickness, infill, orientation, and postprocessing operation. They found that infill and postprocessing had the greatest influence on roughness [31]. Thrimurthulu et al. [32] simultaneously optimized surface roughness and build time, as a function of slice thickness and build deposition orientation. Both parameters influenced roughness. Wang et al. [33] studied the effects of: layer thickness, deposition style, support style, deposition orientation in the Z direction (build angle), deposition orientation in the X direction (raster angle), and build location on the tensile strength, dimensional accuracy, and surface roughness of printed parts. They observed that layer thickness was the most influential parameter.

The aim of the present paper is to define a geometrical model for surface roughness in lateral walls, in FDM printing processes. The model considers the different print orientations, with simulated results being compared to experimental results. To do so, cylindrical samples are printed with different print orientations of between 0° and 85°, in PLA. Roughness is measured along the generatrix of the samples, by means of a contact roughness meter. Then, the results from the model are compared to the experimental results for different print orientation angles.

2. Materials and Methods

2.1. Geometrical Model

A geometrical model was defined to calculate roughness in lateral walls, for parts with different print orientations. Two assumptions were made (Figure 1):

- The shape of the cross-section of the filaments after deposition is a rectangle with rounded edges, with a semicircle at each side [19].
- There is no overlapping of adjacent filaments due to material diffusion, since processing temperatures are not excessively high.

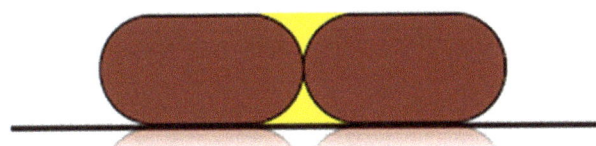

Figure 1. Schematic of the cross-section of two adjacent deposited filaments with print orientation angle of 0° (the horizontal line corresponds to the printing bed).

Considering these assumptions, arithmetical mean height Ra values were calculated for each print orientation, according to the following procedure:

1. The geometry of two deposited filaments (one on top of the other) is drawn for each print orientation studied, using the Solid Woks 2017 software (Dassault Systèmes Solidworks Corporation, Waltham, MA, USA). The tangent line at the edge of the two filaments is determined and the figure is rotated until the tangent line becomes a horizontal line. Figure 2 shows an example for print orientation angle of 45°.

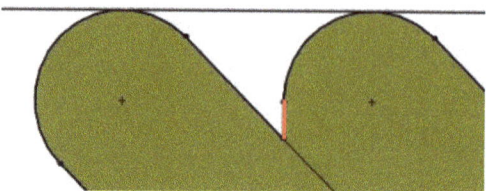

Figure 2. Schematic of the cross-section of two deposited filaments with print orientation angle of 45°.

2. The shape of the edges of the two filaments is considered to be the theoretical roughness profile of the lateral wall of the parts. In order to avoid profiles with negative draft angle from the vertical direction (which are not found in experimental roughness profiles), vertical lines are drawn in the area where the end of one filament adjoins the other filament, if necessary (see red line in Figure 2).
3. The total measurement length of the profiles was defined as the distance between the centers of the circumferences of the edges of the two layers (Figure 3).

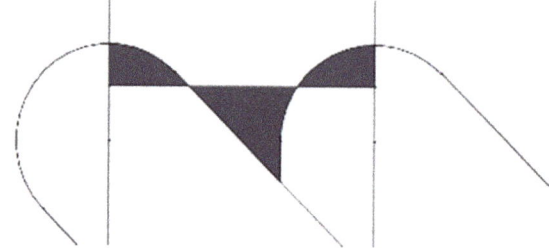

Figure 3. Profile for print orientation angle of 45°, with the areas highlighted in grey.

4. The center line of the profiles was found with Solid Works, taking into account the mean value theorem for integrals. The center line divides a profile function into two parts, so that the areas contained by the profile above and below the center line are equal (Figure 3). The first mean value theorem for integrals says that for all continuous functions in the area [a, b] a point c exists within the interval [a, b], which makes the area below the function equal to its image at point c for all the interval length, according to Equation (1).

$$(b-a) \cdot f(c) = \int_a^b f(x)\, dx. \tag{1}$$

5. The arithmetical mean height roughness parameter Ra (in μm) was calculated according to Equation (2).

$$Ra = \frac{1}{L} \int_0^L |f(x)|\, dx \tag{2}$$

where L is the measurement length in mm, and $f(x)$ is the discrete function that defines the roughness profile, in mm.

In order to compare the simulated results of the model with the experimental results obtained with a contact roughness meter, the geometry of the roughness meter tip was added to the ideal geometry of the layers. Its cross-section was assumed to be an isosceles rectangle triangle of 1 mm height, with sharp edges.

Two different cases were found:

(a) For print orientation angles lower or equal to 45°, the tip leans on two surfaces, and a new profile is obtained which shows shallower valleys than the previous one (Figure 4).

Figure 4. Schematic of the printed layers with the roughness tip, for print orientation angles higher than 45°.

(b) For print orientation angles higher than 45°, the tip leans on one of the two sides of the profile. Moreover, it is not able to reach the lowest part of the profile (Figure 5). The modified valleys have the same depth as the original ones, but the shape of the profile changes.

Figure 5. Representation of the roughness tip with the printed layers, for print orientation angle higher than 45°.

New simulated Ra values were calculated from the modified profiles.

2.2. Printing Process

A double extruder Sigma printer from BCN3D Technologies (Barcelona, Spain) was used. Cylindrical PLA samples were printed, of 12.7 mm diameter and 25.4 mm height, according to a height-to-width ratio of 2.

Printing parameters are provided in Table 1 (Appendix A).

Table 1. Printing parameters of the experimental tests.

Parameter	Values
Layer height (mm)	0.25
Infill ratio (%)	50
Nozzle diameter (mm)	0.4
Printing speed (mm/s)	60
Printing temperature (°C)	205
Print orientation angle (°)	From 5 to 85

Layer height is the thickness of each deposited layer. Infill ratio is the amount of solid material within the volume of a printed structure. Infill type was rectangular in all cases, with raster angle 0°. Air gap is the space between filaments, and depends on the infill ratio used. Shells are the layers that are printed around the infill area. No shell was printed in this case.

Print orientation angle and build angle are complimentary angles. They are shown in Figure 6.

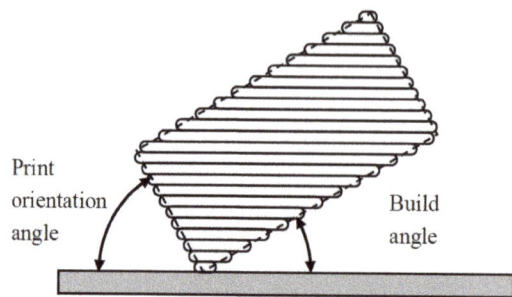

Figure 6. Schematic of a printed part with the print orientation angle and the build angle.

2.3. Roughness Measurement

Roughness was measured in a contact Taylor Hobson Talysurf 2 roughness meter (AMETEK Inc., Berwyn, PA, USA), with two different Gaussian filters of cut-off 8 mm and 2.5 mm respectively. Several roughness parameters were taken into account: arithmetical mean height, Ra, maximum height of the profile, Rz, kurtosis, Rku, skewness, Rsk, and mean width of the profile elements, Rsm.

Measurement direction coincides with one generatrix of the cylinders, specifically the one that is placed opposite the printing supports. As an example, the blue lines in Figure 7 show the measuring direction of two specimens with different print orientation angles.

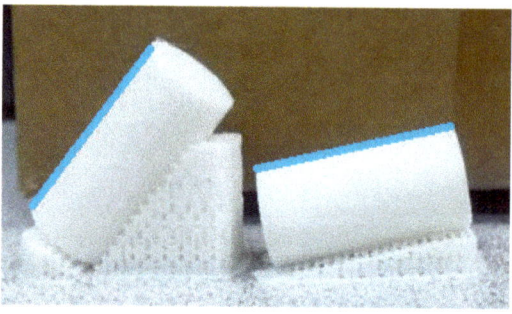

Figure 7. Printed specimens with the measurement direction highlighted in blue.

If a print orientation angle of 0° were considered, there would be no need to use printing supports. Thus, roughness would be measured along any generatrix of the specimen.

3. Results

3.1. Roughness Profiles

As an example, Figure 8 presents experimental roughness profiles for different print orientation angles. A print orientation angle of 5° (Figure 8a) corresponds to a regular profile, with the typical shape obtained in lateral walls when layers have no inclination, in FDM processes. The profile shows rounded peaks and sharp valleys, and the peak width corresponds to the layer height employed. As the angle increases, similar profiles are obtained, for example for a print orientation angle of 55° (Figure 8b). For a print orientation angle of 80°, a sawtooth shape is observed for the profile. For a print orientation angle of 85°, the profile becomes more irregular, combining high peaks for the filament edges with a transition flat area between consecutive peaks. The distance between peaks increases. At a print orientation angle of 90°, the layers would be parallel to the direction in which roughness is measured. For this reason, the theoretical roughness value would be zero.

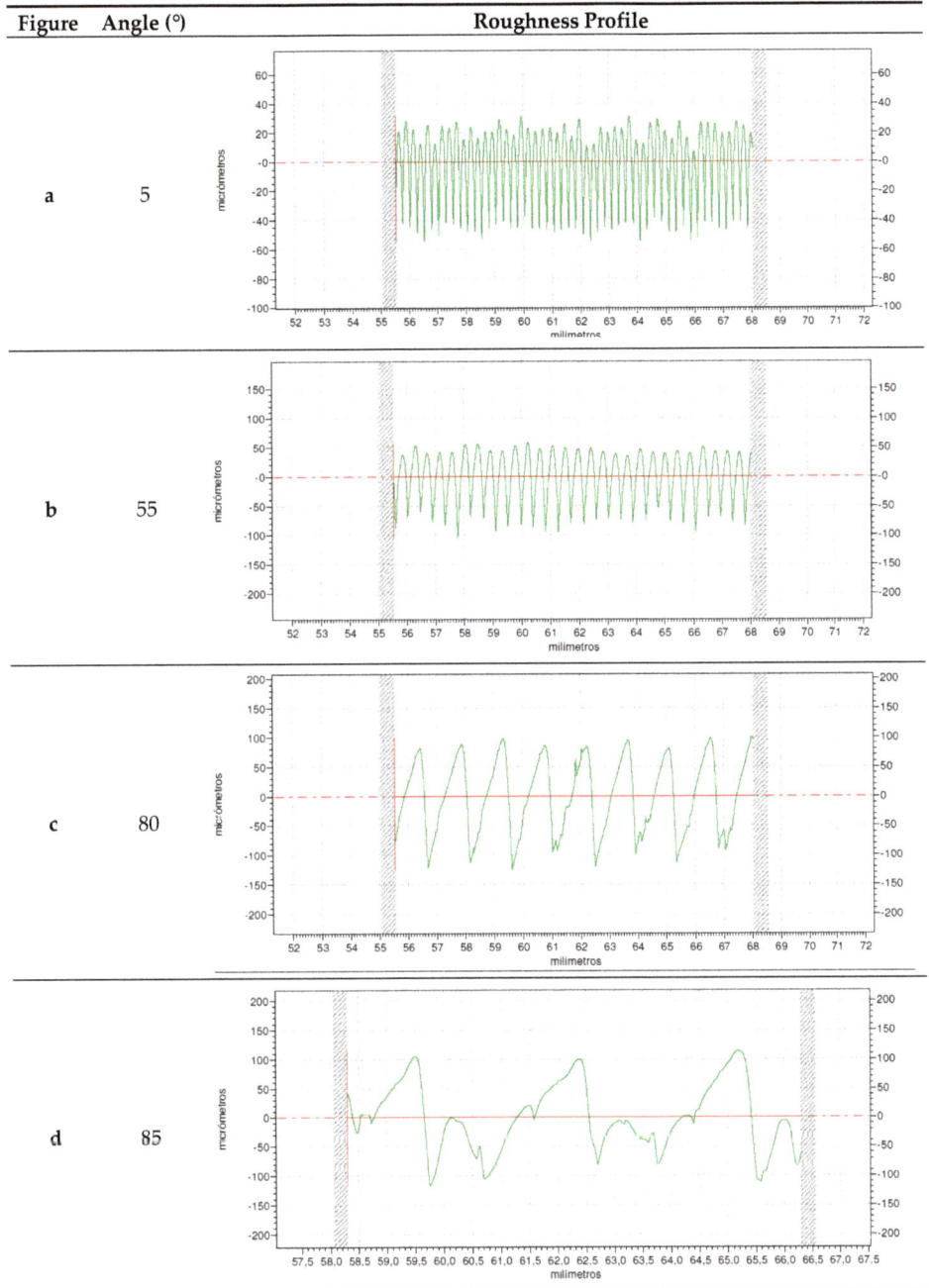

Figure 8. Roughness profiles for print orientation angle of: (**a**) 5°, (**b**) 55°, (**c**) 80°, and (**d**) 85°.

Figure 9 shows a picture (plan view) of a sample manufactured with print angle of 85°.

Figure 9. Plan view of a sample with print orientation angle of 85°.

As print orientation angle increases, the stair-stepping effect becomes more evident. It can be observed that the high inclination of layers leads to a greater distance between crests, with wide plateaus that provide lower roughness values. In addition, the measured roughness profile in this case is more irregular than the rest of the profiles (Figure 8d), causing greater discrepancy between experimental and simulated roughness values.

3.2. Roughness Values

Figure 10 presents the simulated Ra results, considering the tip geometry or not, as well as the measured roughness with cut-off of either 8 mm or 2.5 mm. According to ISO 4288 standard [31], a cut-off value of 2.5 mm is recommended for Ra values between more than 2 µm and 10 µm, and a cut-off value of 8 mm is recommended for Ra values higher than 10 µm. Error bars correspond to ± standard deviation values.

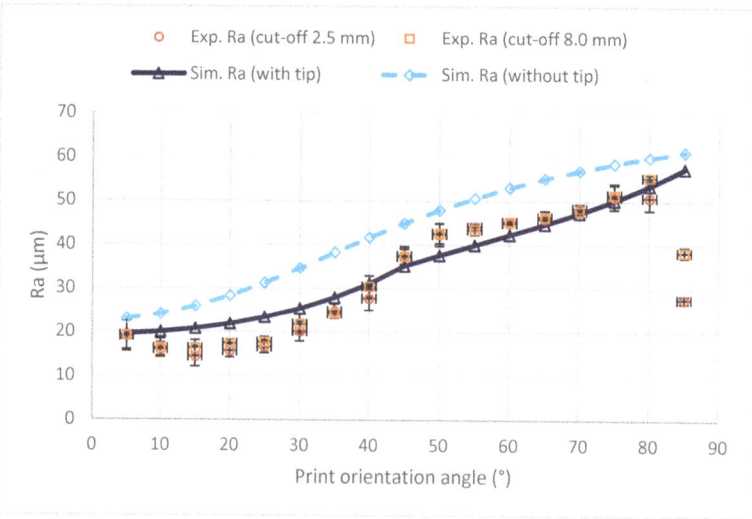

Figure 10. Arithmetical mean height (Ra) vs. print orientation angle.

In all cases, as expected, the roughness results simulated with the tip were lower than those simulated without the tip, since the tip reduces the valley depth of the profile. As a general trend, the experimental values agree with the simulated values with tip up to a print orientation angle of 80°. The results agree with those of Reddy et al. [4], who found that Ra decreases with build angle, which is the complimentary angle of the print orientation angle. They found maximum Ra values of 50 μm for build angles of 10° (printing angle of 80°). However, in the present work, at 85° the experimental roughness decreases significantly with respect to 80°. Such decrease is more important for the cut-off of 2.5 mm than for the cut-off of 8 mm. This suggests that the abrupt transition from high simulated roughness values at the print orientation angle of 80° to the zero simulated roughness value at the print orientation angle of 90° is more gradual in the experimental tests.

In order to analyse the shape of the roughness profiles at high print orientation values, Table 2 provides the experimental values of other roughness parameters, Rz, Rsk, Rku, and Rsm, measured with a cut-off of 8 mm.

Table 2. Rz, Rsk, Rku, and Rsm values.

Print Angle (°)	Mean Value Rz (μm)	Standard Deviation Rz (μm)	Mean Value Rsk	Standard Deviation Rsk	Mean Value Rku	Standard Deviation Rku	Mean Value Rsm (μm)	Standard Deviation Rsm (μm)
50	161.211	17.050	−0.433	0.091	2.032	0.092	388.588	2.344
55	169.910	19.409	−0.372	0.039	1.875	0.097	437.863	1.618
60	187.277	7.030	−0.201	0.033	1.742	0.040	501.911	1.214
65	183.679	12.237	−0.121	0.050	1.748	0.033	593.597	2.417
70	196.639	3.787	−0.258	0.012	1.872	0.002	728.450	3.314
75	225.699	20.530	−0.153	0.103	1.788	0.058	963.619	3.645
80	237.129	10.838	0.035	0.104	1.791	0.017	1428.000	4.048
85	211.161	27.924	0.050	0.137	2.312	0.034	1226.810	8.924

Rz increases with print orientation angle, as expected, up to 80°, and then decreases at print orientation angle of 85°. Skewness shows negative values up to 70°, corresponding to higher valleys than peaks (Figure 8b). At 75° and 80° skewness values are close to zero, corresponding to symmetric profiles (Figure 8c). At 85°, skewness has a positive value, with higher peaks than valleys (Figure 8d). Kurtosis is lower than 3 in all cases, pointing out that the peaks are sharper than those corresponding to a normal distribution of heights. At print orientation angle of 85°, the highest Rku value is obtained of 2.312, corresponding to rounder peaks and valleys. Parameter RSm, mean width of the profile elements, increases with print orientation angle, since the effective distance between layers increases. However, at 85° the parameter decreases, because small roughness peaks are measured in the gaps between adjacent peaks (Figure 8d).

4. Discussion

The proposed model allows simulating Ra values to be obtained in lateral walls of FDM printed parts. Unlike other models, which take into account overlapping among adjacent deposited filaments [15,17], the present model makes the assumption that printing temperature is low enough to avoid overlapping. It also assumes that the shape of the cross-section of the deposited filament is rectangular with rounded edges [19,20].

Experimental Ra values are similar to simulated ones at low print orientation angles, and they increase with print orientation angle as reported by Reddy et al. [7]. However, at high angles above 80°, the experimental roughness values are lower than the simulated ones. This suggests a gradual decrease in the experimental roughness between 80° and 90°. At 90°, the printing direction would be parallel to the measuring direction and, for this reason, the experimental roughness values would be close to zero.

At low print orientation angles, regular profiles are obtained with round peaks and sharp valleys, which are typical of FDM processes [34]. At a high print orientation angle of 85°, the distance between consecutive peaks increases, leading to a flat area or gap. In this case, not only the arithmetical mean height of the profile Ra decreases but the maximum height of profile Rz and the mean width of the profile Rsm. Skewness parameter Rsk becomes positive and kurtosis parameter Rku increases, noting the change in the profile shape [35].

In the future, a similar methodology using the mean value theorem for integrals, can will be applied to calculate simulated Ra in other manufacturing processes, either additive manufacturing processes or subtractive processes, provided that the theoretical geometry of the roughness profile can be obtained.

5. Conclusions

This paper presents a geometrical model for the simulation of roughness profiles obtained with different print orientation angles in FDM processes, in order to determine the mean height of the roughness profile, Ra. In addition, experimental tests were performed. The main conclusions of the paper are as follows:

- Use of the mean value theorem for integrals allows calculating Ra from the geometrical model of the roughness profile in a simple way. This methodology is also valid in case the assumptions of the model need to be varied, or even for other manufacturing processes.
- At low print orientation angles, regular profiles are obtained, in which peak amplitude corresponds to layer height. At high print orientation angles, peak width increases, with a flat area or gap between consecutive peaks.
- As a general trend, both simulated and experimental amplitude roughness values increase with print orientation angle, as the stair-stepping effect is accentuated. However, simulated roughness results decrease abruptly (simulated roughness would be zero at 90° because the roughness measurement direction coincides with the direction of the printed layers), while experimental results show a more gradual decrease starting at around 85°.
- At a high print orientation angle of 85°, skewness parameter Sku becomes positive, kurtosis parameter Rku increases, and the mean width of the profile Rsm shows a slight decrease with respect to 80°, thus noticing the change in the shape of the roughness profile.

Author Contributions: conceptualization, I.B.-C. and R.D.-L.; methodology, I.B.-C. and R.D.-L.; software, A.D.-F.; validation, I.B.-C. and A.D.-F.; formal analysis, I.B.-C. and A.D.-F.; investigation, I.B.-C., A.D.-F. and R.D.-L.; resources, I.B.-C.; data curation, I.B.-C., A.D.-F. and R.D.-L.; Writing—original draft preparation, I.B.-C. and R.D.-L.; Writing—review and editing, I.B.-C.; visualization, A.D.-F.; supervision, I.B.-C.; project administration, I.B.-C.; funding acquisition, I.B.-C.

Funding: This research was funded by the Spanish Ministry of Industry, Economy and Competitiveness, grant number DPI2016-80345R.

Acknowledgments: The authors thank Ramón Casado-López for his help with experimental tests.

Conflicts of Interest: The authors declare no conflict of interest.

Appendix A

List of printing parameters

[profile]

layer_height = 0.25

wall_thickness = 1.2

retraction_enable = True

solid_layer_thickness = 1.2

fill_density = 50

print_speed = 60

print_temperature = 205

print_temperature2 = 205

print_temperature3 = 0

print_temperature4 = 0

print_temperature5 = 0

print_bed_temperature = 65

support = Everywhere

platform_adhesion = Raft

support_dual_extrusion = First extruder

wipe_tower = False

wipe_tower_volume = 50

ooze_shield = False

filament_diameter = 2.85

filament_diameter2 = 2.85

filament_diameter3 = 0

filament_diameter4 = 0

filament_diameter5 = 0

filament_flow = 100

nozzle_size = 0.4

retraction_speed = 40

retraction_amount = 6.8

retraction_dual_amount = 3

retraction_min_travel = 1.5

retraction_combing = No Skin

retraction_minimal_extrusion = 0

retraction_hop = 0.08

bottom_thickness = 0.2

layer0_width_factor = 100

object_sink = 0

overlap_dual = 0.15

travel_speed = 200

bottom_layer_speed = 35

infill_speed = 35

solidarea_speed = 35

inset0_speed = 35

insetx_speed = 35

cool_min_layer_time = 5

fan_enabled = True

skirt_line_count = 2

skirt_gap = 2

skirt_minimal_length = 150.0

fan_full_height = 0.5

fan_speed = 85

fan_speed_max = 100

cool_min_feedrate = 10

cool_head_lift = False

solid_top = True

solid_bottom = True

fill_overlap = 15

perimeter_before_infill = True

support_type = Lines

support_angle = 20

support_fill_rate = 50

support_xy_distance = 0.6

support_z_distance = 0.15

spiralize = False

simple_mode = False

brim_line_count = 5

raft_margin = 3.0

raft_line_spacing = 3.0

raft_base_thickness = 0.3

raft_base_linewidth = 1.0

raft_interface_thickness = 0.28

raft_interface_linewidth = 0.6

raft_airgap_all = 0.0

raft_airgap = 0.22

raft_surface_layers = 2

raft_surface_thickness = 0.15

raft_surface_linewidth = 0.4

fix_horrible_union_all_type_a = True

fix_horrible_union_all_type_b = False

fix_horrible_use_open_bits = False

fix_horrible_extensive_stitching = False

plugin_config = (lp1

    (dp2

    S'params'

    p3

    (dp4

    sS'filename'

    p5

    S'RingingRemover.py'

    p6

    sa.

object_center_x = −1

object_center_y = −1

References

1. Singh, D.; Singh, R.; Boparai, K.S. Development and surface improvement of FDM pattern based investment casting of biomedical implants: A state of art review. *J. Manuf. Process.* **2018**, *31*, 80–95. [CrossRef]
2. Kwok, S.W.; Goh, K.H.H.; Tan, Z.D.; Tan, S.T.M.; Tjiu, W.W.; Soh, J.Y.; Ng, Z.J.G.; Chan, Y.Z.; Hui, H.K.; Goh, K.E.J. Electrically conductive filament for 3D-printed circuits and sensors. *Appl. Mater. Today* **2017**, *9*, 167–175. [CrossRef]
3. Tian, X.; Liu, T.; Yang, C.; Wang, Q.; Li, D. Interface and performance of 3D printed continuous carbon fiber reinforced PLA composites. *Compos. Part A Appl. Sci. Manuf.* **2016**, *88*, 198–205. [CrossRef]
4. Gibson, I.; Rosen, D.; Stucker, B. *Additive Manufacturing Technologies: 3D Printing, Rapid Prototyping, and Direct Digital Manufacturing*; Springer: New York, NY, USA, 2014; ISBN 9781493921126.
5. Ngo, T.D.; Kashani, A.; Imbalzano, G.; Nguyen, K.T.Q.; Hui, D. Additive manufacturing (3D printing): A review of materials, methods, applications and challenges. *Compos. Part B Eng.* **2018**, *143*, 172–196. [CrossRef]
6. Pérez, M.; Medina-Sánchez, G.; García-Collado, A.; Gupta, M.; Carou, D. Surface quality enhancement of fused deposition modeling (FDM) printed samples based on the selection of critical printing parameters. *Materials* **2018**, *11*, 1382. [CrossRef] [PubMed]
7. Reddy, V.; Flys, O.; Chaparala, A.; Berrimi, C.E.; V, A.; Rosen, B. Study on surface texture of Fused Deposition Modeling. *Procedia Manuf.* **2018**, *25*, 389–396. [CrossRef]
8. Peng, T.; Yan, F. Dual-objective Analysis for Desktop FDM Printers: Energy Consumption and Surface Roughness. In Proceedings of the Procedia CIRP, Gulf of Naples, Italy, 18–20 July 2018; Volume 69, pp. 106–111.
9. Kovan, V.; Tezel, T.; Topal, E.S.; Camurlu, H.E. Printing Parameters Effect on Surface Characteristics of 3D Printed Pla Materials. *Mach. Technol. Mater.* **2018**, *12*, 266–269.
10. You, D.H. Optimal printing conditions of PLA printing material for 3D printer. *Trans. Korean Inst. Electr. Eng.* **2016**, *65*, 825–830. [CrossRef]

11. Altan, M.; Eryildiz, M.; Gumus, B.; Kahraman, Y. Effects of process parameters on the quality of PLA products fabricated by fused deposition modeling (FDM): Surface roughness and tensile strength. *Mater. Test.* **2018**, *60*, 471–477. [CrossRef]
12. Mohamed, O.A.; Masood, S.H.; Bhowmik, J.L. Characterization and dynamic mechanical analysis of PC-ABS material processed by fused deposition modelling: An investigation through I-optimal response surface methodology. *Meas. J. Int. Meas. Confed.* **2017**, *107*, 128–141. [CrossRef]
13. Luis Pérez, C.J. Analysis of the surface roughness and dimensional accuracy capability of fused deposition modelling processes. *Int. J. Prod. Res.* **2002**, *40*, 2865–2881. [CrossRef]
14. Pandey, P.M.; Reddy, N.V.; Dhande, S.G. Improvement of surface finish by staircase machining in fused deposition modeling. *J. Mater. Process. Technol.* **2003**, *132*, 323–331. [CrossRef]
15. Ahn, D.; Kweon, J.H.; Kwon, S.; Song, J.; Lee, S. Representation of surface roughness in fused deposition modeling. *J. Mater. Process. Technol.* **2009**, *209*, 5593–5600. [CrossRef]
16. Boschetto, A.; Giordano, V.; Veniali, F. Modelling micro geometrical profiles in fused deposition process. *Int. J. Adv. Manuf. Technol.* **2012**, *61*, 945–956. [CrossRef]
17. Ding, D.; Pan, Z.; Cuiuri, D.; Li, H.; Van Duin, S.; Larkin, N. Bead modelling and implementation of adaptive MAT path in wire and arc additive manufacturing. *Robot. Comput. Integr. Manuf.* **2016**, *39*, 32–42. [CrossRef]
18. Kaji, F.; Barari, A. Evaluation of the Surface Roughness of Additive Manufacturing Parts Based on the Modelling of Cusp Geometry. *IFAC-PapersOnLine* **2015**, *48*, 692–697. [CrossRef]
19. Hodgson, G.; Ranellucci, A.; Moe, J. *Slic3r Manual—Flow Math*; Aleph Objects: Loveland, CO, USA, 2016.
20. Jin, Y.A.; Li, H.; He, Y.; Fu, J.Z. Quantitative analysis of surface profile in fused deposition modelling. *Addit. Manuf.* **2015**, *8*, 142–148. [CrossRef]
21. Buj-Corral, I.; Petit-Rojo, O.; Bagheri, A.; Minguella-Canela, J. Modelling of porosity of 3D printed ceramic prostheses with grid structure. *Procedia Manuf.* **2017**, *13*, 770–777. [CrossRef]
22. Wang, P.; Zou, B.; Ding, S. Modeling of surface roughness based on heat transfer considering diffusion among deposition filaments for FDM 3D printing heat-resistant resin. *Appl. Therm. Eng.* **2019**, *161*, 114064. [CrossRef]
23. Bottini, L.; Boschetto, A. Interference fit of material extrusion parts. *Addit. Manuf.* **2019**, *25*, 335–346. [CrossRef]
24. Popescu, D.; Zapciu, A.; Amza, C.; Baciu, F.; Marinescu, R. FDM process parameters influence over the mechanical properties of polymer specimens: A review. *Polym. Test.* **2018**, *69*, 157–166. [CrossRef]
25. Domingo-Espin, M.; Puigoriol-Forcada, J.M.; Garcia-Granada, A.A.; Llumà, J.; Borros, S.; Reyes, G. Mechanical property characterization and simulation of fused deposition modeling Polycarbonate parts. *Mater. Des.* **2015**, *83*, 670–677. [CrossRef]
26. Knoop, F.; Schoeppner, V. Mechanical and Thermal Properties of Fdm Parts Manufactured with Polyamide 12. In Proceedings of the Solid Freeform Fabrication Symposium, Austin, TX, USA, 10–12 August 2015.
27. Uddin, M.S.; Sidek, M.F.R.; Faizal, M.A.; Ghomashchi, R.; Pramanik, A. Evaluating Mechanical Properties and Failure Mechanisms of Fused Deposition Modeling Acrylonitrile Butadiene Styrene Parts. *J. Manuf. Sci. Eng.* **2017**, *139*, 081018. [CrossRef]
28. Chacón, J.M.; Caminero, M.A.; García-Plaza, E.; Núñez, P.J. Additive manufacturing of PLA structures using fused deposition modelling: Effect of process parameters on mechanical properties and their optimal selection. *Mater. Des.* **2017**, *124*, 143–157. [CrossRef]
29. Sood, A.K.; Ohdar, R.K.; Mahapatra, S.S. Experimental investigation and empirical modelling of FDM process for compressive strength improvement. *J. Adv. Res.* **2012**, *3*, 81–90. [CrossRef]
30. McLouth, T.D.; Severino, J.V.; Adams, P.M.; Patel, D.N.; Zaldivar, R.J. The impact of print orientation and raster pattern on fracture toughness in additively manufactured ABS. *Addit. Manuf.* **2017**, *18*, 103–109. [CrossRef]
31. Chaudhari, M.; Jogi, B.F.; Pawade, R.S. Comparative Study of Part Characteristics Built Using Additive Manufacturing (FDM). In Proceedings of the Procedia Manufacturing, Tirgu Mures, Romania, 4–5 October 2018.
32. Thrimurthulu, K.; Pandey, P.M.; Reddy, N.V. Optimum part deposition orientation in fused deposition modeling. *Int. J. Mach. Tools Manuf.* **2004**, *44*, 585–594. [CrossRef]
33. Wang, C.C.; Lin, T.W.; Hu, S.S. Optimizing the rapid prototyping process by integrating the Taguchi method with the Gray relational analysis. *Rapid Prototyp. J.* **2007**, *13*, 304–315. [CrossRef]

34. Ibrahim, D.; Ding, S.; Sun, S. Roughness Prediction For FDM Produced Surfaces. In Proceedings of the International Conference Recent Treads in Engineering & Technology (ICRET'2014), Batam, Indonesia, 13–14 February 2014.
35. Li, Y.; Linke, B.S.; Voet, H.; Falk, B.; Schmitt, R.; Lam, M. Cost, sustainability and surface roughness quality—A comprehensive analysis of products made with personal 3D printers. *CIRP J. Manuf. Sci. Technol.* **2017**, *16*, 1–11. [CrossRef]

© 2019 by the authors. Licensee MDPI, Basel, Switzerland. This article is an open access article distributed under the terms and conditions of the Creative Commons Attribution (CC BY) license (http://creativecommons.org/licenses/by/4.0/).

Article

Elastic Asymmetry of PLA Material in FDM-Printed Parts: Considerations Concerning Experimental Characterisation for Use in Numerical Simulations

Ma-Magdalena Pastor-Artigues [1,*], Francesc Roure-Fernández [1], Xavier Ayneto-Gubert [1], Jordi Bonada-Bo [1], Elsa Pérez-Guindal [2] and Irene Buj-Corral [3]

[1] Department of Strength of Materials and Structural Engineering (RMEE), Barcelona School of Industrial Engineering (ETSEIB), Universitat Politècnica de Catalunya–Barcelona Tech (UPC), 08028 Barcelona, Spain; francesc.roure@upc.edu (F.R.-F.); javier.ayneto@upc.edu (X.A.-G.); jordi.bonada@upc.edu (J.B.-B.)
[2] Department of Strength of Materials and Structural Engineering (RMEE), Vilanova i la Geltrú School of Engineering (EPSEVG), Universitat Politècnica de Catalunya–Barcelona Tech (UPC), 08800 Vilanova i la Geltrú, Spain; elsa.perez@upc.edu
[3] Department of Mechanical Engineering (EM), Barcelona School of Industrial Engineering (ETSEIB), Universitat Politècnica de Catalunya–Barcelona Tech (UPC), 08028 Barcelona, Spain; irene.buj@upc.edu
* Correspondence: m.magdalena.pastor@upc.edu; Tel.: +34-93-401-6732

Received: 31 October 2019; Accepted: 15 December 2019; Published: 18 December 2019

Abstract: The objective of this research is to characterise the material poly lactic acid (PLA), printed by fused deposition modelling (FDM) technology, under three loading conditions—tension, compression and bending—in order to get data that will allow to simulate structural components. In the absence of specific standards for materials manufactured in FDM technology, characterisation is carried out based on ASTM International standards D638, D695 and D790, respectively. Samples manufactured with the same printing parameters have been built and tested; and the tensile, compressive and flexural properties have been determined. The influences of the cross-sectional shape and the specimen length on the strength and elastic modulus of compression are addressed. By analysing the mechanical properties obtained in this way, the conclusion is that they are different, are not coherent with each other, and do not reflect the bimodular nature (different behaviour of material in tension and compression) of this material. A finite element (FE) model is used to verify these differences, including geometric non-linearity, to realistically reproduce conditions during physical tests. The main conclusion is that the test methods currently used do not guarantee a coherent set of mechanical properties useful for numerical simulation, which highlights the need to define new characterisation methods better adapted to the behaviour of FDM-printed PLA.

Keywords: FDM; PLA; mechanical properties; bimodulus materials; standards; finite element analysis (FEA)

1. Introduction

Additive manufacturing (AM) technologies allow for converting virtual models into physical models in a quick and easy way by means of tool-free processes. Different polymeric materials are being produced for 3D Printing (3DP) with a wider range of properties. 3DP has many applications in sectors such as automotive, electronics or medical [1]. Aliphatic polyesters, in particular poly lactic acid (PLA), are suitable materials for in vivo applications because of their biocompatibility, biodegradability, good mechanical strength and processability. PLA is the most researched and used aliphatic biodegradable polyester. It is a leading biomaterial for numerous applications in both medicine and industry, and the ability to adapt its properties for specific applications makes the market capacity of PLA products very broad, which has catalysed an extensive and growing amount of research aimed at the use of this material in innovative forms and applications [2–7].

If PLA-printed parts are to be usable as real industrial or biomechanical components, their structural and mechanical reliability has to be proved by means of strength and stiffness verifications. These will be done by classical strength of materials calculations or by finite element (FE) simulations. In either case, it is essential to have a coherent set of mechanical properties of the material under the different service conditions (tension, compression, bending, torsion, etc.). Flexural strength and tensile strength are two of the most commonly used values for comparing plastic materials. Compressive strength gives a good indication of short-term load capacities. Rigidity is expressed by the modulus of elasticity in tension and flexion. Reference data on the mechanical properties of PLA are available in the literature [2,8–14], with the tensile test being the most common of the tests performed to characterise the mechanical behaviour of PLA [15–21].

However, information of the bulk material behaviour is not always useful for carrying out numerical simulations to check the proper in-service behaviour of a fused deposition modelling (FDM)-printed component. To simulate the behaviour of a FDM-printed component, it is previously necessary to characterise the material in the same way that it is in the component. Many works start from the existing standards for polymeric materials to characterise also their FDM-printed versions. The ASTM standards D638 (tensile properties), D695 (compressive properties) and D790 (flexural properties) are widely used for this purpose.

The objective of this work was to obtain the mechanical properties of parts printed on PLA for use in numerical simulations, and to contrast the procedures for such purposes based on the use of ASTM standards. The results obtained considering an isotropic behaviour were compared for the different types of loads in the standards (tension, compression and bending). Nevertheless, when considering the bimodular behaviour of the material [22–25], inconsistencies were observed between the results obtained in the different tests. Differences in compression behaviour were also observed depending on the shape and proportions of the samples.

It is concluded that the currently accepted approach, based on characterisation according to the above standards (or the equivalent ISO standards), is not suitable for this purpose. It is necessary to define new characterisation methods that take into account of the bimodularity of the material and ensure a consistent set of mechanical characteristics for numerical simulation. This will be the next step of this research.

2. Test Methods

The PLA samples were 3D-printed with FDM technology, in a BCN3D Sigma v17 machine (BCN3D Technologies, Barcelona, Spain). It is an FDM desktop 3D printer with an independent dual extruder.

Specifications and properties of the PLA filament, provided by the manufacturer of the machine, are shown in Table 1.

Table 1. Filament specifications and properties

Specification/Parameter	Value
Diameter (mm)	2.85
Specific gravity (ISO 1183) (g/cm^3)	1.24
Tensile strength at yield (ISO 527) (MPa)	70
Strain at yield (ISO 527) (%)	5
Strain at break (ISO 527) (%)	20
Tensile modulus (ISO 527) (MPa)	3120
Melting temperature (ISO 11357) (°C)	115 ± 35
Glass transition temperature (ISO 11357) (°C)	57
Molecular weight (g/mol)	1.598×10^5

The slicing software was Cura 0.1.5. All the specimens were manufactured with constant printing parameters, which are provided in Table 2.

Table 2. Printing parameters.

Parameter	Value
Infill ratio (%)	90
Nozzle diameter (mm)	0.4
Printing temperature (°C)	205
Printing speed (mm/s)	50
Printing pattern	Rectangular
Raster angle (°)	45
Layer height (mm)	0.1

Anisotropy as well as the influence of different printing parameters, like infill percentage or printing speed, were beyond the scope of this study.

ASTM (American Society for Testing and Materials) and ISO (International Organization for Standardization) mechanical testing standards are widely used to determine mechanical properties of plastics. These test procedures assume the material is continuous and homogeneous, although not necessarily isotropic. They do not include particular considerations for additive manufacturing [26,27]. Forster [28] reviews existing procedures for testing polymers and analyses their feasibility for additive manufacturing processes. It should be mentioned; however, that there are technical committees working on developing new standards for additive manufacturing (ISO/TC261, ASTM F42).

ASTM D638-02a Standard Test Method for Tensile Properties of Plastics [29], ASTM D695-02a Standard Test Method for Compressive Properties of Rigid Plastics [30] and ASTM D790-02 Flexural Properties of Unreinforced and Reinforced Plastics and Electrical Insulating Materials [31] were followed in the tests carried out.

As regards the shape and dimensions of the specimens tested, the recommendations of the standards were followed. For the tensile and bending tests, the options were quite clear; however, in the case of the compression test, two cross-section shapes were possible, and the specimen length depended on the mechanical property to be obtained.

2.1. Specimens

The shape and dimensions of the specimens were defined in accordance with the standards (Figures 1–3) as above mentioned, and six samples of each series were manufactured: One set of specimens for tensile tests, another set for bending tests, and four sets for compression tests, where two different cross-section shapes and two different specimen lengths were possible.

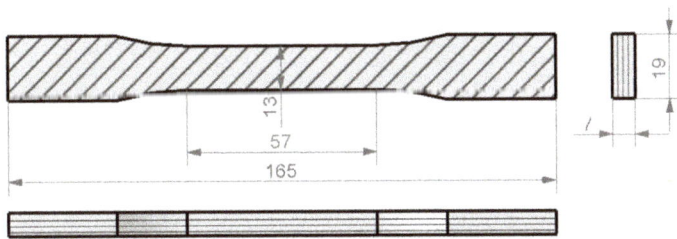

Figure 1. Drawing of the tensile test specimen (dimensions in mm).

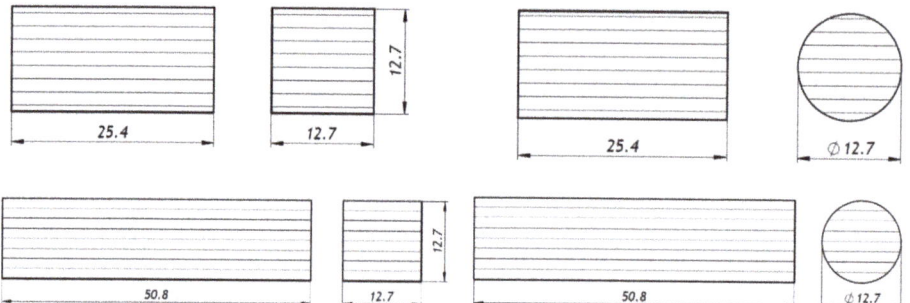

Figure 2. Drawing of the compression test specimens (dimensions in mm).

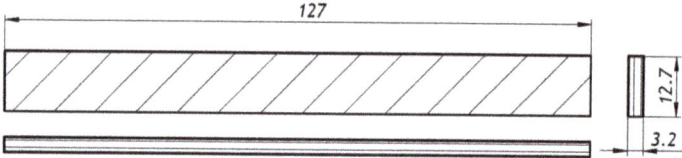

Figure 3. Drawing of the three-point bending test specimen (dimensions in mm).

As mentioned above, the printing parameters were kept constant (Table 2). The layers were oriented in the direction of the stresses. Figure 4 shows the print raster and the direction of the layers.

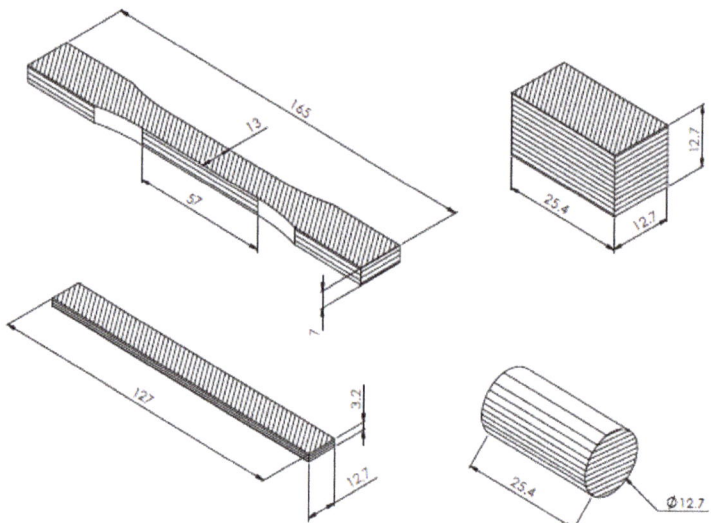

Figure 4. Structure of specimens with the raster and layers.

Figure 5 shows the 36 PLA 3D-printed samples grouped in six sets.

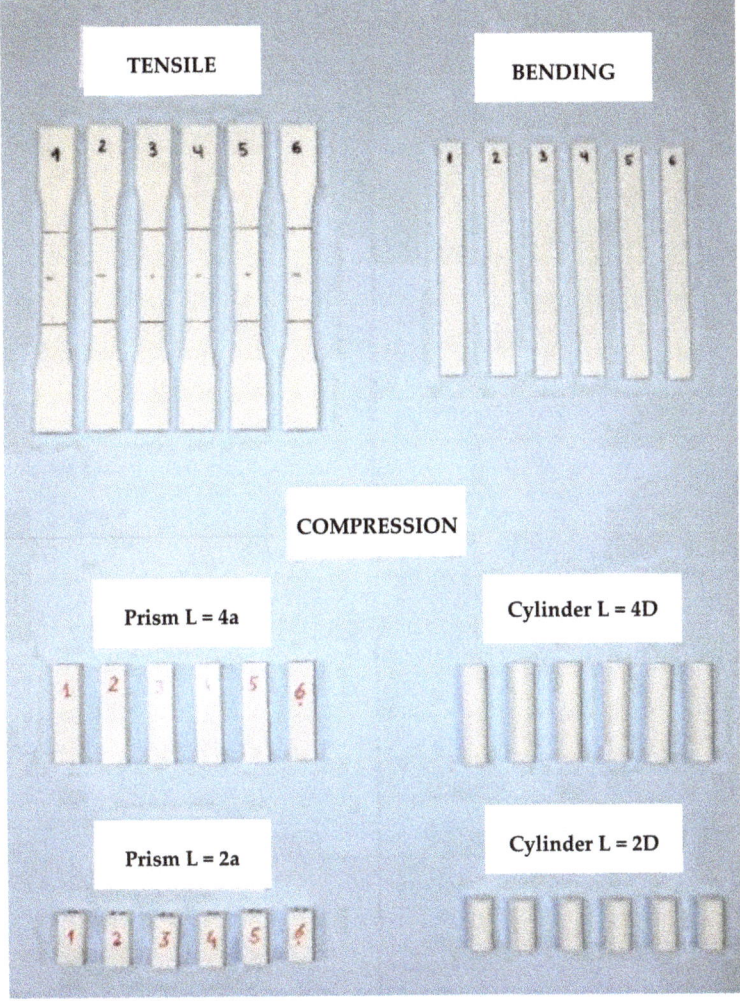

Figure 5. Poly Lactic Acid (PLA) samples for mechanical testing.

All test specimens were weighed on a KERN 400-55N precision scale (KERN & SOHN GmbH, Balingen, Germany), and measures were taken by means of a calliper to obtain the actual dimensions of each one of them. Tensile, compression and three-point bending tests were performed. All of them were carried out by means of an INSTRON machine model 3366 (INSTRON®, MA, USA) with the necessary equipment for each test.

2.2. Experimental Tests

2.2.1. Tensile Test

According to the ASTM D638 standard [29], the speed of testing was set at 5 mm/min and the load–extension curve of the specimen was recorded. Longitudinal (INSTRON 2630-102) and transverse (INSTRON I3575-250M-ST) strain measuring devices (extensometers) were attached to the specimen in order to determine the Poisson's ratio (Figure 6) (INSTRON®, MA, USA).

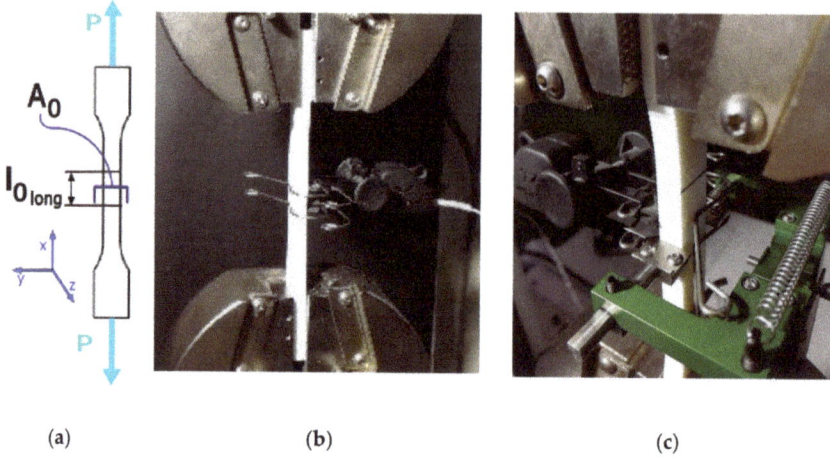

Figure 6. Tensile test assembly. Tensile test diagram (**a**); the longitudinal (**b**) and transverse (**c**) extensometers fitted on the specimen are shown.

From the data recorded during the test, the values of stress (σ), strain (ε) and modulus of elasticity (E) were calculated (Equation (1)).

$$\left\{\sigma = \frac{P}{A_0} \quad \varepsilon = \frac{\Delta l}{l_0} \quad E = \frac{\sigma}{\varepsilon} \right., \tag{1}$$

where:
- P = tensile load;
- A_0 = initial cross-sectional area;
- Δl = increment of distance between gauge marks;
- l_0 = initial specimen gauge length.

2.2.2. Compression Test

In the case of the compression test, the ASTM D695 standard [30] specifies that the test specimen shall be in the form of a right cylinder or prism (square), whose length is twice its width or diameter. However, when the modulus of elasticity and offset yield-stress are desired, the test specimen shall be of such dimensions that the slenderness ratio (λ) is in the range from 11 to 16:1. In this case, the preferred specimen sizes were 12.7 (a) by 12.7 (a) by 50.8 mm (L) (prism), or 12.7 in diameter (D) by 50.8 mm (L) (cylinder) (Figure 7). In literature consulted concerning this test, the length of the specimens was usually twice its principal width or diameter. Exceptionally [32–35] long-length specimens were tested.

$$\lambda = \frac{L}{i} = \frac{L}{\sqrt{\frac{I}{A}}} = \begin{cases} \frac{L}{\sqrt{\frac{\pi D^4}{64}}} = \frac{4L}{D} \in [11,16] \Rightarrow L \in [2.75D, 4D] & \text{for cylinder} \\ \frac{L}{\sqrt{\frac{\frac{1}{12}a^4}{a^2}}} = \frac{2\sqrt{3}L}{a} \in [11,16] \Rightarrow L \in [3.2a, 4.6a] & \text{for prism (square)} \end{cases}, \tag{2}$$

where:
- i = least radius of gyration;
- I = moment of inertia;
- A = area of the cross-section.

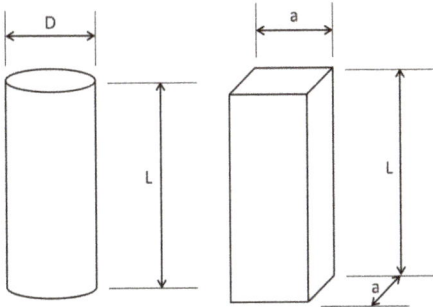

Figure 7. Dimensions of compression test specimens.

Limiting the slenderness ratio (λ) to the range 11 to 16:1 avoids 1) the influence of the end conditions on the results and 2) the buckling of the sample in the elastic range of the test.

Four series of specimens were built, called short specimens (L = 2a, L = 2D) and long specimens (L = 4a, L = 4D), whose length values were in the range given by Equation (2).

The speed of testing was set at 1.3 mm/min for short specimens (L = 2a, L = 2D) in accordance with paragraph 9 of the standard, and at 2.6 mm/min for long specimens (L = 4a, L = 4D), thus preserving the constant strain rate in all tests.

The test setup is shown in Figure 8.

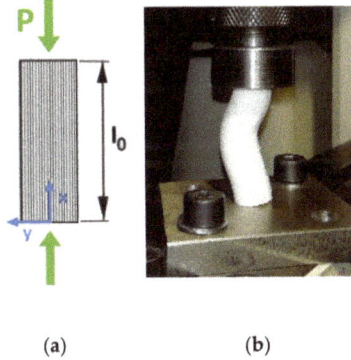

(a) (b)

Figure 8. Compression test diagram (**a**); long specimen at the end of the test (**b**).

In a similar way to the tensile test, the values of stress (σ), strain (ε) and modulus of elasticity (E) were calculated (Equation (1)).

2.2.3. Three-Point Bending Test

The three-point bending test was performed in accordance with ASTM D790 [31]. Figure 9 illustrates the test setup. The dimensions of the specimens can be seen in Figure 3.

The machine was set for the rate of crosshead motion (R) calculated according to Equation (3) taken from the standard:

$$R = \frac{ZL^2}{6h} = \frac{0.01 \times 52^2}{6 \times 3.2} \approx 1.4 \ mm/min, \qquad (3)$$

where:

L = support span (52 mm);

h = depth of beam (3.2 mm);
Z = rate of strain of the outer fibre (mm/mm/min), where Z shall be equal to 0.01.

The test ended when the deformation reached 5%, which was equivalent to a vertical displacement (δ) of 7 mm (Procedure A).

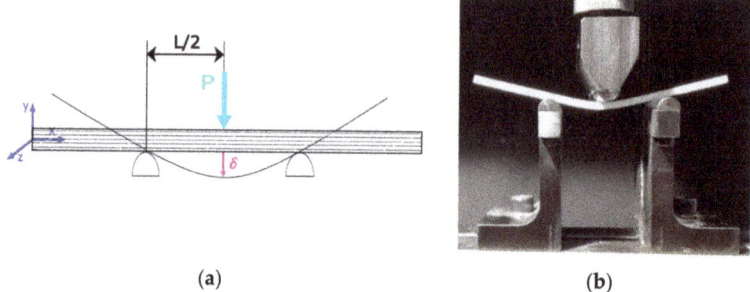

Figure 9. Bending test diagram (**a**); specimen under loading (**b**).

The flexural stress (σ_f) and flexural strain (ε_f) were calculated according to Equation (4). The flexural modulus of elasticity (E_B) is calculated from Equation (5).

$$\left\{\sigma_f = \frac{3PL}{2bh^2} \quad \varepsilon_f = \frac{6\delta h}{L^2}\right., \tag{4}$$

where:
P = load at a given point on the load–deflection curve (N);
b = width of beam tested (mm)

2.3. Numerical Test

Finite Element (FE) analyses were carried out previously to reproduce mechanical testing on ABS 3D-printed parts [36,37]. When using experimental data in numerical models, it is important to ascertain under what conditions the mechanical properties were obtained [34].

This section presents a first approximation of a finite element model of the compression test. A three-dimensional model is shown that does not include the internal structure of the material. The software used was ANSYS® Academic Research Mechanical, Release 19.1 [38].

- The simplified hypotheses assumed in the model were:
- Nominal dimensions were used to define the geometry of the sample
- The behaviour of the material was assumed to be linear, elastic and isotropic
- Neither the porosity of the material nor the manufacturing process was simulated. Figure 10 shows an enlarged image of the surface of one of the specimens, where the porosity and internal structure of the material is seen.
- Only a quarter of the specimen was modelled due to the symmetry of the analysis (Figure 11).

The features of the analysis were:

- Finite element Solid 186—3D, 20-node—was used for prismatic and cylindrical specimens;
- Material data (E,ν) found from experimental tests were introduced;
- The analysis was geometrically non-linear (GNA). Through an iterative process (Newton–Raphson), an equilibrium of forces was reached at each load step. The loading process was controlled by displacements.

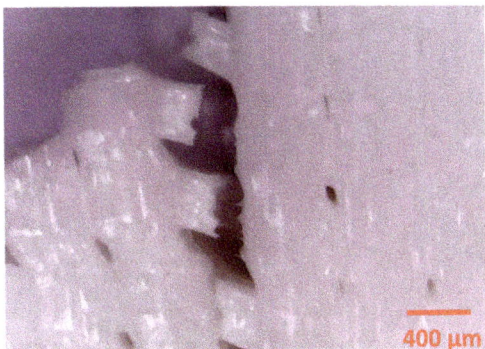

Figure 10. Enlargement of the surface area of a sample, where the porous nature of the material can be seen.

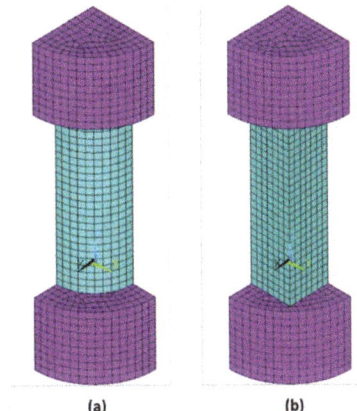

Figure 11. Finite element model for the cylindrical (**a**) and prismatic (**b**) short specimens. The purple elements correspond to the steel plates of the testing equipment.

The analysis reproduced the conditions of the compression test. The steel plates of the test setup were also simulated, and a contact between the specimen and the steel was introduced with a coefficient of friction of approximately 0.4 [39]. To define the boundary conditions, the bottom line of the bottom steel plate was fixed with zero displacement in all directions, and a negative vertical displacement was defined in the top steel plate. Symmetry boundary conditions were introduced at the two planes of symmetry.

3. Results

3.1. Results of Experimental Tests

3.1.1. Tensile Test Results

Figure 12 shows the stress–strain curves, overlapped for all specimens. Figure 13 illustrates the tensile failure of a sample after the test.

The modulus of elasticity (E), strength (σ), elongation (ε) and Poisson's ratio (ν) were derived from the data recorded during the test and subsequent calculations following the recommendations of the standard (Equation (1)).

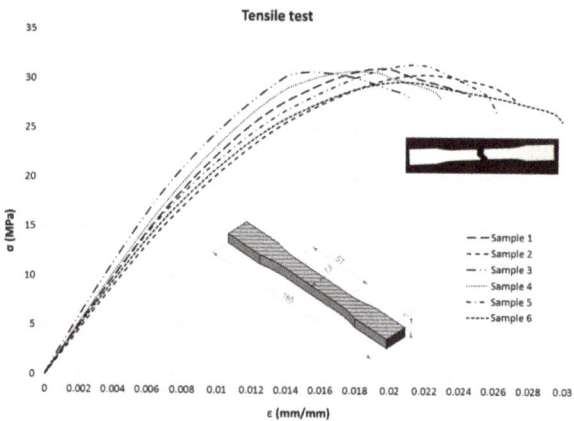

Figure 12. Tensile stress–strain curves.

Figure 13. Aspect of brittle break of a tensile specimen by the minimum cross-section—7 ×19 (mm^2).

The regression line of the initial linear part of the stress–strain curve was plotted, and the slope was taken as an E value (Figure 14).

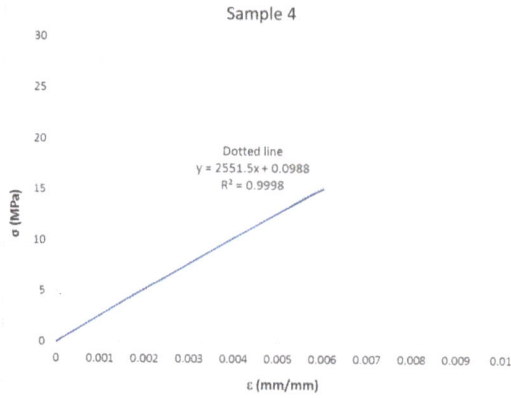

Figure 14. Regression line (dotted line) taken for calculation of modulus of elasticity (E).

The Poisson's ratio (ν) was calculated from the data recorded by the two extensometers (linear and transverse). Figure 15 shows both deformations as a function of the applied load. With the slopes of these two straight lines, ν was calculated by dividing the values by each other.

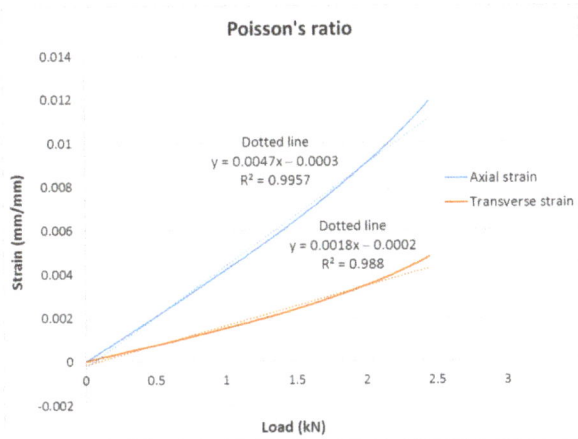

Figure 15. Regression lines (dotted lines) taken to calculate the Poisson's ratio. The continuous lines represent the strain measured by the extensometers as a function of the applied load.

Values of modulus of elasticity (E), tensile strength (σ), elongation at tensile strength (ε) and Poisson's ratio (ν) derived from tensile tests are shown in Table 3, including mean and standard deviation (SD).

Table 3. Tensile properties obtained from tensile tests (ASTM D638).

Specimen	Modulus of Elasticity E (MPa)	Tensile Strength σ (MPa)	Elongation at σ ε (mm/mm)	Poisson's Ratio ν
1	2200.7	31.13	0.019	-
2	2063.6	30.52	0.022	-
3	2776.4	30.78	0.015	-
4	2551.5	30.84	0.018	0.38
5	2394.0	31.55	0.021	0.37
6	2346.7	29.82	0.021	0.36
Mean	2388.8	30.77	0.019	-
SD	252.8	0.59	0.002	-

3.1.2. Compression Test Results

Analogously to the methodology followed for the tensile tests, from the force-displacement data recorded during the test, the stress–strain curve was obtained. The graphs were adjusted by doing the toe compensation indicated by the Annex A1 of the ASTM D 695 standard [30]. It consisted of ignoring the initial "toe" region of the stress–strain curve, and obtaining the corrected zero point of the stress axis by intersecting the prolongation of the linear region of the curve with the stress axis.

The modulus of elasticity (E), strength (σ) and elongation (ε) were determined from the data recorded during the test and subsequent calculations (Equation (1)).

It can be seen that, apparently, the strength increased in the plastic range of the material. This was because, as a result of compression, transformations occurred in the material, porosity was reduced, volume decreased and density increased. When the internal structure of the material changed, the load capacity changed. However, this did not affect the calculated parameters.

Values of modulus of elasticity (E), compressive yield point strength (σ) and elongation at compressive yield point strength (ε) are shown in Tables 4–6, including mean and standard deviation (SD). Since there were four series of test specimens in this case, it was deemed convenient to show the results separately (i.e., a table for each magnitude).

Table 4. Modulus of elasticity from compression test (ASTM D695).

Specimen	E (MPa)			
	Prism (L = 2a)	Cylinder (L = 2D)	Prism (L = 4a)	Cylinder (L = 4D)
1	921.2	1076.1	1397.7	1472.4
2	799.6	1077.6	1458.4	1463.8
3	724.2	1099.2	1251.7	1516.2
4	816.4	1113.5	1453.0	1563.0
5	802.7	1097.6	1355.3	1564.5
6	802.4	1123.3	1304.9	1549.4
Mean	811.1	1097.9	1370.2	1521.5
SD	63.2	18.9	82.4	45.0

Table 5. Compressive yield point strength from compression test (ASTM D695).

Specimen	σ (MPa)			
	Prism (L = 2a)	Cylinder (L = 2D)	Prism (L = 4a)	Cylinder (L = 4D)
1	30.06	41.17	32.57	42.49
2	25.97	40.66	39.79	41.56
3	24.25	42.58	28.83	43.64
4	27.57	42.82	40.10	45.52
5	26.58	41.78	31.71	45.14
6	26.79	42.13	31.10	44.27
Mean	26.87	41.86	34.02	43.77
SD	1.92	0.83	4.76	1.53

Table 6. Elongation at compressive yield point strength from compression test (ASTM D695).

Specimen	ε (mm/mm)			
	Prism (L = 2a)	Cylinder (L = 2D)	Prism (L = 4a)	Cylinder (L = 4D)
1	0.041	0.043	0.026	0.042
2	0.039	0.043	0.035	0.038
3	0.039	0.044	0.029	0.041
4	0.038	0.043	0.036	0.042
5	0.038	0.043	0.031	0.043
6	0.039	0.042	0.032	0.033
Mean	0.039	0.043	0.031	0.040
SD	0.001	0.000	0.004	0.004

Figure 16a,b shows the compressive response at medium levels of deformation for short specimens. For high levels of deformation, the long specimens failed by global buckling (Figure 16c,d), and the failure pattern of the short specimens was a barrel shape (Figure 17).

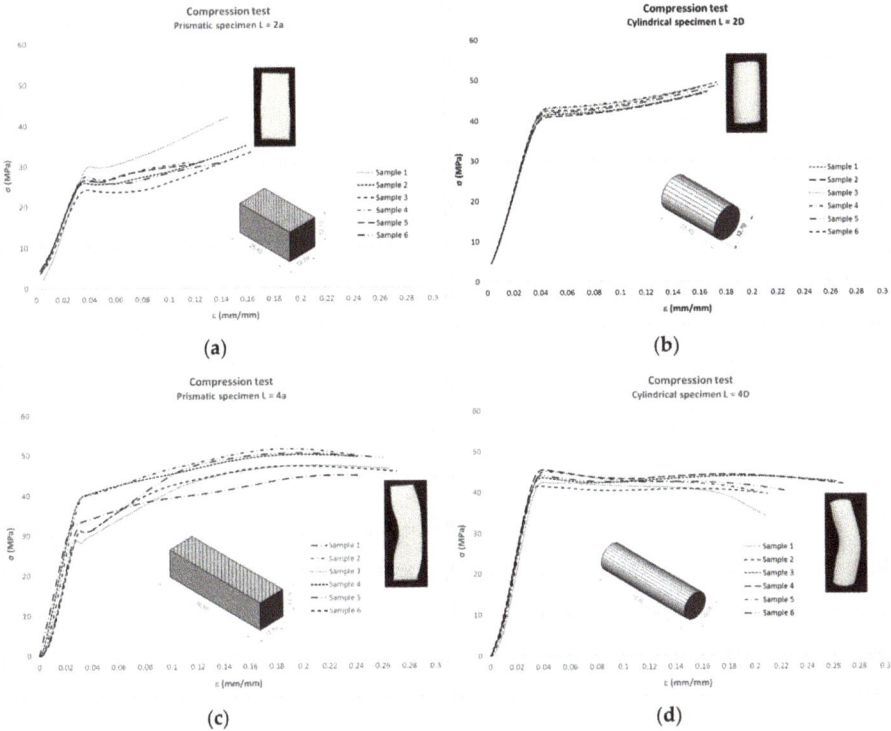

Figure 16. Compressive stress–strain diagrams: prismatic short shape (**a**), cylindrical short shape (**b**), prismatic long shape (**c**) and cylindrical long shape (**d**).

Figure 17. Samples tested until different levels of deformation.

3.1.3. Bending Test Results

Stress–strain curves, overlapped for all specimens, are shown in Figure 18. The maximum value of σ_f was taken as flexural strength, and ε_f was the respective deformation.

Values of the tangent modulus of elasticity (E_B), flexural stress (σ_f) and flexural strain (ε_f) are shown in Table 7, including mean and standard deviation (SD). Flexural stress (σ_f) and flexural strain (ε_f) were calculated using Equation (4). According to the standard, the modulus of elasticity in bending (E_B) was calculated by means of Equation (5):

$$E_B = \frac{L^3 \frac{P}{\delta}}{4bh^3} = \frac{PL^3}{48\delta I},\qquad(5)$$

where:

I is the moment of inertia of the cross-section (mm^4).

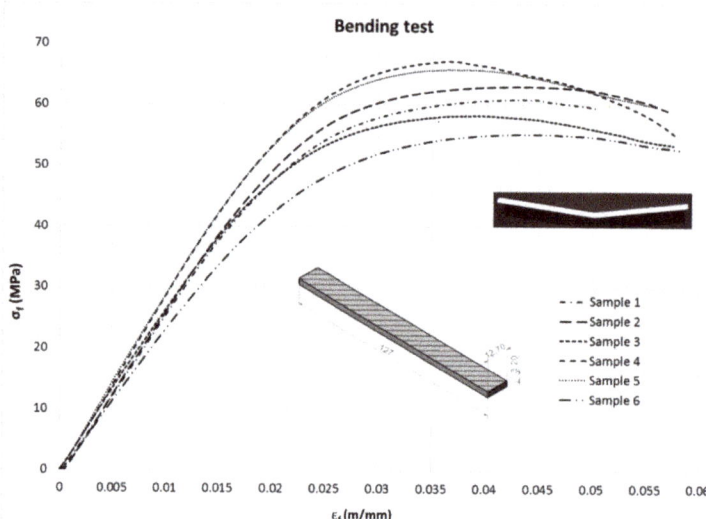

Figure 18. Flexural stress versus flexural strain curves.

Table 7. Flexural properties from three-point bending test (ASTM D790).

Specimen	E_B (MPa)	σ_f (MPa)	ε_f (mm/mm)
1	2603.0	60.92	0.042
2	2657.4	63.19	0.043
3	2572.1	58.34	0.039
4	2900.2	67.26	0.036
5	2859.0	65.87	0.036
6	2265.4	55.22	0.043
Mean	2642.8	61.80	0.040
SD	229.0	4.57	0.003

3.2. Numerical Testing Results (FE Analysis)

Figures 19–22 show the normal stress plots for each of the four specimen types: short prismatic (Figure 19), short cylindrical (Figure 20), long prismatic (Figure 21) and long cylindrical (Figure 22).

There were no significant differences in the influence of the shape of the cross-section.

It can be perceived that longer samples exhibited a quasi-uniform distribution of stresses over a wide intermediate length, while shorter samples had hardly a uniform distribution in the intermediate cross-section, which means that the boundary conditions at the ends of the specimen had more influence on shorter samples. This effect was more notorious in the prismatic rather than the cylindrical specimens.

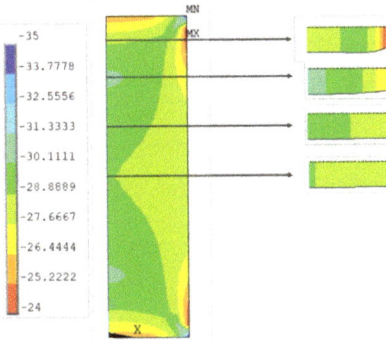

Figure 19. Distribution of normal compressive stresses (N/mm^2) for the middle section of the symmetry plane. Short prismatic specimen.

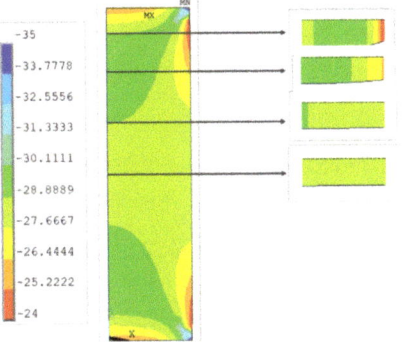

Figure 20. Distribution of normal compressive stresses (N/mm^2) for the middle section of the symmetry plane. Short cylindrical specimen.

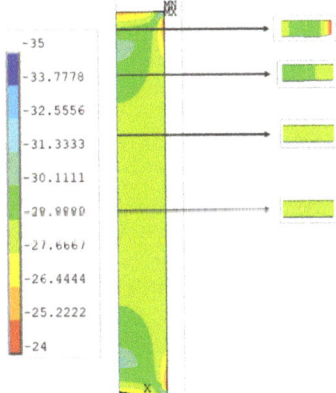

Figure 21. Distribution of normal compressive stresses (N/mm^2) for the middle section of the symmetry plane. Long prismatic specimen.

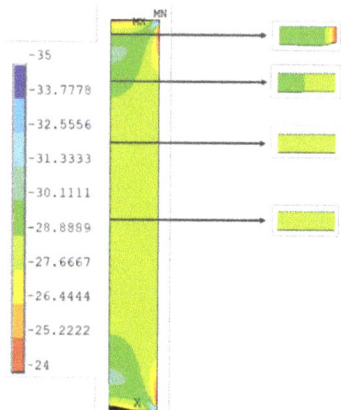

Figure 22. Distribution of normal compressive stresses (N/mm^2) for the middle section of the symmetry plane. Long cylindrical specimen.

4. Discussion of Results

4.1. Experimental Tensile, Compression and Bending Tests

In order to compare the results between the three types of tests, in the case of the compression test, the average of the long specimens was taken, given that the standard recommends this when determining the modulus of elasticity.

The second, third and fourth columns of Table 8 summarize the modulus of elasticity (E), strength (σ) and elongation (ε) found for each type of test, respectively, while the last two columns capture mechanical properties of processed PLA by injection [2,22]. Figures 23–25 show the results in bar graph format.

Table 8. Summary table of the mechanical properties of the three tests.

Test	E (MPa)	σ (MPa)	ε (mm/mm)	Farah et al. [2]	*Song et al. [22]
Compression	1445.9	38.89	0.036	-	3200 ÷ 3880, 70.80 ÷ 71.94, –
Tensile	2388.8	30.77	0.019	3700, 65.6, 0.04	3430 ÷ 3680, 29.79 ÷ 46.76, –
Flexural	2642.8	61.80	0.040	-	-

(*) depending on the strain rate.

It can be seen that the highest modulus of elasticity was the flexural modulus, with a value of 2640 MPa, followed by the tensile modulus with 2390 MPa. This was not a significant difference; however, the value of the modulus of elasticity in compression was 1445 MPa, approximately 40% lower than the previous two. This scenario would be anomalous from the point of view of a bimodular behaviour, as it should be an intermediate value between them. In effect, using the calculation method described in [23], given the elastic modules of 2515 MPa in tension and 1445 MPa in compression, the modulus in three-point bending should be around 1836 MPa (the effect of shear deformation in the specimen was taken into account), considerably lower than the observed value of 2575 MPa.

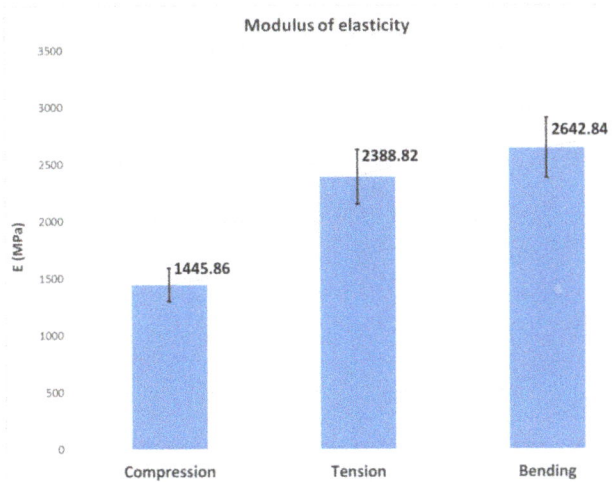

Figure 23. Bar graph with the mean value of the E variable.

This apparent anomaly was investigated by re-evaluating the values of the elastic modules in order to find out the causes of the discrepancy. The following corrections were made to the data sets of the three tests:

- The nominal definitions of stress (force per unit of initial area) and strain (increase in length per unit of the initial length of the extensometer) were changed to their true values (true stress—force per unit of area of the deformed section; true strain—logarithmic strain), assuming the constant volume hypothesis (the volume of the material remains constant during deformation), in order to consider the effect of the variation of the cross section during the tensile and compression tests.
- The non-linearity in the initial phase of the stress–strain curve in the compression test was corrected for the long samples (L = 4D and L = 4a) by performing the toe compensation indicated in annex A1 of standard D695-02a. The parallelism between the faces of the samples was critical. This effect was more pronounced in short samples than in long samples.
- Calculations of the initial elastic modulus were carried out for the three types of tests (tension, compression and bending) considering the linear part of the stress–strain curves (strain from 0 to 0.004 mm/mm).

Table 9 lists the new elastic modulus values found after these adjustments.

Table 9. Summary of re-evaluated modulus of elasticity.

Specimen	E (MPa)			
	Tension	Compression L = 4D	Compression L = 4a	Bending
1	2497	1405	1405	2455
2	2282	1362	1375	2666
3	2845	1485	1225	2679
4	2599	1507	1441	2699
5	2478	1452	1353	2601
6	2388	1459	1276	2349
Mean	2515	1445	1346	2575
SD	194	53	81	142

The new values obtained showed less dispersion, both between samples of the same test and between the different types of cross-sections in the compression tests of long specimens.

As can be seen, the values of the tensile and bending modules were now more similar to those shown in Figure 23, although the mean bending value was slightly higher than the mean tensile value.

However, the results obtained from the three standard tests were still inconsistent with bimodular behaviour. This inconsistency is mainly attributed to differences in sample geometry, which also generates process differences, and to different strain rates in the standard tests. Since these are uniaxial tests, and the print layers are very thin, the effect of anisotropy was considered to be minor.

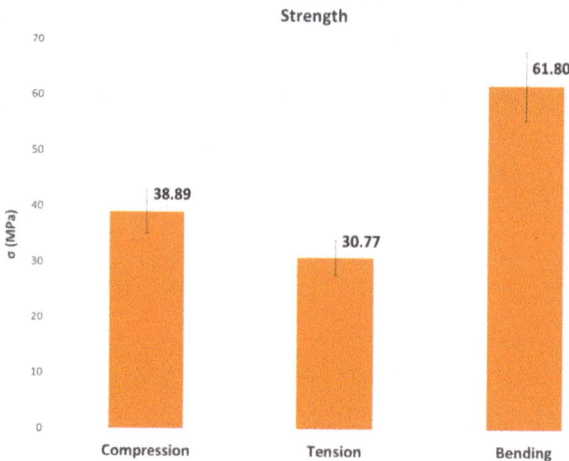

Figure 24. Bar graph with the mean value of the σ variable.

The most remarkable thing about graphics related to strength is that the flexural value was considerably higher than the other two, 60% higher than the compressive strength and twice as high as the tensile strength, which is clearly the lowest.

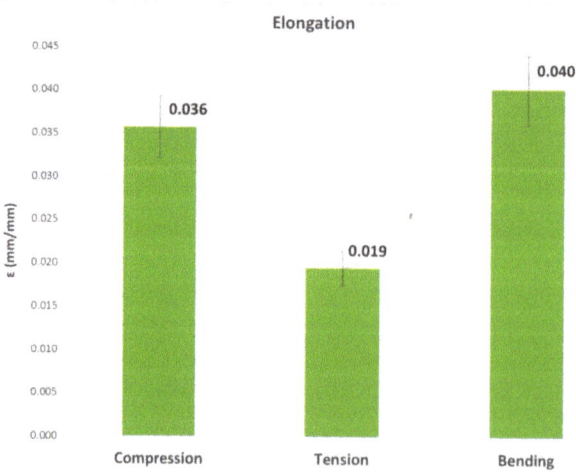

Figure 25. Bar graph with the mean value of the ε variable.

Finally, the graphs of the elongation values were compared. As above, the highest value corresponded to the bending test, followed by compression, although they did not show such a significant difference. The elongation for the tensile test was the lowest.

These results are shown here in summary:

$$E_{bending} \approx E_{tensile} \gg E_{compression}$$

$$\sigma_{bending} \gg \sigma_{compression} > \sigma_{tensile}$$

$$\varepsilon_{bending} \gtrsim \varepsilon_{compression} \gg \varepsilon_{tensile}$$

The results obtained from the tests of the four series of compression specimens, two lengths and two cross-section shapes, are analysed in the following Section.

4.2. Specific Analysis of the Compression Test

As can be seen in Figure 26, the modulus of elasticity of long specimens was considerably higher regardless of the cross-section shape. This result is consistent with [34]: Young's modulus decreases when the diameter:length ratio increases. In the case of cylindrical specimens, the value for long specimens was 40% higher, being 70% higher in prismatic specimens. On the other hand, the cylinder shape also had a higher value within each length. Thus, the highest value was that of long-cylinder specimens.

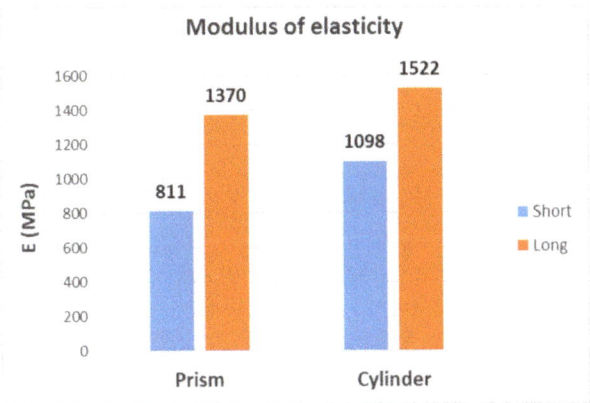

Figure 26. Comparison of modulus of elasticity in compression tests.

In the last row of Table 4, the dispersions of the values of the compressive modulus of elasticity (in %) were calculated for each type of sample. All values are within the acceptable range of variation when the modulus of elasticity is measured, as this is a parameter that tends to present higher dispersion than other ones (e.g., breaking stresses). Some conclusions can be drawn from Table 4:

(a) It was observed that the rectangular samples (2a and 4a) showed higher dispersion (in %) than the cylindrical ones (2D and 4D). Therefore, it is recommended the use of cylindrical specimens to determine the modulus E.
(b) The difference between the mean value of the modulus of elasticity obtained from samples 4D and 4a was 11.0 %, whereas the difference between the mean value obtained from the 2D and 2a samples was 35.3%.

By combining reasoning a) and b), it is concluded that the most appropriate sample type to determine the modulus E is the L = 4D type.

In addition to the dispersion inherent in a compression test, the influence of two other factors was identified and analysed:

(c) The flexibility of the testing machine caused an increase in the displacement between the compression plates, which implied obtaining an elastic modulus lower than the real modulus of the material. Although the compression test standard ASTM D695-02a did not provide for correction for this effect, the actual modulus E can be calculated from the measured modulus E' by using the following Equation:

$$E = \frac{L}{\frac{L}{E'} - \frac{A}{K_M}}, \tag{6}$$

where:
L is the length of the specimen;
A is the area of the cross-section;
K_M is the rigidity of the testing machine.

Table 10 shows the values initially determined for each type of specimen together with the values achieved after making the correction for the flexibility of the testing machine, as well as the variation (in %).

Table 10. Influence of the flexibility of the testing machine.

Specimen	E' (Measured) (MPa)	E (Real) (MPa)	Variation (%)
L = 2a	811.1	938	16
L = 2D	1097.9	1283	17
L = 4a	1370.2	1547	13
L = 4D	1521.5	1690	11

It was observed that the effect of the flexibility of the testing machine was less for long specimens (4a and 4D). This is an additional reason for using long specimens to determine the modulus E.

(d) Friction between the sample and the load plates of the machine limited the effect of Poisson at the ends of the sample, and caused a barrel shape, which altered the assumed uniform distribution of stresses and deformations, thus increasing the apparent elastic modulus. The barrel shape was observed in some of the specimens (see Figure 17). To verify the influence of this effect on the Young's modulus, two finite element simulations of the compression tests were performed. In the first one, it was assumed that there was no friction, and in the second one, the friction completely blocked the sliding between both surfaces in contact. The sample stiffness variation (which is proportional to the Young's modulus) is shown in Table 11, where the percentage compression stiffness variation is expressed with regard to the frictionless model.

Table 11. Variation of the stiffness with respect to the model without friction, obtained by FEA.

Specimen	Variation in Stiffness (%)
L = 2a	4.17
L = 2D	3.56
L = 4a	2.08
L = 4D	1.78

It was noticed that the influence of the coefficient of friction between the specimen and the test machine plates was low, being lower for long specimens (4a and 4D). This is an additional argument for using long specimens to determine the modulus E.

In the values of compressive strength (Figure 27), it is observed that cylindrical samples had higher strength values, regardless of the length of the sample. The differences between them were not as noticeable as in the case of the modulus of elasticity.

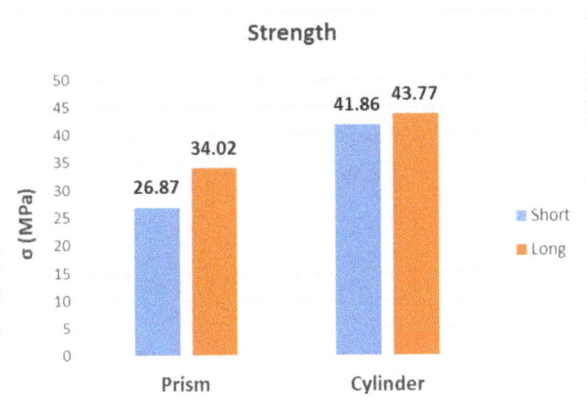

Figure 27. Comparison of compressive strength.

Finally, the elongation (Figure 28) obtained for cylindrical samples was also slightly higher than that obtained for prismatic samples. In addition, higher values were found in the short samples than in the long ones. However, in the case of elongation, the differences were less meaningful than in the case of the modulus of elasticity and strength.

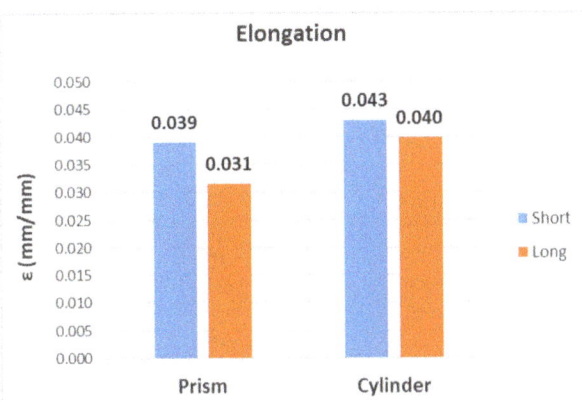

Figure 28. Comparison of elongation in compression tests.

These results are summarized in the following expressions:

$$E_{Cylinder} > E_{Prism} \text{ and } E_{Long} \gg E_{Short}$$

$$\sigma_{Cylinder} > \sigma_{Prism} \text{ and } \sigma_{Long} \approx \sigma_{Short}$$

$$\varepsilon_{Cylinder} > \varepsilon_{Prism} \text{ and } \varepsilon_{Long} < \varepsilon_{Short}$$

5. Conclusions

The mechanical properties of PLA manufactured by FDM were determined under tensile, compressive and flexural stresses. The results obtained are quite consistent, considering the low dispersion of results within each group of specimens and in comparison with available data.

However, the results obtained in this work show that PLA has a double asymmetry in its tensile and compressive behaviours: On the one hand the asymmetry in strength, and on the other hand asymmetry in the constitutive behaviour, which suggests the need to treat this material by means of a bimodular elasticity model.

Nevertheless, characterisation based on standard tests presents significant difficulties when its purpose goes beyond its application to quality control tasks, such as numerical simulation.

The behaviour of 3D-printed materials is highly sensitive to process factors and, in the case of polymeric materials, also to the effect of different strain rates applied in each test. This makes it difficult to achieve a consistent characterisation of the elastic constants among the various types of standard tests available today, even when taking into account that the different dimensions and shapes of the specimens in each test can cause process differences that affect in the measured properties.

For this reason, it is necessary to define a new characterisation procedure that allows obtaining a consistent set of elastic constants, with the minimum number of tests possible, especially adapted to bimodular materials. This result would be of great interest for carrying out simulations of the structural behaviour of 3D printed parts. Recently, Mazzanti et al. [40] concluded there is no recognised international standard governing the characterisation of the tensile, compressive or flexural properties of 3D-printed materials. The current standards are those used for the characterisation of bulk polymeric materials. In this case, the geometric characteristics are standardized through the concepts of stress and strain, but in the case of 3D printing, this is difficult because the specimen is actually a structure, not a material.

With respect to the compression test, it has been demonstrated that there is less variability of results in cylindrical specimens than in prismatic specimens (probably a result of the manufacturing process, since at the corners of the prismatic shapes the printing head can deposit excess material). In addition, when determining the modulus of elasticity, it is confirmed that, following the recommendation of the ASTM D695 standard, the longer specimens provide results that are more consistent.

The need to perform compression tests to characterise the elastic modulus to compression should be reconsidered. This could be consistently deduced using a bimodular model from the bending test, through the flexural modulus, or also from the shear test, through the transverse modulus of elasticity G, thus avoiding the uncertainties associated with the compression test.

The simulation of the compression test shows that the model is sensitive to the boundary conditions applied at the ends of the specimen (friction, etc.). Despite the introduction of geometric non-linearity in the analysis (GNA), this is not enough to correctly reproduce the actual behaviour of the material in a coherent way.

Author Contributions: Conceptualization, M.-M.P.-A., F.R.-F. and X.A.-G.; methodology, M.-M.P.-A and F.R.-F.; software, J.B.-B.; formal analysis, X.A.-G.; resources, E.P.-G. and I.B.-C.; writing—original draft preparation, M.-M.P.-A.; writing—review and editing, F.R.-F., X.A.-G., J.B.-B., I.B.-C. and E.P.-G.; supervision, M.-M.P.-A. All authors have read and agreed to the published version of the manuscript.

Funding: This research has been funded by the Spanish Ministry of Economy, Industry and Competitiveness; Grant Number DPI2016-80345-R.

Acknowledgments: The authors thank Marina Blasco, Bachelor of Mechanical Engineering, for her important contribution to the work. They also thank Ramón Casado-López and Francesc-Joaquim García-Rabella for their help with experimental tests.

Conflicts of Interest: The authors declare no conflict of interest.

References

1. Wohlers. *Wohlers Report 2014: Additive Manufacturing and 3D Printing State of the Industry, Annual Worldwide Progress Report*; Wohlers Associates Inc.: Fort Collins, CO, USA, 2014.
2. Farah, S.; Anderson, D.G.; Langer, R. Physical and mechanical properties of PLA, and their functions in widespread applications-A comprehensive review. *Adv. Drug Deliv. Rev.* **2016**, *107*, 367–392. [CrossRef] [PubMed]
3. Chen, L.; He, Y.; Yang, Y.; Niu, S.; Ren, H. The research status and development of additive manufacturing technology. *Int. J. Adv. Manuf. Technol.* **2017**, *89*, 3651–3660. [CrossRef]
4. Srivastava, V.K. A Review on Advances in Rapid Prototype 3D Printing of Multi-Functional Applications. *Sci. Technol.* **2017**, *7*, 4–24.
5. Chiulan, I.; Frone, A.N.; Brandabur, C.; Panaitescu, D.M. Recent Advances in 3D Printing of Aliphatic Polyesters. *Bioengineering* **2018**, *5*, 2. [CrossRef] [PubMed]
6. Dizon, J.R.C.; Espera, A.H., Jr.; Chen, Q.; Advincula, R.C. Mechanical characterization of 3d-printed polymers (Review). *Addit. Manuf.* **2018**, *20*, 44–67. [CrossRef]
7. Popescu, D.; Zapciu, A.; Amza, C.; Baciu, F.; Marinescu, R. FDM process parameters influence over the mechanical properties of polymer specimens: A review. *Polym. Test* **2018**, *69*, 157–166. [CrossRef]
8. Tymrak, B.M.; Kreiger, M.; Pearce, J.M. Mechanical properties of components fabricated with open-source 3-D printers under realistic environmental conditions. *Mater. Des.* **2014**, *58*, 242–246. [CrossRef]
9. Chacón, J.M.; Caminero, M.A.; García-Plaza, E.; Núñez, P.J. Additive manufacturing of PLA structures using fused deposition modelling: Effect of process parameters on mechanical properties and their optimal selection. *Mater. Des.* **2017**, *124*, 143–157. [CrossRef]
10. Innofil. Comparable data sheet Innofil3D filaments. In *Overview Mechanical Properties of Printed Test Specimens*; Basf: Emmen, the Netherlands; Available online: https://www.innofil3d.com/wp-content/uploads/2016/06/Pro1_Comparison_Sheet_small.pdf (accessed on 6 May 2019).
11. 3Faktur. *FDM/FFF Materials: ABS and PLA*; 3Faktur GmbH: Jena, Germany; Available online: https://3faktur.com/en/3d-printing-materials-technologies/fdm-materials-pla-and-abs/ (accessed on 6 May 2019).
12. MakerBot. *PLA and ABS Strength Data*; MakerBot Industries LLC: New York, NY, USA; Available online: http://download.makerbot.com/legal/MakerBot_R__PLA_and_ABS_Strength_Data.pdf (accessed on 6 May 2019).
13. SD3D. *PLA Technical Data Sheet*; SD3D Printing, Inc.: San Diego, CA, USA; Available online: https://www.sd3d.com/wp-content/uploads/2017/06/MaterialTDS-PLA_01.pdf (accessed on 22 July 2017).
14. Ultimaker. *PLA Technical Data Sheet*; Ultimaker BV: Utrecht, the Netherlands; Available online: https://ultimaker.com/download/74599/UM180821%20TDS%20PLA%20RB%20V10.pdf (accessed on 6 May 2019).
15. Lanzotti, A.; Grasso, M.; Staiano, G.; Martorelli, M. The impact of process parameters on mechanical properties of parts fabricated in PLA with an open-source 3-D printer. *Rapid Prototyp. J.* **2015**, *21*, 5. [CrossRef]
16. Ahmed, M.; Islam, M.; Vanhoose, J.; Rahman, M. Comparisons of Elasticity Moduli of Different Specimens Made Through Three Dimensional Printing. *3D Print Addit. Manuf.* **2017**, *4*, 105–109. [CrossRef]
17. Cantrell, J.T.; Rohde, S.; Damiani, D.; Guranani, R.; DiSandro, L.; Anton, J.; Young, A.; Jerez, A.; Steinbach, D.; Kroese, C.; et al. Experimental characterization of the mechanical properties of 3D-printed ABS and polycarbonate parts. *Rapid Prototyp. J.* **2017**, *23*, 811–824. [CrossRef]
18. Ferreira, R.T.L.; Amatte, I.C.; Dutra, T.A.; Bürger, D. Experimental characterization and micrography of 3D printed PLA and PLA reinforced with short carbon fibers. *Compos. Part B Eng.* **2017**, *124*, 88–100. [CrossRef]
19. Johnson, G.A.; French, J.J. Evaluation of Infill Effect on Mechanical Properties of Consumer 3D Printing Materials. *Int. J. Eng. Technol. Innov.* **2018**, *3*, 179–184.
20. Seol, K.S.U.; Zhao, P.; Shin, B.C.; Zhang, S.U. Infill Print Parameters for Mechanical Properties of 3D Printed PLA Parts. *J. Korean Soc. Manuf. Process Eng.* **2018**, *17*, 9–16. [CrossRef]
21. Subramaniam, S.R.; Samykano, M.; Selvamani, S.K.; Ngui, W.K.; Kadirgama, K.; Sudhakar, K.; Idris, M.S. Preliminary investigations of polylactic acid (PLA) properties. In *AIP Conference Proceedings*; API: College Park, MD, USA, 2019.
22. Song, Y.; Li, Y.; Song, W.; Lee, K.; Lee, K.Y.; Tagarielli, V.L. Measurements of the mechanical response of unidirectional 3D-printed PLA. *Mater. Des.* **2017**, *123*, 154–164. [CrossRef]

23. Chamis, C.C. Analysis of the three-point-bend test for materials with unequal tension and compression properties. In *NASA Technical Note TN D-7572*; National Aeronautics and Space Administration: Washington, DC, USA, 1974.
24. Vgenopoulos, D.; Sweeney, J.; Grant, C.A.; Thompson, G.P.; Spencer, P.E.; Caton-Rose, P.; Coates, P.D. Nanoindentation analysis of oriented polypropylene: Influence of elastic properties in tension and compression. *Polymer* **2018**, *151*, 197–207.
25. Huang, T.; Pan, Q.X.; Jin, J.; Zheng, J.L.; Wen, P.H. Continuous constitutive model for bimodulus materials with meshless approach. *Appl. Math. Model.* **2019**, *66*, 41–58. [CrossRef]
26. Kuznetsov, V.; Solonin, A.N.; Urzhumtsev, O.D.; Schilling, R.; Tavitov, A.G. Strength of PLA Components Fabricated with Fused Deposition Technology Using a Desktop 3D Printer as a Function of Geometrical Parameters of the Process. *Polymers* **2018**, *10*, 313. [CrossRef]
27. Zhang, Q.; Mochalin, V.N.; Neitzel, I.; Hazeli, K.; Niu, J.; Kontsos, A.; Zhou, J.G.; Lelkes, P.I.; Gogotsi, Y. Mechanical properties and biomineralization of multifunctional nanodiamond-PLLA composites for bone tissue engineering. *Biomaterials* **2012**, *33*, 5067–5075. [CrossRef]
28. Forster, A.M. Materials Testing Standards for Additive manufacturing of Polymer Materials: State of the Art and Standards Applicability. In *Additive Manufacturing Materials*; White, L., Ed.; Nova Science Publishers: Hauppauge, NY, USA, 2015.
29. ASTM D638-02a. *Standard Test Method for Tensile Properties of Plastics*; ASTM: West Conshohocken, PA, USA, 2003.
30. ASTM D695-02a. *Standard Test Method for Compressive Properties of Rigid Plastic*; ASTM: West Conshohocken, PA, USA, 2002.
31. ASTM D790-02. *Flexural Properties of Unreinforced and Reinforced Plastics and Electrical Insulating Materials*; ASTM: West Conshohocken, PA, USA, 2002.
32. Bagheri, A.; Buj, I.; Ferrer, M.; Pastor, M.M.; Roure, F. Determination of the Elasticity Modulus of 3D-Printed Octet-Truss Structures for Use in Porous Prosthesis Implants. *Materials* **2018**, *11*, 2420. [CrossRef]
33. Brischetto, S.; Ferro, C.G.; Maggiore, P.; Torre, R. Compression Tests of ABS Specimens for UAV Components Produced via the FDM Technique. *Technologies* **2017**, *5*, 20. [CrossRef]
34. Dhôte, J.X.; Comer, A.J.; Stanley, W.F.; Young, T.M. Investigation into compressive properties of liquid shim of aerospace bolted joints. *Compos. Struct.* **2014**, *109*, 224–230. [CrossRef]
35. Lammens, N.; Kersemans, M.; DeBaere, I.; VanPaepegem, W. On the visco-elasto-plastic response of additively manufactured polyamide-12 (PA-12) through selective laser sintering. *Polym. Test* **2017**, *57*, 149–155. [CrossRef]
36. Baich, L.J. Impact of Infill Design on Mechanical Strength and Production Cost in Material Extrusion Based Additive Manufacturing. Ph.D. Thesis, Youngstown State University, Youngstown, OH, USA, 2016.
37. Sayre III, R. A Comparative Finite Element Stress Analysis of Isotropic and Fusion Deposited 3D Printed Polymer. Ph.D. Thesis, Rensselaer Polytechnic Institute, New York, NY, USA, 2014.
38. ANSYS. *Academic Research Mechanical, Release 19.1*; ANSYS Inc.: Canonsburg, PA, USA, 2019.
39. Bainbridge, J. 3D Printing Filament Properties. Available online: https://github.com/superjamie/lazyweb/wiki/3D-Printing-Filament-Properties (accessed on 4 June 2019).
40. Mazzanti, V.; Malagutti, L.; Mollica, F. FDM 3D Printing of Polymers Containing Natural Fillers: A Review of their Mechanical Properties. *Polymers* **2019**, *11*, 1094. [CrossRef]

© 2019 by the authors. Licensee MDPI, Basel, Switzerland. This article is an open access article distributed under the terms and conditions of the Creative Commons Attribution (CC BY) license (http://creativecommons.org/licenses/by/4.0/).

Article

Dimensional and Geometrical Quality Enhancement in Additively Manufactured Parts: Systematic Framework and A Case Study

Natalia Beltrán, David Blanco *, Braulio José Álvarez, Álvaro Noriega and Pedro Fernández

Department of Construction and Manufacturing Engineering, University of Oviedo, Pedro Puig Adam St., E.D.O.5, 33203 Gijon, Asturias, Spain; nataliabeltran@uniovi.es (N.B.); braulio@uniovi.es (B.J.Á.); noriegaalvaro@uniovi.es (Á.N.); pedrofa@uniovi.es (P.F.)
* Correspondence: dbf@uniovi.es

Received: 31 October 2019; Accepted: 27 November 2019; Published: 28 November 2019

Abstract: In order to compete with traditional manufacturing processes, Additive Manufacturing (AM) should be capable of producing medium to large batches at industrial-degree quality and competitive cost-per-unit. This paper proposes a systematic framework approach to the problem of fulfilling dimensional and geometric requirements for medium batch sizes of AM parts, which has been structured as a three-step optimization methodology. Firstly, specific work characteristics are analyzed so that information is arranged according to an Operation Space (factors that could have an influence upon quality) and a Verification Space (formed by quality indicators and requirements). Standard process configuration leads to characterization of the standard achievable quality. Secondly, controllable factors are analyzed to determine their relative influence upon quality indicators and the optimal process configuration. Thirdly, optimization of part dimensional and/or geometric definition at the design level is performed in order to improve part quality and meet quality requirements. To evaluate the usefulness of the proposed framework under quasi-industrial condition, a case study is presented here which is focused on the dimensional and geometric optimization of surgical-steel tibia resection guides manufactured by Laser-Power Bed Fusion (L-PBF). The results show that the proposed approach allows for part quality improvement to a degree that matches the initial requirements.

Keywords: additive manufacturing; quality enhancement; process parameters; design optimization

1. Introduction

Additive Manufacturing (AM) is defined as "the process of joining materials to make objects from 3D model data, usually layer upon layer, as opposed to subtractive manufacturing fabrication methodologies" according to ISO 17296-1 [1] and ASTM 2792-12 [2]. This definition encompasses a wide variety of processes used to manufacture three-dimensional objects by means of vertical stacking of bi-dimensional layers. Most of the technologies involved have gained maturity in recent years. This has allowed AM applications to evolve from prototype manufacturing to small-batch-size production. Nevertheless, according to Gartner´s hype cycle, consistent adoption of AM in manufacturing operations will still take 5 to 10 years of development [3]. There are many factors that influence AM's difficulties to match the requirements of medium and high batch-size production. Some are related to working volumes and production rates of machines at the current state of development. Others reflect the difficulty of producing parts with similar mechanical behavior to those obtained by traditional manufacturing processes. Finally, dimensional and geometric quality deficiencies of AM parts have also been highlighted as common disadvantages [4,5], which explains their relevance as research subjects during the last decade [6–10]. Quality improvement is a *sine qua non* condition for the generalized industrial adoption of AM processes, since cost-per-unit reduction would not be enough by itself. Research in the field of dimensional and/or geometric quality improvement of AM parts could be grouped according to three different approaches: error analysis, error prevention and error correction.

Error analysis pays attention to the influence of process parameters on the dimensional or geometrical accuracy of parts. The usual error analysis approach involves comparing the theoretical values of the design parameters and their correspondent values measured upon the manufactured part [11–13]. Consequently, these works provide useful information regarding the expected results of an AM process in terms of quality and accuracy. Research under this approach shows a wide variety of factors influencing the accuracy of AM parts. This is related to the variety of physical principles and implementations available, which lead to specific research solutions for each individual case. These studies also show a lack of uniformity regarding the indicators used to compare the results of modifying process configuration. Researchers sometimes use indicators with no real meaning from an industrial point of view [11,13]. Tests are also frequently carried out upon test specimens and geometries designed *ad hoc* [11,12].

Approaches based on error prevention aim to establish the optimal process configuration to manufacture parts with the highest achievable quality. This goal is commonly based on the analysis of possible error sources and their relative influence upon quality, so that their effects could be minimized by means of parameters adjustment. Once again, the variety of physical principles and configuration parameters used in different AM processes hinders the adoption of a unified methodology for error prevention. Deposition speed or scanning energy are among the factors considered under this approach, but factors related to practical decisions like part location and orientation [6,14] are also used. This last category aims to provide useful recommendations on the best way to place the parts on the working volume. Finally, internal part model parameters, like raster angle or layer thickness [15], are also analyzed. This approach could explore the possibilities of a given technology to improve quality by acting upon influence factors. Nevertheless, this could also be its main limitation, since once the optimal configuration has been determined and errors still exceed tolerance requirements, there is no margin for further improvement.

Finally, error correction approaches act upon accuracy by working on the strategies used to convert 3D geometry into a series of flat layers (slicing), on the generation of material deposition paths and tool trajectories, or directly upon the CAD model. Therefore, this group is formed by those approaches that intend to overpass the limitations derived from the combination of process, technology and geometry to improve part dimensional or geometrical quality. This global objective could be achieved by different solutions. Some works just compensate deviations from the theoretical values [16]; others aim to compensate mechanical errors in the machine [17]; some works elaborate complex models to compensate the influence of different parameters upon the overall quality [18]. In recent years, there has been a tendency to apply machine-learning methods to provide error correction in AM [10,19,20]. Some works focus on compensating in-plane shape deformation [20,21]. Other approaches work with the 3D geometry, mainly compensating thermal deformations that have previously been modelled via finite elements (FE) simulation [22] or by means of virtual manufacturing models [23,24]. These types of works build "predictive" mathematical models that could be based on experimental data [10,20] or build from theoretical models [21,23]. Although Artificial Neural Networks (ANN) are frequently used for building predictive models [10,21,22], alternative mathematical modelling is also used for this purpose (e.g., Gaussian process multi-task learning [20] or particle swarm optimization [25]. The information provided by predictive models could be later used to change input parameters in order to fulfil tolerances.

In sum, the objective of improving dimensional and geometric quality in AM parts is frequently addressed without a recognizable methodology or a standardized procedure, mainly due to the variety of processes and influence factors. This situation is even more pronounced since most of the research has been carried upon "laboratory specimens", neglecting the relevance of industrial tolerances or the problems derived from medium to large batch productions. Additionally, most of the research has been carried out upon parts designed *ad hoc*, with different levels of complexity, that greatly differ between studies. These artefacts frequently consist of a collection of basic geometries (planes, cylinders, spheres) arranged in one unique part. There are also examples of dimensional quality comparison between biological-type parts (usually bones) in the fields of surgical reconstruction. Organizations like the American National Institute of Standards and Technology (NIST) have even proposed their own round-robin artefacts for establishing repeatability or reproducibility values for a particular AM technology or machine [26]. Accordingly, there are huge differences between quality indicators: some could be considered as artificial indicators, since they are calculated through complex mathematical expressions that ponder a series of individual parameters; others are usually referred to as "volumetric errors" although they lack a standardized physical meaning. Both types of indicators are useful to provide an impression of the overall manufacturing accuracy or for comparison between different process configurations, machines or technologies. Nevertheless, they tend to ignore the fact that dimensional tolerances in industry are limited to Features of Size (FoS) [27], as they are related to fit purposes. Finally, another frequent problem is that error is sometimes assumed to be linear, whereas, in AM processes, there are many factors that could have a non-linear influence upon error, like volumetric shrinkage of thermoplastic in Material Extrusion (ME) processes.

In the present work, a systematic framework for the minimization of dimensional and geometrical errors and tolerance fulfilment in AM parts is presented. This methodology has been specially developed to be applied upon FoS and medium to large production batches. In the following sections, a description of the framework will be presented. The systematic approach is structured in three consecutive steps: Work Analysis, Process Optimization and Design Optimization. The methodology has been evaluated using a case study under quasi-industrial conditions, which is the dimensional and geometric optimization of surgical-steel tibia resection guides manufactured by L-PBF.

2. Systematic Framework Description

The proposed framework has been designed to be used when production of a given part simultaneously fulfils two conditions:

- At least one of its features is a Feature of Size. Consequently, part geometry must include at least one cylinder or two parallel opposite planes [27] affected by a dimensional tolerance;
- Batch size and part-added value justify a proportional investment of test specimens and optimization effort.

Therefore, parts with features used for fitting purposes would be candidates for optimization via the proposed approach, whereas parts with features affected only by general tolerances would not be worth of such optimization efforts. Similarly, small batch sizes would not justify the effort of a systematic optimization. In these cases, alternative improvement strategies (e.g., trial-and-error) should be considered. Three consecutive stages have been proposed for the optimization: Work Analysis, Process Optimization and Design Optimization (Figure 1).

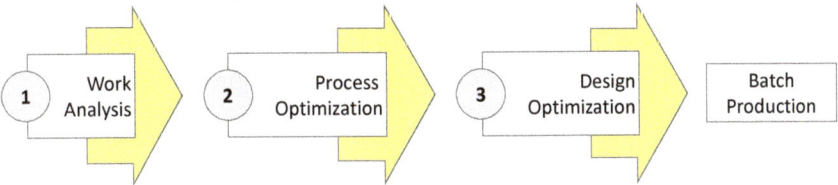

Figure 1. Steps of proposed approach.

2.1. Work Analysis

Firstly, a preliminary analysis of the work to be done would be carried out, with the objective of achieving a full description of the problem and evaluating lack-of-quality issues. This implies collecting all the information regarding part, process, production and equipment to elaborate an initial problem statement, defining an operational space and performing an initial quality characterization (Figure 2). Input information should be collected and structured according to three categories—production requirements, design specifications and process characteristics:

- Production requirements encompass information about expected production rates, batch size, maximum profitable manufacturing cost-per-unit and all additional requirements—like expected strength or hardness—that would not be related to geometrical requirements;
- Design specifications would comprise the geometrical information of the 3D CAD file, along with dimensional and geometric tolerances requirements. Information about shapes and dimensions should be taken into account even if they are not directly affected by tolerances. Part material should also be included in this category;
- Process characteristics should include all data regarding AM processes and equipment within the scope of the problem. Depending on each particular situation, process and/or machine could be previously set or included in the problem (which implies that process would be considered as an additional factor). At the process level, staff in charge of the optimization must analyze the information about the fundamentals of the process, including physical principle, range of materials or common manufacturing defects. At the equipment level, attention should be paid to the characteristics of machines, like workspace dimensions, axes speed, operation limits, or appropriate ranges for configuration parameters (layer thickness, building speed, etc.). Batch size and part-added value justify a proportional investment on test specimens and optimization effort.

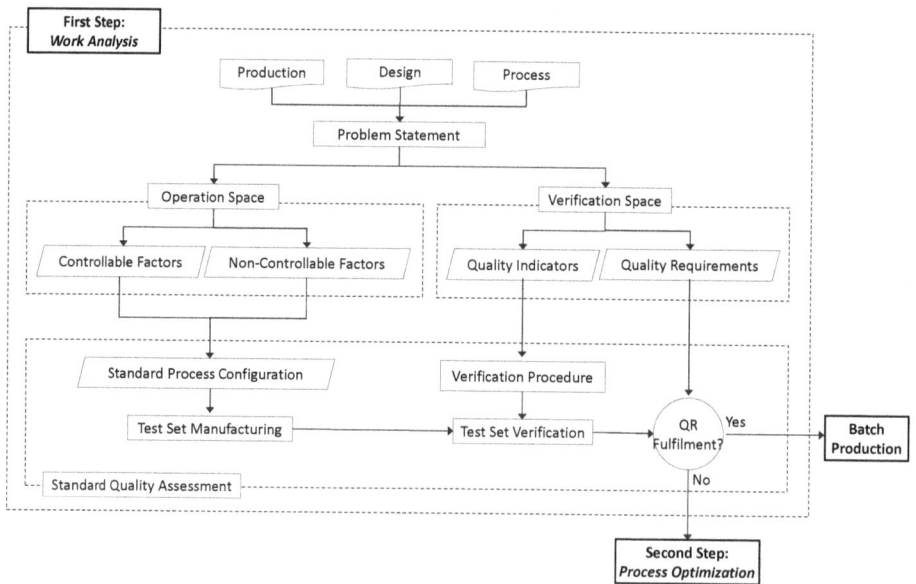

Figure 2. Work Analysis workflow.

Once all the relevant information has been collected, a statement of the problem should be performed. The objective of this task is to define both an Operation Space and a Verification Space.

- The Operation Space would consist of all those factors that could have an influence upon part quality. This means that every single factor whose modification would presumably affect, to a certain degree, the quality of the part, must be included in the Operation Space. Factors could be subject to modification (controllable factors) or not (non-controllable factors). Controllable factors are those that can be modified according to production decisions. This category includes discrete factors (e.g., selecting "glossy" or "matte" finishing in a Material Jetting process) and continuous factors (e.g., nozzle temperature in ME of thermoplastics) that could adopt many different values within a certain range. On the other hand, non-controllable factors are those that, having an influence upon part quality, should not be modified. This category would include design decisions of requirements (shape, dimensions, material, etc.) that have been set during the design stage. It also should include production decisions (batch size, process, production machine, etc.) that are not subjected to possible modifications. Process parameters could be also considered non-controllable factors when their values have been set according to material or machine supplier recommendations, workshop procedures or workforce experience. Categorization of influence factors into non-controllable and controllable groups is a key task. Most factors undoubtedly belong to one of these groups, but special attention should be paid to those factors that do not have a significant influence upon quality according to previous know-how, since there is a risk of neglecting their influence upon a particular part or feature, despite their actual significance;

- The Verification Space would be formed by Quality Indicators (QI) and Quality Requirements (QR). QI would be used for evaluating the degree of compliance of the tolerances imposed during the design stage. They would usually match FoS quality requirements, like dimensions (diameter of a cylindrical feature or distance between parallel flat surfaces) or geometrical deviation of controlled features (flatness, parallelism, cylindricity or concentricity). Nevertheless, QI could also be defined as relative differences between those parameters and their optimal values (e.g., the difference between the measured diameter and the middle value of the tolerance interval). During the definition of QI, it would also be necessary to define the verification procedure. This means that an inspection plan should be elaborated, including the materials and methods used for verifying each part and calculating actual values for each QI. Finally, QR are defined as the range of acceptable values that each single QI should adopt to enable batch production.

Once information has been structured into the Operating Space and the Verification Space, the objective of this first step is to determine if a standard process configuration could ensure the fulfilment of QR. In order to check this condition, a test set must be manufactured and verified. This implies that staff in charge of improving part quality must decide which process configuration should be used, by setting all controllable factors. This task should be performed taking into account previous experiences, good practices and workforce know-how.

The size of the test set must be determined in order to properly check QR fulfilment and, simultaneously, minimize the number of test specimens to be manufactured. Robustness of this task will increase with the number of replicas, whereas test size is conditioned by experimental cost. Nevertheless, a minimum of two building trays for each particular manufacturing configuration should be demanded, in order to contemplate experimental error. In accordance, calculation of QI values should also be done by means of arithmetic average values of repeated measurements. Staff are encouraged to consult the available literature regarding Design of Experiments (DOE) and Quality Assessment [28].

Manufactured test specimens would then be measured by means of the verification procedure, so that measured values for QI would be calculated and compared with QR. If the results indicate that those requirements would be fulfilled for an acceptable number of parts (defined as the percentage of valid parts per total production), then parts would be considered suitable for batch production and the procedure would finish. If requirements are not appropriately fulfilled, then the strategy continues through its second step. An intermediate situation could also occur when some of the QR are fulfilled but not all of them. In this case, the efforts during Process Optimization should be focused on those requirements that have not been properly fulfilled.

2.2. Process Optimization

Process Optimization (Figure 3) determines if quality requirements could be fulfilled by just acting upon controllable factors. The complexity of testing the significance on quality of variations in Controllable Factors sharply increases with their number. Checking their influence could be simple when few factors have to be considered, but turns to be extremely complex when the number of factors that should be tested increases. To minimize this problem, factors could be ranked according to their level of significance by means of statistical tools, like Design of Experiments (DOE). This Significance Analysis would be performed as an iterative task to reduce experimental effort, since considering all possible Controllable Factors for DOE could be an inefficient approach when some initial restrictions based on previous know-how have been incorporated into the analysis. Consequently, an initial appreciation of each factor significance could be established based on research papers results, reference books, workforce know-how, etc. This approach would help to fix some factors and reduce experimental effort.

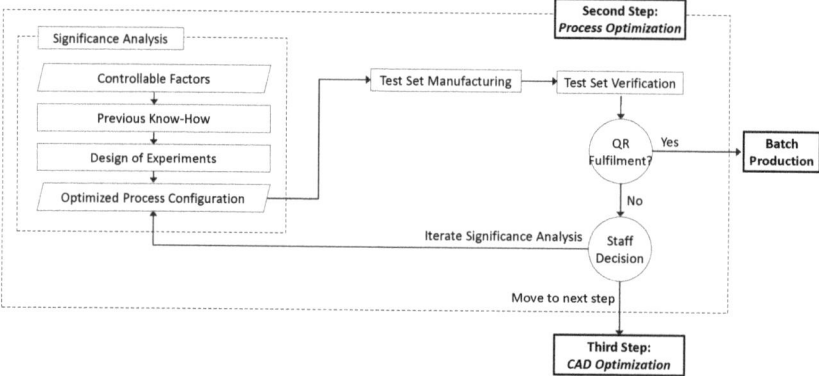

Figure 3. Process Optimization workflow.

For instance, it is widely accepted that layer thickness affects geometric quality in ME, since coarse layers increase the staircase effect of sloped surfaces. Consequently, although layer thickness is usually a controllable factor and could be considered during Process Optimization, it can be assumed that increasing layer thickness would not improve geometric quality, and thus this factor should not be included in the optimization step.

Once the number of factors has been initially reduced by means of previous know-how, further decisions should be sustained by experimental testing and supported by statistical analysis. Experimental designs could demand huge experimental effort when a high number of factors are considered. In these cases, the use of fractional factorial designs is widely recommended. Fractional designs would allow for reducing the experimental error by minimizing the number of experiments (and, consequently, the cost of manufacturing and measuring test specimens) by running a fraction of a full factorial design. This approach has the disadvantage of confounding effects of higher-order interaction, but it will be useful to characterize main effects and low-order interactions at a reduced experimental cost. Accordingly, an analysis of variance could provide an ordered list of factors, reflecting their relative influence upon each QI variance. Factors that show no influence upon QI should no longer be considered for Process Optimization. On the other hand, those factors that show a significant influence upon QI results should be ranked according to their relative importance. This procedure could lead directly to an optimized process configuration if only categorical factors (fan on/fan off) have been considered. Nevertheless, if continuous and discrete factors have been included in the DoE, further research could be demanded. In any case, analysis of variance would provide a regression equation that models how QI behaves according to changes in the influence factors. This equation should be used to optimize a particular QI, whereas multiple response regression optimization methods could be used for simultaneous optimization of multiple QIs.

This optimization effort should lead to a newly optimized process configuration and, once this initial optimization has been established, a new test set should be manufactured in order to check if the QR are fulfilled. A positive result would lead to starting batch production under the optimized configuration, whereas a negative result would lead to revising the significance analysis. If no completely positive result could be achieved, staff should take the decision to finish this step and move onto the Design Optimization step.

2.3. Design Optimization

Part manufacturing after Process Optimization may still produce features that do not fulfil expected QR. Consequently, part errors (deviations between QI values and optimal QR) should be mathematically modeled and design parameters that could have an influence on the results must be identified. It has to be noted that, in this methodology, design factors are limited to those that could be modified at the CAD definition step. This means that inner-part characteristics like layer thickness or wall thickness are considered process factors, since they are defined at the Computer-Aided Manufacturing (CAM) step.

There are clear differences regarding the level of complexity of error modelling for the non-fulfilment of dimensional QR and geometrical QR:

- Dimensional non-fulfilments could be corrected if the observed variability of QI is comparatively lower than the range of acceptable values defined by each correspondent QR. In these cases, Design Optimization searches for a model where the values of design parameters related to QI could be modified to improve quality. This means that optimization could be conducted without modifying the initial parameterization of the part, since only the values of existing parameters would be modified according to the correspondent inverse model;
- Geometrical non-fulfilments, on the other hand, demand the change of part parameterization at the design stage, which could be a more complex task. In this situation, a model of geometric distortion would be necessary. The ideal situation at this point is that analysis of part geometric distortion leads to a recognizable pattern that could be easily parameterized—e.g., if a theoretically cylindrical feature resembles an elliptic cylinder once manufactured, the geometric distortion model should determine the orientation of the major and minor axis, and the correspondent equation coefficients and re-parameterization turns, to be almost direct. Nevertheless, in other situations, modelling the deformed geometry would require methods of higher complexity. E.g., if the deformation does not resemble any common primitive, the theoretical cylinder could be modelled by means of a free-form adjustment based on Non-Uniform Rational Basis Splines (NURBS). Research efforts should be conducted in order to select the most appropriate fitting model, minimizing as much as possible the model complexity and taking into account each particular CAD suite capability for manipulating different types of parameterization. In both cases, original parameterization should be substituted with a new parameterization that follows actual manufactured geometries. Once this objective is achieved, the problem is similar to that described for dimensional non-fulfillments (defining a predictive model and an inverse model) with the difference that, in this case, alternative design parameters are used.

Additionally, it must be taken into account that, although geometrical re-parameterization could influence dimensional results, it is possible that dimensional optimization would not significantly influence geometrical QR fulfilment. This fact leads to the proposal that geometrical optimization, if necessary, should be carried out before dealing with dimensional optimization.

Once an adequate parameterization has been defined, it would be used to predict the most probable value that each QI would adopt in the final part as a function of controllable factors and design factors. Consequently, a new set of test specimens that include variations in the values of those design parameters that influence QR fulfillment would be manufactured and measured, and QI measures would be used to elaborate a "predictive" model.

There are many mathematical models that could be used at this step, ranging from simple linear regression to complex computing systems (response surfaces, artificial neural networks, etc.). The staff in charge of optimization should evaluate, in each case, which would be the best option for building a predictive model, pondering the required experimental effort and model complexity.

Once the predictive model has been made available, the objective is to define a new mathematical model that answers the inverse question: what should the proper values of design parameters be to obtain a QI as close as possible to the optimal QR? This new model should be known as an "inverse" model. The inverse model would provide new values for design parameters, so that a newly optimized design could be obtained. Ideally, new parts manufactured with this optimized design would be closer to the design theoretical objective, fulfilling the desired QR. Figure 4 contains a graphical explanation of this strategy.

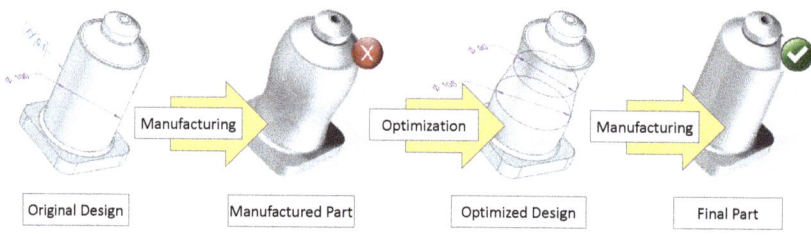

Figure 4. Design Optimization strategy.

The proposed workflow for the Design Optimization step is provided in Figure 5.

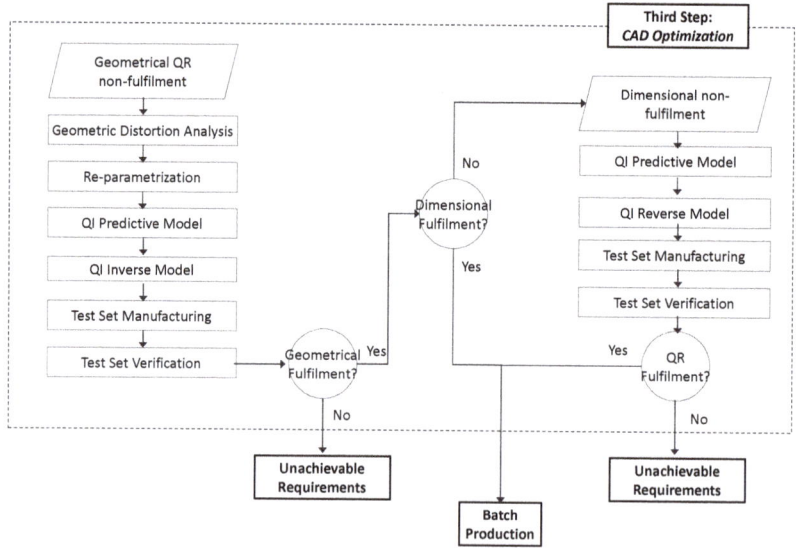

Figure 5. Design Optimization workflow.

2.4. Practical Recommendations

Once the different steps of the methodology have been described, there are some practical recommendations that should be taken into account:

- Optimization steps should be carried out using the minimal number of experiments and also minimizing the number of test specimens that would be manufactured;
- Subjective decisions regarding process configuration should be avoided whenever possible;
- Several tasks could have an iterative application when results suggest that additional research could be worthwhile;
- Factors categorized as non-controllable by following previous know-how or good-practices recommendations could have a more significant influence than previously thought for certain geometries. Accordingly, if QI variance could not be put under control by means of controllable-factor adjustment, such categorizations must be carefully revised.

To evaluate the usefulness of the proposed framework, a case study is presented in the next section.

3. Case Study: Optimization of Surgical-Steel Tibia Resection Guides

Evaluation case studies must fulfil two conditions: the design must contemplate at least one FoS affected by a dimensional tolerance, and optimization effort must be in accordance with potential batch size. Under both premises, several alternatives were considered until a surgical-steel tibia resection guide was finally selected. This part is a metallic insert used for the guidance of resection instruments during knee arthroplasty, a surgical procedure that is carried out approximately 600,000 times a year in Europe according to EUROSTAT statistical reports [29]. Among different alternatives for resection tool guidance, the one considered here (Figure 6) is a bi-component design, consisting of a polyamide customized alignment part (single-use) and a surgical steel insert (multiple-uses).

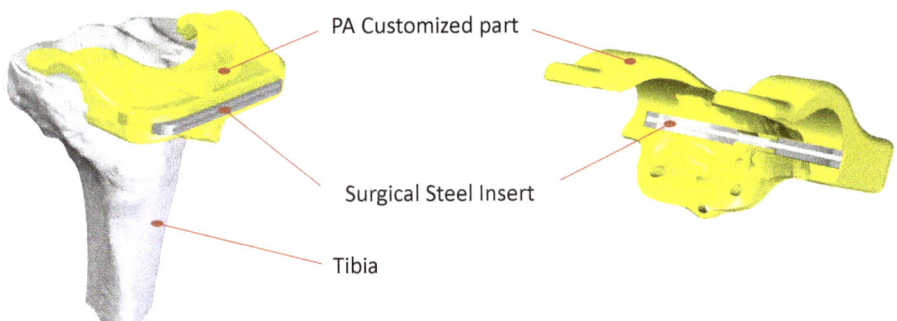

Figure 6. Parts of a bi-component tibia resection guide.

From a functional point of view, metallic inserts must fit into the PA part and become a single element (no relative movement allowed). At the same time, the theoretical resection plane should be accurately oriented with respect to the alignment features. Guidance is achieved by means of a deep and narrow through slot, defined by two parallel opposite flat surfaces and two slightly tapered lateral sides, resembling an arrow slit, to facilitate instrument handling. This geometry can be obtained through traditional manufacturing processes in two stages. Firstly, basic geometry could be manufactured directly by milling a metal plate or a casting preform; secondly, although external geometry could be completed by milling, the slot is so narrow and deep that it should be manufactured by means of Wire EDM (Electro Discharge Machining) or Sinker EDM. The combined manufacturing costs of these processes could range from 100 to 300 € per unit according to TEKNOS experts. Nevertheless, such geometry could be also manufactured by means of metal AM processes at a competitive cost.

In following sections, the consecutive steps of the proposed methodology applied to such geometry will be explained and discussed.

3.1. Step 1: Work Analysis

Work analysis starts by gathering together all available information about production requirements, design specifications and process characteristics. Firstly, inserts must be manufactured in a material suitable for biomedical applications, like surgical stainless steel or titanium, by means of a process capable of achieving good quality. This condition led to selecting L-PBF for manufacturing the knee resection inserts included in this work. This process uses the energy of a focused laser beam (typically Nd-YAG) to locally melt metal powder into a solid part [30]. An EOSINT M270 machine has been chosen to manufacture test specimens. This machine has a 250 × 250 mm working area, whereas building height could reach up to 215 mm. A 200 W Yb-Fiber is used alongside with F-Theta precision lens. Layer thickness could range from 20 to 100 µm, whereas other process parameters (scanning speed or effective power consumption) are material-dependent and established according to the indications of the manufacturer. Parts have been manufactured at PRODINTEC Technological Center. The estimated batch size was established at 1000 units per year, and parts must be manufactured according to the design shown in Figure 7. Tolerances were defined by the research team taking into account part functionality and meaningfulness regarding the objectives of this particular case study, but they are not intended to be applicable to other resection guide designs, since present design values for geometric tolerances values could possibly be too restrictive.

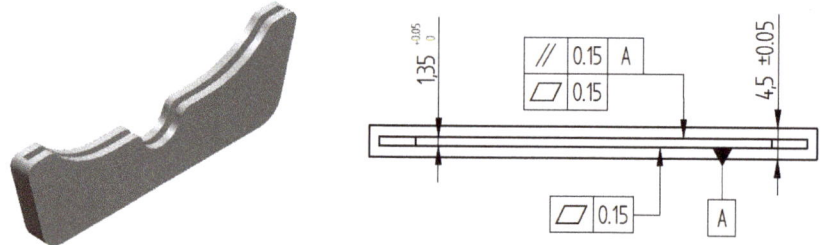

Figure 7. Part design (**a**) Perspective view; (**b**) Main tolerances in mm.

This part includes two FoS affected by tolerances: external width (nominal value 4.5 mm with a symmetric tolerance of ±0.05 mm) and slot width (nominal value 1.35 mm with an asymmetric tolerance of +0 to +0.05 mm). Internal surfaces of the slot are also affected by a 0.15 mm flatness tolerance, whereas parallelism between those surfaces and the external ones has to be under 0.15 mm. Other part features do not demand specific tolerances, so it can be assumed that usual manufacturing quality would be adequate, since non-compliance of general tolerances related to these features would not critically affect instrument guiding during surgery. Verification will be carried out using a DEA Global Image 09-15-08 Coordinate Measurement Machine (CMM). This machine has been calibrated according to EN 10360-2:2001, with a maximum permissible error in length measurement (MPE_E) of $2.2 + 3 \cdot L/1000$ µm (L in mm) and a maximum permissible error in probing repeatability (MPE_P) of 2.2 µm. PC-DMIS metrology software was used to perform verification operations. Temperature in the laboratory during verification procedures is maintained within 20 ± 2 °C. Once information has been gathered and the main features of the problem have been stated, the sequence of tasks continues with the definition of the Operation and Verification Spaces and the subsequent Standard Quality Assessment.

3.1.1. Operation Space

Influence Factors should be grouped in two categories: controllable and non-controllable (Table 1).

Table 1. Operation Space.

Category	Factor–Level 1	Factor–Level 2	Controllable	Non-Controllable
Production				
	Batch Size			X
	Type of process			X
	Equipment			X
Design				
	Geometry			X
	Dimensions			X
	Material			X
Process				
	Process Parameters			
		Layer Thickness	X	
		Volume Rate		X
		Scan Speed		X
		Power		X
		Ambient Parameters		X
		Base Temperature		X
		Part Orientation		X
		Part Location	X	
		Type of Support	X	
Post-Processing				
		Support Removal		X
		Thermal/Heat Treatment	X	
		Sand Blasting	X	

Since L-PBF has been selected as the most appropriate AM process for this work, and production shall be carried out in an EOS M270, neither process type nor machine could be initially considered as controllable factors. These decisions would also condition subsequent ones, since other possible influence factors, like the type of material, should be accordingly limited to those that the process/machine combination can handle. Therefore, since EOSINT M270 uses proprietary materials, possible materials are reduced to stainless steel PH1 or titanium Ti64. Since weight is not a crucial parameter, SS PH1 has been selected as the construction material, and therefore should be considered as a non-controllable factor. Geometry and dimensions have been set at the design stage. Consequently, all features, geometries and dimensions in the CAD have been also considered as non-controllable factors. Exceptions to this rule are those FoS affected by tolerances since, even when not subjected to modifications at this stage, they could be modified during the Design Optimization stage. Layer thickness in this machine ranges from 20 to 100 μm, so it is a controllable factor. Nevertheless, once layer thickness has been selected, volume rate, scanning speed and effective power are also defined according to material technology specifications. This implies that they cannot be modified by the operator and, consequently, they must be considered as non controllable factors. This is also applicable to ambient parameters like nitrogen atmosphere (1.5% oxygen), building platform model, or base temperature (40°). Part orientation has a great influence upon processing time, which usually implies that parts are oriented so that minimum Z travelling is required. Nevertheless, issues regarding manufacturing of slot surfaces were taken into account to avoid excessive overhanging, which would require support structures inside the slot. This concern involved selecting a vertical orientation for the slot and, consequently, orientation has been labelled as a non-controllable factor.

On the other hand, the location of parts within the workspace can be modified by the operator with minimal restrictions (like minimum allowed space between adjacent parts), so it should be considered as a controllable factor. Support structure type is also a controllable factor. Finally, post processing operations could also have an influence upon quality. A support removal operation is unavoidable if support structures are used. Other post-processing operations, like sand blasting or thermal treatments, are optional, and so they should be considered as controllable factors.

3.1.2. Verification Space

QI would be related to part FoS, affected by dimensional and geometric tolerances. They would also be ranked according to their relevance to part functionality.

Accordingly, the distance between parallel surfaces of the slot (DS) has been considered as the most relevant QI, since it critically affects resection instrument performance during surgery. An excessive distance would cause noticeable clearance between the slot and the resection instrument, whereas an insufficient distance would make its movement difficult. The second most relevant, Flatness of Slot parallel surfaces (FSR for the Rear surface and FSF for the Frontal surface) could also have an influence during resection, while Parallelism between these Slot surfaces (PS) must be controlled in order to allow uniform behavior of the instrument with independence in its orientation during resection. Finally, Distance between External surfaces (DE) has a relatively lower relevance, since its insertion in the alignment part would be favored by PA flexibility. QR have been defined as the acceptable range of values that each QI should adopt. These limits have been established during the design stage, as reflected in Table 2.

Table 2. Quality indicators with their correspondent quality requirements sorted by relevance.

Priority Order	QI	Lower Limit (mm)	Upper Limit (mm)
1	DS	1.350	1.400
2	FSR	-	0.150
3	FSF	-	0.150
4	PS	-	0.150
5	DE	4.450	4.550

3.1.3. Standard Quality Assessment

Standard Quality Assessment implies manufacturing a test set and checking if the QI values measured during verification fulfil QR. In order to manufacture the test set, all controllable factors must be revised, and a Standard Process Configuration defined. This task should be done by taking into account existent know-how, which should include the literature or research works, supplier recommendations or personnel's previous experience. Consequently, layer thickness was set at 20 µm, according to the recommended value for SS PH1. Layer thickness selection determines other variables that, like Volume Rate, fixed at 1.8 mm^3/s, are included in the technology files provided by the manufacturer. Part location could also affect quality but, since the objective is to manufacture medium-to-high batches, it is necessary to accommodate the maximum possible number of units in each tray. Accordingly, it was decided that parts within the test set would be distributed along the whole work area. A lightweight supporting structure was selected, since it is the easiest to remove and minimizes removal cost. The same criterion was used to decide that no thermal processing would be applied for the standard configuration. On the other hand, since sand blasting is used to minimize the effect of metallic projections upon surface quality, this post-processing operation was included as part of the process. Once the values and alternatives for all these factors have been defined, Process Configuration could be considered complete.

Regarding the Verification Procedure, it has to be noted that QI could be calculated using just four planes adjusted to slot parallel surfaces and external parallel surfaces. To digitize each internal surface, a regular grid with 284 points was used. The distance between adjacent points is 1.4 mm in the same column and 1.6 mm in the same row. Due to slot restricted accessibility, a spherical-end stylus probe with 0.7 mm diameter and 20 mm length was used. In the case of external surfaces, regular grids with 171 digitized points were used. The distance between adjacent points was 2.2 mm in the same column and 2.26 mm in the same row. Complete digitizing of each part in this work (including alignment routine) was repeated thrice, and, each time, four planes were adjusted to each set of digitized points (Figure 8). Consequently, QI was calculated thrice, and average values were obtained.

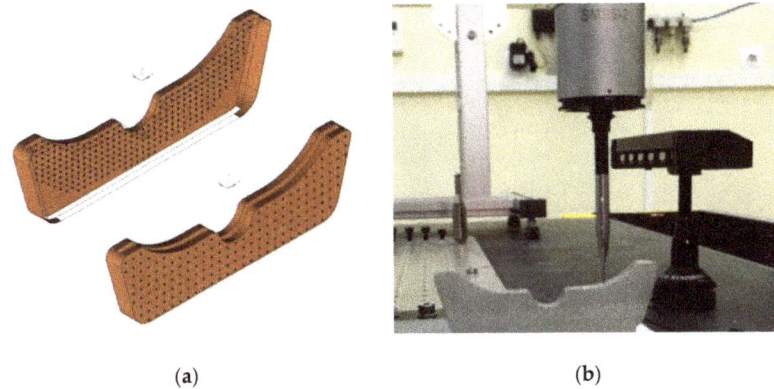

Figure 8. Verification Procedure. (**a**) Distribution of points on the internal and external surfaces; (**b**) CMM measurement of a test specimen.

The test set used to check part quality under standard process configuration was arranged as a series of sixteen test specimens, manufactured in four independent trays (four units each). These parts were located on the corners of the tray, so that the effect of part location upon QI could be observed through results' analysis. Once the specimens were manufactured, parts were removed by mechanical means and sand blasted before CMM verification at the laboratory. Table 3 collects the average values of the three measurements performed for each QI and each part.

Table 3. Standard Quality Assessment results.

Item	Tray	Location	DS (mm)	FSR (mm)	FSF (mm)	PS (mm)	DE (mm)
1	01	1	1.359	0.135	0.266	0.278	
2	01	2	1.378	0.188	0.113	0.193	4.504
3	01	3	1.398	0.121	0.378	0.382	4.572
4	01	4	1.405	0.160	0.331	0.406	4.561
5	02	1	1.450	0.140	0.275	0.374	4.555
6	02	2	1.400	0.177	0.109	0.219	4.477
7	02	3	1.397	0.120	0.103	0.209	4.467
8	02	4	1.440	0.317	0.127	0.425	4.600
9	03	1	1.429	0.123	0.397	0.435	4.568
10	03	2	1.396	0.125	0.167	0.194	4.479
11	03	3	1.381	0.141	0.170	0.189	4.453
12	03	4	1.495	0.295	0.137	0.469	4.572
13	04	1	1.485	0.338	0.257	0.636	4.668
14	04	2	1.359	0.120	0.190	0.201	4.482
15	04	3	1.424	0.116	0.076	0.193	4.489
16	04	4	1.405	0.112	0.128	0.148	4.487
Average (mm)			1.413	0.171	0.201	0.309	4.527
Standard Deviation (mm)			0.040	0.076	0.103	0.140	0.060
QR Fulfilment			43.7%	62.2%	43.7%	6.25%	56.2%

Results clearly indicate that QR would not be fulfilled under standard manufacturing conditions. DS average value (1.413 mm) indicates that slots tend to be wider than expected (13 μm wider than the correspondent QR upper limit). Additionally, DS standard deviation (0.040 mm) indicates an unexpectedly high variability. This indicates that the process is not under control and standard configuration would not allow for batch production. Similar conclusions can be derived from the analysis of the other QR, since none of them are fulfilled under standard conditions. Geometric indicators (PD, FSD and FSF) clearly exceed the desired limits, while simultaneously presenting very high values for the standard deviation. Finally, DE achieves the desired quality in nine out of sixteen parts. Nevertheless, DE standard deviation (0.060 mm) clearly indicates that an unacceptable percentage of parts would not fulfil this condition during batch manufacturing.

Consequently, Standard Quality Assessment reveals that QR are far from being fulfilled. This means that the methodology should move onto the second step: Process Optimization.

3.2. Step 2: Process Optimization

The objective of Process Optimization is to work exclusively upon process configuration parameters to fulfil quality requirements. This means that the level of significance that variations in controllable factors exert upon variation in QI values must be established. At this stage, controllable factors have been reduced to Layer Thickness, Part Location, Type of Support, Thermal Treatment and Sand Blasting. A DOE considering five parameters could be performed at this point to gain a statistical assessment of each factor's relative influence upon QI. Nevertheless, an analysis of factor influence likeness has been previously performed, to evaluate if the number of factors could be reduced in the first iteration of Process Optimization

- Layer Thickness value was set during the Standard Quality Assessment at the minimum achievable level (20 µm). Delgado et al. [31] have not found a statistically significant effect of Layer Thickness upon dimensional error in L-PBF for test specimens with non-sloped surfaces (like the one analyzed in the present work). On the other hand, Nguyen et al. [32] reported an increase in dimensional accuracy with decreasing layer thickness, and this result seems to have been confirmed by Maamoun et al. [33]. Although it is sometimes difficult to compare results, no research has been found supporting the hypothesis that thicker layers could improve dimensional quality. Accordingly, since the minimum achievable value for Layer Thickness has already been used for the Standard Process Configuration, this factor should not be included during the Process Optimization step;
- Sand blasting is the usual finishing process in L-PBF, since it is intended to remove metallic projections that appear as spikes on part surfaces. Consequently, sand blasting is presumed to reduce the unevenness of surfaces. This means that suppressing sand blasting would probably cause an increment in flatness and parallelism deviations and also could affect dimensional quality. These arguments led to the consideration of sand blasting as an unavoidable step;
- Part location within the working area could influence quality results, and this hypothesis could be tested using data from the previous step. Analyzing how DS measures are related to part location within the trays, the Pearson Correlation provides a p-value of 0.721. This value reveals an extremely low probability of a linear relationship between DS and part location for the standard process configuration. Additionally, none of the other indicators (FSR, FSF, PS and DE) show any correlation with part location according to their p-values. This means that variability in QI cannot be explained by part location within the tray, and, thus, modifying location would not allow for fulfilment of QR;
- Lightweight supporting structures were preferred during the Standard Quality Assessment because part removal was easier and did not demand subsequent machining. Nevertheless, manual removal of support could have an influence upon observed lack of quality [34], so Type of Support should be included as a possible influence factor in the Process Optimization step;
- Similarly, thermal stresses accumulated during manufacturing operations (in-layer and layer-upon-layer) could provoke a noticeable distortion in the part, after they are released from support [35]. Accordingly, thermal treatment should be included in the Process Optimization step.

As a result of the analysis, only two factors were left for Process Optimization: Type of Support and Thermal Treatment. In order to check if those factors have a real influence upon Quality Indicators, it was decided that two additional trays (Tray 05 and Tray 06) should be manufactured using solid support (instead of a lightweight one) and applying a thermal treatment before releasing parts from the tray. Thermal treatment was intended to release thermal residual stresses of parts and was carried out following the recommendations of the material supplier (EOS) and manufacturer (PRODINTEC), by maintaining the tray at a 482 °C during four hours in a Nabertherm oven. Then, the parts and tray were left to cool at room temperature before being taken to a sawing operation. Once each part was released from the tray, it was taken to a Computer Numerical Control (CNC) milling machine to complete support removal. An *ad hoc* designed jig was used to prevent part deformation during milling. When the overall process had finished, parts were verified with the CMM using the same procedure as for trays 01 to 04. Comparisons between values of QI calculated from trays 01 to 04 (lightweight support/no thermal treatment) and 05 to 06 (solid support/thermal treatment) are provided in Figure 9.

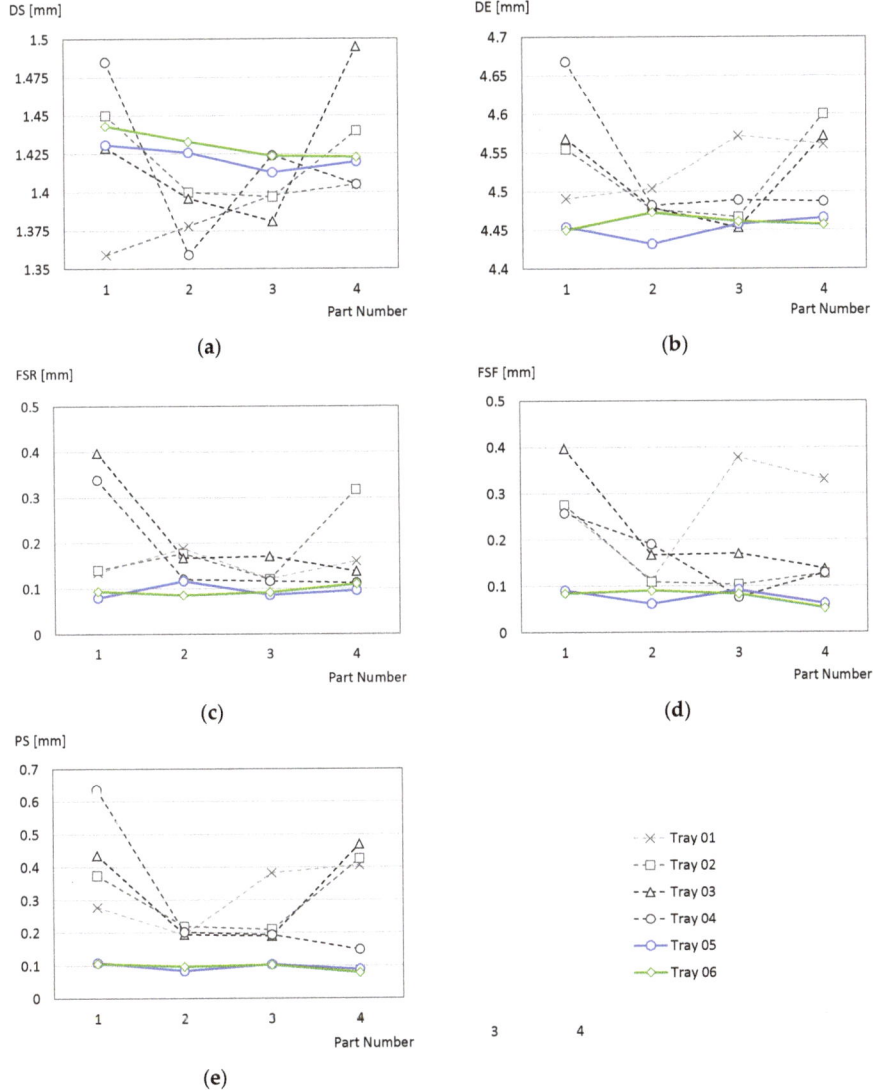

Figure 9. Comparison of QI measures for trays 01 to 06: (**a**) *DS*; (**b**) *DE*; (**c**) *FSR*; (**d**) *FSF*; (**e**) *PS*.

Results indicate that the variability observed during the first step was related to the factors included in this Process Optimization step, since the QI measured for parts from trays 05 and 06 are clearly more uniform. This result is especially remarkable in the case of QI derived from geometric tolerances, since all the parts in the new trays fulfil QR for *FSR*, *FSF* and *PS*.

In the case of dimensional requirements, none of the new parts fulfil QR for *DS*, whereas only six parts fulfil the requirement for *DE* (although *DE* values are particularly close to the lower acceptable value). Nevertheless, the most important fact about measured dimensions is that variability seems to be significantly lower than that observed during the first step. Standard deviation results for dimensional QI pointed to the possibility of fulfilling QR by means of an optimization of design parameters.

Table 4 provides the measurement results of these eight specimens. The conclusion of this analysis is that the resection guide must be manufactured using solid support structures and that a thermal treatment, like the one described above, must be applied, and both elements have been thereafter incorporated to the Optimized Process Configuration. Manufacturing using this configuration directly allows the fulfilment of geometrical QRs, but it is still clearly unsuccessful in the case of the dimension of the slot and external width.

Table 4. Optimized Quality Assessment results.

ID	*DS* (mm)	*FSR* (mm)	*FSF* (mm)	*PS* (mm)	*DE* (mm)
Average	1.427	0.095	0.077	0.096	4.456
Standard Deviation	0.009	0.013	0.016	0.011	0.012
QR Fulfilment	0%	100%	100%	100%	75%

Reaching this point, no additional QI improvement could be reasonably achieved by means of process parameters without acting upon design. Consequently, the methodology moved to the third step: Design Optimization.

3.3. Step 3: Design Optimization

Design Optimization can affect both dimensional and geometric tolerances, and the level of complexity required depends on the results observed during the previous stages. In this case study, geometric QR have been fulfilled via Process Optimization, so Design Optimization must focus on dimensional QR. Dimensional optimization implies building a mathematical model capable of accurately predicting the value that each QI would reach, as a function of those controllable factors whose significance has been considered relevant within the scope of the problem. In this case, study of both *DS* and *DE* present low variability after the Process Optimization step (*DS* standard deviation is 9 μm and *DE* standard deviation is 12 μm). This suggests that a unique linear compensation of designed theoretical values for all the parts within a tray could be applied. In the simplest formulation, the average deviation of both QI with respect to correspondent theoretical dimensions could be calculated and the design parameters modified accordingly, assuming linear behavior of results.

Nevertheless, although this could be the optimal approach for *DE*, it would not be equally recommended for *DS*. Quality Requirement for *DE* has a 100 μm range; consequently, uniform compensation should reasonably get most of the parts within QR. However, the Quality Requirement for *DS* has a 50 μm range. In order to achieve further improvements of *DS*, it was decided that the level of significance of remaining controllable factors (part location on the tray, with respect to *X* and *Y* axis) upon *DS* measures variability has to be verified. Although these factors have not shown significance for parts manufactured under Standard Process Configuration, the reduction in variability achieved via Optimized Process Configuration could have modified this circumstance.

Accordingly, a two-level full-factorial 2^2 DOE has been defined. Possible curvature effects have been taken into account by including two central points. This design allows for using data from trays 05 and 06. Part location has been coded according to a virtual *XY* origin ideally placed at the geometric center of the tray.

Parts located on the left side of the tray have been coded as $X = -1$, whereas those on the right side have been coded as $X = 1$. Similarly, parts located closer to the door have been coded as $Y = -1$, whereas those far from the door have been coded as $Y = 1$. Central locations have been coded $X = 0$ and $Y = 0$. Table 5 contains the structure of experiments for the DOE and the measured values for QI.

Table 5. Design of experiments (DOE) structure and results.

ID	X	Y	DS (mm)	FSR (mm)	FSF (mm)	PS (mm)	DE (mm)
1	−1	−1	1.431	0.080	0.091	0.108	4.454
2	−1	1	1.426	0.117	0.062	0.084	4.432
3	0	0	1.421	0.078	0.057	0.070	4.458
4	0	0	1.423	0.089	0.082	0.089	4.453
5	1	−1	1.413	0.086	0.092	0.103	4.458
6	1	1	1.420	0.096	0.062	0.088	4.466
7	−1	−1	1.443	0.094	0.084	0.105	4.450
8	−1	1	1.433	0.085	0.090	0.097	4.473
9	0	0	1.428	0.088	0.077	0.080	4.452
10	0	0	1.428	0.092	0.083	0.101	4.455
11	1	−1	1.424	0.092	0.083	0.101	4.461
12	1	1	1.423	0.110	0.052	0.078	4.457

DS results (Table 5) have been processed using Minitab 17 statistical software to obtain variance analysis. Results are reflected in Table 6.

Table 6. DS Variance Analysis.

Source	DF	Adj SS	Adj MS	f-Value	p-Value
Model	5	0.000592	0.000118	23.11	0.001
Blocks	1	0.000169	0.000169	32.93	0.001
Linear	2	0.000361	0.000181	35.24	0.000
X	1	0.000351	0.000351	68.51	0.000
Y	1	0.00001	0.00001	1.98	0.209
2-Way Interactions	1	0.000055	0.000055	10.76	0.017
X*Y	1	0.000055	0.000055	10.76	0.017
Curvature	1	0.000007	0.000007	1.37	0.286
Error	6	0.000031	0.000005		
Lack-of-Fit	4	0.000029	0.000007	7.19	0.126
Pure Error	2	0.000002	0.000001		
Total	11	0.000623			
Model Summary					
S	R-sq	R-sq(adj)	R-sq(pred)		
0.0022638	95.06%	90.95%	73.96%		

Analysis of variance points out that the location along the X axis has a significant influence upon DS values. Additionally, although the Y location appears non-significant, the interaction of X and Y has also been found to be significant. This means that variance in DS could be modelled by considering the location of parts within the manufacturing tray, with respect to both X and Y axes, since a linear relationship between them could not be discarded. Figure 10 contains an explicative Pareto chart of standardized effects, where the relative significance of factors A (X), B (Y) and interaction $A*B$ ($X*Y$) can be observed. Additionally, the interaction plot for DS helps to explain the effect of each factor upon DS. Parts located to the left tend to have a wider slot than parts located to the right. Parts located closer to machine door are expected to have wider slots than parts located far from the door when the left side of the tray is analyzed, but this behavior is flipped (wider slots for distant parts) when parts manufactured on the right area of the tray are analyzed.

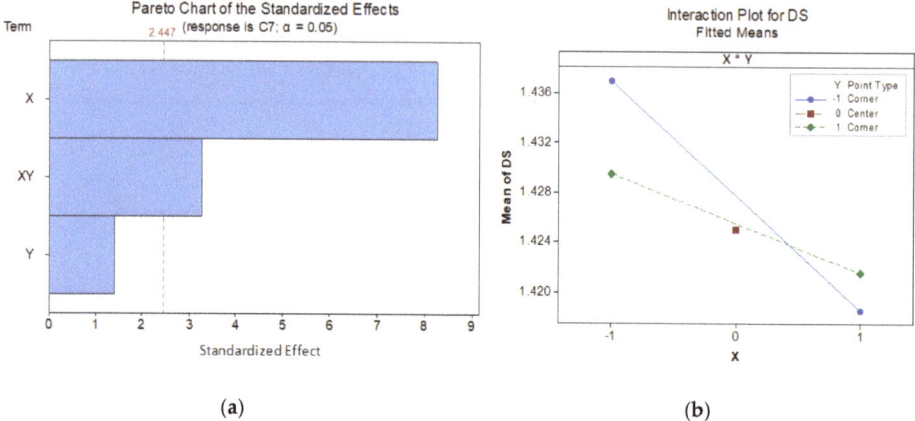

Figure 10. *DS* statistical graphs: (**a**) Pareto chart of the standardized effects; (**b**) Interaction plot.

This behavior illustrates the significance of $X*Y$ interaction, and indicates that modelling *DS* variability would have to include both X and Y locations as parameters.

Additionally, values for center points indicate that a linear relationship between location and DS should be expected, since there is no evidence of curvature. Although there was no need to analyze $X*Y$ significance upon *DE*, this task could be carried out without additional experiments. Consequently, an additional analysis of variance has been performed for *DE* using data from Table 5. Results indicate that neither X nor Y have any significance on DE variability. This result implies that observed variability cannot be explained by means of part location within tray, so these factors should not be taken into account when building a predictive model for DE. Figure 11 provides a Pareto chart for *DE*.

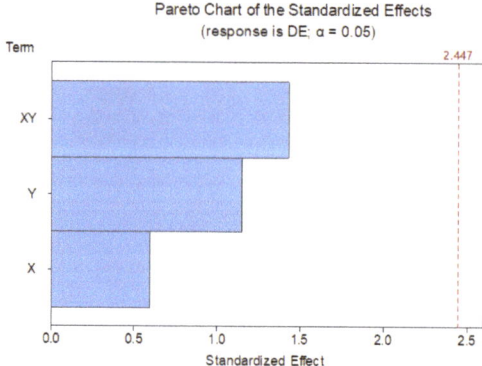

Figure 11. Pareto chart of the standardized effects for *DE*.

In sum, although the part location according to X and Y axes should be taken into account when predicting *DS* variability (values would differ significantly for different locations), they should not be considered for *DE* (values would be similar, independent of part location). Accordingly, no predictive model is required for *DE*. Instead, a simple linear compensation of the average values of correspondent QI will be used in the present case study.

3.3.1. Predictive and Inverse Models

To elaborate the predictive model for DS, some considerations must be given:

- Design values for DS will be thereafter denoted as DS_D, whereas the correspondent values measured after part manufacturing will be denoted DS_M. Equivalent notation will be used for DE_D and DE_M;
- Modifying DS_D could not only affect DS_M, but also DE_M and, vice versa, modifying DE_D could have and influence upon DS_M;
- Predictive model should include design factor (DS_D) as part of its parameterization. To achieve better results, the set of data that would be used to adjust such models should also include variations in design factors.

Cross-influence of DS_D and DE_D has been addressed in this case study by focusing on the predictive model for DS_M, and calculating DE_D compensation from average values of DE_M obtained from those parts used to construct the DS_M model. Since the distribution of DS_M with part location have previously shown no curvature, a simple polynomial expression has been used for predictive model $DS_M = f_1(X, Y, DS_D)$ Consequently, DS_M values should be predicted as a function of X location, Y location and DS_D, being a_1, a_2, a_3 and a_4 coefficients that minimize the adjustment error of such a function (1).

$$DS_M = a_1 + a_2 \times X + a_3 \times Y + a_4 \times DS_D \tag{1}$$

To calculate these coefficients, a new tray arrangement was defined so that the theoretical distance between slot parallel surfaces could also be taken into account. Consequently, two additional trays were defined and manufactured: the first one includes six parts, four located at the corners of the tray and the other two at the center, with a nominal DE of 1.350 mm. The second one follows the same distribution, but the width of the slot has been reduced to 1.250 mm. These limits have been selected according to previous results, which show that DE_M tends to be noticeably higher than DE_D. Consequently, optimized values for this parameter could be reasonably expected to be smaller than the initial ones and probably within the 1.350 to 1.250 mm range. Parts were thereafter measured and results can be found in Table 7.

Table 7. DS values used for predictive model construction.

ID	X (mm)	Y (mm)	DS_D (mm)	DS_M (mm)
1	−77.25	−90.58	1.25	1.311
2	−77.25	90.58	1.25	1.307
3	−12.25	0	1.25	1.329
4	12.25	0	1.25	1.331
5	77.25	−90.58	1.25	1.319
6	77.25	90.58	1.25	1.306
7	−77.25	−90.58	1.35	1.437
8	−77.25	90.58	1.35	1.430
9	−12.25	0	1.35	1.424
10	12.25	0	1.35	1.426
11	77.25	−90.58	1.35	1.418
12	77.25	90.58	1.35	1.421

Model coefficients have been calculated by means of a least square iterative method and results are provided in Table 8.

Table 8. Predictive model coefficients.

Coefficient	Estimated Value
a_1	-4.32500×10^{-2}
a_2	-3.25450×10^{-5}
a_3	-2.89799×10^{-5}
a_4	1.08833×10^{-0}

Once the predictive model was defined, an inverse model was constructed. This inverse model should provide an optimized design value for DS (DS_O), so that the correspondent measured value DS_M of the manufactured part is as close as possible to the theoretical (optimal) value for DS (DS_T). This objective could be achieved by means of an optimization problem (2). Note that, in the original design, DS_T and DS_D were equivalent, whereas, after optimization, each part would have a different DS_O.

$$minerror = \min(DS_M - DS_T)^2 = \min(f(X, Y, DS_D) - DS_T)^2 \quad (2)$$

Consequently, the value that minimizes such functions should be an optimized DS_D (denoted as DS_O for clarification purposes). This problem could be efficiently solved by means of an interior-point method, taking advantage of the easily calculable derivatives of the function. In the present work, MATLAB *fmincon* command has been used to determine the optimal design values for DS.

Finally, DE compensation value was calculated from DE_M results obtained from Table 7 parts, so that the average value reflects possible variations derived from DE_D modification. Accordingly, a DE_M average value of -0.043 mm has been calculated and linear compensation should provide an optimized value of 4.532 for DE_O.

3.3.2. Optimized Design Quality Assessment

To evaluate fulfilment of QR once the design was optimized, an inverse model was used to design a verification tray with nine parts. This tray included six positions that had already been used to elaborate the prediction model and four additional intermediate ones (never used before). According to the predictive and inverse models, the design dimensions of slots were different for every individual part, whereas DE_O is unique (4.543 mm) for all parts. Table 9 includes design values and measured results.

Table 9. Optimized Design Quality Assessment results.

ID	X (mm)	Y (mm)	DS_O (mm)	DS_M (mm)	DE_O (mm)	DE_M (mm)
1	−77.25	−90.58	1.298	1.374	4.543	4.507
2	−77.25	90.58	1.303	1.378	4.543	4.508
3	−38.625	−45.29	1.301	1.383	4.543	4.506
4	−38.625	45.29	1.303	1.380	4.543	4.503
5	0	0	1.303	1.371	4.543	4.488
6	38.625	−45.29	1.303	1.374	4.543	4.508
7	38.625	45.29	1.305	1.380	4.543	4.514
8	77.25	−90.58	1.303	1.381	4.543	4.514
9	77.25	90.58	1.308	1.387	4.543	4.521

As can be observed, QR have been fulfilled for every manufactured part in both DE and DS. To illustrate the evolution of QI from the Standard Process Configuration to the last step of the optimization procedures, Figures 12 and 13 provide two comparative histograms of DS_M and DE_M.

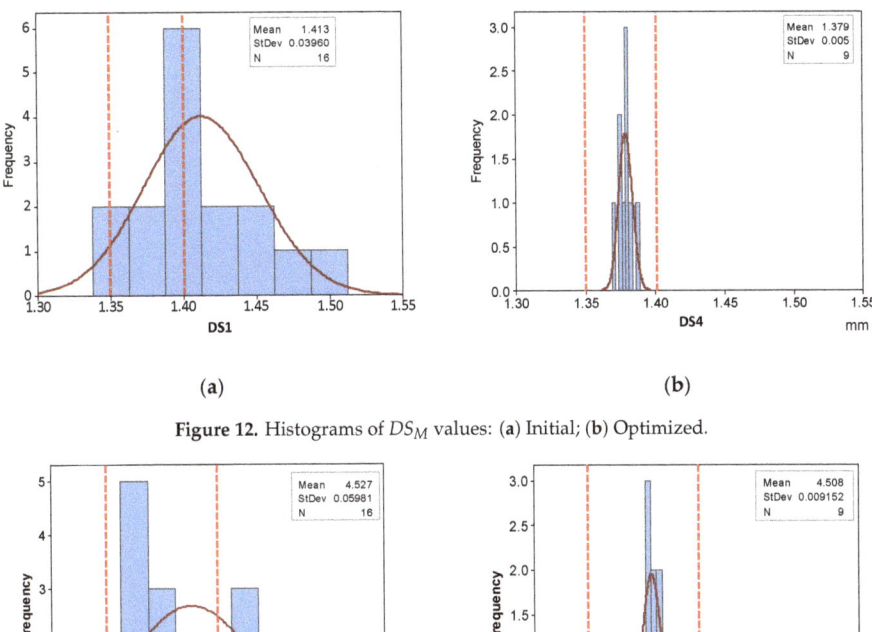

Figure 12. Histograms of DS_M values: (a) Initial; (b) Optimized.

Figure 13. Histograms of DE_M values: (a) Initial; (b) Optimized.

It can be observed that improvement was achieved by two effects: centering the average value of slot width with respect to the limits of the correspondent QR, and reducing initial variability to the extent that an extremely high percentage of parts should fulfil required quality. In fact, standard deviation within the verification tray has been reduced to only 5 µm, whereas its value calculated for the Optimized Process Quality Assessment was 9 µm. QR for DE has also been achieved via Design Optimization but, since the same dimensional compensation has been used, standard deviation presents a similar value (9 µm).

According to these results, batch production could commence once the implications of Process and Design optimization steps have been incorporated into production configuration.

4. Discussion

The proposed framework allows for combining the advantages of different approaches analyzed in the literature review. Some of the works related to the error analysis and error prevention [6,7,14] could be incorporated under our approach to the Process Optimization stage. In fact, once consolidated, their conclusions on factors' influence upon part quality could be part of an intensive knowledge base to simplify decision-making procedures in AM. Similarly, the mathematical methods proposed for error prediction used in works devoted to error correction [17,18,22] could be easily incorporated into the Design Optimization stage. However, the definition of this framework makes it unnecessary to apply optimization to the whole part, but instead is focused on FoS affected by dimensional tolerances and their related geometrical tolerances. A global compensation of geometric distortions [22,23] could be researched, to see if it is preferable to our proposal of a specific compensation, but there is a risk of orientating efforts to models that, despite being able to improve overall dimensional and geometric quality, failed to fulfil a specific tolerance. Nevertheless, both approaches should not be considered exclusive, since it is possible to anticipate that, in the future, machine learning and artificial intelligence approaches [10,21] will both be incorporated to machine control systems. The proposed framework would check if specific QR have been fulfilled once optimization procedures have been incorporated into machine technology and, in case embedded optimization rules are still not enough to match tolerances, it would provide a methodology that manufacturers could easily follow. The degree of complexity of the tasks included in the proposed framework is highly dependent on each particular part's design and process characteristics, but it encourages taking advantage of available knowledge to simplify experimental effort. Models capable of accurately predicting dimensional and geometric errors grow in complexity with the number of factors contemplated, so an approach that delays simulation efforts until problem complexity has been reduced (minimizing factors) will probably be more useful to end-use manufacturers in the short term. Further research should be done in order to provide detailed rules on how some decisions should be adopted. Quantitative evaluation of the possible consequences of opting for one process configuration alternative could help staff to reach higher levels of objectivity while minimizing subjective decisions. Moreover, defining a model that describes all available possibilities in terms of statistical analysis and mathematical modelling, while simultaneously helping staff to decide which of these tools should be preferable for a particular situation, would be of great use.

5. Conclusions

The achieved results support the usefulness of a systematic framework for dimensional and geometric quality enhancement of additively manufactured parts. Work Analysis has permitted a reasonable understanding of the role of different influence factors, grouped according to controllable and non-controllable categories, to define both the Operating Space and the standard process configuration. An initial evaluation of QR fulfilment by means of a verification procedure and a test set provides an idea of the differences between measured QI values and QR objectives. Sorting controllable factors according to their relative influence upon QI values helps to simplify the Operational Space and drives testing to the most promising configurations. This contributes to a reduction in experimental effort and helps save costs and time. A balance between know-how and experimental effort should allow for an improvement in part quality that could eventually make further development unnecessary. Nevertheless, once the process's capabilities have been exhausted, the possibility of working upon the design parameters, or even upon the parameterization model itself, could take part quality to a higher level. The application of this framework to a quasi-industrial case, involving dimensional and geometrical optimization of surgical-steel tibia resection guides, helps explain the proposed workflow in more detail. In fact, L-PBF manufacturing of these inserts has an approximate cost-per-unit of 37 €, according to PRODINTEC, with an approximate production rate of 1.6 units per hour (based on 50 specimen trays). This cost is truly competitive with the conventional manufacturing alternatives, ranging from 100 to 300 €. These results help to reinforce the idea that the proposed framework could contribute to the global objective of AM quality improvement.

Author Contributions: Conceptualization, N.B. and D.B. methodology, N.B., D.B. and Á.N.; validation, B.J.Á. and P.F.; investigation, N.B., B.J.Á. and P.F.; data curation, B.J.Á., Á.N and P.F.; writing—original draft N.B., D.B. and B.J.Á.; project administration, D.B.; funding acquisition, D.B.

Funding: This research was funded by the Asturian Institute for Economic Development (IDEPA) and ArcelorMittal as part of the RIS3 strategy, grant number SV-PA-15-RIS3-4.

Acknowledgments: The authors wish to thank IDEPA and ArcelorMittal for their support and also acknowledge Pablo Suarez from Teknos Biomedical Engineering and Ignacio Dosil from PRODINTEC for their advice.

Conflicts of Interest: The authors declare no conflict of interest. The funders had no role in the design of the study; in the collection, analyses, or interpretation of data; in the writing of the manuscript, or in the decision to publish the results.

References

1. *Additive Manufacturing—General Principles—Part 1: Terminology*; ISO/DIS 17296-1; International Organization for Standardization: Genève, Switzerland, 2014.
2. *Standard Terminology for Additive Manufacturing Technologies*; ASTM F2792-12; ASTM International: West Conshohocken, PA, USA, 2012.
3. Basiliere, P.; Shanler, M. *Hype Cycle for 3D Printing*; Gartner Inc.: Stamford, CT, USA, 2019.
4. Wohlers, T. *Wohlers Report 2012—Additive Manufacturing and 3D Printing State of the Industry: Annual Worldwide Progress Report*; Wohlers Associates: Fort Collins, CO, USA, 2012.
5. Feenstra, F. Additive Manufacturing: SASAM Standardisation Roadmap 2014. Available online: https://www.rm-platform.com/downloads2/send/2-articles-publications/607-sasam-standardisation-roadmap-2014 (accessed on 7 September 2019).
6. Masood, S.H.; Rattanawong, W.; Iovenitti, P. Part build orientations based on volumetric error in fused deposition modeling. *Int. J. Adv. Manuf. Tech.* **2000**, *16*, 162–168. [CrossRef]
7. Noriega, A.; Blanco, D.; Alvarez, B.J.; Garcia, A. Dimensional Accuracy Improvement of FDM Square Cross-Section Parts Using Artificial Neural Networks and an Optimization Algorithm. *Int. J. Adv. Manuf. Tech.* **2013**, *69*, 2301–2313. [CrossRef]
8. Boschetto, A.; Bottini, L. Triangular mesh offset aiming to enhance Fused Deposition Modeling accuracy. *Int. J. Adv. Manuf. Tech.* **2015**, *80*, 99–111. [CrossRef]
9. Armillotta, A.; Bellotti, M.; Cavallaro, M. Warpage of FDM parts: Experimental tests and analytic model. *Robot. CIM Int. Manuf.* **2018**, *50*, 140–152. [CrossRef]
10. Ikeuchi, D.; Vargas-Uscategui, A.; Wu, X.; King, P.C. Neural Network Modelling of Track Profile in Cold Spray Additive Manufacturing. *Materials* **2019**, *12*, 2827. [CrossRef] [PubMed]
11. Chang, D.Y.; Huang, B.H. Studies on profile error and extruding aperture for the RP parts using the fused deposition modelling process. *Int. J. Adv. Manuf. Technol.* **2011**, *53*, 1027–1037. [CrossRef]
12. Brajlih, T.; Valentan, B.; Balic, J.; Drstvensek, I. Speed and accuracy evaluation of additive manufacturing machines. *Rapid Prototyp. J.* **2011**, *17*, 64–75. [CrossRef]
13. El-Katatny, I.; Masood, S.H.; Morsi, Y.S. Error analysis of FDM fabricated medical replicas. *Rapid Prototyp. J.* **2010**, *16*, 36–43. [CrossRef]
14. Byun, H.S.; Lee, K.H. Determination of the optimal build direction for different rapid prototyping processes using multi-criterion decision making. *Robot. CIM Int. Manuf.* **2006**, *22*, 69–80. [CrossRef]
15. Lee, B.H.; Abdullah, J.; Khan, Z.A. Optimization of rapid prototyping parameters for production of flexible ABS object. *J. Mater. Process. Technol.* **2005**, *169*, 54–61. [CrossRef]
16. Kumar, V.V.; Tagore, G.R.N.; Venugopal, A. Some investigations on geometric conformity analysis of a 3-D freeform objects produced by rapid prototyping (FDM) process. *Int. J. Appl. Res. Mech. Eng.* **2011**, *1*, 82–86.
17. Tong, K.; Joshi, S.; Lehtinet, E.A. Error compensation for fused deposition modeling (FDM) machine by correcting slice files. *Rapid Prototyp. J.* **2008**, *14*, 4–14. [CrossRef]
18. Sood, A.K. Study on Parametric Optimization of Used Deposition Modelling (FDM) Process. Ph.D. Thesis, National Institute of Technology, Rourkela, India, 2011.
19. Baturynskaa, I.; Semeniuta, O.; Martinsen, K. Optimization of process parameters for powder bed fusion additive manufacturing by combination of machine learning and finite element method: A conceptual framework. *Procedia CIRP* **2018**, *67*, 227–232. [CrossRef]
20. Zhu, Z.; Answer, N.; Huang, Q.; Mathieu, L. Machine learning in tolerancing for additive manufacturing. *CIRP Ann. Manuf. Technol.* **2018**, *67*, 157–160. [CrossRef]
21. Shen, Z.; Shang, X.; Li, Y.; Bao, Y.; Zhang, X.; Dong, X.; Wan, L.; Xiong, G.; Wang, F.Y. PredNet and CompNet: Prediction and High-Precision Compensation of In-Plane Shape Deformation for Additive Manufacturing. In Proceedings of the 2019 IEEE 15th International Conference on Automation Science and Engineering (CASE), Vancouver, BC, Canada, 22–26 August 2019. [CrossRef]

22. Chowdhury, S.; Anand, S. Artificial Neural Network Based Geometric Compensation for Thermal Deformation in Additive Manufacturing Processes. In Proceedings of the ASME 2016 11th International Manufacturing Science and Engineering Conference (MSEC 2016), Blacksburg, VA, USA, 27 June–1 July 2016. [CrossRef]
23. Paul, R. Modeling and Optimization of Powder Based Additive Manufacturing (AM) Processes. Ph.D. Thesis, University of Cincinnati, Cincinnati, OH, USA, 2013.
24. Moroni, G.; Syam, W.P.; Petrò, S. Towards early estimation of part accuracy in additive manufacturing. *Procedia CIRP* **2014**, *21*, 300–305. [CrossRef]
25. Liu, X.; Liu, L.; Zhao, Y.; Ma, J.; Meng, L.I.; Mao, L. Optimization of Forming Accuracy of Additive Manufacturing Complex Parts Based on PSO. In Proceedings of the 14th IEEE Conference on Industrial Electronics and Applications (ICIEA), Xi'an, China, 19–21 June 2019. [CrossRef]
26. Moylan, S.; Brown, C.U.; Slotwinski, J. Recommended Protocol for Round Robin Studies in Additive Manufacturing. *J. Test. Eval.* **2016**, *44*, 1009–1018. [CrossRef]
27. General Tolerances. *Part 1: Tolerances for Linear and Angular Dimensions without Individual Tolerance Indications*; ISO 2768-1; International Organization for Standardization: Genève, Switzerland, 1989.
28. Montgomery, D.C. *Design and Analysis of Experiments*, 9th ed.; John Wiley & Sons: Hoboken, NJ, USA, 2017.
29. Surgical Operations and Procedures Performed in Hospitals (Eurostat). Available online: http://ec.europa.eu/eurostat/statistics-explained/images/0/06/Surgical_operations_and_procedures_Health2015B.xlsx (accessed on 21 September 2019).
30. Gibson, I.; Rosen, D.W.; Stucker, B. *Additive Manufacturing Technologies. Rapid Prototyping to Direct Digital Manufacturing*; Springer: New York, NY, USA, 2010.
31. Delgado, J.; Ciurana, J.; Rodríguez, C.A. Influence of process parameters on part quality and mechanical properties for DMLS and SLM with iron-based materials. *Int. J. Adv. Manuf. Technol.* **2012**, *60*, 601–610. [CrossRef]
32. Nguyen, B.; Luu, D.N.; Nai, S.M.L.; Zhu, Z.; Chen, Z.; Wei, J. The role of powder layer thickness on the quality of SLM printed parts. *Arch. Civ. Mech. Eng.* **2018**, *18*, 948–955. [CrossRef]
33. Maamoun, A.H.; Xue, Y.F.; Elbestawi, M.A.; Veldhuis, S.C. Effect of Selective Laser Melting Process Parameters on the Quality of Al Alloy Parts: Powder Characterization, Density, Surface Roughness, and Dimensional Accuracy. *Materials* **2018**, *11*, 2343. [CrossRef]
34. Järvinen, J.-P.; Matilainen, V.; Li, X.; Piili, H.; Salminen, A.; Mäkelä, I.; Nyrhilä, O. Characterization of effect of support structures in laser additive manufacturing of stainless steel. *Phys. Procedia* **2014**, *56*, 72–81. [CrossRef]
35. Mercelis, P.; Kruth, J.P. Residual stresses in selective laser sintering and selective laser melting. *Rapid Prototyp. J.* **2006**, *12*, 254–265. [CrossRef]

 © 2019 by the authors. Licensee MDPI, Basel, Switzerland. This article is an open access article distributed under the terms and conditions of the Creative Commons Attribution (CC BY) license (http://creativecommons.org/licenses/by/4.0/).

Article

Mechanical Properties of 3D-Printing Polylactic Acid Parts subjected to Bending Stress and Fatigue Testing

J. Antonio Travieso-Rodriguez [1,*], Ramon Jerez-Mesa [2], Jordi Llumà [3], Oriol Traver-Ramos [1], Giovanni Gomez-Gras [4] and Joan Josep Roa Rovira [3]

[1] Mechanical Engineering Department, Escola d'Enginyeria de Barcelona Est, Universitat Politècnica de Catalunya, Avinguda d'Eduard Maristany, 10–14, 08019 Barcelona, Spain; oriol.traver@upc.edu
[2] Engineering Department, Faculty of Sciences and Technology, Universitat de Vic—Universitat Central de Catalunya, C. Laura, 13 Vic, 08500 Barcelona, Spain; ramon.jerez@uvic.cat
[3] Materials Science and Metallurgical Engineering Department, Universitat Politècnica de Catalunya, Escola d'Enginyeria de Barcelona Est, Avinguda d'Eduard Maristany, 10–14, 08019 Barcelona, Spain; jordi.lluma@upc.edu (J.L.); joan.josep.roa@upc.edu (J.J.R.R.)
[4] Industrial Engineering Department, IQS School of Engineering, Universitat Ramon Llull, Via Augusta, 390, 08017 Barcelona, Spain; giovanni.gomez@iqs.url.edu
* Correspondence: antonio.travieso@upc.edu

Received: 25 October 2019; Accepted: 19 November 2019; Published: 22 November 2019

Abstract: This paper aims to analyse the mechanical properties response of polylactic acid (PLA) parts manufactured through fused filament fabrication. The influence of six manufacturing factors (layer height, filament width, fill density, layer orientation, printing velocity, and infill pattern) on the flexural resistance of PLA specimens is studied through an L27 Taguchi experimental array. Different geometries were tested on a four-point bending machine and on a rotating bending machine. From the first experimental phase, an optimal set of parameters deriving in the highest flexural resistance was determined. The results show that layer orientation is the most influential parameter, followed by layer height, filament width, and printing velocity, whereas the fill density and infill pattern show no significant influence. Finally, the fatigue fracture behaviour is evaluated and compared with that of previous studies' results, in order to present a comprehensive study of the mechanical properties of the material under different kind of solicitations.

Keywords: additive manufacturing; 3D printing; fused filament fabrication; flexural properties; fatigue; PLA

1. Introduction

Manufacturing through fused filament fabrication (FFF) or 3D-printing is a phenomenon that has drastically changed the way manufacturing is understood, mainly during the last decade [1]. The interest comes from the clear advantages that this group of technologies presents with respect to traditional manufacturing technologies; that is, great freedom of design and innovation capacities, a stronger connection between design and manufacturing, or the ability to manufacture unique pieces [2]. In addition, additive manufacturing (AM) systems have been easily implemented in domestic or low-scale manufacturing environments as a cheap and easy manufacturing technology.

Regardless of the rapid expansion of AM, the problem related to the identification and prediction of the mechanical behaviour and physical characteristics of the final pieces has been the main handicap for its application in industrial environments or final pieces. This difficulty lies in the fact that the parameters to be defined during the manufacturing process are numerous and interact with one another; and, on the other hand, because of the anisotropy of the material, caused by the high influence of the filament orientations in the manufacturing space [3]. Furthermore, anisotropy also originated thanks to the difference between the bonding forces between strands of the same layer (intralayer) and between layers (interlayer) [4]. For these reasons, the orientation of the layers is a key parameter to be defined when taking into account the work conditions of the piece.

According to Bellehumeur et al. [5], the mechanical resistance of parts is the result of the addition of three factors: the resistance of the filaments, the resistance of the union between filaments of the same layer, and the resistance of the union between layers. The inherent resistance of the filaments mainly depends on the mechanical properties of the raw material and the strength of the joints depends on the cohesion between filaments. This is proportional to the thermal energy of the filaments when they come into contact when being placed. The union is a local sinter in which polymer chains are shared. This process is applicable to all joints, between layer threads of both the same layer and different ones.

The authors Gurrala and Regalla [6], Gray et al. [7], and Zhong et al. [8] agree that the orientation of the layers must be coincident with the directions of the expected service loads to optimize the mechanical properties. In contrast, in compression forces, owing to the buckling effect, the fibres tend to bend. Therefore, the fibres should be oriented perpendicular to the load in this case [9].

This same effect of the orientation of the layers on the mechanical properties of the workpieces, has also been observed in other processes of AM, as in the technology of laminated object manufacturing (LOM), according to Olivier et al. [10]; selective sintering by laser, as reported Ajoku et al. [11]; or stereolithography presented by Quintana et al. [12].

Another parameter with great influence on the mechanical properties is the height of the layer. When the layers have a lower height, the parts show an overall better cohesion between layers, because the contact surface is greater and the empty space between filaments is smaller. This effect improves the transport mechanism of thermal energy, favouring the welding between wires, as found in the work of [9]

On the other hand, the thickness or width of the extruded filament is also a parameter that significantly influences the mechanical behaviour. It has a great impact on the transport mechanisms of thermal energy, which will affect the cohesion of the threads, according to the study proposed by Wang et al. [13].

The printing strategy determines the paths of the machine head in the creation of the piece. Within this context, the printed pieces are composed of two characteristic zones: the contour and infill. The outline is the skin that delimits the piece and corresponds to the outer perimeters. The infill is the one formed by the trajectories that the nozzle follows to fill the empty space that remains inside the contour, as depicted in Figure 1.

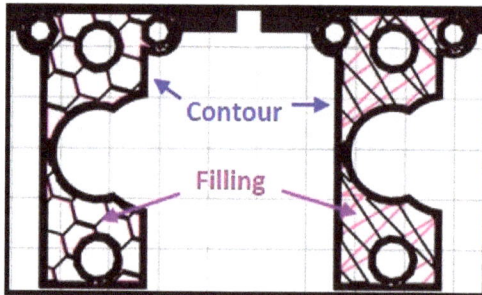

Figure 1. Section of a piece printed with two types of fill patterns. Left: honeycomb, right: linear.

Generally, in each layer, the contour is first performed followed by the internal filling with the selected printing strategy. Each one provides different mechanical properties. In the present work, the influence of several patterns shall be studied, as well as different infill densities, to assess their impact on the workpiece flexural behaviour.

The printing velocity is also a modifiable parameter. It can be defined for each printing zone, being independent for the contours, fills, and upper and lower layers. The velocity will be a parameter of study in this work since it has influence in the process of melting and solidification of the filaments. In addition, it affects the rate of extruded material.

Considering the aforementioned base of knowledge about FFF, this paper aims to study the influence of the manufacturing parameters on the mechanical properties of pieces made of polylactic acid (PLA) manufactured by FFF. Specifically, the flexural mechanical properties of these parts are evaluated. The results obtained are also compared with the those obtained in a previous study by Gómez-Gras et al. [14] and Jerez-Mesa et al. [15], performed on the same material subjected to a different loading mode. The main novelty delivered by this paper is that it contributes to the enrichment of mechanical behavioural data regarding PLA material. So far, an extensive study about bending properties and their direct comparison to fatigue performance linked to process parameters has not been found in the literature. For this reason, the results presented in this paper complement other results regarding tensile or fatigue properties, presented by authors in previous references, as presented above. The makers and users of FFF machines often ask about the best way to manufacture their parts. The answer should be that printing parameters should be chosen according to the expected part behaviour; this paper contributes to enriching that answer.

2. Materials and Methods

In this paper, the flexural mechanical properties of PLA are assessed. The influence of the manufacturing parameters in these properties will also be analysed. Therefore, the first experimental stage explained in this paper comprises a series of four-point bending tests performed on prismatic test specimens, following the American Society for Testing and Materials (ASTM) D6272-2 standard [16].

To better understand the influence of the significant parameters, different images of the fractured areas were taken and subsequently analysed. In addition, to complement the fractography, a micro scratch test was performed, which helped to better understand the fracture mechanism of the pieces.

In a second experimental stage, a fatigue Whöler curve generated through flexural fatigue tests was drawn to analyse whether the best conditions obtained in the four-point bending tests also derive in good fatigue properties.

2.1. Four-Point Bending Tests

2.1.1. Specimens Manufacture

The design of the specimens used in the study was done with SOLIDWORKS® Research Edition 2019 software (Dassault Systèmes, Vélizy-Villacoublay, France) and the models were filleted with Slic3r software (GNU Affero General Public License) [17]. Subsequently, they were manufactured in the domestic 3D printer, Pyramid 3Dstudio XL Single Extruder. Their geometry is shown in Figure 2, with dimensions according to the standard that governs the bending test. All manufactured specimens were submitted to a quality control, in which they were weighed and measured with a calliper. Therefore, they had to be validated before testing from a dimensional and constructive point of view. The resulting lengths, widths, and weights were statistically processed, and those specimens whose descriptors were out of the ±2% were considered not to comply and were immediately discarded.

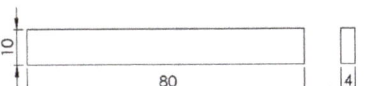

Figure 2. Test specimen's geometry: 80 mm × 10 mm × 4 mm, according to the D6272-02 ASTM standard.

The material used in the manufacture of the specimens, as discussed above, is PLA. It is a biodegradable thermoplastic. The choice of PLA as the study material was based on the fact that it is the most used material in domestic 3D printing. In this case, the selected filament was manufactured by Fillamentum Company from the Czech Republic. It has a diameter of 3 mm and its extrusion temperature is around 210 °C. The technical information provided by the manufacturer is indicated in Table 1.

Table 1. Mechanical properties of polylactic acid (PLA).

Mechanical Property	Value
Yield strength	60 MPa
Elongation at break	6%
Tensile modulus	3600 MPa
Flexural strength	83 MPa
Flexural modulus	3800 MPa

2.1.2. Taguchi Experimental Design

To carry out the four-point bending study, the design of experiments (DOE) technique was used. The design consists of the combination of the printing parameters that are considered most influential in mechanical behaviour. Six parameters are included in the study, and three levels of each one are defined (Table 2). They were selected taking into account the bibliography studied, as well as the experience of previous work of the research group.

Table 2. Parameters and levels used in design of experiments (DOE).

Parameter	Level		
	1	2	3
Filament width (mm)	0.3	0.4	0.6
Layer height (mm)	0.1	0.2	0.3
Fill density (%)	25	50	75
Printing velocity (mm/s)	20	30	40
Layer orientation	X- axis	Y- axis	Z- axis
Infill pattern	Linear	Rectilinear	Honeycomb

Filament width: Determined by the diameters of the extrusion nozzles: 0.3, 0.4, and 0.6 mm. It defines the volume and surface of the extruded threads, as well as the welding surface between filaments (Figure 3A).

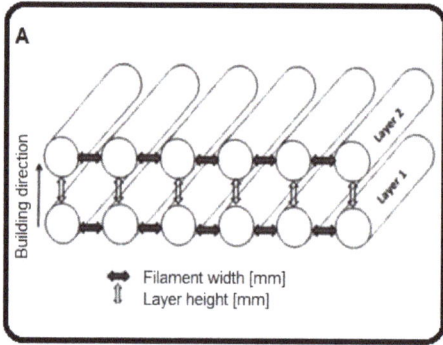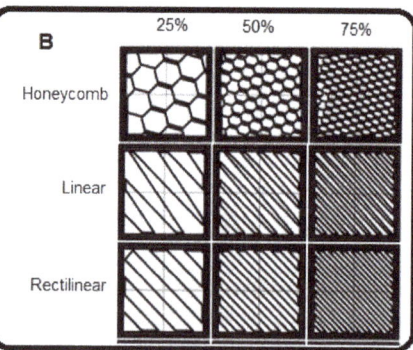

Figure 3. Schematic representation of the parameters used in the study: (A) filament width and layer height, (B) infill pattern and fill density.

Layer height: Describes the thickness of each layer and, therefore, the number of layers the printed piece will have. It affects the volume and surface of the threads, as well as the welding between layers. The manufacturing time is inversely proportional to the layer height. Thinner layers imply more layers to print and a longer production time (Figure 3A).

Fill density: Represents the amount of material that is deposited within the contours. It avoids relative movements between contours and gives robustness to the pieces. It also determines the distance between the inner threads and affects material consumption (Figure 3B).

Fill pattern: Defines the trajectories that the nozzle follows to fill the empty space within the contour. Each pattern will create a different interior geometry producing different mechanical behaviours (Figure 3B).

Orientation: The specimens will be printed in the direction of the three coordinate axes: X, Y, and Z, as shown in Figure 4. In this way, the stacking of the layers will be done in three different ways and their behaviour can be studied. Normally, the stacking direction is the most determinant factor in mechanical behaviour [18].

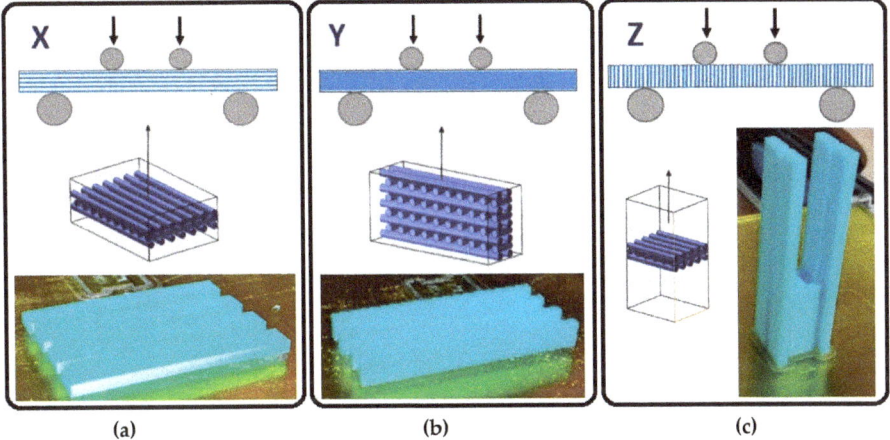

Figure 4. The orientation of the layers' stacking, in the manufactured specimens. (**a**) X-axis oriented; (**b**) Y-axis oriented; (**c**) Z-axis oriented.

Printing velocity: It determines the extrusion and deposition of the threads' velocity. The velocity is defined for each part of the piece (inner, external perimeters, inner threads, and so on) to optimize the manufacturing time. In this study, the same velocity was defined for all parts of the piece to homogenize its structure.

In this study, a Taguchi L27 DOE was used. This method has been applied successfully in other studies concerning the mechanical properties of FFF pieces [14] Table 3 shows an orthogonal matrix with a specific combination of parameters used. The influence of these separately as well as their interaction will be studied.

Table 3. Orthogonal matrix of Taguchi L27 for the DOE.

N°	Filament Width [mm]	Layer Height [mm]	Infill Density (%)	Printing Velocity [mm/s]	La Y-axiser Orientation	Infill
1	0.3	0.1	25	20	X-axis	Rectilinear
2	0.3	0.1	50	30	Y-axis	Linear
3	0.3	0.1	75	40	Z-axis	Honeycomb
4	0.3	0.2	25	30	Y- axis	Honeycomb
5	0.3	0.2	50	40	Z- axis	Rectilinear
6	0.3	0.2	75	20	X- axis	Linear
7	0.3	0.3	25	40	Z- axis	Linear
8	0.3	0.3	50	20	X- axis	Honeycomb
9	0.3	0.3	75	30	Y- axis	Rectilinear
10	0.4	0.1	25	30	Z- axis	Linear
11	0.4	0.1	50	40	X- axis	Honeycomb
12	0.4	0.1	75	20	Y- axis	Rectilinear
13	0.4	0.2	25	40	X- axis	Rectilinear
14	0.4	0.2	50	20	Y- axis	Linear
15	0.4	0.2	75	30	Z- axis	Honeycomb
16	0.4	0.3	25	20	Y- axis	Honeycomb
17	0.4	0.3	50	30	Z- axis	Rectilinear
18	0.4	0.3	75	40	X- axis	Linear
19	0.6	0.1	25	40	Y- axis	Honeycomb
20	0.6	0.1	50	20	Z- axis	Rectilinear
21	0.6	0.1	75	30	X- axis	Linear
22	0.6	0.2	25	20	Z- axis	Linear
23	0.6	0.2	50	30	X- axis	Honeycomb
24	0.6	0.2	75	40	Y- axis	Rectilinear
25	0.6	0.3	25	30	X- axis	Rectilinear
26	0.6	0.3	50	40	Y- axis	Linear
27	0.6	0.3	75	20	Z- axis	Honeycomb

The rest of the parameters that affect the conception of the test specimens remained constant.

2.1.3. Experimental Setup

The tests were carried out on the Microtest EM2/20 universal electromechanical machine, with a capacity of 20 kN, displacement of 300 mm, and a speed range 0–160 mm/min. The force acquisition was performed with a load cell of 500 N and a precision of 0.03 N.

The test consists of placing the specimen of a rectangular cross section over two supports and loading it at two points by means of two loading rollers; each at an equal distance from the adjacent support point. The specimen is bent at a constant speed, until the external fibres break, or until the maximum deformation of the external fibres reaches a 5% elongation. The parameters used in the experiment are described in the D6272-02 ASTM standard; that is, a support span of 64 mm and a load span of 21.3 mm (Figure 5).

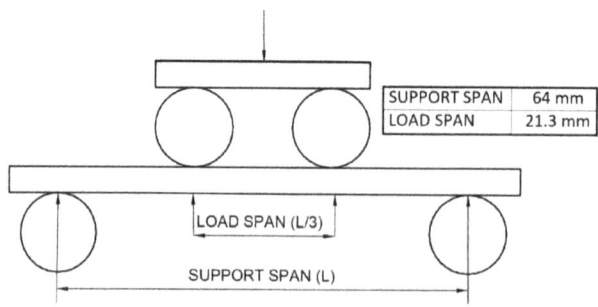

Figure 5. Diagram of the four-point bending test method, according to the D6272-02 ASTM standard.

The deflection value will be obtained through image processing. High-definition video capture is planned for all tests. That way, the displacement will be obtained through image processing, by following a marker painted on the lower fibre of the specimen. The displacement will be determined to calculate the overall deflection (Figure 6). On the other hand, the force applied by the loading rollers will be measured with a load cell. The objective of data processing is to create the stress–strain curve of the specimens [19]. From the obtained curve, the following results will be extracted: Young's modulus (E), elastic limit ($Rp_{0.2}$), maximum strength (σ_{max}), and maximum deformation (ε).

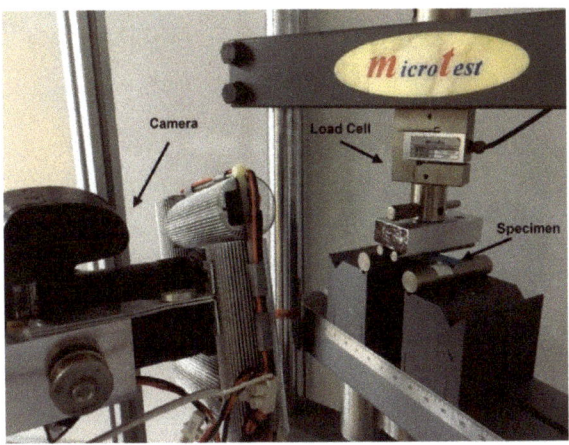

Figure 6. The installation used to perform the four-point bending tests.

The test method used contemplates two different types, which differ in the test speed according to the behaviour of the test piece.

Type A. Used in test specimens that break with little deflection.

Type B. Used in the test specimens that absorb large deflections during the test.

The Type A test will end when breakage is detected in the outer fibres of the test pieces, and the Type B test will end when specimens break or the deflection D = 10.9 mm, according to measurements of the specimens and the parameters used.

A previous experimental testing was performed to validate the adequacy of the described method. From these experiments, it was detected that specimens printed in the direction of the Z-axis do not admit deflection, and present brittle failure, while the specimens printed in the direction of the X- and Y-axes admit large deflections. The summary of the test types can be seen in Table 4.

Table 4. Test parameters.

Concept	Test Type A	Test Type B
Specimen's orientation	Z	X and Y
Test speed	1.9 mm/min	19 mm/min
End of test	When break appears in the external fibres	When breaks or deflection = 10.9 mm

2.1.4. Data Analysis

The data analysis was processed by following the steps described as follows:

1. Separation of the frames of the High Definition videos of each test. The camera used registered the image at approximately 60 fps. The tests lasted between 45 s and 2 min, so, in each of the 108 tests, between 2700 and 7200 frames were processed.

2. Calculation of the specimen's deflection through the frames. Position markers were painted on the outer fibre of the specimen, where the maximum deflection occurs, and on the static rollers (Figure 7A). The difference between the final position and the initial one, between the most displaced marker of the specimen and the markers on the static rollers, is considered the maximum deflection (Figure 7B). This analysis was performed through a self-designed MATLAB® code (version 2018) with image processing functions.

The calculation of the stress that is generated in the specimen at each moment by means of Equation (1) is as follows:

$$S = \frac{PL}{bd^2}, \qquad (1)$$

where

S = Stress in the outermost fibre (MPa)
P = Applied load (N)
L = Distance between support rollers (64 mm)
b = Width of the specimen (10 mm)
d = Thickness of the specimen (4 mm)

3. Analysis of the stress–strain curve obtained to extract the study parameters.

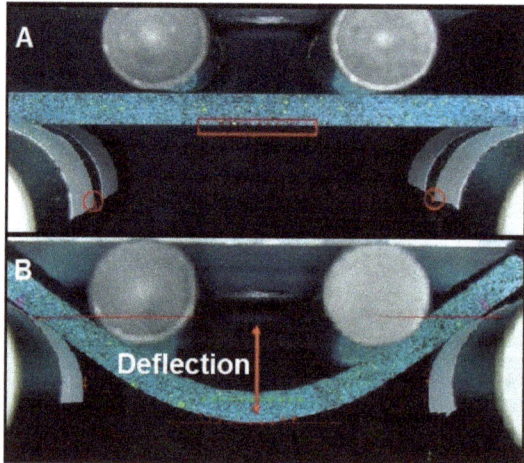

Figure 7. Schematic representation of the data collection process during the tests. (**A**) Initial position of markers; (**B**) final position (red crosses) and initial position (green crosses) of the markers.

2.2. Fractography and Scratch Test

In order to analyse the influence of the parameters that were significant, a SMZ-168 MOTIC stereo microscope was used to observe the fractures surfaces. The most interesting fracture phenomena were photographed with a MOTICAM 2300 camera. Both equipment were manufactured by Motic®, Xiamen, China.

Also, micro scratch tests were conducted in a scratch tester unit (CSM-Instruments, Needham, MA, USA) (Figure 8A) using a sphere-conical diamond indenter with a radius of 200 µm. Tests were done under a linearly increasing load, from 0 to 70 N, at a loading rate of 10 mm·min^{-1} and in an interval length of 5 mm, according to the ASTM C1624-05 standard [20]. Figure 8B shows the two different scratches per specimen that were carried out in order to observe the reproducibility of the induced damage. Furthermore, the micro scratch tests were conducted in the longitudinal and transversal printing direction to observe the main plastic deformation mechanisms induced. Surface damage induced during scratch tests was observed by a desktop scanning electron microscopy (SEM) Phenom XL from ThermoFisher Scientific (Waltham, MA, USA) (Figure 8C).

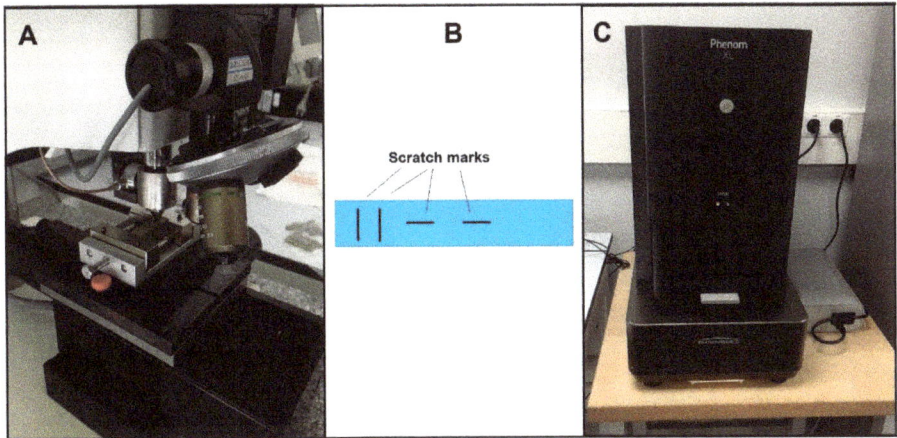

Figure 8. Micro scratch test. (**A**) scratch tester unit; (**B**) specimen; (**C**) scanning electron microscopy (SEM) ThermoFisher Scientific Phenom XL.

2.3. Fatigue Test

To complete this study, it is proposed to analyse, in a second experimental stage, how cylindrical specimens behave when manufactured through the optimal parameter set found in the previous study, subjected to a rotating fatigue test. This will also allow the comparison with other values previously obtained for the same material using other printing conditions [14].

The rotating bending fatigue test consists of applying a variable bending moment on a cylindrical test piece of known dimensions that rotates on its own axis. In this way, alternative tensile and compressive stresses are generated in the external fibres in each rotation. The test was carried out on printed cylindrical specimens like the one shown in Figure 9. For the fabrication of the fatigue specimens, the same 3D printer was used.

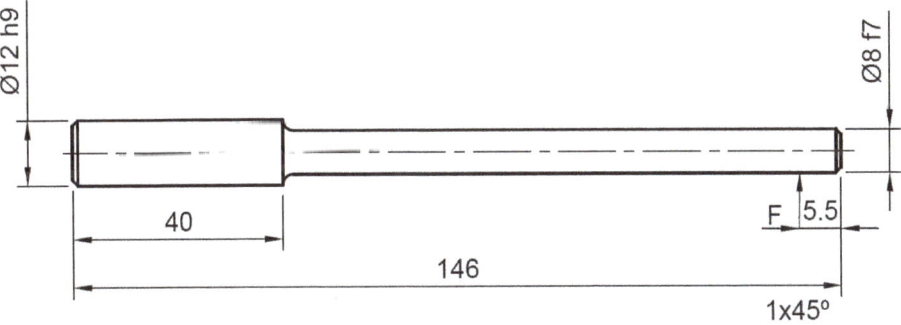

Figure 9. Dimensions of the test specimens used in the fatigue test.

3. Results and Discussion

3.1. Four-Point Bending Test

Table 5 shows the results, for each printing configuration, of the stress-strain curve as the average results of the five repetitions and their standard deviation.

Table 5. Average results and standard deviations of the material properties. E: Young's modulus, $Rp_{0.2}$: yield strength, σ_{max}: maximum strength, ε: maximum deformation, Std: standard deviation for each property.

#	E (GPa)	Std	$Rp_{0.2}$ (MPa)	Std	σ_{max} (MPa)	Std	ε	Std
1	2.36	0.18	53.8	3.19	64.2	8.18	4.72	1.16
2	3.06	0.07	83.5	0.95	96.0	2.98	4.90	0.64
3	1.79	0.03	11.8	1.74	11.8	1.74	0.70	0.13
4	2.74	0.03	69.7	4.10	79.0	4.97	4.68	1.10
5	1.23	0.10	7.92	1.58	7.96	1.58	0.81	0.24
6	2.71	0.03	60.1	3.09	80.8	2.36	5.85	0.50
7	0.59	0.05	6.71	1.76	6.7	1.76	1.20	0.22
8	2.78	0.11	60.6	3.45	64.1	4.43	3.37	0.32
9	2.81	0.06	65.1	3.61	79.2	6.11	4.91	0.59
10	2.29	0.29	37.1	4.04	37.1	4.04	1.64	0.05
11	3.34	0.19	67.9	3.16	83.7	4.53	4.57	0.17
12	3.69	0.08	95.3	4.26	120.0	1.38	5.34	0.20
13	2.41	0.07	50.2	6.97	72.3	8.23	5.72	0.18
14	3.45	0.33	85.0	3.67	104.6	2.16	4.98	0.17
15	2.07	0.21	26.2	3.34	26.1	3.34	1.49	0.41
16	3.19	0.06	73.4	1.15	83.8	3.87	4.09	0.36
17	1.20	0.09	10.6	1.60	10.6	1.60	1.02	0.13
18	1.44	0.27	26.7	3.25	35.7	4.11	5.09	0.74
19	3.61	0.07	87.4	2.53	95.5	7.35	3.73	0.72
20	3.02	0.27	43.4	3.64	43.5	3.64	1.50	0.12
21	3.23	0.02	70.8	3.51	93.1	4.52	5.09	0.22
22	2.33	0.25	21.3	3.22	21.4	3.22	0.53	0.60
23	2.85	0.19	63.4	6.54	86.4	3.27	5.36	0.64
24	3.70	0.14	90.8	2.28	109.5	4.70	5.11	0.95
25	1.90	0.08	44.0	5.16	60.4	4.43	6.21	0.28
26	2.96	0.15	75.2	3.49	86.7	8.68	4.54	1.56
27	2.30	0.15	25.3	6.61	25.4	6.61	0.91	0.61

An analysis of variance (ANOVA) was performed on the dataset included in the Taguchi experimental array, for each parameter that describes the mechanical behaviour of the evaluated specimens. To validate the statistical relevance of the parameters included in the model, the p-value associated with the ANOVA was compared to a significance level of 5%.

One of the first observations derived from the experimental testing is that specimens printed in the Z-axis direction presented fragile failure, as their failure mode was governed by the lower resistance between layers deposited vertically, thus with a lower neck growth area between them. For that reason, the elastic limit ($Rp_{0.2}$) associated with these specimens was by default considered equal to their maximum strength (σ_{max}). This approach was necessary to perform the statistical analysis, and allows the brittle behaviour to be included in the statistical analysis.

Alongside the yield limit and the maximum strength, the Young's modulus and maximum deformation were considered as response variables to analyse the influence of the different parameters in the statistical study. The following subsections describe the influence of the different parameters on the considered mechanical properties.

3.1.1. Young's Modulus

As a predictable result, the specimens oriented along the Z-axis direction present the lowest rigidity of all, owing to their described brittle behaviour, and thus can be orientation defined as the most influential parameter (Figure 10A). The highest deformation module in the elastic regime is defined by an orientation of the fibres along the Y-axis direction, because of the different pattern deposited in this direction with regards to the X–axis orientation.

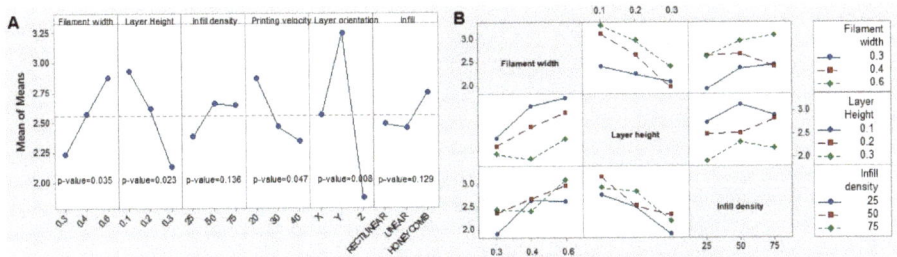

Figure 10. Main effects of (**A**) means and (**B**) interactions on Young's modulus.

On the other hand, an increase in the value of Young's modulus occurs when the filament width increases, probably because of the higher inertia of the single filaments that restrict bending. This effect of higher inertia of the surface is also achieved by decreasing the layer height, as it derives in a higher value of Young's modulus. This effect could be related to the fact that porosity is decreased by a lower layer height (and, complementarily, stiffness is increased). Following the same line, the printing velocity proves to increase the stiffness of the specimen as it is lower, probably again by the increase of the overall stiffness.

Of all the tested parameters, both the fill density and the infill pattern had a negligible impact (p-value of the ANOVA test > 5%) and no clear trend, which seems to disagree with the previous analysis. However, it must be considered that the small size of the specimens was derived in a lack of filling, and the geometry was composed basically of boundary layers that have relegated the infill to a second plane in this experimental campaign.

Figure 10B shows that no significant interaction among parameters is observed, as the p-values of them are all greater than 0.05.

3.1.2. Yield Strength

Figure 11 shows the influence of the printing parameters on the elastic limit. Again, the layer height and the infill pattern do not show a significant influence. The effect of the other parameters on the response follows the same pattern as in the case of Young's modulus. The most influential parameter again is the printing orientation. With the Y-axis orientation, the highest elastic limit is achieved, while the Z-axis orientation shows the lowest one. In addition, with the X-axis orientation, an intermediate value is achieved with respect to the other printing orientations. The layer height has an influence somewhat higher than that of the filament width, but in the opposite way; as the layer height decreases or the filament width increases, the elastic limit increases. Although the printing velocity has low relevance, a trend is observed: when the velocity decreases, the elastic limit increases.

When analysing the interactions between the different parameters, it is concluded that there is no significant interaction, as the p-values in each case are much higher than 0.05. The same happens for the rest of the parameters. This is positive because it means that the influence of the parameters on the response is independent of each other, at least in the ranges of values analysed.

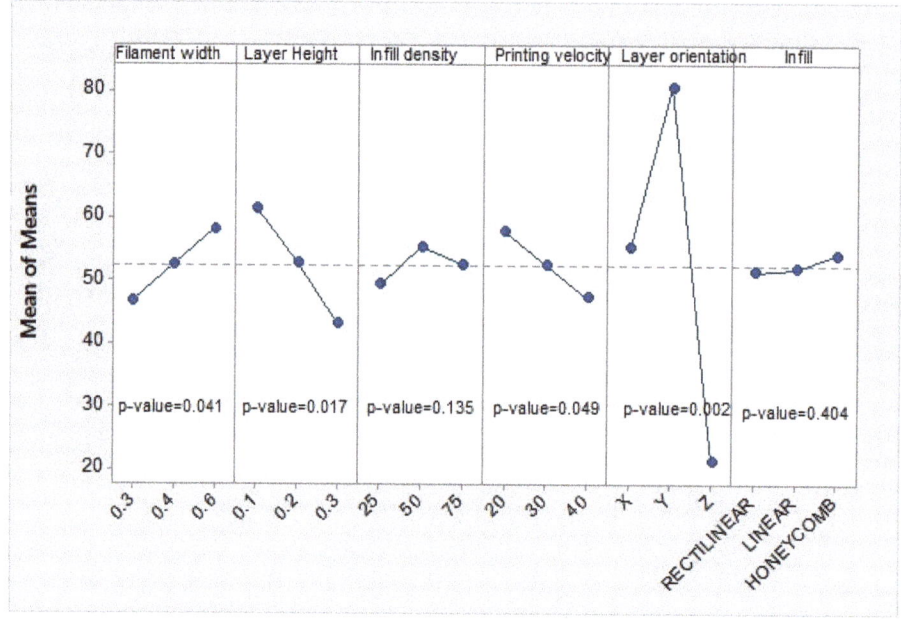

Figure 11. Main effects of means for yield strength.

3.1.3. Maximum Strength

The behaviour of the parameters follow the same pattern as the elastic limit case (Figure 12). The layer orientation is still the parameter with the greatest influence on the mean value, followed by the layer height, filament width, and printing velocity, with less influence. Fill density and infill pattern do not have a statistically significant influence.

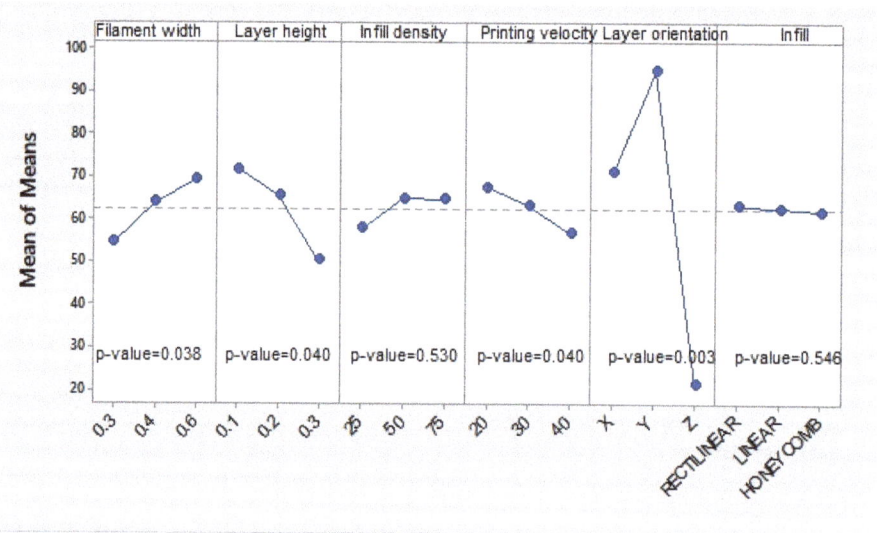

Figure 12. Main effects of means for maximum strength.

3.1.4. Maximum Deformation

Figure 13 reveals that the only significant parameter is orientation. The X-axis and Y-axis orientations cause the greatest elongation and the Z-axis orientation causes the smallest one. Filament width, layer height, fill density, and printing velocity do not present any pattern or proportionality. The honeycomb fill pattern produces the least effect. Regarding the signal S–N, the only robust parameter is again the orientation.

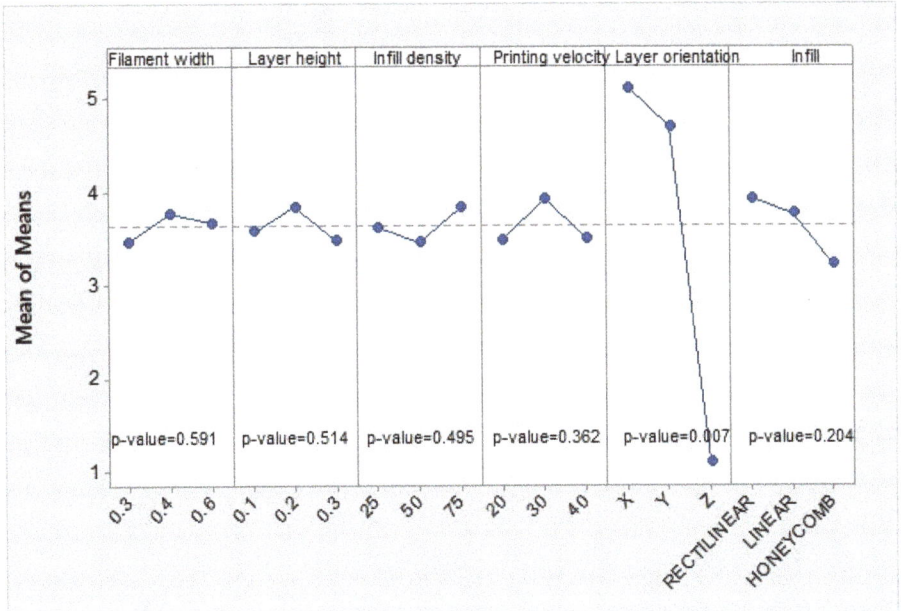

Figure 13. Main effects of means for maximum deformation.

3.1.5. Summary

In Table 6, a summary of the analysis of the influence of each parameter under study on the different mechanical properties studied can be seen. More green checks indicate that the factor is more influential on the response. Three checks indicates that p-value < 0.01, two checks indicate that 0.01 < p-value < 0.04, one check indicates that 0.04 < p-value < 0.05. The red cross is assigned to the parameters that are not statistically significant (p-value > 0.05). The orientation is the most influential parameter in the zone of both the elastic and plastic behaviour of the pieces tested. The layer height and the filament width are also parameters that influence all of the properties studied, except for the maximum deformation. The same thing happens with printing velocity, but to a lesser extent. In Table 7, the optimum levels of each parameter are shown.

Table 6. Significance value of the parameters with respect to the answers.

Factor	Elastic Properties		Plastic Properties	
	Young's Modulus (E)	Yield Strengt ($Rp_{0.2}$)	Maximum Strength (σ_{max})	Maximum Deformation (ε)
Layer orientation	✓✓✓	✓✓✓	✓✓✓	✓✓✓
Layer Height	✓✓	✓✓	✓✓	✗
Filament width	✓✓	✓	✓✓	✗
Printing velocity	✓	✓	✓	✗
Infill density	✗	✗	✗	✗
Infill pattern	✗	✗	✗	✗

Table 7. Parameters' level to maximize the response.

Factor	Young's Modulus (E)	Yield Strength ($Rp_{0.2}$)	Maximum Tension (σ_{max})	Maximum Deformation (ε)
Filament width	0.6 mm	0.6 mm	0.6 mm	0.2 mm
Layer Height	0.1 mm	0.1 mm	0.1 mm	0.2 mm
Infill density	75%	75%	75%	75%
Printing Velocity	20 mm/s	20 mm/s	20 mm/s	30 mm/s
Layer Orientation	Y-axis	Y-axis	Y-axis	X-axis
Infill pattern	Honeycomb	Honeycomb	Honeycomb	Rectilinear

Of all parameters, the lack of influence of infill density deserves a special mention. This observation has already been made by other authors, such as Admed & Susmel (2019) [21] and Andrzejewska et al. (2017) [22]. These authors explain that the mechanical properties of PLA specimens with a 100% infill density depend on three main aspects, namely, the mechanical properties of filaments, the bonding forces between layers, and bonding forces between filaments of the same layer. Decreasing the infill density derives in the loss of bonding strength between filaments of the same layer, regardless of the distance between filaments in the same layer, which is the direct effect of infill density reduction. That is, the effect of changing infill density is more conspicuous when reducing from 100% to any other value, hence the lack of relevance of decreasing it from 75% to 25%. Furthermore, we could add a second fact explaining the lack of influence of infill density on the results, which could be related to the fact that bending specimens are of reduced dimensions, meaning that their mechanical behaviour is governed by their skin and, to a much lesser extent, the infill, which is only comprised by a few layers.

The manufacturing orientation plays a vital role in defining the flexural behaviour of specimens, as stress is normal to the specimen section, and the orientation of the bonding area between filaments shall define the way in which the material processes the stress. This result contrasts with that obtained when the specimens are subjected to fatigue tests, where layer height is the most influential parameter owing to the fact that the limiting factor here is the prevention of crack propagation, and not bearing stress itself [9].

3.2. Fractography and Scratch Test

It was already noted that the main factor that determines the strength of the specimens is the orientation of the stacking layers. In Figure 14, the outlooks of the fracture section of some printed specimens in the different directions of the coordinate axes are compared. The specimens printed along the X-axis direction (Figure 14A) or along the Y-axis direction (Figure 14B) have a slight ductile behaviour with high elongation, good flexural strength, and high rigidity. This reaction is caused by the filaments being aligned with the main stress direction. The resistance depends on the strength of the intra-layer bond and the strength of the filaments. The stiffness and flexural strength are slightly higher in the Y-axis orientation specimens. The reason is once again the arrangement of the layers. Although the two orientations have the filaments parallel to the direction of the stresses, the specimens printed in the X-axis direction can become delaminated between layers when they are bent. The delamination is produced by the breakage of the weak interlayer bonds. The specimen becomes flexible and is unable to withstand the bending stress, although the intralayer bonds remain intact.

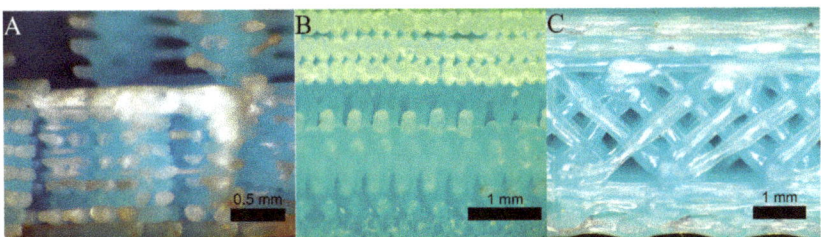

Figure 14. Breaking section, specimens with orientation in the (**A**) X-, (**B**) Y-, and (**C**) Z-direction.

In Figure 14C, the section of the rupture of a test specimen printed in Z-axis orientation is shown. These specimens have fragile behaviour and little deformation, low resistance to bending, and low rigidity. This is caused because the layers are oriented perpendicularly to the stresses generated in the specimen during the bending test. For this reason, failure has occurred in the weak interlayer weld without the affection of filament integrity.

The second most influential factor is the layer height, followed by the filament width. The smaller the layer height and the larger the filament width, the stiffer and more resistant to bending is the test specimen. This is directly related to the compactness of the threads and the welding between threads (Figure 15).

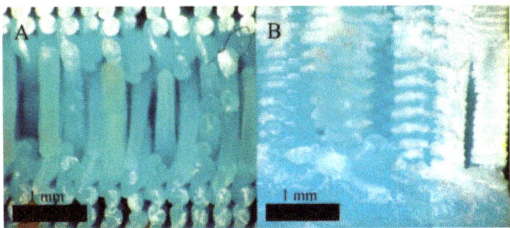

Figure 15. (**A**) Test specimen with a layer height of 0.3 mm and filament width of 0.3 mm. Test specimen 9_2. Microscopic photography. (**B**) Test specimen with a layer height of 0.1 mm and filament width of 0.6 mm. Test specimen 21_3. Microscopic photography.

Figure 15A shows the extreme case with the maximum layer height of 0.3 mm and minimum filament width of 0.3 mm. With these dimensions, the threads are cylindrical and produce low compaction and weak welding owing to the scarce contact surface between threads. On the other hand, as the layer height decreases and the filament width increases, the threads have a flat shape, with a larger welding surface. Figure 15B shows the optimal case with a minimum layer height of 0.1 mm and maximum filament width of 0.6 mm. In summary, the welding surface of the threads, where the micro-welds are produced between the chains of the polymer deposited at the beginning and those of the filament that is then deposited on it, is determinant in the mechanical behaviour. The greater the welding surface, the greater the rigidity and strength of the piece.

Figure 16 shows the micro scratch test tracks in both the (A) perpendicular and (B) parallel direction to the filaments, on the same piece printed in the X-axis direction shown in Figure 14A. It can be seen how, up to the tested force (70 N), the material deforms ductilely without cracking in the base material, as the indenter moves. It also looks like the burrs produced by the extruder are torn. The fact that there are no disclosures between filaments implies that the adhesion between them in the same layer is enough to resist the efforts applied during the test.

Figure 16. Micro scratch test: (**A**) perpendicular to the printing direction; (**B**) parallel to the printing direction.

The graph in Figure 17 shows the results of the micro scratch tests: (A) perpendicular and (B) parallel to the direction of the filaments in the range of test forces. The values of normal force, friction force, penetration depth, residual depth, and friction coefficient are clearly observed. While the value of the friction coefficient measured in the perpendicular test shows oscillations, owing to the abrasive wear of the burrs (see arrows in Figure 17A), in the parallel test, its value remains almost constant. It could be possible to sense that these pieces are not showing remarkable wear adhesive.

On the other hand, these burrs left during the extrusion process form channels on the piece surface. If it is true that this worsens the surface roughness of the pieces, they could be useful for retaining lubricant adhered to the sides of the burr ridges; more taking into account that they do not increase their friction coefficient too much, as shown in Figure 17.

3.3. Fatigue Test

The parameter that marks the difference between both curves in Figure 18 is the layer height, being 0.1 mm for the results of this study and 0.3 mm for the referenced study [14].

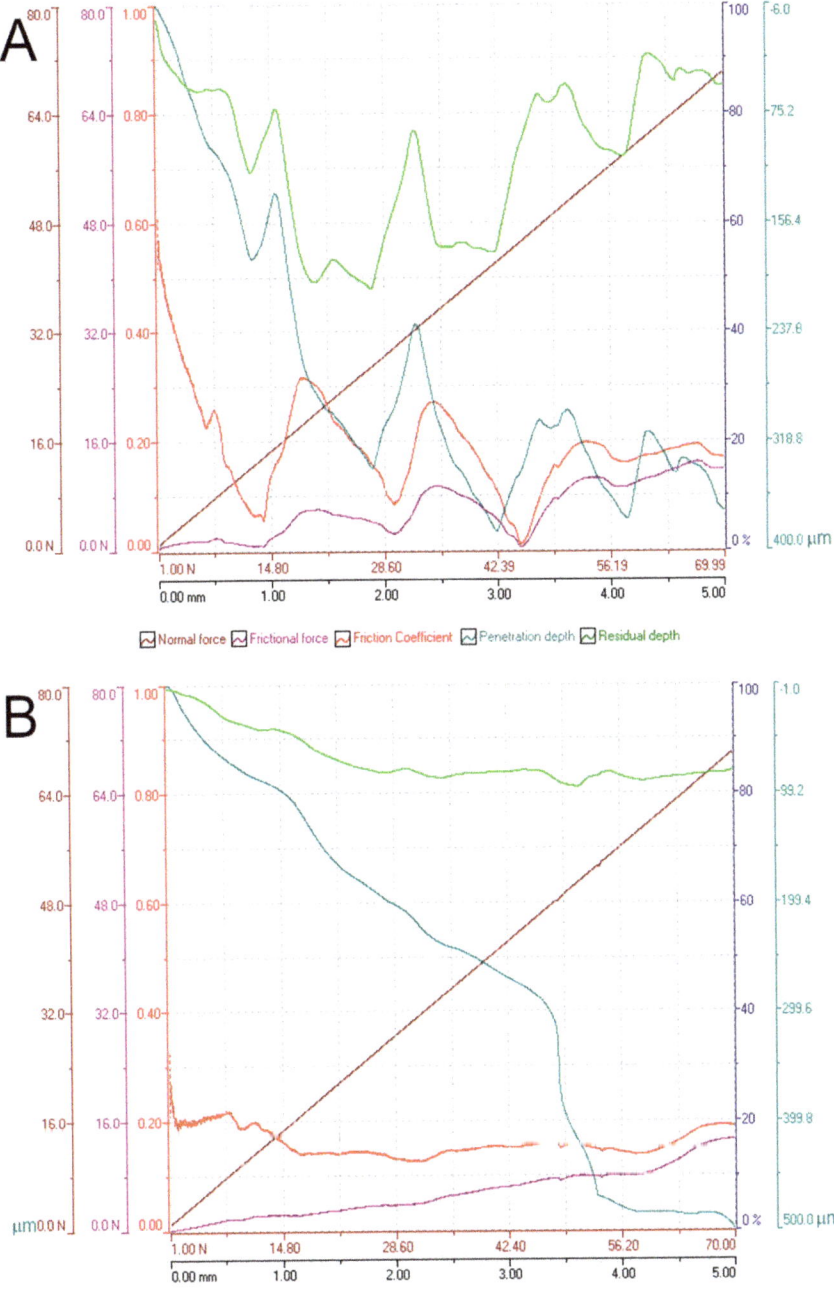

Figure 17. Micro scratch test results: (**A**) perpendicular to the printing direction; (**B**) parallel to the printing direction.

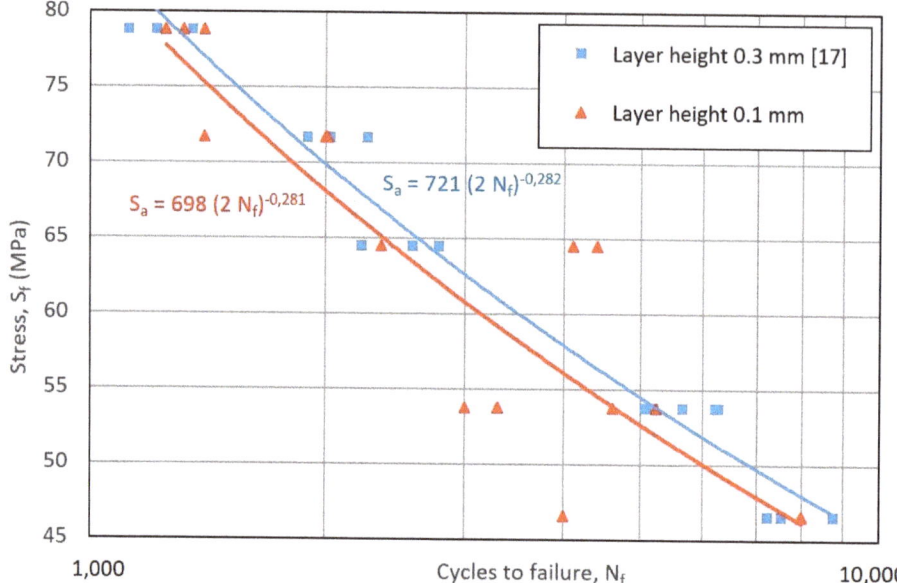

Figure 18. Wöhler curve for the results obtained in this study and those obtained in the work of [14].

Although the authors of [14] find that layer height is slightly significant—and although it seems that the following assumption holds: the higher the height layer value, the greater the improvement detected regarding resistance—this cannot be assured, as the errors calculated for the multiplicative factor and the exponent in both equations mean that they can be the same.

Therefore, although a dependence on the layer height is insinuated, the current data do not allow it to assert it.

4. Conclusions

The influence of the layer orientation, layer height, filament width, printing velocity, fill density, and infill pattern on the flexural performance of PLA specimens was studied through a Taguchi DOE. The following conclusions can be extracted:

1. The orientation of the stacking of the layers is the most influential parameter in the rigidity, in the flexural resistance, and in the maximum deformation.
2. The layer height and the filament width had a great significance in stiffness and flexural strength and no influence on maximum deformation.
3. Printing velocity had a small, but significant effect on rigidity and flexural strength and no influence on maximum deformation.
4. The fill density and infill pattern had no effect on the studied mechanical properties.
5. The orientation of stacking layers in Y, the layer height of 0.1 mm, the filament width of 0.6 mm and the printing velocity of 20 mm/min was the optimal combination obtained that will allow maximizing rigidity and flexural resistance.
6. The printing direction in Y-axis showed the best mechanical behaviour owing to its resistance depending on the strong intralayer bond.
7. The large filament width, the small layer height, and the low printing velocity formed test specimens with better compaction and better welding between wires, and generated a better rigidity and resistance to bending.
8. It could not be ensured that higher layer height improves fatigue life.

9. Depending on the mechanical property to enhance, the combination of optimal parameters to use is different.

Author Contributions: Conceptualization, J.A.T.-R. and R.J.-M.; Methodology, R.J.-M., G.G.-G. O.T.-R. And J.J.R.R.; Software, J.L. And O.T.-R.; Validation, J.L., J.J.R.R. and R.J.-M.; Formal Analysis, G.G.-G. and R.J.-M.; Investigation, J.A.T.-R., G.G.-G. and O.T.-R.; Resources, J.A.T.-R. and J.L.; Data Curation, J.L. J.J.R.R. and R.J.-M.; Writing-Original Draft Preparation, J.A.T.-R.; Writing-Review & Editing, J.L., G.G.-G. and R.J.-M.; Visualization, J.L. And J.J.R.R.; Supervision, J.A.T.-R. And R.J.-M.; Project Administration, J.A.T.-R., Funding Acquisition, J.A.T.-R., R.J.-M. and J.L.

Funding: This research did not receive any specific grant from funding agencies in the public, commercial, or not-for-profit sectors.

Acknowledgments: J.J. Roa would like to acknowledge the Serra Hunter program of the *Generalitat de Catalunya*.

Conflicts of Interest: The authors declare no conflict of interest.

Abbreviations

AM	additive manufacturing
FFF	fused filament fabrication
PLA	polylactic acid
DOE	design of experiments
ANOVA	analysis of variance

References

1. Go, J.; Schiffres, S.N.; Stevens, A.G.; Hart, A.J. Rate limits of additive manufacturing by fused filament fabrication and guidelines for high-throughput system design. *Addit. Manuf.* **2017**, *16*, 1–11. [CrossRef]
2. Fontrodona, J.; Blanco, R. Estado Actual y Perspectivas de la Impresión en 3D. Artículos de Economía Industrial. 2014. Available online: http://empresa.gencat.cat/web/.content/19_-_industria/documents/economia_industrial/impressio3d_es.pdf (accessed on 4 July 2019).
3. Domingo-Espin, M.; Puigoriol-Forcada, J.M.; Garcia-Granada, A.A.; Llumà, J.; Borros, S.; Reyes, G. Mechanical property characterization and simulation of fused deposition modeling Polycarbonate parts. *Mater. Des.* **2015**, *83*, 670–677. [CrossRef]
4. Bellini, A.; Güçeri, S. Mechanical characterization of parts fabricated using fused deposition modeling. *Rapid Prototyp. J.* **2003**, *9*, 252–264. [CrossRef]
5. Bellehumeur, C.; Li, L.; Sun, Q.; Gu, P. Modeling of Bond Formation Between Polymer Filaments in the Fused Deposition Modeling Process. *J. Manuf. Process.* **2004**, *6*, 170–178. [CrossRef]
6. Gurrala, P.K.; Regalla, S.P. Part strength evolution with bonding between filaments in fused deposition modelling: This paper studies how coalescence of filaments contributes to the strength of final FDM part. *Virtual Phys. Prototyp.* **2014**, *9*, 141–149. [CrossRef]
7. Gray, R.W.; Baird, D.G.; Helge Bøhn, J. Effects of processing conditions on short TLCP fiber reinforced FDM part. *Rapid Prototyp. J.* **1998**, *4*, 14–25. [CrossRef]
8. Zhong, W.; Li, F.; Zhang, Z.; Song, L.; Li, Z. Short fiber reinforced composites for fused deposition modelling. *Mater. Sci. Eng. A* **2001**, *301*, 125–130. [CrossRef]
9. Sood, A.K.; Ohdar, R.K.; Mahapatra, S.S. Parametric appraisal of mechanical property of fused deposition modelling processed parts. *Mater. Des.* **2010**, *31*, 287–295. [CrossRef]
10. Olivier, D.; Travieso-Rodriguez, J.A.; Borros, S.; Reyes, G.; Jerez-Mesa, R. Influence of building orientation on the flexural strength of laminated object manufacturing specimens. *J. Mech. Sci. Technol.* **2017**, *31*, 133–139. [CrossRef]
11. Ajoku, U.; Saleh, N.; Hopkinson, N.; Hague, R.; Erasenthiran, P. Investigating mechanical anisotropy and end-of-vector effect in laser-sintered nylon parts. *Proc. Inst. Mech. Eng. B J. Eng. Manuf.* **2006**, *220*, 1077–1086. [CrossRef]
12. Quintana, R.; Choi, J.W.; Puebla, K.; Wicker, R. Effects of build orientation on tensile strength for stereolithography-manufactured ASTM D-638 type I specimens. *Int. J. Adv. Manuf. Technol.* **2010**, *46*, 201–215. [CrossRef]

13. Wang, T.; Xi, J.; Jin, Y. A model research for prototype warp deformation in the FDM process. *Int. J. Adv. Manuf. Technol.* **2007**, *33*, 1087–1096. [CrossRef]
14. Gomez-Gras, G.; Jerez-Mesa, R.; Travieso-Rodriguez, J.A.; Lluma-Fuentes, J. Fatigue performance of fused filament fabrication PLA specimens. *Mater. Des.* **2018**, *140*, 278–285. [CrossRef]
15. Jerez-Mesa, R.; Travieso-Rodriguez, J.A.; Lluma-Fuentes, J.; Gomez-Gras, G.; Puig, D. Fatigue lifespan study of PLA parts obtained by additive manufacturing. *Procedia Manuf.* **2017**, *13*, 872–879. [CrossRef]
16. ASTM D6272–02. *Standard Test Method for Flexural Properties of Unreinforced and Reinforced Plastics and Electrical Insulating Materials*; Bending, F.-P., Ed.; ASTM International: West Conshohocken, PA, USA, 2002.
17. Hodgson, G. Slic3r Manual—Infill Patterns and Density. Manualslic3rorg, 2017. Available online: http://manual.slic3r.org/ (accessed on 4 July 2019).
18. Puigoriol-Forcada, J.M.; Alsina, A.; Salazar-Martín, A.G.; Gomez-Gras, G.; Pérez, M.A. Flexural Fatigue Properties of Polycarbonate Fused-deposition Modelling Specimens. *Mater. Des.* **2018**, *141*, 414–425. [CrossRef]
19. Morales-Planas, S.; Minguella-Canela, J.; Lluma-Fuentes, J.; Travieso-Rodriguez, J.A.; García-Granada, A.A. Multi Jet Fusion PA12 manufacturing parameters for watertightness, strength and tolerances. *Materials* **2018**, *11*, 1472. [CrossRef] [PubMed]
20. ASTM C1624-05. *Standard Test Method for Adhesion Strength and Mechanical Failure Modes of Ceramic Coatings by Quantitative Single Point Scratch Testing*; ASTM International Standards: West Conshohocken, PA, USA, 2005.
21. Ahmed, A.A.; Susmel, L. Static assessment of plain/notched polylactide (PLA) 3D-printed with different infill levels: Equivalent homogenised material concept and Theory of Critical Distances. *Fatigue Fract. Eng. Mater. Struct.* **2019**, *42*, 883–904. [CrossRef]
22. Andrzejewska, A.; Pejkowski, Ł.; Topoliński, T. Tensile and Fatigue Behavior of Additive Manufactured Polylactide. *3D Print. Addit. Manuf.* **2019**, *6*, 1–9. [CrossRef]

 © 2019 by the authors. Licensee MDPI, Basel, Switzerland. This article is an open access article distributed under the terms and conditions of the Creative Commons Attribution (CC BY) license (http://creativecommons.org/licenses/by/4.0/).

Article

An Experimental and Numerical Analysis of the Compression of Bimetallic Cylinders

Ana María Camacho [1,*], Álvaro Rodríguez-Prieto [1], José Manuel Herrero [1], Ana María Aragón [1], Claudio Bernal [1], Cinta Lorenzo-Martin [2], Ángel Yanguas-Gil [2] and Paulo A. F. Martins [3]

1. Department of Manufacturing Engineering, Universidad Nacional de Educación a Distancia (UNED), 28040 Madrid, Spain; alvaro.rodriguez@invi.uned.es (Á.R.-P.); jherrero74@alumno.uned.es (J.M.H.); amaragon@invi.uned.es (A.M.A.); cbernal@ind.uned.es (C.B.)
2. Applied Materials Division, Argonne National Laboratory, 9700 Cass Ave, Lemont, IL 60439, USA; lorenzo-martin@anl.gov (C.L.-M.); ayg@anl.gov (Á.Y.-G.)
3. Instituto de Engenharia Mecânica, Instituto Superior Técnico, Universidade de Lisboa, Av. Rovisco Pais, 1049-001 Lisboa, Portugal; pmartins@tecnico.ulisboa.pt
* Correspondence: amcamacho@ind.uned.es; Tel.: +34-913-988-660

Received: 5 November 2019; Accepted: 5 December 2019; Published: 7 December 2019

Abstract: This paper investigates the upsetting of bimetallic cylinders with an aluminum alloy center and a brass ring. The influence of the center-ring shape factor and type of assembly fit (interference and clearance), and the effect of friction on the compression force and ductile damage are comprehensively analyzed by means of a combined numerical-experimental approach. Results showed that the higher the shape factor, the lower the forces required, whereas the effect of friction is especially important for cylinders with the lowest shape factors. The type of assembly fit does not influence the compression force. The accumulated ductile damage in the compression of bimetallic cylinders is higher than in single-material cylinders, and the higher the shape factor, the lower the damage for the same amount of stroke. The highest values of damaged were found to occur at the middle plane, and typically in the ring. Results also showed that an interference fit was more favorable for preventing fracture of the ring than a clearance fit. Microstructural analysis by scanning electron microscopy revealed a good agreement with the finite element predicted distribution of ductile damage.

Keywords: metal forming; bi-metallic; cylinders; compression; finite elements; experimentation; microscopy

1. Introduction

In recent years, there has been a considerable growth in the use of multi-material components due to their advantages over single-material components regarding the possibility of tailoring physical properties, improving stiffness and strength, reducing overall weight, and saving the number of parts and the assembly costs in mechanical systems made of multiple components. Reductions in weight can, for example, be achieved through the combination of materials with lower densities than the original ones. Significant cost savings in electric power systems used in modern hybrid and electric vehicles can also be obtained by combining materials with different electrical and thermal conductivities. Improvements in the surface integrity of components exposed to extreme conditions are also of special interest; namely, in applications subjected to high friction contacts (enhancing the tribological properties) or high corrosion environments, such as those existing in marine and chemical industries [1]. In fact, the range of potential applications of multi-material components is so wide that they can also be found in the production of high denomination coins for security and aesthetic reasons.

The interest in multi-material components is scientifically and socially recognized by its inclusion in different work programs of the EU Horizon 2020, in which the manufacturing of multi-materials by additive manufacturing (for research, transport, customized goods, or biomaterials), and the combination of commercial materials into multi-material components for industrial applications [2] were selected as key research topics. The importance of additive manufacturing is confirmed by the growing number of publications in the field, which are focused on both directed energy deposition (DED) and powder bed fusion (PBF) [3] based techniques. Laser engineering net shaping (LENS) [4–6] is a powder DED based process; laser metal deposition (LMD) [7] is a wire DED based process; and selective laser melting (SLM) [8] is one of the most promising PBF processes for fabricating multi-material components.

Still, there are limitations in the use of additive manufacturing that are similar to those found in the fabrication of multi-material components by welding (e.g., friction stir welding, and laser and explosive welding) [5]. In fact, joining of dissimilar materials suffers from the risk of formation of brittle intermetallic metallurgical structures, and thermal heating-cooling cycles give rise to residual stresses, distortions, and geometric inaccuracies [3,9]. Table 1 summarizes the main problems associated with the production of multi-material components by additive manufacturing, welding, and forming.

Table 1. Main limitations in the fabrication of multi-material components by means of additive manufacturing, welding, and metal forming.

Main Limitations [1]	Additive Manufacturing (DED and PBF)	Welding (Friction Stir Welding, Laser and Explosive Welding)	Metal Forming (Extrusion, Rolling, Upsetting)
Materials compatibility	X	X	-
Formation of brittle intermetalics	X	X	-
Microstructure thermal effects	X	X	-
Distortion	X	X	-
Residual stresses	X	X	-
Delamination	X	X	X
Formability limits	-	-	X

[1] X: limitation associated to category of processes.

As seen in Table 1, metal forming successfully overcomes most of the difficulties that are found in the production of multi-material components by additive manufacturing and welding. The main problems are due to formability issues and to the risk of delamination because thermal effects do not play a role in the cold metal forming based process that is considered in this paper.

Despite this, formability studies on multi-material components made from commercial materials by means of metal forming are not very widespread in literature. Studies are mostly limited to bimetallic components made of two different metallic alloys, such as the publications on the extrusion of bi-metallic components [10–12] and on the combination of forming and joining to produce bimetallic bearing bushings [13].

Coin minting of bimetallic disks is probably the most well-known application in the field [14,15] and the technology was recently thrown to a higher level of complexity by the development of new bi-material collection coins with a polymer composite center and a metallic ring to generate innovative aesthetics and incorporate advanced holographic security features [16]. Finite element modelling was utilized to investigate the influence of the initial clearance between the polymer center and the metallic ring on the mechanics of coin minting and performance of the resulting force fit joint.

Other researchers like Essa et al. [17] discussed the possibility of producing bimetallic components or preforms by upsetting, after concluding that some geometries with a good interfacial contact between the center and the ring can be successfully employed as preforms for further processing. A similar conclusion was made by Misirly et al. [18] after analyzing the open die forging of bimetallic cylinders with steel rings and brass and pure copper centers, and observing that pure copper prevents the formation of cavities at the center-ring interfaces.

A recently publish work by Cetintav et al. [1] on the compression of trimetallic cylinders with aluminum centers and steel, copper, and brass rings, focused on the improvements in mechanical properties and weight reduction that result from the utilization of multi-material components.

More recently, Wernicke et al. [19] developed a new type of hybrid gear made from aluminum and steel to obtain significant weight reductions and locally adapted mechanical properties without the need of performing subsequent heat-treatment processes.

In the meantime, there have also been other investigations in the field aimed at analyzing the deformation mechanics and predicting the compression forces in multi-material components. This is the case of Plancak et al. [20], who developed two special purpose analytical models to calculate the compression force and validated their predictions against experimental tests performed on bimetallic cylinders with centers and rings made from different commercial steels. These models were later improved by Gisbert et al. [21] to include shear friction.

Under these circumstances, this paper aims to analyze the formability of bimetallic cylindrical billets produced by compression by means of a numerical and experimental based investigation. Compression forces and accumulation of ductile damage were analyzed by means of a work plan including different shape factors and two assembly fits between the center and the ring (interference and clearance). Scanning electron microscopy (SEM) observations were included to identify the major defects and to correlate the location of these defects with the finite element predicted distribution of ductile damage after compression.

The problems of delamination included in Table 1 will not be addressed because these are mainly found when the compression forces are not applied perpendicular to the contact surfaces between the different materials to be joined, as in case of extrusion and rolling [17]. This is not the case in the present investigation.

2. Materials and Methods

2.1. Materials and Experimental Work Plan

The bimetallic cylindrical test samples utilized in the investigation have an aluminum alloy UNS A92011 center and a brass UNS C38500 ring (Figure 1).

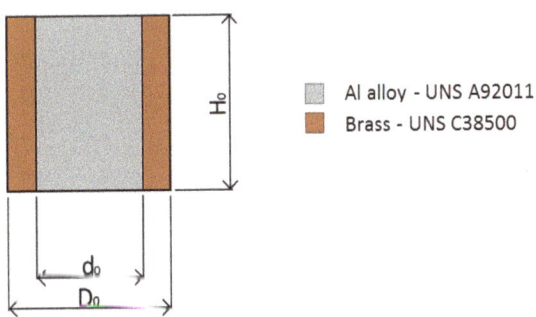

Figure 1. Bimetallic cylindrical test samples and notation utilized in the paper.

The aluminum center and the brass rings were machined from commercial rods with 12 and 15 mm diameters, respectively. Both materials were utilized in their as-supplied conditions and their chemical compositions are listed in Table 2.

Table 2. Chemical compositions of the aluminum alloy UNS A92011 [22] and brass UNS C38500 [23].

Material	Al (wt.%)	Cu (wt.%)	Fe (wt.%)	Si (wt.%)	Zn (wt.%)	Pb (wt.%)
UNS A92011	92.0	5.5	0.7	0.4	-	-
UNS C38500	-	58.0	-	-	39.0	3.0

The physical and mechanical properties of both materials are included in Table 3.

Table 3. Physical and mechanical properties of the aluminum alloy UNS A92011 [22] and brass UNS C38500 [23].

Property	UNS A92011	UNS C38500
Density (kg/m^3)	2840	8470
Hardness (HB)	110	90–160
Youngs' modulus (GPa)	70–72.5	90–100
Elongation A	6–12	15–25
Yield point (MPa)	125–230	220–350
UTS (MPa)	275–310	350–500

The density of brass is three times higher than that of the aluminum alloy but its beta metallurgical phase, which is very appropriate for applications with extreme contact pressures, limits its ductility in cold forming. The overall rigidity of brass is also higher than that of the aluminum alloy because the latter has a smaller yield stress and a smaller ultimate tensile strength (UTS), meaning that it requires less energy to be plastically deformed. The experimental work plan is summarized in Table 4 and made use of cylindrical test samples with different height to diameter ratios, H_0/d_0 (previously designated as the "shape factor") and two different types of assembly fit. The assembly fit (P1i) corresponds to test samples in which the center was mounted into the ring with interference. For this purpose, the center was pushed into the ring using the universal testing machine that was also used in the compression tests. The assembly fit (P2i) corresponds to test samples in which the center was mounted into ring with a clearance of 0.1 mm in order to ensure easy sliding between the two parts.

Table 4. Summary of the experimental work plan [1,2].

Group (Assembly Fit)	Sample	D_0 (mm)	d_0 (mm)	H_0 (mm)	H_0/d_0
Interference	P1a	12	8	8	1.00
	P1b	12	8	10	1.25
	P1c	12	8	12	1.50
	P1d	12	8	14	1.75
	P1e	12	8	16	2.00
Clearance	P2a	12	8	8	1.00
	P2b	12	8	10	1.25
	P2c	12	8	12	1.50
	P2d	12	8	14	1.75
	P2e	12	8	16	2.00

[1] The dimensional parameters (D_0, d_0, H_0) are defined in Figure 1. [2] a,b,c,d,e: denotes the shape factor (H_0/d_0) of the sample.

Figure 2 shows the bi-metallic cylindrical test samples (notation according to Table 4) before compression. Two samples were prepared for each testing condition.

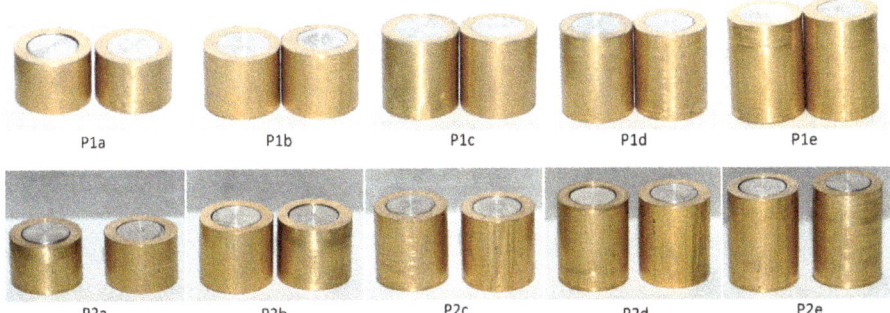

Figure 2. Bimetallic cylindrical test samples before compression. Notation in accordance with Table 4.

2.2. Equipment and Experimental Procedure

The compression of the bimetallic cylindrical test samples was performed in a universal testing machine Hoytom HM-100kN (Hoytom HM-100kN, Hoytom, S.L., Leioa, Spain) with control software Howin 32 RS (version 3.11, Hoytom, S.L., Leioa, Spain). A precision cut-off machine Mecatome P100 (Mecatome P100, PRESI, Brié et Angonnes, France) was utilized to prepare the test samples for analysis and micrographic observation after compression.

The experimental procedure consisted of the following steps:

1. Before compression, the samples and the compression die platens were properly cleaned with ethanol.
2. The samples were then placed in the center of the lower die platen.
3. Compression was performed for each sample. Repeatability of the testing conditions (ram speed, pre-contact force and end of testing) was ensured by the control software of the universal testing machine. A pre-contact force of 50 N was utilized, after which the ram moved at a constant speed of 1 mm/s until reaching a compression force of 90 kN. The tests were performed at room temperature under dry, lubricated conditions.
4. After testing, the samples were cut along their axial cross-sectional plane with the precision cut-off machine.

2.3. Finite Element Modeling

Finite element simulations were carried out with the commercial finite element computer program DEFORM 3D. The compression die platens were modelled as rigid objects and the bimetallic cylinders were modelled as an assembly between two plastically deformable objects (center and ring). The center and ring were discretized by means of approximately 11,000 tetrahedral elements. The detail of the initial finite element meshes is provided in Figure 3a.

The center and ring materials (aluminum alloy UNS A92011 and brass UNS C38500) were assumed to be isotropic and their flow curves (true stress–true strain curves) are disclosed in Figure 4.

Friction was modelled by means of the law of constant friction. As explained by Essa et al. [17], it is not possible to accurately define the frictional conditions prevailing at the center-ring contact interface. But previous research in multi-material upsetting [1,17,18], also lead to the conclusion that variations of the friction factor in the range of 0 to 0.5 do not influence the overall deformation of multi-material components.

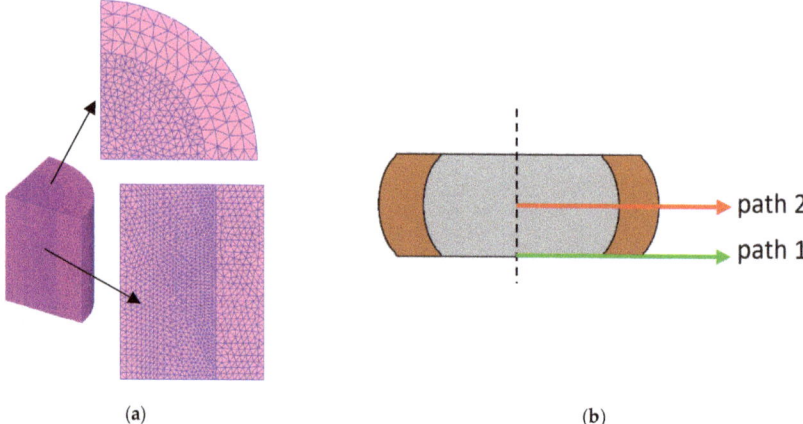

(a)   (b)

Figure 3. Finite element modelling of the compression of bimetallic cylinders: (**a**) Detail of the initial mesh; (**b**) Identification of the two paths (path 1 in green and path 2 in red) that will be later utilized in the presentation to analyze ductile damage.

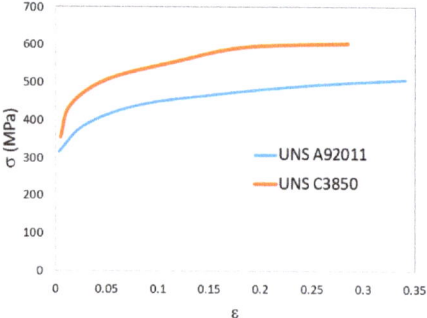

Figure 4. Flow curves of the aluminum alloy UNS A92011 and brass UNS C3850.

Therefore, taking into consideration a previous study performed by the authors [24], which points to the same above-mentioned conclusion, it was decided to use a friction factor equal to 0.08 along the center-ring contact interface. The same study was utilized to define a value of 0.12 at the contact interfaces between the specimen and the upper and lower die platens.

The accumulation of ductile damage D was modelled by means of the Cockcroft–Latham criterion [25]. According to this criterion, fracture is supposed to occur when the accumulated ductile damage reaches a critical value D_{crit}, for a given temperature and strain rate loading condition

$$D_{crit} = \int_0^{\bar{\varepsilon}_f} \sigma^* d\bar{\varepsilon}, \quad (1)$$

where σ^* is the maximum principal stress, $\bar{\varepsilon}$ is the equivalent strain, and $\bar{\varepsilon}_f$ is the equivalent strain at fracture.

The ductile damage distributions included in this paper are based on a normalized version of the Cockcroft–Latham criterion (1),

$$C = \int_0^{\bar{\varepsilon}} \frac{\sigma^*}{\bar{\sigma}} d\bar{\varepsilon}, \quad (2)$$

where $\bar{\sigma}$ is the effective stress. The values of strain and stress were calculated at the centre of each tetrahedral element, and therefore, the values of the normalized accumulated ductile damage C were also accumulated at the centre of the elements.

Damage distribution along the two paths shown in Figure 3b were calculated during post-processing of results. Path 1 was taken at the contact interface between the deformed cylinders and lower die platen, and path 2 was taken from the middle plane of the cylinders after compression.

2.4. Scanning Electron Microscopy (SEM)

Microstructural observation and analysis of the test samples with interference (P1a to P1e) were carried out with a high-resolution scanning electron microscope of the Center for Nanoscale Materials (CNM) at Argonne National Laboratory. The equipment utilized was a Hitachi S-4700-II (Hitachi, Krefeld, Germany), with an electron dispersive spectroscopic (EDS) detector, Bruker XFlash 6160 (Bruker, Billerica, MA, USA).

P1a to P1e samples were microstructurally characterized because damage was observed but fracture did not occur. This approach allows one to assess the formability of this multi-material sample prepared with interference, so more useful information was obtained from the defects found in the microstructural characterization.

The results obtained from these observations were compared with the finite element predictions of accumulated ductile damage. Further validation of the finite element computations was performed by comparing the numerical and experimental force-displacement evolutions. This is displayed in the following section.

3. Results and Discussion

3.1. Compression Forces

Figure 5 shows the bimetallic cylindrical test samples after compression. As seen, the influence of the assembly fit only provides visible differences for the samples with a shape factor $H_0/d_0 = 2$ because of cracking in the samples where the center was mounted in the ring with clearance.

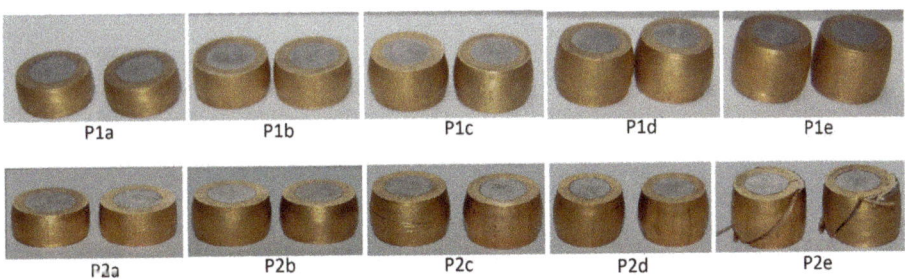

Figure 5. Bimetallic cylindrical test samples after compression. Failure by cracking is observed in sample P2e (Table 4).

The lack of visible differences in the other test samples with smaller shape factors H_0/d_0 was further confirmed by the finite element predicted evolution of force with displacement shown in Figure 6. In fact, the force-displacement evolution is only sensitive to the shape factors H_0/d_0, as in case of single-material (solid) cylinders—the higher the shape factor, the lower the compression forces. In other words, there is no influence of the type of assembly fit on the force-displacement evolution for the test samples with shape factors $H_0/d_0 = 1$, 1.25, 1.5, and 1.75.

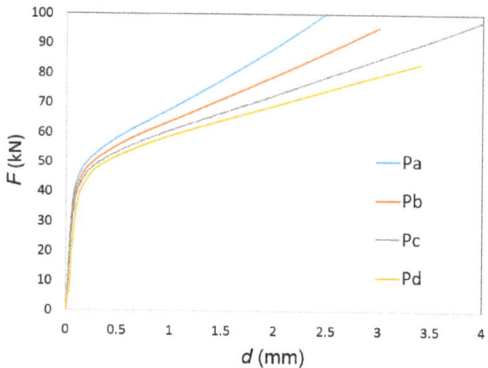

Figure 6. Finite element predicted evolution of the force with displacement for the compression of bimetallic cylindrical test samples with different shape factors H_0/d_0 (Pa: 1.00, Pb: 1.25, Pc: 1.50, Pd: 1.75).

Figure 7a shows a comparison between the experimental and finite element predicted force-displacement evolution for the entire set of test samples included in Table 4.

The first conclusion to be taken from these results is the lack of influence of the type of assembly fit on the experimental evolution of the force with displacement, as it had been previously observed in finite elements.

The second conclusion is that the reason why finite elements are not able to predict the drop in force after cracking of the two test samples P2e is because modelling did not take crack propagation into consideration.

The third conclusion is that the overall agreement between experimental and numerical prediction of the force-displacement evolution improves as the shape factor H_0/d_0 increases. In the compression of single-material (solid) cylinders, this discrepancy was attributed to the influence of friction, which becomes more important and leads to more significant deviations, as the cylinders reduce their height—typically when the shape factor goes below 0.5 [26,27].

This type of influence was also observed in the compression of bimetallic cylinders, especially for test samples P1a-P2a and P1b-P2b with the lowest shape factors. Improvements of the numerical estimates would require tuning the friction factor for each shape factor H_0/d_0 in order to match the experimental results. This was not carried out because the actual differences between numerical and experimental results were considered not relevant for the overall aims and objective of the investigation.

The finite element distribution of effective strain after 3.5 mm displacement of the upper die platen is shown in Figure 7b. Effective strain values were obtained at the center of each tetrahedral element and interpolated between old (distorted) and new meshes during remeshing procedures [28]. As seen, the effective strain values are higher for small shape factors H_0/d_0 and the distribution is more homogeneous for high shape factors H_0/d_0. This result is interesting because it goes against the expected conclusion that cracks would be triggered in test sample P2e because its overall level of effective strain and its overall level of inhomogeneity would be the highest.

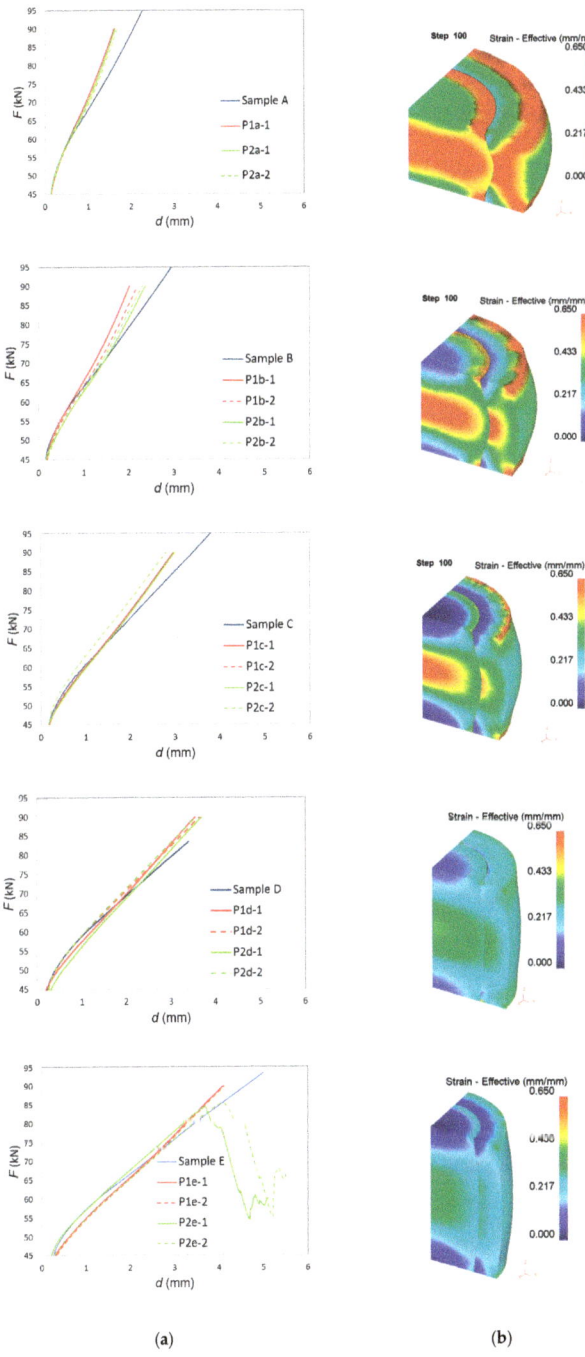

Figure 7. (**a**) Experimental and finite element predicted evolution of force with displacement for the entire set of test cases included in Table 4; (**b**) finite element predicted evolution of effective strain after 3.5 mm displacement of the upper die platen.

3.2. Ductile Damage

Figure 8 shows the cylindrical test samples cut along their axial cross-sectional planes, after compression. As seen, some of the samples with clearance fit (P2i) showed a permanent joint between the center and the ring after the compression—one sample for each shape factor. On the contrary, two of the samples with interference fit (P1a and P1c) showed separation between the center and the ring after compression. The latter result was attributed to the appearance of internal voids at the contact interface, as reported by Cetintav et al. [1], who previously observed the existence of such voids for height reductions of 30%.

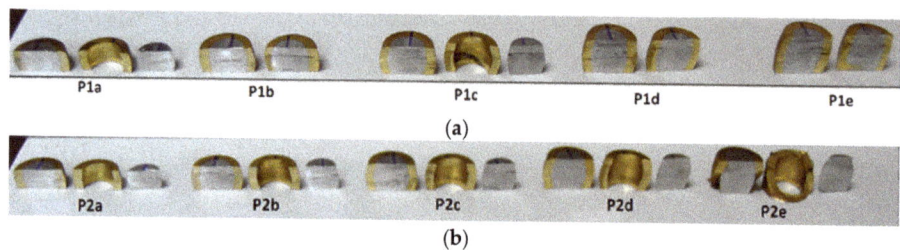

Figure 8. Cross section of bimetallic cylindrical test samples mounted with (**a**) interference fit (P1i) and (**b**) clearance fit (P2i), after compression.

Because Essa et al. [17], also claimed the occurrence of voids when the ratio between the center and the ring diameters was higher than 0.6, it is not possible to claim a general design rule for obtaining bimetallic cylinders with permanent joints between the center and the ring after compression.

The distribution of accumulated ductile damage in both bimetallic and single-material cylinders made of the aluminum alloy UNS A92011 is disclosed in Figure 9. As seen, damage is higher in bimetallic cylinders and a discontinuity is also observed in the center-ring contact interface.

The accumulated ductile damage along paths 1 and 2 (Figure 3) is plotted in Figure 10. Results show that the higher the shape factor, the lower the damage in the center, especially for sample Pe, where damage is almost negligible due to the limited amount of inhomogeneous material flow (small amount of barreling).

Moreover, the mostly damaged region was found to occur at the middle plane of the outer ring surface, as previously claimed by Silva et al. [29]. The only exception is sample Pa, with the lowest shape factor, in which the highest value of damage, and therefore, the most critically damaged region, were also found at the intersection between the center-ring interface and the die platens (refer to path 1 in Figure 10). As expected, the cylinder center does not experience a significant amount of damage due to a nearly homogeneous material flow.

Now, focusing our attention of the test samples with the highest shape factor (samples Pe), one concludes that cracking in both samples P2e after a displacement of 3.7 mm is not compatible with the fact that the Pe samples are those presenting the smallest amounts of accumulated ductile damage. In fact, the finite element predicted damage was below 0.07 (Figure 11), and therefore, is more compatible with the absence of cracking observed in samples P1e than with the existence of cracks in samples P2e.

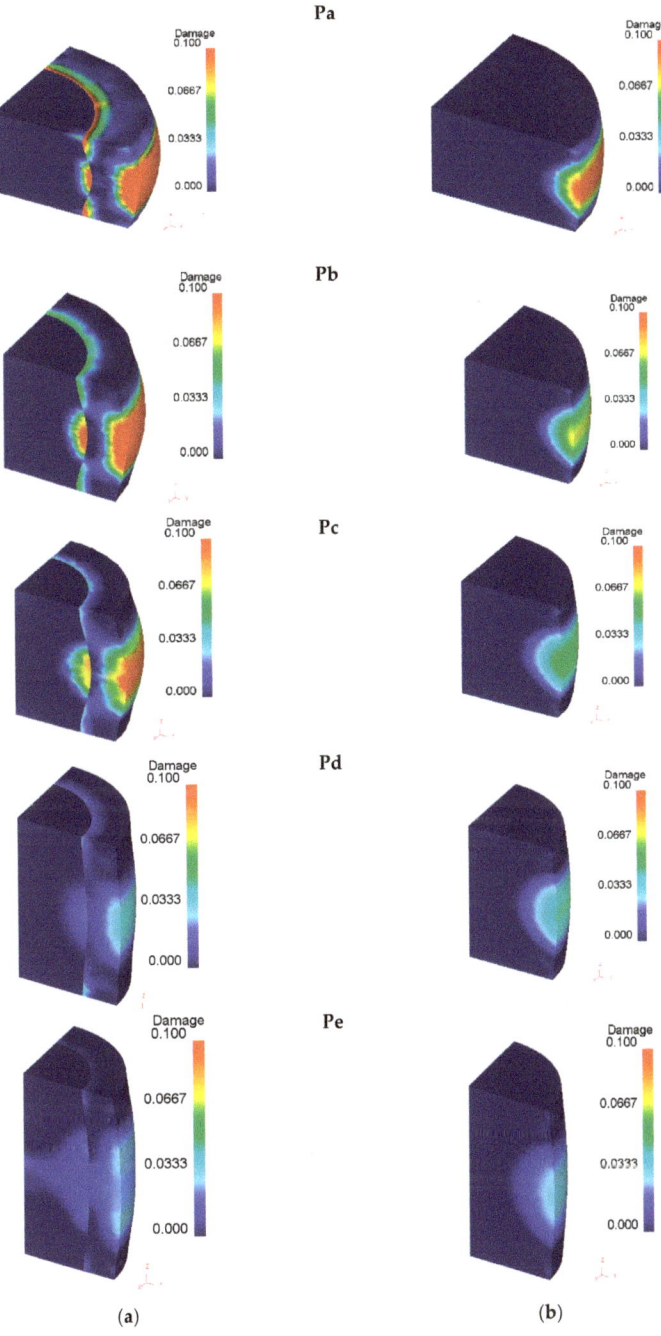

Figure 9. Finite element distribution of accumulated ductile damage in (**a**) bimetallic cylindrical test samples and (**b**) single-material cylindrical test samples made from the aluminum alloy UNS A92011 after 3.5 mm displacement of the upper die platen.

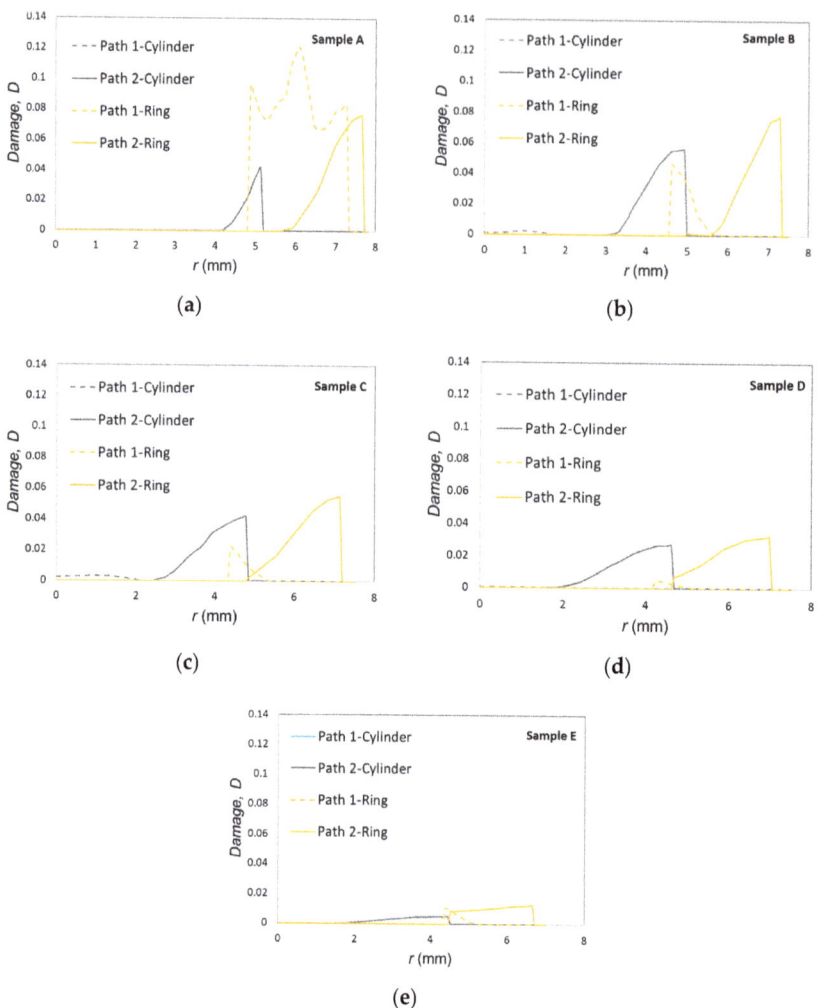

Figure 10. Finite element accumulated ductile damage as a function of the radial distance from the symmetry axis for paths 1 and 2 (Figure 4), after 3.5 mm displacement of the upper die platen: (**a**) Pa; (**b**) Pb; (**c**) Pc; (**d**) Pd; (**e**) Pe.

Despite the above-mentioned contradictory results, cracking in Figure 11 can only be explained by a combination between the maximum accumulated damage at the outer ring and at the intersection between the contact interface and the die platens. This explanation is not straightforwardly evident, but the experimental results allow concluding that the type of assembly fit (P1 versus P2) plays a key role on the development of cracks. In particular, interference fit is more favorable to preventing cracking of the ring.

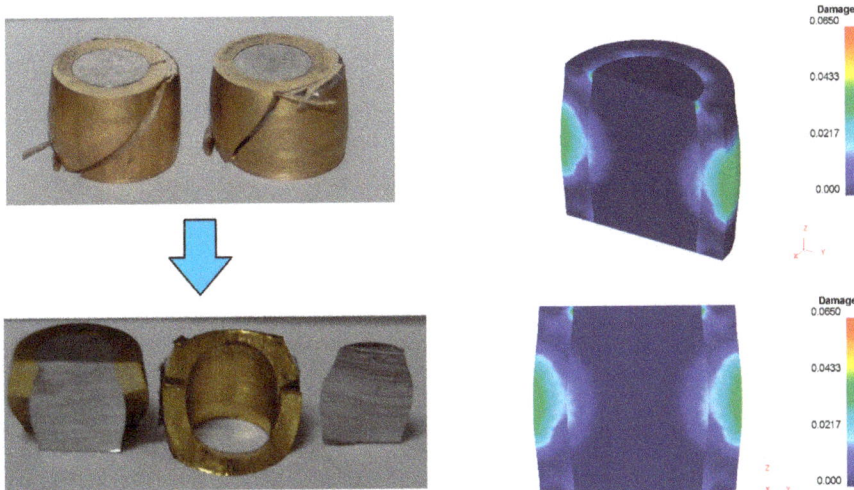

Figure 11. Cross section of the sample P2e showing failure by cracking and the corresponding finite element prediction of ductile damage (after 3.7 mm displacement of the upper die platen).

3.3. Microstructural Observations

Figure 12 shows different defects (voids, cracks, and microcracks) found in the center and ring in the tests samples with interference fit (P1i). The surface observed corresponds to the middle plane because it is the most damaged region, as described in Section 3.2.

Table 5 exhibits a comparison between the defects detected by SEM and the finite element predictions of ductile damage. The comparison is very good.

Finally, Figure 13 shows, graphically, the locations of major presence of defects (maximum damage) observed by SEM and the damage location range predicted by the finite element analysis.

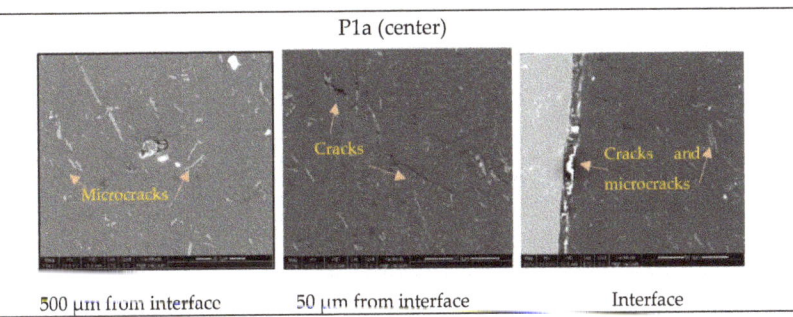

Figure 12. *Cont.*

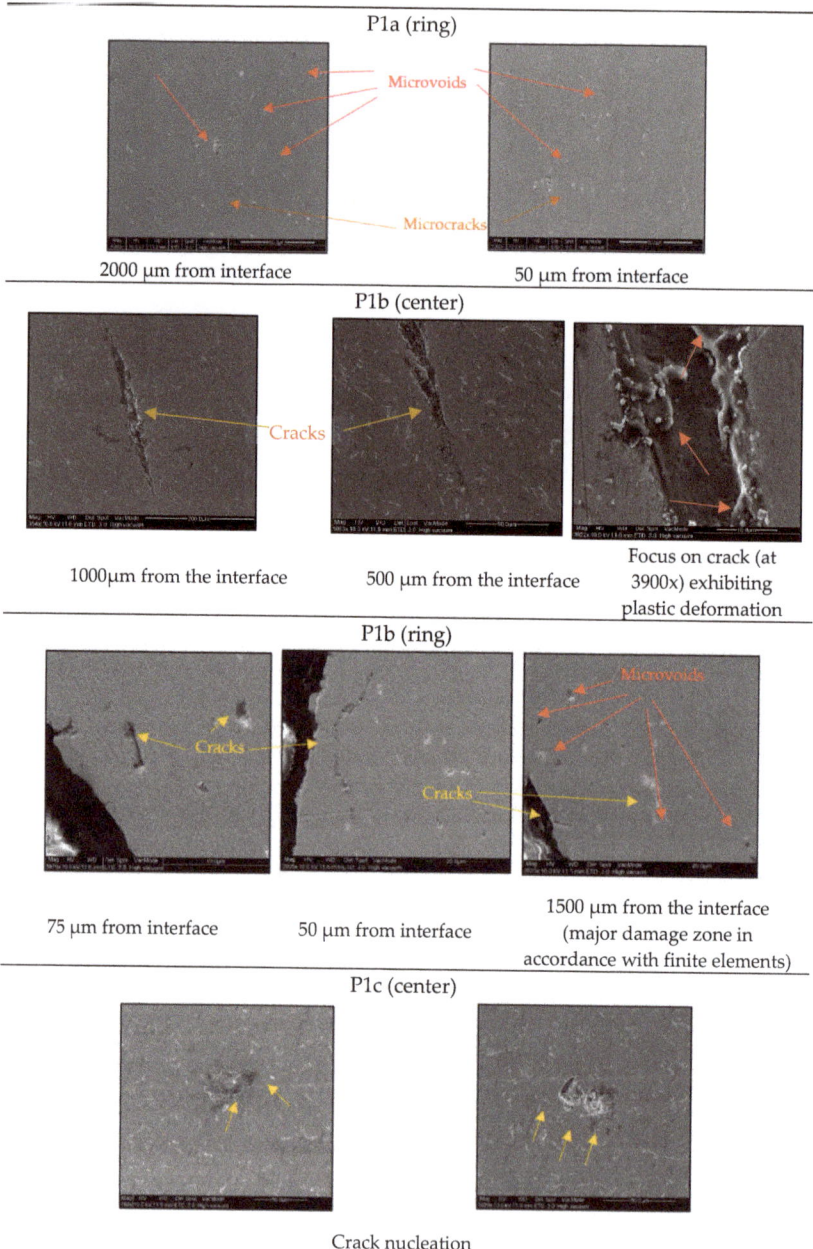

Figure 12. Cont.

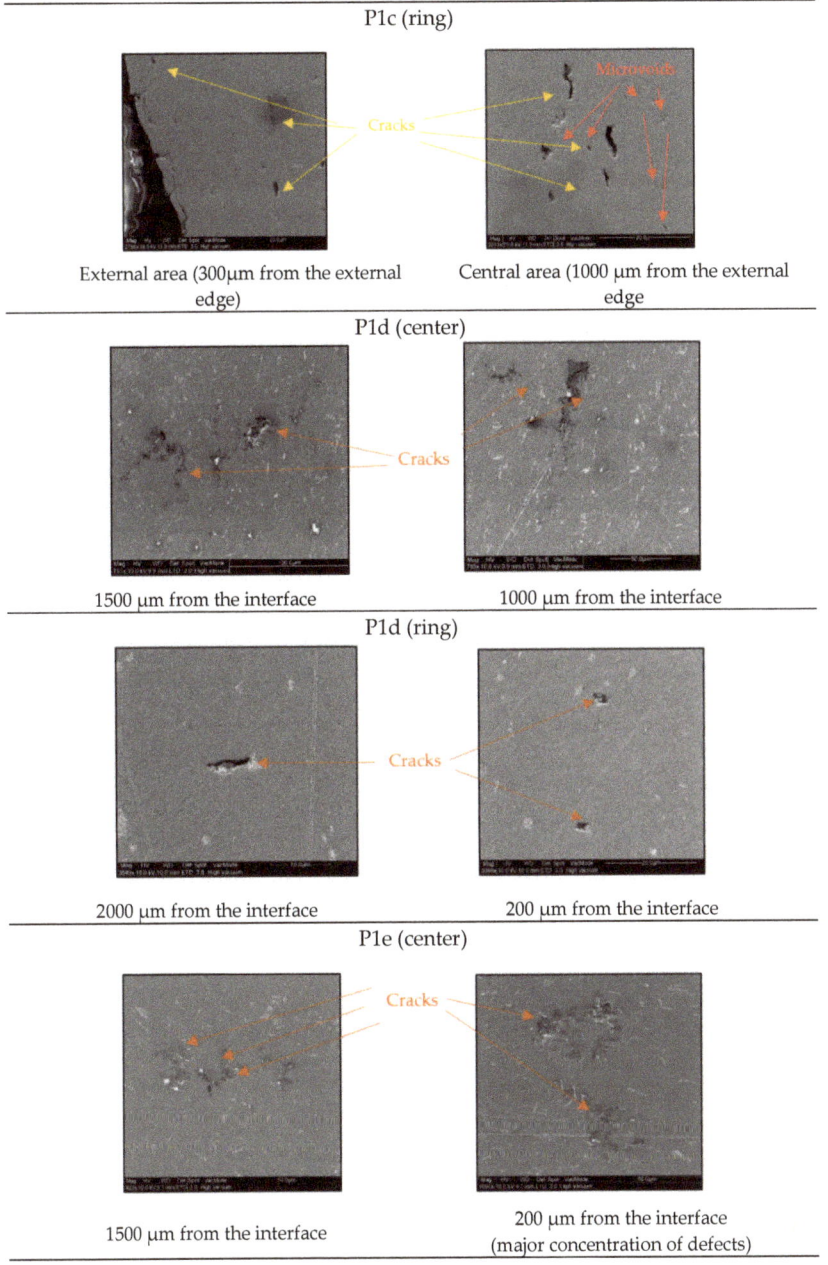

Figure 12. Microstructural observations in the center, and ring of the test samples' interference fit (P1i).

Table 5. Comparison between SEM observations and finite element predictions of ductile damage.

Sample	SEM Observation	Finite Element Prediction of Ductile Damage
P1a	**Centre**: Concentration of defects (cracks, microcracks, microvoids) at 500 μm distance to the interface ($r \cong 5.20$ mm) and in close agreement with the location of the maximum finite element prediction of damage (5.10 mm).	**Centre**: Accumulation of ductile damage for radial distances (r) between 4.15 and 5.25 mm (Figure 10). Maximum damage ($D = 0.04$) located at a radial distance of = 5.10 mm.
	Ring: Microcracks and microvoids located at 2000 μm distance from the interface ($r \cong 7.20$ mm) and close to the maximum damage predicted by finite elements.	**Ring**: Accumulation of ductile damage for radial distances (r) between 5.66–7.75mm (Figure 10). Maximum damage ($D = 0.078$) located at a radial distance of 7.66 mm.
P1b	**Centre**: Concentration of defects (cracks, microcracks, microvoids) at 500–1000 μm distance to the interface ($r \cong 5.20$mm) and in close agreement with the location of the maximum finite element predicted damage.	**Centre**: Accumulation of ductile damage for radial distances (r) between 3.33 and 5 mm (Figure 10). Maximum damage ($D = 0.055$) located at a radial distance between 4.66 and 4.90 mm.
	Ring: Microcracks and microvoids located at 2000μm from the interface (r $\cong$7.00 mm) and close to the maximum damage predicted by finite elements.	**Ring**: Accumulation of ductile damage for radial distances (r) between 5.66 and 7.33 mm (Figure 10). Maximum damage ($D = 0.078$) located at a radial distance between 7.1 and 7.33 mm.
P1c	**Centre**: Concentration of microcracks in the central area (between $r = 3$ and 5 mm).	**Centre**: Accumulation of ductile damage for radial distances (r) between 2.50 and 4.90 mm (Figure 10). Maximum damage ($D = 0.04$) located at a radial distance of 4.70 mm.
	Ring: Concentration of microcracks and cracks at 300-1000 μm distance from the interface ($r \cong 7.10$ mm).	**Ring**: Accumulation of ductile damage for radial distances (r) between 4.66-7.25 mm (Figure 10). Maximum damage ($D = 0.055$) located at a radial distance of 7 and 7.10 mm.
P1d	**Centre**: Cracks at 1000-2000 μm distance from the interface ($r \cong 2.50$ mm–3.5 mm).	**Centre**: Accumulation of ductile damage for radial distances (r) between 2.00 and 4.66mm (Figure 12). Maximum damage ($D = 0.03$) at a radial distance of 4.66 mm.
	Ring: Microcracks at 200–2000 μm distance from the interface ($r \cong 5$–6.80 mm).	**Ring**: Accumulation of ductile damage for radial distances (r) between 4.66-7 mm (Figure 10). Maximum damage ($D = 0.03$) located at a radial distance of 6.33 and 7 mm.
P1e	**Centre**: Microcracks observed at 200–1500 μm distance from the interface ($r \cong 4.60$ mm).	**Centre**: Accumulation of ductile damage for radial distances (r) between 1.66 and 4.5mm (Figure 10). Maximum damage ($D = 0.005$) at a radial distance between 3.33 and 4.33mm.
	Ring: No relevant defects were observed.	**Ring**: Accumulation of ductile damage for radial distances (r) between 4.4 and 6.66mm (Figure 10). Maximum damage ($D = 0.01$) located at a radial distance of 6 and 6.66 mm.

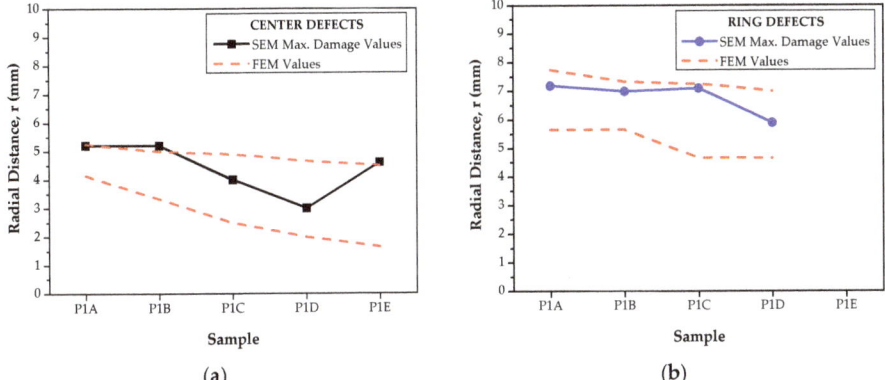

Figure 13. Representation of the locations with major presences of defects (maximum damage) observed by SEM, and the damage location range predicted by the finite element analysis. (**a**) Center; (**b**) ring.

4. Conclusions and Future Work

This paper looked at the compression of bimetallic cylinders from a combined damage and microstructural point of view. The cylinders were made from an aluminum alloy UNS A92011 center and a brass UNS C38500 ring with various height to diameter ratios ("shape factor ratios") and difference assembly fit tolerances (by interference and clearance). Ductile damage predictions were obtained from finite element modelling with a commercial finite element (FE) program, whereas the microstructural observations were carried out with a scanning electron microscope (SEM).

The comparison between the experimental and numerical predicted forces showed that the shape factor ratio influences the force-displacement evolution in a similar way to what is commonly found in the compression of single-material cylinders. Differences between experimental and numerical results were more significant for test samples P1a-P2a and P1b-P2b, having the lowest shape factors, due to variations in friction for large amounts of height reduction. In contrast, the type of assembly fit did not influence the overall force-displacement evolutions.

SEM observations of voids, microcracks, and cracks revealed a general good agreement with the finite element estimates of ductile damage. In particular, SEM observations detected voids, microcracks, and cracks in specific areas of the maximum predicted damage. Numerical simulations also showed that ductile damage is higher in bimetallic cylinders than in single-material cylinders made from the aluminum alloy UNS A92011, and that there is a discontinuity in ductile damage distribution at the contact interface between the center and the core.

The overall results allow concluding that the higher the shape factor, the lower the damage, for the same amount of displacement of the upper die platen. This is due to differences in material flow inhomogeneity that favor the test samples with larger shape factors, but it is in clear contradiction to the fact that cracking was only found in test samples P2e having the largest shape factors. The explanation for this discrepancy was attributed to the type of assembly fit. In particular, results show that mounting the center in the ring with an interference fit prevents the occurrence of cracking. This is the most favorable condition for application in multi-material forming by compression. The amount of interference-fit needs to be addressed in future work, together with conducting complementary microstructural analysis in order to find relations between microstructure and material flow.

Author Contributions: Conceptualization, A.M.C., C.B., and Á.R.-P.; formal analysis, A.M.C., J.M.H., Á.R.-P., A.M.A., and C.L.-M.; funding acquisition, A.M.C., Á.R.-P., and Á.Y.-G.; investigation, A.M.C., Á.R.-P., J.M.H., and C.B.; methodology, A.M.C., J.M.H., and Á.R.-P.; project administration, A.M.C.; resources, A.M.C., C.L.-M., and Á.Y.-G.; supervision, A.M.C. and Á.R.-P.; validation, A.M.C. and Á.R.-P.; writing—original draft, A.M.C. and Á.R.-P.; writing—review and editing, A.M.C., Á.R.-P., Á.Y.-G., and P.A.F.M.

Funding: This research was funded by the Annual Grants Call of the E.T.S.I.I. of UNED through the projects of references 2014-ICF04, 2015-ICF04, and 2019-ICF04. A mobility grant for junior researchers was also granted by MES (Manufacturing Engineering Society) to Álvaro Rodríguez-Prieto.

Acknowledgments: The authors would like to take this opportunity to thank the Research Group of the UNED "Industrial Production and Manufacturing Engineering (IPME)" for the support provided during the development of this work. We also acknowledge to Center for Nanoscale Materials (CNM), supported by the US Department of Energy, Office of Science and Office of Basic Energy Sciences under contract number DE-AC02-06CH11357. Paulo Martins would like to thank the support of by Fundação para a Ciência e a Tecnologia of Portugal and IDMEC under LAETA-UID/EMS/50022/2019.

Conflicts of Interest: The authors declare no conflict of interest.

References

1. Cetintav, I.; Misirli, C.; Can, Y. Upsetting of Tri-Metallic St-Cu-Al and St-Cu60Zn-Al Cylindrical Billets. *Int. J. Mech. Aerosp. Ind. Mechatron. Manuf. Eng.* **2016**, *10*, 1632–1635.
2. European Comission Horizon 2020 Work Programme 2018-2020 5.ii. Nanotechnologies, Advanced Materials, Biotechnology and Advanced Manufacturing and Processing. Available online: https://ec.europa.eu/research/participants/data/ref/h2020/wp/2018-2020/main/h2020-wp1820-leit-nmp_en.pdf (accessed on 30 October 2019).
3. DebRoy, T.; Wei, H.L.; Zuback, J.S.; Mukherjee, T.; Elmer, J.W.; Milewski, J.O.; Beese, A.M.; Wilson-Heid, A.; De, A.; Zhang, W. Additive manufacturing of metallic components—Process, structure and properties. *Prog. Mater. Sci.* **2018**, *92*, 112–224. [CrossRef]
4. Onuike, B.; Heer, B.; Bandyopadhyay, A. Additive manufacturing of Inconel 718—Copper alloy bimetallic structure using laser engineered net shaping (LENS^{TM}). *Addit. Manuf.* **2018**, *21*, 133–140. [CrossRef]
5. Onuike, B.; Bandyopadhyay, A. Functional bimetallic joints of Ti6Al4V to SS410. *Addit. Manuf.* **2020**, *31*, 100931. [CrossRef]
6. Sahasrabudhe, H.; Harrison, R.; Carpenter, C.; Bandyopadhyay, A. Stainless steel to titanium bimetallic structure using LENS^{TM}. *Addit. Manuf.* **2015**, *5*, 1–8. [CrossRef]
7. Reichardt, A.; Dillon, R.P.; Borgonia, J.P.; Shapiro, A.A.; McEnerney, B.W.; Momose, T.; Hosemann, P. Development and characterization of Ti-6Al-4V to 304L stainless steel gradient components fabricated with laser deposition additive manufacturing. *Mater. Des.* **2016**, *104*, 404–413. [CrossRef]
8. Zhang, M.; Yang, Y.; Wang, D.; Song, C.; Chen, J. Microstructure and mechanical properties of CuSn/18Ni300 bimetallic porous structures manufactured by selective laser melting. *Mater. Des.* **2019**. [CrossRef]
9. Mukherjee, T.; Zuback, J.S.; Zhang, W.; DebRoy, T. Residual stresses and distortion in additively manufactured compositionally graded and dissimilar joints. *Comput. Mater. Sci.* **2018**, *143*, 325–337. [CrossRef]
10. Knezevic, M.; Jahedi, M.; Korkolis, Y.P.; Beyerlein, I.J. Material-based design of the extrusion of bimetallic tubes. *Comput. Mater. Sci.* **2014**, *95*, 63–73. [CrossRef]
11. Berski, S.; Dyja, H.; Banaszek, G.; Janik, M. Theoretical analysis of bimetallic rods extrusion process in double reduction die. *J. Mater. Process. Technol.* **2004**, *153–154*, 583–588. [CrossRef]
12. Alcaraz, J.L.; Gil-Sevillano, J. An analyisis of the extrusion of bimetallic tubes by numerical simulation. *Int. J. Mech. Sci.* **1996**, *38*, 157–173. [CrossRef]
13. Behrens, B.A.; Goldstein, R.; Chugreeva, A. Thermomechanical processing for creating bi-metal bearing bushings. In *Thermal Processing in Motion 2018*; ASM International: Spartanburg, SC, USA, 2018; pp. 15–21.
14. Leitão, P.J.; Teixeira, A.C.; Rodrigues, J.M.C.; Martins, P.A.F. Development of an industrial process for minting a new type of bimetallic coin. *J. Mater. Process. Technol.* **1997**, *70*, 178–184. [CrossRef]
15. Barata, M.J.M.; Martins, P.A.F. A study of bi-metal coins by the finite element method. *J. Mater. Process. Technol.* **1991**, *26*, 337–348. [CrossRef]
16. Afonso, R.M.; Alexandrino, P.; Silva, F.M.; Leitão, P.J.; Alves, L.M.; Martins, P.A. A new type of bi-material coin. *Proc. Inst. Mech. Eng. Part B J. Eng. Manuf.* **2019**, *233*, 2358–2367. [CrossRef]
17. Essa, K.; Kacmarcik, I.; Hartley, P.; Plancak, M.; Vilotic, D. Upsetting of bi-metallic ring billets. *J. Mater. Process. Technol.* **2012**, *212*, 817–824. [CrossRef]
18. Misirli, C.; Cetintav, I.; Can, Y. Experimental and fem study of open die forging for bimetallic cylindrical parts produced using different materials. *Int. J. Mod. Manuf. Technol.* **2016**, *8*, 69–74.

19. Wernicke, S.; Gies, S.; Tekkaya, A.E. Manufacturing of hybrid gears by incremental sheet-bulk metal forming. *Procedia Manuf.* **2019**, *27*, 152–157. [CrossRef]
20. Plancak, M.; Kacmarcik, I.; Vilotic, D.; Krsulja, M. Compression of bimetallic components–analytical and experimental investigation. *Ann. Fac. Eng. Hunedoara–Int. J. Eng.* **2012**, *X-2*, 157–160.
21. Gisbert, C.; Bernal, C.; Camacho, A.M. Improved Analytical Model for the Calculation of Forging Forces during Compression of Bimetallic Axial Assemblies. *Procedia Eng.* **2015**, *132*, 298–305. [CrossRef]
22. Aluminium Alloys—Aluminium 2011 Properties, Fabrication and Applications. Available online: https://www.azom.com/article.aspx?ArticleID=2803#4 (accessed on 21 October 2019).
23. Brass Alloys—Brass CZ121 Properties, Fabrication and Applications. Available online: https://www.azom.com/article.aspx?ArticleID=2822# (accessed on 21 October 2019).
24. Herrero, J.M. Estudio del Comportamiento a Compresión de Componentes Cilíndricos bi-Metálicos Mediante Técnicas Numéricas y Experimentales. Bachelor's Thesis, Departamento de Ingeniería de Construcción y Fabricación, UNED, Madrid, Spain, 2017.
25. Cockcroft, M.; Latham, D. Ductility and the workability of metals. *J. Inst. Met.* **1968**, *96*, 33–39.
26. Kopp, R.; Wiegels, H. *Einführung in die Umformtechnick*; Verlag Mainz: Aachen, Germany, 1999; ISBN 978-3-86073-821-6.
27. Marín, M.M.; Camacho, A.M.; Bernal, C.; Sebastián, M.A. Investigations on the influence of the shape factor and friction in compression processes of cylindrical billets of AA 6082-T6 aluminum alloy by numerical and experimental techniques | Investigaciones sobre la influencia del factor de forma y del rozamien. *Rev. Metal.* **2013**, *49*, 200–212. [CrossRef]
28. Scientific Forming Technologies. *DEFORM v11.2 User's Manual*; Scientific Forming Technologies Corporation: Columbus, OH, USA, 2017.
29. Silva, C.M.A.; Alves, L.M.; Nielsen, C.V.; Atkins, A.G.; Martins, P.A.F. Failure by fracture in bulk metal forming. *J. Mater. Process. Technol.* **2015**, *215*, 287–298. [CrossRef]

© 2019 by the authors. Licensee MDPI, Basel, Switzerland. This article is an open access article distributed under the terms and conditions of the Creative Commons Attribution (CC BY) license (http://creativecommons.org/licenses/by/4.0/).

Article

Study of the Degree of Cure through Thermal Analysis and Raman Spectroscopy in Composite-Forming Processes

Juan A. García-Manrique *, Bernabé Marí, Amparo Ribes-Greus, Llúcia Monreal, Roberto Teruel, Llanos Gascón, Juan A. Sans and Julia Marí-Guaita

Institute of Design for Manufacturing and Automated Production, Universitat Politècnica de València (UPV), Camí de Vera s/n, 46022 València, Spain; bmari@fis.upv.es (B.M.); aribes@ter.upv.es (A.R.-G.); lmonreal@mat.upv.es (L.M.); r.teruel@upvnet.upv.es (R.T.); llgascon@mat.upv.es (L.G.); juasant2@upvnet.upv.es (J.A.S.); juliasetze@gmail.com (J.M.-G.)
* Correspondence: jugarcia@mcm.upv.es; Tel.: +34-963-877-622

Received: 8 November 2019; Accepted: 28 November 2019; Published: 2 December 2019

Abstract: The curing of composite materials is one of the parameters that most affects their mechanical behavior. The inspection methods used do not always allow a correct characterization of the curing state of the thermosetting resins. In this work, Raman spectroscopy technology is used for measuring the degree of cure. The results are compared with conventional thermal gravimetric analysis (TGA), differential scanning calorimetry (DSC), and scanning electron microscope (SEM). Carbon fiber specimens manufactured with technologies out of autoclave (OoA) have been used, with an epoxy system Prepreg System, SE 84LV. The results obtained with Raman technology show that it is possible to verify the degree of polymerization, and the information is complementary from classical thermal characterization techniques such as TGA and DSC; thus, it is possible to have greater control in curing and improving the quality of the manufactured parts.

Keywords: prepeg; carbon fiber; Raman spectroscopy

1. Introduction

There is currently a great demand from the aerospace, wind, nautical and automobile industries for the development of new high-performance composites. The cure characterization of the parts affects both the forming processes and the repair–maintenance processes. Cure time is generally calculated very conservatively to ensure complete curing of the part before removing it from the mold. However, this practice greatly slows manufacturing times, and in some cases, it damages the part. The "in situ" control techniques of the mold filling and cure monitoring provide enough information to reduce the injection and curing times. Biomedical applications are used to increase the thickness of the part's fillers; however, a high degree of cure could be difficult to achieve, reducing its mechanical properties and the biocompatibility [1].

The manufacturing processes of composite materials can be divided into two large families: the wet layup and the dry layup processes (see Figure 1). Wet processes are those where the resin and the fabric have been mixed and fabrics already impregnated with the resin are supplied. The chemical polymerization reaction is "frozen" by the low supply temperature (−20 °C) and its reaction will be activated by heat. These materials are formulated with an excess of resin that must be evacuated during the fiber-compaction phase. When the resin is heated, its viscosity decreases and can flow through the thickness of the part. Thanks to this resin displacement through the thickness, the layers that form the laminate will be joined together. Autoclave, mechanical press, or atmospheric pressure can be used for compaction. This paper analyzes the forming processes compacted only with atmospheric pressure, which are known as VBO (vacuum bag only) or OoA (out of autoclave). These processes are those that have an easier processing; however, sometimes they are very expensive due to the large amount of labor they need, the cost of raw materials, etc.

On the other hand, regarding the dry layup forming processes, the impregnation phase of the fabrics takes place inside the mold. Therefore, the manufacturing process has a filling phase, a melt front advance, injection points, vents, etc. Depending on the type of injection or the stiffness of the countermold, it will define the different forming processes that make up the family of Liquid Composite Molding (LCM). These processes have great advantages over wet processes because they allow manufacturing with the same quality but at a significantly lower cost. Depending on the resin used, they may need an oven curing stage where the resin cures. However, these require a greater knowledge regarding the behavior of materials and manufacturing techniques than wet processes.

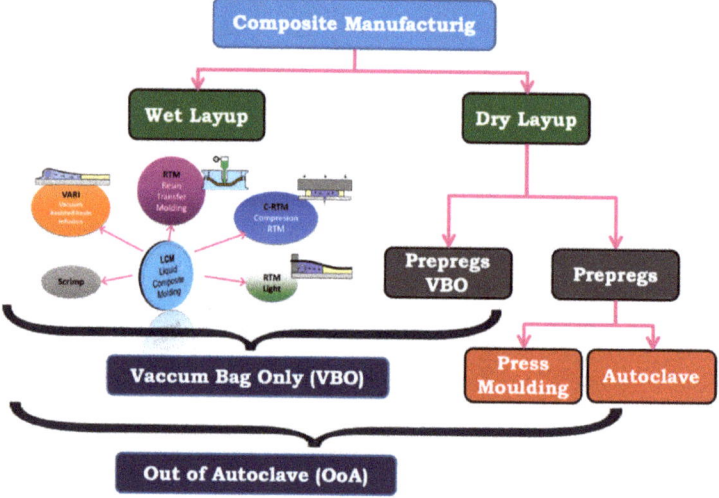

Figure 1. General classification of composite manufacturing process.

The objective of any composite manufacturing process is to impregnate the fabric with resin in the most efficient way and evacuate all the pores or voids that reduce the mechanical properties of the parts. In this context, the simulation of the formation and displacement of the pores has been an important research topic in recent years. An extensive review on this topic can be found in (Pillai, 2004 [2]; Park and Lee, 2011 [3]). Autoclave curing processes (OoA, out of autoclave) and vacuum bag-only compression (VBO, vacuum bag only) processes are often considered equivalent, but there are some differences between them. OoA processes, as the name implies, are all processes that heal without using an autoclave, but it is possible to apply a compaction pressure as large as necessary, for example, using a hot press for curing. However, concerning compaction processes with vacuum bag only, the compaction pressure is limited to atmospheric pressure. In both cases, the cost reduction with respect to autoclave manufacturing is important, and production times can be greatly reduced. In both cases, the manufacture of components out of autoclave with the same quality as those obtained with the autoclave is a major technological challenge. In this context, it is necessary to know the behavior of the materials during the forming process, and numerical models adapted to the different material scales are needed. In all forming processes, either wet (prepregs) or dry (LCM), numerical methods that include capillary effects and/or resin–air interface modeling should be used. The most advanced models in these techniques are based on numerical schemes by biphasic finite elements [4] that use the Stokes or Darcy's flows models adapted to the biphasic behavior.

Nowadays, these models cannot solve the full geometry of the part on a macro scale, so reduced schemes are used for its numerical calculation. The equations to be solved are very similar for dry path processes where the flow to be studied is in-plane XY; as for the wet path, processes where the flow to be studied is basically through the thickness Z. The most traditional manufacturing process of high-performance composite parts is prepregs. These materials are made of fiber and resin and are supplied in sheets. To achieve the maximum mechanical properties, the laminate must undergo a compaction phase and a heat phase for resin curing. The material compaction is necessary to achieve the maximum fiber volumetric fraction, and in addition, the excess of resin with which the prepregs are manufactured must be evacuated. This excess of resin flows in through the thickness direction and is necessary to get the bond between each layer of the laminate. Maximum compaction is achieved using expensive autoclaves that are only accessible to certain high value-added industry such as aerospace or racing vehicles.

The prepreg forming process (OoA) basically consists of three stages: layup, vacuum closing, and curing. First, the prepregs are cut in the required number of layers and sizes to obtain the thickness and shape of the part. Preimpregnated layers are stacked manually or using automated techniques (automatic tape layup). Intermediate compaction steps are performed to improve the layers compaction by applying vacuum at room temperature (see Figure 2). Once the part has reached the required thickness, it is compressed under vacuum, and the curing cycle begins. During the compaction phase, physical phenomena occur that are normally neglected. These phenomena include capillarity, drainage, inhibition, or pore mobility [5].

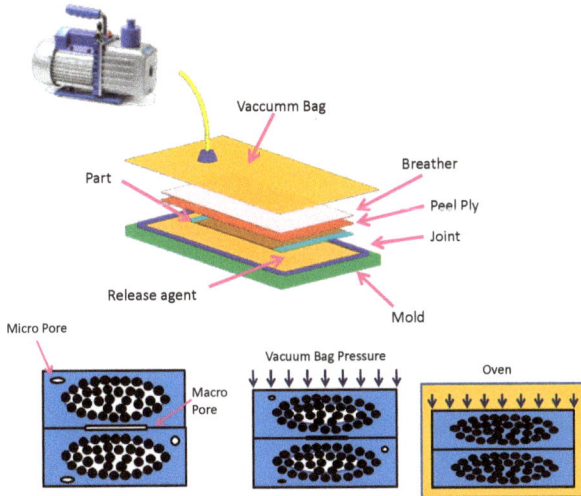

Figure 2. Prepreg technology.

Therefore, the resin flows through the fabric and fills both the spaces between the tows and inside of the tows. The distances or gaps between the tows are around tenths of a millimeter, while the spaces inside the tows are 10–15 μm. Then, the problem needs a double-scale approximation, including the meso and the micro scale. Many authors have tried to solve this problem by modifying the permeability [6] and the subsequent treatment of flow equations as a traditional LCM problem. There is a difference of several magnitudes of order between the value of the permeability inside the tows (micro scale) and the space between the tows (macro or meso scale). As a result, the numerical simulation of the double-scale process is extremely complex. There are simplified models that estimate the saturation of the fabrics, but the work done so far does not consider many of the physical phenomena that occur inside the fabrics, resulting in them being inadequate for most applications of industrial interest.

In this sense, it is worth highlighting the work carried out by Professor Veronic Michaud of the Ecole Polytechnique de Lausanne [7], who proposed a numerical model for the behavior of air permeability during the curing phase. Other researchers proposed the phenomenological characterization of the process, such as the research group of Professor S.G. Advani at the University of Delaware (USA) [8]. This approach requires the experimental data to have quantitative and qualitative results [9].

Composite manufacturers in the aerospace industry and other large industries, such as the automobile and wind industries, are looking for manufacturing processes out of autoclave (OoA) that can achieve 2% content in pores with less expensive and more efficient equipment. Autoclaves are used when the highest quality is needed in the final part, with a pore content of less than 2% and high glass transition temperatures (Tg). Aerospace autoclaves normally operate at 120 to 230 degrees Celsius within a nitrogen atmosphere at 7 bar pressure. It has been shown that processing with a preimpregnated vacuum bag only produces high porosity composite laminates, due to the air and moisture trapped in layers and between layers, which cannot be evacuated, and the lower (atmospheric) pressure cannot sufficiently compact. Moisture in prepreg can cause pores when processed only with the vacuum bag, but when processed in an autoclave, the higher pressure causes moisture to condense, suppressing pore growth. The first preimpregnated OoA was designed in the early 1990s for initial curing at low temperature (approximately 60 °C), followed by post-curing at high temperature (approximately 110 °C) [10].

The thermosetting resins used in the manufacture of high-performance composite (HPC) are initially processed in a liquid state, and by an exothermic curing reaction the polymer chains are crosslinked, forming covalent bonds of high hardness and chemical stability. When these properties are mixed with the appropriate fiber, they become an excellent composite material. The composite materials analyzed in this work are composed of a high modulus epoxy resin and long carbon fiber (HPC). The curing of this type of resin can be carried out by chemical methods, thermal methods, or a combination of both. The most commonly used method is the last one, since it allows greater control over the quality of the curing and reduces the manufacturing time. In all curing processes, not only the time and temperatures, but also the reaction rate must be controlled. There is an optimum in terms of cure temperature, which should coincide with the time when the reaction rate is higher. A higher temperature causes degradation of the material, rather than an increase in the reaction rate. The classic methods of resin characterization are differential scanning calorimetry (DSC) and dynamic mechanical analysis (DMA). These methods provide excellent information on the glass transition temperature, the heat generated by the chemical reaction, the cure rate, and the degree of cure. However, all these methods have the disadvantage of being destructive methods, and therefore can only be applied on a laboratory scale.

The cure reaction as a function of the manufacturing variables is a challenge for the analysis and design of processing operations. In addition, the physical properties of composite materials strongly depend on their microstructure and are directly related with the failure of the fibers or delaminations. Calorimetry will be used for the macrokinetic analysis of cure reactions. The cure reaction of epoxy resin will be analyzed also by means of thermal characterization using differential scanning calorimetry (DSC) and thermogravimetry (TGA). The vibrational properties of the composites were studied by Raman spectroscopy. The degree of cure of three samples of composite material with different levels of cure has been measured. The Raman spectroscopy vibrational method has been used and compared with those from thermal gravimetric analysis (TGA) and differential scanning calorimetry (DSC). It has been observed that complementary information can be obtained from classical trials, which allows progress toward non-destructive quality control in CFRP composites.

2. Materials and Methods

The coupons manufacturing begins with the consolidation stage. At this stage, the trapped air between the prepreg sheets is removed by vacuum sealing with a bag. This vacuum is applied for approximately 15 min at room temperature. Due to the stiffness of the prepreg layers, each of these consolidations are applied to 3–5 layers, depending on the total thickness of the part. In any case, the first layer is consolidated individually to ensure a good surface finish.

The SE 84LV is used for the manufacture of the coupons and can be cured at low temperature (85 °C) or, for faster molding of components, at 120 °C. Once the entire compaction sequence has been carried out, the mold is introduced into the oven (in-house manufacture) where the viscosity is reduced. The compaction pressure generates the flow of the resin, which mainly flows through the thickness of the part, and the resin excess is trapped in the absorption blanket. The polymerization reaction is activated by heat, and the oven must be programmed to control the resin reaction rate. The oven must control the temperature to avoid significant gradients inside the mold; therefore, the temperature rise ramp must be slow (approximately 1 degree/min). The oven temperature should be held for 10 h at 85 °C, and the tolerance must not exceed ±5 °C. Once the part is totally cured, quality control is recommended to ensure that no air has been trapped inside the part; the methods used are ultrasonic, stereography, thermography, etc.

A thermogravimetric study (TGA) of the samples has been carried out in order to simulate and optimize the percentage of volumetric fraction for the resin, fiber, and residue. Once the process conditions have been determined, the samples are introduced into a muffle furnace and the non-isothermal thermogravimetric multi-rate experiments are performed. Mettler-Toledo TGA/SDTA 851 (Mettler-Toledo, Columbus, OH, USA) equipment has been used. The samples weigh 10–11 mg and are heated in an alumina holder with a capacity of 70 µL. A heating curve from 30 °C to 1000 °C has been programmed with a rate of 10°/min in a controlled flow of oxygen atmosphere (50 mL/min) to simulate the real manufacturing conditions. Each experiment has been repeated three times, and finally, the characterization was carried out with the software STAR® 9.10 from Mettler-Toledo. In accordance with ASTM D3171-11 [4], "Standard methods for constituent content of composite materials", the weight percentage of each constituent is obtained from the following expressions:

$$P1(\%) = \frac{P_1}{P_0} \cdot 100 \quad (1)$$

$$P2(\%) = \frac{P_2}{P_0} \cdot 100 \quad (2)$$

$$P3(\%) = \frac{P_3}{P_0} \cdot 100 \quad (3)$$

$$\text{Residue }(\%) = P1 - P2 - P3 \quad (4)$$

$$P_0 = \text{initial mass} \quad (5)$$

$$P_1 = \text{final mass} \quad (6)$$

$$P_3 = \text{carbon mass} \quad (7)$$

$$P_4 = \text{resin mass} \quad (8)$$

The degree of resin cure of the specimens has been measured by differential scanning calorimetry for later comparison with the results obtained from Raman spectroscopy. A Mettler Toledo DSC 822 DSC (Mettler-Toledo, Columbus, OH, USA) with samples of 4–6 mg weight was used and sealed in aluminium pans with a capacity for 40 µL. The programmed heating curve was from 25 to 300 °C with a heating rate of 10 °C/min for dynamic DSC scanning. The total heat of the reaction, HT, is obtained from the total area in the heat generation versus the time line graph. The degree of cure and the rate of the degree of cure will be determined in the isothermal scanning experiment. The measurements were made at temperatures from 125 to 145 °C with increments of 5 °C and a rate of heating of 10 degrees/min.

The isothermal curing curve is calculated from the total area enclosed by these exothermic curves. The samples are cooled and heated again (from 25 to 300 °C at 10 °C/min) after each isothermal scan to obtain the residual heat of the reaction, RH. This value is obtained from the area under the exothermic peak in the resulting curve. To carry out Raman spectroscopy tests, a backscattering geometry by a confocal HORIBA Jobin-Yvon LabRam high-resolution micro Raman spectrometer (HORIBA Jobin-Yvon, NJ, USA) has been used. The test characteristics were 1200 grooves/mm grating at a 100-µm slit and 50× objective, in combination with a thermoelectrically cooled multichannel CCD (charge-coupled device) detector (spectral resolution below 3 cm^{-1}). A solid-state laser with a power of 50 mW emitting at 532.12 nm has been used.

3. Results and Discussion

The degree of cure of three composite samples with different characteristics has been analyzed. Sample 1 is a non-cured sample, which is used as the reference. It is a carbon prepreg sheet with any degree of cure, corresponding to a sample at the beginning of the manufacturing process. Sample 2 has been semi-cured out of autoclave, and Sample 3 was completely cured also in OoA. Figure 3 shows the FESEM micrographs for the three studied samples with different degrees of cure. The effect of the cure time is evident in these micrographs, the resins covering the carbon fibers disappear with the evolution of the cure reaction. According to Figure 3, the diameter of carbon fibers was found to be about 6 µm.

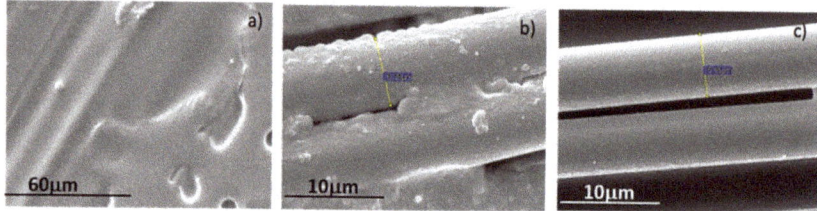

Figure 3. FESEM micrographs of the samples with different degree of cure: (**a**) Non-cured, (**b**) Semi-cured, (**c**) Cured.

3.1. Derivative Thermogravimetric (DTG)

Derivative thermogravimetric (DTG) is a type of thermal analysis in which the rate of material weight changes upon heating is plotted against temperature. It is used to simplify reading the weight versus temperature thermogram peaks, which occur close together. The results of the DTG analysis for the semi-cured (Sample 2) and cured (Sample 3) samples are shown in Figures 4 and 5, respectively.

In both figures, three peaks corresponding to three phases were observed by the dynamic thermogravimetric studies in oxidative conditions, as specified in Table 1.

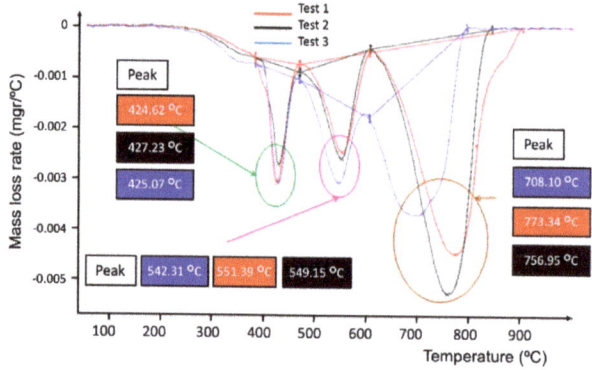

Figure 4. Derivative thermogravimetric (DTG) for semi-cured composite samples (Sample 2).

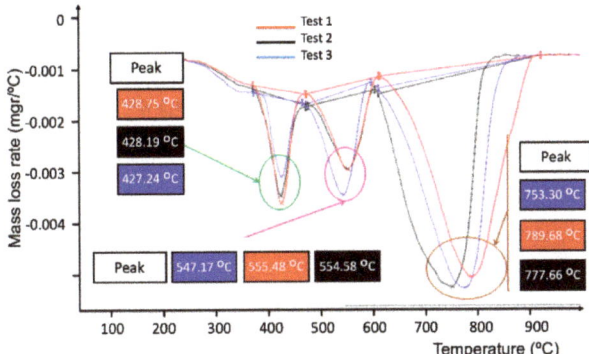

Figure 5. DTG for cured composite samples (Sample 3).

Table 1. Thermal degradation temperature. Samples 2 and 3.

Sample Number	Tpeak1	Tpeak2	Tpeak3
Sample 2	427 °C	551 °C	773 °C
Sample 3	430 °C	553 °C	790 °C

It can be noticed that in Figure 4, the first phase starts at about 425 °C, the second phase reaches a maximum rate near 550 °C, and finally, the third phase is near 790 °C. The third phase corresponds with the degradation of the residue. For both samples 2 and 3, the initial (lost) masses are shown in Table 2 and the mass losses are presented in Table 3.

Table 2. Initial mass for Samples 2 and 3.

Scan	P_0 (mg) Sample 2	P_0 (mg) Sample 3
C1	1.1655	1.1276
C2	1.1630	1.1634
C3	1.0930	1.1214

Table 3. Mass loss in the tests of Samples 2 and 3.

Sample 2			Sample 3		
Tc = 427 °C	Tc = 551 °C	Tc = 773 °C	Tc = 430 °C	Tc = 553 °C	Tc = 790 °C
Mass Loss P_1 (mg)	Mass Loss P_2 (mg)	Mass Loss P_3 (mg)	Mass Loss P_1 (mg)	Mass Loss P_2 (mg)	Mass Loss P_3 (mg)
0.4103	0.0681	0.6859	0.4098	0.2497	0.4648
0.4192	0.2413	0.5009	0.4174	0.5555	0.1862
0.3805	0.0359	0.6755	0.3855	0.0276	0.7073

Table 4 displays the fraction percentages for Samples 2 and 3, after applying the cure phases from Equations (1)–(4).

Table 4. Fraction percentages for Samples 2 and 3.

Fraction	Sample 2			Sample 3		
	Rep. 1 (%)	Rep. 2 (%)	Rep. 3 (%)	Rep. 1 (%)	Rep. 2 (%)	Rep. 3 (%)
P1	35	36	34	36	35	34
P2	5	20	3	22	47	2
P3	58	43	61	41	16	63
Residue	2	1	2	1	2	1

The analysis of the decomposition process of the composites suggests that the early degradation step is due to the removal of the most volatile components and low molecular mass species, with a rate of about 35% for the three replicas of both samples. The second stage of the thermal decomposition would correspond to the scission of the polymer network, and the later decomposition stage would be related to the thermal degradation of the carbon fibers [11]. Table 5 displays the average and standard deviation values for the four studied fractions. The high standard deviations found for fractions P2 and P3 indicate a non-uniform curing process. This behavior is in good agreement with the composite internal double scale, as represented in Figure 6. The gap between fibers inside and between tows are around 5 μm and 500 μm respectively, and then the cure kinetic is different.

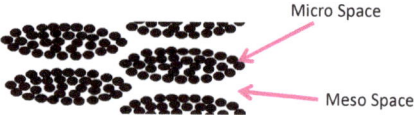

Figure 6. Micro and meso spaces between and inside tows.

Table 5. Standard deviation and average for Samples 2 and 3.

Fraction	Sample 2		Sample 3	
	Average	Standard Deviation	Average	Standard Deviation
P1	35.0	1.0	35.0	1.0
P2	9.3	9.3	23.7	22.5
P3	54.0	9.6	40.0	23.5
Residue	0.7	0.6	1.3	0.6

Differential scanning calorimetry (DSC) was used to characterize the degree of cure of the epoxy carbon composites. Figure 7 shows the exothermic heat curves of the epoxy system for Sample 1 at a heating rate of 10 °C/min. The area under the curve, which is calculated from the extrapolated baseline at the end of the reaction, was used to assess the total heat of the cure, giving a value of HT = 135 J/g. In general, the curing reaction of a thermosetting resin can spread over a wide temperature range. However, considering the sensitivity and response limitation to heat changes, the isothermal DSC analysis is usually run in a moderate curing range of temperature [12]. According to the dynamic curing DSC curves displayed in Figure 6, the drop on the weight is a Ti = 125 °C and it reaches a minimum at Tpeak = 145 °C. This means that the temperature for optimum curing is in the range 125–145 °C.

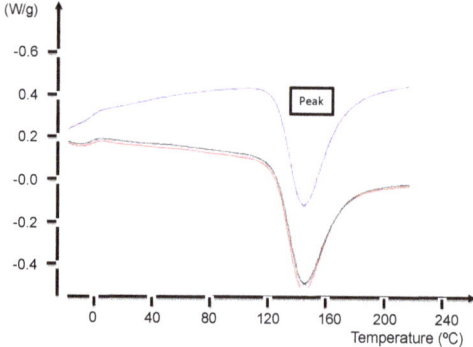

Figure 7. DSC measurements for Sample 1: exothermic heat curves at a rate of 10 °C/min.

The isothermal degree of cure was obtained from isothermal DSC curves (Figure 8). As expected, when the curing temperature increased, the time required to reach the minimum value became shorter.

The isothermal heat of cure, ΔHi, was derived from the total area under the curve. The residual heat of reaction (ΔHR) is obtained by a cooling and heating process for each T_{iso} value. The values of ΔHi and ΔHR are shown in Table 6 at the T_{iso} temperature.

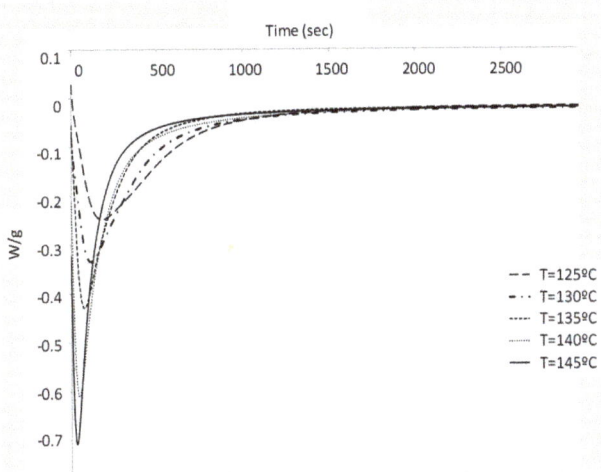

Figure 8. Isothermal DSC curves for composite samples.

Table 6. ΔH$_i$ and ΔH$_R$ for five isothermal experiments.

T_{iso} (°C)	ΔH$_i$ (J/g)	ΔH$_R$ (J/g)
125	132.70	2.87
130	134.50	3.12
135	118.26	3.50
140	135.99	3.62
145	123.60	6.12

Then, the degree of cure at T_{iso} can be calculated through two equivalent expressions:

$$\alpha_{Hi} = \frac{\Delta H_i}{\Delta H_T} \quad (9)$$

$$\alpha_{Hi} = \frac{\Delta H_i}{\Delta H_T}. \quad (10)$$

Table 7 shows the results obtained for the degree of cure, α, for each. As can be seen, there is no clear trend on the behavior of reaction heats, which suggests that a non-uniform process is occurring. This circumstance has also been observed in the TGA experiments, so it is interpreted that the information of both results is complementary and can help the interpretation of physical phenomena to be modeled with the cure equations.

Table 7. Values of DOC for Sample 2.

T_{iso} (°C)	α_{HR}	α_{Hi}
125	0.98	0.98
130	0.98	0.99
135	0.97	0.88
140	0.97	1.00
145	0.95	0.92

3.2. Raman Spectroscopy

Raman scattering measurements of the three samples (1, 2, and 3) were performed under the same conditions: two accumulations of 120 s of exposure time and a neutral 0.6 density filter. The use of an attenuator has been revealed to be necessary to avoid the radiation damage on the composite samples. The Raman spectrum shown for each sample is obtained by calculating the average of the three points of one fiber after checking that several fibers gave similar results. In order to obtain the optimal conditions to perform the measurements, different combinations of exposure time and attenuators have been studied. Once the tests are finished, no significant changes are observed in the surface of the samples measured by Raman spectroscopy.

Figure 9 shows the Raman spectra for the three composite fiber samples. The Raman spectra have been represented vertically to clarify their visualization. All samples have a peak located around 1352, 1585 and 1620 cm^{-1}. The peak located at 1585 cm^{-1} is a consequence of ordered or graphitic carbon (known as G band), while the peaks located at 1352 cm^{-1} (D band) and 1620 cm^{-1} (D' band) are assigned to disordered carbon atoms, which is usually explained by the double-resonance Raman mechanism in carbon [12,13]. With the present spectral resolution of our Raman spectrometer (below 3 cm^{-1}), the D' band appears as a shoulder of the G band and cannot be separated. Therefore, only the values of the D and G bands were used in our calculations.

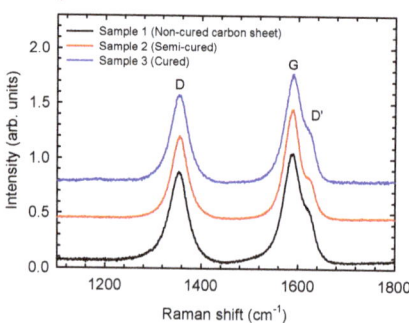

Figure 9. Raman spectra of three analyzed composite samples as a function of their degree of cure. The Raman spectra have been vertically shifted to improve the comparison among them.

The ratio between the intensities of both D and G bands gives us an idea of the crystalline order of the sample. The crystalline size, L_a, is derived from the Knight formula [14,15],

$$L_a(\text{nm}) = (2.4 \times 10^{-10})\, \lambda^4 \left(\frac{I_D}{I_G}\right)^{-1} \tag{11}$$

where λ is the laser line wavelength in nanometer units, and I_D and I_G are the intensities of the D and G bands, respectively [16,17]. The values for the crystalline order and crystallite sizes for the composite samples with different cure degrees are displayed in Table 8.

Table 8. Crystalline order and crystallite size derived from the Knight formula.

Sample	I_D (a.u.)	I_G (a.u.)	I_D/I_G	L_a (nm)
1 (Semi-cured)	0.73	0.98	0.74	25.8
2 (Cured)	0.79	0.98	0.81	23.9
3 (Carbon sheet)	0.81	0.97	0.84	23.0

The results of Raman spectroscopy indicate that the crystalline order increases in the cured samples, while the crystallite size decreases with the cure. These results are consistent with the results obtained with DTG studies. As can be seen, the degree of cure is directly related to an increase in temperature peaks, as shown in DTG studies, which results in a rise of the crystalline order.

4. Conclusions

Thermal analysis and Raman spectroscopy technologies provide relevant information for the modeling of the kinetic behaviour of the resins in composite materials. The control of the degree of curing of composite parts is fundamental for the optimization of the mechanical properties. The degree of polymerization of carbon–epoxy composites was investigated through several techniques such as FESEM, TGA, DSC, and Raman spectroscopy. FESEM micrographs reveal that the surface of the carbon fibers is better defined as the cure time becomes longer, due to the homogeneous redistribution of covering resins. TGA and DSC experiments confirm that the thermal characteristics of cured samples depend on the applied cure process. Raman spectroscopy offers an assessment of the crystalline order and crystallite sizes through the ratio between the intensities of D and G bands, which corresponds to the disordered and ordered phases of graphitic carbon, respectively. The results show that both Raman spectroscopy and thermal analysis are useful and complementary techniques for evaluating the degree of cure in composite materials. The technologies studied in this article are easily extrapolated to other materials or different applications, such as biomaterial for tissue engineering or biocomposites. Nowadays, many biotissues are manufactured with thermoset resins that need proper curing to achieve maximum mechanical properties. The problems associated with correct curing such as shrinkage or surface defects can be easily improved with these techniques.

Author Contributions: Conceptualization, J.A.G.-M. and L.G.; methodology, A.R.-G., L.M. and R.T.; validation, all authors; formal analysis, B.M., J.A.S. and J.M.-G.; writing—Review and editing, all authors.

Funding: This research was funded by Spanish Ministerio de Ciencia, Innovación y Universidades through the grants ENE2016-77798-C4-2-R, FIS2017-83295-P and RED2018-102612-T and CENTRO DE EXCELENCIA EN NANOFIBRAS LEITAT CHILE, 13CEI2-21839. J.A.S. also acknowledges Ramón y Cajal Fellowship for financial support (RYC-2015-17482).

Conflicts of Interest: The authors declare no conflict of interest. The funders had no role in the design of the study; in the collection, analyses, or interpretation of data; in the writing of the manuscript, or in the decision to publish the results.

References

1. Flury, S.; Hayoz, S.; Peutzfeldt, A.; Hüsler, J.; Lussi, A. Depth of cure of resin composites: Is the ISO 4049 method suitable for bulk fill materials. *Dent. Mater.* **2012**, *28*, 521–528. [CrossRef] [PubMed]
2. Pillai, K.M. Modeling the Unsaturated Flow in Liquid Composite Molding Processes: A Review and Some Thoughts. *J. Compos. Mater.* **2004**, *38*, 2097–2118. [CrossRef]
3. Park, C.H.; Lee, W. Modeling void formation and unsaturated flow in liquid composite molding processes: A survey and review. *J. Reinf. Plast. Compos.* **2011**, *30*, 957–977. [CrossRef]
4. Gascón, L.; García, J.A.; Lebel, F.; Ruiz, E.; Francois, T. Numerical prediction of saturation in dual scale fibrous reinforcements during Liquid Composite Molding. *Compos. Part A Appl. Sci. Manuf.* **2015**, *77*, 275–284. [CrossRef]
5. Dong, A.; Zhao, Y.; Zhao, X.; Yu, Q. Cure Cycle Optimization of Rapidly Cured Out-Of-Autoclave Composs. *Materials* **2018**, *11*, 421. [CrossRef] [PubMed]
6. Cender, T.; Gangloff, J.; Simacek, P.; Advani, S.G. Void reduction during out-of-autoclave thermoset prepreg composite processing. In Proceedings of the Conference: SAMPE Seattle, Seattle, WA, USA, 2–5 June 2014.
7. Sequeira-Tavares, S.; Caillet-Bois, N.; Michaud, V.; Månson, J.-A. Non-autoclave processing of honeycomb sandwich structures: Skin through thickness air permeability during cure. *Compos. Part A Appl. Sci. Manuf.* **2010**, *41*, 646–652. [CrossRef]
8. Simacek, P.; Neacsu, V.; Advani, S.G. A Phenomenological Model for Fiber Tow Saturation of Dual Scale Fabrics in Liquid Composite Molding. *Polym. Compos.* **2010**, *31*, 1881–1889. [CrossRef]
9. Yun, M.; Sas, H.; Simacek, P.; Advani, S.G. Characterization of 3D fabric permeability with skew terms. *Compos. Part A Appl. Sci. Manuf.* **2017**, *97*, 51–59. [CrossRef]

10. Grunenfelder, L.K.; Nutt, S.R. Void formation in composite prepregs—effect of dissolved moisture. *Compos. Sci. Technol.* **2010**, *70*, 2304–2309. [CrossRef]
11. Régnier, N.; Fontaine, S. Determination of the Thermal Degradation Kinetic Parameters of Carbon Fibre Reinforced Epoxy Using TG. *J. Therm. Anal. Calorim.* **2001**, *64*, 789. [CrossRef]
12. Katagiri, G.; Ishida, H.; Ishitani, A. Raman spectra of graphite edge planes. *Carbon* **1988**, *26*, 565–571. [CrossRef]
13. Hardisa, R.; Jessopb, J.L.P.; Petersa, F.E.; Kesslerc, M.R. Cure kinetics characterization and monitoring of an epoxy resin using DSC, Raman spectroscopy, and DEA. *Compos. Part A Appl. Sci. Manuf.* **2013**, *49*, 100–108. [CrossRef]
14. Farquharson, S.; Smith, W.W.; Rigas, E.J.; Granville, D. Characterization of polymer composites during autoclave manufacturing by Fourier transform Raman spectroscop. *Opt. Methods Ind. Process.* **2001**, *4201*, 103–111. [CrossRef]
15. Cançado, L.G.; Takai, K.; Enoki, T.; Endo, M.; Kim, Y.A.; Mizusaki, H.; Jorio, A.; Coelho, L.N.; Magalhães-Paniago, R.; Piment, M.A. General equation for the determination of the crystallite size La of nanographite by Raman spectroscopy. *Appl. Phys. Lett.* **2006**, *88*, 163106. [CrossRef]
16. Ivanova, B.; Spiteller, M. A quantitative solid-state Raman spectroscopic method for control of fungicides. *Analyst* **2012**, *137*, 3355–3364. [CrossRef] [PubMed]
17. Chen, K.; Wu, T.; Wei, H.; Wu, X.; Li, Y. High spectral specificity of local chemical components characterization with multichannel shift-excitation Raman spectroscopy. *Sci. Rep.* **2015**, *5*, 13952. [CrossRef] [PubMed]

© 2019 by the authors. Licensee MDPI, Basel, Switzerland. This article is an open access article distributed under the terms and conditions of the Creative Commons Attribution (CC BY) license (http://creativecommons.org/licenses/by/4.0/).

Article

Deformation-Assisted Joining of Sheets to Tubes by Annular Sheet Squeezing

Luis M. Alves, Rafael M. Afonso, Frederico L.R. Silva and Paulo A.F. Martins *

Instituto de Engenharia Mecânica, Instituto Superior Técnico, Universidade de Lisboa, Av. Rovisco Pais, 1049-001 Lisboa, Portugal; luisalves@tecnico.ulisboa.pt (L.M.A.); rafael.afonso@tecnico.ulisboa.pt (R.M.A.); frederico.rocha.e.silva@tecnico.ulisboa.pt (F.L.R.S.)
* Correspondence: pmartins@tecnico.ulisboa.pt; Tel.: +351-21-841-9006

Received: 28 October 2019; Accepted: 25 November 2019; Published: 26 November 2019

Abstract: This paper is built upon the deformation-assisted joining of sheets to tubes, away from the tube ends, by means of a new process developed by the authors. The process is based on mechanical joining by means of form-fit joints that are obtained by annular squeezing (compression) of the sheet surfaces adjacent to the tubes. The concept is different from the fixing of sheets to tubes by applying direct loading on the tubes, as is currently done in existing deformation-assisted joining solutions. The process is carried out at room temperature and its development is a contribution towards ecological and sustainable manufacturing practices due to savings in material and energy consumption and to easier end-of-life disassembly and recycling when compared to alternative processes based on fastening, riveting, welding and adhesive bonding. The paper is focused on the main process parameters and special emphasis is put on sheet thickness, squeezing depth, and cross-section recess length of the punches. The presentation is supported by experimentation and finite element modelling, and results show that appropriate process parameters should ensure a compromise between the geometry of the mechanical interlocking and the pull-out strength of the new sheet–tube connections.

Keywords: joining; forming; sheet–tube connections; experimentation; modelling and simulation

1. Introduction

In recent years, there has been a growing utilization of deformation-assisted joining processes driven by an increasing demand of assembling lightweight components. Deformation-assisted joining processes are classified into three distinct groups according to their operating principles:

(i) Based on mechanical, hydraulic, and magnetic loading, such as clinching [1,2], self-pierce riveting [3], sheet-bulk compression [4], hydraulic forming [5], and electro-magnetic forming [6];
(ii) Based on solid-state welding, such as friction stir welding [7], friction spot welding [8] and explosive welding [9];
(iii) Based on fusion welding combined with plastic deformation, such as resistance spot and projection welding [10] and weldbonding [11].

Some of the above-mentioned processes have been extensively investigated for joining sheets and tubes made from similar or dissimilar materials. The state-of-the-art reviews by Mori et al. [12] and Groche et al. [13] provide detailed information on the most significant developments and applications of deformation-assisted joining processes to connect sheets and tubes.

Despite the progresses in sheet-to-sheet and tube-to-tube connections, the joining of sheets-to-tubes is still preferentially accomplished by means of conventional fastening and welding technologies. Little progress has been made in the application of deformation-assisted joining to sheet–tube connections, aside from recent developments involving the utilization of form-fit mechanical joints based on compression beads formed by local buckling [14], on the combination of tube compression beads with flaring [15], and on the combination of partial sheet-bulk compression of tubes with upsetting [16] or flaring [17] (Figure 1a–c).

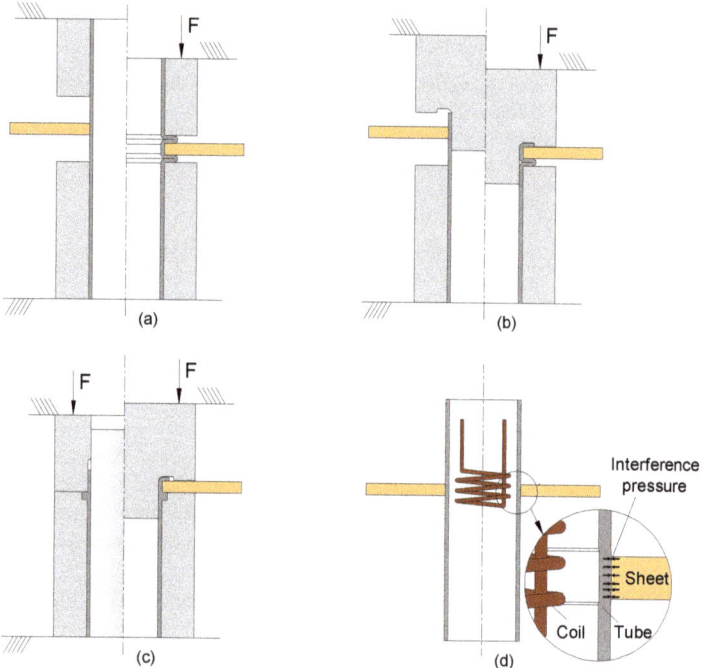

Figure 1. Deformation-assisted joining of sheets to tubes by mechanical joining. (**a**) Form-fit joints produced by tube compression beads formed by local buckling. (**b**) Form-fit joints produced by a combination of tube compression beads and flaring. (**c**) Form-fit joints produced by a combination of partial sheet-bulk compression of tubes and flaring. (**d**) Force-fit joints produced by the pressure that remains after elastic recovery from magnetic loading.

The utilization of force-fit mechanical joints that exclusively rely on the pressure that remains on the contact interfaces after elastic recovery from mechanical [18], hydraulic [5], or magnetic loading [6] is not a feasible solution for the fixing of sheets to tubes due to the limited contact surfaces provided by the small thickness sheets that are commonly used in the industry (Figure 1d). Similar constraints apply to new developments for the joining of thick blocks (or plates) to tubes by compressing the upper end of a connecting block flange [19,20] because of the need to produce the connecting flange by machining.

Regarding the connection of sheets to tubes, which is the main objective of this paper, it is worth mentioning that all the previously mentioned deformation-assisted joining processes based on mechanical form-fit joints [14–17] involve multiple operations. This undercuts their advantages in material consumption, energy requirements, and end-of-life recycling with disadvantages related to efficiency and costs that are better ensured by conventional technologies based on fastening, welding, and adhesive bonding. Besides the limitations resulting from existing processes based on mechanical form-fit joints being carried out in multiple operations, there is also the risk of failure by cracking in the compression beads produced by local buckling. This reduces their applicability to tubes with low fracture toughness [21].

All of the above-mentioned problems prompted the authors to develop a new process for fixing sheets to tubes at room temperature that is based on an entirely new mechanical joining concept in which loading is applied on the sheet surface instead of the tube itself [22]. The process is schematically shown in Figure 2, and consists of squeezing (compressing) the annular surface of the sheet adjacent to the tube in order to ensure the material from the sheet flows inwards and is shaped as a form-fit joint with good mechanical interlocking between the sheet and the tube.

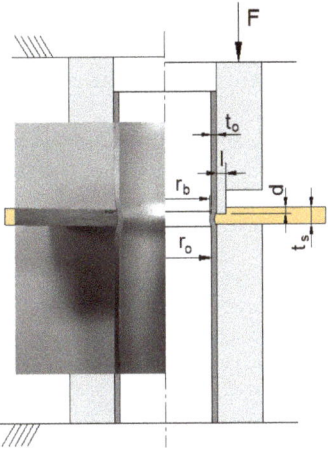

Figure 2. A representation of the deformation-assisted joining of sheets to tubes by annular sheet squeezing at the open and closed positions [22]. A photograph of a longitudinal cross section of a test specimen is enclosed.

The new joining process is performed in a single punch stroke and requires both the sheet and the tube to have some degree of ductility to plastically shape the form-fit joint. The optimum operating conditions require the material of the tube to have a lower elastic modulus than that of the sheet, so that the pressure remains on the contact interface after producing the form-fit joint as a result of the more pronounced elastic recovery of the tube in the direction of the sheet. Otherwise, the resulting form-fit joint may end up slightly loose as in the case of sheet–tube connections made from dissimilar materials (e.g., metals and polymers or composites) with very different elastic modulus. Another requirement of the new proposed joining process is the necessity of the sheet strength being similar or higher to that of the tube to allow for easy shaping of the inner tube bead that is needed to produce the form-fit mechanical joint.

Finally, it is worth mentioning that the new joining process circumvents the previously-mentioned difficulties resulting from the utilization of tubular materials with low fracture toughness.

This paper is focused on the main process parameters of deformation-assisted joining of sheets to tubes by annular sheet squeezing, hereafter designated as 'mechanical joining of sheets to tubes'. Special emphasis is put on the sheet thickness t_s, squeezing depth d, and cross-section recess length l of the punches due to their influence on the inner tube bead shape of the form-fit joint and on the quality and performance of the mechanical joint between the sheet and the tube. The presentation includes results from experimentation and finite element modelling, and from destructive pull-out tests that were carried out to determine the maximum force that the new joints can withstand before failing.

2. Materials and Methods

2.1. Materials and Flow Curves

The work on the mechanical joining of sheets to tubes by annular sheet squeezing was carried out in two different aluminum alloys. The sheets were made of aluminum AA5754-H111 with a 5 mm thickness and the tubes were made of aluminum AA6063-T6 with an outer radius of 16 mm and a 1.5 mm wall thickness.

The flow curves of the two materials were obtained by combining tensile and stack compression tests [23] performed in a hydraulic testing machine with a cross-head speed of 5 mm/min. The tensile tests allowed characterization of the stress response of the materials for small values of strain, whereas the stack compression tests were utilized to determine the stress response for larger values of strain, beyond plastic instability in tension, following a procedure similar to that of Silva et al. [4]. The flow curves of the AA5754-H111 sheets and AA6063-T6 tubes are shown in Figure 3.

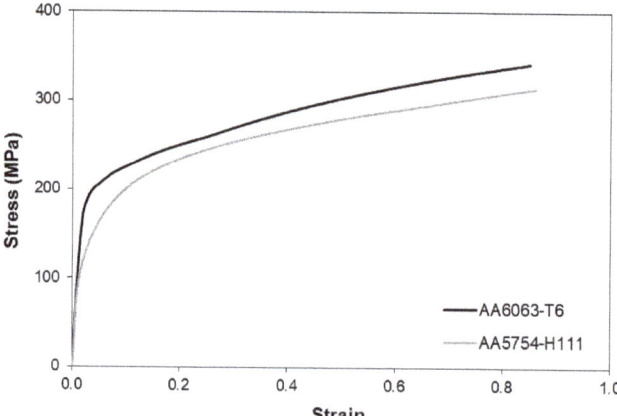

Figure 3. Flow curves of the two aluminum alloys utilized in the experiments.

The reason for using two materials with similar flow curves and nearly identical elasticity moduli (with differences within the range of 68 to 68.9 GPa) was to allow for the study of the performance of the form-fit joints alone and independently of the interfacial pressure, which develops on the sheet–tube contact surface in case of form-fit joints made from materials with a different elasticity modulus.

2.2. Experimental Tests

In their original paper on the mechanical joining of sheets to tubes by annular sheet squeezing [22], the authors put emphasis on the influence of the cross-section recess length l of the punch on plastic material flow inside the sheet thickness. The squeezing depth d was kept constant and a deformation-zone geometry parameter $\Delta = t_s/l$, defined as the ratio of the sheet thickness t_s to the cross-section recess length l, was introduced to characterize material flow and identify the deformation modes associated with acceptable and unacceptable joints.

In a subsequent paper [24], the authors focused on the complementary work plan by analyzing the influence of the squeezing depth d and keeping the cross-section recess length l of the punch at a fixed value. The investigation allowed the characterization of the physics behind material separation at the cross-section recess corner of the punch and to reach a better understanding of the influence of the squeezing depth d on the pull-out destructive strength of the joints.

The work plan giving support to this paper takes the combined influence of d and l into account. The goal is to understand how changes in both variables at the same time will influence material flow and the overall pull-out performance of the joints so that a procedure can be reached to determine the combination of squeezing depth d and cross-section recess length l that is capable of ensuring the best form-fit joint for a supplied set of geometries and materials.

Such a procedure has not been addressed in previous papers and is of paramount importance when producing form-fit joints in dissimilar materials in which the strength of the tube is similar or slightly higher than that of the sheet.

The overall methodology utilized in the experimental tests consisted on the following steps:

1. The tube and the sheet were first cut to the required sizes;
2. A hole with a diameter equal to the outer tube diameter was drilled in the sheet;
3. The tube and the sheet were then placed in the tooling system installed in a hydraulic testing machine INSTRON SATEC 1200 kN;
4. The hydraulic testing machine moved the punch downwards in order to locally compress the sheet up to a total squeezing depth
5. Steps 1 to 4 were repeated several times in order to vary the sheet thickness t_s, the cross-section recess length of the punch, and the squeezing depth d according to the values listed in Table 1. All the tests were performed at room temperature;
6. After finishing the tests, selected samples were cut along their axial cross-sectional planes in order to analyze their mechanical form-fit joints, to take photographs, and to perform measurements;
7. Finally, another set of selected samples was subjected to destructive pull-out tests in order to evaluate the overall performance of the joints.

Table 1. The experiments on the mechanical joining of sheets to tubes by annular sheet squeezing. Notation is in accordance with Figure 2.

r_0 (mm)	t_0 (mm)	t_s (mm)	l (mm)	d (mm)
14.5	1.5	2.5, 5	0.5–2.5	1.0–4.0

Table 1 presents a summary of the experiments on mechanical joining together with the major geometrical specifications of the sheets and tubes that were utilized in the investigation. The influence of the sheet thickness t_s was taken into consideration by extending the experimental work to sheets with a smaller thickness ($t_s = 2.5$ mm) than the nominal supplied thickness $t_s = 5$ mm.

Figure 4 presents a schematic representation of the experimental setup that was utilized to evaluate the performance of the new form-fit joints. The joints were subjected to destructive pull-out tests in which the sheet was detached from the tube, and the objective was the determination of the maximum force F that the joints can withstand before failing.

2.3. Finite Element Modelling

The in-house finite element computer program I-form was utilized to analyze the mechanical joining of sheets to tubes by annular sheet squeezing. The computer program is based on the finite element flow formulation [25], and its implementation follows the extension of the rigid-plastic Markov's principle [26] of minimum plastic work to include incompressibility and contact between deformable bodies,

$$\Pi = \int_V \bar{\sigma}\dot{\bar{\varepsilon}}\, dV + \frac{1}{2}K \int_V \dot{\varepsilon}_v^2\, dV - \int_{S_T} T_i u_i\, dS + \int_{S_f} \left(\int_0^{|u_r|} \tau_f du_r \right) dS + \frac{1}{2}K_1 \sum_{c=1}^{N_c} (g_n^c)^2 + \frac{1}{2}K_2 \sum_{c=1}^{N_c} (g_t^c)^2 \quad (1)$$

In the first term of Functional (1), the symbols $\bar{\sigma}$ and $\dot{\bar{\varepsilon}}$ denote the effective stress and the effective strain rate, respectively,

$$\bar{\sigma} = \sqrt{\frac{3}{2}\sigma'_{ij}\sigma'_{ij}} \quad \dot{\bar{\varepsilon}} = \sqrt{\frac{2}{3}\dot{\varepsilon}_{ij}\dot{\varepsilon}_{ij}} \qquad (2)$$

where σ'_{ij} is the deviatoric stress tensor and $\dot{\varepsilon}_{ij}$ is the strain rate tensor.

In the second term, the symbol $\dot{\varepsilon}_v$ is the volumetric strain rate, given by

$$\dot{\varepsilon}_v = \delta_{ij}\dot{\varepsilon}_{ij} \qquad (3)$$

where δ_{ij} is the Kronecker delta and K is a large positive number utilized to impose incompressibility in volume V by means of a penalty factor.

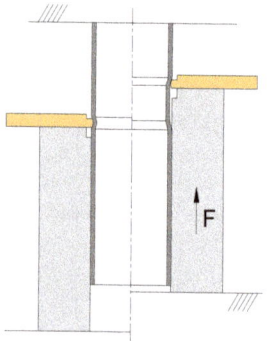

Figure 4. A representation of the experimental pull-out destructive setup before (left) and during testing (right).

The third term of Functional (1) makes use of the surface tractions T_i and velocities u_i on the surface S_T, whereas the fourth term takes care of the frictional effects along the contact interface S_f between the sheet and tube with the tools. In this work, tools were assumed as rigid bodies, and τ_f and u_r denote the friction shear stress and relative sliding velocity of the sheet and tube. The friction shear stress was modelled according to the law of constant friction,

$$\tau_f = m\,k \qquad (4)$$

where m is the friction factor between the sheets and tubes, taken as 0.1 after checking the predicted forces that best matched the experimental results. The symbol k denotes the flow shear stress.

The fifth and sixth terms account for the contact between the sheet and tube modelled as deformable bodies along their contact interfaces defined by means of N_c pairs extracted from the faces of the elements that were utilized in their discretization. The symbols g_n^c and g_t^c denote the normal and tangential gap velocities in the contact pairs, which are penalized by large numbers K_1 and K_2 to avoid penetration. A more detailed look into the numerical implementation of Functional (1) in the finite element computer program I-form can be found in reference [27].

Figure 5 shows in detail the finite element model before and after the sheet is mechanically joined to the tube. The finite element models utilized in the numerical modelling of the process made use of rotational symmetry conditions and required discretization of the longitudinal cross-section of the sheets and tubes by means of approximately 20,000 and 800 quadrilateral elements, respectively. The sheet and the tube were modelled as deformable bodies, whereas the tools were modelled as rigid objects and discretized by means of linear contact-friction elements. Several remeshings were carried out to avoid excessive element distortion during annular sheet squeezing.

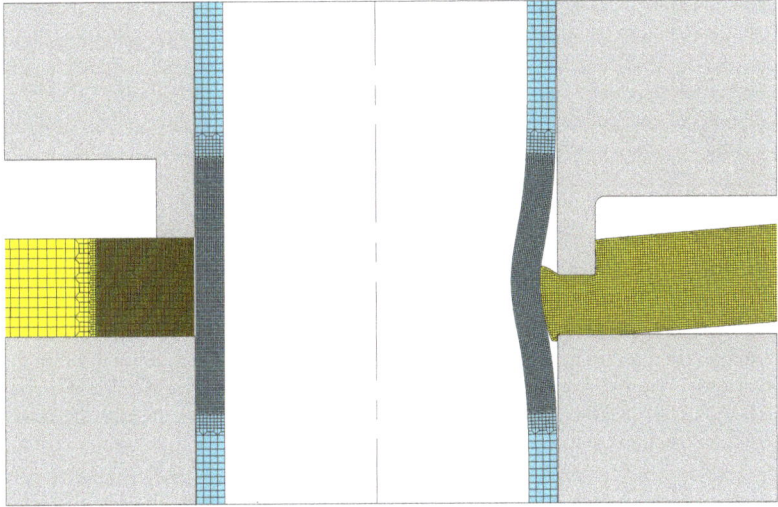

Figure 5. Element meshes before and after fixing the sheet to the tube using mechanical joining by annular sheet squeezing ($l = 2$ mm and $d = 2$ mm and $t_s = 5$ mm).

3. Results

Figure 6 shows the finite element computed reduction of the inner tube radius $R = (r_0 - r_b)/r_0$ as a function of the cross-section recess length l of the punch for four different values of the squeezing depth d. Experimental measurements of R for a squeezing depth $d = 2$ mm and a cross recess length $l = 2$ mm are included to assess the validity and reliability of the finite element estimates.

The first conclusion derived from Figure 6 is that small values of d lead to small amounts of material being displaced against the tube, and therefore, to the development of less pronounced inner tube bead geometries of the form-fit joints. In contrast, large values of d give rise to large amounts of material being squeezed against the tube and to more pronounced inner tube bead geometries of the form-fit joints. This is graphically shown in the two schemes that are included in Figure 6.

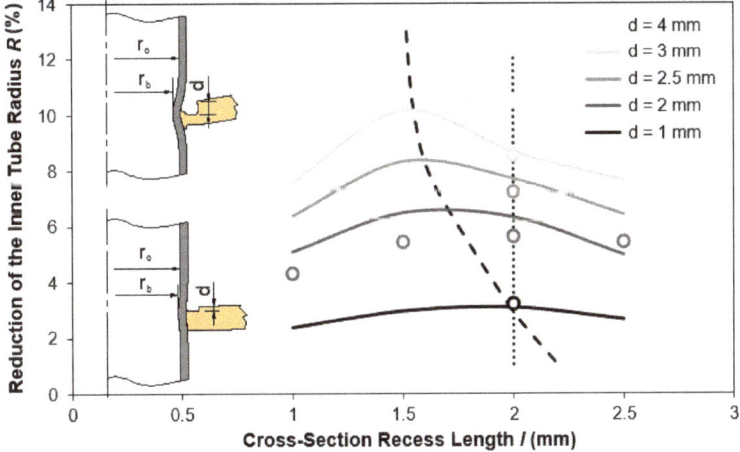

Figure 6. Reduction of the inner tube radius R as a function of the cross-section recess length l of the punch for different values of the squeezing depth d.

As seen in Figure 6, a typical $R(l)$ evolution passes through a peak corresponding to the cross-section recess length l that provides the maximum reduction R_{max} of the inner tube radius for a given squeezing depth d. The dashed curve passing through all these peaks defines the correlation between l and d that maximizes the inner tube bead geometry of the form-fit joints.

To the left of the dashed curve, there is a decrease in the amount of sheet material being squeezed as the cross-section recess length l diminishes. This leads to smaller values of R and to the development of less pronounced inner tube bead geometries of the form-fit joints. At the limit, there will be no form-fit joint.

Contrary to what one would expect, to the right of the dashed curve, the $R(l)$ evolutions should increase were it not for the squeezed sheet material starting to moving outwards instead of inwards.

This last conclusion is confirmed by the computed evolution of the normalized radial velocity v_r/v_p, where v_p is the vertical punch velocity, shown in Figure 7b. As seen in case of $d = 2$ mm, when the cross-section recess length of the punch increases from $l = 2$ mm to $l = 2.5$ mm, there is a shift of the neutral point (NP), corresponding to the transition between inward and outward material flow towards the left corner of the punch, meaning that more squeezed sheet material will start flowing outwards (refer to the black arrows included in Figure 7b).

Another consequence of the squeezed sheet material starting to flow outwards is the occurrence of bending. Bending gives rise to form-fit joints with a lack of perpendicularity between the sheet and the tube, as shown in the bottom experimental and numerically predicted cross-sections that are included in Figure 7a.

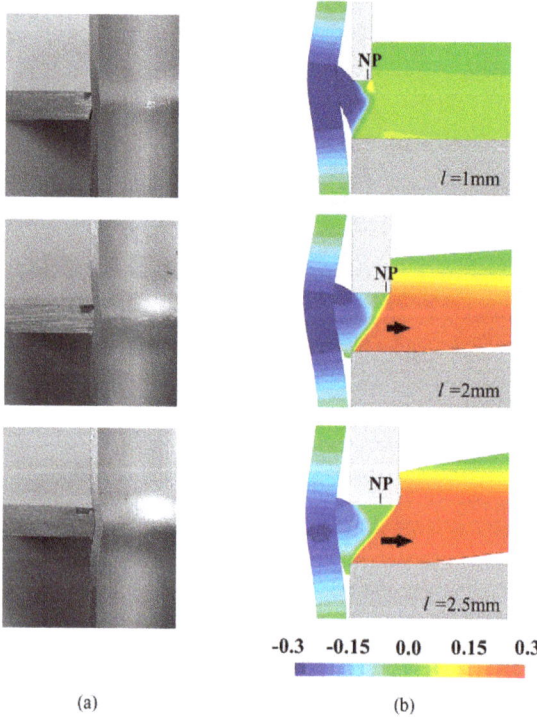

Figure 7. Influence of the cross-section recess length l of the punch on the form-fit joints for a squeezing depth $d = 2$ mm. (**a**) Experimental cross-sections of the form-fit joints; (**b**) Finite element predicted cross-sections of the form-fit joints with the distribution of normalized radial velocity v_r/v_p.

4. Discussion

4.1. Pull-Out Destructive Forces

The results obtained in Section 3 showed that both the cross-section recess length l and the squeezing depth d of the punch play a key role in the geometry of the inner tube bead of the form-fit joints.

One question that naturally arises from Figure 6 is whether the correlation between l and d that maximizes the inner tube bead geometry of the form-fit joints is capable of ensuring the maximum pull-out forces that the joints can safely withstand before failing.

To answer this question, the authors took the form-fit joint with $l = 2$ mm and $d = 2$ mm lying close to the correlation line that maximizes the inner tube bead geometry and compared its pull-out destructive force with those obtained for other joints that were obtained by varying d along the dotted vertical line of Figure 6.

The results from experimental tests and numerical simulations are shown in Figure 8 and allow to conclude that the maximum pull-out destructive force F is not obtained for the process operating conditions that maximize the inner tube bead geometry of the form-fit joints.

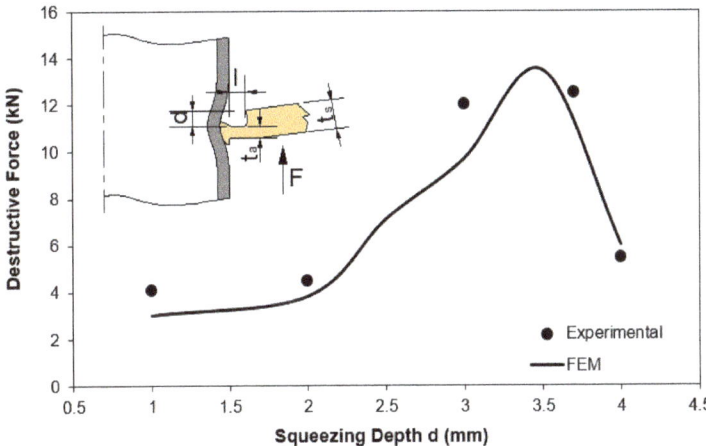

Figure 8. Influence of the squeezing depth d on the pull-out destructive force F of sheet–tube connections obtained with a punch having a cross-section recess length $l = 2$ mm.

Whilst at a first glance, this result may seem easy to explain because large values of d should give rise to larger inner tube beads of the form-fit joints, there is an additional phenomenon that also needs to be considered. Otherwise, it is not possible to fully understand the experimental and numerical results included in Figure 8, because the increase of the pull-out force F with d is only monotonic up to a peak, after which the force drops sharply.

The additional phenomenon that needs to be considered for understanding the evolution of the pull-out force F with d is the failure mechanism. In particular, the change in mechanism when the active sheet thickness t_a left below the cross-section recess length of the punch (refer to the inset included in Figure 8) becomes very small.

Taking, for example, the pull-out force vs. displacement evolutions for the form-fit joints with $d = 3$ mm and $d = 4$ mm retrieved from Figure 8, it is easy to observe two totally different separation mechanisms (Figure 9a). One mechanism, leading to higher forces, is similar to tube extrusion while the other mechanism, leading to lower forces and a sharp drop at the end, is related to shearing.

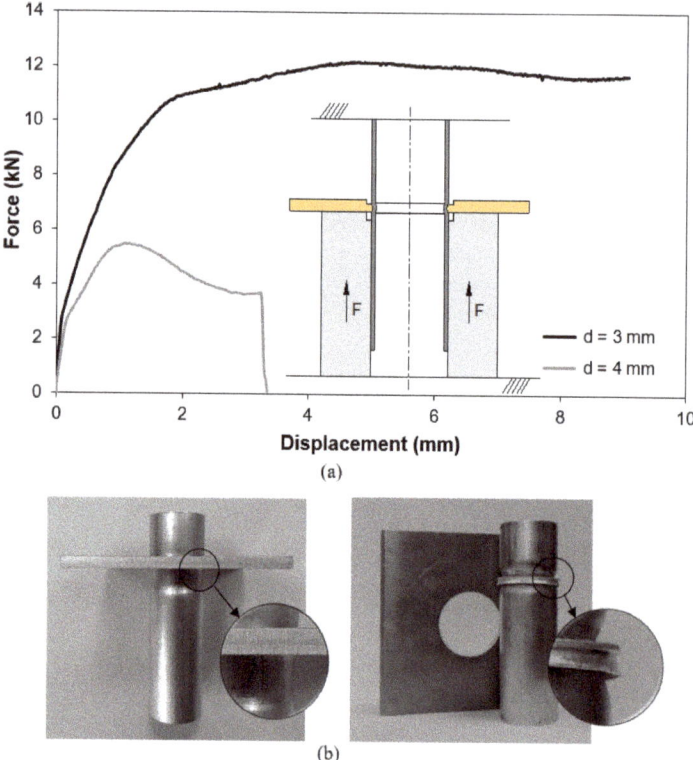

Figure 9. Pull-out tests. (**a**) Experimental evolution of the force with displacement for destructive pull-out tests performed in two different form-fit joints that were produced with a punch having a cross-section recess length $l = 2$ mm. (**b**) Photographs of the two different tests specimens after failure by extrusion (left) and shearing (right).

In case of the first mechanism, observed in the pull-out test of the form-fit joint with $d = 3$ mm, the sheet acts as a floating die and the tube is forced to plastically deform in order to reduce its inner radius from r_0 to r_b. The maximum pull-out force F attained in the destructive test is approximately equal to 12 kN.

The second mechanism, observed in the pull-out test of the form-fit joint with $d = 4$ mm, is typical of shearing along the small active sheet thickness t_a that was left below the cross-section recess length of the punch. The maximum pull-out force F associated to this mechanism is smaller than that of tube extrusion.

The photographs included in Figure 9b illustrate the differences between the two above-mentioned failure mechanisms. The conclusion to be taken from these tests is that the design of a joint to be produced by a punch having a cross-section length $l = 2$ mm must consider a squeezing depth $d = 3 - 3.5$ mm to ensure high pull-out destructive forces.

In connection to this, it is worth mentioning that the concept of maximum pull-out force utilized by the authors refers to the force that prevents collapse by avoiding detachment of the two components and not to the force that is needed to produce the first relative movement between the sheet and the tube.

4.2. Sheet Thickness

The other question one may have about the mechanical joining of sheets to tubes by annular sheet squeezing is whether the process works for smaller sheet thicknesses than that utilized in the previous sections. To answer this question authors decided to produce sheet–tube joints using the same tube geometry but reducing the sheet thickness from $t_s = 5$ mm to $t_s = 2.5$ mm. The photograph included in Figure 10a shows that the process is still feasible for sheets with smaller thicknesses.

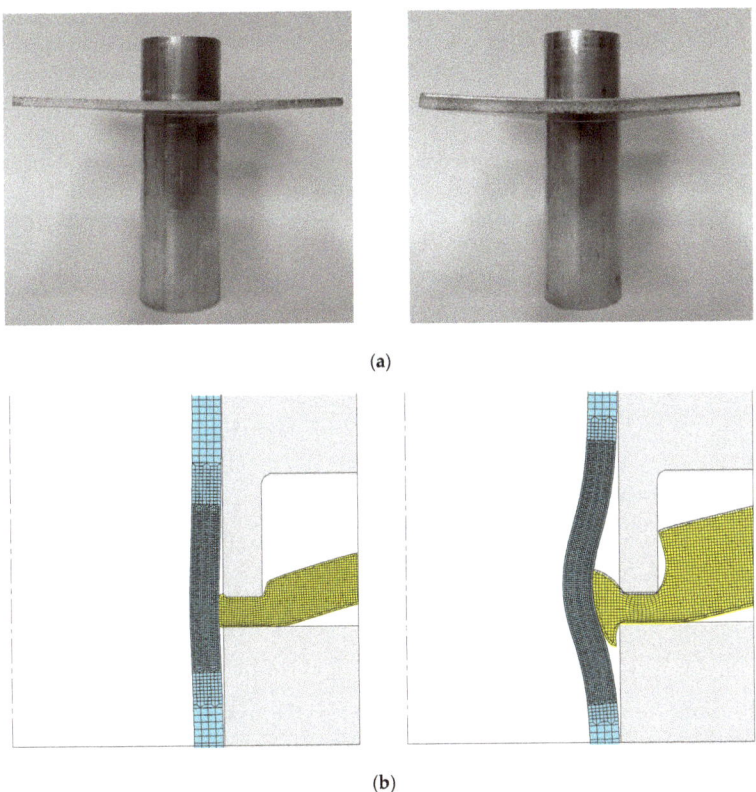

Figure 10. Mechanical joining of sheets to tubes by annular sheet squeezing using a punch with a cross section recess length $l = 2$ mm and two different thicknesses $t_s = 2.5$ mm and $t_s = 5$ mm. (a) Photographs of sheets and tubes after being joined. (b) Finite element predicted cross section at the end of stroke for the two test cases shown in (a) using a squeezing depth $d = 1$ mm (left) and $d = 3.5$ mm (right).

In view of the above, it is interesting for the readers to know which values of the cross-section recess length l and squeezing depth d of the punch were utilized to produce the form-fit joint with a sheet thickness $t_s = 2.5$ mm. For this purpose, it is important to observe in Figure 8 that the maximum pull-out force of a form-fit joint produced with a punch having a cross-section recess length $l = 2$ mm is obtained for a squeezing depth $d = 3.5$ mm. Larger values of d, leading to active sheet thicknesses $t_a < 1.5$ mm, provide smaller pull-out forces because the failure mechanism changes from extrusion to shearing.

So, in order to use the same punch ($l = 2$ mm) for comparison purposes, and prevent failure by shearing in case $t_a < 1.5$ mm, it was decided to use a squeezing depth $d = 1$ mm for obtaining the form-fit joint with a sheet thickness $t_s = 2.5$ mm.

Figure 10b shows a detail of the computed finite element cross sections for the form-fit joints obtained for the two different sheet thicknesses. As seen, the magnitude of bending is significant in both cases, namely in the sheet with smaller thickness due to its lower stiffness. However, the phenomenon can be easily avoided by diminishing the gap between the cross-section recess length and the remaining flat surface of the punch so that the total amount of bending is limited.

5. Conclusions

Deformation-assisted joining of sheets to tubes by annular sheet squeezing is based on mechanical joining by means of form-fit joints. The amount of sheet material to be squeezed and the final shape and volume of the inner tube beads of the form-fit joints is controlled by the cross-section recess length l of the punch, the squeezing depth d and by the geometry of the punch and sheet, namely the sheet thickness t_s.

By varying these parameters, it is possible to change the plastic flow of the squeezed sheet material from a predominantly inward into a combination of inward and outward. In particular, it is possible to define a combination of parameters capable of ensuring maximum shapes and volumes of the form-fit joints for given values of d or l. However, the maximum shapes and volumes of the form-fit joints do not necessarily provide the maximum pull-out destructive forces because there is a minimum active sheet thickness t_a below which the failure mechanism changes from extrusion to shearing and the forces drop significantly.

In connection to this, it is also shown that the new mechanical joining process is not limited to thick sheets because what needs to be fulfilled is a combination of parameters capable of ensuring a sound mechanical joint for values of the active sheet thickness t_a above the critical threshold of failure by shearing.

Author Contributions: L.M.A. and R.M.A. designed and fabricated the tools. L.M.A., R.M.A. and F.L.R.S. performed the experimental and numerical simulation work. L.M.A and P.A.F.M. developed the finite element computer program and analyzed the results. P.A.F.M. supervised the overall research work and wrote the article with the collaboration of all the other authors.

Funding: This research was funded by Fundação para a Ciência e a Tecnologia of Portugal under LAETA - UID/EMS/50022/2019 and PDTC/EMS-TEC/0626/2014.

Acknowledgments: The authors would like to acknowledge the support provided by IDMEC and Fundação para a Ciência e a Tecnologia of Portugal.

Conflicts of Interest: The authors declare no conflict of interest. The funders had no role in the design of the study; in the collection, analyses, or interpretation of data; in the writing of the manuscript, or in the decision to publish the results.

References

1. Chen, C.; Zhao, S.; Han, X.; Zhao, X.; Ishida, T. Experimental investigation on the joining of aluminum alloy sheets using improved clinching process. *Materials* **2017**, *10*, 887. [CrossRef] [PubMed]
2. Chen, C.; Zhao, S.; Han, X.; Wang, Y.; Zhao, X. Investigation of flat clinching process combined with material forming technology for aluminum alloy. *Materials* **2017**, *10*, 1433. [CrossRef] [PubMed]
3. Jäckel, M.; Grimm, T.; Niegsch, R.; Drossel, W.G. Overview of current challenges in self-pierce riveting of lightweight materials. *Proceedings* **2018**, *3*, 384. [CrossRef]
4. Silva, D.F.M.; Silva, C.M.A.; Bragança, I.M.F.; Nielsen, C.V.; Alves, L.M.; Martins, P.A.F. On the performance of thin-walled crash boxes joined by forming. *Materials* **2018**, *11*, 1118. [CrossRef] [PubMed]
5. Marre, M.; Rautenberg, J.; Tekkaya, A.E.; Zabel, A.; Biermann, D.; Wojciechowski, J.; Przybylski, W. An experimental study on the groove design for joints produced by hydraulic expansion considering axial or torque load. *Mater. Manuf. Process.* **2012**, *27*, 545–555. [CrossRef]
6. Psyk, V.; Risch, D.; Kinsey, B.L.; Tekkaya, A.E.; Kleiner, M. Electromagnetic forming-A review. *J. Mater. Process. Technol.* **2011**, *211*, 787–829. [CrossRef]

7. Serio, L.M.; Palumbo, D.; Filippis, L.A.C.D.; Galietti, U.; Ludovico, A.D. Effect of friction stir process parameters on the mechanical and thermal behavior of 5754-H111 aluminum plates. *Materials* **2016**, *9*, 122. [CrossRef]
8. Zhang, B.; Chen, X.; Pan, K.; Wang, J. Multi-objective optimization of friction stir spot-welded parameters on aluminum alloy sheets based on automotive joint loads. *Metals* **2019**, *9*, 520. [CrossRef]
9. Mahmood, Y.; Dai, K.; Chen, P.; Zhou, Q.; Bhatti, A.A.; Arab, A. Experimental and numerical study on microstructure and mechanical properties of Ti-6Al-4V/Al-1060 Explosive Welding. *Metals* **2019**, *9*, 1189. [CrossRef]
10. Cortés, V.H.V.; Guerrero, G.A.; Granados, I.M.; Hernández, V.H.B.; Zepeda, C.M. Effect of retained austenite and non-metallic inclusions on the mechanical properties of resistance spot welding nuggets of low-alloy TRIP steels. *Metals* **2019**, *9*, 1064. [CrossRef]
11. Lim, Y.C.; Park, H.; Jang, J.; McMurray, J.W.; Lokitz, B.S.; Keum, J.K.; Wu, Z.; Feng, Z. Dissimilar materials joining of carbon fiber polymer to dual phase 980 by friction bit joining, adhesive bonding, and weldbonding. *Metals* **2018**, *8*, 865. [CrossRef]
12. Mori, K.; Bay, N.; Fratini, L.; Micari, F.; Tekkaya, A.E. Joining by plastic deformation. *CIRP Ann. Manuf. Technol.* **2013**, *62*, 673–694. [CrossRef]
13. Groche, P.; Wohletz, S.; Brenneis, M.; Pabst, C.; Resch, F. Joining by forming-A review on joint mechanisms, applications and future trends. *J. Mater. Process. Technol.* **2014**, *214*, 1972–1994. [CrossRef]
14. Alves, L.M.; Dias, E.J.; Martins, P.A.F. Joining sheet panels to thin-walled tubular profiles by tube end forming. *J. Clean. Prod.* **2011**, *19*, 712–719. [CrossRef]
15. Alves, L.M.; Martins, P.A.F. Single-stroke mechanical joining of sheet panels to tubular profiles. *J. Manuf. Process.* **2013**, *15*, 151–157. [CrossRef]
16. Alves, L.M.; Afonso, R.M.; Silva, C.M.A.; Martins, P.A.F. Joining tubes to sheets by boss forming and upsetting. *J. Mater. Process. Technol.* **2018**, *252*, 773–781. [CrossRef]
17. Alves, L.M.; Afonso, R.M.; Silva, C.M.A.; Martins, P.A.F. Joining sandwich composite panels to tubes. *J. Mater. Des. Appl.* **2019**, *233*, 1472–1481. [CrossRef]
18. Hao, J.W.; Pan, H.L.; Tang, S.H. Research of elastomeric expansion for tube-to-tubesheet joint. In Proceedings of the ASME Pressure Vessels and Piping Division Conference, Vancouver, BC, Canada, 23–27 July 2006; pp. 1–5.
19. Wrobel, N.; Rejek, M.; Krolczyk, G.; Hloch, S. *Testing of Tight Crimped Joint Made on a Prototype Stand*; Advances in Manufacturing, Lecture Notes in Mechanical Engineering; Springer: Cham, Switzerland, 2017; pp. 497–507. ISBN 978-3-319-68619-6.
20. Rejek, M.; Wróbel, N.; Królczyk, J.; Królczyk, G. Designing and testing cold-formed rounded connections made on a prototype station. *Materials* **2019**, *12*, 1061. [CrossRef]
21. Sizova, I.; Sviridov, A.; Bambach, M. Avoiding crack nucleation and propagation during upset bulging of tubes. *Int. J. Mater. Form.* **2017**, *10*, 443–451. [CrossRef]
22. Alves, L.M.; Silva, F.L.R.; Afonso, R.M.; Martins, P.A.F. Joining sheets to tubes by annular sheet squeezing. *Int. J. Mach. Tools Manuf.* **2019**, *143*, 16–22. [CrossRef]
23. Alves, L.M.; Nielsen, C.V.; Martins, P.A.F. Revisiting the fundamentals and capabilities of the stack compression test. *Exp. Mech.* **2011**, *51*, 1565–1572. [CrossRef]
24. Alves, L.M.; Silva, F.L.R.; Afonso, R.M.; Martins, P.A.F. A new joining by forming process for fixing sheets to tubes. *Int. J. Adv. Manuf. Technol.* **2019**, *104*, 3199–3207. [CrossRef]
25. Tekkaya, A.E.; Martins, P.A.F. Accuracy, reliability and validity of finite element analysis in metal forming: A user's perspective. *Eng. Comput.* **2009**, *26*, 1026–1055. [CrossRef]
26. Markov, A.A. On variational principles in theory of plasticity. *Prikl. Mat. Mek.* **1947**, *II*, 339–350.
27. Nielsen, C.V.; Zhang, W.; Alves, L.M.; Bay, N.; Martins, P.A.F. *Modelling of Thermo-Electro-Mechanical Manufacturing Processes with Applications in Metal Forming and Resistance Welding*; Springer: London, UK, 2013; ISBN 978-1-4471-4643-8.

© 2019 by the authors. Licensee MDPI, Basel, Switzerland. This article is an open access article distributed under the terms and conditions of the Creative Commons Attribution (CC BY) license (http://creativecommons.org/licenses/by/4.0/).

Article

On The Influence of Rotary Dresser Geometry on Wear Evolution and Grinding Process

Leire Godino [1],*, Jorge Alvarez [2], Arkaitz Muñoz [1] and Iñigo Pombo [1]

[1] Faculty of Engineering Bilbao, University of the Basque Country (UPV/EHU), Plaza Ingeniero Torres Quevedo, 1, 48013 Bilbao, Bizkaia, Spain; arkaitzmu97@gmail.com (A.M.); inigo.pombo@ehu.eus (I.P.)
[2] Ideko S. Coop., Arriaga Kalea, 2, 20870 Elgoibar, Gipuzkoa, Spain; jalvarez@ideko.es
* Correspondence: leire.godino@ehu.eus; Tel.: +34-946-014822

Received: 22 October 2019; Accepted: 20 November 2019; Published: 22 November 2019

Abstract: Dressing is a critical issue for optimizing the grinding process. Dresser tool and dresser parameters must be designed according to the grinding wheel material, shape, or even the dimensional and geometrical tolerances of the workpiece and its surface roughness. Likewise, one of the problematic issues of dressers is the wear that they suffer. In order to tackle this issue, the present work characterized the wear of two rotary dressers by analysing the wear behaviour depending on the pit radius of the dressers while studying the influence of the wear on ground surfaces. This work showed that the rotary dresser with a higher pit radius presents wear that is approximately 28% higher than the dresser with a half pit radius.

Keywords: OD grinding; CBN; dressing; rotary dresser; wear; CVD diamond

1. Introduction

The grinding process is a very important machining process characterized by the high added value of the ground parts. Its capacity for manufacturing advanced materials of poor machinability while achieving high-quality surfaces is made possible due to the combination of grinding wheel design, new advances in abrasives and an optimal dressing process. In recent years, the use of super-abrasive grinding wheels, primarily CBN (cubic boron nitride), has become increasingly extensive in the industry due to their thermal stability and the absence of chemical affinity with ferrous materials. Moreover, CBN grinding wheels are suitable when high removal rates are required and are used in creep feed grinding and high-speed grinding applications. Due to the different bond used with CBN grinding wheels, vitrified grinding wheels present more advantages than the other resin bond, metallic bond or electroplated grinding wheels. These advantages are primarily related to the ability to transport the coolant, the efficiency of chip removal and the capacity for dressing using rotatory dressing tools. Moreover, CBN grinding wheels are not the only widely used wheel in industrial applications, with conventional grinding wheels also being used due to their versatility and low cost in comparison with super-abrasives. The latest advances make them more competitive with regard to abrasive grain shape and crystalline structure.

One of the most important aspects of the grinding process is dressing, which is used in order to regenerate the abrasive capacity of the grinding wheel and the initial shape of the abrasive tool. Among the various types of dressers, these can be classified as static and rotary. The present work was focused on the rotary dressing tool, which consists of a cylindrical body with a single layer of inserted or deposited diamond particles. Different designs of rotary dressers can be found depending on the shape and size of the diamonds and their location, as well as their bonding chemical compositions. The main advantage of using this tool is the high dressing speed that can be achieved in comparison with a stationary single or multiple diamond tool, along with the possibility of generating a complex profile on the grinding wheel surface [1]. In addition, the wear level suffered by the dresser is relatively low [2]. Taking this fact into account, these types of dressers are used for dressing CBN and diamond grinding wheels with vitreous bond and conventional profile grinding wheels. The main advantages of rotary dressing tools are the high dressing speed and the possibility of generating complex profiles in the grinding wheel [1]. Moreover, this type of dressing tool provides the process with a high level of repeatability and precision, reducing processing costs and the number of rejected parts.

However, the complexity of the process is high, particularly when compared with the one carried out using a stationary diamond tool. In this regard, specific works focused on the influence of particular dressing parameters on the characteristics of the wheel surface can be found in the specialized literature. In [3], the first attempts were made to analyse the influence of dressing parameters on the wheel performance when using roller diamond tools. The Schmitt diagram is currently used in the industry to design dressing processes. In [2], a more in-depth analysis was conducted on the state of the art regarding grinding wheel conditioning, including the rotary dressing process. This work shows that a considerable amount of research has focused on the influence of the dressing parameters on grinding wheel performance, including the influence of the radial feed of the dressing process on the roughness of the wheel surface, theoretical pathways of dressing grains, influence of dressing speed ratios on the radial dressing force and effective grinding wheel roughness or the influence of the depth of dressing cut on specific normal grinding force. In addition, some studies on diamond tool wear were included, particularly [4] and the studies of Linke and Klocke [5,6], which focused on analysing diamond tool wear for stationary dressers.

Recent works regarding the analysis of the rotary dressing process are worth mentioning. For instance, in [7], the authors analysed the performance of various diamond tools under the same dressing conditions for small grinding wheels for internal grinding. In [8], a new type of rotary diamond tool was presented. In this case, the geometry and density of abrasive grits was completely monitored. These authors evaluated the performance of the new dresser against conventional dressers in terms of force and wear of the abrasives. The results were positive and the forces for the new dresser were found to be approximately 50% lower than that of the classic dressers, while the wear was also improved, although no quantitative values were reported. In [9], an analysis was presented regarding the performance of a roller dresser for the case of the micro-grinding of a titanium alloy. The authors focused their study on analysing the influence of the overlap ratio (U_d) on the ground surface quality. The main conclusion reached by the authors was that extremely high values of overlap ratios were suitable for achieving high surface quality in micro-grinding processes. Finally, Palmer et al. [10] analysed the characteristics of the dressed wheel surface when using roller dressers.

Although one of the most relevant issues in dressing is the wear of the dresser, very little information can be found in the specialized literature regarding the wear of rotary dressers. This fact is very important when using this kind of dressing tool for two main reasons. The first is related to the characteristics of the dressed surface. In the case of rotary dressing tools, excessive wear of the dressing tool implies changes in the tool geometry and hence the tool sharpness parameter. This could imply changes in a_d, or even in U_d, and in the characteristics of the dressed surface. In Figure 1, the influence of a worn dressing tool on a plane grinding wheel is shown, with a smaller a_d being achieved for the worn rotary dresser.

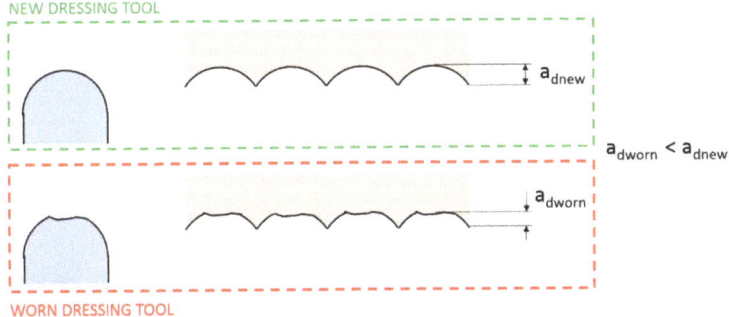

Figure 1. Influence of the worn rotary dressing tool on the wheel surface.

The second reason is related to the geometry of the dressed wheel for non-plane profiles in the wheel. Depending on the geometry of the dressing tool, a dressing path is programmed in the machine. Once the rotary diamond tool has lost its shape due to wear after several dressing passes, either the dressing path or the rotary tool must be changed. If the wear suffered by the rotary diamond tool is not controlled, several ground parts could be rejected. Figure 2 shows the effect of grinding a profile wheel with a worn rotary dressing tool. In the left part of the image, the theoretical wheel profile is shown in green, and also points to the programmed path of the rotary dresser. In contrast, in the right part of the image, the path of the worn rotary dressing tool is shown. In this case, the real wheel profile is going to be larger than theoretical profile. Thus, when the part is ground, of the removed material is higher than programmed ones. This effect leads to rejection of the part.

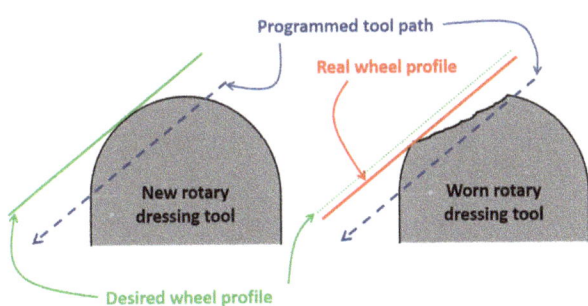

Figure 2. Influence of the worn rotary dressing tool on the wheel profile.

In order to tackle these problems, the present work constitutes a preliminary approach toward characterizing the wear of rotary dressers. From research point of view, there is not a unique parameter to define the wear of rotary dressers. In contrast, the volumetric wear of grinding wheels is defined by G-ratio and the wear of stationary dressers is quantified using dressing wear ratio, G_d, applied by Shi and Akemon [11] for stationary blade diamond tools. Therefore, the aim of the present work was to define a wear parameter to quantify the wear suffered by the rotary dressing tool. A new parameter, termed "wear parameter, W_d", was presented. This parameter allowed the characterization of rotary dresser wear in order to be comparable to the stationary blade dressers. To this end, one of the objectives was to develop a systematic methodology for analysing and characterizing the wear suffered by a rotary dressing tool. The proposed methodology included the development of specific software (in Python) to measure the rotary diamond tool wear, the proposal of a parameter to measure such wear and an analysis of the grinding wheel behaviour, paying particular attention to the consumed power and surface roughness. This methodology was used to analyse the wear suffered by different geometry rotary dressing tools when dressing plane profile CBN grinding wheels with a vitreous bond.

For this purpose, the employed experimental setup and methodology was first presented. Second, the results are analysed and the W_d is defined. Finally, the main conclusions drawn from this work is presented.

2. Materials and Methods

This study examined the wear of rotary dressing tools and its influence on ground workpieces. To this end, dressing and grinding tests were combined on a cylindrical grinding machine (DanobatGroup, Elgoibar, Spain). The study consisted of the analysis of two rotary dressers, varying the diamond pattern and the dresser geometry, i.e., the tip radius. RIG 52035 and RIG 52034 were manufactured by TYROLIT. For simplicity, the two rotary dressers are referred to as RIG 35 and RIG 34, respectively. Table 1 lists the main characteristics of the rotary dressers. In both cases, CVD (cultivated diamond) diamonds were inserted. In the case of RIG35, the interlayer was positioned. In contrast, in RIG34 dresser, the CVD diamonds were aligned. Regarding to the geometry of the rotary dressers, RIG35 presented a pit radius of 0.5 mm, whereas the pit radius of RIG34 was 0.25 mm. As previously mentioned, rotary dresser geometry is essential when profile grinding wheels are dressed. Any variation in the rotary dresser geometry is copied on the wheel surface.

In the present work, a CBN grinding wheel was used to conduct dressing and grinding tests. In order to distinguish between the influence of the wear on the ground workpiece surface and the grinding wheel shape, a straight grinding wheel was used. Thus, the influence of dresser wear on the wheel shape was not taken into account in this analysis, and only ground surface quality was analysed. The nomenclature of the used grinding wheel was CBN170N100V (UNESA S.L, Hernani, Spain), which was a medium hard wheel, presenting a vitreous bond and high density of abrasive grains. The external diameter of the wheel was Ø450 mm, the width was 10.3 mm, and the grain size was approximately Ø170 μm, corresponding to a finishing grinding wheel. Furthermore, plunge grinding tests were carried out on a cylindrical workpiece of hardened steel (AISI 52100). The medium hardness of the workpiece was approximately 54 HRC, while the external diameter of the parts was Ø80 mm.

Table 1. Characteristics of rotary dressing tools.

	RIG 35	RIG 34	Common Characteristics	
Diamond pattern			Diamond type	CVD
			External diameter [mm]	120
Pit radius [mm]	0.5	0.25	Internal diameter [mm]	40
Width [mm]	4.26	3.88	Internal angle	35°

The present study involved dressing and grinding, the analysis of the wear suffered by the rotary dresser, and the influence of wear on the grinding wheel surface and hence on ground workpiece surfaces. Accumulative dressing tests and plunge grinding tests were conducted using the same cylindrical grinding machine, *DANOBAT FG600S* (© DanobatGroup, Elgoibar, Spain), as shown in Figure 3. Moreover, CBN grinding wheel was used at a cutting speed of 50 m/s, and a water-based coolant with a concentration of 3.2% was used at a pressure of 13 bar. Furthermore, to conduct a complete analysis of the process during both dressing and grinding, real power consumption was measured using a *Load Control UPC* (© Load Controls Incorporated, Sturbridge, MA, USA) power meter, and a *USB-6008* data acquisition card from *National Instruments*. Additionally, in order to quantify the influence of dresser geometry on wear evolution and the effect of wear on the subsequent grinding process, a new methodology was developed, which is described in the following section. The complete approach was validated through experimental grinding tests under industrial grinding conditions, as detailed below.

Figure 3. Dressing and grinding test set up.

Table 2 lists the dressing and grinding parameters. Moreover, the two dresser tools presented a similar dressing overlap ratio in order to compare the influence of the pit radius in the wear of the rotary dresser and in the ground surface. The overlap ratio is the relation between the effective width of the dressing tool and the feed per wheel revolution ($U_d = b_d/f_d$). A high value of dressing overlap ratio leads to a smooth grinding wheel surface, but the grinding forces and the specific energy are high. In contrast, a low value of dressing overlap ratio generates a sharper surface, with more cutting edges, thereby decreasing grinding forces. The range of values for the dressing process is U_d = 2–20 [12]. For the present study, a smooth wheel surface was required. Thus, the dressing overlap ratio was U_d = 2.64 for RIG 35 and U_d = 2.017 for RIG 34. Additionally, the smoothness of the grinding wheel surface depends not only on the dressing overlap ratio, but also on the dressing sharpness ratio, which is defined as the relation between the dressing depth and the effective width of the dresser ($\gamma_d = a_d/b_d$). This parameter represents the influence of the dresser shape on the wheel surface. In the present study, in both cases rotary dressers were studied, varying the pit radius. Thus, γ_d = 0.05 for RIG 35 and γ_d = 0.07 for RIG 34.

Table 2. Dressing and grinding parameters.

Dressing Parameters		Grinding Parameters	
Cutting speed v_s [m/s]	50	Cutting speed v_s [m/s]	50
Dresser speed n_d [rpm]	3453	Feed rate v_f [mm/min]	1.194
Dressing depth a_d [µm]	10	Depth of cut a_e [µm/rev]	8
Speed ratio between dresser and wheel	0.434	Speed ratio between wheel and workpiece	80
Dressing feed rate v_{fd} [mm/min]	100	Removed material in diameter [µm]	300

Accumulative dressing tests were carried out for each dressing tool, removing a total of 49,415 mm^3 of wheel volume. The first step was to measure the new surface of the rotary dresser. The topography of new rotary dresser was characterized using a Confocal microscope Leica DCM3D® (Leica microsystems AG, Wetzlar, Germany). Once the starting surface was characterized, the dressing test was carried out. Moreover, in Table 2, fine-dressing parameters are built. The CBN grinding wheel was continuously dressed, removing 12,354 mm^3 of abrasive material. Immediately after the dressing process, the plunge grinding test was carried out. The specific rate of material removal during the grinding process was Q'_w = 5 mm^3/mm s. After both the dressing and grinding tests, the worn surface of the rotary dressing tool and the ground surface were analysed. First, the dresser topography was measured again using a confocal microscope. The measurement after each dressing tests allowed for analysing the evolution of the wear during a complete test. Second, in order to observe the effect of dressing on the ground workpiece, the roughness of the ground surface is measured using a portable surface roughness tester (Taylor Hobson, Leicester, United Kingdom).

This step was completed a total of four times, following the same methodology, with 12,354 mm^3 of abrasive material being removed at each step. Table 3 details the range of workpiece material removed at each step. The complete test was finished after dressing a total of 49,415 mm^3. Similarly, during both dressing and grinding tests, the power consumption was measured in order to analyse the influence of dressing with a worn rotary dressing tool on the efficiency of the process. The last step involved the analysis of the data obtained during the tests. Power was readily analysed, while the topography data had to be processed and analysed to characterize the wear, which was a complex process. Thus, in the present work, a methodology for quantifying dresser wear was proposed, which is detailed hereafter.

Table 3. Range of material removed from the workpiece for each step of a given test.

Complete Test			
Step 1	Step 2	Step 3	Step 4
0-12354 mm^3	12,354–24,707 mm^3	24,707–37,061 mm^3	37,061–49,415 mm^3

A Methodology for Quantifying Wear in Rotary Dressers

As a first step, it was necessary to accurately establish the geometry of the brand-new rotary dresser. To this end, specific tooling was designed for the dresser to take measurements on a Leica DCM3D® confocal microscope (Leica microsystems AG, Wetzlar, Germany), as shown in Figure 4. The tooling system, together with the first measurement of the dresser surface, were necessary to set the references required to quantify the wear parameters.

Figure 4. Initial characterization of the geometry of the rotary dresser on the confocal microscope.

Using this equipment, dresser topographies were obtained. Each state of wear was measured in four different zones along the profile of the disk, separated by 90 degrees. This is shown in the first image of Figure 5. The complete measured area was 2.546 × 8.477 mm^2 and 2.808 mm in height, with a height resolution of 12 µm. The blue light was used in order to avoid dresser surface brightness. Three-dimensional (3D) profiles were then extracted, as seen in the second image. Profile comparison was carried out by slicing the 3D geometry and obtaining 2D curves. Five curves were obtained for each 3D profile. For this purpose, the topography layer of the LeicaMap® (Leica microsystems AG, Wetzlar, Germany) was used, as shown in the third and fourth images of Figure 5. At this stage, it is important to note that the reference must be set at the diamond and not at the bonding, since the latter will suffer more pronounced wear. Therefore, the intensity layer (shown in the third image of Figure 5) of the data was used as a reference because, on this layer, the infiltrated diamonds can be clearly observed and slices were made to coincide with CVD diamonds. Thus, only the wear of the diamond was taken into account, avoiding the influence of the bond. Moreover, this layer used the same scale as that used by the topography layer from which the 2D profiles were extracted. It is important to note that the rotary dresser profiles could not be obtained using a stylus profilometer due to the shape of the dresser surface and the abrasive surface. Therefore, a confocal microscope is the best option to analyse this kind of surface.

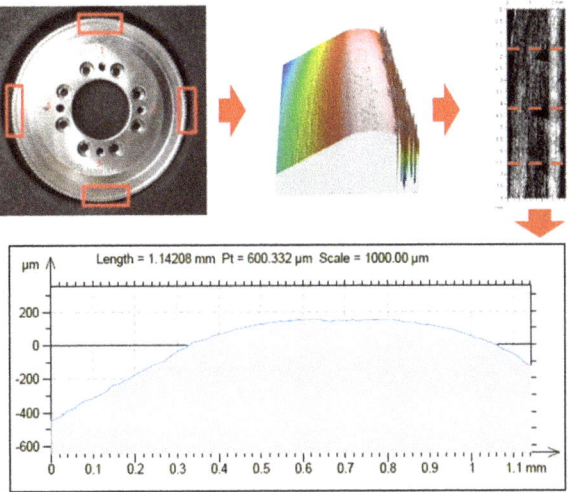

Figure 5. Rotary dresser measurement to characterize the wear.

Once the 2D profiles are available, it is possible to compare different states of wear of the dresser. Various geometrical parameters can be selected for comparison. In this work, two auxiliary parameters (namely worn area and contact length) and one main wear indicator (wear height, h_d) were defined. To obtain these parameters, profiles at different wear stages must be overlapped while maintaining stable references. In order to do so, a Python app was developed.

The first utility developed to assist profile comparison was used for profile smoothing. Any possible measurement defect, or even the presence of noise on the signal, was filtered using a low band pass filter. After some experimentation, the order of the filter was set at 6, the sample frequency at 15 Hz, and the cut-off frequency at 1.5 Hz. The second step involved overlapping the profiles at different wear stages. Errors were removed using a best-fitting technique on the profiles. In order to apply the best-fitting technique, reference points were set. These points did not suffer wear during dressing because they were not in contact with the grinding wheel. In Figure 6a, it is shown that points 1 and 2 are the reference points of corresponding new and worn profiles. Moreover, in Figure 6b, the two overlapping profiles show the wear suffered by the rotary dresser tool.

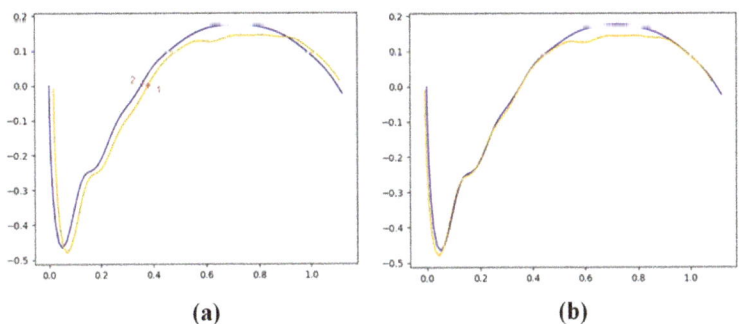

Figure 6. (a) Incorrect overlapping of profiles 1 and 2; (b) solution after applying best-fitting technique.

The contact area was defined to correspond to the limits set by the points where deviations were more important in value. Subsequently, the worn area and the contact zone can be effectively quantified, as shown in Figure 7a. The worn area can later be used to estimate the total volume of dresser worn during the operation. Further, maximum and average values of the wear height parameter h_d can be obtained at this stage, as shown in Figure 7b.

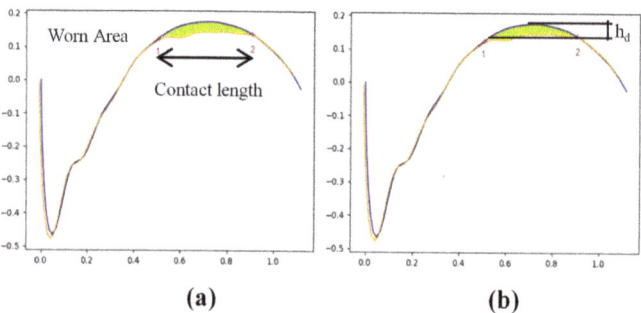

Figure 7. (a) Contact area and zone of the dresser where wear concentrates; (b) definition of wear height h_d.

3. Results and Discussion

In this section, the wear of the rotary dressers with CVD infiltrated diamonds was analysed. First, the influence of dresser geometry on the wear was studied, after which the influence of dresser wear on the surface quality of the workpiece after dressing was assessed. The majority of works that have analysed dresser wear used the wear volume (V_d), the wear of removed grinding wheel (V_w), and thus, the dressing wear ratio (G_d), in order to quantify the wear of the dressers, to compare different wear states, and to determine dresser life. Almost all the studies have been conducted for stationary dressers, namely single point or blade diamond tools [11]. In contrast, in the present work, dressers of differing geometry were analysed, and these parameters were not appropriate for making the comparison. Therefore, wear height was used due to the dissimilarities in pit radius and hence the wear volume of both the studied rotary dressers, as Figure 8a shows. For a given h_d, V_d can be approximately half the value for RIG 34 in comparison with RIG 35. Moreover, the height is the parameter that affects dressing precision and, as a consequence, the quality of the grinding process. If the rotary dresser decreases in diameter due to the height wear, the wheel does not achieve the designed dimensions and shape, obtaining larger dimensions. The error is translated to the workpiece during grinding, as removing more than the corresponding material would cause the workpiece to be rejected. Thus, it was necessary to analyse the evolution of h_d during the process.

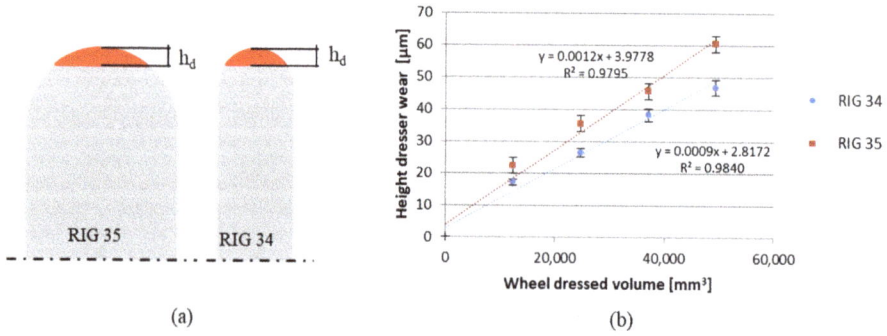

Figure 8. (a) Dresser comparison and (b) wear height evolution during a complete test.

Figure 8b represents the evolution of an accumulative dresser wear with the workpiece removed from the material. A linear increase in wear was shown with V_w for both rotary dressers. RIG35 presented a slightly higher slope than RIG34, achieving higher wear at the end of the tests. When 49,415 mm³ of grinding wheel was removed, the rotary dresser RIG 35 presented a level of wear that was 28% higher than RIG 34. Thus, RIG 35, which had a greater pit radius, showed a higher tendency towardswear than RIG 34. In this regard, it is necessary to highlight that if the comparison had been made through V_d, the wear difference between RIG 34 and RIG 35 would have been 145%, and also higher for RIG35. This result indicates that the volumetric parameters are valid for comparing dressers with the same geometry when the compared volume is equivalent. Thus, the parameter for determining the wear of two rotary dressers that differ in geometry—in this case, different pit radius or even diameter—must be the wear height.

Moreover, dressing volumetric ratio, i.e., the relation between the removed volume of the wheel and the worn volume of the rotary dressing tool ($G_d = V_w/V_d$), is also used to determine dressing process efficiency in the case of stationary dressers. However, as mentioned, this is not a valid parameter for comparing rotary dressers of differing geometry. In this regard, the present work measured efficiency through the relation between removed volume of the wheel and the height of the wear of the rotary dressing tool, defining the dressing wear parameter ($W_d = V_w/h_d$). High values of W_d imply that the rotary dresser suffers less wear when dressing a higher quantity of wheel volume.

Regarding the case studied here, in Figure 9, the rotary dresser with lower pit radius, RIG34, presents a higher value of W_d than RIG35. Thus, when using a rotary dresser with a half pit radius, the W_d was approximately 31% higher in the case under study. Thus, RIG 34 was more efficient than RIG 35, increasing dresser life and hence becoming a more economic process due to the high cost of CVD rotary dressers. With the wear values obtained, both h_d and W_d, it can be confirmed that the rotary dresser with smaller pit radius presented less wear in height, as well as higher dressing efficiency for the studied dressing and grinding parameters.

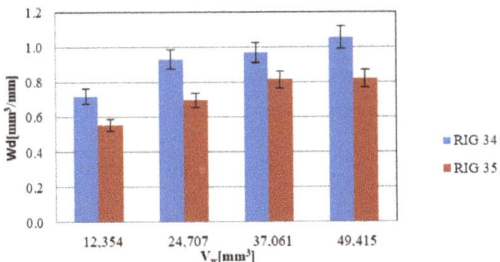

Figure 9. Dressing wear parameters.

Figure 10 plots the mean value of power consumption during grinding after each dressing step. To analyse the power, only the three last steps of the test, from 24,707 mm³ until the end of the test, were taken into account. Thus, the first contact, in which the dresser wear presented a transitory behaviour, was avoided. Thus, from 24,707 mm³ to the end of the test, power consumption during grinding decreased for both rotary dressers. However, when comparing the two rotary dressers, power consumption tended to decrease more markedly after dressing with RIG35 in comparison with RIG34. If the grinding wheel was dressed using the RIG35, the power consumption during grinding was approximately 18% higher than if the wheel was dressed using the RIG34 rotary dresser. This could be due to the influence of the loss of sharpness ratio on the wheel surface.

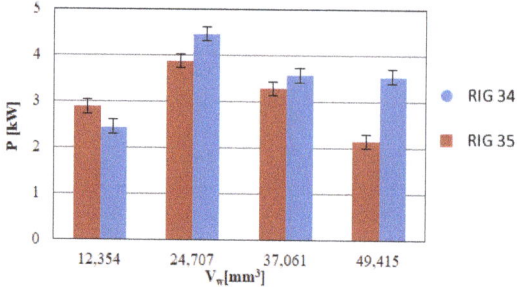

Figure 10. Power consumption during grinding.

The initial sharpness ratio was $\gamma_d = 0.05$ for RIG35 and $\gamma_d = 0.07$ for RIG34. Thus, RIG 34 was sharper, leading to a smoother wheel surface with more active (but shallower) cutting edges, as shown in the upper part of Figure 11a. When dressers were worn, the sharpness ratio decreased. The resulting wheel surface had fewer (and deeper) cutting edges than at the beginning of the tests. This helps to remove material during grinding, consuming less power. However, this is not the case for any dresser wear state, that is, it only occurs if the dresser is slightly worn. Analysing Figure 9a and taking into account the shape of rotary dressers, for the case studied here, RIG34 presented lower power consumption during grinding because the dressing process generated a greater quantity of (more shallow) cutting edges. Likewise, for RIG 35, in the last state it can be observed that the power consumption was similar. From this point, the power increased because the wear of the dresser was too high.

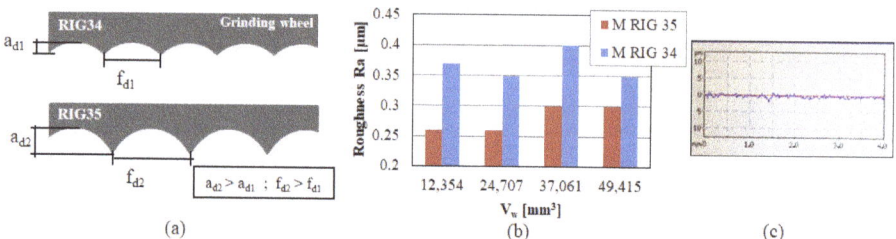

Figure 11. (a) Influence of sharpness ratio and the shape of the rotary dresser on the wheel surface, (b) roughness of workpiece after grinding and (c) the real Ra profile corresponding to 0.26 μm.

Finally, the influence of dressing on the quality of the workpiece surface was analysed. To this end, the roughness of ground workpieces was measured. In Figure 11c, the real profile generated by the rotary dresser tool RIG 35 is shown. During a complete test, higher Ra was achieved for RIG 34 than for RIG35. Thus, RIG35 led to smoother ground surfaces with the used parameters. If the evolution of the roughness during the test is analysed, different behaviour is shown in both cases, as displayed in Figure 11b. For RIG35, first, slightly lower values of Ra were measured (approximately 0.26 μm), and 0.3 μm were observed by the end of the test. In contrast, RIG34 did not present a tendency toward roughness, and the values varied from 0.35 to 0.4. In the last studied state, the difference in roughness between the two surfaces was lower than 16%. In any case, the values of Ra obtained were lower than 0.4 μm. Thus, a good surface quality was achieved despite the wear of rotary dressers.

4. Conclusions

After conducting a complete analysis of worn dressers and studying the influence of worn dressers on the ground workpiece surface, the obtained results were carefully analysed. From a discussion of the results, the following conclusions can be drawn:

1. It was demonstrated that dressing wear ratio, G_d, is not an adequate parameter to quantify the wear of rotary dressers. If wear volumes are compared, the results indicate that RIG35 presented wear approximately 145% higher than RIG34. In contrast, if heights, h_d, are compared, the difference is around 28%.
2. Dressing wear parameter ($W_d = V_w/h_d$) was defined in order to compare rotary dressers of different pit radius and diamond density.
3. With regard to the height of the wear of rotary dressers, h_d, a linear increase of was shown during a complete test. Moreover, the dresser with a higher pit radius presented a wear about 60 μm higher than that with a half pit radius.
4. Regarding to W_d, results demonstrated a more efficient dressing process for lower pit radius. Thus, in the present study, when using a rotary dresser with a half pit radius, the W_d was approximately 31% higher.
5. Regarding the grinding process, the rotary dresser with the higher pit radius presented relatively lower power consumption. Dressing with higher pit radius led to a wheel surface with fewer, albeit deeper, cutting edges. Thus, the material was removed easily from the workpiece, consuming less power.
6. Finally, grinding after dressing with both pit radius led to good surface quality, with an Ra lower than 0.4 μm.

These results suggest that for the range of dresser life studied here, the wear suffered was around 60 μm in height for RIG35 and 47 μm for RIG34, while the roughness of the ground workpiece did not undergo any significant changes. Moreover, it is necessary to bear in mind that this analysis was carried out for a plane grinding wheel, so the influence of the geometry lost in the wheel profile was not analysed. In this preliminary approach, only the plane grinding wheels were tested. Thus, it will be of interest to tackle the problem of non-plane grinding wheels in future works.

Author Contributions: Conceptualization, J.A.; Data curation, A.M.; Funding acquisition, I.P.; Investigation, I.P.; Methodology, L.G.; Software, A.M.; Supervision, J.A.; Writing—original draft, L.G.; Writing—review & editing, I.P.

Funding: The authors gratefully acknowledge the funding support received from the Spanish Ministry of Economy and Competitiveness and the FEDER operation program for funding the project "Scientific models and machine-tool advanced sensing techniques for efficient machining of precision components of Low Pressure Turbines" (DPI2017-82239-P).

Acknowledgments: The authors gratefully acknowledge the Basque Digital Innovation Hub (BDIH). Experiments have been carried out in facilities of the Digital Grinding Innovation Hub, part of the Basque Digital Innovation Hub (BDIH) initiative of the Basque Government.

Conflicts of Interest: The authors declare no conflict of interest. The founding sponsors had no role in the design of the study; in the collection, analyses, or interpretation of data; in the writing of the manuscript, and in the decision to publish the results.

References

1. Klocke, F.; Soo, S.; Karpuschewski, B.; Webster, J.A.; Novovic, D.; Elfizy, A.; Axinte, D.A.; Tönissen, S. Abrasive machining of advanced aerospace alloys and composites. *CIRP Ann.-Manuf. Technol.* **2015**, *64*, 581–604. [CrossRef]
2. Wegener, K.; Hoffmeister, H.W.; Karpuschewski, B.; Kuster, F.; Hahmann, W.C.; Rabiey, M. Conditioning and monitoring of grinding wheels. *CIRP Ann.-Manuf. Technol.* **2011**, *60*, 757–777. [CrossRef]
3. Schmitt, R. *Abrichten von Schleifscheiben mit diamantbestückten Rollen*; TU Braunschweig: Braunschweig, Germany, 1968; p. 93.
4. Brinksmeier, E.; Çinar, M. Characterization of Dressing Processes by Determination of the Collision Number of the Abrasive Grits. *CIRP Ann.-Manuf. Technol.* **1995**, *44*, 299–304. [CrossRef]
5. Linke, B. Dressing process model for vitrified bonded grinding wheels. *CIRP Ann.-Manuf. Technol.* **2008**, *57*, 345–348. [CrossRef]
6. Klocke, F.; Linke, B. Mechanisms in the generation of grinding wheel topography by dressing. *Prod. Eng.* **2008**, *2*, 157–163. [CrossRef]

7. Daneshi, A.; Jandaghi, N.; Tawakoli, T. Effect of dressing on internal cylindrical grinding. *Procedia CIRP.* **2014**, *14*, 37–41. [CrossRef]
8. Spampinato, A.; Axinte, D.; Butler-Smith, P.; Novovic, D. On the performance of a novel dressing tool with controlled geometry and density of abrasive grits. *CIRP Ann.-Manuf. Technol.* **2017**, *66*, 337–340. [CrossRef]
9. Kadivar, M.; Azarhoushang, B.; Shamray, S.; Krajnik, P. The effect of dressing parameters on micro-grinding of titanium alloy. *Precis. Eng.* **2018**, *51*, 176–185. [CrossRef]
10. Palmer, J.; Ghadbeigi, H.; Novovic, D.; Curtis, D. An experimental study of the effects of dressing parameters on the topography of grinding wheels during roller dressing. *J. Manuf. Process.* **2018**, *31*, 348–355. [CrossRef]
11. Shih, A.J.; Clark, W.I.; Akemon, J.L. Wear of the blade diamond tools in truing vitreous bond grinding wheels. *Wear.* **2001**, *250*, 593–603. [CrossRef]
12. Rowe, W.B. Chapter 4. Grinding wheel dressing. In *Principles of Modern Grinding Technology*; William Andrew: Burlington, MA, USA, 2009; pp. 59–78.

© 2019 by the authors. Licensee MDPI, Basel, Switzerland. This article is an open access article distributed under the terms and conditions of the Creative Commons Attribution (CC BY) license (http://creativecommons.org/licenses/by/4.0/).

Article

Repairing Hybrid Mg–Al–Mg Components Using Sustainable Cooling Systems

David Blanco [1,†], Eva María Rubio [1,*], Marta María Marín [1] and Joao Paulo Davim [2]

1. Department of Manufacturing Engineering, Industrial Engineering School, Universidad Nacional de Educación a Distancia (UNED), St/Juan del Rosal 12, E28040 Madrid, Spain; dblanco78@alumno.uned.es (D.B.); mmarin@ind.uned.es (M.M.M.)
2. Department of Mechanical Engineering, University of Aveiro, 3810-193 Aveiro, Portugal; pdavim@ua.pt
* Correspondence: erubio@ind.uned.es; Tel.: +34-913-988-226
† Programa de Doctorado en Tecnologías Industriales.

Received: 5 December 2019; Accepted: 13 January 2020; Published: 15 January 2020

Abstract: This paper focused on the maintenance or repair of holes made using hybrid Mg–Al–Mg components by drilling, using two sustainable cooling techniques (dry machining and cold compressed air) and taking surface roughness on the inside of the holes as the response variable. The novelty of the work is in proving that the repair operations of the multi-material components (magnesium–aluminum–magnesium) and the parts made of aluminum and magnesium (separately) but assembled to form a higher component can be done simultaneously, thus reducing the time and cost of the assembly and disassembly of this type of component. The study is based on a design of experiments (DOE) defined as a product of a full factorial 2^3 and a block of two factors (3 × 2). Based on our findings, we propose that the analyzed operations are feasible under sustainable conditions and, in particular, under dry machining. Also, the results depend on the machining order.

Keywords: hybrid components; light alloys; magnesium; aluminum; drilling; dry machining; cold compressed air; lubrication and cooling systems; arithmetical mean roughness; Ra; average maximum height; Rz; repair and maintenance operations

1. Introduction

Today, energy efficiency and sustainability play an increasingly important role in the development of new materials and their applications, especially in the transport industry, due to the pollution generated by the vehicles or their different parts in many stages of their life cycle.

Therefore, in order to reduce such contamination, it is necessary to approach the problem in a global way, analyzing all the factors that can have an influence during the manufacturing, the use, the maintenance, the repair, and the recycling of each part.

Although new forms of energy are being investigated to propel vehicles, these new alternative energies need to be greatly improved to match the level of development achieved with fossil fuels.

Until these new kinds of sustainable energy become competitive, in the transport sector, it is still essential to decrease the weight of vehicles to reduce the quantity of fossil fuels used, and, as a consequence, the pollution associated with its consumption. This is particularly true for the aeronautic and aerospace sectors.

To achieve this goal, research focuses on the use of the combination of different materials to create new multi-material components with better global properties than those of the original individual materials. These materials are called hybrid components, and the number of studies associated with them grew exponentially in recent years. Rubio and collaborators [1] affirmed that, among the possible combinations of materials used to form hybrid components with structural uses, the metal–polymer and metal–metal combinations are two of the most used, with aluminum alloys being most used to form these types of hybrid materials. Some of the main applications of these combination materials in the automotive sectors are car sleeper roofs, door structures, car fronts, tubular components of the exhaust system, and tubular components of the suspension system. In the aeronautic sector, the main applications are in the fuselage of airplanes and helicopters, the wings and rotors, and the other control surfaces (ailerons, flaps, spoilers, aerodynamic brakes, slats, and horizontal and vertical stabilizers).

The use of lightweight structural materials is widespread in various industries, particularly the aeronautics and automotive industries, since the weight of aircraft and cars is directly related to their consumption and pollution. A 10% reduction in the weight of an automobile can lead to improvements in fuel efficiency (by 8%), acceleration, and braking performance, and can reduce CO and hydrocarbon (HC) emissions by 4.5%, and NOx emissions by 8.8% [2]. In the civil aircraft industry, the weight of a Boeing 747–400 is approximately 183,500 kg, and the estimated fuel savings for an airplane is, over short distances, 117–134 kg of kerosene per kilogram of reduced weight and, over long distances, 172–212 kg of kerosene per kilogram of reduced weight [3].

Within the industries mentioned above, light alloys are used widely thanks to their excellent weight/mechanical strength ratio. Also, in recent years, combinations of light alloys (hybrid components) began to be used. Among them, it is possible to highlight the combination of aluminum and magnesium alloys. Both types of alloys can be combined with each other or with other lightweight and resistant materials to create hybrid materials, thereby extending the boundaries of the material's property space [4–8]. Also, magnesium and aluminum present other interesting advantages in relation to sustainability; for example, both are easy to recycle [9,10].

Regarding magnesium, several recycling options exist. Firstly, when the scrap is of sufficient quality, it can be reprocessed to obtain parts with good specifications. Secondly, magnesium scrap can be used to produce other metallic materials. Finally, magnesium can be recycled for use as a raw material in the production of fertilizers [11].

Aluminum has 95% recyclability at the end of its useful life. Also, recycling saves 95% of the energy used in its initial production. Moreover, according to the Environmental Product Declarations (EPD), it was shown that aluminum contains an average of 39% recycled aluminum and 61% primary aluminum. In addition to its recyclability, aluminum has an excellent carbon footprint. The EPD states that, for its different types (anodized and lacquered, with and without a thermal break), the values of CO_2 range between 10.3 and 11.8 kg of CO_2 per kg of aluminum, and recycling can achieve exemptions of between 3.0 and 4.0 kg of CO_2 per kg of aluminum.

The primary non-renewable energy used in the manufacture of aluminum products is another indicator of the impact they generate. As with the previous rate, considering all aluminum types included in the EPD, the primary energy is 420 MJ and 324 MJ per kg of aluminum. Therefore, aluminum recycling provides energy savings of between 49 and 38 MJ [9].

On the other hand, the geometric, dimensional shape and surface requirements are very strict in the aeronautics or aerospace sectors, which makes these parts types expensive and sometimes difficult or even impossible to keep stock of them ready for when it is necessary to maintain or repair damaged parts. Therefore, it is important to guarantee that it is possible to carry out efficient and sustainable repair or maintenance operations, thereby extending the lifetime of these parts and improving sustainability [12–14].

As the density of magnesium (1740 kg/m^3) is the lowest among structural metals (being two-thirds that of aluminum and one-quarter that of steel), its alloys are interesting candidates for combination with heavier alloys to reduce weight. However, magnesium has low machinability and high flammability (especially as powder or chips). In fact, the magnesium flame temperature and its alloys can reach 3100 °C and, once the fire starts, it is difficult to extinguish since there is continuous combustion of nitrogen, carbon dioxide, and water [15]. Also, molten magnesium reacts violently with water. For these reasons, it is necessary to study the behavior of magnesium when it is mechanized along with other materials (forming hybrid components), testing if such combinations are suitable for manufacturing, repair, and maintenance. For example, steel produces sparks during machining at cutting speeds between 200 to 300 m/min [16], and this can be extremely dangerous if magnesium is present.

Therefore, when magnesium-containing hybrid components are going to be mechanized, it is necessary to take certain security measures regarding the lubricants or coolant systems employed. Depending on the material or materials with which magnesium is combined, different lubrication/cooling techniques can be used [17,18] both individually (dry machining [19–31], minimum quantity lubrication (MQL) [30–42], solid lubrication [43–47], cryogenic cooling [48–55], gaseous cooling [56–61], nanofluids [62–66], and sustainable cutting fluids [67–72]) and in combination [73–76]. Some of these techniques were tested in several previous works with the intention of better describing the behavior of the individual materials (especially aluminum [19–23], titanium [30,31,39], and magnesium [16,28,77–104]). From the point of view of the optimization of the costs of the process and its sustainability, the ideal would be (1) to be able to completely eliminate lubricants or coolants and carry out dry machining, (2) to test more recently developed techniques (such as machining with minimum quantity of lubricant, cold compressed air or cryogenic refrigerants), and (3) to develop new lubricants or refrigerants compatible with magnesium.

The machining of this type of hybrid component results in an increase in process instability due to the different properties of the materials that form them and their particular cutting characteristics [12]. This makes it necessary to determine the best cutting parameters for each combination of materials, especially when strict design requirements must be reached. Although the literature contains an important number of experimental works with regard the machining of hybrid components, only some of them addressed hybrid components based on magnesium. Most studies focused on machining processes [105–113] and tried to find the optimal combination of machining conditions and lubrication/cooling systems by means of experimental tests, taking the surface roughness required in a particular industry sector or application as a response variable. Others dealt with friction and wear between contact materials [114,115], the effects of pre-treatments on the adhesion of hybrid materials [116], innovative techniques for forming these types of components [117,118], or identification of some of the most prominent issues.

The scope of this work is to prove that repair and maintenance operations can fix holes made in pieces of hybrid components based on magnesium and aluminum, not only in an efficient way, but also sustainably. Therefore, considering the above, this experimental study focused on the drilling of hybrid Mg–Al–Mg components. For decades, aluminum and magnesium were used (separately) in the aeronautical sector due to their good weight/mechanical properties. Thus, given that the density of magnesium is two-thirds that of aluminum, it was hypothesized that they could be used together, reducing weight by replacing aluminum with magnesium where possible.

The drilling process is used regularly in this sector and, thus, it was well analyzed; for only the assembly of the wing to the fuselage, thousands of holes are required [119,120].

The repair and maintenance operations were selected as they represent an additional challenge versus those of manufacturing, since, as previously mentioned, the lack of parts in aeronautical stock means having to do the repair in the shortest possible time to reduce the costs associated with the downtime of the aircraft. Also, two cooling systems (dry machining and cold compressed air) were tested to analyze the sustainability of the process. Surface roughness was chosen as a response variable as it is one of the most widespread in the literature, thereby allowing a better contrast of the results obtained; the required values for the sector are also standardized (0.8 µm < Ra < 1.6 µm) [121].

The novelty of the work is in proving that the repair operations of multi-material components (magnesium–aluminum–magnesium) and the parts made of aluminum and magnesium (separately) but assembled to form a higher component can be repaired simultaneously. This approach saves time and reduces cost.

2. Methodology

As this work is part of a broader research project that involves different geometries, material combinations, cutting conditions, tools, and lubrication/cooling systems, the methodology is similar to that followed in other previous works [105–108] and is based on the guidelines given by Montgomery [122].

2.1. Pre-Experimental Planning

Here, we report the findings of an experimental study addressing the repair or maintenance of holes made in parts of Mg–Al–Mg hybrid components using sustainable cooling systems. The study focuses on the aeronautical sector. As the response variable, we chose surface roughness since it is commonly used as a reference of quality in aeronautical components and is also used to evaluate the efficiency of the machining processes; thus, there are several works in the literature with which to make comparisons.

To determine the factors, levels, and range of their values, it should be noted that these are repair operations; hence, the depth of cut must be as small as possible to maintain the dimensional design requirements. On the other hand, since these are two non-ferrous alloys with similar machinability characteristics, it would be sufficient to test a unique type of tool. Moreover, since two cooling systems (dry machining and cold compressed air) were tested, and only a single pre-drilled part with eight holes through the Mg–Al–Mg combination was available, it was necessary to adapt the remaining factors and levels to the number of holes. Therefore, we decided to use the feed rate and the spindle speed (and two levels for each) as factors.

Additionally, as it was thought that the surface of the drilled holes could be damaged, not only by the cutting process but also due to friction caused by the chips inside of them (due to the accumulation along the mechanized length), and seeing that a similar factor was taken into account in other works [19–26], it was decided to include two additional factors related to the location relative to the insert (with three levels, one for each one of the stacks: Mg–Al–Mg) and related to the location relative to the specimen, that is, each hole (with two levels, one at the entry of the holes and one at the exit holes), where measurement of the surface roughness was taken.

2.2. Experimental Design

Considering everything explained in the pre-experimental planning, the depth of cut and the type of tool do not affect the design of experiments (DOE) since they only have one level. For the feed rate, f (mm/rev), the spindle speed, N (rpm), and the type of cooling system, C, two levels were taken for each, i.e., ($f1, f2$), ($N1, N2$), and ($C1, C2$), respectively. In the same way, for the additional factors where the surface roughness was measured, i.e., location relative to the insert, LRI, and location relative to the specimen, LRS, three ($LRI1, LRI2, LRI3$) and two ($LRS1, LRS2$) levels were taken, respectively. Table 1 describes the factors and levels selected for this experimental analysis.

Table 1. Factors and levels.

Factors	Levels
Feed rate, f (mm/rev)	$f1, f2$
Spindle speed, N (rpm)	$N1, N2$
Type of cooling system, C	$C1, C2$
Location relative to the insert, LRI	$LRI1, LRI2, LRI3$
Location relative to the specimen, LRS	$LRS1, LRS2$

The surface roughness was taken as the response variable, and the average roughness values (Ra) and the average maximum height (Rz) were measured in the different zones defined by the factors location relative to the insert (LRI) and location relative to the specimen (LRS). The factors and levels are shown in Table 1, and a DOE, as a product of a full factorial 2^3 and a block of two factors (3 × 2), was defined with a total of eight experimental re-drills and 24 measurements of the surface roughness, which provided a total of 48 values (24 of *Ra* and 24 of *Rz*). Also, the design was randomized to reduce the influence of non-considered variables [122] (Table 2).

Table 2. Experimental design: product of a full factorial 2^3 and a block of two factors (3 × 2).

No.	f*	N**	C	LRI	LRS	No	f*	N**	C	LRI	LRS	No.	f*	N**	C	LRI	LRS
1	f1	N1	C1	LRI1	LRS1	1	f1	N1	C1	LRI2	LRS1	1	f1	N1	C1	LRI3	LRS1
1	f1	N1	C1	LRI1	LRS2	1	f1	N1	C1	LRI2	LRS2	1	f1	N1	C1	LRI3	LRS2
2	f1	N2	C1	LRI1	LRS1	2	f1	N2	C1	LRI2	LRS1	2	f1	N2	C1	LRI3	LRS1
2	f1	N2	C1	LRI1	LRS2	2	f1	N2	C1	LRI2	LRS2	2	f1	N2	C1	LRI3	LRS2
3	f2	N1	C1	LRI1	LRS1	3	f2	N1	C1	LRI2	LRS1	3	f2	N1	C1	LRI3	LRS1
3	f2	N1	C1	LRI1	LRS2	3	f2	N1	C1	LRI2	LRS2	3	f2	N1	C1	LRI3	LRS2
4	f2	N2	C1	LRI1	LRS1	4	f2	N2	C1	LRI2	LRS1	4	f2	N2	C1	LRI3	LRS1
4	f2	N2	C1	LRI1	LRS2	4	f2	N2	C1	LRI2	LRS2	4	f2	N2	C1	LRI3	LRS2
5	f1	N1	C2	LRI1	LRS1	5	f1	N1	C2	LRI2	LRS1	5	f1	N1	C2	LRI3	LRS1
5	f1	N1	C2	LRI1	LRS2	5	f1	N1	C2	LRI2	LRS2	5	f1	N1	C2	LRI3	LRS2
6	f1	N2	C2	LRI1	LRS1	6	f1	N2	C2	LRI2	LRS1	6	f1	N2	C2	LRI3	LRS1
6	f1	N2	C2	LRI1	LRS2	6	f1	N2	C2	LRI2	LRS2	6	f1	N2	C2	LRI3	LRS2
7	f2	N1	C2	LRI1	LRS1	7	f2	N1	C2	LRI2	LRS1	7	f2	N1	C2	LRI3	LRS1
7	f2	N1	C2	LRI1	LRS2	7	f2	N1	C2	LRI2	LRS2	7	f2	N1	C2	LRI3	LRS2
8	f2	N2	C2	LRI1	LRS1	8	f2	N2	C2	LRI2	LRS1	8	f2	N2	C2	LRI3	LRS1
8	f2	N2	C2	LRI1	LRS2	8	f2	N2	C2	LRI2	LRS2	8	f2	N2	C2	LRI3	LRS2

*f (mm/rev); ** N (rpm).

2.3. Performing the Experiment

Before carrying out the re-drilling tests, it was necessary to collect the specimens of the hybrid parts, the tools, and the cooling systems, as well as introduce the parameter values into the machine tools and establish cutting conditions and data collecting protocols. Next, the machining operations were carried out and, finally, photographs and videos of the trials were taken for subsequent analysis.

2.4. Statistical Analysis of the Data

Once the machining process was finished, the arithmetical mean roughness (*Ra*) and average maximum height (*Rz*) were measured. The data were statistically analyzed, including an analysis of variance (ANOVA) to identify the influential factors for surface roughness variation and the interactions among them.

2.5. Conclusions

The main conclusions extracted from the descriptive analysis of the obtained results and their statistical analysis were established.

3. Applications and Results

3.1. Materials

The hybrid component specimen was made of three parallelepiped plates of dimensions 50 × 50 × 15 mm pre-drilled with eight holes of 8 mm in diameter. The three plates were mechanically fixed so that they could be easily disassembled to take the measurements inside the holes. The plates placed above and below were of magnesium alloy (UNS M11917), and the other (placed between them) was of aluminum alloy (UNS A92024). The chemical composition of both materials is given in Table 3.

Table 3. Chemical composition of the materials used for the manufacturing specimens.

UNS M11917 (AZ91D)	UNS A92024 (AA2024 T351)
Al 8.30–9.70%	Al 90.7–94.7%
Cu ≤ 0.03%	Cr ≤ 0.1%
Fe ≤ 0.005%	Cu 3.8–4.9%
Mg 90%	Fe ≤ 0.5%
Mn ≥ 0.13%	Mg 1.2–1.8%
Ni ≤ 0.002%	Mn 0.3–0.9%
Si ≤ 0.1%	Si ≤ 0.5%
Zn 0.35–1%	Ti ≤ 0.15%
–	Zn ≤ 0.25%

These materials were selected because the authors had previous experience in their machining, both independently [12–14,16–23,25,28–32,77–79,82–94] and together [105–108]. Also, there were interesting works of other researchers in the literature, thus allowing comparisons to be drawn [15,24,26,27,80,81,95–104].

3.2. Tools

As the target of this study was to analyze the feasibility of carrying out the repair and maintenance operations in an efficient and sustainable way, a single level for the depth of cut factor, d, was taken (d = 0.5 mm, using of a 9-mm-diameter drill bit). On the other hand, keeping the depth of cut at a low level also helps to keep the cutting temperature low and, therefore, to keep the magnesium temperature far from its ignition temperature.

The tools used in the trials were helical drill bits of high performance. They were made of a high-speed-steel (known as Cobalt Steel, HSSE, or HSS-E), obtained by powder metallurgy (PM) (Figure 1). The tools, with reference HSS-E-PM A1 1257, were purchased from Garant (Hoffmann Iberia, San Fernando de Henares, Madrid, Spain).

Figure 1. Helical drill bits HSS-E-PM A1 1257 manufactured by Garant [123].

Their dimensions were 9 mm of diameter, 81 mm of helical length, and 131 mm of total length. In addition, their special geometry allowed self-centering and optimal chip evacuation.

3.3. Machines and Equipment

The trials were carried out in a Tongtai TMV510 machining center (Tongai Machine &Tool Co., Luzhu Dist, Kaohsiung City, Taiwan) equipped with a Control Numeric Computer (CNC) Fanuc (Fanuc Iberia, Castelldefels, Barcelona, Spain) (Figure 2a). A drilling cycle was programmed that made the tool penetrate 10 mm, and then return to evacuate the generated chips. The same sequence was repeated until the tool crossed the entire width of the piece formed by the three stacks of Mg–Al–Mg. This was done so that the accumulated chips inside the holes did not scratch the surface or stop or hinder the tool inside the piece. Cold compressed air (CCA) was used as the cooling system, implementing a Vortec Cold Air Gun (Vortec, Cincinnati, Ohio, USA) (Figure 2b). The roughness measurements were taken using a Mitutoyo Surftest SJ 401 roughness tester (Figure 2c) with the following settings: measuring range, 800 µm; resolution, 0.000125 µm; transverse length, 25 mm; cut off, 0.8 mm; scan rate, 4 mm (N = 5); the standard ISO 1997 [124] was used.

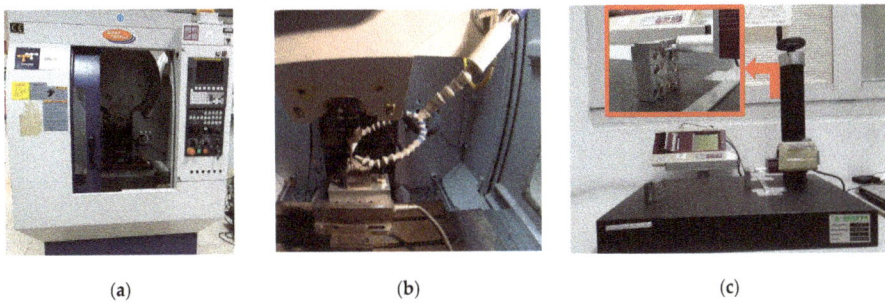

Figure 2. (**a**) Tongtai TMV510 machining center; (**b**) details of the Vortec Cold Air Gun during the trials; (**c**) Mitutoyo Surftest SJ 401 roughness tester.

3.4. Experimental Tests

The design of experiments, the materials, tools, machines, and equipment used in the trials, and the parameter value ranges are given in Table 4.

Table 4. Factors, levels, and values. CCA—cold compressed air.

Factors	Level Values
Feed rate, f (mm/rev)	$f1 = 0.05$; $f2 = 0.10$
Spindle speed, N (rpm)	$N1 = 500$; $N2 = 1200$
Type of cooling system, C	$C1$ = CCA; $C2$ = dry
Location relative to the insert, LRI	$LRI1$ = Mg; $LRI2$ = Al; $LRI3$ = Mg
Location relative to the specimen, LRS	$LRS1$ = specimen entry zone; $LRS2$ = specimen exit zone

The locations of the measurement zones of the surface roughness are shown in Figure 3. $LRI1$ denotes the first magnesium plate, $LRI2$ denotes the aluminum plate, and $LRI3$ denotes the second (and last) magnesium plate. LRS took into account the location of the measuring zone inside each hole after re-drilling it, and its levels were defined as $LRS1$ (the specimen entry zone) and $LRS2$ (the specimen exit zone). Figure 4 provides a graphical summary of the experimental set-up.

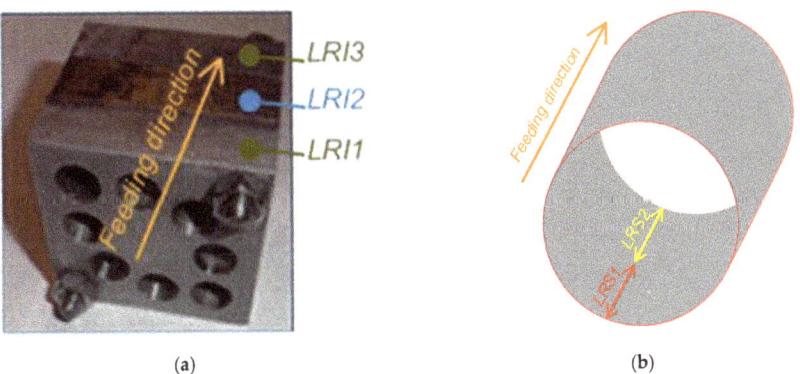

Figure 3. The concrete locations of the measurement zones of the surface roughness: (**a**) location relative to the insert, LRI; (**b**) location relative to the specimen, LRS.

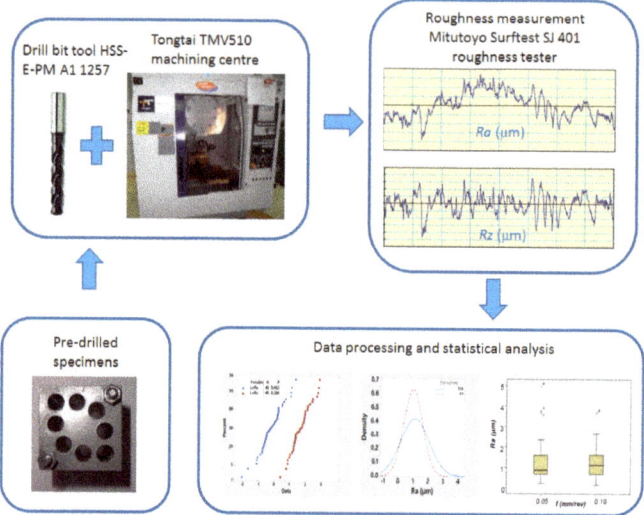

Figure 4. The experimental set-up.

3.5. Analysis and Discussion of the Results

After performing the eight re-drilling tests, surface roughness measurements were made in each hole in the three plates (entry and exit zones). The values of the arithmetical mean roughness (Ra) and the average maximum height (Rz) were calculated (in micrometers) and are given in Table 5.

Table 5. The arithmetical mean roughness (Ra) and the average maximum height (Rz) obtained during the measurement tests.

No.	f (mm/rev)	N (rpm)	C	LRI	LRS	Ra (µm)	Rz (µm)
1	0.05	500	CCA	LRI1	LRS1	0.35	2.70
1	0.05	500	CCA	LRI1	LRS2	0.81	4.30
2	0.05	1200	CCA	LRI1	LRS1	1.06	6.10
2	0.05	1200	CCA	LRI1	LRS2	1.23	7.20
3	0.10	500	CCA	LRI1	LRS1	1.27	8.50
3	0.10	500	CCA	LRI1	LRS2	1.31	7.00
4	0.10	1200	CCA	LRI1	LRS1	1.03	6.40
4	0.10	1200	CCA	LRI1	LRS2	1.33	7.20
5	0.05	500	Dry	LRI1	LRS1	1.22	6.90
5	0.05	500	Dry	LRI1	LRS2	0.19	2.20
6	0.05	1200	Dry	LRI1	LRS1	0.61	4.60
6	0.05	1200	Dry	LRI1	LRS2	0.73	5.70
7	0.10	500	Dry	LRI1	LRS1	0.73	5.40
7	0.10	500	Dry	LRI1	LRS2	0.48	3.60
8	0.10	1200	Dry	LRI1	LRS1	0.98	7.00
8	0.10	1200	Dry	LRI1	LRS2	0.62	4.00
1	0.05	500	CCA	LRI2	LRS1	2.94	16.80
1	0.05	500	CCA	LRI2	LRS2	3.00	16.20

Table 5. Cont.

No.	f (mm/rev)	N (rpm)	C	LRI	LRS	Ra (µm)	Rz (µm)
2	0.05	1200	CCA	LRI2	LRS1	1.31	7.20
2	0.05	1200	CCA	LRI2	LRS2	3.95	18.00
3	0.10	500	CCA	LRI2	LRS1	1.20	8.70
3	0.10	500	CCA	LRI2	LRS2	2.11	11.50
4	0.10	1200	CCA	LRI2	LRS1	0.13	1.50
4	0.10	1200	CCA	LRI2	LRS2	2.91	15.20
5	0.05	500	Dry	LRI2	LRS1	1.24	6.80
5	0.05	500	Dry	LRI2	LRS2	1.30	6.80
6	0.05	1200	Dry	LRI2	LRS1	1.88	11.20
6	0.05	1200	Dry	LRI2	LRS2	0.51	4.00
7	0.10	500	Dry	LRI2	LRS1	0.87	5.10
7	0.10	500	Dry	LRI2	LRS2	0.33	2.40
8	0.10	1200	Dry	LRI2	LRS1	1.66	8.40
8	0.10	1200	Dry	LRI2	LRS2	1.70	9.80
1	0.05	500	CCA	LRI3	LRS1	0.48	4.00
1	0.05	500	CCA	LRI3	LRS2	0.59	3.40
2	0.05	1200	CCA	LRI3	LRS1	0.64	4.00
2	0.05	1200	CCA	LRI3	LRS2	0.56	3.50
3	0.10	500	CCA	LRI3	LRS1	0.56	3.20
3	0.10	500	CCA	LRI3	LRS2	0.78	5.10
4	0.10	1200	CCA	LRI3	LRS1	1.13	6.10
4	0.10	1200	CCA	LRI3	LRS2	0.92	5.90
5	0.05	500	Dry	LRI3	LRS1	0.34	2.40
5	0.05	500	Dry	LRI3	LRS2	0.67	6.50
6	0.05	1200	Dry	LRI3	LRS1	0.53	3.50
6	0.05	1200	Dry	LRI3	LRS2	0.63	3.80
7	0.10	500	Dry	LRI3	LRS1	0.46	2.80
7	0.10	500	Dry	LRI3	LRS2	0.42	3.60
8	0.10	1200	Dry	LRI3	LRS1	0.59	3.80
8	0.10	1200	Dry	LRI3	LRS2	0.53	3.90

Initially, a descriptive method was used to analyze the Ra and Rz values. The obtained results are separated into Tables 6 and 7. Tables 6 and 7 give the values of Ra and Rz, respectively, in each plate (both in the entry and exit zones of the holes).

Table 6. Values of Ra in each plate at the entry and at the exit zones of the holes.

No.	F (mm/rev)	N (rpm)	C	Ra (µm)					
				LRI1		LRI2		LRI3	
				LRS1	LRS2	LRS1	LRS2	LRS1	LRS2
1	0.05	500	CCA	0.35	0.81	2.94	3.00	0.48	0.59
2	0.05	1200	CCA	1.06	1.23	1.31	3.95	0.64	0.56
3	0.10	500	CCA	1.27	1.31	1.20	2.11	0.56	0.78
4	0.10	1200	CCA	1.03	1.33	0.13	2.91	1.13	0.92
5	0.05	500	Dry	1.22	0.19	1.24	1.30	0.34	0.67
6	0.05	1200	Dry	0.61	0.73	1.88	0.51	0.53	0.63
7	0.10	500	Dry	0.73	0.48	0.87	0.33	0.46	0.42
8	0.10	1200	Dry	0.98	0.62	1.66	1.70	0.59	0.53

Table 7. Values of Rz in each plate at the entry and exit zones of the holes.

No.	F (mm/rev)	N (rpm)	C	Rz (µm)					
				LRI1		LRI2		LRI3	
				LRS1	LRS2	LRS1	LRS2	LRS1	LRS2
1	0.05	500	CCA	2.70	4.30	16.80	16.20	4.00	3.40
2	0.05	1200	CCA	6.10	7.20	7.20	18.00	4.00	3.50
3	0.10	500	CCA	8.50	7.00	8.70	11.50	3.20	5.10
4	0.10	1200	CCA	6.40	7.20	1.50	15.20	6.10	5.90
5	0.05	500	Dry	6.90	2.20	6.80	6.80	2.40	6.50
6	0.05	1200	Dry	4.60	5.70	11.20	4.00	3.50	3.80
7	0.10	500	Dry	5.40	3.60	5.10	2.40	2.80	3.60
8	0.10	1200	Dry	7.00	4.00	8.40	9.80	3.80	3.90

From the Ra and Rz values given in Tables 6 and 7, the graphics of Figure 5 were drawn. Figure 5 shows the normal distribution of Ra (left column) and Rz (right column) with respect to feed rate, f (mm/rev), (a) Ra and (b) Rz; spindle speed, N (rpm), (c) Ra and (d) Rz; type of cooling system, C, (e) Ra and (f) Rz; location relative to the insert, LRI, (g) Ra and (h) Rz; and location relative to the specimen, LRS, (i) Ra and (j) Rz.

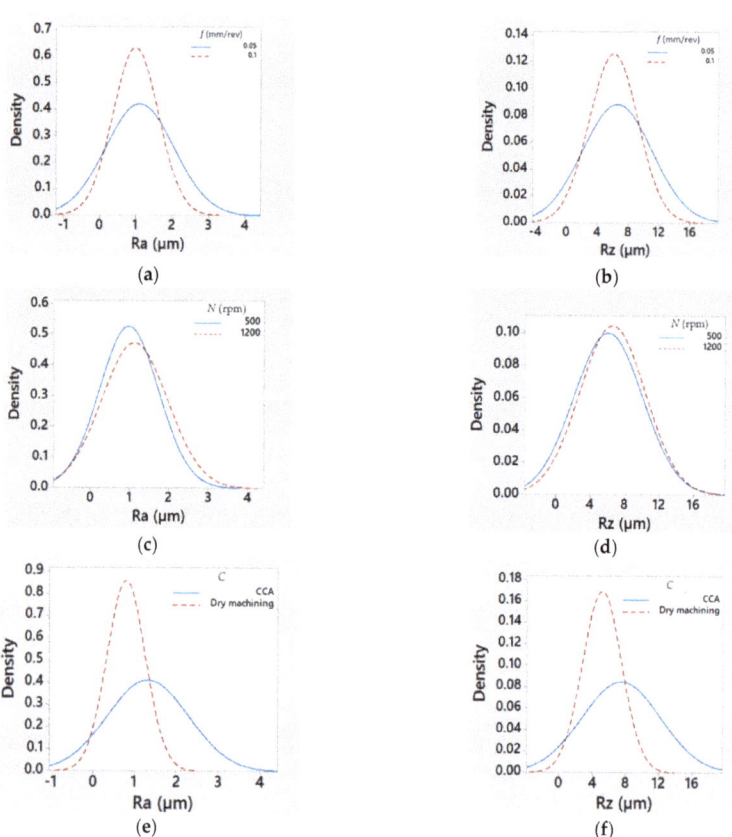

Figure 5. Cont.

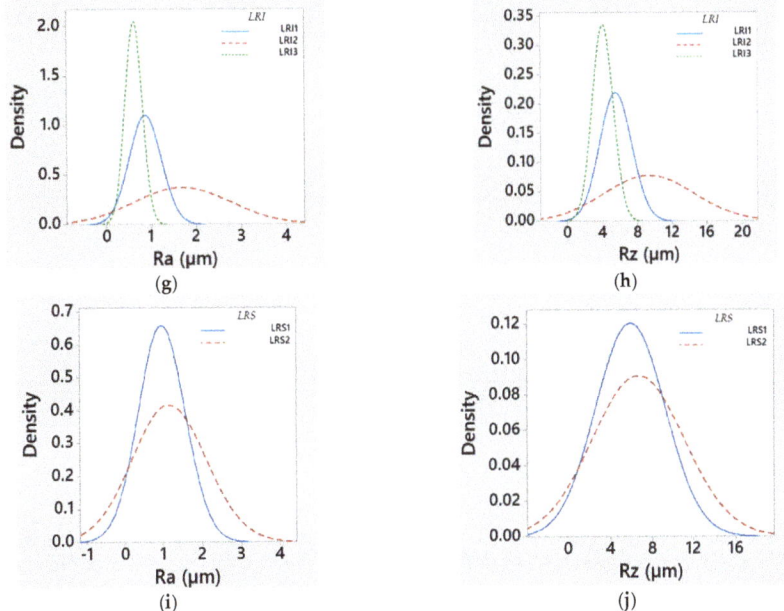

Figure 5. Normal distribution of Ra (µm) and Rz (µm), respectively, with respect to (**a**,**b**) feed rate, f (mm/rev); (**c**,**d**) spindle speed, N (rpm); (**e**,**f**) type of cooling system, C; (**g**,**h**) location regarding insert, LRI; (**i**,**j**) location regarding specimen, LRS.

Taking into account that a good behavior of the results is considered when the obtained values are concentrated in the interval [0.8 µm; 1.6 µm] given by the standard [121], a first approach to the analysis can be made by observing data collected in Tables 6 and 7 and graphics from Figure 5. Thus, it was possible to affirm that Ra and Rz had a similar behavior for the cutting parameters; however, they were perhaps slightly better for high feed rates (f = 0.10 mm/rev) and dry machining and they were very similar for both tested values of the spindle speed (perhaps slightly better for low values N = 500 rpm). Regarding the location relative to the insert, in both magnesium plates, Ra and Rz were better than in the aluminum plate. Also, when comparing the results of the first and the last magnesium plates, the results were lower for the latter ($LRI3$). However, $LRI1$ was considered as better since the surface roughness values were closer to the standard values used in the aeronautic sector (between 0.8 µm and 1.6 µm) [121]. Finally, regarding the location relative to the specimen, the results were lower at the entry of the holes than at the exit.

The values from Tables 6 and 7 are plotted in Figures 6 and 7, respectively, revealing possible combinations of parameters that could be used for repairing hybrid parts by re-drilling. Figure 8 plots the Ra and Rz values obtained in the trials and collected in Table 5. All of them were inside of the usual upper and lower limits given in the chart of conversion relations between Ra and Rz, according to DIN 47 [125].

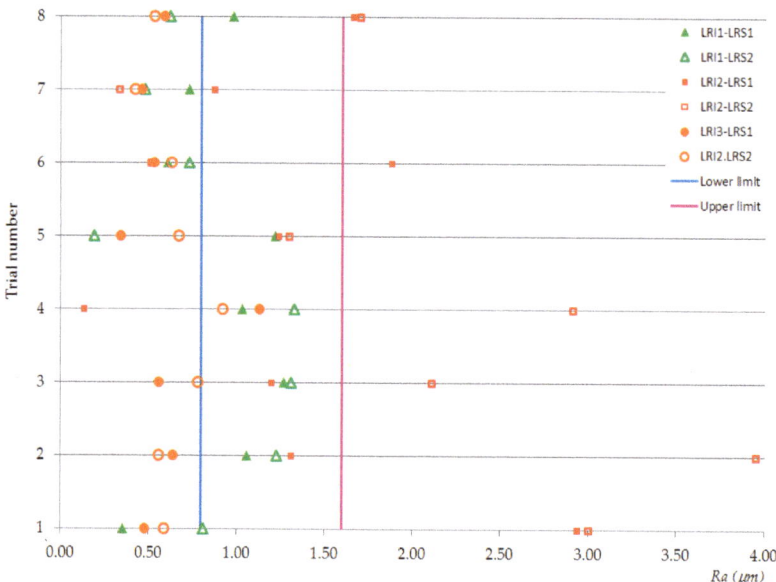

Figure 6. Graphic representation of the *Ra* values.

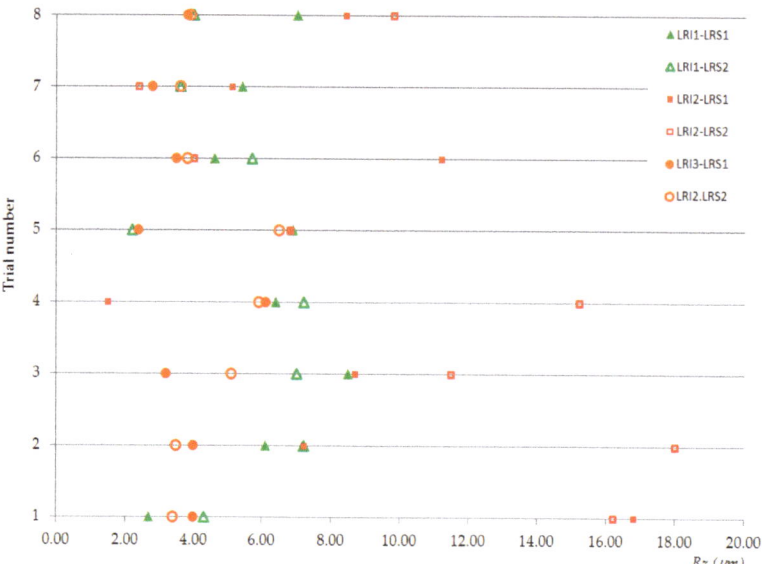

Figure 7. Graphic representation of the *Rz* values.

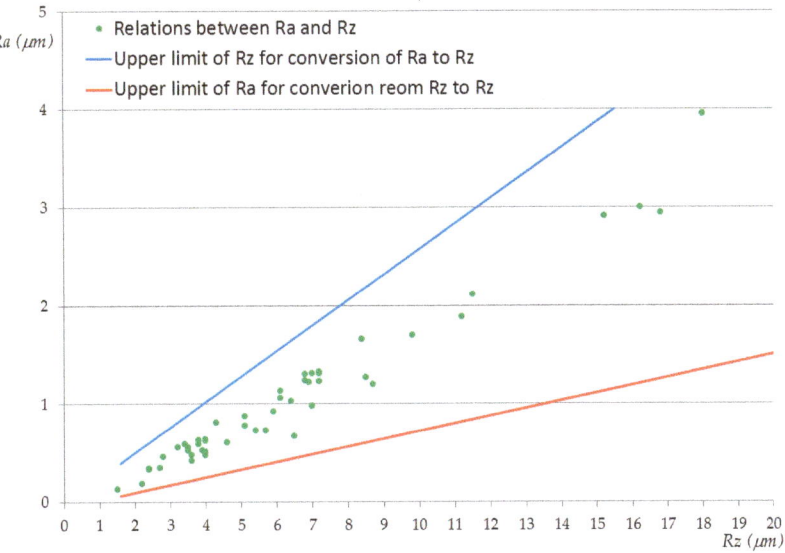

Figure 8. The relationship between *Ra* and *Rz* values.

Observing the *Ra* and *Rz* values in Figures 6 and 7, we see that, for the aluminum plate, the roughness values were higher than for the magnesium plate, especially at the exit of the holes and when using cold compressed air as the cooling system. Therefore, it seems reasonable to select aluminum as the more critical material when establishing the cutting parameters. As the results were better at the entry of the holes, it would perhaps be possible to improve the results by modifying the geometry of hybrid component, for example, by searching for the adequate proportions of the thicknesses of the combined materials. On the other hand, in the second plate of magnesium, most of the surface roughness values (except the obtained one for $f = 0.05$ mm/rev, $N = 1200$ rpm, and $C = CCA$) were lower than those established in the design requirement standards. Therefore, roughness values might be improved by drilling halfway, turning the hybrid component, and then continuing to drill from the opposite side.

Reviewing the surface roughness values at the entry of the aluminum holes (Table 6 column *LRI2*, *LRS1*), it can be seen that test numbers 2, 3, 5, and 7 presented values within the range of the values given by the standard (0.8 µm < *Ra* < 1.6 µm) [121]. Among them, test numbers 2 and 3 were within the same range for the first magnesium plate; test number 5 could be an option if the magnesium plates were about half as thick, since the *LRI1* had values within such an interval; test number 7 indicated room for improvement because the roughness value of the aluminum was close to the lower limit of the roughness required in the aeronautic sector (0.8 µm). In fact, comparing the roughness values obtained in test number 7 with those from test number 8, we propose that a new parameter combination is possible by selecting a spindle speed equal to 1000 rpm or near this value; this would also increase the feed speed, decrease the machining time, and, consequently, improve the efficiency of the process.

In addition, an ANOVA was performed to identify the factors that influence the variation of the response variables, *Ra* and *Rz*. To apply an ANOVA, it is necessary that the variables meet three conditions: (1) each data group must be independent, (2) the results obtained for each group must follow a normal distribution (although a breach of this assumption is supported when the distribution is symmetric), and (3) the variances of each data group must not differ significantly (homoscedasticity).

Using the data extracted directly from the experiment, the *Ra* and *Rz* values did not follow a normal distribution (Shapiro–Wilk test p-value < 0.05). Therefore, the data were processed using logarithmic transformation, maintaining its order but softening the effect of outliers.

By this approach, normally distributed Ln*Ra* and Ln*Rz* values (Shapiro–Wilk test *p*-value > 0.05) were obtained (Figure 9). In addition, the condition of homoscedasticity was also fulfilled (Levene statistic, *p*-value > 0.05), and independent data groups had a similar number of cases (Table 8). In the analysis, interactions up to the third order were considered, and successive iterations were performed until all values were significant. In each iteration, the statistically less significant effect was excluded if it had a *p*-value greater than 0.05. Tables 9 and 10 give the outcome of the first and the last ANOVA over Ln*Ra*, and Tables 11 and 12 collect the outcome of the first and the last ANOVA over Ln*Rz*, respectively.

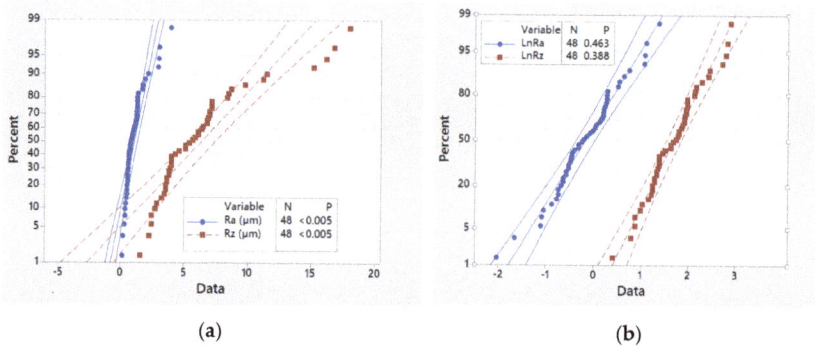

Figure 9. Probability plots: (**a**) *Ra* and *Rz*; (**b**) Ln*Ra* and Ln*Rz*.

Table 8. Homogeneity test of variances for factors f and N and response variables Ln*Ra* and Ln*Rz*.

	f (mm/rev)		N (rpm)	
	Levene Statistic	Significance	Levene Statistic	Significance
Ln*Ra*	0.30	0.59	0.58	0.46
Ln*Rz*	0.21	0.65	0.13	0.72

Table 9. Outcome of the first iteration for the ANOVA over Ln*Ra*.

Source	DF *	Sum of Squares	Mean Square	F-value	$p > F$
Corrected Model	23	13.037	0.567	1.415	0.202
Intercept	1	1.486	1.486	3.709	0.066
LRI	2	4.983	2.491	6.217	0.007
N	1	0.306	0.306	0.764	0.391
f	1	0.010	0.010	0.025	0.875
C	1	1.882	1.882	4.697	0.040
LRI × N	2	0.365	0.183	0.456	0.639
LRI × f	2	1.766	0.883	2.203	0.132
LRI × C	2	0.070	0.035	0.087	0.917
$N × f$	1	0.002	0.002	0.006	0.938
$N × C$	1	0.306	0.306	0.764	0.391
$f × C$	1	0.003	0.003	0.008	0.929
LRI × $N × f$	2	0.368	0.184	0.459	0.638
LRI × $N × C$	2	0.801	0.401	1.000	0.383
LRI × $f × C$	2	0.906	0.453	1.130	0.340
$N × f × C$	1	0.576	0.576	1.437	0.242
LRI × $N × f × C$	2	0.694	0.347	0.865	0.434
Error	24	9.617	0.401	–	–
Total	48	24.141	–	–	–
Corrected Total	47	22.654	–	–	–

* DF, degrees of freedom.

Table 10. Outcome of the last iteration for the ANOVA over LnRa.

Source	DF *	Sum of Squares	Mean Square	F-value	p > F
Corrected Model	5	6.935a	1.387	3.706	0.007
Intercept	1	1.486	1.486	3.971	0.053
LRI	2	4.983	2.491	6.656	0.003
C	1	1.882	1.882	5.029	0.030
Error	42	15.720	0.374	–	–
Total	48	24.141	–	–	–
Corrected Total	47	22.654	–	–	–

* DF, degrees of freedom.

Table 11. Outcome of the first iteration for the ANOVA over LnRz.

Source	DF *	Sum of Squares	Mean Square	F-value	p > F
Corrected Model	23	8.726a	0.379	1.600	0.130
Intercept	1	137.220	137.220	578.50	0.000
LRI	2	3.673	1.836	7.742	0.003
C	1	0.963	0.963	4.061	0.055
f	1	0.009	0.009	0.039	0.845
N	1	0.150	0.150	0.633	0.434
LRI × C	2	0.190	0.095	0.402	0.674
LRI × f	2	1.027	0.514	2.165	0.137
LRI × N	2	0.172	0.086	0.363	0.699
C × f	1	0.014	0.014	0.057	0.813
C × N	1	0.232	0.232	0.978	0.332
f × N	1	0.011	0.011	0.048	0.828
LRI × C × f	2	0.490	0.245	1.034	0.371
LRI × C × N	2	0.840	0.420	1.770	0.192
LRI × f × N	2	0.381	0.190	0.803	0.460
C × f × N	1	0.311	0.311	1.310	0.264
LRI × C × f × N	2	0.263	0.131	0.553	0.582
Error	24	5.693	0.237		
Total	48	151.639			
Corrected Total	47	14.419			

* DF, degrees of freedom.

Table 12. Outcome of the last iteration for the ANOVA over LnRz.

Source	DF *	Sum of Squares	Mean Square	F-value	p > F
Corrected Model	5	4.826	0.965	4.226	0.003
Intercept	1	137.22	137.220	600.78	0.000
LRI	2	3.673	1.836	8.040	0.001
C	1	0.963	0.963	4.217	0.046
Error	42	9.593	0.228		
Total	48	151.63			
Corrected Total	47	14.419			

* DF, degrees of freedom.

Taking into account the results shown in Tables 10 and 12, we conclude that the most influential factors, in both cases, were the location relative to the insert (LRI) and the type of cooling system (C), and that there were no interactions among factors with influence.

Considering the re-drilling surface roughness variability of hybrid Mg–Al–Mg components explained by the statistically significant effect obtained from the ANOVA, the percentage of variability attributed to each factor is shown in Table 13, and the contribution of each effect was obtained as the percentage of the sum of squares values of each significant effect relative to the sum of squares of all significant effects.

Table 13. Percentage variability of the statistically significant effects obtained from ANOVA.

Source	Ra		Rz	
	Sum of Squares	Variability Percentage	Sum of Squares	Variability Percentage
LRI	5.0	72.6%	3.7	79.2%
C	1.9	27.4%	1.0	20.8%
Total	6.9	100%	4.6	100%

4. Conclusions

This work focused on the maintenance and repair of holes made of hybrid Mg–Al–Mg components by drilling, using two sustainable cooling techniques: dry machining and cold compressed air. The aeronautic and aerospace sectors were selected as relevant applications. In such sectors, the pieces have strict design requirements for surface roughness (0.8 µm < Ra < 1.6 µm). Ra and Rz were taken as response variables. From our analyses, we propose that Ra and Rz values have similar behaviors and they exhibit the following characteristics:

- They are better for high feed rates and dry machining.
- They are very similar for both tested values of spindle speed, although perhaps slightly better for low values.
- They are, regarding the location relative to the insert, better in both magnesium plates than in the aluminum plate; also, when comparing the results of the first and the last magnesium plate, the results were lower for $LRI3$, yet the values for $LRI1$ were considered better since the surface roughness values were closer to the aeronautic industry standard.
- They display, for the location regarding specimen, better behavior at the entry of the holes than at the exit.
- They are higher in the plate of aluminum than in the magnesium one, particularly at the exit of the holes and, in a more pronounced way, using cold compressed air as a cooling system. Therefore, aluminum is considered a more valuable material when selecting the cutting parameters. Also, as the results are better (lower values) at the entry of the holes than at the exit, it will perhaps be possible to improve the results by modifying the geometry of the hybrid component, for example, by searching for the adequate thicknesses among the different combined materials.
- They are lower, in most cases, in the second plate of magnesium than the established standard. Therefore, the process can be influenced by the drilling direction, and it could be improved by drilling halfway, turning the part, and drilling again from the opposite side.

In addition, from the ANOVA analysis, we found that the factors that influence the response variables Ra and Rz are location relative to the insert and type of cooling system, with percentages of influence of 72.6% and 27.4%, respectively, for Ra, and 79.2% and 20.8%, respectively, for Rz.

With this work, we showed that it is possible to simultaneously repair magnesium–aluminum–magnesium multi-material components and parts made of aluminum and magnesium (separately) but assembled to form a higher component using sustainable cooling systems (dry machining). This approach reduces the time and cost associated with the assembly and disassembly of these types of components during maintenance or repair.

In conclusion, we propose three ways to optimize (or at least improve) the process: (1) using different parameters values (for example, higher values of the spindle speed that increase the efficiency of the process); (2) designing a hybrid component with new proportions of the thicknesses of the materials combined; (3) applying other drilling sequences (e.g., firstly drilling halfway and then turning the part and drilling from the opposite side).

Author Contributions: E.M.R., M.M.M., and D.B. contributed to the conceptualization, methodology, and formal analysis. D.B. performed the investigation. E.M.R. and M.M.M. managed the project resources. E.M.R., M.M.M., and D.B. prepared the original draft of the manuscript. E.M.R., M.M.M., D.B., and J.P.D. reviewed and edited the manuscript. E.M.R., M.M.M., D.B., and J.P.D. contributed to data visualization. E.M.R., M.M.M., and J.P.D. supervised the study. E.M.R. and M.M.M. were responsible for the funding acquisition and project administration. All authors read and agreed to the published version of the manuscript.

Funding: This work was partly funded by grants from the Ministerio de Ciencia, Innovación, y Universidades, and the Industrial Engineering School-UNED (RTI2018-102215-B-I00, REF2019-ICF05, and REF2019-ICF08), Spain.

Acknowledgments: The authors thank the Industrial Production and Manufacturing Engineering (IPME) Research Group and the Industrial Engineering School-UNED (Projects REF2019-ICF05 and REF2019-ICF08).

Conflicts of Interest: The authors declare no conflicts of interest.

References

1. Rubio, E.M.; Blanco, D.; Marín, M.M.; Carou, D. Analysis of the latest trends in hybrid components of lightweight materials for structural uses. In Proceedings of the 8th Manufacturing Engineering Society International Conference 2019 (MESIC 2019), Madrid, Spain, 19–21 June 2019.
2. Lee, D.; Morillo, C.; Oller, S.; Bugeda, G.; Oñate, E. Robust design optimization of advance hybrid (fiber-metal) composite structures. *Compos. Struct.* **2013**, *99*, 181–192. [CrossRef]
3. DRL. *Innovation Report 2011*; Institute of Composites Structures and Adaptative Systems: Braunschweig, Germany, 2011.
4. Ashby, M. Hybrid Materials to Expand the Boundaries of Material-Property Space. *J. Am. Ceram. Soc.* **2011**, *94*, 3–14. [CrossRef]
5. Ashby, M.F.; Bréchet, Y.J.M. Design hybrid mate. *Acta Mater.* **2003**, *51*, 5801–5821. [CrossRef]
6. Boyer, R.R.; Cotton, J.D.; Mohaghegh, M.; Schafrik, R.E. Materials considerations for aerospace applications. *Mrs Bull.* **2015**, *40*, 1055–1066. [CrossRef]
7. Taub, A.I.; Luo, A.A. Advanced lightweight materials and manufacturing processes for automotive applications. Materials considerations for aerospace applications. *Mrs Bull.* **2015**, *40*, 1045–1054. [CrossRef]
8. Montemayor, L.; Chernow, V.; Greer, J.R. Materials by design: Using architecture in material design to reach new property spaces. *Mrs Bull.* **2015**, *40*, 1122–1129. [CrossRef]
9. Available online: https://www.asoc-aluminio.es/info-aea (accessed on 14 January 2020).
10. Available online: https://www.aluminum.org/ (accessed on 14 January 2020).
11. Berrio, L.F.; Echeverry, M.; Correa, A.A.; Robledo, S.M.; Castaño, J.G.; Echeverría, F. Development of the magnesium alloy industry in Colombia—An opportunity. *Dyna* **2017**, *84*, 55–64.
12. Sáenz de Pipaón, J.M. Diseño y Fabricación de Probetas de Componentes Híbridos con Aleaciones de Magnesio para Ensayos de Mecanizado. Ph.D. Thesis, UNED (Universidad Nacional de Educación a Distancia), Madrid, Spain, 2013.
13. Saa, A.J. Estudio Experimental Basado en la Rugosidad Superficial para la Selección de Herramientas y Condiciones de Corte en Operaciones de Refrentado en Seco a Baja Velocidad de Piezas de Magnesio. Ph.D. Thesis, UNED (Universidad Nacional de Educación a Distancia), Madrid, Spain, 2015.
14. Carou, D. Estudio Experimental para Determinar la Influencia de la Refrigeración/Lubricación en la Rugosidad Superficial en el Torneado Intermitente a Baja Velocidad de Piezas de Magnesio. Ph.D. Thesis, UNED (Universidad Nacional de Educación a Distancia), Madrid, Spain, 2013.
15. Dreizin, E.L.; Berman, C.H.; Vicenzi, E.P. Condensed-phase modifications in magnesium particle combustion in air. *Combust. Flame* **2000**, *122*, 30–42. [CrossRef]
16. Carou, D.; Rubio, E.M.; Davim, J.P. Analysis of ignition risk in intermittent turning of UNS M11917 magnesium alloy at low cutting speeds based on the chip morphology. *Proc. Inst. Mech. Eng. Part B J. Eng. Manuf.* **2015**, *229*, 365–371. [CrossRef]
17. Benedicto, E.; Carou, D.; Rubio, E.M. Technical, Economic and Environmental Review of the Lubrication/Cooling Systems used in Machining Processes. *Procedia Eng.* **2017**, *184*, 99–116. [CrossRef]
18. Rubio, E.M.; de Agustina, B.; Marín, M.M.; Bericua, A. Cooling systems based on cold compressed air: A review of the applications in machining processes. *Proc. Eng.* **2015**, *132*, 413–418. [CrossRef]
19. Saá, A.J.; Agustina, B.; Marcos, M.; Rubio, E.M. Experimental study of dry turning of UNS A92024-T3 aluminium alloy bars based on surface roughness. In Proceedings of the AIP Conference Proceedings, Alcoy, Spain, 17–19 June 2009; Volume 1181, pp. 151–158.

20. Agustina, B.; Saá, A.; Marcos, M.; Rubio, E.M. Analysis of the machinability of aluminium alloys UNS A97050-T7 and UNS A92024-T3 during short dry turning tests. In *Advanced Materials Research*; Trans Tech Publications: Bäch, Switzerland, 2011; Volume 264, pp. 931–936.
21. Agustina, B.; Rubio, E.M. Experimental study of cutting forces during dry turning processes of UNS A92024-T3 aluminium alloys. In Proceedings of the 4th Manufacturing Engineering Society International Conference 2011 (MESIC 2011), Cadiz, Spain, 21–23 September 2011; pp. 1–6.
22. Agustina, B.; Rubio, E.M. Analysis of cutting forces during dry turning processes of UNS A92024-T3 aluminium bars. *Adv. Mater. Res.* **2012**, *498*, 25–30. [CrossRef]
23. Agustina, B.; Rubio, E.M.; Sebastian, M.A. Surface roughness model based on force sensors for the prediction of the tool wear. *Sensors* **2014**, *4*, 6393–6408. [CrossRef] [PubMed]
24. Arokiadass, R.; Palaniradja, K.; Alagumoorthi, N. Effect of process parameters on surface roughness in end milling of Al/SiCp MMC. *Int. J. Eng. Sci. Technol.* **2011**, *13*, 276–284.
25. Rubio, E.M.; Camacho, A.M.; Sánchez, J.M.; Marcos, M. Surface roughness of AA7050 alloy turned bars, analysis of the influence of the length of machining. *J. Mater. Process. Technol.* **2005**, *162*, 682–689. [CrossRef]
26. Batista, M.; Sánchez-Carrilero, M.; Rubio, E.M.; Marcos, M. Cutting Speed and Feed Based Analysis of Chip Arrangement in the Dry Horizontal Turning of UNS A92024 Alloy. *Ann. DAAAM Proc.* **2009**, *20*, 967–968.
27. Bisker, J.; Christman, T.; Allison, T.; Goranson, H.; Landmesser, J.; Minister, A.; Plonski, R. DOE Handbook. In *Primer on Spontaneous Heating and Pyrophoricity*; U.S. Department of Energy: Washington, DC, USA, 1994; pp. 1–68.
28. Rubio, E.M.; Sáenz de Pipaón, M.J.; Villeta, M.; Sebastián, M.A. Experimental study for improving repair operations of pieces of magnesium UNS M11311 obtained by dry turning. In Proceedings of the 12th CIRP Conference on Modelling of Machining, San Sebastian, Spain, 7–8 May 2009; pp. 819–826.
29. Carou, D.; Rubio, E.M.; Davim, J.P. Discontinuous cutting: Failure mechanisms, tools materials and temperature study—A review. *Rev. Adv. Mater. Sci.* **2014**, *38*, 110–124.
30. Rubio, E.M.; Bericua, A.; de Agustina, B.; Marín, M.M. Analysis of the surface roughness of titanium pieces obtained by turning using different cooling systems. In Proceedings of the 12th CIRP Conference on Intelligent Computation in Manufacturing Engineering, Naples, Italy, 18–20 July 2018.
31. Carou, D.; Rubio, E.M.; Agustina, B.; Marín, M.M. Experimental study for effective and sustainable repair and maintenance of bars made of Ti-6Al-4V alloy application to the aeronautic industry. *J. Clean. Prod.* **2017**, *164*, 465–475. [CrossRef]
32. Carou, D.; Rubio, E.M.; Davim, J.P. A note on the use of the minimum quantity lubrication (MQL) system in turning. *Ind. Lubr. Tribol.* **2015**, *67*, 256–261. [CrossRef]
33. Khan, A.M.; Jamil, M.; Mia, M.; Pimenov, D.Y.; Gasiyarov, V.R.; Gupta, M.K.; He, N.; Khan, A.M. Multi-Objective Optimization for Grinding of AISI D2 Steel with Al_2O_3 Wheel under MQL. *Materials* **2018**, *11*, 2269. [CrossRef]
34. Mia, M.; Dey, P.R.; Hossain, M.S.; Arafat, M.T.; Asaduzzaman, M.; Shoriat, M.; Tareq, S.M. Taguchi S/N based optimization of machining parameters for surface roughness, tool wear and material removal rate in hard turning under MQL cutting condition. *Measurement* **2018**, *122*, 380–391. [CrossRef]
35. Mia, M.; Rifat, A.; Tanvir, M.F.; Gupta, M.K.; Hossain, M.J.; Goswami, A. Multi-objective optimization of chip-tool interaction parameters using Grey-Taguchi method in MQL-assisted turning. *Measurement* **2018**, *129*, 156–166. [CrossRef]
36. Mia, M. Mathematical modeling and optimization of MQL assisted end milling characteristics based on RSM and Taguchi method. *Measurement* **2018**, *121*, 249–260. [CrossRef]
37. Sharma, A.K.; Tiwari, A.K.; Dixit, A.R. Effects of Minimum Quantity Lubrication (MQL) in machining processes using conventional and nanofluid based cutting fluids: A comprehensive review. *J. Clean. Prod.* **2016**, *127*, 1–18. [CrossRef]
38. Singh, G.R.; Gupta, M.; Mia, M.; Sharma, V. Modeling and optimization of tool wear in MQL-assisted milling of Inconel 718 superalloy using evolutionary techniques. *Int. J. Adv. Manuf. Technol.* **2018**, *97*, 481–494. [CrossRef]
39. Singh, G.R.; Pruncu, C.I.; Gupta, M.K.; Mia, M.; Khan, A.M.; Jamil, M.; Pimenov, D.Y.; Sen, B.; Sharma, V.S. Investigations of Machining Characteristics in the Upgraded MQL-Assisted Turning of Pure Titanium Alloys Using Evolutionary Algorithms. *Materials* **2019**, *12*, 999–1016. [CrossRef]

40. Maruda, R.; Feldshtein, E.; Legutko, S.; Krolczyk, G. Analysis of Contact Phenomena and Heat Exchange in the Cutting Zone Under Minimum Quantity Cooling Lubrication conditions. *Arab. J. Sci. Eng.* **2016**, *41*, 661–668. [CrossRef]
41. Maruda, R.K.; Krolczyk, G.; Wojciechowski, S.; Zak, K.; Habrat, W.; Nieslony, P. Effects of extreme pressure and anti-wear additives on surface topography and tool wear during MQCL turning of AISI 1045 steel. *J. Mech. Sci. Technol.* **2018**, *32*, 1585–1591. [CrossRef]
42. Mia, M.; Morshed, M.S.; Kharshiduzzaman, M.; Razi, M.; Mostafa, M.; Rahman, S.; Ahmad, I.; Hafiz, M.; Kamal, A. Prediction and optimization of surface roughness in minimum quantity coolant lubrication applied turning of high hardness steel. *Measurement* **2018**, *118*, 43–51. [CrossRef]
43. Scharf, T.W.; Prasad, S.V. Solid lubricants: A review. *J. Mater. Sci.* **2013**, *48*, 511–531. [CrossRef]
44. Nageswara, D.; Vamsi, P. The influence of solid lubricant particle size on machining parameters in turning. *Int. J. Mach. Tools Manuf.* **2008**, *48*, 107–111. [CrossRef]
45. Wenlong, S.; Jianxin, D.; Hui, Z.; Pei, Y.; Jun, Z.; Xing, A. Performance of a cemented carbide self-lubricating tool embedded with MoS2 solid lubricants in dry machining. *J. Manuf. Process.* **2011**, *13*, 8–15. [CrossRef]
46. Maheshwera, U.; Ratnam, Y.; Reddy, R.; Kumar, S. Measurement and Analysis of Surface Roughness in WS2 Solid Lubricant Assisted Minimum Quantity Lubrication (MQL) Turning of Inconel 718. *Procedia CIRP* **2016**, *40*, 138–143.
47. Varma, J.; Patel, C. A review of effect of solid lubricant in hard turning of alloy steel. *Int. J. Adv. Res. Technol.* **2013**, *2*, 12–15.
48. Mia, M.; Gupta, M.K.; Lozano, J.A.; Carou, D.; Pimenov, D.Y.; Królczyk, G.; Khan, A.M.; Dhar, N.R. Multi-objective optimization and life cycle assessment of eco-friendly cryogenic N2 assisted turning of Ti-6Al-4V. *J. Clean. Prod.* **2019**, *210*, 121–133. [CrossRef]
49. Mia, M. Multi-response optimization of end milling parameters under through-tool cryogenic cooling condition. *Measurement* **2017**, *111*, 134–145. [CrossRef]
50. Hong, S.; Broomer, M. Economical and ecological cryogenic machining of AISI 304 austenitic stainless steel. *Clean Prod. Process.* **2000**, *2*, 157–166. [CrossRef]
51. Ghosh, R.; Zurecki, Z.; Frey, J.H. Cryogenic Machining with Brittle Tools and Effects on Tool Life. In *International Mechanical Engineering Congress and Exposition*; ASME: New York, NJ, USA, 2003; pp. 1–9.
52. Islam, A.K.; Mia, M.; Dhar, N.R. Effects of internal cooling by cryogenic on the machinability of hardened steel. *Int. J. Adv. Manuf. Technol.* **2017**, *90*, 11–20. [CrossRef]
53. Kamata, Y.; Obikawa, T. High speed MQL finish-turning of Inconel 718 with different coated tools. *J. Mater. Process. Technol.* **2007**, *192*, 281–286. [CrossRef]
54. Lu, T.; Jawahir, I. Metrics-based Sustainability Evaluation of Cryogenic Machining. *Procedia CIRP* **2015**, *29*, 520–525. [CrossRef]
55. Jawahir, I.; Attia, H.; Biermann, D.; Duflou, J.; Klocke, F.; Meyer, D.; Newman, S.; Pusavec, F.; Putz, M.; Rech, J.; et al. Cryogenic manufacturing processes. *CIRP Ann.* **2016**, *65*, 713–736. [CrossRef]
56. Sartori, S.; Ghiotti, A.; Bruschi, S. Temperature effects on the Ti6Al4V machinability using cooled gaseous nitrogen in semi-finishing turning. *J. Manuf. Process.* **2017**, *30*, 187–194. [CrossRef]
57. Sun, S.; Brandt, M.; Palanisamy, S.; Dargusch, M.S. Effect of cryogenic compressed air on the evolution of cutting force and tool wear during machining of Ti–6Al–4V alloy. *J. Mater. Process. Technol.* **2015**, *221*, 243–254. [CrossRef]
58. Shokrani, A.; Dhokia, V.; Newman, S. Environmentally conscious machining of difficult-to-machine materials with regard to cutting fluids. *Int. J. Mach. Tools Manuf.* **2012**, *57*, 83–101. [CrossRef]
59. Najiha, M.; Rahman, M.; Yusoff, A. Environmental impacts and hazards associated with metal working fluids and recent advances in the sustainable systems: A review. *Renew. Sustain. Energy Rev.* **2016**, *60*, 1008–1031. [CrossRef]
60. Hosseini, A.; Shabgard, M.; Pilehvarian, F. On the feasibility of a reduction in cutting fluid consumption via spray of biodegradable vegetable oil with compressed air in machining Inconel 706. *J. Clean. Prod.* **2015**, *108*, 90–103.
61. Hosseini, T.; Shabgard, M.; Pilehvarian, F. Application of liquid nitrogen and spray mode of biodegradable vegetable cutting fluid with compressed air in order to reduce cutting fluid consumption in turning Inconel 740. *J. Clean. Prod.* **2015**, *108*, 90–103.

62. Jamil, M.; Khan, A.; Hegab, H.; Gong, L.; Mia, M.; Gupta, M.; He, N. Effects of hybrid Al2O 3-CNT nanofluids and cryogenic cooling on machining of Ti–6Al–4V. *Int. J. Adv. Manuf. Technol.* **2019**, *102*, 3895–3909. [CrossRef]
63. Sen, B.; Mia, M.; Gupta, M.K.; Rahman, M.A.; Mandal, U.K.; Mondal, S.P. Influence of Al2O3 and palm oil–mixed nano-fluid on machining performances of Inconel-690: IF-THEN rules–based FIS model in eco-benign milling. *Int. J. Adv. Manuf. Technol.* **2019**, *103*, 3389–3404. [CrossRef]
64. Khan, A.M.; Jamil, M.; Salonitis, K.; Sarfraz, S.; Zhao, W.; He, N.; Mia, M.; Zhao, G. Multi-Objective Optimization of Energy Consumption and Surface Quality in Nanofluid SQCL Assisted Face Milling. *Energies* **2019**, *12*, 710–732. [CrossRef]
65. Singh, R.K.; Sharma, A.K.; Dixit, A.R.; Tiwari, A.K.; Pramanik, A.; Mandal, A. Performance evaluation of alumina-graphene hybrid nano-cutting fluid in hard turning. *J. Clean. Prod.* **2017**, *162*, 830–845. [CrossRef]
66. Gupta, M.K.; Jamil, M.; Wang, X.; Song, Q.; Liu, Z.; Mia, M.; Hegab, H.; Khan, A.M.; Collado, A.; Pruncu, C.I.; et al. Performance Evaluation of Vegetable Oil-Based Nano-Cutting Fluids in Environmentally Friendly Machining of Inconel-800 Alloy. *Materials* **2019**, *12*, 2792–2812. [CrossRef] [PubMed]
67. Mijanovic, K.; Sokovic, M. Ecological aspects of the cutting fluids and its influence on quantifiable parameters of the cutting processes. *J. Mater. Process. Technol.* **2001**, *109*, 181–189.
68. Skerlos, S.J.; Hayes, K.F.; Clarens, A.F.; Zhao, F. Current Advances in Sustainable Metalworking Fluids Research. *Int. J. Sustain. Manuf.* **2008**, *1*, 180–202. [CrossRef]
69. Erhan, S.; Sharma, B.; Perez, J. Oxidation and low temperature stability of vegetable oil-based lubricants. *Ind. Crop. Prod.* **2006**, *24*, 292–299. [CrossRef]
70. Shashidhara, Y.; Jayaram, S. Vegetable oils as a potential cutting fluid-An evolution. *Tribol. Int.* **2010**, *43*, 1073–1081. [CrossRef]
71. Lawal, S.; Choudhury, I.; Nukman, Y. Application of vegetable oil-based metalworking fluids in machining ferrous metals—A review. *Int. J. Mach. Tools Manuf.* **2012**, *52*, 1–12. [CrossRef]
72. Xavior, M.; Adithan, M. Determining the influence of cutting fluids on tool wear and surface roughness during turning of AISI 304 austenitic stainless steel. *J. Mater. Process. Technol.* **2009**, *209*, 900–909. [CrossRef]
73. Gupta, M.; Mia, M.; Singh, G.R.; Pimenov, D.; Sarikaya, M.; Sharma, V. Hybrid cooling-lubrication strategies to improve surface topography and tool wear in sustainable turning of Al 7075-T6 alloy. *Int. J. Adv. Manuf. Technol.* **2019**, *101*, 55–69. [CrossRef]
74. Sanchez, J.; Pombo, I.; Alberdi, R.; Izquierdo, B.; Ortega, N.; Plaza, S.; Martinez-Toledano, J. Machining evaluation of a hybrid MQL CO2 grinding technology. *J. Clean. Prod.* **2010**, *18*, 1840–1849. [CrossRef]
75. Su, Y.; He, N.; Li, L.; Iqbal, A.; Xiao, M.H.; Xu, S.; Qiu, B.G. Refrigerated cooling air cutting of difficult-to-cut materials. *Int. J. Mach. Tools Manuf.* **2007**, *47*, 927–933. [CrossRef]
76. Zhang, C.; Zhang, S.; Yan, X.; Zhang, Q. Effects of internal cooling channel structures on cutting forces and tool life in side milling of H13 steel under cryogenic minimum quantity lubrication condition. *Int. J. Adv. Manuf. Technol.* **2016**, *83*, 975–984. [CrossRef]
77. Sáenz de Pipaón, M.J.; Rubio, E.M.; Villeta, M.; Sebastián, M.A. Influence of cutting conditions and tool coatings on the surface finish of workpieces of magnesium obtained by dry turning. In Proceedings of the 19th International DAAAM Symposium, Trnava, Slovakia, 22–25 October 2008; pp. 604–605.
78. Carou, D.; Rubio, E.M.; Lauro, J.P.; Davim, J.P. Experimental investigation on surface finish during intermittent turning of UNS M11917 magnesium alloy under dry and near dry machining conditions. *Measurement* **2014**, *56*, 136–154. [CrossRef]
79. Carou, D.; Rubio, E.M.; Lauro, J.P.; Davim, J.P. Experimental investigation on finish intermittent turning of UNS M11917 magnesium alloy under dry machining. *Int. J. Adv. Manuf. Technol.* **2014**, *75*, 1417–1429. [CrossRef]
80. Ozsváth, P.; Szmejkál, A.; Takács, J. Dry milling of magnesium-based hybrid materials. *Transp. Eng.* **2008**, *36*, 73–78. [CrossRef]
81. Prakash, S.; Palanikumar, K.; Mercy, J.L.; Nithyalakshmi, S. Evaluation of surface roughness parameters (Ra, Rz) in drilling of MDF composite panel using Box-Behnken experimental design (BBD). *Int. J. Des. Manuf. Technol.* **2011**, *5*, 52–62.
82. Rubio, E.M.; Valencia, J.L.; Saá, A.J.; Carou, D. Experimental study of the dry facing of magnesium pieces based on the surface roughness. *Int. J. Precis. Eng. Manuf.* **2013**, *14*, 995–1001. [CrossRef]

83. Rubio, E.M.; Villeta, M.; Carou, D.; Saá, A.J. Comparative analysis of sustainable cooling systems in intermittent turning of magnesium pieces. *Int. J. Precis. Eng. Manuf.* **2014**, *15*, 929–940. [CrossRef]
84. Rubio, E.M.; Valencia, J.L.; de Agustina, B.; Saá, A.J. Tool selection based on surface roughness in dry facing repair operations of magnesium pieces. *Int. J. Mater. Prod. Technol.* **2014**, *48*, 116–134. [CrossRef]
85. Sáenz de Pipaón, J.M.; Rubio, E.M.; Villeta, M.; Sebastián, M.A. Analysis of the chips obtained by dry turning of UNS M11311 magnesium. In Proceedings of the 3rd Manufacturing Engineering Society International Conference 2009, Alcoy, Spain, 17–19 June 2009; pp. 33–38.
86. Sáenz de Pipaón, J.M.; Rubio, E.M.; Villeta, M.; Sebastián, M.A. Selection of the cutting tools and conditions for the low speed turning of bars of magnesium UNS M11311 based on the surface roughness. In *Innovative Production Machines and Systems*; Whittles Publishing: Cambridge, UK, 2010; pp. 174–179.
87. Villeta, M.; de Agustina, B.; Sáenz de Pipaón, J.M.; Rubio, E.M. Efficient optimisation of machining processes based on technical specifications for surface roughness: Application to magnesium pieces in the aerospace industry. *Int. J. Adv. Manuf. Technol.* **2012**, *60*, 1237–1246. [CrossRef]
88. Rubio, E.M.; Valencia, J.L.; Carou, D.; Saá, A.J. Inserts selection for intermittent turning of magnesium pieces. In *Applied Mechanics and Materials*; Trans Tech Publications: Bäch, Switzerland, 2012; Volume 217, pp. 1581–1591.
89. Rubio, E.M.; Villeta, M.; Saá, A.J.; Carou, D. *Analysis of Main Optimization Techniques in Predicting Surface Roughness in Metal Cutting Processes*; Trans Tech Publications: Bäch, Switzerland, 2012; Volume 217, pp. 2171–2182.
90. Rubio, E.M.; Villeta, M.; Agustina, B.; Carou, D. Surface roughness analysis of magnesium pieces obtained by intermittent turning. In *Materials Science Forum*; Trans Tech Publications: Bäch, Switzerland, 2014; Volume 773, pp. 377–391.
91. Carou, D.; Rubio, E.M.; Lauro, C.H.; Davim, J.P. The effect of minimum quantity lubrication in the intermittent turning of magnesium based on vibration signals. *Measurement* **2016**, *94*, 338–343. [CrossRef]
92. Carou, D.; Rubio, E.M.; Lauro, C.H.; Brandão, L.C.; Davim, J.P. Study based on sound monitoring as a means for superficial quality control in intermittent turning of magnesium workpieces. *Procedia CIRP* **2017**, *62*, 262–268. [CrossRef]
93. Carou, D.; Rubio, E.M.; Davim, J.P. Chapter 5. Machinability of magnesium and its alloys: A review. In *Traditional Machining Processes*; Springer: Berlin/Heidelberg, Germany, 2015; pp. 133–152.
94. Berzosa, F.; de Agustina, B.; Rubio, E.M.; Davim, J.P. Feasibility Study of Hole Repair and Maintenance Operations by Dry Drilling of Magnesium Alloy UNS M11917 for Aeronautical Components. *Metals* **2019**, *9*, 740–755. [CrossRef]
95. Basmacı, G.; Taskin, A.; Koklu, U. Effect of tool path strategies and cooling conditions in pocket machining of AZ91 magnesium alloy. *Indian J. Chem. Technol.* **2019**, *26*, 139.
96. Bruschi, S.; Bertolini, R.; Ghiotti, A.; Savio, E.; Guo, W.; Shivpuri, R. Machining-induced surface transformations of magnesium alloys to enhance corrosion resistance in human-like environment. *CIRP Ann.* **2018**, *67*, 579–582. [CrossRef]
97. Danilenko, B.D. Selecting the initial cutting parameters in machining magnesium alloys. *Russ. Eng. Res.* **2009**, *29*, 316–319. [CrossRef]
98. Giraud, E.; Rossi, F.; Germain, G.; Outeiro, J.C. Constitutive Modelling of AZ31B-O Magnesium Alloy for Cryogenic Machining. *Procedia CIRP* **2013**, *8*, 522–527. [CrossRef]
99. Nasr, M.N.A.; Outeiro, J.C. Sensitivity Analysis of Cryogenic Cooling Machining of Magnesium Alloy AZ31B-O. *Procedia CIRP* **2015**, *31*, 264–269. [CrossRef]
100. Outeiro, J.C.; Rossi, F.; Fromentin, G.; Poulachon, G.; Germain, G.; Batista, A.C. Process Mechanics and Surface Integrity Induced by Dry and Cryogenic Machining of AZ31B-O Magnesium Alloy. *Procedia CIRP* **2013**, *8*, 487–492. [CrossRef]
101. Pimenov, D.; Erdakov, I. ANN Surface Roughness Optimization of AZ61 Magnesium Alloy Finish Turning: Minimum Machining Times at Prime Machining Costs. *Materials* **2018**, *11*, 808.
102. Rashid, R.; Sun, S.; Wang, G.; Dargusch, M. Experimental investigation of laser assisted machining of AZ91 magnesium alloy. *Int. J. Precis. Eng. Manuf.* **2013**, *14*, 1263–1265. [CrossRef]
103. Salahshoor, M.; Guo, Y.B. Cutting mechanics in high speed dry machining of biomedical magnesium-calcium alloy using internal state variable plasticity model. *Int. J. Mach. Tools Manuf.* **2011**, *51*, 579–591. [CrossRef]

104. Tonshoff, H.; Winkler, J.; Tonshoff, H. The influence of tool coatings in machining of magnesium. *Surf. Coat. Technol.* **1997**, *94*, 610–616. [CrossRef]
105. Rubio, E.M.; Sáenz de Pipaón, J.M.; Valencia, J.L.; Villeta, M. Design, Manufacturing and Machining Trials of Magnesium Based Hybrid Parts. In *Machining of Light Alloys: Aluminium, Titanium and Magnesium*; CRC-Press: Boca Raton, FL, USA, 2018.
106. Rubio, E.M.; Villeta, M.; Valencia, J.L.; Sáenz de Pipaón, J.M. Experimental Study for Improving the Repair of Magnesium–Aluminium Hybrid Parts by Turning Processes. *Metals* **2018**, *8*, 59. [CrossRef]
107. Rubio, E.M.; Villeta, M.; Valencia, J.L.; Sáenz de Pipaón, J.M. Cutting Parameter Selection for Efficient and Sustainable Repair of Holes Made in Hybrid Mg–Ti–Mg Component Stacks by Dry Drilling Operations. *Materials* **2018**, *11*, 1369. [CrossRef]
108. Rubio, E.M.; Blanco, D.; Marín, M.M.; Saenz de Pipaon, J.M. Analysis of the surface roughness of Mg-Al-Mg hybrid components obtained by drilling using different cooling systems. In Proceedings of the 13th CIRP Conference on Intelligent Computation in Manufacturing Engineering, Naples, Italy, 17–19 July 2019.
109. Sanz, C.; Fuentes, E.; Gonzalo, O.; Bengoetxea, I.; Obermair, F.; Eidenhammer, M. Advances in the ecological machining of magnesium and magnesium-based hybrid parts. *Int. J. Mach. Mach. Mater.* **2008**, *4*, 302–319. [CrossRef]
110. Satheesh, J.; Tajamul, P.; Madhusudhan, T.H. Optimal machining conditions for turning of AlSiC Metal Matrix Composites using ANOVA. *Int. J. Innov. Res. Sci. Eng. Technol.* **2013**, *2*, 6171–6176.
111. Sokolowsky, J.H.; Szablewski, D.; Kasprzak, W.; Ng, E.G.; Dumitrescu, M. Effect of tool cutter immersion on Al-Si bi-metallic materials in high-speed milling. *J. Achiev. Mater. Manuf. Eng.* **2006**, *17*, 15–20.
112. Vilches, F.J.T.; Hurtado, L.S.; Fernandez, F.M.; Bermudo, C. Analysis of the chip geometry in dry machining of aeronautical aluminum alloys. *Appl. Sci.* **2017**, *7*, 132. [CrossRef]
113. Rafai, N.H.; Islam, M.N. An investigation into dimensional accuracy and surface finish achievable in dry turning. *Mach. Sci. Technol.* **2009**, *13*, 571–589. [CrossRef]
114. Fletcher, D.I.; Kapoor, A.; Steinhoff, K.; Schuleit, N. Theoretical analysis of steady-state texture formation during wear of a bi-material surface. *Wear* **2001**, *251*, 1332–1336. [CrossRef]
115. Arokiasamy, S.; Anand Ronald, B. Experimental investigations on the enhancement of mechanical properties of magnesium-based hybrid metal matrix composites through friction stir processing. *Int. J. Adv. Manuf. Technol.* **2017**, *93*, 493–503. [CrossRef]
116. Troconis, B.C.R.; Frankel, G.S. Effects of Pretreatments on the Adhesion of Acetoacetate to AA2024-T3 Using the Blister Test. *Corrosion* **2014**, *70*, 483–495. [CrossRef]
117. Vicario, I.; Crespo, I.; Plaza, L.M.; Caballero, P.; Idoiaga, I.K. Aluminium Foam and Magnesium Compound Casting Produced by High-Pressure Die Casting. *Metals* **2016**, *6*, 24. [CrossRef]
118. Taub, A. Automotive materials: Technology trends and challenges in the 21st century. *MRS Bull.* **2006**, *31*, 336–343. [CrossRef]
119. Zhang, L.; Dhupia, J.S.; Wu, M.; Huang, H. A Robotic Drilling End-Effector and Its Sliding Mode Control for the Normal Adjustment. *Appl. Sci.* **2018**, *8*, 1892–1910. [CrossRef]
120. Ralph, W.C.; Johnson, W.S.; Toivonen, P.; Makeev, A.; Newman, J.C. Effect of various aircraft production drilling procedures on hole quality. *Int. J. Fatigue* **2006**, *28*, 943–950.
121. The American Society of Mechanical Engineers. *Surface Texture: Surface Roughness, Waviness and Lay*; ASME: New York, NY, USA, 2010.
122. Montgomery, D.C. *Design and Analysis of Experiments*; John Wiley & Sons, Inc.: New York, NY, USA, 2005.
123. Available online: https://www.hoffmann-group.com (accessed on 14 January 2020).
124. *ISO 4287:1997 Geometrical Product Specifications (GPS)—Surface Texture: Profile Method—Terms, Definitions and Surface Texture Parameters*; International Organization for Standardization: Geneva, Switzerland, 1997.
125. Available online: https://www.kometgroup.com/en/komet/komet/ (accessed on 14 January 2020).

© 2020 by the authors. Licensee MDPI, Basel, Switzerland. This article is an open access article distributed under the terms and conditions of the Creative Commons Attribution (CC BY) license (http://creativecommons.org/licenses/by/4.0/).

Article

An Approach to Sustainable Metrics Definition and Evaluation for Green Manufacturing in Material Removal Processes

César Ayabaca [1,2,*] and Carlos Vila [1]

1 Department of Mechanical Engineering and Materials, Universitat Politècnica de València, 46022 València, Spain; carvipas@upv.es
2 Department of Mechanical Engineering, Faculty of Mechanical Engineering, Escuela Politécnica Nacional, Quito 170524, Ecuador
* Correspondence: ceaysar1@doctor.upv.es; Tel.: +34-61-7710-407

Received: 31 October 2019; Accepted: 9 January 2020; Published: 14 January 2020

Abstract: Material removal technologies should be thoroughly analyzed not only to optimize operations but also to minimize the different waste emissions and obtain cleaner production centers. The study of environmental sustainability in manufacturing processes, which is rapidly gaining importance, requires activity modeling with material and resource inputs and outputs and, most importantly, the definition of a balanced scorecard with suitable indicators for different levels, including the operational level. This paper proposes a metrics deployment approach for the different stages of the product life cycle, including a conceptual framework of high-level indicators and the definition of machining process indicators from different perspectives. This set of metrics enables methodological measurement and analysis and integrates the results into aggregated indicators that can be considered for continuous improvement strategies. This approach was validated by five case studies of experimental testing of the sustainability indicators in material removal operations. The results helped to confirm or modify the approach and to adjust the parameter definitions to optimize the initial sustainability objectives.

Keywords: green manufacturing; sustainability metrics; cleaner product life cycle; material removal processes

1. Introduction

The analysis of industrial manufacturing processes from the sustainability point of view started during the early 1980s in order to meet the sustainable development concept that arose in the 1970s as a result of a general worry about the global environment due to pollution and the consumption of energy and raw materials. In 1983, the United Nation (UN)'s World Commission on the Environment and Development, known as the Brundtland Commission, prepared a formal report entitled "Our Common Future", in which the concept of sustainable development was defined as "development that meets the current needs of people without compromising the ability of future generations to meet theirs" [1]. Global movements and policies were generated in order to find common strategies that could be applied worldwide [2,3].

Subsequently, in the area of product manufacturing, many research projects were initiated to generate proposals for process improvement and the optimization of the consumption of resources and raw materials [4–6] as well as rules and regulations for waste management for cleaner production methods [7]. At the UN summit in 2015, the 17 goals for sustainable development for 2015 to 2030 were laid down, including goal number 9 (Industry, Innovation, and Infrastructure) which is focused on building resilient infrastructures, promoting inclusive and sustainable industrialization, and fostering innovation [8]. Among others, we could list the most important international initiatives that encourage a better understanding of sustainability:

(a) United Nations, Sustainable Development Goals.
(b) ISO 26000: 2010 Social Responsibility [9].
(c) United Nations Global Compact (Global Compact) [10].
(d) Guide for Organisation for Economic Co-operation and Development (OECD) multinationals. (Global Reporting Initiative) [11].

Since the end of the last century, researchers, technicians, managers, and environmentalists have recommended three economic, environmental, and social dimensions for evaluating economic, environmental, and social aspects, which are known as sustainability dimensions [12].

We must underline the contribution of Zackrisson et al. [13], who studied the relationship between performance measurement systems and sustainability. Although the research was done in Swedish manufacturing companies, they found that at the shop floor level, 90% of the indicators have at least an indirect relationship with one or more of the economic, environmental, or social dimensions, while 26% of the indicators are indirectly related to the environmental dimension.

In materials and manufacturing engineering research topics, indicators are being proposed to measure the sustainability of industrial processes. Reich-Weiser et al. defined a general set of metrics for sustainable manufacturing [14], while Shuaib et al. [15] and Jayal et al. [16] designed a framework and a model for sustainable manufacturing, respectively. Singh et al. developed an expert system for the performance evaluation of small and medium enterprises [17]. With a greater focus on machining technologies, we find the work of Rajurkar et al., which explored how to ensure sustainability and optimize non-traditional machining processes [18].

Since machining technologies pollute and consume energy and raw materials, it is understandable that more research actions are needed. This work is part of the research into the design of green manufacturing activities within the product life cycle, focusing on machining technologies. This work will propose a framework and a list of indicators that will help to get data and information from manufacturing activities as part of the Life Cycle Assessment. The paper is structured into six sections. The Section 2 reviews the state of the art of green manufacturing applied to machining and presents our vision of a green manufacturing activities model. The Section 3 describes the framework for defining the metrics from the point of view of materials, parts, and processes during the product life cycle. The Section 4 describes the validation experiments, while the conclusions and future work are presented in the last two sections.

2. State of the Art in Green Machining Operations

2.1. Analysis of Previous Works of Sustainability in Industrial Manufacturing Processes

The aim of this work was to carry out several experiments to calculate different types of sustainable indicators with different materials, processes, and machining centres. Many studies have been carried out on sustainable manufacturing and, in the literature, we can find many focused on specific disciplines, such as machining technologies. Peralta et al. [19] showed the trends emerging in the last 15 years in the sustainability of machining processes from the point of view of the triple bottom line and the three general dimensions: economic, ecological, and equity. Bhanot et al. [20] presented a statistically validated study that proposed a comprehensive sustainability framework for the manufacturing domain to strengthen the enablers and mitigate barriers based on the responses of researchers and industry professionals.

Eastwood et al. [21] developed a sustainable assessment methodology to both improve the accuracy of the existing approaches in identifying the sustainability impact of a product and to assist manufacturing decision makers using unit process modeling and life cycle inventory techniques. The proposed methodology can quantify sustainability metrics by aggregating information from the process level, where various metrics require different aggregation methods, from the manufacturing process to the manufacturing system level.

The work of Garretson et al. [22] facilitated standards development efforts by harmonizing the terms used to describe production processes. A set of 47 terms focusing on process characterization and describing sustainable production was generated, although terms unique to individual production processes were omitted. The terms were organized into six categories to define the overall concepts: Scope, Boundary, Material, Measurement, Model, and Flow. Then, definitions of the terms were derived from: (a) the literature in sustainable manufacturing and chemical and process industries, (b) process characterization and planning, (c) organization standards, and life cycle assessment and management. The reported terms and definitions are not unique to sustainable production.

Helleno et al. [23] proposed a conceptual method of integrating a new group of sustainability indicators into the Value Stream Mapping (VSM) tool to assess manufacturing processes. The method was applied in three case studies, and the results demonstrated that the proposed method identified different levels of manufacturing process sustainability and thus enabled the development of improved scenarios.

Kluczek [24] introduced an "improvement scenario" in a company producing heating devices, between existing and new processes, based on an approach that can be applied to perform the sustainability assessment of manufacturing processes, requiring less detailed data, time, and expert knowledge but still providing a company-level analysis.

Latif et al. [25] developed an interactive model to determine the sustainability index based on user responses; the model is able to provide suggestions to improve sustainability, as well as carbon footprint reduction, and can assist industry to identify its shortcomings in achieving sustainability, can determine the carbon footprint reduction potential, and can compare the sustainability index as a benchmark measure.

Moldavska and Welo [26] analyzed the different definitions of sustainable manufacturing (SM) and identified the current understanding using an inductive content analysis of definitions published in a variety of academic journals. It is proposed that the findings can serve as a foundation for the development of a common language in both the research field and industrial practice.

In the literature reviewed, Winroth et al. [27] compiled the existing Sustainable Framework of Indicators, in which sustainability assessments can be carried out at different levels within an organization.

The hierarchical dimension of the activities (global, national, corporate, or factory), and the functional dimension (product, supplier, production, logistics, and customer) are shown in Figure 1a. This analysis allows us to select the production indicators at the factory level, at which in each industrial process there are proposals for the measurement of sustainability, among which manufacturing companies use the key performance indicators (KPIs) for the control and monitoring of their processes for continuous improvement.

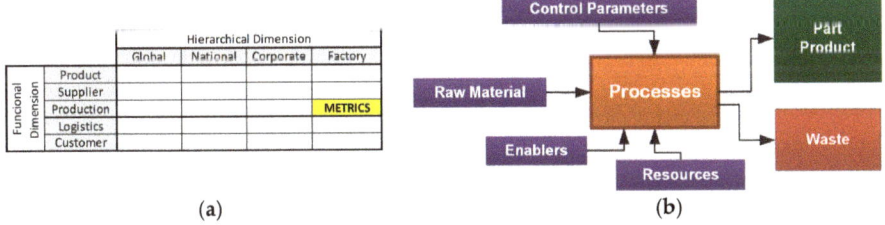

Figure 1. Dimensions selection sustainability indicators modeling. (**a**) Criteria for Metrics Selection [27]. (**b**) Generic Process Flow Diagram [28].

For example, Linke et al. [28] proposed a generic process diagram: resources (raw material, energy, auxiliary materials, etc.) and enablers (the machine, the worker, the tools, etc.) are considered as inputs to the process, as shown in Figure 1b.

The process can be evaluated through performance parameters, indicators, or metrics, which can be related not only to outputs but also to resources, enablers, and inputs in order to give feedback during the manufacturing phase of the product life cycle. The final quality of the manufactured part or the assembled product will be achieved by controlling activities but also by obtaining data and information from the metrics. The waste output generated from the process will be in the form of physical material or environmental pollution.

From the perspective of manufacturing throughout the product life cycle, industrial processes can be analyzed by considering stages such as the extraction of raw materials, manufactured materials, product manufacturing, shipping, distribution, use, recycling, and the final disposal of the product. It is important to emphasize that a certain percentage of the recycled material can be integrated back into the material manufacturing process, while the rest must be used in other applications.

Vila et al. [29] proposed a framework for defining a structured set of metrics that are customizable for operations in different manufacturing technologies. Although the research work was applied to AISI 1018 material turning operations in order to analyze the surface integrity of the part, the contribution established the relationships between the machining parameters of the turning process and the final properties of the manufactured part, such as roughness, microhardness, and other parameters. This was the first attempt to link general sustainable metrics with technological-related metrics. For this reason, several perspectives were proposed. Table 1 shows Product, Process, and Resources (PPR) perspectives and some activity indicators that can be defined during the product life cycle. High-level general indicators aligned with sustainable objectives are initially created for each activity.

Table 1. Product, material, and resources activities as general performance indicators.

PPR Perspective	Activity	Generic Indicator	Units	Sustainable Objective
Material	Raw Materials Extraction	Material class	Options	↗
		Material properties	Value	↗
	Manufacture	% Recycled content	%	↗
		Weight	Kg	↘
Product	Transport and Distribution	Region where it is manufactured.	Km, CO_2	↘
		Transportation to end user.	M	↘
	Use	Energy need throughout its useful life	kWh	↘
	Recycling	% Recycling	%	↗
	End Disposition	% Burned	%	↘
		% Spell	%	↘
Process	Product Manufacture	Energy of the manufacturing process.	kWh	↘
		Useful product lifetime of the product.	s	↗
		Energy needed for assembly.	kWh	↘

Note: Objectives symbol's meaning ↗ maximize; ↘ minimize.

With this general view of manufacturing sustainable metrics, the next step was to explore previous works in order to define a sustainable scorecard indicator for machining or material removal techniques.

2.2. Dimensions of the Sustainability Metrics for Machining Processes

In the review of the state of the art, we can find some contributions that define sustainable metrics and show how important it is to define them at different levels of the product life cycle. One of the most interesting works was done by Bhanot et al. [30], in which they analyzed the complex interdependences between parameters that affect the result of a metal cutting process when seeking sustainable objectives. From the literature and from other previous works, we can highlight the manufacturing aspects to measure.

In manufacturing, it is critical to guarantee the competitiveness of each activity from many points of view, and a balanced optimization between the economic, environmental, and social dimensions must be obtained. Therefore, it is necessary to define not only the technological metrics, but also the sustainability metrics aligned with the manufacturing process. For example, for each dimension, we can list some aggregated metrics:

(a) Economic Dimension: Surface Roughness, Material Removal Rate (MRR), Tool Life per Edge, Production Rate per Edge, Production Cost per Component, Process and Production Management.
(b) Environmental Dimension: Coolant Consumption, Carbon Emission, Energy Consumption, Cutting Temperature, Recyclable Waste Production, Non-Recyclable Waste Production, Waste Management.
(c) Social Dimension: Individual Productivity, Relations with Other Workers, Worker Skills, Rotation Flexibility at Work, Punctuality at Work, Senior Management Support, Total Satisfaction, Suspicious Work Environment, Degree of Support from Authorities, Compliance with Worker Requirements.

The indicators in the economic and environmental dimensions can be defined through analytic expressions, and they can be quantitatively evaluated using data mining and process calculations. In the social dimension, the indicators are mainly evaluated qualitatively. Some of these indicators are introduced in Table 2. According to the product life cycle phase and for different PPR perspectives—Process, for example—we can find technical metrics associated with the economic dimension (Material Removal Rate), the environment dimension (Cutting Temperature), or the social dimension (Worker Skills).

Table 2. Generic indicators for sustainable machining. PPR: Product, Process, and Resources.

PPR Perspective	Phase	Generic Indicator	Acronyms	Units	Sustainable Dimension
Product	Use End Disposition	Surface Roughness	R_a	μ	Economic
		Refrigerant Consumption	R_c	m^3	Environmental
		Carbon Emissions	C_e	CO_2	Environmental
Process	Product Manufacture	Material Remove Rate	MRR	m^3/s	Economic
		Tool Life per Edge	T.L/edge	Min	Economic
		Production Rate per Edge	PR/edge	Units	Economic
		Production Cost per Component	PC/edge	€/part	Economic
		Energy Consumption	Ec	kWh	Environmental
		Cutting Temperature	Ct	°	Environmental
		Worker Productivity	Wp	%	Social
		Relations with Other Workers	Rw	%	Social
		Worker Skills	Ws	%	Social
		Rotation Flexibility at Work	Rf	%	Social
		Punctuality at Work	Pw	%	Social
		Senior Management Support	Sms	%	Social
		Total Satisfaction	Ts	%	Social
		Auspicious Work Environment	Awe	%	Social
		Support from Authorities	Sfa	%	Social
		Worker Requirements	Wr	%	Social

The appropriate selection of sustainable indicators allows for the diagnosis of continuous improvement plans in industrial processes, especially in manufacturing processes. However, it is still difficult to define social metrics in manufacturing activities, as Ayabaca and Vila presented in their work and, moreover, to analyze the data [31].

Bhanot et al. [32] presented a study on a machining group in which the interdependencies of different sustainable machining parameters were examined in the context of milling and turning processes.

In order to ensure competitiveness in the manufacturing field, there must be a balance between the economic, environmental, and social dimensions. Gupta et al. [33] presented an experimental investigation that compared empirical and experimental results, which was complemented by a desirability optimization technique, to study the impact on cutting forces, surface roughness, tool wear, surface topography, microhardness, and surface chemical composition in turning the aerospace material titanium (grade-2) alloy, considering Minimum Quantity Lubrication (MQL) conditions.

Hegab et al. [34] developed and discussed a sustainability assessment algorithm for machining processes. The four life cycle stages (pre-manufacturing, manufacturing, use, and post-use) are included in the proposed algorithm. Energy consumption, machining costs, waste management, environmental impact, and personal health and safety are used to express the overall sustainability assessment index. Kadam et al. [35] analyzed the surface integrity in high-speed machining of Inconel 718, and the results show that a good surface finish and residual stresses in compressive regimes can be ensured in the high-speed machining range with low MRR in a water-vapor machining environment, this also being feasible at high MRR in dry cutting.

Benedicto et al. [36] presented a comprehensive analysis of the use of cutting fluids and their main alternatives in machining, focusing on the economic, environmental, and technical dimensions. Zhao et al. [37] reviewed a critical assessment of energy consumption in a machining system at the process, machine, and system levels. Machine tool power demands in different machine states with different components were also discussed, and the predictive methods of energy consumption at different levels were summarized. Energy consumption reduction strategies to achieve sustainable manufacturing were also discussed.

Abbas et al. [38] presented an extensive study of the effectiveness of using different cooling and lubrication techniques when turning AISI 1045 steel. Three multi-objective optimization models were employed to select the optimal cutting conditions. The results offer a clear guideline for selecting the optimal cutting conditions based on different scenarios: MQL nanofluid compared to dry and flood approaches.

Ali et al. [39] found that the tool path strategy has a significant influence on the end outcomes of face milling, considering the surface topography with respect to different cutter path strategies and the optimal cutting strategy for the material Al 2024. Li et al. [40] evaluated the cutting performance of cutting tools in the high-speed machining (HSM) of AISI 4340 by using tools coated with TiN/TiCN/TiAlN multi-coating, TiAlN + TiN coating, TiCN + NbC coating, and AlTiN coating, respectively. A TiN/TiCN/TiAlN multi-coated tool is the most suitable for the high-speed milling of AISI 4340 due to the lower cutting force, lower cutting temperature, and high diffusion resistance of the material. Gupta et al. [41] discussed the features of two innovative techniques for machining an Inconel-800 superalloy by plain turning while considering some critical parameters, reducing the amount of cutting fluid while using sustainable methods. Near dry machining (NDM) will be possible and will solve the problem of chemical components in the fluids being harmful to human health.

In recent research, Gamage et al. [42] used a Taguchi design of experiments and analysis of variance (ANOVA) to identify the significant parameters that optimize the process energy consumption of wire electro-discharge machining (WEDM) of the superalloys Inconel-718 and Ti64Al4V. The results indicate that the preferred parameters to minimize the specific energy consumption are workpiece thickness, wire material, wire diameter, and pulse-OFF time. The reduction of carbon emissions corresponds to the non-working energy consumption of the machines, which is also calculated.

Gunda et al. [43] presented a novel technique for the generation of machining techniques—namely, high-pressure minimum quantity solid lubricant (HP-MQSL) and an experimental setup, with an aim of improving process performance and eliminating the use of cutting fluids in machining operations.

Lu and Jawahir [44] presented a sustainability evaluation methodology for manufacturing processes based on cryogenic machining processes which involves a metrics-based Process Sustainability Index (ProcSI) evaluation. This helps to decide the best cutting conditions from the sustainable manufacturing viewpoint.

Pusavec et al. [45] presented an experimental study of the sustainable high-performance machining of Inconel 718 with the development of performance-based predictive models for dry, near-dry (MQL), cryogenic, and cryolubrication (cryogenic þ near-dry) machining processes using the response surface methodology (RSM). The models developed in the first part of the paper are used in the second part for process evaluation and optimization, to determine the optimum machining conditions for an overall process performance improvement.

Goindi and Sarkar [46] presented a review of all aspects of dry machining, including the sustainability aspects of machining, especially focusing on three research objectives: (1) identifying the areas where dry machining has been successfully adopted and where it has not been possible to do so, (2) reporting on the research work carried out and various alternative solutions provided by the researchers in the area of dry machining, and (3) finding gaps in the current knowledge and suggesting some directions for further work to make dry machining more sustainable, profitable, and adaptable to product manufacturing. Shin et al. [47] presented a component-based energy-modeling methodology to implement the online optimization needed for real-time control in a milling machine. Models that can predict energy up to the tool path level at specific machining configurations are called component models.

Um et al. [48] proposed an approach for deriving an energy estimation model from general key performance indicators of the sustainability of machine tools in the laser welding process of an automotive assembly line and the milling process of an aircraft part manufacturer. ANOVA and RSM are widely used for optimizing cutting parameter tools. Zhang et al. [49] proposed using the Pareto diagram to calculate multi-objective optimization, although this is difficult when there are more than two objectives. This proposal lists and characterizes all the 128 scenarios of sustainable machining operations, considering seven objectives that include energy, cost, time, power, shear force, tool life, and surface finish. The results show that all the scenarios can be converted into a simple objective situation that has a single solution or a set of contradictory bi-objective cases that can be represented on a simple Pareto front.

The use and storage of the calculated indicators, together with modern systems of data acquisition and information management in real time, will strengthen the implementation of advanced manufacturing systems, or what is called Industry 4.0. Activities such as team maintenance and specific service requirements can be planned and adjusted in real time. Gao et al. [50] reviewed the historical development of prognosis theories and techniques and projected their future growth in the emerging cloud infrastructure.

2.3. Sustainable Metrics for Manufacturing Processes

The first contribution of the research after the review of the state of the art is shown in Figure 2, which gives the Sustainability Approach to Manufacturing Industrial Processes and summarizes the stages of the product life cycle: inputs, enablers, manufactured parts, and waste. This framework was outlined after a deep analysis of manufacturing industrial processes and activities. The activities model was sketched using ICAM Definition (IDEF) methodologies [51] and Unified Modeling Language (UML). Figure 2 presents the top level from the hierarchical model that is later deployed in different layers with more detail of activities and customized for different manufacturing technologies.

The activity model includes, for a manufacturing process, the inputs, which can be either raw material or a geometrical preform from previous manufacturing technology. The input of controls depends on the manufacturing technique, and it is represented here as a Manufacturing Process Group Technique (MPGT). For each manufacturing technology group (i), we can find different variations or techniques (j). The evaluation of the technological process parameters, according to the manufacturing process plan (control), will assure the quality of the final product or the quality of an individual part.

The controls parameters will depend on the technological process that is applied, while the performance parameters, which can be also introduced, define the process efficiency.

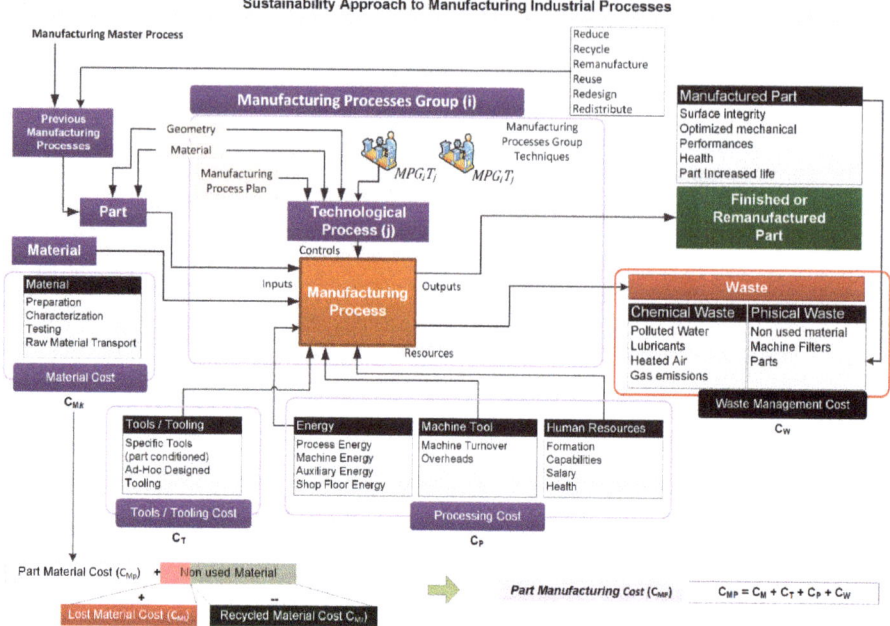

Figure 2. Activity model of a generic industrial manufacturing processes.

The model describes the general enablers and resources that are grouped into tools and tooling, energy consumption, machine tools, and human resources. For each enabler and resource, we can analyze the different issues and therefore define specific metrics. For example, the model shows countable metrics such as the machine-consumed energy or uncountable metrics such as works formation.

Finally, an important issue is to define the economic impact considering all the activities, inputs, controls, resources, enablers, and outputs. The costs of part manufacturing can be calculated considering the material costs (C_M), the cost of tools (C_T), the costs of the process (C_P), and the costs of waste management (C_W).

For the final output, the analysis of all the metrics defined for inputs, controls, enablers, and resources, for each product life cycle phase and from different perspectives, will help to establish whether the result meets the technical, functional, and sustainability requirements.

With this activity model, the next step is to validate it for a manufacturing process and technology that, in our research, will be material removal or machining processes and technologies.

2.4. Sustainability Metrics for Machining Operations

At the factory level, a manufacturing technology or technique requires one to define correctly the process plan so that the processes, equipment, people, etc., in the production system perform a specific function. Manufacturing process engineering requires a clear and complete description of the information associated with the process and the exchange of data to be of such a quality that forecasts of the results can be obtained. The modeling of systems and processes is expected to standardize the process—what is done, what is controlled, what resources are required—and define the products generated.

For the improvement of industrial quality, a knowledge of industrial processes is required. In this case, we will analyze the process in which it is defined: function, inputs, outputs, resources, and controls that allow measuring performance, as well as the emissions generated, which are evaluated in the context of sustainability.

The second contribution of this research is the matching of the previous general model in a specific model for machining processes and technologies and the definition of metrics in each phase for different machining activities and detailed operations. The activity model for machining is shown in Figure 3, and it represents the material inputs, cutting tool preparation, and the different resources used for machining: cutting fluids, compressed air, energy consumption, facilities consumption, machine tool use, and repayment and human resources.

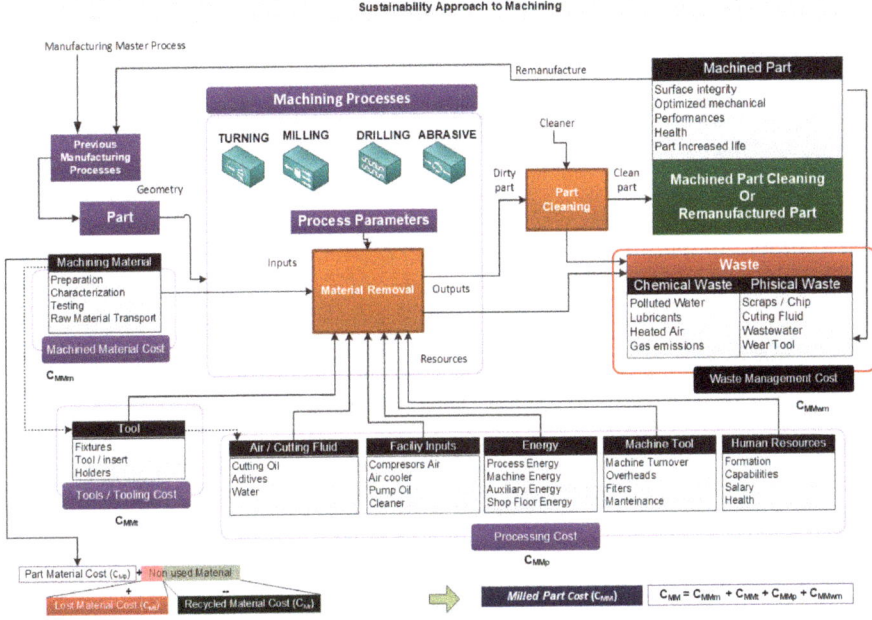

Figure 3. Activity model for machining operations and metrics definition.

For machining processes and technologies, the activity model was defined considering the following basic issues:

- MAIN ACTIVITY: material removal or machining. The most common machining processes and technologies in industrial shop floors are turning, milling, drilling, and grinding, although this activity model can be similar to advanced machining processes and technologies such as chemical machining, electrochemical machining, thermal machining (laser cutting), or advanced mechanical machining (water cutting).
- INPUTS: materials to obtain the final part; in this case, the preform to be machined. In this manufacturing process, we consider the material characterization, testing, preparation, and transport as inputs.
- RESOURCES: cutting tools, compressed air, cutting fluids, facility inputs, energy, machine tools, and human resources. These resources are different depending on the individual operation defined in the macro and micro manufacturing process plan. For example, regarding the consumed energy, we will define the metrics for machine tool consumed energy and cutting fluids consumed energy, as well as the compressed air requirement for the operation and other auxiliary systems.
- CONTROLS: technological instructions and process indicators defined in the micro manufacturing process plan, which ensure the efficiency and effectiveness of the process in order to obtain the final

product. Apart from these, we introduce indicators that can be evaluated from the sustainability perspective, considering economic, environmental, and social dimensions.

- OUTPUTS: The final machined part must be cleaned at the end of the process, since it generally uses cutting fluids with chemical agents. However, the most important issue is that the process generates removed material in chips that we must manage and recycle.
- WASTE: Although it is desired to minimize the total waste, depending on the number of different machining phases, we can divide this metric for each one. We consider scrap or residuum generated by the production process—this can be physical (chips, raw material details, or broken cutting tools), chemical (used cutting fluids mixed with microscopic chips or wastewater) or air pollution, due to gas emissions.

Nevertheless, this activity model is not enough if we want to design a balanced scorecard that includes sustainability indicators. It is obvious that a product's design will have a great influence on how it is manufactured and what materials, processes, and systems are used, as one of the specialists in green manufacturing, Dornfeld [52], indicated. These contributions to sustainability indicators provide the basis for acquiring data during the product life cycle. We propose four main phases for the product life cycle, which include design, manufacturing, use, and end of life, as shown in Figure 4, in order to locate the proposed activity model.

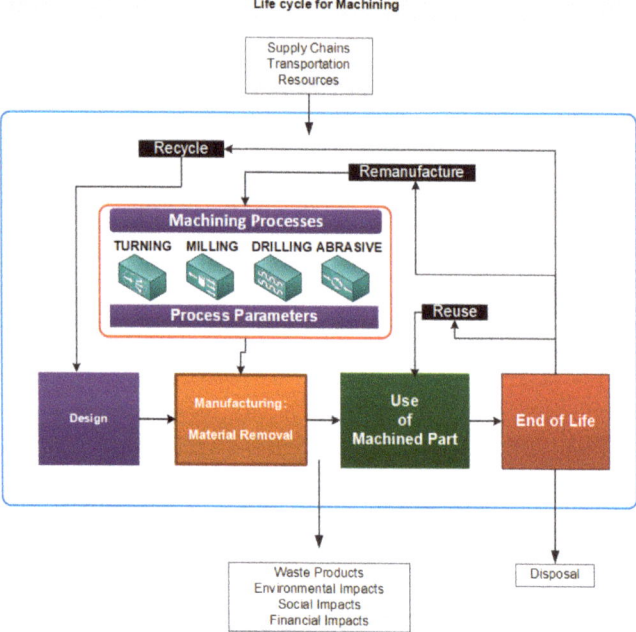

Figure 4. Main product life cycle phases where the activity model is positioned.

The approach to these four phases will allow us to define indicators in each one from the PPR perspective

1. Design. This phase includes raw material management and product design and development stages. To design indicators, we consider not only materials flowing from mining but also from recycled products and cause–effect actions on next phases in engineering activities.

2. Manufacturing. In our research, we consider material removal, machining, processes, and technologies, and the proposed activity model (Figure 3) is incorporated into this phase and we will mainly present indicators here.
3. Use. This third phase is related to product use and service. Thus, we will focus on individual part maintenance or spare parts.
4. End of life. This phase supposes the disposal or the recycling of products once obsolescence is reached and, therefore, individual part disposal or recycling according to companies' sustainability strategy.

Supply chains and materials transport are sometimes considered a separate life cycle phase, but from our point of view, transport and distribution happen throughout the product's life cycle. The complete supply chain is also an integral part of the product life cycle, as these supply chains must produce, deliver, and collect a finished good for use or at the end of its life.

With the activity model for machining and the product life cycle framework, we define the sustainability metrics in our research.

3. A Framework for Sustainability Machining Metrics

The proposal includes the definition of grouped indicators from the PPR perspective. In this work, we will not include resource indicators, and the Product perspective will be subdivided into Materials and Parts. The name of the grouped indicator will start with the PPR name (Material, Part, or Process) followed by another name related to what we want to measure, as shown in Figure 5.

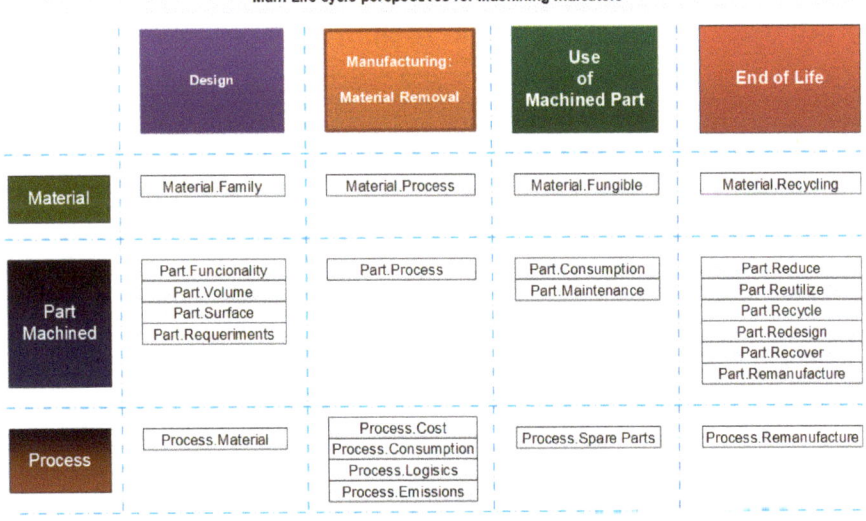

Figure 5. Machining indicators from different PPR perspectives along the product life cycle.

In order to describe the indicators, we have divided the proposal into phases. Table 3 shows how the specific information is organized in this section, and the following tables contain detailed information about the indicators per phase and per PPR perspective.

Table 3. General and specific information on sustainable machining indicators.

Phase	General Information	Detailed Information
Design	Definitions of Material, Product, and Process in the stage of design (Table 4)	In Table 8, specific information and definitions of Material, Product, and Process in the design stage, manufacturing, use and end of life.
Manufacturing	Definitions of Material, Product, and Process in the stage of manufacturing (Table 5)	
Use	Definitions of Material, Product, and Process in the use stage (Table 6)	
End of Life	Definitions of Material, Product, and Process in the end-of-life Stage (Table 7)	

The definition of the indicators can be used to create a balanced scorecard aligned with a company's sustainability strategy. The first approach to specific indicators is shown in Tables 4–8, and they include both quantitative and qualitative indicators that can be used all along the product life cycle and its name and a short description.

For the first product life cycle phase, Design, the design considerations are strengthened by the search for new materials that have high performances for parts and have allowed a wide range of manufacturing options. High-performance materials and new mechanical and chemical characteristics are incorporated into the databases in product life cycle management (PLM) platforms and help engineers to make the right decision. Table 4 shows the general definitions for the Design phase.

Table 4. Design phase indicators definition.

Phase	PPR Perspective	Indicator Name	Description
Design	Material	Material.Family	Identifies the materials used for the design of each component of the product
	Product	Part.Functionality	Describes the function of the part and determines the mechanical relationships with the others in the set
		Part.Volume	Indicates the volume of a part. The mathematical definition depends on its geometry
		Part.Surface	Reveals the surface integrity of a part/product
		Part.Requirements	Lists the functional requirements of the part, for example roughness, dimensional tolerances, etc.
	Process	Process.Material	Illustrates design considerations that affect compatibility between manufacturing processes and the selected material group

In machining processes, the relationship between the material, geometry, cutting tool, and machine tool opens the field to the research in manufacturing process optimization and new materials. The operating conditions (process parameters for machining) depend on the efficient performance of these four elements. Table 5 describes, for the manufacturing phase, the indicators for the machining process, including turning, milling, drilling, and boring operations, among others.

Machine tools can be manual, automated, or numerical control-driven. Today, most of them are ready for Industry 4.0 connection, and CAD/CAE/CAM applications are associated with shop floor cells and provide data to PLM platforms. The manufacturer's recommended tool parameters (controls) are based on extensive studies of the process, part material, part geometry, and tool performances. With appropriate sustainable indicators, we can improve them. For example, the energy consumption of the process and its emissions metrics will require experimental measurements on the shop floor to establish the process indicators and online data collection.

Table 5. Manufacturing phase indicators definition.

Phase	PPR Perspective	Indicator Name	Description
Manufacturing	Material	Material.Process	Establishes the design considerations that affect the compatibility between the material selected with the manufacturing process
	Product	Part.Process	Illustrates the design considerations that affect the geometric compatibility of the part with the manufacturing process.
	Process	Process.Cost	Reveals the cost of the process per part/unit
		Process.Consumption	Shows the energy consumption in the manufacture of the part. This indicator contains more detailed indicators for each energy source (W, L/h, etc.)
		Process.Logistics	Describes the material flow, internal and external to the shop floor. This indicator contains more detailed indicators according to the part process plan
		Process.Emissions	Indicates the emissions of solids, liquids, and gases produced in the process. This indicator contains lower-level indicators with different perspectives.

Machined parts used as machine components or consumer product components must meet the design specifications, in which the consumables and the consumption of energy sources for their operation must be considered. Table 6 shows the definitions of metrics from this perspective.

Table 6. Use phase indicators definition.

Phase	PPR Perspective	Indicator Name	Description
Use	Material	Material.Fungible	Indicates the necessary materials or components used by the product during its phase of use
	Product	Part.Consumption	Indicates the consumption of various energy sources used by the product for proper operation (water, electricity, gas, etc.)
		Part.Maintenance	Lists the maintenance actions that must be undertaken during use, mainly those programmed, with an estimate of unscheduled maintenance
	Process	Process.SpareParts	Shows the manufacturing orders that must be issued to maintain the legally established stock of the product during the use phase and after production ends

The end of life of a machined part can be postponed by maintenance and repairs, which may include processes for the recovery of dimensional tolerances, for which an analysis of surface integrity may be necessary. Table 7 shows the main definitions.

The specific information regarding some indicators is shown in Table 8. The table shows the indicator, the simplified name, acronym, units, goal, and the possible source of the information. The indicator units depend on the specific variables that are measured, and the objective may be seeking the maximum ($\nearrow$) or the minimum ($\searrow$). The sources of information can be standards or databases of materials, machines, tools, and consumables, which can be taken as a reference for the analyzed process.

Table 7. End-of-life phase indicators definition.

Phase	PPR Perspective	Indicator Name	Description
End of Life	Material	Material.Recycling	Identifies the amount of material that can be recycled for each of the parts/components used in the product
	Product	Part.Reduce	Identifies the parts or components that can be removed without damaging the proper functioning of the product
		Part.Reutilize	Identifies the parts or components that can be reused as components of another new product
		Part.Recycle	Identifies the parts or components that can be recycled and included as part of the base material as spare parts without damaging the proper functioning of the product
		Part.Redesign	Identifies parts or components that are likely to be redesigned to minimize the environmental impact of the assembly
		Part.Recover	Identifies the parts or components that can be recovered as spare parts without damaging the proper functioning of the product
		Part.Remanufacture	Identifies the parts or components that can be re-passed through a new manufacturing process and incorporated into a new product.
	Process	Process.Remanufacturing	Establishes the ability of the manufacturing process to form materials from a product in the end-of-life phase

Table 8. Specific information regarding the Indicators from the main life cycle perspective.

Phase	PPR Perspective	Indicator Definition	Expression	Units [Example]	Goal	Source of Information
Design	Material	Material.Family	M.F.	kg	↗	Standard/Databases
	Product	Part.Functionality	P.F	Functionality	↗	Guides
		Part.Volume	P.Vm	mm^3	↗	@
		Part.Surface	P.Sf	µ; mm^2	↗	@
		Part.Requirements	P.R	# Requirements	↗	@
	Process	Process.Material	P.Mc	Machining Operations	↘	Guides
Manufacturing	Material	Material.Process	M.Pc.	Machining strategy	↘	Standard/Guides
	Product	Part.Process	P.P	Machining path	↘	Guides
	Process	Process.Cost	P.C	€/unit	↘	@
		Process.Consumption	P.Co	Kg; €	↘	@
		Process.Logistics	P.L	s; m; €	↘	@
		Process.Emissions	P.E	Kg CO_2	↘	@
Use	Material	Material.Fungible	M.Fu	Kg	↘	Database
	Product	Part.Consumption	P.C	kW/h	↘	Standards; @
		Part.Maintenance	P.M	OEE	↘	Guides; @
	Process	Process.Spare Parts	P.Rec	# orders	↘	Guides; @
End of Life	Material	Material.Recycling	M.R	Recycled Kg/Kg components	↗	Guides, Standard; @
	Product	Part.Reduce	P.Redu	parts/unit	↗	Guides; @
		Part.Reutilize	P.Reu	parts/product	↗	Guides; @
		Part.Recycle	P.Rec	parts/product	↗	Guides; @
		Part.Redesign	P.Reds	parts/product	↗	Guides; @
		Part.Recover	P.Rep	parts/product	↗	Guides; @
		Part.Remanufacture	P.Ref	parts/product	↗	Guides; @
	Process	Process.Remanufacturing	P.RMfg	%	↗	Guides; @

Note: ↗ maximize; ↘ minimize; @ various information sources; # number of.

4. Experimental Development, Results, and Discussion

In order to validate the metrics definition, a set of experiments were designed. The experiments had the objective of acquiring data for some indicators and to analyze how variations on manufacturing process plan parameters could affect the sustainable indicators.

The experiments were included in a green product life cycle management initiative and, for example, advanced Computer Aided Design and Manufacturing (CAD/CAM) tools were used to prepare the design of experiments for different indicators. It should also be noted that the proposal managed high-level and low-level indicators and indicators in different phases that can be validated on the shop floor. For example, machining time can be determined through the design phase indicators with CAM applications, and the real operation time can be measured in the machine tool numerical control and then compared.

The experiments are summarized in Table 9 and, for each one, the indicators from the PPR perspective are evaluated. For each experiment, a basic machining technology was tested with previous machining simulation arrangements.

Table 9. Experimental validation of sustainable machining indicators. MRR: Material Removal Rate.

Phase	Test	PPR Perspective	Method	Machining Process	Material	Metric Evaluated	Goal
DESIGN	#1	Process	Simulation CAD/CAM Autodesk	Milling: Surface Facing	AISI1045	MRR	↗
						Machining Time	↘
	#2	Process	Simulator CAD/CAM 3DExperience	Milling: Concave Surfaces	AISI1045	Machining Strategies	↗
				Milling: Convex Surface	AISI1045	Machining Strategies	↗
MANUFACTURING	#3	Part	Measurements and tests	Turning Straight Turning	AISI1018	Roughness	↘
						Microhardness	↗
						Surface Metallographic	↗
						Mechanical Performance	↗
						Plastic Deformation	↘
	4#4	Part	Measurements between two machining centers	Milling: Surface Facing	AISI1045	Roughness (Machine Tool A)	↘
						Roughness (Machine Tool B)	↘
						Power Consumption (Machine Tool A)	↘
						Power Consumption (Machine Tool B)	↘
	#5	Process	Measurements in Social Dimension	Turning, Milling	Various	16 Sustainability Indicators	↗

Note: ↗ maximize; ↘ minimize.

In the following subsections, the five experiments are briefly described to show how to obtain data for the defined sustainable indicators.

4.1. Test #1. Material Removal Rate (MRR) and Machining Time

The objective of this experiment was to set the minimum processing time and the highest possible MRR in a planning process. The design of the experiments considered cutting directions of 0°, 45°, and 90°, and the material was AISI1045 in the Gentiger machining tool (Taichung City, Taiwan), with the cutting tool Mitsubishi VPX300R 4004SA32SA and LOGU1207080PNER-M (MP6120) inserts (Tsukuba, Japan). The cutting diameter $\varnothing = 40$ mm, the number of flutes $Z_c = 4$, and the main cutting angle $K = 90°$. The operation was done in all cases without cutting fluids.

The CAD/CAM Inventor HSM 2019 application was used to find the 27 possible combinations of the Taguchi method to get the combination of four parameters (ABCD) that reaches the highest MRR and the minimum processing time. In this experiment, A is the direction of the cutting trajectory pass p_d, B is the depth of cut a_p, C is the cutting speed v_c, and D is the feed rate per tooth f_z. The time is expressed in min:s.

The indicator, shown in Table 9, revealed that the highest MRR material removal rate obtained for the roughing operations was MRR = 10.1 cm^3/min. In this operation, the control parameters were p_d = (0° or 45° or 90°), a_p = 1.2, v_c = 1671, and f_z = (0.10 or 0.12 or 0.08).

For the finishing machining operation, the highest material removal rate obtained was MRR = 6.1 cm^3/min. In this operation, the control parameters were p_d = (0° or 45° or 90°); a_p = 0.8; v_c = 1671; and f_z = (0.12 or 0.08 or 0.10).

Finally, the minimum machining time obtained for the finishing operation was t_{min} = 2:25 min:s. The process parameters were p_d = 0°, a_p = 0.8, v_c = 1671, and f_z = 0.12.

4.2. Test #2 Machining Strategies on Concave and Convex Surfaces

The objective of this experiment was to find the machining strategy with the shortest machining time among the possible options of the cutting trajectories strategy. The surface geometry was specially designed for the test, and the machined material was AISI1045.

The experimental setup used a high-performance Gentiger machining center, and the cutting tools for surface machining were the Mitsubishi Ball Mill VQ4SVBR0600 (cutting diameter Ø 6 mm) for roughing surface milling and the Mitsubishi VQ4SVBR0300 (cutting diameter Ø 3 mm) for the finishing operations from Tsukuba production centre (Japan). Both were done with the recommended cutting fluids, which encouraged us to optimize its use for sustainability reasons. All the experiments were prepared with the CAM application of the 3DEXPERIENCE 2019 platform. In this case, 21 different options were achieved.

Several evaluations were made on the part that had a concave and convex surface. The test analyzed different options for tool path generation and trajectory strategies in three-dimensional surface milling with all the possible combinations and simulations [53]. For example, with the *style pocketing* surface machining option and the *back and forth* strategy, it was found that the total operation time was reduced by 10%, while only a 3% time reduction could be achieved for the rest of the options compared to the longest one.

In this experiment, we could obtain qualitative information about the cutting trajectory strategy and quantitative information about the operation time and, therefore, the energy consumption.

4.3. Test #3. Roughness, Microhardness, Plastic Deformation

The objective of this experiment was to find the effect of the machining parameters on the surface quality of the part, which was a quality requirement. The part was a machine axis, and the main machining operations were turning. The part material was AISI1018, and the machine tool was the ROMI numerical control lathe (Sao Paulo, Brasil).

The turning operation cutting tool used was the SANDVIK DNMG 15 06 08-PM4325, and the operations were done with cutting fluids. The machining microprocess plan with trajectory strategies was prepared with the CAD/CAM application SolidCAM 2018 [29].

Mathematical relationships were found to predict these properties and recommendations for the use of different parameters. This test used experimental data from a turning process to determine the influence of the machining parameters in the surface quality, such as the depth of cut, yield strength, plastic deformation, and roughness. The analytical indicator helped us to decide which combination of parameters can be accepted or rejected according to the requirements and dimensional tolerances of the part. Experimental equations were obtained for the selected process, machine, material, and tool.

4.4. Test #4: Roughness and Power Consumption

The objective of this experiment was to reproduce the machining process plan to build the same part in two different machining centers (A and B) that were geographically distributed. The experiment was reproduced with exactly the same machining process parameters, and the aim was to determine whether the minimum roughness and the minimum power consumption would be obtained with the same parameters [54].

The experiment was carried out in a high-performance Gentiger Machining Center (Taichung City, Taiwan) (A) and in a high performance Deckel Maho Machining Center (Pfronten, Germany) (B).

The part material was AISI1045, and the cutting tool was the Mitsubishi VPX300R 4004SA32SA, with LOGU1207080PNER-M inserts (MP6120). The cutting diameter $Ø = 40$ mm, the number of flutes $Zc = 4$, and the main cutting angle $K = 90°$. The operation was dry machining, without cutting fluids.

Although both experiments had the same machining cutting parameters, it was discovered that the machine has an important influence on the result and, therefore, on the micro process plan.

Power consumption. In Machine #A, the minimum power consumption was 2.79 kWh, with the conditions of a pass direction of 90°, a cutting depth of 1.0 mm, a cutting speed of 180 m/min, and a feed per tooth of 0.1 mm/tooth. In Machine #B, the minimum power consumption was 4.88 kWh, with the cutting conditions being a pass direction of 0°, a cutting depth of 0.8 mm, a cutting speed of 140 m/min, and a feed per tooth of 0.08 mm/tooth.

Roughness. In Machine #A, the minimum value of $R_a = 0.55$ μm, with the cutting conditions of a pass direction of 0°, a cutting depth of 0.8 mm, a cutting speed of 210 m/min, and a feed per tooth of 0.12 mm/tooth. In Machine #B, the minimum value of $R_a = 0.83$ μm, with the cutting conditions of a pass direction of 45°, a cutting depth of 0.8 mm, a cutting speed of 210 m/min, and a feed per tooth of 0.08 mm/tooth.

The conclusion was that although we had twin machine tools with similar performances, we have to slightly customize the process plan to reach the indicator objective.

4.5. Test #5. Social Dimension Analysis

For this experiment, the evaluation of the social dimension in machining was the main objective. The experiment was carried out on a shop floor with numerical control machine tools that perform machining processes on various parts at the same time.

An assessment questionnaire was designed and completed by workers, shop floor officers, and middle managers, obtaining the minimum number of answers to validate the method.

The indicators were calculated by the grey relational theory. There were 16 indexes analyzed: Worker Productivity, Relations with Other Workers, Worker's Skill Level, Flexibility of Job Rotation, Punctuality, Top Management Support on Various Issues, Job Satisfaction, Conducive Working Environment, Awareness of Sustainable Manufacturing Initiatives, Technological Upgrades, Financial Support (loans, etc.), Required Product Quality, and Waste Management [31].

The obtained results when evaluating the sustainability indicators in the social dimension, after the application of the Plan, Do, Check, Act (PDCA) continuous improvement cycle were 9.21% higher than the initial evaluation after the implementation of the improvements.

Some improvements were implemented after the analysis of the initial evaluation and the final evaluation of this sustainability dimension, which is one of the most difficult to get information and data analysis for. Fortunately, it helped to implement objective indicators on the machining shop floor.

5. Conclusions

This paper's contributions can be highlighted in three main areas of interest that have been presented, from the indicators' definition to the shop floor in machining operations.

The first one is the general activity model of industrial manufacturing processes that can be deployed in more detailed activities to identify indicators, where needed, for the manufacturing phase of the product life cycle management.

The second one is the customization of the activity model for material removal and machining processes. In this model, we detected the manufacturing machining process inputs, controls, resources, and enablers in order to define the general, technical, and sustainability indicators. These indicators can be defined for different product life cycle phases in an organized way, which is why we defined the PPR life cycle phases matrix.

The different experiments carried out provided skills, data, and information about the applied indicators in the manufacturing and materials engineering discipline. They gave real case studies for validating the metrics' definition.

Apart from the manufacturing phase metrics' definition, the use of manufacturing authoring applications within a PLM platform in the design stage can simulate the manufacturing process and help to predict its behavior under the required conditions. Computer-aided manufacturing simulation software has different simulation levels that can help to define and validate machining strategies or manufacturing cell activities to ensure good product quality, achieving sustainable strategies.

In the comparative study of the two machining centers, the lowest roughness and the lowest energy consumption were obtained with different machining parameters. The experiment was carried out on the same material and the same tool, and it was determined that sustainability indicators must be established for each machining center.

When considering surface roughness and the power consumed as the variables to find the best cutting conditions, it was determined that each machining center has its own operating parameters for these conditions. These parameters are related to each other, and the value depends on the material selected, the tool, the machine center, and the lubrication.

The manufacturing parameters that can be tested with virtual manufacturing help to minimize the iterative process when fixing the indicators' objective values. In other cases, it will be more difficult, and we will need to do shop floor measures, as shown in the last experiment.

Finally, sustainability indicators should be evaluated in the product design stage for the best results, and the characteristics of the manufactured part can best be predicted by including the sustainability criteria in the product life cycle management (PLM) platforms.

6. Future Work

The research plans aligned with this proposal, and we learned multiple lessons, including several key actions. Firstly, we suggest a proposal to incorporate indicators' reports into product life cycle management platforms, in which the sustainability alternatives proposed in the design stage can be evaluated, as part of digital twins' implementation for Industry 4.0 demonstrators. Secondly, further research should carry out experiments to determine the influence of mooring in milling processes and its influence on the quality of the part and the sustainability indicators. Finally, future research should focus on developing a system for evaluating sustainability indicators that can quantify the increase in the indicators when the product is improved, while the options or alternatives are analyzed in the design and manufacturing stages.

Author Contributions: C.A. conceptualization and writing—review and editing, analyzed the results, and prepared the draft manuscript. C.V. supervised the research, reviewed the analysis of the results, and wrote the final manuscript. All authors have read and agreed to the published version of the manuscript.

Funding: This research was funded by the Escuela Politécnica Nacional (Ecuador) Research Project: PIS 16-15, the Universitat Politècnica de València UPV (Spain) and the Carolina Foundation (Spanish Government Scholarships) Call 2017.

Acknowledgments: The authors would like to express their gratitude for the support provided by the Escuela Politécnica Nacional (Ecuador), the Universitat Politècnica de València (Spain) and the Carolina Foundation (Spanish Government) with the corresponding grants.

Conflicts of Interest: The authors declare no conflict of interest.

References

1. Brundtland, G.H. Our Common Future—Call for Action. *Environ. Conserv.* **1987**, *14*, 291–294. [CrossRef]
2. Blewitt, J. *Understanding Sustainable Development*; Routledge: Abingdon upon Thames, UK, 2012.

3. Berke, P.R.; Conroy, M.M. Are We Planning for Sustainable Development? *J. Am. Plan. Assoc.* **2000**. [CrossRef]
4. De Ron, A.J. Sustainable production: The ultimate result of a continuous improvement. *Int. J. Prod. Econ.* **1998**, *56–57*, 99–110. [CrossRef]
5. Aarseth, W.; Ahola, T.; Aaltonen, K.; Økland, A.; Andersen, B. Project sustainability strategies: A systematic literature review. *Int. J. Proj. Manag.* **2017**, *35*, 1071–1083. [CrossRef]
6. Jansen, L. The challenge of sustainable development. *J. Clean. Prod.* **2003**. [CrossRef]
7. Vieira, L.C.; Amaral, F.G. Barriers and strategies applying Cleaner Production: A systematic review. *J. Clean. Prod.* **2016**, *113*, 5–16. [CrossRef]
8. United Nations. *The Sustainable Development Goals Report*; United Nations: New York, NY, USA, 2017.
9. ISO Central Secretariat; Frost, R. ISO 26000 Social Responsibility Basics. *ISO Focus* **2011**.
10. United Nations. United Nations Global Compact Guide to Corporate Sustainability. *Int. J. Proj. Manag.* **2014**. [CrossRef]
11. Global Reporting Initiative (GRI). GRI and ISO 26000: How to use the GRI Guidelines in conjunction with ISO 26000. *Design* **2011**.
12. Elkington, J. Towards the Sustainable Corporation: Win-Win-Win Business Strategies for Sustainable Development. *Calif. Manag. Rev.* **1994**. [CrossRef]
13. Zackrisson, M.; Kurdve, M.; Shahbazi, S.; Wiktorsson, M.; Winroth, M.; Landström, A.; Almström, P.; Andersson, C.; Windmark, C.; Öberg, A.E.; et al. Sustainability Performance Indicators at Shop Floor Level in Large Manufacturing Companies. *Procedia CIRP* **2017**, *61*, 457–462. [CrossRef]
14. Reich-Weiser, C.; Vijayaraghavan, A.; Dornfeld, D.A. Metrics for Sustainable Manufacturing. In Proceedings of the ASME International Manufacturing Science and Engineering Conference, MSEC2008, Evanston, IL, USA, 7–10 October 2008; Volume 1, pp. 327–335.
15. Shuaib, M.; Seevers, D.; Zhang, X.; Badurdeen, F.; Rouch, K.E.; Jawahir, I.S. Product sustainability index (ProdSI): A metrics-based framework to evaluate the total life cycle sustainability of manufactured products. *J. Ind. Ecol.* **2014**, *18*, 491–507. [CrossRef]
16. Jayal, A.D.; Badurdeen, F.; Dillon, O.W.; Jawahir, I.S. Sustainable manufacturing: Modeling and optimization challenges at the product, process and system levels. *CIRP J. Manuf. Sci. Technol.* **2010**, *2*, 144–152. [CrossRef]
17. Singh, S.; Olugu, E.U.; Musa, S.N. Development of Sustainable Manufacturing Performance Evaluation Expert System for Small and Medium Enterprises. *Procedia CIRP* **2016**, *40*, 608–613. [CrossRef]
18. Rajurkar, K.P.; Hadidi, H.; Pariti, J.; Reddy, G.C. Review of Sustainability Issues in Non-Traditional Machining Processes. *Procedia Manuf.* **2017**, *7*, 714–720. [CrossRef]
19. Peralta Álvarez, M.E.; Marcos Bárcena, M.; Aguayo González, F. On the sustainability of machining processes. Proposal for a unified framework through the triple bottom-line from an understanding review. *J. Clean. Prod.* **2017**, *142*, 3890–3904. [CrossRef]
20. Bhanot, N.; Rao, P.V.; Deshmukh, S.G. An integrated approach for analysing the enablers and barriers of sustainable manufacturing. *J. Clean. Prod.* **2017**, *142*, 4412–4439. [CrossRef]
21. Eastwood, M.D.; Haapala, K.R. A unit process model based methodology to assist product sustainability assessment during design for manufacturing. *J. Clean. Prod.* **2015**, *108*, 54–64. [CrossRef]
22. Garretson, I.C.; Mani, M.; Leong, S.; Lyons, K.W.; Haapala, K.R. Terminology to support manufacturing process characterization and assessment for sustainable production. *J. Clean. Prod.* **2016**, *139*, 986–1000. [CrossRef]
23. Helleno, A.L.; de Moraes, A.J.I.; Simon, A.T.; Helleno, A.L. Integrating sustainability indicators and Lean Manufacturing to assess manufacturing processes: Application case studies in Brazilian industry. *J. Clean. Prod.* **2017**, *153*, 405–416. [CrossRef]
24. Kluczek, A. An Overall Multi-criteria Approach to Sustainability Assessment of Manufacturing Processes. *Procedia Manuf.* **2017**, *8*, 136–143. [CrossRef]
25. Latif, H.H.; Gopalakrishnan, B.; Nimbarte, A.; Currie, K. Sustainability index development for manufacturing industry. *Sustain. Energy Technol. Assess.* **2017**, *24*, 82–95. [CrossRef]
26. Moldavska, A.; Welo, T. The concept of sustainable manufacturing and its definitions: A content-analysis based literature review. *J. Clean. Prod.* **2017**, *166*, 744–755. [CrossRef]
27. Winroth, M.; Almström, P.; Andersson, C. Sustainable production indicators at factory level. *J. Manuf. Technol. Manag.* **2016**, *27*, 842–873. [CrossRef]

28. Linke, B.; Das, J.; Lam, M.; Ly, C. Sustainability indicators for finishing operations based on process performance and part quality. *Procedia CIRP* **2014**, *14*, 564–569. [CrossRef]
29. Vila, C.; Ayabaca, C.; Díaz-Campoverde, C.; Calle, O. Sustainability analysis of AISI 1018 turning operations under surface integrity criteria. *Sustainability* **2019**, *11*, 4786. [CrossRef]
30. Bhanot, N.; Rao, P.V.; Deshmukh, S.G. An Assessment of Sustainability for Turning Process in an Automobile Firm. *Procedia CIRP* **2016**, *48*, 538–543. [CrossRef]
31. Ayabaca, C.; Vila, C. Assessment of the Sustainability Social Dimension in Machining Through Grey Relational Analysis. In Proceedings of the 2018 IEEE International Conference on Engineering, Technology and Innovation, ICE/ITMC, Stuttgart, Germany, 17–20 June 2018.
32. Bhanot, N.; Rao, P.V.; Deshmukh, S.G. Sustainable Manufacturing: An Interaction Analysis for Machining Parameters using Graph Theory. *Procedia Soc. Behav. Sci.* **2015**. [CrossRef]
33. Gupta, M.K.; Sood, P.K.; Singh, G.; Sharma, V.S. Sustainable Machining of Aerospace Material-Ti (grade-2) alloy: Modelling and Optimization. *J. Clean. Prod.* **2017**. [CrossRef]
34. Hegab, H.A.; Darras, B.; Kishawy, H.A. Towards sustainability assessment of machining processes. *J. Clean. Prod.* **2018**, *170*, 694–703. [CrossRef]
35. Kadam, G.S.; Pawade, R.S. Surface integrity and sustainability assessment in high-speed machining of Inconel 718—An eco-friendly green approach. *J. Clean. Prod.* **2017**, *147*, 273–283. [CrossRef]
36. Benedicto, E.; Carou, D.; Rubio, E.M. Technical, Economic and Environmental Review of the Lubrication/Cooling Systems Used in Machining Processes. *Procedia Eng.* **2017**, *184*, 99–116. [CrossRef]
37. Zhao, G.Y.; Liu, Z.Y.; He, Y.; Cao, H.J.; Guo, Y.B. Energy consumption in machining: Classification, prediction, and reduction strategy. *Energy* **2017**, *133*, 142–157. [CrossRef]
38. Abbas, A.T.; Benyahia, F.; El Rayes, M.M.; Pruncu, C.; Taha, M.A.; Hegab, H. Towards optimization of machining performance and sustainability aspects when turning AISI 1045 steel under different cooling and lubrication strategies. *Materials* **2019**, *12*, 3023. [CrossRef]
39. Ali, R.A.; Mia, M.; Khan, A.M.; Chen, W.; Gupta, M.K.; Pruncu, C.I. Multi-response optimization of face milling performance considering tool path strategies in machining of Al-2024. *Materials* **2019**, *12*, 1013. [CrossRef]
40. Li, Y.; Zheng, G.; Cheng, X.; Yang, X.; Xu, R.; Zhang, H. Cutting performance evaluation of the coated tools in high-speed milling of aisi 4340 steel. *Materials* **2019**, *12*, 3266. [CrossRef]
41. Gupta, M.K.; Pruncu, C.I.; Mia, M.; Singh, G.; Singh, S.; Prakash, C.; Sood, P.K.; Gill, H.S. Machinability investigations of Inconel-800 super alloy under sustainable cooling conditions. *Materials* **2018**, *11*, 2088. [CrossRef]
42. Gamage, J.R.; DeSilva, A.K.M.; Chantzis, D.; Antar, M. Sustainable machining: Process energy optimisation of wire electrodischarge machining of Inconel and titanium superalloys. *J. Clean. Prod.* **2017**, *164*, 642–651. [CrossRef]
43. Gunda, R.K.; Reddy, N.S.K.; Kishawy, H.A. A Novel Technique to Achieve Sustainable Machining System. *Procedia CIRP* **2016**, *40*, 30–34. [CrossRef]
44. Lu, T.; Jawahir, I.S. Metrics-based sustainability evaluation of cryogenic machining. *Procedia CIRP* **2015**, *29*, 520–525. [CrossRef]
45. Pusavec, F.; Deshpande, A.; Yang, S.; M'Saoubi, R.; Kopac, J.; Dillon, O.W.; Jawahir, I.S. Sustainable machining of high temperature Nickel alloy—Inconel 718: Part 1—Predictive performance models. *J. Clean. Prod.* **2014**, *81*, 255–269. [CrossRef]
46. Goindi, G.S.; Sarkar, P. Dry machining: A step towards sustainable machining—Challenges and future directions. *J. Clean. Prod.* **2017**, *165*, 1557–1571. [CrossRef]
47. Shin, S.J.; Woo, J.; Rachuri, S. Energy efficiency of milling machining: Component modeling and online optimization of cutting parameters. *J. Clean. Prod.* **2017**, *161*, 12–29. [CrossRef]
48. Um, J.; Gontarz, A.; Stroud, I. Developing energy estimation model based on sustainability KPI of machine tools. *Procedia CIRP* **2015**, *26*, 217–266. [CrossRef]
49. Zhang, T.; Owodunni, O.; Gao, J. Scenarios in multi-objective optimisation of process parameters for sustainable machining. *Procedia CIRP* **2015**, *26*, 373–378. [CrossRef]
50. Gao, R.; Wang, L.; Teti, R.; Dornfeld, D.; Kumara, S.; Mori, M.; Helu, M. Cloud-enabled prognosis for manufacturing. *CIRP Ann.* **2015**, *64*, 749–772. [CrossRef]

51. Menzel, C.; Mayer, R.J. The IDEF Family of Languages. In *Handbook on Architectures of Information Systems*; Springer: Berlin/Heidelberg, Germany, 1998.
52. Dornfeld, D.A. Moving towards green and sustainable manufacturing. *Int. J. Precis. Eng. Manuf.-Green Technol.* **2014**, *1*, 63–66. [CrossRef]
53. Vila, C.; Ayabaca, C.; Gutiérrez, S.; Meseguer, A.; Torres, R.; Yang, X. Analysis of Different Tool Path Strategies for Free Form Machining with Computer Aided Surface Milling Operations. In Proceedings of the 8th Manufacturing Engineering Society International Conference MESIC 2019, Madrid, Spain, 19–21 June 2019; Barajas, C., Caja, J., Calvo, R., Maresca, P., Palacios, T., Eds.; ETSI Aeronáutica y del Espacio, UPM: Madrid, Spain, 2019; p. 137.
54. Ayabaca, C.; Vila, C.; Abellán-Nebot, J.V. Comparative study of Sustainability Metrics for Face Milling AISI 1045 in different Machining Centers. In Proceedings of the 8th Manufacturing Engineering Society International Conference MESIC 2019, Madrid, Spain, 19–21 June 2019; Barajas, C., Caja, J., Calvo, R., Maresca, P., Palacios, T., Eds.; ETSI Aeronáutica y del Espacio, UPM: Madrid, Spain, 2019; p. 133.

© 2020 by the authors. Licensee MDPI, Basel, Switzerland. This article is an open access article distributed under the terms and conditions of the Creative Commons Attribution (CC BY) license (http://creativecommons.org/licenses/by/4.0/).

Review

Sustainable Lubrication Methods for the Machining of Titanium Alloys: An Overview

Enrique García-Martínez [1,2], Valentín Miguel [1,2,*], Alberto Martínez-Martínez [2], María Carmen Manjabacas [1,2] and Juana Coello [1,2]

1. High Technical School of Industrial Engineers of Albacete, University of Castilla-La Mancha, 02071 Albacete, Spain; Enrique.GMartinez@uclm.es (E.G.-M.); mcarmen.manjabacas@uclm.es (M.C.M.); juana.coello@uclm.es (J.C.)
2. Regional Development Institute, Science and Engineering of Materials, University of Castilla La Mancha, 02071 Albacete, Spain; alberto.martinez@uclm.es
* Correspondence: valentin.miguel@uclm.es; Tel.: +34-967-599-200

Received: 29 October 2019; Accepted: 19 November 2019; Published: 22 November 2019

Abstract: Titanium is one of the most interesting materials in modern manufacturing thanks to its good mechanical properties and light weight. These features make it very attractive for use in the aeronautical and aerospace industries. Important alloys, such as Ti6Al4V, are extensively used. Nevertheless, titanium alloys present several problems in machining processes. Their machinability is poor, affected by low thermal conductivity, which generates very high cutting temperatures and thermal gradients in the cutting tool. Lubricants and cutting fluids have traditionally been used to solve this problem. However, this option is unsustainable as such lubricants represent a risk to the environment and to the health of the operator due to their different chemical components. Therefore, novel, sustainable and green lubrication techniques are necessary. Dry machining is the most sustainable option. Nevertheless, difficult-to-machine materials like titanium alloys cannot be machined under these conditions, leading to very high cutting temperatures and excessive tool wear. This study is intended to describe, analyse and review the non-traditional lubrication techniques developed in turning, drilling and milling processes since 2015, including minimum quantity of lubricant, cryogenic lubrication, minimum quantity of cooling lubrication or high-pressure coolant. The aim is to provide a general overview of the recent advances in each technique for the main machining processes.

Keywords: titanium alloys; sustainable lubrication; cryogenic lubrication; MQL

---

1. Introduction

Because of their mechanical properties, titanium and titanium alloys are one of the most commonly used materials in manufacturing processes in certain key industries such as the aeronautical, aerospace and medical ones. These alloys exhibit high strength while also being very light [1], and also have very high wear and corrosion resistance. These unique characteristics, along with their possessing the highest strength to weight ratio [2], make these alloys very attractive for such industries. Titanium and titanium alloys offer the best mechanical features possible for applications in which low weight is required.

Nevertheless, pure titanium and titanium alloys present several problems for machining processes, which mean they are classified as difficult-to-machine materials. Their low modulus of elasticity and extreme strength at high temperature [3], and inferior thermal conductivity generate long ductile chips and relatively large contact length between chip and cutting tool in machining processes. Thus, very high temperatures are reached and aggressive thermal gradients appear in the cutting tool.

Their low thermal conductivity and high heat capacity play a critical role in the heat dissipation process [4], which finally causes tool wear and a rapid reduction in tool life.

Furthermore, the plastic deformation of the material, friction and high chemical affinity of the titanium and the cutting tool materials produce built up edge (BUE), affecting the geometry of the tool and impairing the surface integrity of the final machined part.

The combination of all these conditions reduces the machinability of titanium alloys, the improvement of which is one of the greatest challenges recently addressed by a large number of researchers. Such researchers are striving to provide new machining strategies for this material, as well as to determine optimal cutting conditions in the machining processes.

Ti6Al4V is the most significant titanium alloy from the point of view of industrial applications. Table 1 shows the distribution of the most frequently used titanium alloys in the literature we have reviewed. Ti6Al4V alloy is estimated to account for 50% of global titanium metal production, and 80% of this corresponds to the aerospace and medical industries [2]. Ti6Al4V is an $\alpha - \beta$ titanium alloy, in which the amount of beta stabilizer added is about 4–6%. These $\alpha - \beta$ alloys can obtain different mechanical properties by heat treatment and their characteristics are optimal for application with warm temperatures (400 °C) [5].

Table 1. Proportion and composition of titanium and titanium alloys in the reviewed literature.

Material	Number of Studies	Proportion (%)	Composition (%)							
			Al	V	Mo	Cr	Fe	Sn	Zr	Nb
Titanium	3	5.08	-	-	-	-	-	-	-	-
Ti6Al4V	49	83.05	6	4			0.25			
Ti5553	3	5.08	5	5	5	3	0.5			
TC17	1	1.69	5		4	4		2	2	
Ti6Al7Nb	1	1.69	6							7
Ti aluminide	2	3.39								

Ti553 alloy is a relatively modern alloy, which is gaining prominence in applications in the automotive, chemical and medical industries because of its excellent properties, such as its good strength-to-weight ratio and high hardness at heating state [4]. The machinability of this alloy has not been extensively studied in comparison with Ti6Al4V, and thus the optimal conditions for the machining of this material are still far from determined.

Titanium aluminide (TiAl) is an intermetallic chemical compound of titanium and aluminium as base metals and other elements in small proportions [6]. This material achieves superior properties compared to traditional titanium alloys, such as excellent heat, corrosion and oxidation resistance. Moreover, it is very light, thanks to its high proportion of aluminium. However, it exhibits poor ductility and low fracture toughness [5].

TC17 is an alloy with high toughness with some applications in gas turbine engine components. Although Ti6Al7Nb alloy has similar properties as Ti6Al4V, no general applications are found with it.

1.1. Traditional Lubrication Approach in Titanium Machining Processes

Flood lubrication with abundant quantities of lubricant has traditionally been the focus of attempts to overcome the machinability problems of titanium alloys [7]. The lubricant, coolant and cutting fluids (CFs) industrially used are estimated to account for about 17% of the total manufacturing costs of the final part [8], while the cutting tool costs are only 4% of total machining costs [9]. The cost factor, along with the environmental and health risks associated with the use of these cutting fluids, are the main reasons for attempting to reduce their use. Cutting fluids used as cooling agents contain environmentally hazardous and harmful chemical elements [6] and may cause diseases, such as respiratory problems, asthma and cancer.

The functions of the cutting fluids include lubrication and cooling, which can be carried out in parallel. In short, the objectives pursued by the application of these fluids is to improve the dissipation of the heat generated in the cutting area, which is very high in the machining of titanium alloys (up to 1000 °C), and to reduce the friction between the chip and the cutting tool.

The better or worse performance of the cutting fluid depends on the machining process, as well as the cutting conditions and the fluid characteristics, such as density, viscosity and specific heat. According to Lin et al. [10], the properties of the cutting fluid may affect the cutting conditions when turning Ti6Al4V, depending on the lubrication method. Three main types of cutting fluids exists: mineral, semi-synthetic and synthetic. Synthetic and semi-synthetic fluids are aqueous based fluids in which the good heat conduction of water combined with the oil properties enhances the performance of the cutting fluid.

1.2. Need for Sustainability

As previously explained, cutting fluids are damaging for the environment and human health, apart from involving high additional costs in the machining processes. There has been some efforts for using sustainable cutting fluids based on vegetable oils like sesam, coconut, sunflower, palm and others [11]. Some applications of this kind of fluids increases the cutting tool life a 170% in drilling of materials difficult to cut as AISI 316 steel. Coconut oil improves significantly the efficiency of machining processes and with the palm oil better results for Ti6Al4V alloy have been obtained compared to traditional flood lubrication [11,12]. Nevertheless the sustainability of this kind of fluids are controversial nowadays, specially palm oil.

Thus, the next qualitative advance in cutting fluids consists of the use of synthetic fluids made out of a liquid vegetable base and the addition of nanoparticles as Al_2O_3, MoS_2, diamond and graphene [11]. The synthetics fluids reduce the cutting forces and temperature and improve the surface finishing after machining. Moreover, some of these components act in an efficient way in applications in which friction at high pressure appears [13]. Nevertheless the cost of these lubricants is much higher than traditional ones [12].

For these reasons, in recent years, different lubrication and cooling techniques have been developed with the goal of reducing the use of cutting fluids and improving the machinability of titanium alloys under environmentally-friendly conditions [7,12,14–16]. Figure 1 shows the evolution of the number of papers related to the principal sustainable lubrication techniques.

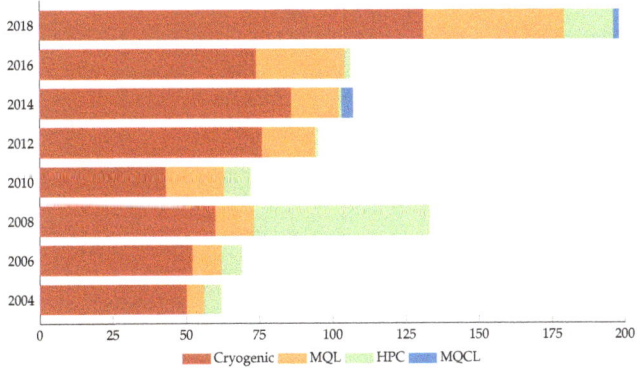

Figure 1. Evolution of principal sustainable machining techniques for titanium alloys.

These modern techniques provide different solutions for the cooling and lubrication problem, reducing the quantity of lubricant, as in the case of the minimum quantity of lubricant technique (MQL), or substituting the harmful lubricant for another green substance that can provide the cooling action, as in the case of cryogenic lubrication with liquid nitrogen, LN_2. This paper reviews the main sustainable techniques, such as minimum quantity of lubricant (MQL), cryogenic lubrication, minimum quantity cooling lubrication (MQCL) and high-pressure cooling (HPC), based on the recent literature since 2015.

Although dry machining can be considered a sustainable technique due to the lack of any lubricant, it cannot be applied in titanium and titanium alloy in most cases because of the high temperatures reached, which cause excessive tool wear and considerably reduce the quality of the machined parts.

The minimum quantity lubrication method (MQL) involves the atomization of cutting fluid droplets mixed in the air and directed at the cutting interface of the tool. The MQL technique is comparatively less hazardous for the environment than flood lubrication because it consumes very little toxic and non-biodegradable cutting fluid and needs less energy to pump the atomized fluid at higher flow rates and pressure [14].

The MQL strategy limits the cutting temperature by reducing the friction force tool-chip during the cutting process, thanks to the effective lubrication directly applied on the rake face of the tool. Compared to traditional flood lubrication, MQL requires much less than 1 L/h of lubricant [17] which is a great decrease compared to over 100 L/h adopted flood lubrication [12], although specific equipment is needed to atomize the lubricant droplets. In addition, vegetable-based oils are extensively used in this lubrication method [8].

Although air contributes to refrigerating the cutting interface and to dissipating heat by using the convective heat transfer mechanism, it has been proven for difficult-to-machine materials, like titanium alloys and nickel-based materials, that the MQL technique is not completely suitable because of its low heat transfer capacity [10] and its role depends on the specific machining process and cutting conditions. Typically, the cooling action of the air is not sufficient to reduce the cutting temperature, being the main drawback of MQL lubrication. Moreover, Mathew et al. [6] demonstrated a low efficiency of the MQL lubrication for deep hole drilling. However, it has been extensively demonstrated that MQL method is able to reduce cutting forces and to improve tool life compared to dry machining, and, depending on cutting conditions, compared to traditional flood lubrication [1]. Mathew et al. [6] reported MQL improvement on Ti aluminide drilling compared to flood lubrication based on cutting forces thanks to better penetration of the lubricant when eliminated chips obstruct its passage. Benjamin et al. [17] explain that the MQL method is able to overcome the formation of a vapor blanket in the cutting zone, which inhibits the effective lubrication effect of flood cooling in operations with low machinability materials.

As a result of the low quantity of lubricant applied in this technique, metallic chips generated during the process are almost dry and can be easily recycled [1].

Definitively, the MQL method solves the friction problem between the chip and the cutting tool by means of effective lubrication with a low quantity of lubricant compared to flood lubrication, which makes it a sustainable, environmentally friendly lubrication technique. Nevertheless, MQL does not enhance the cooling effect efficiently and is unable to rapidly dissipate the heat generated during the machining process.

The MQL approach has been developed to improve the cooling action. Many authors have demonstrated the improvement of mixing oil with water as a mode of transport and cooling source [10]. This technique, called Oil on Water (OoW) MQL, uses water droplets to transport a small quantity of lubricant. Specific equipment is required.

With this improvement, lubricant oil performs the lubrication effect, reducing the friction coefficient between chip and tool, while water, when evaporated, provides a cooling effect, which is more effective than supplied air flow. Briefly, when the water and oil droplet reach the objective surface, water evaporates, decreasing the cutting temperature, and oil remains, forming the lubricant film that reduces the friction force [14].

Another variant whose purpose is to improve heat dissipation in the MQL method is the minimum quantity of cooling lubrication technique (MQCL) [15]. Figure 2 shows the schematic diagrams for MQL and MQCL set ups. In this method, MQL is combined with sub-zero cooling, provided by low temperature air to enhance heat transfer from the tool-chip interface [17]. This system uses temperatures below 0 °C, but not as low as those achieved in cryogenic lubrication. For both MQL and MQCL procedures, the droplet size of the lubricant must be controlled and always greater than 5–10 µm. If the particles are smaller than that value, they can remain in the air and might cause healthy problems in the machine laborer [15,18]. Pervaiz et al. [8] performed their investigation on turning of Ti6AL5V titanium alloy by using vegetable-oil based lubricant at a flow rate of 60–100 mL/h and supplied air at −4 °C. They found that this MQCL cooling strategy is a successful substitute for conventional flood cooling methods, mostly at high cutting speeds.

Benjamin et al. [17] carried out an investigation on milling of Ti6Al4V alloy under the MQCL approach, using refined palm oil as lubricant at a flow rate of 350 mL/h and a cold temperature of −10 °C. They found surface roughness improvements compared to the MQL technique.

Nevertheless, the optimal conditions of this technique are still under development; Maruda et al. [19] establishes the threshold of useful conditions in terms of droplets diameters and emulsion mass flow and state that the number of droplets that arrives to the part surface are influenced by the volumetric air flow, and the device-part distance. Anyway, some research is being done regarding to improve the performance index of this kind of lubrication, i.e., the use of extreme pressure-antiwear additives [20].

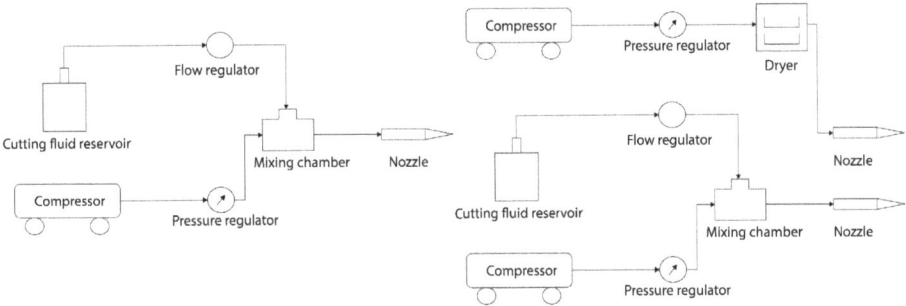

Figure 2. Schematics of MQL and MQCL systems.

The cryogenic lubrication method has been extensively studied in recent years. In cryogenic lubrication, liquid nitrogen LN_2 is usually supplied by means of small-diameter nozzles at a temperature range between −194 °C and −200 °C [16]. The significant cooling action produced is the greatest advantage of this strategy, compared to the MQL method. Cryogenic lubrication is able to effectively control the cutting temperature, reducing the heat generated in the cutting zone, which represents a substantive improvement on the machining of titanium alloys thanks to the high thermal gradients that appear in the machining operations.

Although this technique has been widely studied in the literature on turning, milling and drilling processes, it is not clear under what conditions the improvements are achieved and whether it is economically profitable, due to the expensive cost of liquid nitrogen [15]. The price of liquid nitrogen is a limiting factor for the application of this lubrication technique since large quantities of nitrogen are required, as well as special equipment. Moreover, in many industries, the savings that can be achieved by increasing the life of the cutting insert or reducing the cutting oil is not sufficient compared to the cost of liquid nitrogen. Nonetheless, LN_2 is a green lubricant, being harmless for the environment and human health, which makes this technique attractive. A variant of the deployment of liquid nitrogen is the use of dry ice or supercritical CO_2 [14]. The differences lie in the temperatures. While liquid nitrogen boils at −178 °C, carbon dioxide boils at −78 °C [21]. In addition, the preservation methods are different. LN_2 must be kept at atmospheric pressure and low temperatures in insulated containers, needing specific equipment, while CO_2 needs to be preserved at room temperature and high pressure, about 55 bar.

The cooling effect of CO_2 is due to the expansion of the gas when it comes out through the nozzle. A pressure drop takes place, resulting in a phase change into liquid and solid due to the Houle-Thompson effect, which promotes the cooling of the cutting area [22].

Shokrani et al. [2] found that cryogenic lubrication with LN_2 on the milling of the Ti6Al4V alloy reduced surface roughness by 40% in comparison to flood cooling, while tool life incremented almost three times. Nevertheless, Dix et al. [23] found, using the Finite Element Method (FEM), that cryogenic cooling produced high torque and force in the axial direction on drilling in comparison to dry machining.

Bordin et al. [24], reported that cryogenic cooling is the optimal solution and most suitable method for machining titanium parts in case of medical applications, because it minimizes the further severe cleaning requirements thanks to the lack of oil lubricant.

High-pressure cooling (HPC) is a technique whose principle is not based on the reduction of the quantity of lubricant, but on its efficient use, through effective penetration in the contact zone between the tool and the chip and in the flank face. In comparison with flood cooling, in which the average flow rate of lubricant is about 1.7 L/min [12], HPC utilizes a greater quantity of coolant. Ezugwu et al. [25] performed their study by using flow rates between 18.5 L/min and 24 L/min, while Mia et al. [26] utilized a flow rate of 6 L/min at a pressure of 8 MPa. For these reasons, HPC should not be classified as a green, environmentally friendly lubrication technique.

Laser-assisted machining is a thermally assisted machining process in which the method of operation is notably different from cryogenic lubrication and which has been gaining popularity in recent years [27]. In this modern technique, the specific area of the workpiece is preheated before the cutting process to reduce the flow stress and enable chip formation. This method, effective for machining processes of difficult-to-cut materials, such as titanium alloys, requires specific equipment and high precision in the control of laser variables [28]. The power of the laser and its movement velocity are critical parameters to promote improvements in the reduction of the cutting force. Nevertheless, diffusion can be increased thanks to the temperature reached by the material (about 500 °C). Therefore, the wear processes involved can be accelerated. In addition, thermal shocks can be produced by the temporal delay existing between the material overheated and the arrival of the cutting tool. Tool damage would appear if the laser variables are not properlly controlled [29].

2. Literature Review Methodology

This study is strictly focused on the machining processes of titanium and titanium alloys, considering the importance of these materials in current industrial manufacturing processes. The main objective of this review is to analyse the state of the art of the principal modern lubrication techniques that have been widely developed over approximately the last 20 years.

The literature includes many reviews on these lubrication techniques. However, they either give a limited explanation of each method or focus on only one of them, analysing a number of studies on one machining process.

The purpose of this paper is to provide firstly a general overview of each technique, and then a detailed analysis of each one, addressing the principal machining processes (turning, milling and drilling), based on the most novel works found in the published literature, and always with regard to titanium and titanium alloys. For this reason, this work review s from 2015 to 2019. It has been conducted using the Web of Science search engine.

Machining has been used as the general topic for the research. Terms such as MQL, HPC, cryogenic, MQCL, or the same terms written in full have been included in the search for papers related to each technique and filtering from the year 2015. All the papers reviewed are classified in Table 2, according to the type of process, material used, technique evaluated, number of citations and impact of the journal. The number of papers for each technique is shown in Figure 3.

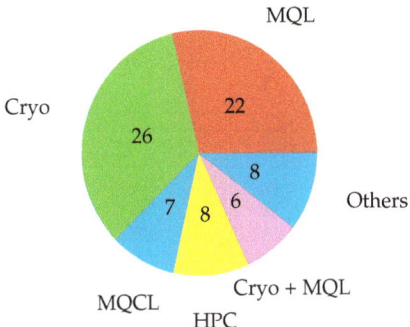

Figure 3. Number of papers on different techniques. Minimum quantity of lubricant (MQL), cryogenics (Cryo), high pressure coolant (HPC), minimum quantity of cooling lubrication (MQCL).

Cryogenic and minimum quantity of lubricant are the two most extensively studied techniques, while high-pressure cooling is only used in turning processes. Other techniques, such as dry machining, ultrasonic vibration assisted machining, machining with specific inserts or electrostatic high-velocity solid lubricant machining have been classified together.

A statistical analysis of these papers has been developed with the aim of obtaining the main parameters studied for each of the principal machining processes, such as cutting forces, surface integrity, cutting temperature, tool wear, chip morphology, friction, specific energy or power consumption. The principal variables studied in each paper on drilling, milling and turning are shown in Tables 3–5.

Tables 3–5 reveal that the most commonly researched aspects are cutting forces, tool wear and surface integrity of the machined parts, while parameters such as friction coefficient or power consumption are rarely analysed, Figure 4. Cutting temperature is an important variable that is the subject of less research than cutting forces or tool wear due to the difficulty of its being accurately measured in the cutting area.

With the aim of providing a more detailed overview of the state of the art, the highest-ranked papers on drilling, milling and turning have been filtered, firstly taken into account the number of citations and the impact index of the journal. In this way, a number around 15 papers have been managed for turning and milling processes; for drilling, all papers have been carefully analysed as the few number of total papers published. After that, a second filtering step has been applied based on a detailed reading, selecting a balanced quantity of papers about each main lubrication technique. This led to take into account eight manuscripts for turning and milling processes.

For the selection of these papers, a content analysis has been made that considers the influence of the lubrication process on the technological capacity of the processes in terms of surface roughness, cutting forces, temperature and tool life. In addition, it has been evaluated qualitatively the discussion and conclusions carried out by the own authors in comparison with traditional lubrication.

The main strenghts of the review methodology applied herein are the relevant and complete classification of involved papers based on the machining process, the lubrication technique used and the different variables analysed and researched by the authors. This permits to locate in an easy way the points of interest of each paper to be considered as background for specific researchings. Likewise, based on some representative papers, the different lubrication techniques and their effect have been carefully analysed providing the grounded fundamentals of them.

Table 2. Classification of reviewed literature (June 2019).

Reference	Year of Publication	Studied Material	Machining Process	Analysed Lubrication Technique	Number of Citations	Impact
[6]	2017	Ti aluminide	Drilling	MQL	9	1
[30]	2016	Ti6Al4V	Drilling	Cryogenic/MQL	9	1
[31]	2016	Ti6Al4V	Drilling	Oil on water MQL	0	
[32]	2018	Ti aluminide	Drilling	MQL	0	2
[1]	2015	Ti6Al4V	Milling	Cryogenic/MQL/Cryogenic + MQL	15	3
[2]	2016	Ti6Al4V	Milling	Cryogenic LN_2	7	3
[8]	2017	Ti6Al4V	Milling	Cryogenic/MQL/Cryogenic + MQL	12	3
[17]	2017	Ti6Al4V	Milling	MQL and MQCL	5	1
[22]	2017	Ti6Al4V	Milling	Cryogenic CO_2/MQL/Cryogenic + MQL	4	
[27]	2015	Ti6Al4V	Milling	Laser-assisted	27	1
[29]	2016	Ti6Al4V	Milling	Laser-assisted	10	2
[33]	2015	Ti6Al4V	Milling	Laser-assisted	5	2
[34]	2017	Ti6Al4V	Milling	Cryogenic	3	2
[35]	2017	Ti6Al4V	Milling	Cryogenic LN_2	2	2
[36]	2017	Ti6Al4V	Milling	Cryogenic/Dry	0	
[37]	2016	Ti6Al4V	Milling	Cryogenic/Dry	3	3
[38]	2018	Ti6Al4V	Milling	MQL+nanoparticles	1	2
[39]	2016	Ti6Al4V	Milling	Cryogenic/Dry	2	3
[40]	2018	Ti6Al4V	Milling	Cryogenic + MQL	0	2
[41]	2017	Ti6Al4V	Milling	Cryogenic	0	2
[42]	2018	Ti6Al4V	Milling	Ultrasonic vibration-assisted	1	3
[43]	2019	Ti6Al4V	Milling	Cryogenic LN_2	0	2
[44]	2018	Ti6Al4V	Milling	Laser-assisted	0	1
[3]	2016	Ti6Al4V	Turning	HPC	18	2
[4]	2018	Ti5553	Turning	MQL/HPC	2	1
[8]	2016	Ti6Al4V	Turning	MQCL	13	2
[10]	2015	Ti6Al4V	Turning	Oil on water MQL/MQCL	19	2
[21]	2016	Ti6Al4V	Turning	Cryogenic/MQL/Cryogenic + MQL	14	
[25]	2018	Ti6Al4V	Turning	HPC	0	1
[26]	2017	Ti6Al4V	Turning	HPC	9	1
[45]	2017	Ti6Al4V	Turning	Cryogenic LN_2	13	2
[46]	2016	Ti6Al4V	Turning	HPC	12	2
[47]	2017	Ti Grade 2	Turning	MQL	10	1
[48]	2017	Ti6Al4V	Turning	Gaseous nitrogen	7	1
[49]	2017	Ti6Al4V	Turning	Cryogenic + ethanol	6	
[50]	2017	Ti Grade 2	Turning	MQL (R-H vortex tube)	7	2
[51]	2017	Ti6Al4V	Turning	Dry/MQL	6	1
[52]	2017	Ti6Al4V	Turning	MQCL	7	3
[53]	2018	Ti6Al4V	Turning	MQL	6	2
[54]	2018	Ti6Al4V	Turning	MQL	5	2
[55]	2017	Ti6Al4V	Turning	Dry	3	
[56]	2017	Ti6Al4V	Turning	Cryogenic LN_2 (single and dual)	5	3
[57]	2019	Ti6Al4V	Turning	Cryogenic LN_2	3	1
[58]	2018	Ti6Al4V	Turning	Specific insert	3	1
[59]	2016	Ti6Al4V	Turning	Cryogenic/MQL/sub zero	2	
[60]	2017	Ti6Al4V	Turning	Cryogenic LN_2	3	2
[61]	2018	Ti5553	Turning	Cryogenic/MQL/HPC	3	2
[62]	2018	Ti6Al4V	Turning	Cryogenic LN_2	2	2
[63]	2018	Ti6Al4V	Turning	Cryogenic LN_2	2	2
[64]	2017	Ti6Al4V	Turning	Cryogenic LN_2/N_2	2	2
[65]	2017	Ti6Al4V	Turning	MQL	2	2
[66]	2019	Ti6Al4V ELI	Turning	MQCL	1	1
[67]	2018	Ti6Al4V	Turning	Sub-zero HPC	0	1
[68]	2018	Ti6Al7Nb	Turning	MQL	1	2
[69]	2017	Ti6Al4V	Turning	Nanolubrication	1	2
[70]	2018	TC17	Turning	MQL/oil on water MQL	0	1
[71]	2019	Ti5553	Turning	Cryogenic LN_2/CO_2	0	1
[72]	2019	Pure titanium	Turning	Cryogenic + MQL	0	2
[73]	2018	Ti6Al4V	Turning	MQL/U-CMQL	0	2
[74]	2018	Ti6Al4V	Turning	MQL and MQCL	0	2
[75]	2019	Ti6Al4V ELI	Turning	HPC	0	2
[76]	2018	Ti6Al4V	Turning	Cryogenic LN_2/CO_2	0	3
[77]	2017	Ti6Al4V	Turning	EHVSL	0	3
[78]	2018	Ti6Al4V	Turning	Cryogenic LN_2 (Internal cooling)	0	4
[79]	2018	Ti6Al4V	Turning	Laser-assisted	0	

Table 3. Main variables studied in papers on drilling.

Reference	Cutting Force	Cutting Temperature	Surface Integrity	Chip Morphology	Specific Cutting Energy	Tool Wear	Tool Life	Friction Coefficient	Power Consumption
[6]	x			x		x			
[30]	x		x			x			
[31]						x	x		
[32]		x				x			

Table 4. Main variables studied in papers on milling.

Reference	Cutting Force	Cutting Temperature	Surface Integrity	Chip Morphology	Specific Cutting Energy	Tool Wear	Tool Life	Friction Coefficient	Power Consumption
[1]	x					x			
[2]			x		x	x	x		x
[9]	x			x		x			
[17]		x	x	x		x		x	
[22]						x	x		
[27]		x				x	x		
[29]	x	x							
[33]	x	x				x			
[34]	x			x		x		x	
[35]		x				x			
[36]	x		x			x			x
[37]					x	x			x
[38]	x	x	x			x			
[39]			x			x	x		
[40]	x		x						
[41]	x		x						
[42]						x			x
[43]			x			x	x		
[44]	x				x				

Table 5. Main variables studied in papers on turning.

Reference	Cutting Force	Cutting Temperature	Surface Integrity	Chip Morphology	Specific Cutting Energy	Tool Wear	Tool Life	Friction Coefficient	Power Consumption
[3]			x			x	x		
[4]	x		x	x					
[8]	x		x			x			
[10]	x	x	x	x		x			
[21]			x		x	x	x		
[25]			x	x		x	x		
[26]	x	x	x	x		x			
[45]	x		x						
[46]	x	x		x					
[47]	x		x			x			
[48]			x			x			
[49]	x			x		x			
[50]	x		x			x			x
[51]	x		x			x	x		x
[52]	x		x			x			
[54]			x	x					
[55]	x					x			
[56]		x	x	x	x	x	x		
[57]	x	x	x		x				
[58]		x	x	x		x			
[59]	x	x		x		x	x		
[60]	x		x			x			
[61]	x	x		x		x			
[62]				x					
[63]	x		x	x					
[64]						x			
[65]	x		x			x			
[66]		x	x	x		x			
[67]	x	x		x		x			
[68]	x		x	x	x	x		x	
[69]			x			x			x
[70]	x					x			
[71]	x	x	x	x	x	x			
[72]	x		x			x			x
[73]	x		x	x		x			
[74]						x	x		
[75]	x			x					
[76]			x						
[77]	x		x			x			
[78]		x							
[79]	x								

Nevertheless, the methodology is conditioned to the use of a single search engine that, althought is the most important one, might set aside some significant papers. Besides, sometimes, the keywords in the papers could not be well identified according to our searching criteria.

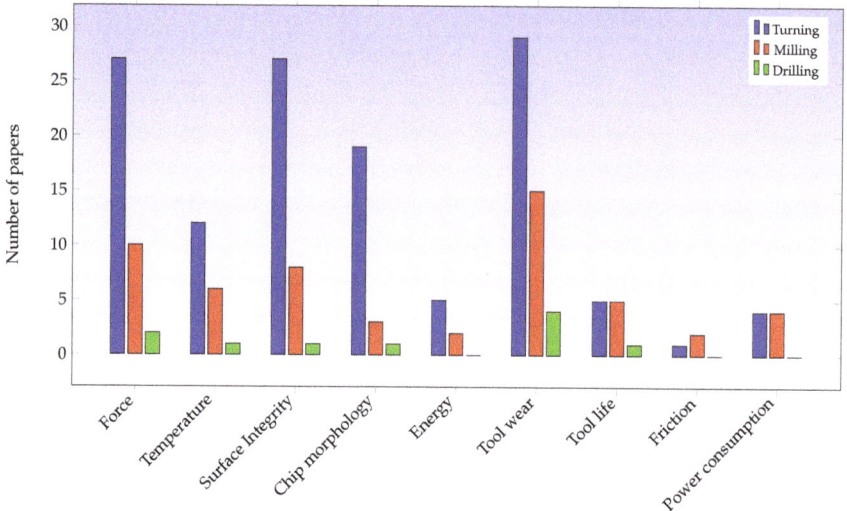

Figure 4. Main criteria mentioned in papers on turning, milling and drilling.

2.1. Sustainability in Milling Processes

A detailed review of the most significant sustainable techniques in milling processes was conducted. The papers selected are shown in Table 6.

With this selection of milling research papers, a general overview can be obtained about the state of art of each of the main sustainable techniques.

Shokrani et al. [2] investigated the application of cryogenic lubrication with liquid nitrogen LN_2 in the milling process of Ti6Al4V alloy, compared to dry and flood lubrication. They performed the experimental investigation by supplying cryogenic nitrogen around the cutting tool at $-197\,°C$, 1.5 bar of pressure and with a flow rate of 0.4 L/min, but without submerging the workpiece in LN_2. For the experiments, they used a TiN-TiAlN coated solid carbide end mill and tested three cutting velocities (30 m/min, 115 m/min and 200 m/min), three feed rates (0.03 mm/tooth, 0.065 mm/tooth, and 0.1 mm/tooth) and three depths of cut (1 mm, 3 mm and 5 mm).

This study examined the potential of the cryogenic technique to improve surface roughness and tool life, thanks to decelerating thermally induced wear, while reducing power and energy consumption. The authors found that, on average, surface roughness was 30.42% lower in comparison with dry and flood machining. They also found that the environment (dry, flood or cryogenic) has a 21.5% contribution on final surface roughness and a 73% contribution in power consumption.

Park et al. [1] obtained poor results for the application of only cryogenic LN_2, both by the internal and external method. The cutting force at the beginning of the process was lower than for flood lubrication, but a significant increase was obtained at the final pass. In addition, the criterion of tool rupture was achieved in both cases. The reduction of tool life is explained by the lack of effective lubrication due to the deep axial depth of cut with a large contact area between the tool and chip, causing excessive adhesion on the tool.

Table 6. Selected papers on milling.

Reference	Journal	Title	Year of Publication
[34]	The International Journal of Advanced Manufacturing Technology	Effect of cryogenic treatment on the microstructure and the wear behavior of WC-Co end mills for machining of Ti6Al4V titanium alloy	2017
[17]	Tribology International	On the benefits of sub-zero air supplemented minimum quantity lubrication systems: An experimental and mechanistic investigation on end milling of Ti6Al4V alloy	2017
[22]	The 50th CIRP Conference on Manufacturing Systems. Procedia CIRP	Investigation of the Influence of CO_2 Cryogenic Coolant Application on Tool Wear	2017
[9]	International Journal of Precision Engineering and Manufacturing	Milling of titanium alloy with cryogenic cooling and minimum quantity lubrication (MQL)	2017
[2]	Machining Science and Technology	Comparative investigation on using cryogenic machining in CNC milling of Ti6Al4V titanium alloy	2016
[1]	Journal of Mechanical Science and Technology	The effect of cryogenic cooling and minimum quantity lubrication on end milling of titanium alloy Ti6Al4V	2015
[29]	International journal of advanced manufacturing technology	Determination of optimal laser power according to the tool path inclination angle of a titanium alloy workpiece in laser-assisted machining	2016
[27]	Wear	Tool life and wear mechanisms in laser assisted milling Ti6Al4V	2015

These authors also investigated the end milling process of Ti6Al4V alloy, comparing different techniques, such as flood coolant lubrication, MQL mixture with Hexagonal Boron Nitride nano-particles and the combination of internal cryogenic and Nano-MQL. They conducted the experiments at cutting speeds of 72 m/min for flood lubrication and 86 m/min for the other techniques, feed rate of 0.1 mm/tooth and depth of cut of 24.5 mm. They used a tool coated with Aluminium Chromium Nitride (AlCrN).

They found that the application of Nano-MQL always reduces the cutting forces in comparison with flood cooling, obtaining the best results for the combination of Nano-MQL and internal cryogenic. In addition, tool life was enhanced by up to 32%, the reasons being the effective lubrication supplied by Nano-MQL and the reduction of cutting temperature thanks to cryogenic lubrication.

In another study, Park et al. [9] analysed face milling and end milling of Ti6Al4V alloy under dry, flood and MQL. They used an uncoated insert and the conditions for the face milling experiments were cutting speed of 47.7 m/min, 76.4 m/min, 100 m/min and 120 m/min, feed rate of 0.15 mm/rev and depth of cut of 2 mm. For the end milling experiments, the cutting speeds were 72 m/min and 90 m/min, the feed rate was 0.1 mm/rev and the depth of cut was 1.5 mm. They tested two MQL methods: the application of a vegetable oil at the tool-chip interface at 5 bars and 3 mL/min and MQL mixture with exfoliated graphite nano-platelets (MQLN). They showed that MQLN is most effective at a high cutting speed because nano-particles play the role of a lubricant reducing the friction between the tool and work material.

They found MQL reduced cutting force compared to dry and flood machining in all cases, obtaining the best performance in combination with cryogenic conditions.

Similar results were obtained by Tapoglou et al. [22]. These authors reported that the best performing cryogenic method was CO_2 in combination with the MQL technique, being better than MQL and the combination of CO_2 and air when they used a single insert. When they used five inserts, they found that at 70 m/min and 80 m/min, the combination of MQL and cryogenic CO_2 improved tool life by up to 29% and 32%, respectively, compared to MQL. Nevertheless, the application of cryogenic CO_2 at maximum velocity reduced tool life in comparison with the MQL technique, thus not being an efficient lubrication method.

Similarly, Benjamin et al. [17] focused their study on comparing MQL and MQL with sub-zero cooling on end milling of Ti6Al4V alloy. They used refined palm oil as lubricant with a flow rate of 350 mL/h and a pressure of 2 bar. The cold stream was supplied by a Vortex tube at −10 °C and 6 bar. The cutting conditions were cutting speed between 90 m/min and 150 m/min and feed rate from 0.025 mm/rev to 0.075 mm/rev.

They found that lower temperatures were achieved under the MQCL technique in comparison with MQL with cold air, which increased the viscosity of the palm oil, improving its lubricating properties.

In addition, with MQCL, the tool life was increased from 11.06 min to 15.9 min and flank wear decreased by 19%.

Park et al. [9] examined what increase in flow rate is needed in the MQL technique as cutting velocity increases. They showed that an optimal flow rate exists for each cutting condition that improves tool life.

Celik et al. [34] studied the influence of cryogenic treatment of WC-Co end mills for machining Ti6Al4V alloy. They analysed the behaviour of coated and uncoated cryogenically treated end mills over 12 h, 24 h and 36 h, developing the experimental tests under dry condition. They found that 36 h cryogenically treatment reduced the friction coefficient and friction forces. In the milling process, an increment of cutting forces was demonstrated in each pass due to toll wear. Increasing the cryogenic treatment times from 12 to 36 h positively affected the coated and uncoated tools, except for the AlTiN sample, in which the coating substrate interfacial adhesion bond was weakened.

AlCrN coated tools with 36 h of cryogenic treatment showed the best performance for both tool wear and cutting forces, while AlTiN coated tools exhibited lower performance with the treatment. In conclusion, it can be said that cryogenic treatment is not an effective method to improve milling process performance when working with an AlTiN coated tool.

Bermhingham et al. [27] studied tool life and wear in laser assisted milling of Ti6Al4V. They performed their face milling experiments under dry conditions, MQL, flood coolant, LAM and a combination of LAM with MQL with the aim of analysing the influence of the environment in tool life. Using a PVD coated (TiN) mill, they tested three laser power levels (50 W, 100 W and 150 W) to analyse the influence of overheating the workpiece material.

They found that laser-assisted milling increased tool life at some cutting speeds and decreased it at others. LAM at high power (150 W) was found to produce the lowest tool life of any test, which means that overheating reduces tool performance due to thermal shock. Compared to the rest of lubrication techniques, laser-assisted milling generated an improvement in tool life at low velocity. For the combination of LAM with MQL, no failure appeared at 69 m/min after 28 min of testing. It was verified that for high cutting velocities, laser-assisted milling has a detrimental effect on tool life due to thermal shock, while the combination of LAM with MQL is capable of improving tool life as the cooling action of MQL slows the rate of thermal wear processes.

Sim et al. [29] analysed the optimal laser power and inclination angle for laser-assisted milling of Ti6Al4V alloy. Using FEM, they studied the influence of the milling rotation angle in relation to the inclination angle of the workpiece surface with respect to the horizontal surface. They found that for inclination angles greater than 30°, the preheated temperature decreased in line with the increase of the inclination angle in the tool path until 75°, from which a new increment took place, meaning that the minimum value occurs for 75°.

In the experimental tests, they found that the cutting force decreased depending on the increase in the path inclination angle. They showed that the preheating temperature decreased as the tool path inclination angle increased, but the preheating temperature increased when the tool path inclination angle was 75° or more, which corresponds to the simulated results.

2.2. Sustainability in Turning Processes

There now follows a detailed review of the most important sustainable techniques in milling processes. The papers selected are shown in Table 7.

Table 7. Selected papers on turning.

Reference	Journal	Title	Year of Publication
[3]	The International Journal of Advanced Manufacturing Technology	High-pressure coolant on flank and rake surfaces of tool in turning of Ti6Al4V: investigations on surface roughness and tool wear	2016
[47]	Journal of Cleaner Production	Sustainable machining of aerospace material- Ti (grade 2) alloy: modelling and optimization	2017
[48]	Journal of Manufacturing Processes	Temperature effects on the Ti6Al4V machinability using cooled gaseous nitrogen in semi-finishing turning	2017
[45]	The International Journal of Advanced Manufacturing Technology	Study of surface roughness and cutting forces using ANN, RSN, and ANOVA in turning of Ti6Al4V under cryogenic jets applied at flank and rake faces of coated WC tool	2017
[49]	CIRP Journal of Manufacturing Science and Technology	Increasing efficiency of Ti-alloy machining by cryogenic cooling and using ethanol in MRF	2017
[8]	The International Journal of Advanced Manufacturing Technology	An experimental investigation on effect of minimum quantity cooling lubrication (MQCL) in machining titanium alloy (Ti6Al4V)	2016
[10]	The International Journal of Advanced Manufacturing Technology	Tool wear in Ti6Al4V alloy turning under oils on water cooling comparing with cryogenic air mixed with minimal quantity lubrication	2015
[21]	48th CIRP Conference on Manufacturing Systems – CIRP CMS 2015. Procedia CIRP	Investigation of cooling and lubrication strategies for machining high-temperature alloys	2016

This selection of papers on turning provides a general overview on the state of the art of each principal sustainable technique.

Mia et al. [3] performed their experimental study on turning of Ti6Al4V alloy by applying high-pressure coolant technique on the flank and rake surfaces of the tool. They compared the difference in surface roughness and tool wear between machining at dry and HPC conditions, using a coated carbide tool (TiN, WC and Co).

They found that double jet action significantly improved the heat transfer because the coolant is able to reach the point of highest temperature by overcoming the chip obstruction and removing it from the cutting zone. Using ANOVA analysis, they found that the environment lubrication has an 18% influence on the final surface roughness. HPC reduced the contact length between chip and tool and inhibited the probability of BUE formation, improving the tribological interaction. Similar results were obtained by Busch et al. [21].

Nevertheless, although HPC decreases the surface roughness at high cutting speed, at low cutting speed the effect was contrary, due to the sliding of chips over the tool surfaces. If coolant output pressure is not sufficient to break the chip and cutting velocity is low, the adhesiveness of the chip is enhanced.

Pervaiz et al. [8] studied the effect of minimum quantity cooling lubrication (MQCL) on Ti6Al4V turning. They used a vegetable oil-based MQL system at flow rates between 60 mL/h and 100 mL/h. The cooling action was added by pressurized air at 0.5 MPa and −4 °C and an uncoated cutting insert was used.

It was found that the application of different oil flow rates under MQCL had no impact on surface roughness. At 90 m/min and lower feed levels, the cutting force showed a decreasing trend with the increment of lubricant flow rate due to effective lubrication. Nevertheless, at the highest feed level, a higher flow rate increased the cutting force. Pervaiz et al. suggested that increased cooling of the workpiece helps maintain material hardness without thermal softening. Similar conclusions were reached by Lin et al. [10] when they analysed the turning process of Ti6Al4V under MQCL conditions. They found that although the lowest air temperature of −26 °C provided a better cooling effect by reducing the cutting temperature, the lowest cutting forces and surface roughness were obtained for −16 °C of temperature. This is due to the reduction of Ti6Al4V temperature that enharden the material. However, both temperatures provided better results than for dry and wet machining conditions.

Sartori et al. [48] conducted their research on semi-finishing turning of Ti6Al4V ELI using cooled gaseous nitrogen, comparing its effects to wet condition and cryogenic liquid nitrogen lubrication. The cutting parameters were cutting speed of 80 m/min, feed rate of 0.2 mm/rev and depth of cut of 0.25 mm. A TiAlN coated tungsten carbide tool was used. Gaseous nitrogen was applied at 2.5 bar in a range of temperatures between 0 °C and −150 °C. For the wet condition, a water emulsion with 5% of semi-synthetic cutting fluid was utilized. For cryogenic tests, liquid nitrogen was supplied at 15 bar and −196 °C.

They found that LN_2 lubrication increased flank wear by 30% with respect to the wet condition, while cooled N2 always improved tool life. The best results took place at −100 °C, with a reduction of flank wear of 26% and 43% with respect to wet and LN_2 cooling. Less difference was found with an N2 temperature between 0 °C and −100 °C, which does not justify the cost involved in reducing the temperature. Crater wear was considerably reduced by N2 at −75 °C, while the greatest reduction took place at −150 °C. The lowest tool wear was found at −100 °C, with a reduction of 43% compared to LN_2, for which the thermal power required to achieve a low temperature is almost 10 times the thermal power required by gaseous nitrogen. However, the improvements were already evident from −50 °C.

Liquid nitrogen was found to ensure the best surface roughness of the machined part. However, N2 conditions at −100 °C and −150 °C were also close to this result.

Nevertheless, good results were obtained by Krishnamurthy et al. [49] in their study. They found that cryogenic treatment allowed a reduction of 300 N in cutting forces in comparison with dry machining, which corresponds to a 25% reduction. During the cryogenic treatment, chip segments became susceptible to fracture, which was detected by the segmentation of the chip. Using a Charpy V-notch impact test, they found that less energy is needed to break Ti6Al4V alloy at cryogenic conditions. Nevertheless, in these conditions, the shear strain of the machined piece was lower, which is an indicator of the high flow stress of Ti6Al4V at this temperature.

Krishnamurthy et al. [49] performed turning of Ti6Al4V alloy by cryogenic cooling and using ethanol in metal removal fluid (MRF). They performed turning tests under different conditions, such as dry machining, flooded machining using water-based an MRF mixture with vegetable oil and flooded machining using the same MRF with an addition of 10% of volume of ethanol, both with a flow rate of 8.3 mL/s. Finally, they performed dry machining after having immersed Ti6Al4V workpieces in a bath of LN_2 for 20 min.

Applying ethanol, a further reduction of 65% in cutting forces was generated and the best roughness surface was obtained, thanks to a reduction in the coefficient of friction at the tool-workpiece interface.

Mia et al. [45], in other different study, evaluated surface roughness and cutting forces in the turning of Ti6Al4V under cryogenic lubrication, applied at flank and rake faces and using two cutting inserts (SNMM 120408 and SNMG 1240408).

Using an ANOVA, they found that, under cryogenic conditions, the greatest variation in cutting force came from the cutting velocity, followed by the feed rate, while the tool showed an insignificant impact. For the surface roughness, the variations came from the cutting tool, SNMM insert, associated with lower cutting forces, producing the lowest surface roughness with a reduction of 50% in comparison with the other insert.

Gupta et al. [47] studied the turning process of Ti grade 2 under MQL conditions in comparison with dry machining. By using Box-Behnkens response surface methodology and ANOVA, they obtained the most influent parameters in the machining process. The turning tests were conducted under a cutting velocity from 200 m/min to 300 m/min, feed rate from 0.1 mm/rev to 0.2 mm/rev, depth of cut of 1 mm and approach angle from 60° to 90°, by using a CBN insert, coated with TiN. For MQL conditions, the flow rate was fixed at 300 mL/h, with an output pressure of 5 bar.

They verified that with the MQL lubrication condition, lower cutting temperatures were achieved due to effective lubrication, producing lower forces than for dry cutting, for which the no lubrication condition resulted in greater chip-tool contact length. Gupta et al. found that an increase of the cutting speed from 200 m/min to 250 m/min was beneficial and cutting forces, surface roughness and tool wear were reduced thanks to an initial increase in cutting temperature, which is able to soften the workpiece. However, the increment of cutting velocity from 250 m/min to 300 m/min, generated a negative effect on cutting forces, tool wear and surface roughness because MQL lubrication does not sufficient cooling capacity to evacuate heat.

Lin et al. [10] analysed the turning process of Ti6Al4V under the oils on water MQL and the MQCL approach. Oils on water (OoW) is a lubrication method in which a thin layer of oil is transported by droplets of water, providing effective lubrication and cooling effect. They tested two forms of spraying the droplets, from the interior of the cutting tool with a specific insert (IOoW) and from the outside of the cutting tool (EOoW). Based on chip morphology, cutting temperatures, forces, surface roughness and tool wear, they analysed the differences between OoW techniques and MQCL, for which MQL was mixed with cryogenic air between $-16\,^{\circ}\text{C}$ and $-26\,^{\circ}\text{C}$.

Under EOoW conditions, the effects of three different spraying locations (rake face, flank face, and rake and flank faces) were studied. Under IOoW experiments, the effects of small (1.2 L/h) and large (2.4 L/h) amounts of water were studied. Two different lubricants were used, fatty alcohol and synthetic ester, with different lubricity and cooling abilities. The turning tests were conducted under a cutting speed of 70 m/min, 90 m/min and 110 m/min, feed rate of 0.25 mm/rev and depth of cut of 1 mm, using coated carbide inserts.

From these EOoW turning tests, they found the lowest forces for flank face spraying location because the chips were much shorter due to the air direction of the shingle. Nevertheless, the highest temperature was achieved for this configuration due to the lack of lubricant on the rake face. The lowest temperature was obtained from the double air direction. EOoW on the flank face provided the best surface roughness thanks to better chip breaking.

In the case of internal oil on water, it was found that the larger amount of lubrication reduced cutting forces and cutting temperatures thanks to effective cooling capacity. However, poorer surface roughness and greater tool wear was obtained in comparison with the lower quantity of lubricant. Lin et al. suggested the explanation that when a small amount of water is used, the mixture of oil and water takes the desirable configuration, oil forms a thin layer around the water droplet and provides an effective lubrication. In the case of using a large amount of oil, the configuration is inverse. Water forms a thin layer around the oil droplets, reducing the lubrication properties.

In addition, Lin et al. reported that lubricant properties had an influence on the IOoW method but not on the EOoW method.

2.3. Sustainability in Drilling Processes

A detailed review of the main sustainable techniques in milling processes was conducted. The papers selected are shown in Table 8.

Table 8. Selected papers on drilling.

Reference	Journal	Title	Year of Publication
[32]	Materials and Manufacturing Processes	*Temperature rise in workpiece and cutting tool during drilling of titanium aluminide under sustainable environment*	2018
[6]	Journal of Cleaner Production	*Environmentally friendly drilling of intermetallic titanium aluminide at different aspect ratio*	2017
[30]	Precision Engineering	*Micro-drilling of Ti6Al4V alloy: The effects of cooling/lubricating*	2016
[31]	16th Machining Innovations Conference for Aerospace Industry	*Effect of water oil mist spray (WOMS) cooling on drilling of Ti6Al4V alloy using Ester oil based cutting fluid*	2016

This selection of turning papers provides a general overview on the state of art of each principal sustainable technique.

Perçin et al. [30] studied the effect of MQL and cryogenic lubrication on micro-drilling processes of Ti6Al4V. They compared cutting parameters, such as cutting forces, torque or tool wear, by developing micro-drilling tests under dry, wet, MQL and cryogenic conditions. The experiments were conducted under five different spindle speeds and five feed rates. The depth of the holes was established as a fixed 3 mm. They used a 700 µm diameter uncoated tungsten carbide micro-drill. For the MQL test, lubricant was supplied under low pressure, lower than 3 bar, because higher pressures broke the tool.

They found that, under cryogenic conditions, the cutting force was always greater than for dry and wet conditions, while under MQL, for the greatest feed rates, the force was lower than for the dry condition. The improvement was also seen for the lowest cutting speed in comparison with dry cutting.

The use of lubrication reduced cutting force and torque. At the beginning of the hole, torque was higher for wet and MQL lubrication because the lubricant made it more difficult for the chip to be released. However, with the cutting time, wet and MQL conditions provided better performance in the drilling process due to effective lubrication reducing the torque.

Nevertheless, the best quality for surface roughness was obtained for both wet and MQL lubrication. The results for the cryogenic conditions were worse than those for the dry condition. In addition, they found that the increment of cutting speed reduced surface micro-hardness due to thermal softening. Cryogenic lubrication conditions reduced softening, obtaining greater hardness. For this reason, cryogenic lubrication provided the greatest surface micro-hardness. It also ensured the minimum tool wear, while MQL generated less wear compared to dry cutting due to less abrasion.

Mathew et al. [32] investigated the drilling process of titanium aluminide under MQL lubrication. The experiments were conducted under a cutting speed of 72 m/min and feed rate of 0.01 mm/rev with a depth of cut of 10 mm (low aspect ratio) and 37.5 mm (high aspect ratio). A 4 mm diameter solid carbide drill (TiAlN coated) was used.

They found that in the high aspect ratio (HAR) method, in the absence of coolant for dry cutting, the temperature generated on the workpiece and the cutting tool was significantly higher than for MQL lubrication. This means that MQL provided effective lubrication to evacuate the heat.

Built-up edge was formed in both cases, but, for the MQL condition, there was found to be less BUE formation due to the reduction of chip adhesion on the tool. During LAR tests, the probability of BUE forming is reduced. Mathew et al. analysed tool wear based on tool roughness after 4 holes, comparing it to the initial roughness, reporting that MQL lubrication reduces tool wear significantly, providing as good results as wet machining. Similar results were presented by Nandgaonkar et al. [31]. They developed their experimental investigation on drilling Ti6Al4V alloy under oil on water MQL approach by using a 8 mm diameter solid carbide twist drill coated with TiAlN, finding that 43 holes of 20 mm depth could be made under dry conditions, while, under oil on water MQL conditions, 55 holes could be completed. This means that tool life was improved by 27% thanks to effective lubrication which reduces chip-tool contact length and cutting temperature. The authors reported that tool wear was reduced by 66% under oil on water MQL conditions.

In another study in the same experimental conditions [6], Mathew et al. analysed the evolution of cutting force and torque on drilling titanium aluminide at different aspect ratios. They found that for low aspect ratio (LAR), MQL lubrication provided a reduction in cutting force compared to wet lubrication and dry machining. Torque under MQL and wet conditions was always lower and more stable than under dry drilling.

However, different results were obtained for HAR drilling. In this case, the thrust force achieved under wet conditions was lower than for MQL lubrication, mainly at the end of the drilling process. The authors explained that for HAR drilling the air pressure is not sufficient to evacuate the chip, in addition to the amount of lubricant being insufficient to ensure the evacuation of heat, which causes higher tool wear.

In conclusion, these papers propose MQL as an efficient solution for drilling titanium aluminide at low aspect ratio, while it must be optimized for high aspect ratio drilling, for which the performance index is lower than for wet lubrication.

3. Conclusions

Titanium and titanium alloys are widely used in industry for aeronautical and biomedical applications, thanks to their good mechanical properties and low density. Ti6Al4V is the most important alloy, representing 50% of titanium production, although in recent years interest has been growing in titanium aluminides. Based on the literature review, the following conclusions can be drawn:

- Due to the low modulus of elasticity and thermal conductivity, as well as the high chemical affinity with the cutting inserts, the machinability of titanium alloys is extremely poor, exhibiting several difficulties for the machining processes.
- Sustainable lubrication methods are necessary to replace flood lubrication in machining processes, so that machinability is increased and is more environmentally friendly. Techniques such as cryogenic lubrication, minimum quantity of lubricant (MQL), minimum quantity cooling lubrication (MQCL) or laser-beam assisted machining (LBM) have been analysed, focusing on turning, milling and drilling processes.
- The cryogenic lubrication technique has been found to be a potential substitute for traditional flood lubrication. Many authors have concluded that it is an effective method of reducing surface roughness on machined parts, obtaining significant improvements compared to flood lubrication, up to 65% on turning processes [49]. Nevertheless, no clear conclusions have been reached on the performance of LN_2 cryogenic lubrication when tool wear and tool life is analysed. A number of authors have obtained tool life improvements under certain cutting conditions on milling and turning processes, but many others have found that tool wear is greatly increased under cryogenic conditions due to the increment of titanium hardness at low temperatures, which also increases the cutting force compared to those obtained under traditional lubrication conditions, although they are smaller than for dry machining. A solution to the hardness increment proposed by various authors is cryogenic treatment of the tool before machining, without affecting the part to be machined, obtaining good results.
- The minimum quantity of lubrication (MQL) method has been proposed by many authors as a good solution to reduce oil quantity in machining processes, achieving the same or better results as for flood lubrication. Cutting forces in milling and turning processes can be reduced by using the MQL technique, although for drilling processes, it has been found to have a great dependence of the cutting conditions and hole aspect ratios, because the difficulty of eliminating the chips in deep hole drilling increases the forces. For low aspect ratios, increments up to 27% on tool life have been obtained on drilling processes [32].
- The minimum quantity cooling lubrication (MQCL) technique has been found to be a good combination of MQL and low temperature supplement. It can efficiently combine the friction reduction effect of MQL and temperature reduction effect of the cooling agent. The application of cooling air or gaseous N2 at temperatures from −4 °C to −26 °C has been found to be a good solution to reduce cutting forces and surface integrity on turning processes. Nevertheless, the combination of MQL and cryogenic LN_2 has been found to be a less efficient method than MQL, due to increased cutting forces.
- Laser-beam assisted machining (LAM) has been suggested as a potential method for reducing cutting forces in milling operations, thanks to decreased titanium flow stress. Nevertheless, the process parameters are difficult to control and the use of a high power laser has been shown to cause excessive wear on the tool. Some authors have combined LBM with MQL with good results.

Finally, it has been found that the most widely studied criteria for the analysis of machining processes are cutting forces, surface roughness and tool wear, but important phenomena, such as friction, have not been sufficiently studied. For this reason, future research should focus its efforts on the characterisation of the friction phenomenon between the tool and the chip, which is closely related to cutting process performance.

In addition, the analysis should be extended to other materials that are gaining prominence in applications within the industrial sector, such as titanium aluminides, for which the machining processes have been the subject of limited study, especially under environmentally sustainable working conditions.

Author Contributions: This paper is to establish the starting point of the doctoral thesis of E.G.-M. under the supervision of V.M. All authors are active members of the research group Science and Engineering of Materials at the Castilla-La Mancha University in Spain. E.G.-M., J.C., A.M.-M. and M.C.M. have done the literature research and have selected and classified the papers according to the aims persecuted in the manuscript by a first reading of them. E.G.-M. has analysed the selected papers in detail and has written the manuscript in a continuous discussion process with the rest of the authors. V.M. has directed and planning the manuscript, and has supervised the writing of the paper and the relevance of the references considered in it. Conceptualization, V.M.; methodology, V.M., M.C.M., E.G.-M.; formal analysis, V.M., M.C.M., J.C., A.M.-M. and E.G.-M.; investigation, V.M., E.G.-M.; resources, V.M., A.M.-M., M.C.M., J.C.; data curation, E.G.-M.; writing—original draft preparation, V.M., A.M.-M., E.G.-M., J.C., M.C.M.; writing—review and editing, V.M. and E.G.-M.; visualization, E.G.-M.; supervision, V.M.; project administration, V.M.; funding acquisition V.M., A.M.-M., J.C., M.C.M.

Funding: This research received no external funding.

Conflicts of Interest: The authors declare no conflict of interest.

References

1. Park, K.H.; Yang, G.D.; Suhaimi, M.A.; Lee, D.Y.; Kim, T.G.; Kim, D.W.; Lee, S.W. The effect of cryogenic cooling and minimum quantity lubrication on end milling of titanium alloy Ti6Al4V. *J. Mech. Sci. Technol.* **2015**, *29*, 5121–5126. [CrossRef]
2. Shokrani, A.; Dholia, V.; Newman, S.T. Comparative investigation on using cryogenic machining in CNC milling of Ti6Al4V titanium alloy. *Mach. Sci. Technol.* **2016**, *20*, 465–494. [CrossRef]
3. Mia, M.; Khan, M.A.; Dhar, N.R. High-pressure coolant on flank and rake surfaces of tool in turning of Ti6Al4V: Investigations on surface roughness and tool wear. *Int. J. Adv. Manuf. Technol.* **2016**, *90*, 1825–1834. [CrossRef]
4. Kaynak, Y.; Gharibi, A.; Yilmaz, U.; Köklü, U.; Aslantas, K. A comparison of flood cooling, minimum quantity lubrication and high pressure coolant on machining and surface integrity of titanium Ti-5553 alloy. *J. Manuf. Process.* **2016**, *34*, 503–512. [CrossRef]
5. Davim, J.P. *Machining of Titanium Alloys*; Materials Forming, Machining and Tribology; Springer: Berlin/Heidelberg, Germany, 2014.
6. Mathew, N.T.; Vijayaraghavan, L. Environmentally friendly drilling of intermetallic titanium aluminide at different aspect ratio. *J. Clean. Prod.* **2017**, *141*, 439–452. [CrossRef]
7. Ali Osman, K.; Özgür, H.; Seker, U. Application of minimum quantity lubrication techniques in machining process of titanium alloy for sustainability: A review. *Int. J. Adv. Manuf. Technol.* **2019**, *100*, 2311–2332. [CrossRef]
8. Pervaiz, S.; Rashid, A.; Deiab, I.; Nicolescu, C.M. An experimental investigation of effect of minimum quantity cooling lubrication (MQCL) in machining titanium alloy (Ti6Al4V). *Int. J. Adv. Manuf. Technol.* **2016**, *87*, 1371–1386. [CrossRef]
9. Park, K.H.; Suhaimi, M.A.; Yang, G.D.; Lee, D.Y.; Lee, S.W.; Kwon, P. Milling of titanium alloy with cryogenic cooling and minimum quantity lubrication (MQL). *Int. J. Precis. Eng. Manuf.* **2017**, *18*, 5–14. [CrossRef]
10. Lin, H.; Wang, C.; Yuan, Y.; Chen, Z.; Wang, Q.; Xiong, W. Tool wear in Ti6Al4V alloy turning under oils on water cooling comparing with cryogenic air mixed with minimal quantity lubrication. *Int. J. Adv. Manuf. Technol.* **2015**, *81*, 87–101. [CrossRef]
11. Sen, B.; Mia, M.; Krolczyk, G.M.; Mandal, U.K.; Monal, S.P. Eco-Friendly Cutting Fluids in Minimum Quantity Lubrication Assisted Machining: A Review on the Perception of Sustainable Manufacturing. *Int. J. Precis. Eng. Manuf.-Green Technol.* **2019**, 1–32. [CrossRef]

12. Revuru, R.S.; Posinasetti, N.R.; VSN, V.R.; Amrita, M. Application of cutting fluids in machining of titanium alloys-a review. *Int. J. Adv. Manuf. Technol.* **2017**, *91*, 2477–2498. [CrossRef]
13. Miguel, M.; Martínez, A.; Coello, J.; Avellaneda, F.J.; Calatayud, A. A new approach for evaluating sheet metal forming based on sheet drawing test. Application to TRIP 700 steel. *J. Mater. Process. Technol.* **2013**, *213*, 1703–1710. [CrossRef]
14. Pervaiz, S.; Anwar, S.; Qureshi, I.; Ahmed, N. Recent Advances in the Machining of Titanium Alloys using Minimum Quantity Lubrication (MQL) Based Techniques. *Int. J. Precis. Eng. Manuf.-Green Technol.* **2019**, *6*, 133–145. [CrossRef]
15. Krolczyk, G.M.; Maruda, R.W.; Krolczyk, J.B.; Wojciechowski, S.; Mia, M.; Nieslony, P.; Budzik, G. Ecological trends in machining as a key factor in sustainable production—A review. *J. Clean. Prod.* **2019**, *218*, 601–615. [CrossRef]
16. Gupta, K.; Laubscher, R.F. Sustainable machining of titanium alloys: A critical review. *J. Eng. Manuf.* **2017**, *231*, 2543–2560. [CrossRef]
17. Benjamin, D.M.; Sabarish, V.N.; Hariharan, M.V.; Raj, D.S. On the benefits of sub-zero air supplemented minimum quantity lubrication systems: An experimental and mechanistic investigation on end milling of Ti6Al4V alloy. *Tribol. Int.* **2017**, *119*, 464–473. [CrossRef]
18. Maruda, R.W.; Stanislaw Legutko, E.F.; Krolczyk, G.M. Research on emulsion mist generation in the conditions of minimum quantity cooling lubrication (MQCL). *Tehnicki Vjesnik* **2015**, *22*, 1213–1218.
19. Maruda, R.W.; Krolczyk, G.M.; Feldshtein, E.; Pusavec, F.; Szydlowski, M.; Legutko, S.; Sobczak-Kupiec, A. A study on droplets sizes, their distribution and heat exchange for minimum quantity cooling lubrication (MQCL). *Int. J. Mach. Tools Manuf.* **2015**, *100*, 81–92. [CrossRef]
20. Maruda, R.W.; Krolczyk, G.M.; Nieslony, P.; Wojciechowski, S.; Michalski, M.; Legutko, S. The influence of the cooling conditions on the cutting tool wear and the chip formation mechanism. *J. Manuf. Process.* **2016**, *24*, 107–115. [CrossRef]
21. Busch, K.; Hochmuth, C.; Pause, B.; Stoll, A.; Wertheim, R. Investigation of cooling and lubrication strategies for machining high-temperature alloys. *Procedia CIRP* **2016**, *41*, 835–840. [CrossRef]
22. Topoglou, N.; Lopez, M.I.A.; Cook, I.; Taylor, C.M. Investigation of the Influence of CO_2 Cryogenic Coolant Application on Tool Wear. *Procedia CIRP* **2017**, *63*, 745–749. [CrossRef]
23. Dix, M.; Wertheim, R.; Schmidt, G.; Hochmuth, C. Modeling of drilling assisted by cryogenic cooling for higher efficiency. *CIRP Ann. Manuf. Technol.* **2014**, *63*, 73–76. [CrossRef]
24. Bordin, A.; Bruschi, S.; Ghiotti, A.; Bariani, P.F. Analysis of tool wear in cryogenic machining of additive manufactured Ti6Al4V alloy. *Wear* **2015**, *328–329*, 88–99. [CrossRef]
25. Ezugwu, E.O.; da Silva, R.B.; Bonney, J.; Costa, E.S.; Sales, W.F.; Machado, A.R. Evaluation of Performance of Various Coolant Grades When Turning Ti6Al4V Alloy with Uncoated Carbide Tools Under High Pressure Coolant Supplies. *J. Manuf. Sci. Eng.* **2018**, *141*, 014503. [CrossRef]
26. Mia, M.; Dhar, N.R. Effects of duplex jets high-pressure coolant on machining temperature and machinability of Ti6Al4V superalloy. *J. Mater. Process. Technol.* **2018**, *252*, 688–696. [CrossRef]
27. Bermingham, M.J.; Sim, W.M.; Kent, D.; Gardiner, S.; Dargusch, M.S. Tool life and wear mechanisms in laser assisted milling Ti6Al4V. *Wear* **2015**, *322–323*, 151–163. [CrossRef]
28. Lee, C.M.; Kim, D.H.; Baek, J.T.; Kim, E.J. Laser Assisted Milling Device: A Review. *Int. J. Precis. Eng. Manuf.-Green Technol.* **2016**, *3*, 199–208. [CrossRef]
29. Sim, M.S.; Lee, C.M. Determination of optimal laser power according to the tool path inclination angle of a titanium alloy workpiece in laser-assisted machining. *Int. J. Adv. Manuf. Technol.* **2016**, *83*, 1717–1724. [CrossRef]
30. Perçin, M.; Aslantas, K.; Ucun, I.; Kaynak, Y.; Çicek, A. Micro-drilling of Ti6Al4V alloy: The effects of cooling/lubricating. *Precis. Eng.* **2016**, *45*, 450–462. [CrossRef]
31. Nandgaonkar, S.; Gupta, T.V.K.; Joshi, S. Effect of Water Oil Mist Spray (WOMS) Cooling on Drilling of Ti6Al4V Alloy Using Ester Oil Based Cutting Fluid. *Procedia Manuf.* **2016**, *6*, 71–79. [CrossRef]
32. Mathew, N.T.; Lazmanan, V. Temperature rise in workpiece and cutting tool during drilling of titanium aluminide under sustainable environment. *Mater. Manuf. Process.* **2018**, *33*, 1765–1774. [CrossRef]
33. Hedberg, G.K.; Shin, Y.C.; Xu, L. Laser-assisted milling of Ti6Al4V with the consideration of surface integrity. *Int. J. Manuf. Technol.* **2015**, *79*, 1645–1658. [CrossRef]

34. Celik, O.N.; Sert, A.; Gasan, H.; Ulutan, M. Effect of cryogenic treatment on the microstructure and the wear behavior of WC-Co end mills for machining of Ti6Al4V titanium alloy. *Int. J. Adv. Manuf. Technol.* **2017**, *95*, 2989–2999. [CrossRef]
35. Wang, F.; Hou, B.; Wang, Y.; Liu, H. Diffusion thermodynamic behavior of milling Ti6Al4V alloy in liquid nitrogen cryogenic cooling. *Int. J. Adv. Manuf. Technol.* **2017**, *95*, 2783–2793. [CrossRef]
36. Masood, I.; Jahanzaib, M.; Wasim, A. Sustainability Assessment for Dry, Conventional and Cryogenic Machining in Face Milling of Ti6Al4V. *J. Eng. Technol.* **2017**, *36*, 309–320.
37. Shokrani, A.; Dhokia, V.; Newman, S.T. Energy conscious cryogenic machining of Ti6Al4V titanium alloy. *J. Eng. Technol.* **2016**, *232*, 1690–1706.
38. Li, M.; Yu, T.; Zhang, R.; Yang, L.; Li, H.; Wang, W. MQL milling of TC4 alloy by dispersing graphene into vegetable oil-based cutting fluid. *Int. J. Adv. Manuf. Technol.* **2018**, *99*, 1735–1753. [CrossRef]
39. Masood, L.; Jahanzaib, M.; Haider, A. Tool wear and cost evaluation of face milling grade 5 titanium alloy for sustainable machining. *Adv. Prod. Eng. Manag.* **2016**, *11*, 239–250. [CrossRef]
40. Bai, X.; Li, C.; Dong, L.; Yin, Q. Experimental evaluation of the lubrication performances of different nanofluids for minimum quantity lubrication (MQL) in milling Ti6Al4V. *Int. J. Adv. Manuf. Technol.* **2019**, *101*, 2621–2632. [CrossRef]
41. Wang, F.; Wang, Y.; Liu, H. Tool wear behavior of thermal-mechanical effect for milling Ti6Al4V alloy in cryogenic. *Int. J. Adv. Manuf. Technol.* **2017**, *94*, 2077–2088. [CrossRef]
42. Zheng, K.; Liao, W.; Dong, Q.; Sun, L. Friction and wear on titanium alloy surface machined by ultrasonic vibration-assisted milling. *J. Braz. Soc. Mech. Sci. Eng.* **2018**, *40*, 411. [CrossRef]
43. Shokrani, A.; Newman, S.T. Cutting Tool Design for Cryogenic Machining of Ti6Al4V Titanium Alloy. *Materials* **2019**, *12*, 477. [CrossRef] [PubMed]
44. Woo, W.S.; Lee, C.M. A study on the optimum machining conditions and energy efficiency of a laser-assisted fillet milling. *Int. J. Precis. Eng. -Manuf.-Green Technol.* **2018**, *5*, 593–604. [CrossRef]
45. Mia, M.; Khan, M.A.; Dhar, N.R. Study of surface roughness and cutting forces using ANN, RSM, and ANOVA in turning of Ti6Al4V under cryogenic jets applied at flack and rake faces of coated WC tool. *Int. J. Adv. Manuf. Technol.* **2017**, *93*, 975–991. [CrossRef]
46. Mia, M.; Khan, M.A.; Dhar, N.R. High-pressure coolant on flank and rake surfaces of tool in turning of Ti6Al4V: Investigations forces, temperature, and chips. *Int. J. Adv. Manuf. Technol.* **2016**, *90*, 1977–1991.
47. Gupta, M.K.; Sood, P.K.; Singh, G.; Sharma, V.S. Sustainable machining of aerospace material- Ti(grade 2) alloy: Modeling and optimization. *J. Clean. Prod.* **2017**, *147*, 614–627. [CrossRef]
48. Sartori, S.; Ghiotti, A.; Bruschi, S. Temperature effects on the Ti6Al4V machinability using cooled gaseous nitrogen in semi-finishing turning. *J. Manuf. Process.* **2017**, *30*, 187–194. [CrossRef]
49. Krishnamurthy, G.; Bhowmick, S.; Altenhof, W.; Alpas, A.T. Increasing efficiency of Ti-alloy machining by cryogenic cooling and using ethanol in MRF. *CIRP J. Manuf. Sci. Technol.* **2017**, *18*, 159–172. [CrossRef]
50. Singh, G.; Sharma, V.S. Analyzing machining parameters for commercially pure titanium (Grade 2), cooled using minimum quantity lubrication assisted by a Ranque-Hilsch vortex tub. *Int. J. Adv. Manuf. Technol.* **2017**, *88*, 2921–2928. [CrossRef]
51. Faga, M.G.; Priarone, P.C.; Robiglio, M.; Settineri, L.; Tebaldo, V. Technological and Sustainability Implications of Dry, Near-Dry and Wet Turning of Ti6Al4V. *Int. J. Precis. Eng. Manuf.-Green Technol.* **2017**, *4*, 129–139. [CrossRef]
52. Pervaiz, S.; Deiab, I.; Rashib, A.; Nicolescu, M. Minimal quantity cooling lubrication in turning of Ti6Al4V: Influence on surface roughness, cutting force and tool wear. *J. Eng. Manuf.* **2017**, *231*, 1542–1558. [CrossRef]
53. Hegab, H.; Umer, U.; Deiab, I.; Kishawy, H. Performance evaluation of Ti6Al4V machining using nano-cutting fluids under minimum quantity lubrication. *Int. J. Adv. Manuf. Technol.* **2018**, *95*, 4229–4241. [CrossRef]
54. Hegab, H.; Kishawy, H.A.; Gadallah, M.H.; Umer, U.; Deiab, I. On machining of Ti6Al4V using multi-walled carbon nanotubes-based nano-fluid under minimum quantity lubrication. *Int. J. Adv. Manuf. Technol.* **2018**, *97*, 1593–1603. [CrossRef]
55. Caggiano, A.; Napolitano, F.; Teti, R. Dry turning of Ti6Al4V: Tool wear curve reconstruction based on cognitive sensor monitoring. *Procedia CIRP* **2017**, *62*, 209–214. [CrossRef]
56. Mia, M.; Dha, N.R. Influence of single and dual cryogenic jets on machinability characteristics in turning of Ti6Al4V. *J. Eng. Manuf.* **2017**, *233*, 711–726. [CrossRef]

57. Mia, M.; Gupta, M.K.; Lozano, J.A.; Carou, D.; Pimenov, D.Y.; Królczyk, G.; Khan, A.M.; Dha, N.R. Multi-objective optimization and life cycle assessment of echo-friendly cyogenic N_2 assisted turning of Ti6Al4V. *J. Clean. Prod.* **2019**, *210*, 121–133. [CrossRef]
58. Rao, C.M.; Rao, S.S.; Herbert, M.A. Development of novel cutting tool with a micro-hole pattern on PCD insert in machining of titanium alloy. *J. Manuf. Process.* **2018**, *36*, 93–103. [CrossRef]
59. Boswell, B.; Islam, M.N. Sustainable cooling method for machining titanium alloy. *Mater. Sci. Eng.* **2016**, *114*, 12–21. [CrossRef]
60. Ayed, Y.; Germain, G.; Melsio, A.P.; Kowalewski, P.; Locufier, D. Impact of supply conditions of liquid nitrogen on tool wear and surface integrity when machining the Ti6Al4V titanium alloy. *Int. J. Adv. Manuf. Technol.* **2017**, *93*, 1199–1206. [CrossRef]
61. Kaynak, Y.; Gharibi, A.; Ozkutuk, M. Experimental and numerical study of chip formation in orthogonal cutting of Ti-5553 alloy: The influence of cryogenic, MQL, and high pressure coolant supply. *Int. J. Adv. Manuf. Technol.* **2018**, *94*, 1411–1428. [CrossRef]
62. Zhao, W.; Gong, L.; Ren, F.; Li, L.; Xu, Q.; Khan, A.M. Experimental study on chip deformation of Ti6Al4V titanium alloy in cryogenic cutting. *Int. J. Adv. Manuf. Technol.* **2018**, *96*, 4021–4027. [CrossRef]
63. Kim, D.Y.; Kim, D.M.; Park, H.W. Predictive cutting force model for a cryogenic machining process incorporating the phase transformation of Ti6Al4V. *Int. J. Adv. Manuf. Technol.* **2018**, *9*, 1293–1304. [CrossRef]
64. Sartori, S.; Taccin, M.; Pavese, G.; Ghiotti, A.; Bruschi, S. Wear mechanisms of uncoated and coated carbide tools when machining Ti6Al4V using LN_2 and cooled N_2. *Int. J. Adv. Manuf. Technol.* **2017**, *95*, 1255–1264. [CrossRef]
65. Revuru, R.S.; Zhang, J.Z.; Posinasetti, N.R.; Kidd, T. Optimization of titanium alloys turning operation in varied cutting fluid conditions with multiple machining performance characteristics. *Int. J. Adv. Manuf. Technol.* **2017**, *95*, 1451–1463. [CrossRef]
66. Rahman, S.S.; Ashraf, M.Z.I.; Amin, A.N.; Bashar, M.S.; Ashik, M.F.K.; Kamruzzaman, M. Turning nanofluids for improved lubrication performance in turning biomedical grade titanium alloy. *J. Clean. Prod.* **2019**, *206*, 108–196. [CrossRef]
67. Kirsch, B.; Bastein, S.; Hasse, H.; Aurich, J.C. Sub-zero cooling: A novel strategy for high performance cutting. *CIRP Ann.-Manuf. Technol.* **2018**, *67*, 95–98. [CrossRef]
68. Lauro, C.H.; Brandao, L.C.; Filho, S.L.M.R.; Davim, J.P. Behaviour of a biocompatible titanium alloy during orthogonal micro-cutting employing green machining techniques. *Int. J. Adv. Manuf. Technol.* **2018**, *98*, 1573–1589. [CrossRef]
69. Mahboob Ali, M.A.; Azmi, A.I.; Khalil, A.N.M.; Leong, K.W. Experimental study on minimal nanolubrication with surfactant in the turning of titanium alloys. *Int. J. Adv. Manuf. Technol.* **2017**, *92*, 117–127. [CrossRef]
70. Zhang, P.; Wang, Y.; Liu, W. The HSC machining mechanism for TC17 under multimedia mixed minimum quantity lubrication. *Int. J. Adv. Manuf. Technol.* **2018**, *95*, 341–353. [CrossRef]
71. Kaynak, Y.; Gharibi, A. Cryogenic Machining of Titanium Ti-5553 Alloy. *J. Manuf. Sci. Eng.* **2018**, *141*, 041012. [CrossRef]
72. Singh, G.; Pruncu, C.I.; Gupta, M.K.; Mia, M.; Khan, A.M.; Jamil, M.; Pimenov, D.Y.; Sen, B.; Sharma, V.S. Investigations of Machining Characteristics in the Upgraded MQL-Assisted Turning of Pure Titanium Alloys Using Evolutionary Algorithms. *Materials* **2019**, *12*, 999. [CrossRef] [PubMed]
73. Yan, L.; Zhang, Q.; Yu, J. Effects of continuous minimum quantity lubrication with ultrasonic vibration in turning of titanium alloy. *Int. J. Adv. Manuf. Technol.* **2018**, *98*, 827–837. [CrossRef]
74. Iqbal, A.; Biermann, D.; Abbas, H.; al-Ghamdi, K.A.; Metzger, M. Machining β-titanium alloy under carbon dioxide snow and micro-lubrication: A study on tool deflection, energy consumption and tool damage. *Int. J. Adv. Manuf. Technol.* **2018**, *97*, 4195–4208. [CrossRef]
75. Slodki, B.; Zebala, W.; Struzikiewicz, G. Turning Titanium Alloy, Grade 5 ELI, With the Implementation of High Pressure Coolant. *Materials* **2019**, *12*, 768. [CrossRef] [PubMed]
76. Sartori, S.; Pezzato, L.; Dabalà, M.; Enrici, T.M.; Mertens, A.; Ghiotti, A.; Bruschi, S. Surface Integrity Analysis of Ti6Al4V After Semi-finishing Turning Under Different Low-Temperature Cooling Strategies. *J. Mater. Eng. Perform.* **2018**, *27*, 4810–4818. [CrossRef]
77. Gunda, R.K.; Narala, S.K.R. Electrostatic high-velocity solid lubricant machining system for performance improvement of turning Ti6Al4V alloy. *J. Eng. Manuf.* **2017**, *233*, 118–131. [CrossRef]

78. Kesriklioglu, S.; Pfefferkorn, F.E. Prediction of tool-chip interface temperature in cryogenic machining of Ti6Al4V: Analytical modeling and sensitivity analysis. *J. Therm. Sci. Eng. Appl.* **2018**, *11*, 011003. [CrossRef]
79. Habrat, W.; Krupa, K.; Laskowski, P.; Sieniawski, J. Experimental analysis of the cutting force components in laser-assisted turning of Ti6Al4V. *Adv. Manuf. Eng. Mater.* **2018**, 237–245._26. [CrossRef]

© 2019 by the authors. Licensee MDPI, Basel, Switzerland. This article is an open access article distributed under the terms and conditions of the Creative Commons Attribution (CC BY) license (http://creativecommons.org/licenses/by/4.0/).

Article

Study of Drilling Process by Cooling Compressed Air in Reinforced Polyether-Ether-Ketone

Rosario Domingo *, Beatriz de Agustina and Marta María Marín

Department of Construction and Manufacturing Engineering, Universidad Nacional de Educación a Distancia (UNED), C/Juan del Rosal 12, E–28040 Madrid, Spain; bdeagustina@ind.uned.es (B.d.A.); mmarin@ind.uned.es (M.M.M.)
* Correspondence: rdomingo@ind.uned.es; Tel.: +34-91-398-6455

Received: 16 December 2019; Accepted: 20 April 2020; Published: 22 April 2020

Abstract: This study is focused on the application of a cooling compressed air system in drilling processes; this environmentally friendly technique allows removing material at very low temperatures, approximately up to −22 °C in the cutting area. The main goals are to find the most improve cutting conditions with less energy consumption, for the drilling of reinforced polyether-ether-ketone with glass fiber at 30% (PEEK-GF30) with cooling compressed air by a Ranque-Hilsch vortex tube, and to find a balance between environmental conditions and adequate process performance. Drilling tests were carried out on plates of PEEK-GF30 to analyze the influence of cutting parameters and environmental temperature (−22, 0 and 22 °C) on variables such as thrust forces, energy and material removed rate by the use of statistical methods; analysis of variance, analysis of means, response surface, and desirability function were employed to identify the optimum region that provides the most improved values of the aforementioned variables. Drill bit diameter was also analyzed to determine the quality of drilled holes. During the drilling processes, force signals were detected by a piezoelectric dynamometer connected to multichannel amplifier and a pyrometer was used to control the temperature. The diameters of the drilled holes were measured by a coordinate measuring machine. Cooling compressed air can be considered an adequate technique to improve the results from an environmental and efficient perspective; in particular, the maximum desirability function was found at a spindle speed of 7000 rpm, a feedrate of 1 mm/rev and a temperature close to −22 °C.

Keywords: drilling; cooling compressed air; thrust force; energy; material removed rate; PEEK-GF30; multi-response optimization; sustainable manufacturing

1. Introduction

Over the past decade, within manufacturing industries there has been an increasing interest in researching techniques that allow the performance of processes with high efficiency under sustainable environments. In fact, different systems have been applied in manufacturing facilities to identify the most improved parameters to combine both objectives; in some cases, the effects on the organization of production system by the implementation of lean techniques [1] and in others, the effects on the manufacturing process where the improved selection of parameters can reduce CO_2 emissions [2]. The machining is the most used manufacturing process in the industry due to its versatility [3], so studies in this field can be relevant. In this context, several methods have been applied to achieve both objectives, efficiency and sustainability, to the most possible extent. One of them is dry machining or machining without the use of any cutting fluid. Krolczyk et al. [4] have carried out an extensive literature review on four ecological methods in machining of difficult to-cut metals, in particular dry machining, minimum quantity lubrication (MQL)/minimum quantity cooling lubrication (MQCL), cryogenic cooling, high-pressure cooling and biodegradable oils; they found that dry machining is the most sustainable procedure with respect to others, despite the high temperature generated during the machining operation, in particular in operations such as drilling; besides, these authors pointed out the non-use of cryogenic cooling in industry despite the benefits of this procedure; this is the combination of high productivity as well as low cost and energy. Sen et al. [5], in their literature review focussed on metal cutting, collected results from life cycle assessment (LCA) models for different cooling techniques, showing that MQL and dry machining have the minimum negative effects respect to techniques such as flood lubrication, MQL, cryogenic cooling with CO_2, cryogenic cooling with LN_2 and cryogenic cooling with MQL. However, machining under such dry conditions causes excessive temperature rise at the interface between workpiece and tool and, in general, an increase in strains. These undesirable effects have been determined by the analysis of forces and strains [6] or considering the influence of tribology [7]; in both cases finite element models (FEMs) were used in the orthogonal cut of titanium alloys, so the application of new cooling techniques can be explored. Cooling technologies [8], such as cryogenic machining with liquid nitrogen and the application of cooling compressed air to the cutting zone during the machining process by means of a Ranque-Hilsch vortex tube have been identified as possible environmentally friendly procedures. These cooling systems are particularly interesting under dry cutting conditions; in this way, cutting fluids are avoided. Goindi and Sarkar [9] identified cooling compressed air obtained by vortex tube separation as a method to analyze into sustainable machining, as a variant of dry machining and also as an element to consider within the MQL. Moreover, the use of cooling compressed air systems in machining processes requires less investment than cryogenic machining and they are easier to employ on an industrial scale. In addition, although the MQL procedure is convenient from the point of view of sustainability, the use of lubricants could not be suitable in polymeric materials due to its possible absorption by the composite.

In fact, the application of cooling compressed air during the machining process is being studied and some researches can show its suitability as environmentally friendly procedure. Jozić et al. [10] developed an experimental study, in which several machining aspects were analyzed in the milling process of a certain steel with this procedure at −34 °C under dry cutting conditions and also with the use of cutting fluid; it was found the most improved solution (less surface roughness, cutting forces, flank wear and more volume of removed material) in case of the application of cooling air during the process. In addition, regarding the milling processes, Perri et al. [11] analyzed the effect of cooling air on the milling tool, using simulation models by FEM and experimental contrast and verifying important difference of temperatures reached respect to the procedure carried out without cooling air. On the other hand, Nor Khairusshima et al. [12] studied the quality and the tool wear in the milling of the carbon fiber reinforced plastic with cooling air at −10 °C; they pointed out that the tool wear and delamination factor were improved at high cutting speeds. Domingo et al. [13] analyzed the effects of cooling air on reinforced and unreinforced polyamides during the tapping, finding that an adequate procedure at −18 °C provided improved results with respect to the values of forces, torques and power reached during the process.

As shown in the aforementioned literature, the cooling compressed air has been also applied in the machining of composites with polymeric matrix and different reinforcements as carbon fiber [12] or glass fiber [13], and an improved performance has been reported. Nevertheless, studies towards the drilling of reinforced polyether-ether-ketone with 30% glass fiber (PEEK-GF30) have not been found in the scientific literature. The novelty in this study is the analysis of its behavior during drilling processes, the most important process in the assembly as the previous operation to riveting and tapping. This material can withstand very low temperatures maintaining a stable behavior [14], therefore the machining of this material with cooling compressed air can be adequate as its mechanical and chemical properties are not expected to modify significantly during the process. Due to its extended use in the industry, its machinability has been studied, finding that with an adequate selection of cutting parameters and tools, it is a material that can be widely used in industry [15].

This environmentally friendly procedure can be contrasted evaluating the main variables of drilling under specific cutting conditions. Variables such as thrust forces should be low to improve process stability, at least in drilling operations; the literature shows that a homogeneous variation of thrust forces does not exist with respect to cutting conditions, at least in different types of composites such as glass fiber reinforced PEEK [16], fibre metal laminates [17], magnesium matrix based silicon carbide and graphene nanoplatelets (Mg/SiC/GNPs), hybrid magnesium matrix composite [18] or wood-based composite of medium density fiberboard [19]. Nevertheless, the energy consumed during the process could be a key factor to stablish the most sustainable procedure, and also the material removed rate (MRR) in order to evaluate the process efficiency. In the study carried out by Davim and Reis [20], the precision of the hole was considered as a variable to measure surface quality, so the diameter of the hole is another variable to study. Although the surface roughness is another variable that determines the hole quality, its influence in inner holes depends on the hole functionality; in this study, the evaluation of roughness surface it is irrelevant as an internal threading operation is expected to approach; its evaluation is important in other further operations such as riveting, however this characteristic can be corrected in other finishing operations, even with the same tool. Thus, the main variable to be accepted as an adequate dimensional accuracy of the holes is their dimension, and moreover it can be influenced by the chip evacuation and thermal expansion of the matrix occurring during the cutting process due to the lack of lubricant, as can be seen in materials based on multilayer metallic and/or composite stacks such as comprising titanium stacks, carbon fibre reinforced plastics (CFRPs) and aluminium [21], functionally graded composite, carbon/epoxy and glass/epoxy composite [22] or carbon fiber-reinforced plastic composite [23]. The reduction of the temperature in the cutting area could avoid this effect.

Taking into account the above-mentioned points, the objectives of this paper are the following: (i) to find the cutting conditions more environmentally friendly, with less energy consumption, in the drilling of reinforced PEEK with glass fiber at 30% with cooling compressed air by vortex tube, and (ii) to establish a balance between environmental conditions and adequate process performance.

2. Materials and Methods

The methodology used in this work combines experimental and statistical procedures.

2.1. Experimental Procedure

Drilling tests were carried out in a CNC vertical machining center, Manga Tongtai TMV-510 (Kaohsiung Hsien, Taiwan). The drilling operations were performed on plates of PEEK-GF30 (Gapi, Bergamo, Italy), with thickness of 6.5 mm, and a coefficient of linear thermal expansion of 30×10^{-6} m/m·K).

The drill bits employed were provided by FMT Tooling Systems Company (Trofa, Portugal); they had a diameter of 6 mm and are made of solid carbide with coating of zirconium oxide. Their material and geometrical characteristics are described in Table 1. Note that, as the point angle of drill was 140°, the corresponding cutting length was 7.592 mm. The selection of cutting tool was made based on the results of the analysis of variables, such as, for example, the energy or CO_2 emissions [24] and preliminary tests. In drilling processes, the use of coating of zirconium oxide provides a good behavior of tool respect to the tool wear, the cutting forces and the chip flow [25]. A design of experiments (DOE) was used to optimize the resources. In particular, a three-level factorial design with three replicates was employed to analyze the influence of the three factors: feedrate, spindle speed and environmental temperature, on the response variables: thrust force, energy required for the drilling operation, material removed rate and drill bit diameter. This was, in total, 81 drilling tests. Later on, in the Section 2.2 statistical procedure, a more detailed description of the statistical analysis method is included.

Table 1. Drill bits characteristics.

	Material	Coating	Point Angle	Flute Helix Angle	Number of Cutting Edges	Nominal Tolerance
Drill bit	Solid carbide	Zirconium oxide	140°	15°	2	h7

Values of spindle speed (N), from 5000 to 7000 rpm and values of feedrate (f) from 0.5 to 1 mm/rev were applied. The drilling tests were carried out at temperatures (T) from −22 to 22 °C. The speed and the feed were chosen taking into account previous tests, information exchanged with the tool supplier, characteristics of the drill bits and high performance. Besides, similar ranges can be found in the literature [15]. Note that the feedrate was very high and this fact allowed quickly increasing the MRR. Firstly, drilling tests were performed at the room temperature that was 22 °C and secondly, drilling tests were carried out with the cooling compressed air system at 0 and −22 °C; thus, three specimens were used, one for each temperature. For the measurement of the temperature, an infrared pyrometer, Optris, was used; in this way, the temperature was monitored along the drilling process. The cooling was achieved by means of a Ranque–Hilsch vortex tube with two outlets of cold fluid, Dual Nozzle CAG Vortec model. This cooling technology allows obtaining the most improved outcomes when the nozzle number is 2, due to a more homogeneous distribution of cooling [26]. In other machining processes, with regard to this technology it is underlined that cutting forces and the power required in the titanium turning could be reduced with the application of vortex tube [27]. Similar findings were obtained by other studies with respect to flank wear and surface roughness towards the machining of aluminum alloys [28].

For the application of the cooling compressed air, it was necessary that the tube received the compressed air with a pressure of 0.8 MPa. To achieve temperatures of 0 and −22 °C, the compressed air had to flow for 8 and 10 min respectively, along vortex tube, before impinging on the drill bit; this step was very important to avoid the formation of ice on the drill bit during the drilling process. The nozzles of vortex tube were positioned at a distance of 30 mm, approximately, to achieve these temperatures.

The thrust force and the torque on the drilling direction were calculated directly during the monitoring of the drilling process. A piezoelectric dynamometer type Kistler 9257B connected to multichannel amplifier type Kistler 5070A was used (see Figure 1). The data were processed by DasyLab software (version 9.0, Measurement Computer, Norton, USA) [29]. The torques were taken to calculate the energy according to Equation (1), derivated of the expression used by Li et al. in the drilling of titanium alloys [30].

$$E = \int_l Ft \times dl + \int_l \left(\frac{2 \times \pi \times To}{f}\right)_z \times dl, \tag{1}$$

where E is the energy required to drill a hole in J, Ft is the thrust force in N, l is the drilling length in m, To is the torque respect Z axis in N·m, and f is the feedrate respect Z axis, in m/rev.

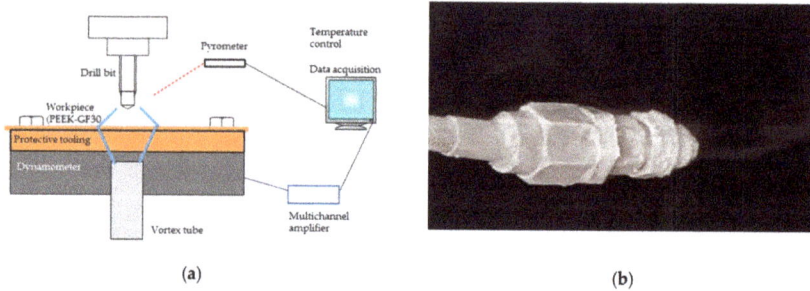

Figure 1. Machining process and data collection: (**a**) Scheme of assembly to drilling process and data capture; (**b**) Outlet of cold fluid after 8 minutes, with ice in the exterior.

In the drilling process, the material removed rate, in mm^3/s, is identified according to Equation (2) [31],

$$MRR = \left(\pi \times Di^2 \times F\right)/(4 \times 60) \tag{2}$$

where Di is the drill bit diameter in mm, F is the feedrate in mm/min.

To control the diameters quality, measurements were taken on 24 points along the circumference of the hole (as considered a suitable number [32]) by a coordinate measuring machine, Mitutoyo BX 303. The least square circle method was used to calculate the diameter measurements. Each measurement was repeated three times. The procedure of the measurement method was developed according to ISO 4291 standard [33]. The surface points were measured using a ball point stylus with a diameter of 1.6 mm. The data obtained were fitted by a Gaussian filter with 50% cut off. The outcomes were treated by Geopak-Win MCC software. Figure 2a shows a scheme of the location of center and circle of measured point after applying least square circle technique. Figure 2b shows the coordinate measurement machine.

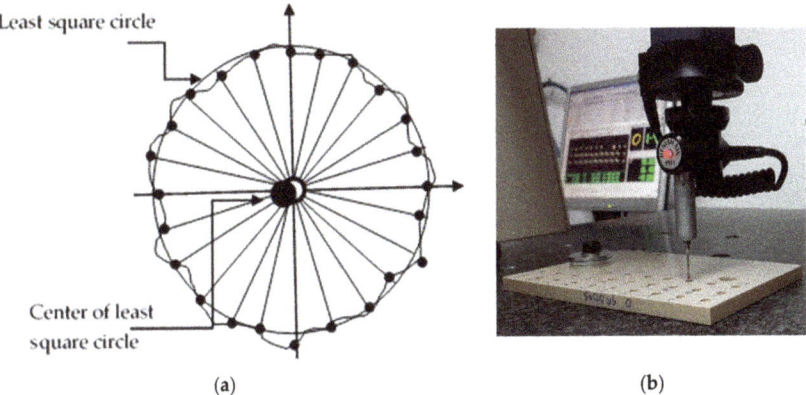

Figure 2. Measurement process of diameters: (**a**) Scheme of determination of center and circle of least square; (**b**) Coordinate measuring machine, during the measurement process of diameters.

2.2. Statistical Procedure

The statistical procedure focused on the application of response surface methodology when several responses should be simultaneously optimized, through a multi-objective method, in particular the desirability function, which allows finding a common objective for variables with different sub-objectives.

The experimental data collected were statistically analyzed, in particular, by the Response Surface with three-level factorial design, 3 × 3 in this case, and using Statgraphics software [34]. The Analysis of Variance (ANOVA) was carried out for the variables thrust forces, energy and MRR, meanwhile the diameter was used to evaluate the quality of the holes. With ANOVA analysis, the significant factors can be identified in a particular confidence interval [33]. The ANOVA of each variable is shown by tables, where the sum of squares or variance of the observations, the degrees of freedom (Df), the mean square, the F-ratio obtained from Fisher–Snedecor distribution and its probability associated (P-value) are represented; values of means squares are calculated dividing the sum of squares by its associated degrees of freedom. In this case, the confidence interval considered was 90%, adequate in manufacturing environments; therefore, a P-value less 0.1 denoted that the factor was significant using an ANOVA analysis. Although a 95% confidence interval is usually common in ANOVA studies, the election of 90% allows increasing the range of significance in the energy variable, which is dependent of other two variables, as it can be observed in Equation (1). Moreover, from a statistical perspective, this interval is adequate [35]. The percentage of contribution of the significant factors to variability was determined by dividing the sum of squares for the factor by sum of squares total. An analysis of means (ANOM) was also carried out to determine if there are differences between the means of each variable for the values considered; this analysis was performed through Tukey-honestly significant difference (HSD) test, which allows identifying what means are different at a confidence interval of 95% [36].

Once an ANOVA study and an ANOM study were developed, regression models were defined based on the surface response [36], and finally, the desirability function was defined. The desirability function, d_i, was defined by Derringer and Suich [37], with different expressions according to the objective of the study variable. In this paper, there are, a variable to maximize, the MRR, and two variables to minimize, the thrust force and the energy. When the variable must be maximized, Equation (3) represents this option [37]:

$$d_i(\hat{y}_i(x)) = \begin{cases} 0 & \text{if } \hat{y}_i(x) < L_i \\ (\hat{y}_i - L_i / U_i - L_i)^s & \text{if } L_i \leq \hat{y}_i(x) \leq U_i \\ 1 & \text{if } \hat{y}_i(x) > U_i \end{cases}, \quad (3)$$

where $\hat{y}_i$ is ith estimated response of the variable (thrust force, energy, MRR), L_i is lower acceptable value, U_i is the upper acceptable value and s is the weight. Equation (4) represents the desirability function when the response must be minimized, being t the weight [37].

$$d_i(\hat{y}_i(x)) = \begin{cases} 1 & \text{if } \hat{y}_i(x) < L_i \\ (U_i - \hat{y}_i / U_i - L_i)^t & \text{if } L_i \leq \hat{y}_i(x) \leq U_i \\ 0 & \text{if } \hat{y}_i(x) > U_i \end{cases}, \quad (4)$$

Therefore $d(\hat{y}_i(x))$ takes, uniquely, a value between 0 and 1 when it transforms a response into free-scale. The global desirability (D) can be determined through a weighted geometric mean by Equation (5) [37]:

$$D = \left(\prod_{i=1}^{n} d_i^{ri}\right)^{1/\sum ri}, \quad (5)$$

where n is the number of variables and ri is the impact value. Thus, D takes also takes values between 0 and 1, and its optimum consists to maximize it. The values of the weights and impact can be fixed by the authors. In case of using Statgraphics software, the values can vary between 1 and 5. This range allows to establish the usual values in manufacturing environments, in particular in machining processes, e.g., in tapping operations [13] or in turning operations [38].

3. Results and Discussion

In Table 2, the experimental results obtained are detailed: thrust forces, energy, MRR and input diameter of holes. The values are the result of calculating the means of three values because the reproducibility of the measurements was very high. Outcomes from forces or energy show the influence of the feedrate, in a way that at higher feedrates, a reduction of the values of these variables were obtained, compared to other experimental data [16], which could be explained by the fact that the flute helix angle of tools improves chip evacuation. A low variability can be observed in data from input diameters; in all tests except in test 21, diameters slightly lower than the nominal drill bit (6 mm) were obtained. The reason could be the improved chip evacuation and the lack of thermal expansion during the drilling tests.

Table 2. Experimental results.

No. Test	T (°C)	N (rpm)	f (mm/rev)	Ft (N)	E (J)	MRR (mm³/s)	Di (mm)
1	−22	5000	0.5	47.12	10.67	1178.10	5.960
2	0	5000	0.5	81.32	12.83	1178.10	5.933
3	22	5000	0.5	59.15	16.28	1178.10	5.987
4	−22	6000	0.5	38.68	15.47	1413.72	5.943
5	0	6000	0.5	85.27	19.54	1413.72	5.913
6	22	6000	0.5	57.05	12.66	1413.72	5.987
7	−22	7000	0.5	31.05	7.54	1649.34	5.950
8	0	7000	0.5	49.26	8.56	1649.34	5.977
9	22	7000	0.5	75.16	19.93	1649.34	5.940
10	−22	5000	0.75	47.72	9.08	1767.15	5.973
11	0	5000	0.75	129.24	13.7	1767.15	5.907
12	22	5000	0.75	58.31	7.82	1767.15	5.953
13	−22	6000	0.75	51.52	7.25	2120.57	5.987
14	0	6000	0.75	96.75	9.8	2120.57	5.930
15	22	6000	0.75	80.4	8.83	2120.57	5.967
16	−22	7000	0.75	26.57	4.85	2474.00	5.933
17	0	7000	0.75	30.70	14.6	2474.00	5.937
18	22	7000	0.75	71.81	10.65	2474.00	5.940
19	−22	5000	1	45.70	5.90	2356.19	5.947
20	0	5000	1	156.66	9.50	2356.19	5.947
21	22	5000	1	70.15	2.91	2356.19	6.000
22	−22	6000	1	47.58	2.67	2827.43	5.967
23	0	6000	1	103.32	8.39	2827.43	5.933
24	22	6000	1	110.36	9.46	2827.43	5.970
25	−22	7000	1	27.69	6.26	3298.67	5.957
26	0	7000	1	56.65	7.69	3298.67	5.960
27	22	7000	1	78.68	7.92	3298.67	5.927

3.1. Analysis of Experimental Results

This subsection is developed to explore the experimental results; the influence of the considered cutting factors (feedrate, spindle speed and environmental temperature) are analyzed.

3.1.1. Thrust Forces

The ANOVA analysis for thrust forces (see Table 3) shows that the significant main factors were environmental temperature (T), spindle speed (N) and feedrate (f), and the significant interactions are T^2, $T \times N$, N^2 and $T^2 \times N$ in a confidence level of 90%. The most influential main factor was the spindle speed, with a contribution of 24.82%, as was expected due to the influence of the incidence of the drill bit on the plate. As can be seen, the contribution of temperature as main factor (first or second order) was 25.31% and taking into account its interactions with spindle speed, $T \times N$ and $T^2 \times N$, their contribution reached 49.12%. Thus, the temperature influence was clear; in fact, in Table 2 it can be observed, as less desirable results were obtained at temperatures close to 0 °C; similar results were approached in the tapping of PA66-GF30 as shown in the study carried out by Domingo et al. [13].

Table 3. Analysis of Variance for Ft.

Source	Sum of Squares	Df	Mean Square	F-Ratio	P-Value
T	913.809	1	913.809	5.18	0.0461
N	6461.76	1	6461.76	36.63	0.0001
f	1524.32	1	1524.32	8.64	0.0148
T^2	5676.25	1	5676.25	32.18	0.0002
T × N	724.941	1	724.941	4.11	0.0701
T × f	338.247	1	338.247	1.92	0.1962
N^2	732.762	1	732.762	4.15	0.0689
N × f	498.843	1	498.843	2.83	0.1235
f^2	22.4396	1	22.4396	0.13	0.7287
T^2 × N	5476.74	1	5476.74	31.05	0.0002
T^2 × f	466.632	1	466.632	2.65	0.1349
T × N^2	29.6117	1	29.6117	0.17	0.6906
T × N × f	3.83645	1	3.83645	0.02	0.8857
T × f^2	52.1043	1	52.1043	0.30	0.5987
N^2 × f	128.633	1	128.633	0.73	0.4131
N × f^2	139.122	1	139.122	0.79	0.3953
Total residual	1763.86	10	176.386		
Total	26,038.0	26			

On the other hand, the ANOM study and the Tukey-HSD test reveal that the pairwise means were significantly different from each other at 95% confidence level, with respect to feedrate (statistically significant differences between pairwise means, 0.5–0.75: −7.67, 0.5–1: −19.19 and 0.5–1: −11.53), spindle speed (statistically significant differences between pairwise means, 5000–6000: 2.71, 5000–7000: 27.53, and 6000–7000: 24.82), and enviromental temperature (statistically significant differences between pairwise means: −22–0: −47.28, −22–22: −33.65, and 0–22: 14.23). This denotes that the thrust force showed different behavior in each of the selected cutting speeds, feedrates and temperatures.

According to the above, thrust forces obtained in drilling processes depended strongly on the enviromental temperature.

3.1.2. Energy

From the ANOVA analysis for energy (Table 4), the factors N^2, f^2, T × N × f, T × f^2 and N × f^2 (not significant) were eliminated in order to achieve a more improved adjustment between the factors. Feedrate and T^2 erre the significant factors at 90% confidence level, with an influence of 9.19% and 6.78%, respectively. Although the effect of these factors was lower than those considered for thrust force, the environmental temperature and cutting conditions reappeared.

Table 4. Analysis of variance for E.

Source	Sum of Squares	Df	Mean Square	F-Ratio	P-Value
T	5.15227	1	5.15227	0.53	0.4783
N	4.47207	1	4.47207	0.46	0.5084
f	**46.5409**	**1**	**46.5409**	**4.78**	**0.0451**
T^2	**34.3523**	**1**	**34.3523**	**3.53**	**0.0800**
T × N	28.49	1	28.49	2.92	0.1079
T × f	7.88941	1	7.88941	0.81	0.3825
N × f	4.45301	1	4.45301	0.46	0.5093
T^2 × N	6.12562	1	6.12562	0.63	0.4402
T^2 × f	7.7748	1	7.7748	0.80	0.3858
T × N^2	2.828	1	2.828	0.29	0.5980
N^2 × f	9.68247	1	9.68247	0.99	0.3347
Total residual	146.171	15	9.74475		
Total	506.569	26			

In this case, from Tukey-HSD test it was obtained that the pairwise means were significantly different from each others at 95% confidence level, with respect to feedrate (statistically significant differences between pairwise means, 0.5–0.75: 4.10, 0.5–1: 6.98 and 0.5–1: 2.86), spindle speed (statistically significant differences between pairwise means, 5000–6000: −0.56, 5000–7000: 0.08, and 6000-7000: 0.67), and temperature (statistically significant differences between pairwise means: −22–0: −3.88, −22–22: −2.97, and 0–22: 0.91). This implies that the energy showed a different behaviour in each of the selected spindle speeds, feedrates and temperatures.

3.1.3. Material Removed Rate

The ANOVA analysis for MRR can be seen in Table 5. Although the data from different temperatures were included in the study, the significant factors were N, f, N × f and f^2, being the most influential factor on the feedrate (f) with a percentage of 44.9%, as was expected despite f^2 having a very low influence. Note that the feedrate considered was very high.

Table 5. Analysis of variance for material removed rate (MRR).

Source	Sum of Squares	Df	Mean Square	F-Ratio	P-Value
T	0.0	1	0.0	0.0	1.0000
N	449,686.0	1	449,686.0	****	0.0000
f	1.79872×10^6	1	1.79872×10^6	****	0.0000
T^2	0.0	1	0.0	0.0	1.0000
T × N	0.0	1	0.0	0.0	1.0000
T × f	0.0	1	0.0	0.0	1.0000
N^2	0.0	1	0.0	0.0	1.0000
N × f	166,550.0	1	166,550.0	****	0.0000
f^2	0.00015	1	0.00015	39,607.88	0.0000
$T^2 \times N$	0.0	1	0.0	0.0	1.0000
$T^2 \times f$	0.0	1	0.0	0.0	1.0000
$T \times N^2$	0.0	1	0.0	0.0	1.0000
T × N × f	0.0	1	0.0	0.0	1.0000
$T \times f^2$	0.0	1	0.0	0.0	1.0000
$N^2 \times f$	0.0	1	0.0	0.0	1.0000
$N \times f^2$	0.0	1	0.0	0.0	1.0000
Total residual	3.78713×10^{-8}	10	3.78713×10^{-9}		
Total	1.14086×10^7	26			

From Tukey-HSD test it was obtained that the pairwise means were significantly different from each other at 95% confidence level, with respect to feedrate (statistically significant differences between pairwise means, 0.5–0.75: −706.85, 0.5–1: −1431.71 and 0.5–1: −706.86) and spindle speed (statistically significant differences between pairwise means, 5000–6000: −353.43, 5000–7000: −706.86, and 6000–7000: −353.43). Obviously, there were no statistically significant differences between pairwise means with respect to temperature because the MRR was independent of it, with differences as: −22–0: 0, −22–22: 0, and 0–22: 0. This implies that the energy showed a different behaviour in each of the selected spindle speeds and feedrates.

3.1.4. Input Diameter

As has been mentioned before, the measurements of input diameters were carried out as quality control evaluation; if measurements were correct the experimental tests could be considered valid. Otherwise, the tests could not be taken out and the experimental data would have to be discarded. In Table 2 can be seen that the greater difference between diameters was 0.093 mm (0.013 mm at 22 °C, 0.070 mm at 0 °C and 0.054 mm at −22 °C), being adequate cooling temperaturesfor different spindle speeds at feedrates of 0.5 mm/rev (Figure 3a), 0.75 mm/rev (Figure 3b) and 1 mm/rev (Figure 3c). In these figures, larger undersized holes could be observed at 0 °C with spindle speed of 5000 and

6000 rpm. However, the difference between them was very low, and it did not seem relevant. Perri et al. [11] found that the effect of the cooling system and the flow air on the displacement of the tool centre point was less intense; this can explain the data obtained in the diameters, and the undersized holes obtained.

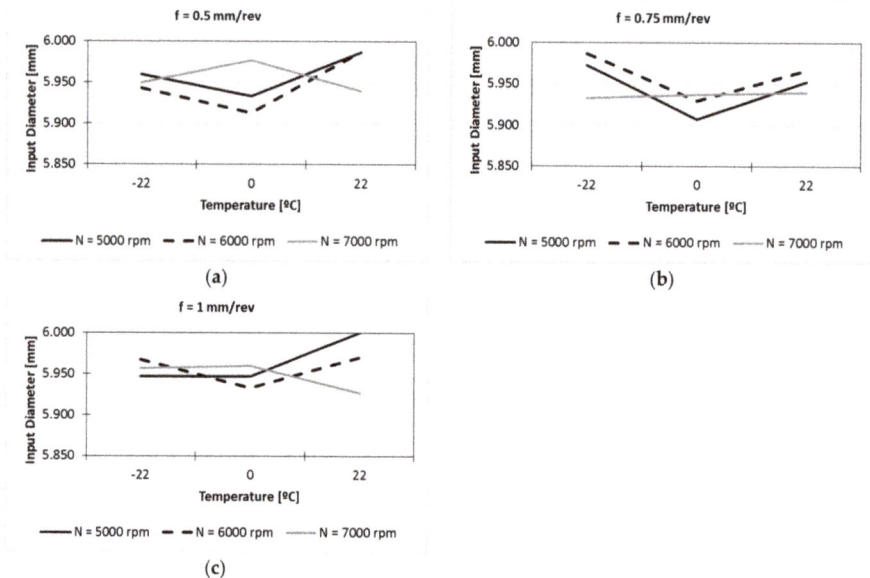

Figure 3. Input diameter executed at different temperatures and spindle speed, and with: (**a**) feedrate of 0.5 mm/rev; (**b**) feedrate of 0.75 mm/rev; (**c**) feedrate of 1 mm/rev.

3.2. Response Surface

Taking into account, uniquely, the significant main factors and interactions, and with a coefficient of determination, R^2, superior to 70% in all of the cases, the regression equations, Equations (6)–(8), are the following:

$$Ft = 83.727 - 5.662 \times T - 0.0415 \times N + 223.276 \times f - 0.455 \times T^2 + \\ + 0.00193 \times T \times N + 0.0000059 \times N^2 + 0.0 \times T^2 \times N, \qquad (6)$$

$$E = -129.475 + 195.038 \times f - 0.01164 \times T^2, \qquad (7)$$

$$MRR = 0.05 - 0.14 \times f + 0.47124 \times N \times f + 0.08 \times f^2, \qquad (8)$$

The effect of the temperature on thrust forces can be seen in Figure 4. The influence of spindle speed and the feedrate was lower at room temperature (Figure 4a) than at 0 °C (Figure 4b). Finally, it is remarkable that, at −22 °C (Figure 4c), the thrust forces were lower and less dependent of cutting conditions and the lowest values were reached. Figure 4 plots the values of Equation (6), also, the evolution of Ft and its trend, with continuous values. The evolution of forces at 0 °C deserves special mention, that showed the force decreased with the increase of spindle speed; the friction seemed to increase at this temperature, though this effect was diminished at higher spindle speeds; in this case, the feedrate more convenient is the lowest value. However, at −22 °C, the friction phenomenon did not seem to be affected, which could be explained because of greater influence of the air pressure at low temperature. Note that, as mentioned in Section 1, PEEK-GF30 maintains a stable behavior at low temperatures [14], which is very important because it prevents the modification of the material properties at these temperatures.

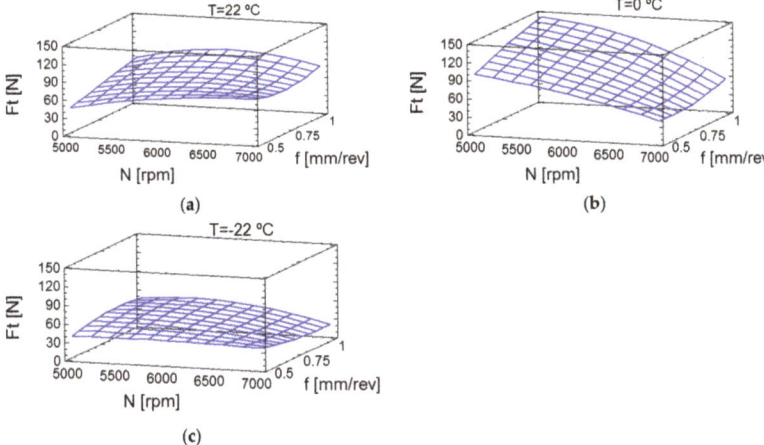

Figure 4. Estimation of response surface for thrust forces: (**a**) Temperature of 22 °C; (**b**) Temperature of 0 °C; (**c**) Temperature of −22 °C.

The estimation of response surface for energy can be seen in Figure 5. At 22 °C (Figure 5a) and 0 °C (Figure 5b) the distribution of the energy obtained was more uniform. At −22 °C (Figure 5c), the highest spindle speeds provided a similar behavior to that in thrust forces at this temperature, despite the greater influence of the torque in the calculation of the energy consumed during drilling [16]. The explanation can be the same as found for the forces, given the dependence of the cutting conditions along the cutting length during drilling (see Equation (1)). To clarify this point, in Figure 5 the values of Equation (7), the evolution of energy and its trend are plotted. Dry machining (at 22 °C) is an option with more energy consumption, maybe because there is a heating in the cutting area.

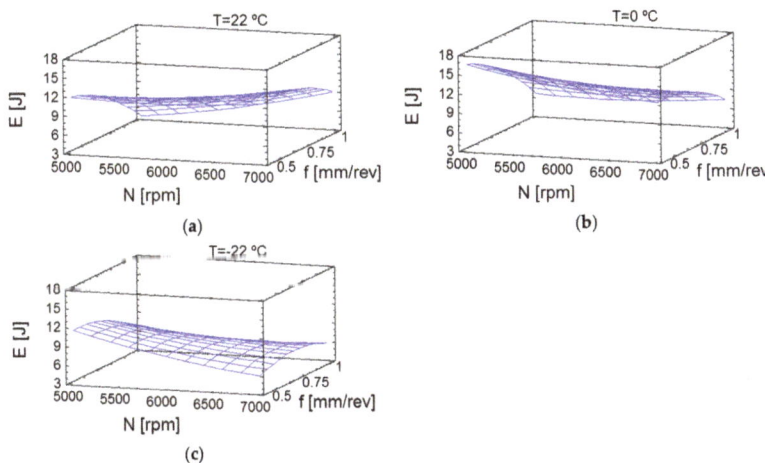

Figure 5. Estimation of response surface for energy: (**a**) Temperature of 22 °C; (**b**) Temperature of 0 °C; (**c**) Temperature of −22 °C.

Figure 6 shows the estimation of response surface for MRR, with the clear and known influence of cutting conditions. This representation of Equation (8) allows the representation of the evolution of MRR, its increase with the cutting conditions, as expected, and its independence of temperature, as can be observed in Equations (2) and (8). In Figure 6, from the last equation MRR values are plotted respect to cutting conditions, spindle speed and feedrate.

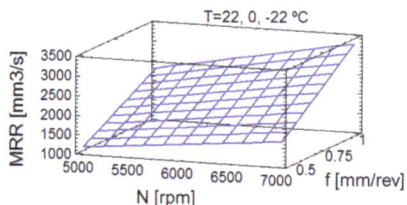

Figure 6. Estimation of response surface for MRR, at 22 °C, 0 °C and −22 °C.

3.3. Multiple Response Surface Optimization

The multiple response surface optimization is represented in Table 6; in this Table, the goal of each variable and the parameters values are shown, lower values of variables (L_i), upper values of variables (U_i), weights considered in the desirability function (s and t), and impact. The values taken for weights and impact avoid that some variables influenced more than others in the results, and besides they are used to optimize the manufacturing processes, in particular in machining operations [13,38]. As can be seen in Table 6, the weights and the impact were the same for each variable, 1 for weights and 3 for impact; in this manner, all variables were compensated. The lower and upper values of each variable were taken from Table 2; the Ft values corresponded to −22 °C, 7000 rpm and 0.75 mm/rev (lowest values) and 0 °C, 5000 rpm and 1 mm/rev (uppest value); the Energy values corresponded to −22 °C, 6000 rpm and 1 mm/rev (lowest value) and 22 °C, 7000 rpm and 0.5 mm/rev (uppest value); and MRR corresponded to any temperature, and 5000 rpm and 0.5 mm/rev (lowest value) and 7000 rpm and 1 mm/rev (uppest value). This indicates the importance of the application of a multi-objective method to seek a common objective when the goals are different and when the minimum and maximum values of each variable correspond to different cutting conditions.

Table 6. Values considered in the desirability function.

Variable	Goal	L_i	U_i	s	t	Impact
Ft [N]	Minimize	26.57	156.66	1	–	3
E [J]	Minimize	2.67	19.93	1	–	3
MRR [mm^3/s]	Maximize	1178.10	3298.67	–	1	3

Figure 7 shows the values of desirability function for thrust force, energy, MRR and Ft-E-MRR; while that the maximum value (1) was achieved for thrust forces and MRR, the minimum value was for the energy, that obviously it was noted in the final value for the combined variable Ft-E-MRR. While with Ft and MRR, the maximum desirability was achieved, the energy showed a value of 0.88. This value can be considered normal due to the dependence of different variables on its calculation (see Equation (1)).

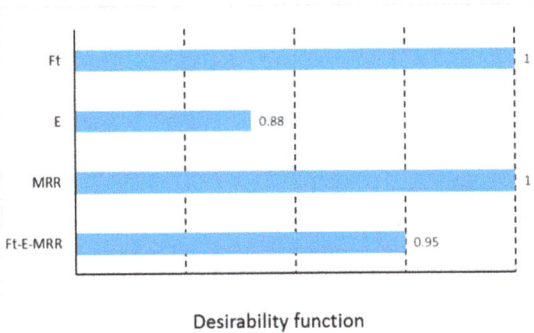

Figure 7. Desirability function.

The optimization allowed finding the maximum desirability at −22 °C, feed of 1 mm/rev and 7000 rpm; under these conditions, the optimum values were 22.93 N for thrust force, 4.2 J for energy and 3298 mm^3/s for MRR. Figure 8 shows the contours of estimated response surface. From this graphic it can be observed that the area of major desirability was located at a temperature of −22 °C and at high spindle speed, considering the thrust force, the energy and the MRR. Note that the optimal parameters allowed machining at high cutting conditions at −22 °C. This possibility can increase the potential use of machining at low temperatures by avoiding thermal expansion of the matrix of composite materials. This balance combines objectives of sustainability and efficiency.

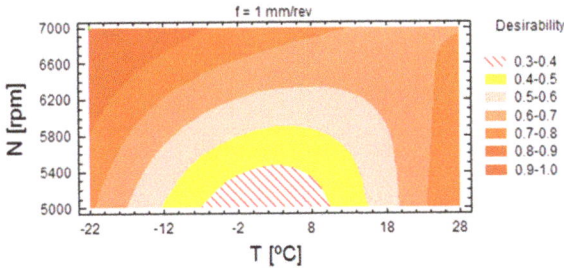

Figure 8. Contours of estimated response surface for f = 1 mm/rev.

4. Conclusions

The drilling of reinforced PEEK plates with drills of coating zirconium oxide and diameter of 6 mm was analyzed at different cutting conditions and environmental temperatures, employing cooling compressed air by a Ranque–Hilsch vortex tube. Experimental data from spindle speed, feedrate, temperature and input diameter of holes were measured, and a statistical study was developed.

The Response Surface methodology with three-level factorial design was applied to optimize, simultaneously, several responses through the desirability function, in order to find the significant factors (spindle, speed, feed rate and temperature) and their interactions on the variables such as thrust force, energy and MRR, the relationships between factors and variables, and the value of desirability function. Thus, the response surface methodology should be optimized.

From the results of this experimental and statistical study, conclusions can be summarized as:

- The environmental temperature (in first, second degree) and its interaction with spindle speed is a significant factor for thrust forces and for energy (in first degree).
- The maximum desirability function was found at highest cutting conditions (7000 rpm and 1 mm/rev) and at −22 °C. The implementation of compressed cooling air provides a balance between the optimal conditions for the analyzed variables.
- Cooling compressed air can be an environmentally friendly procedure that reduces the energy consumption, and besides, it can be compatible with high cutting conditions, which facilitates the high performance of machining processes.
- Finally, oversized holes can be avoided at the conditions used in this study.

Future researches can address lower temperatures, avoiding temperatures close to 0 °C which do not improve the performance with respect to dry conditions at room temperature. Moreover, in future developments, the application of cryogenic drilling on this material can be considered due to the improved outcomes obtained at low temperatures. In addition, the tool wear is proposed to be a factor to analyze.

Author Contributions: Conceptualization, R.D.; methodology, R.D.; validation, B.d.A.; formal analysis, R.D., B.d.A. and M.M.M.; investigation, R.D., B.d.A. and M.M.M.; resources, R.D. and M.M.M.; writing—original draft preparation, R.D. and B.d.A.; writing—review and editing, R.D., B.d.A. and M.M.M.; funding acquisition, R.D. All authors have read and agreed to the published version of the manuscript.

Funding: This research was funded by Spanish Ministry of Science, Innovation and Universities, DPI2014-58007-R and RTI2018-102215-B-I00.

Acknowledgments: The authors also thank the College of Industrial Engineers of UNED for supporting through 2019-ICF08 and 2019-ICF03 projects.

Conflicts of Interest: The authors declare no conflict of interest. The funders had no role in the design of the study; in the collection, analyses, or interpretation of data; in the writing of the manuscript, or in the decision to publish the results.

References

1. Aguado, S.; Alvarez, R.; Domingo, R. Model of efficient and sustainable improvements in a lean production system through processes of environmental innovation. *J. Clean. Prod.* **2013**, *47*, 141–148. [CrossRef]
2. Calvo, L.M.; Domingo, R. Influence of process operating parameters on CO_2 emissions in continuous industrial plants. *J. Clean. Prod.* **2015**, *96*, 253–262. [CrossRef]
3. Groover, M.P. *Fundamentals of Modern Manufacturing: Materials, Processes, and Systems*, 7th ed.; John Wiley & Sons, Inc.: Hoboken, NJ, USA, 2019.
4. Krolczyk, G.M.; Maruda, R.W.; Krolczyk, J.B.; Wojciechowski, S.; Mia, M.; Nieslony, P.; Budzik, G. Ecological trends in machining as a key factor in sustainable production—A review. *J. Clean. Prod.* **2019**, *218*, 601–615. [CrossRef]
5. Sen, B.; Mia, M.; Krolczyk, G.M.; Mandal, U.K.; Mondal, S.P. Eco-Friendly Cutting Fluids in Minimum Quantity Lubrication Assisted Machining: A Review on the Perception of Sustainable Manufacturing. *Int. J. Precis. Eng. Manuf. GT* **2019**. [CrossRef]
6. Alvarez, R.; Domingo, R.; Sebastian, M.A. The formation of saw toothed chip in a titanium alloy: Influence of constitutive models. *Strojniški Vestnik J. Mech. Eng.* **2011**, *57*, 739–749. [CrossRef]
7. Liu, C.; Goel, S.; Llavori, I.; Stolf, P.; Giusca, C.L.; Zabala, A.; Kohlscheen, J.; Paiva, J.M.; Endrino, J.L.; Veldhuis, S.C.; et al. Benchmarking of several material constitutive models for tribology, wear, and other mechanical deformation simulations of Ti6Al4V. *J. Mech. Behav. Biomed.* **2019**, *97*, 126–137. [CrossRef]
8. Wu, Z.; Yang, Y.; Su, C.; Cai, X.; Luo, C. Development and prospect of cooling technology for dry cutting tools. *Int. J. Adv. Manuf. Technol.* **2017**, *88*, 1567–1577. [CrossRef]
9. Goindi, G.S.; Sarkar, P. Dry machining: A step towards sustainable machining—Challenges and future directions. *J. Clean. Prod.* **2017**, *165*, 1557–1571. [CrossRef]
10. Jozić, S.; Bajić, D.; Celent, L. Application of compressed cold air cooling: Achieving multiple performance characteristics in end milling process. *J. Clean. Prod.* **2015**, *100*, 325–332. [CrossRef]

11. Perri, G.M.; Bräunig, M.; di Gironimo, G.; Putz, M.; Tarallo, A.; Wittstock, V. Numerical modelling and analysis of the influence of an air cooling system on a milling machine in virtual environment. *Int. J. Adv. Manuf. Technol.* **2016**, *86*, 1853–1864. [CrossRef]
12. Nor Khairusshima, M.K.; Che Hassan, C.H.; Jaharah, A.G.; Amin, A.K.M.; Md Idriss, A.N. Effect of chilled air on tool wear and workpiece quality during milling of carbon fibre-reinforced plastic. *Wear* **2013**, *302*, 1113–1123. [CrossRef]
13. Domingo, R.; de Agustina, B.; Marín, M.M. A multi-response optimization of thrust forces, torques, and the power of tapping operations by cooling air in reinforced and unreinforced polyamide PA66. *Sustainability* **2018**, *10*, 889. [CrossRef]
14. Yang, X.; Wu, Y.; Wei, K.; Fang, W.; Sun, H. Non-Isothermal Crystallization Kinetics of Short Glass Fiber Reinforced Poly (Ether Ether Ketone) Composites. *Materials* **2018**, *11*, 2094. [CrossRef] [PubMed]
15. Davim, J.P.; Reis, P. Machinability study on composite (polyetheretherketone reinforced with 30% glass fibre–PEEK GF 30) using polycrystalline diamond (PCD) and cemented carbide (K20) tools. *Int. J. Adv. Manuf. Technol.* **2013**, *23*, 412–418. [CrossRef]
16. Domingo, R.; García, M.; Sánchez, A.; Gómez, R. A sustainable evaluation of drilling parameters for PEEK-GF30. *Materials* **2013**, *6*, 5907–5922. [CrossRef]
17. Giasin, K.; Gorey, G.; Byrne, C.; Sinke, J.; Brousseau, E. Effect of machining parameters and cutting tool coating on hole quality in dry drilling of fibre metal laminates. *Compos. Struct.* **2019**, *212*, 159–174. [CrossRef]
18. Abdulgadir, M.M.; Demir, B.; Turan, M.E. Hybrid Reinforced Magnesium Matrix Composites (Mg/Sic/GNPs): Drilling Investigation. *Metals* **2018**, *8*, 215. [CrossRef]
19. Szwajka, K.; Zielińska-Szwajka, J.; Trzepiecinski, T. Experimental Study on Drilling MDF with Tools Coated with TiAlN and ZrN. *Materials* **2019**, *12*, 386. [CrossRef]
20. Davim, J.P.; Reis, P. Damage and dimensional precision on milling carbon fiber-reinforced plastics using design experiments. *J. Mater. Process. Technol.* **2005**, *160*, 160–167. [CrossRef]
21. Shyha, I.S.; Soo, S.L.; Aspinwall, D.K.; Bradley, S.; Perry, R.; Harden, P.; Dawson, S. Hole quality assessment following drilling of metallic-composite stacks. *Int. J. Mach. Tools Manuf.* **2011**, *51*, 569–578. [CrossRef]
22. Köklü, U.; Demir, O.; Avcı, A.; Etyemez, A. Drilling performance of functionally graded composite: Comparison with glass and carbon/epoxy composites. *J. Mech. Sci. Technol.* **2017**, *31*, 4703–4709. [CrossRef]
23. Xia, T.; Kaynak, Y.; Arvin, C.; Jawahir, I.S. Cryogenic cooling-induced process performance and surface integrity in drilling CFRP composite material. *Int. J. Adv. Manuf. Technol.* **2016**, *82*, 605–616. [CrossRef]
24. Domingo, R.; Marín, M.M.; Claver, J.; Calvo, R. Selection of cutting inserts in dry machining for reducing energy consumption and CO_2 emissions. *Energies* **2015**, *18*, 13081–13095. [CrossRef]
25. Schulz, H.; Dörr, J.; Rass, I.J.; Schulze, M.; Leyendecker, T.; Erkens, G. Performance of oxide PVD-coatings in dry cutting operations. *Surf. Coat. Technol.* **2001**, *146–147*, 480–485. [CrossRef]
26. Pinar, A.M.; Uluer, O.; Kırmaci, V. Optimization of counter flow Ranque–Hilsch vortex tube performance using Taguchi method. *Int. J. Refrig.* **2009**, *32*, 1487–1494. [CrossRef]
27. Singh, G.R.; Sharma, V.S. Analyzing machining parameters for commercially puretitanium (Grade 2), cooled using minimum quantity lubrication assisted by a Ranque-Hilsch vortex tube. *Int. J. Adv. Manuf. Technol.* **2017**, *88*, 2921–2928. [CrossRef]
28. Mia, M.; Singh, G.R.; Gupta, K.M.; Sharma, V.S. Influence of Ranque-Hilsch vortex tube and nitrogen gas assisted MQL in precision turning of Al 6061-T6. *Precis. Eng.* **2018**, *53*, 289–299. [CrossRef]
29. Measurement Computing. Available online: www.mccdaq.com/products/DASYLab.aspx (accessed on 20 September 2019).
30. Li, R.; Hegde, P.; Shih, A.J. High-trroughput drilling of titanium alloys. *Int. J. Mach. Tools Manuf.* **2007**, *47*, 63–74. [CrossRef]
31. Kalpakjian, S.; Schmid, S. *Manufacturing Engineering and Technology*, 7th ed.; Pearson: Hoboken, NJ, USA, 2014.
32. Ameur, M.F.; Habak, M.; Kenane, M.; Aouici, H.; Cheikh, M. Machinability analysis of dry drilling of carbon/epoxy composites: Cases of exit delamination and cylindricity error. *Int. J. Adv. Manuf. Technol.* **2017**, *88*, 2557–2571. [CrossRef]
33. ISO 4291. *Methods for the Assessment of Departure from Roundness. Measurement of Variations in Radius*; International Organization for Standardization: Geneva, Switzerland, 1985.
34. Statgraphics. Available online: Statgraphics.net (accessed on 20 June 2019).

35. Montgomery, D.C. *Design and Analysis of Experiments*, 8th ed.; John Wiley & Sons Inc.: New York, NY, USA, 2012.
36. Kutner, M.H.; Natchtsheim, C.J.; Neter, J.; Li, W. *Applied Linear Statistical Models*, 5th ed.; McGraw-Hill Irwin: New York, NY, USA, 2005.
37. Derringer, G.; Suich, R. Simultaneous optimization of several response variables. *J. Qual. Technol.* **1980**, *12*, 214–219. [CrossRef]
38. Kumar, P.; Chauhan, S.R.; Pruncu, C.I.; Gupta, M.K.; Pimenov, D.Y.; Mia, M.; Gill, H.S. Influence of Different Grades of CBN Inserts on Cutting Force and Surface Roughness of AISI H13 Die Tool Steel during Hard Turning Operation. *Materials* **2019**, *12*, 177. [CrossRef] [PubMed]

© 2020 by the authors. Licensee MDPI, Basel, Switzerland. This article is an open access article distributed under the terms and conditions of the Creative Commons Attribution (CC BY) license (http://creativecommons.org/licenses/by/4.0/).

Article

Reusing Discarded Ballast Waste in Ecological Cements

Santiago Yagüe García * and Cristina González Gaya

ETS Ingenieros Industriales, Universidad Nacional de Educación a Distancia (UNED), C/Juan del Rosal, 12, 28040 Madrid, Spain; cggaya@ind.uned.es
* Correspondence: syague14@alumno.uned.es

Received: 13 October 2019; Accepted: 21 November 2019; Published: 25 November 2019

Abstract: Numerous waste streams can be employed in different cement production processes, and the inclusion of pozzolans will, moreover, permit the manufacture of concrete with improved hydraulic properties. Pozzolanic materials can be added to Ordinary Portland Cement (OPC) in the range of 10%–20% by mass of cement. One such example is the phyllosilicate kaolinite (K), and its calcined derivative metakaolin (MK), incorporated in international cement manufacturing standards, due to its high reactivity and utility as a pozzolan. In the present paper, discarded ballast classed as Construction and Demolition Waste (C&DW) is reused as a pozzolanic material. Various techniques are used to characterize its chemical, mineralogical, and morphological properties, alongside its mechanical properties, such as compressive and flexural strength. Discarded ballast in substitution of cement at levels of 10% and 20% produced type II or IV pozzolanic cements that yielded satisfactory test results.

Keywords: cements; ballast waste; cornubianite; mechanical properties; Spain

1. Introduction

The cement industry is one of the main contributors of greenhouse gas emissions, such as CO, CO_2, and NO. In this sector, new strategies are now prioritized in many countries that are trying to surpass the European Union (EU) targets set for 2020, which aspire to 20% lower emission levels than in 1990. Natural additions to cement have been used for the improvement of its properties since the days of antiquity. One of the best known is the addition of pozzalonic material consisting of a type of volcanic pumice, purportedly found around the town of Pozzuoli (Italy) that causes certain hydraulic characteristics in cement, hence the name of pozzalonic cements [1].

A pozzolan material is understood to be a material that generally consists of silica and alumina that, in itself, has no cementitious value when mixed with water. However, when finely ground and in the presence of water, it reacts with portlandite ($Ca(OH)_2$), originating from the hydration of silicates present in the clinker, generating compounds with cementing properties [2]. The normalized artificial additions that present pozzolanic behavior are industrial by-products, such as silica fume and fly ash—the natural pozzolans—and calcined bauxite. The non-normalized materials include paper sludge, calcined sugar cane bagasse ash, and rice husk ash, among others.

Various natural materials that might function as pozzolans have been tested in cement research, and various artificial additions have more recently been undergoing trials, especially industrial waste materials including slags, silica fume, fly ash, etc., which are usually dumped in large volumes in landfill sites [3]. The utility of their addition to cement is two-fold: the elimination of waste and the enhancement of the cement. These additions are understood as technological solutions, designed to improve both the performance of high-strength cements and cement behavior against aggressive environmental agents [4]. Hence, the incorporation of these additions to cement can be seen as an attempt to reduce manufacturing costs, while simultaneously searching for materials that are more respectful towards the environment.

In the current panorama of waste recycling, construction and demolition waste (C&DW) assumes fundamental importance, because it constitutes one of the main waste streams within the EU. The recycling/reutilization rates of C&DW in the EU vary significantly between countries, fluctuating between 5% in Portugal, and 90% in the Netherlands and Estonia. On average, the recycling rate in the 27 countries of the EC is 55% [5].

According to data from the Plan Nacional de Residuos de Construcción y Demolición 2008–2015 (National C&DW Plan in Spain), 40 million tons of C&DW are produced annually, which is equivalent to over 2 kg per person, per day—a higher rate than domestic rubbish. From data provided by the Environment Ministry, 2.5 million tons were recycled in 2018. It implies a C&DW recycling rate of 5.1%—a figure well below the European average—implying that the main means of disposing of these wastes continues to be in regulated or unregulated landfill sites in Spain, as opposed to their recycling or reuse.

The publication, in February 2008, of Royal Decree (RD) 105/2008, in regulation of the production and management of C&DW (BOE of 13 February), implies growing environmental concerns and interest in this matter, that national government and the autonomous regions have been expressing for numerous years. Prior to this point, there was the 2002–2006 Plan Nacional de Residuos Urbanos (PNRU) [National Urban Waste Plan] and, more specific to the type of wastes that are of interest here, the Plan Nacional de Residuos de Construcción y Demolición 2001–2006 (PNRCD) [National C&DW Plan] (BOE 12 July 2001), which approached the treatment, recovery, and recycling of these wastes. These national plans were substituted by the Plan Nacional Integrado de Residuos (PNIR) [National Integrated Waste Plan] for the period 2008–2015 (BOE 26 February 2009), within which, under Point nº 12, it covered the second National C&DW Plan. Among the plans' objectives was the controlled collection and proper management of 95% of C&DW by 2011, the reduction or reuse of 15% of C&DW by 2011, the recycling of 40% of C&DW by 2011, and the exploitation of 70% of all waste construction material packaging, as from 2010.

Among other aspects, the RD 105/2008 introduced the obligation of including a C&DW management study in a building or construction work design project, which had to contain, as a minimum, an estimation of the amount of waste that could be produced, as well as measures for risk prevention, management process, and the exploitation of those waste streams.

Added to the current situation is an increasing demand for fines linked to the environmental restrictions on the exploitation of new quarries that have led to proposals for the reuse of certain waste streams as alternative raw materials. Different studies on the viability of valorizing granite waste and other ornamental rocks have been consulted. In this case, it is the waste from hornfels considered [6–9].

The renewal of a train track is carried out when the ballast does not meet the precise specifications related to wear and alteration of the rocks, which depends on their nature and is controlled by convoys that partially remove and replace the ballast. This control is performed when needed, according to the auscultator train.

In the present work, the addition of a construction waste stream from used ballast is studied for its use as a pozzolan. The need to activate this residue has been assessed using the pozzolanicity test, and the mechanical properties of the mixtures have been obtained by the partial replacement in the Ordinary Portland Cement (OPC) of the waste considered.

2. Materials

The aim was to obtain C&DW from discarded ballast that had been replaced by new ballast. To do so, the process of ballast wear was simulated using an accelerated method, which consisted of wearing down new ballast in a ball mill and collecting the fines. Aggregate materials were employed in this work, with coarse granulometry (gravel) from CANTERAS and CONSTRUCCIONES S.A. (CYCASA), at Aldeavieja, Ávila (Spain). The ballast was taken from the same batches supplied by the quarry for the renewal of the stretch of rail from PB Río Duero - Est. Valladolid C.G., on the Madrid—Segovia—Valladolid High Speed Rail Line.

The sample under study (C) is a hornfels (metamorphic rock) with granoblastic texture, although in many examples the regional schistosity can also be seen. Hornfels is dark rock with a matt shine and opaque colors.

An Ordinary Portland Cement (OPC) type CEM I 42.5 R cement was used by Italcementi group.

3. Methods

The Los Angeles abrasion test [10] was used to monitor wear, in accordance with the conditions specified in annex C of standard UNE ES 13450 [11].

The ballast fragments were worn down in a steel ball mill to obtain the waste product, that was then transported to a laboratory for oven drying for 24 h at 110 °C, to a constant weight (through the removal of humidity).

In accordance with the consulted literature [12,13], the calcination process was followed once the waste had been dried in an electric oven at 600 °C for 2 h (sample CC), with a view to establish the pozzolanic potential of this addition and to improve the activation potential of the waste.

To determine the pozzolanic activity, an accelerated method was used that consisted of placing 1 g of the sample solution in contact with 75 mL of a saturated solution of calcium hydroxide (17.68 mM/L) to 40 °C over 1, 7, 14, 28, and 90 days. The solution underwent vacuum filtration in a Buchner funnel at each age under study. The concentration of calcium ions expressed as calcium oxide or fixed lime was determined in the filtrate by the assay detailed in standard UNE—EN 196-5 [14,15].

The specific surface of the waste was studied using the BET method, through isothermal absorption of nitrogen, and the distribution of particle size (Sympatec Helos 12LA laser diffraction spectrometer, Sympatec, Clausthal-Zellerfeld, Germany).

The solid waste was analyzed through X-ray Fluorescence Spectroscopy (XRF) (to analyze chemical composition, using a Bruker S8 Tiger XDR spectrometer (Bruker, Fremon, CA, USA) with Spectra Plus Quant Express software v1.0.0.13), X-ray Diffraction (XRD) (to analyze mineralogical composition with a SIEMENS D-5000 diffractometer (Anton Parra, Madrid, Spain), working between 3 and 60 degrees with a sweep velocity of 2 degrees per minute), and scanning electron microscopy/energy dispersive X-ray spectroscopy (SEM/EDX) (providing surface data at a microscopic level and surface analysis, using a PHILIPS XL 30 flexible scanning electron microscope (Philips, Leuven Belgié) with a wolfram filament, a BIO-RAD SC 502 disk type sputter target, and an EDAX Energy Dispersive X-ray spectrometer with a DX4i silica/lithium detector and analyzer for chemical analysis, Philips, Leuven, Belgie).

Tests have been carried out on the samples for their resistance to compression and bending according to the UNE EN 196-1 standard [16] (compressive and flexural strength).

4. Results and Discussion

The particle size distribution of the discarded ballast waste was studied, showing a bimodal distribution with two maximum particle sizes from 6 to 15 μm.

BET surface determination provided a value of 1.32 m^2/g, similar to the value obtained for fly ash of 1.40 m^2/g, lower than the value for ceramic tiling of 3.00 m^2/g [17], very much lower than fly ash at 20 m^2/g [18], and higher than the values close to 0.98 m^2/g for ladle furnace slag [18].

The result of the chemical analysis by XRF of the original sample (C) of OPC cement is provided in Table 1.

Table 1. Chemical analysis by X-ray Fluorescence Spectroscopy (XRF) from C sample and Ordinary Portland Cement (OPC) cement.

Oxides (%)	C Sample	OPC Cement
SiO_2	69.64	20.26
Al_2O_3	15.00	4.61
Fe_2O_3	2.52	2.44
MgO	1.60	3.35
Na_2O	3.59	4.14
K_2O	4.04	1.41
P_2O_5	0.17	0.22
TiO_2	0.51	0.14
MnO	0.04	0.03
Chloride (ppm)	290	130
LOI (loss on ignition)	0.52	3.04

In addition, concentrations of zirconium, copper, chrome, cobalt, nickel, strontium, vanadium, zinc, and lead were detected in quantities that were not in excess of 50 ppm. The high content of both silica and aluminum points to good pozzolanic activity.

In turn, the mineralogical composition of the discarded ballast waste, obtained by XRD, indicated the presence of quartz, potassium feldspar, plagioclase (soda-lime feldspar), biotite, and clay-type minerals, such as kaolinite and chlorite with scarce little muscovite, as well as small quantities of hematite, as shown in Figure 1. The components of the cement were tricalcium aluminate, belite, alite, calcite, and ferrite phases.

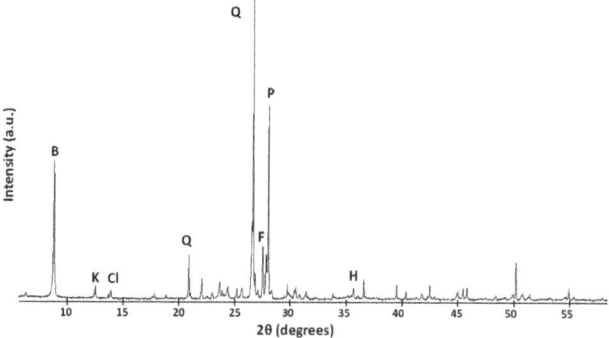

Figure 1. X-ray diffraction by the ballast waste (B = biotite; K = kaolinite; Cl = chlorite; Q = quartz; F = K feldspar; P = Ca, Na feldspar; H = hematite).

4.1. Calcination of Ballast Waste

The waste material was calcined at a temperature of 600 °C for 2 h, in stove at a constant heat to increase its pozzolanic activity. The increase of this activity was due to the loss of structural water in the clay (kaolinite) minerals and in the phyllosilicates (biotite, chlorite).

The result of the mineralogical analysis of the calcined sample (CC) obviously affected the dehydroxylation of the kaolinite, which changed into amorphous metakaolinite, and had an incipient effect on the phyllosilicate structure, which started to lose hydroxyl groups. All other mineral components remained unchanged in the diffractogram.

4.2. Sample Pozzolanicity

The pozzolanicity test on the initial sample and the sample that had been thermally activated at 600 °C/2 h is shown in Figure 2. The improved behavior of sample C and the initial waste is shown, in all cases, except for at 28 days. Thermal activation of the waste was therefore not considered necessary and, henceforth, the initial sample was the only material used in the tests. The differences are so small that thermal activation is not required.

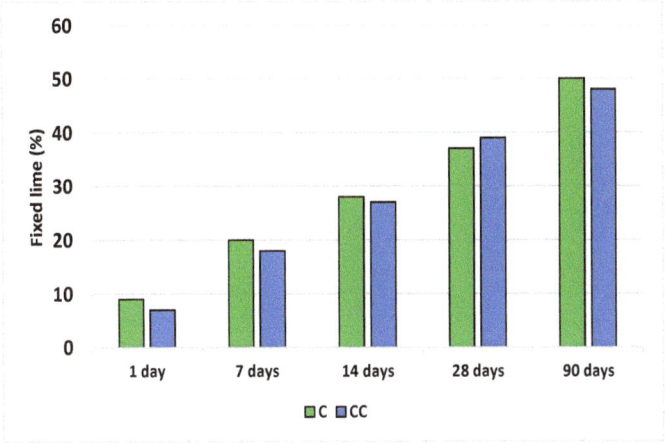

Figure 2. Measures of pozzolanicity of the initial discarded ballast waste (C) and the thermally calcined waste (CC).

4.3. Preparation of Mortars with Discarded Ballast Waste

Standardized CEN sand was used with a granulometry of between 1 and 0.08 mm, which met the requirements specified in standard UNE EN 196-1 [16]. The cement was a CEM 1 42.5 R-type OPC, the composition of which is shown in Table 1 [19].

A mixture of the above components, ballast waste, and Portland cement was used to prepare a mixture that guaranteed the homogeneity of the corresponding mixtures. The cements were differentiated by the substitution of either 10% or 20% by weight of OPC for discarded ballast waste, in an attempt to design type II/A (6/20%) and IV/A (11–35%) cements, in accordance with standard UNE EN 197-1 [14].

4.4. Mechanical Behavior

Compressive strength and flexural tests were performed in accordance with standard UNE EN 196-1 [16]. Prismatic mortar specimens were prepared, measuring 4 cm × 4 cm × 16 cm, with a sand/cement ratio and water/cement ratio of 3/1 and 3/2, respectively. At 24 h after their manufacture, the specimens were demolded and cured, at a temperature of 20 ± 1 °C and a relative humidity of 100%, up until failure.

4.4.1. Mechanical Strength under Compression

All the cements under study presented compressive strengths of over 10 MPa after two days of curing, and greater than or equal to 42.5 MPa over the following 28 days, thereby complying with the mechanical specifications contained in standard UNE EN 197-1 [14] for cements classed as high-strength (42.5 MPa).

From the graph shown in Figure 3, it may be seen that the incorporation of discarded ballast waste in the mortars in no way modified the existing logarithmic tendency (although the trend seems linear, the adaptation to a logarithmic equation meets R^2 better) between compressive strength and curing time, regardless of the substitution level (10% or 20%). Correlation coefficients higher than 0.90 were obtained for R^2.

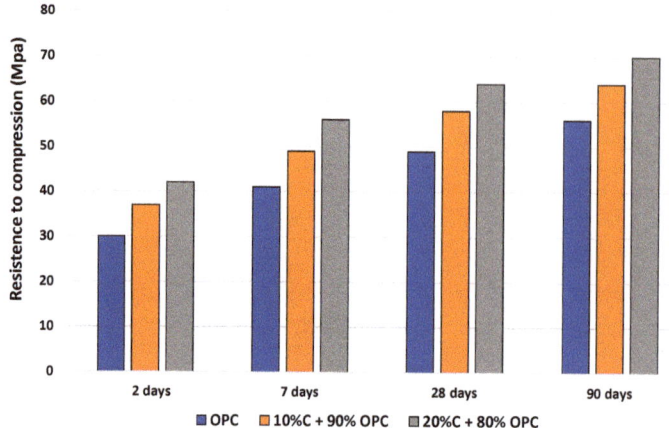

Figure 3. Variation of the compressive strength of the different mortars.

The following expression was used to arrive that figure:

$$y_{OPC} = 6.18 \ln(x), + 41.64, \text{ con } R^2 = 0.945$$

while the two levels of substitution were calculated as follows:

$$y_{10\%+90\%OPC} = 6.39 \ln(x) + 34.88, \text{ with } R^2 = 0.984$$

$$y_{20\%+80\%OPC} = 6.24 \ln(x) + 28.23, \text{ with } R^2 = 0.992$$

It was also noted that the incorporation of the discarded ballast waste implied a significant improvement in performance as the percentage substitution level increased. The weakened performance was close to 11% at 10% C + 90% OPC and around 23% at 20% C + 80% OPC, with regard to the standard OPC specimen, considering a time of 28 days of curing. This tendency to lose strength coincides with the trend noted in conventional mortars by Ramos et al. [20].

In addition, this behavior coincides with the behavior described by Frías et al. [21] and Vardhan et al. [22] in their studies on slate and marble quarry sludge, respectively, in the manufacture of new cements. An addition of 20% seems to improve the conditions of compressive strength compared to one of 10%.

The incorporation of different proportions of SiO_2 and Al_2O_3 improved the structure of the pores, since the ballast waste affects the hydration of the cement and is responsible for the mechanical properties [23]. Fine pores are generated when the Si/Al ratio is high, which increases contact points and resistance. This mechanism is similar to secondary hydration and requires long ages to make it happen.

4.4.2. Flexotraction Strength

Regarding the flexotraction strength of the mortars containing substitutions of ballast waste, a similar tendency was detected to the one observed for compressive strength, with a loss of strength of 10% and 17%, after 28 days of curing in the mixtures with substitution levels of 10% and 20%, respectively, as seen in Figure 4. In this case, the addition of 10% was more favorable in terms of flexotraction than the addition of 20% over time. It can be attributed to the delayed onset of the pozzolanic reaction. It is known that when the pozzolanic reaction begins, the amount of Ca (OH)$_2$ decreases and the microstructure improves, with densification exceeding longer ages [24,25].

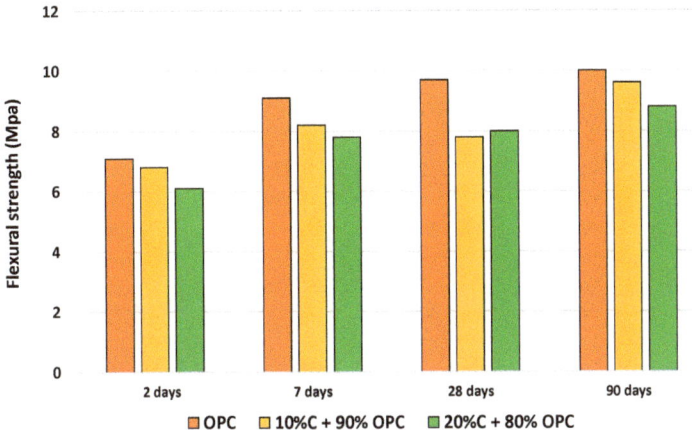

Figure 4. Variation of flexotraction strength of the different mortars.

4.5. Total Porosity and Pore-Size Distribution

Table 2 shows the values of total porosity and the average pore size in the mortars with substitutions of ballast waste at 2 and at 90 days of curing. It is understood that the incorporation of ballast waste generates a slight increase in the total porosity of the mortars with substitutions of 10 and 20%, with a slightly higher increase at substitution levels of 10%, but with values very close to 9% with respect to OPC.

Table 2. Values of the total porosity and the pore diameter average at 2 and 90 days of curing for the mortars studied.

Mortar	Total Porosity (% vol.)		Pore Diameter, Average (µm)	
	2 Days	90 Days	2 Days	90 Days
OPC	13.02	11.98	0.0971	0.0732
10% + 90% OPC	14.52	12.16	0.0989	0.0721
20% + 80% OPC	14.30	12.81	0.1025	0.0742

Table 2, likewise, shows the evolution of the average pore size, with a refinement of the system of pores as the cement hydration process progresses, showing a smaller average size with the age of curing at the higher substitution level (OPC, 10% and 20% of substitution), with respect to the mortars at 2 days.

The above is clear from the SEM images of the mortars under study. Accordingly, SEM-EDAX observations of the grain edges of the mortar specimens (4 cm × 4 cm × 16 cm) with substitution levels of ballast waste at 0%, 10% and 20%, cured over 90 days, have shown that when the cement has no substitution—as shown in Figure 5A—the inter-grain contact is formed by calcium silicate hydrate (CSH) gels and ettringite fibers that give it the slightly porous appearance that can been seen in Figure 5B. In turn, as demonstrated in Figure 5C,D, the inter-grain contact was not observed to be so well defined in the cement with no ballast waste. Nevertheless, as shown in Figure 5E–H, the tendency at levels of substitution of 10% were similar at 20%, although less acute, thus the inter-grain contacts were better defined and substitution levels at 20% were therefore considered unnecessary. The consideration of the results of the previous tests means that the addition is directed towards 10%, which would lead it to be considered a cement type II/A (6/20%) instead of the budget type IV/A (11–35%) in accordance with standard UNE EN 197-1 [14].

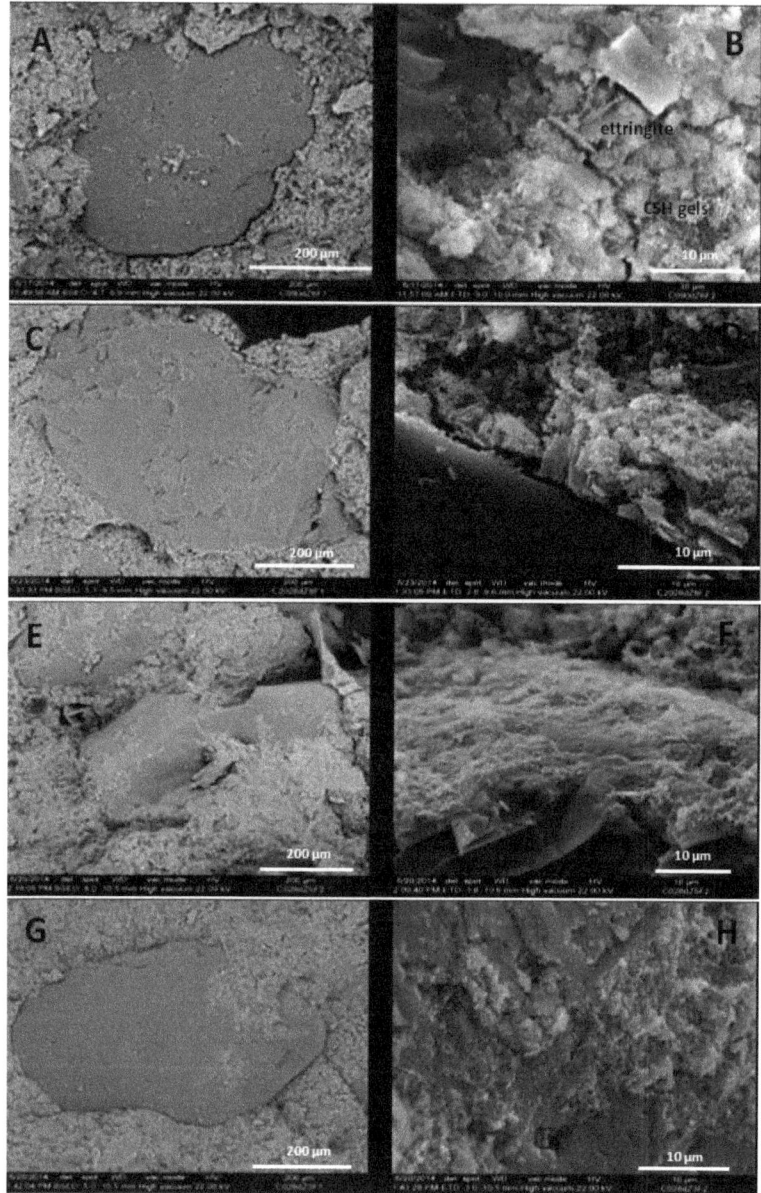

Figure 5. OPC Mortar specimen with no ballast waste: (**A**) magnified image of inter-grain contact; (**B**) grain edge; (**C**) magnified image of inter-grain contact of mortar specimen 20% + 80% OPC; (**D**) grain edge of mortar specimens 10% + 90% OPC; (**E**,**G**) magnified image of inter-grain contact; (**F**,**H**) grain edge.

5. Conclusions

Discarded ballast waste has been used as a pozzolanic addition in cement, to reuse the C&DW instead of sending it to landfill. This type of waste participates in the Circular Economy due to its high amount of silicates, and, when using C&DW, it will be considered as a secondary raw material, as the displacement to waste plants or landfills is unnecessary when taking advantage of its reuse through its management in situ, with consequent economic benefits.

The use of this waste in the same type of facilities eliminates the participation of the necessary track material in its renewal.

The waste does not need any type of activation, neither thermal nor chemical, and can be directly used in mortar mixtures.

The importance of quantifying the addition to the OPC has been investigated at 10% and 20% for ballast waste. Both additions have given good results, especially 10%, for design type II/A (6/20%) cement. In the case of opting for the 20% addition, a type IV/A (11–35%) cement would be obtained, giving good results over time.

The variety of rocks used as ballast warrants further study, looking toward the consideration of materials other than hornfels.

Author Contributions: S.Y.G. contributed materials and performed the experiments and wrote the paper. C.G.G. designed the experiments and analyzed the data.

Funding: This research received no external funding.

Acknowledgments: This paper is based on the ongoing activities that form part of the lead author's Ph.D. thesis, under preparation at the International Doctorate School of the National University of Distance Education (Spain); the authors therefore wish to express their gratitude for the support from that institution. They are also grateful for the assistance kindly provided by Hormigones y Morteros del Río (Aldeavieja, Spain), the Instituto Eduardo Torroja de Madrid, and UAM.

Conflicts of Interest: The authors declare no conflict of interest.

Abbreviations

K	kaolinite
MK	metakaolinite
C&DW	Construction and Demolition Waste
EU	European Union
RD	Real Decreto
PNRU	Plan Nacional de Residuos Urbanos
PNRCD	Plan Nacional de Residuos de Construcción y Demolición
BOE	Boletín Oficial del Estado
OPC	Ordinary Portland Cement
BET method	Brunauer, Emmett and Teller method
XRF	X-ray Fluorescence
XRD	X-ray Diffraction
SEM/EDX	Scanning Electron Microscopy/Energy Dispersive X-ray spectroscopy
LOI	Loss on ignition
CSH	calcium silicate hydrate

References

1. Grist, E.R.; Paine, K.A.; Heath, A.; Norman, J.; Pinder, H. Structural and durability properties of hydraulic lime-pozzolan concretes. *Cem. Concr. Compos.* **2015**, *62*, 212–223. [CrossRef]
2. Velazquez, S.; Monzó, J.M.; Borrachero, M.V.; Payá, J. Assessment of the Pozzolanic Activity of a Spent Catalyst by Conductivity Measurement of Aqueous Suspensions with Calcium Hydroxide. *Materials* **2014**, *7*, 2561–2576. [CrossRef] [PubMed]
3. Paris, J.M.; Roessler, J.G.; Ferraro, C.C.; DeFord, H.D.; Townsend, T.G. A review of waste products utilized as supplements to Portland cement in concrete. *J. Clean. Prod.* **2016**, *121*, 1–18. [CrossRef]
4. Bagheri, M.; Shariatipour, S.M.; Ganjian, E. A review of oil well cement alteration in CO_2 rich environments. *Constr. Build. Mater.* **2018**, *186*, 946–968. [CrossRef]
5. Oficemen Homepage. Available online: www.oficemen.com (accessed on 30 September 2019).

6. Galetakis, M.; Soultana, A. A review on the utilisation of quarry and ornamental stone industry fine by-products in the construction sector. *Constr. Build. Mater.* **2016**, *102*, 769–781. [CrossRef]
7. Bacarji, E.; Toledo Filho, R.D.; Koenders, E.A.B.; Figueiredo, E.P.; Lopes, J. Sustainability perspective of marble and granite residues as concrete fillers. *Constr. Build. Mater.* **2013**, *45*, 1–10. [CrossRef]
8. Medina, G.; Sáez del Bosque, I.F.; Frías, M.; Sánchez de Rojas, M.I.; Medina, C. Mineralogical study of granite waste in a pozzolan/Ca(OH)$_2$ system: Influence of the activation process. *Appl. Clay Sci.* **2017**, *135*, 362–371. [CrossRef]
9. Medina, G.; Sáez del Bosque, I.F.; Frías, M.; Sánchez de Rojas, M.I.; Medina, C. Durability of new recycled granite quarry dust-bearing cements. *Constr. Build. Mater.* **2018**, *187*, 414–425. [CrossRef]
10. AENOR. *Ensayos Para Determinar las Propiedades Mecánicas y Físicas de los Áridos*; Parte 2: Métodos Para la Determinación de la Resistencia a la Fragmentación. UNE EN 1097-2; Asociación Española de Normalización y Certificación: Madrid, Spain, 1999.
11. AENOR. *Áridos Para Balasto*; UNE EN 13450; Asociación Española de Normalización y Certificación: Madrid, Spain, 2003.
12. Vigil de la Villa, R.; Frias, M.; Sánchez de Rojas, M.I.; Vegas, I.; Garcia, R. Mineralogical and morphological changes of calcined paper sludge at different temperatures and retention in furnace. *Appl. Clay Sci.* **2007**, *36*, 279–286. [CrossRef]
13. Frias, M.; Vigil de la Villa, R.; Garcia, R.; Sánchez de Rojas, M.I.; Juan Valdés, A. The Influence of Slate Waste Activation Conditions on Mineralogical Changes and Pozzolanic Behavior. *J. Am. Ceram. Soc.* **2013**, *96*, 2276–2282. [CrossRef]
14. AENOR. *Cemento. Parte I: Composición, Especificaciones y Criterios de Conformidad de los CEMENTOS Communes*; UNE EN 197-1; Asociación Española de Normalización y Certificación: Madrid, Spain, 2011.
15. Khan, K.; Amin, M.N.; Saleem, M.U.; Qureshi, H.J.; Al-Faiad, M.S.; Qadir, M.G. Effect of Fineness of Basaltic Volcanic Ash on Pozzolanic Reactivity, ASR Expansion and Drying Shrinkage of Blended Cement Mortars. *Materials* **2019**, *12*, 2603. [CrossRef] [PubMed]
16. AENOR. *Métodos de Ensayo de Cementos. Parte I. Determinación de Resistencias Mecánicas*; UNE EN 196-1; Asociación Española de Normalización y Certificación: Madrid, Spain, 2005.
17. Sánchez de Rojas, M.I.; Frías, M.; Rodríguez, O.; Rivera, J. Durability of blended cement pastes containing ceramic wastes a pozzolanic addition. *J. Am. Ceram. Soc.* **2014**, *97*, 1543–1551. [CrossRef]
18. Arvaniti, E.C.; Juenger, M.C.G.C.S.; Bernal, A.; Duchesne, J.; Provis, J.L.; Klemm, A.; De Belie, N. Determination of particle size, surface area, and shape of supplementary cementitious materials by different techniques. *Mater. Struct.* **2015**, *48*, 3687–3701. [CrossRef]
19. AENOR. *Métodos de Ensayo de Cementos. Parte 5. ENSAYO de Puzolanicidad Para los Cementos Puzolánicos*; UNE EN 196-5; Asociación Española de Normalización y Certificación: Madrid, Spain, 2011.
20. Ramos, T.; Matos, A.M.; Schmidt, B.; Río, J.O.; Sousa-Coutinho, J. Granitic quarry sludge waste in mortar: Effect on strength and durability. *Constr. Build. Mater.* **2013**, *47*, 1001–1009. [CrossRef]
21. Frías, M.; Vigil, R.; García, R.; de Soto, I.; Medina, C.; Sánchez de Rojas, M.I. Scientific and technical aspect of blended cement matrices containing activated slate wastes. *Cem. Concr. Compos.* **2014**, *48*, 19–25. [CrossRef]
22. Vardhan, K.; Goyal, S.; Siddique, R.; Singh, M. Mechanical properties and microstructural analysis of cement mortar incorporating marble powder as partial replacement building materials. *Constr. Build. Mater.* **2015**, *96*, 615–621. [CrossRef]
23. Johari, M.A.M.; Brooks, J.J.; Kabir, S.; Rivard, P. Influence of supplementary cementitious materials on engineering properties of high strength concrete. *Constr. Build. Mater.* **2011**, *25*, 2639–2648. [CrossRef]
24. Isaia, G.C.; Gastaldini, L.G.; Morases, A.R. Physical and pozzolanic action of mineral additions on the mechanical strength of high performance concrete. *Cem. Concr. Compos.* **2003**, *25*, 69–76. [CrossRef]
25. Bui, D.D.; Hu, J.; Stroeven, P. Particle size effect on the strength of rice husk ash blended gap graded Portland cement concrete. *Cem. Concr. Compos.* **2005**, *27*, 357–366. [CrossRef]

© 2019 by the authors. Licensee MDPI, Basel, Switzerland. This article is an open access article distributed under the terms and conditions of the Creative Commons Attribution (CC BY) license (http://creativecommons.org/licenses/by/4.0/).

Article

Kerf Taper Defect Minimization Based on Abrasive Waterjet Machining of Low Thickness Thermoplastic Carbon Fiber Composites C/TPU

Alejandro Sambruno [1,*], **Fermin Bañon** [1], **Jorge Salguero** [1], **Bartolome Simonet** [2] **and Moises Batista** [1]

1. Department of Mechanical Engineering and Industrial Design, Faculty of Engineering, University of Cadiz, Av. Universidad de Cadiz 10, E-11519 Puerto Real, Cadiz, Spain; fermin.banon@uca.es (F.B.); jorge.salguero@uca.es (J.S.); moises.batista@uca.es (M.B.)
2. Nanotures SL, C. Inteligencia 19, Tecnoparque Agroalimentario, E-11591 Jerez de la Frontera, Cadiz, Spain; bartolome.simonet@nanotures.es
* Correspondence: alejandro.sambruno@uca.es

Received: 27 October 2019; Accepted: 10 December 2019; Published: 13 December 2019

Abstract: Carbon fiber-reinforced thermoplastics (CFRTPs) are materials of great interest in industry. Like thermosets composite materials, they have an excellent weight/mechanical properties ratio and a high degree of automation in their manufacture and recyclability. However, these materials present difficulties in their machining due to their nature. Their anisotropy, together with their low glass transition temperature, can produce important defects in their machining. A process able to machine these materials correctly by producing very small thermal defects is abrasive waterjet machining. However, the dispersion of the waterjet produces a reduction in kinetic energy, which decreases its cutting capacity. This results in an inherent defect called a kerf taper. Also, machining these materials with reduced thicknesses can increase this defect due to the formation of a damage zone at the beginning of cut due to the abrasive particles. This paper studies the influence of cutting parameters on the kerf taper generated during waterjet machining of a thin-walled thermoplastic composite material (carbon/polyurethane, C/TPU). This influence was studied by means of an ANOVA statistical analysis, and a mathematical model was obtained by means of a response surface methodology (RSM). Kerf taper defect was evaluated using a new image processing methodology, where the initial and final damage zone was separated from the kerf taper defect. Finally, a combination of a hydraulic pressure of 3400 bar with a feed rate of 100 mm/min and an abrasive mass flow of 170 g/min produces the minimum kerf taper angle.

Keywords: AWJM (Abrasive waterjet machining); CFRTP (Carbon fiber-reinforced thermoplastics); kerf taper; RSM (response surface methodology); ANOVA (Analysis of variance); C/TPU (carbon/polyurethane)

---

1. Introduction

The use of composite materials in industry has generated a large number of publications and research. Within their wide classification, carbon fiber–reinforced (CFRP) or glass fiber–reinforced (GFRP) polymer matrix composites are the most interesting [1–3]. These materials have an excellent weight-to-mechanical-properties ratio and have been of great importance in recent years, especially in the aerospace and automotive sectors, although others, such as sport, wind energy, and construction, also make use of these composites [4].

However, the type of polymeric matrix used in these materials is thermoset. This generates a series of drawbacks in production and application of these materials, especially in terms of recyclability and processing times [5]. For this reason, there is an alternative to this kind of matrix. In recent years, carbon fibers have been combined with a thermoplastic polymer (CFRTP) to replace thermosets [6,7]. Due to their chemical composition, these polymers have a major advantage over thermosets, as they can be reshaped after curing. In addition, within the wide range of thermoplastic polymers, there are high-performance polymers. These are able to reach large service temperature ranges and achieve excellent impact resistance [8,9]. Also, compared to thermosets, CFRTPs have a high degree of automation in their manufacturing and recyclability. This makes these materials strategic for various industries, such as automotive, aeronautics, civil, or sports [10].

Thermoplastic polyurethane (TPU) stands out. It is an elastic elastomer that can be manufactured by various methods and subsequently machined. According to Biron et al. [11], these polymers have a high performance and a current consumption level. Due to this, they can provide a considerable number of combinations of physical properties that make them extremely flexible materials and adaptable to a multitude of uses [12].

However, due to their complex nature, they are highly difficult to machine in order to obtain their final geometry. The main disadvantage is due to their reduced glass transition temperature. In most conventional technologies, such as milling or drilling, the temperatures generated exceed the glass transition temperature of CFRTP, giving rise to defectology.

Masek et al. [13] carried out a study on milling CFRTPs with different cutting geometries. The thermoplastic matrix used was polyphenylene sulfide (PPS). When undergoing dry machining, i.e., in absence of cooling liquids, the temperatures reached softened the thermoplastic matrix. This boosted the excess removal of the thermoplastic matrix, leaving the reinforcement free in the form of a burr or fraying. This issue is also generated in the machining of composite materials with a thermoset matrix [14]. This, together with the abrasive and adhesive wear generated on the tool, makes conventional processes ineffective when machining CFRTPs.

However, there are non-conventional technologies that make it easier to machine materials that are difficult to machine with conventional technology. Abrasive waterjet machining (AWJM) is now of great interest for the machining of thermoset and thermoplastic composites due to its excellent flexibility and high material removal performance [15,16]. Its main advantage compared to other processes is the reduction of thermal defects during machining. In this sense, the temperatures reached are very low, and CFRTPs and CFRPs can be machined without matrix removal [17–19].

However, AWJM presents defectology due to the inherent nature of the process. The jet, when hitting the material, generates a reduction in kinetic energy. This produces a decrease in its cutting capacity, giving rise to a conical geometry known as a taper [20,21]. This is usually associated with two geometric factors (Figure 1)—Upper width (Wt) and lower width (Wb), giving rise to the angles α (°) and β (°). The sum of these angles gives the taper angle or the kerf taper (KT, Φ).

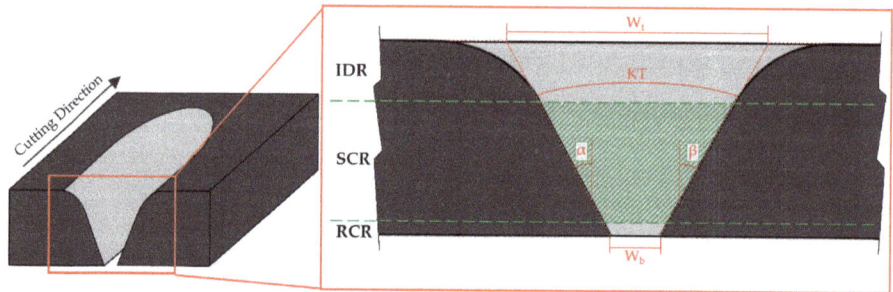

Figure 1. Detail of the cut section where it is appreciated: The initial damage region (IDR), the smooth cutting region (SMC), and the rough cutting region (RCR).

Nevertheless, the methodology for evaluation of conicity used in most articles can generate failures in their measurements. The formation of the zone known as the initial damage region (IDR) is due to the erosive action of abrasive particles, which produce a combination between this defect and the taper angle. In this way, the evaluation of upper width (Wt) can be increased due to this erosive action, giving rise to a dispersion in the evaluated taper, as observed in results shown in references [22–24].

Other researchers [20,24–26], however, have taken into account three regions with different defectology that may appear in the cut section (Figure 1). These are the upper region, called the initial damage region (IDR), where a greater erosive effect is produced due to the impact of the waterjet and abrasive particles on the surface of material; the central region, called the smooth cutting region (SCR), which presents a more homogeneous cut due to the stabilization of the waterjet on the cutting slot; and the lower region, called the rough cutting region (RCR), where the waterjet disperses again, losing a great deal of kinetic energy.

The taper defect is of great importance because it occurs in any material and can result in final geometries that do not have the required dimensions; it especially can result in reduced thicknesses.

Machining a composite material using AWJM creates a series of defects regardless of the matrix the composite is made of. The jet is conditioned by the change of materials that compose the composite. The matrix with which the composite is created will influence, to a greater or lesser extent, the generation of defects after machining. Authors such as Masek et al. [15] and Rao et al. [14] have similar conclusions after machining composites with different matrices. For that reason, due to the fact that CFRTP composition is close to CFRP, similarities can be established in their results.

Dhanawade et al. [27] carried out a study on abrasive waterjet cutting in a thermoset polymeric matrix composite material. The CFRP used was 26 mm thick. In this study, a response surface methodology was carried out in which an ANOVA statistical analysis was performed to determine the influence of the cutting parameters. Dhanawade established that the most influential parameter in the taper angle was hydraulic pressure. An increase in the kinetic energy of the waterjet is produced by increasing this parameter. This increases your cutting capacity and reduces the conicity of the cut. In this way, the kerf taper is directly influenced by the amount of impact of abrasive particles and kinetic energy given by the waterjet.

In addition, an increase in the feed rate of the machine decreases the overlap of abrasive particle impacts, reducing its cutting capacity and thus increasing the taper angle. Kinetic energy is also reduced with an increase in the distance between jet and the material—called the stand-off distance (SOD)—Because it generates a greater dispersion of the jet at exit.

The abrasive mass flow also has a high influence on the conicity generated during cutting. A small increase in this parameter decreases the conicity obtained due to the greater cutting capacity of the jet. However, an excessive increase in the amount of abrasive particles produces a collision between them, rounding their edges and reducing their cutting capacity, which generates a greater angle of conicity.

Similar results were obtained by El-Hofy [17]. In this study, the minimum conicity that can be obtained was indicated by applying high hydraulic pressure combined with a small stand-off distance. The conicity obtained is reduced by increasing the feed rate, contrary to the findings of Dhanawade et al. [27]. This is because, at high pressures, an increase in feed rate generates a smaller upper width (Wt), producing a more constant cut.

The distance between the focusing tube and the surface is a very important parameter for obtaining the proper conicity. This is mainly due to a loss of kinetic energy in the form of dispersion of the jet when leaving the focusing tube. Most studies indicate that a recommended distance is usually 2–3 mm [20,28,29].

Popan et al. [30] studied the influence of the variation of stand-off distances for a thickness of 6 mm. In this study, a reduction in this parameter of up to 0.5 mm reduced the upper cutting width (Wt), thus decreasing the taper. In addition, a reduction in stand-off distance produces a decrease in the radius of zone affected by erosion (IDR) due to the initial impacts of abrasive particles.

Also, the thickness of the material has a fundamental role in the conicity generated by machining. A reduction in material thickness enhances the influence of parameters considered less decisive in large thicknesses. Wong et al. [31] studied waterjet cutting in a thermoset composite material with a 3 mm thickness. In this study, hydraulic pressure and abrasive mass flow (AMF) take second place. In this way, the main parameter that affects the conicity of the cut is the combination of stand-off distance and feed rate. The combination of a high distance with a high feed produces maximum conicity.

In view of the above, more information is required on the influence of cutting parameters on conicity and on the reduction of this defect. In addition, as there is no literature focusing on waterjet machining with abrasive CFRTPs, it is necessary to determine the influence on matrix change [13]. Also, methodology established in most articles, gives rise to considerable errors in the assessment of this defect, by not separating it from the area affected by abrasive particles.

For these reasons, this article proposes the evaluation of taper angle using a new methodology based on image processing. In addition, influence of cutting parameters will be determined by means of an ANOVA statistical analysis, in order to discuss the results obtained in machining of thermoset composite materials.

Finally, by means of a response surface methodology (RSM), a mathematical model will be obtained that predicts the conicity generated in abrasive waterjet cutting of low-thickness thermoplastic matrix composite materials.

2. Materials and Methods

In this article, carbon fiber (Twill 200 g/m^2) was used as reinforcement, and thermoplastic resin (TPU, polyurethane) was used to manufacture the CFRTP composite. This composite was manufactured by a thermoforming process. Table 1 shows its main characteristics, as well as the fiber and matrix, respectively.

Table 1. Fiber, matrix, and carbon fiber–reinforced thermoplastic (CFRTP) characteristics.

Reinforcement	Fiber type	Twill 200 g/m^2
	Fiber thickness	0.25 mm
Matrix	Resin type	TPU (Polyurethane)
	Melting Temperature	145 °C
CFRTP	Ply Orientation	[0°/90°]s
	Number of plies	7
	Composite thickness (t)	2.08 mm
	Fiber/Matrix (%)	74.2/25.8

A three-axis water-jet machine (TCI Cutting, BP-C 3020, Valencia, Spain) was used for the experimentation. The AWJM machine was equipped with an ultra-high capacity pump (KMT, Streamline PRO-2 60, Bd Nauheim, Germany). The water orifice of the machine had a diameter of 0.30 mm. The diameter and length of the focusing tube were 0.8 mm and 94.7 mm, respectively. All trials were carried out by a 120 mesh Indian garnet abrasive material.

In order to carry out the experimental design, a response surface methodology (RSM) was set up. This kind of methodology has already been employed by some authors in several experimental studies of the same order [27,31]. A face-centered composite design (FCD) with a total of 20 trials (8 factorial points–2^3, 6 axial points–2×3, and 6 center points) was established and carried out using Minitab® 18 software (18.1, Minitab, LLC, State College, PA, USA).

A complementary experimental design was carried out. Because of this design, experimentally obtained data can be matched with data predicted by Minitab analytical software (18.1, Minitab, LLC,

State College, PA, USA). This comparison makes it possible to obtain the error generated between these values and confirms the accuracy of the response surface model.

Three main parameters, which include hydraulic pressure, feed rate, and abrasive mass flow, were employed to determine their influence of the kerf taper generated. These parameters were designated based on the limitations of the CNC machine used, as well as the levels most employed in reviewed literature [17,22,24,31,32]. Also, they were converted into three different levels (−1, 0, 1) that represent minimum, central, and maximum values, respectively (Table 2). In order to establish a stand-off distance in accordance with the results of other authors [20,28,29], 2.5 mm was the stand-off distance used in all the tests.

Table 2. Cutting parameters set.

Parameter	Symbol	Units	Level −1	Level 0	Level 1
Hydraulic Pressure	P	bar	1200	2500	3400
Feed Rate	FR	mm/min	100	300	500
Abrasive Mass Flow	AMF	g/min	170	225	340
Stand-off distance	SOD	mm	Fixed at 2.5		

Figure 2 shows the distribution of 20 test. They are machined in a 170 × 25 mm specimen with an 8 mm gap to optimize material consumption. Before machining, a horizontal cut was made at coordinate 0.0 in order to ensure the perpendicularity of each cut with the final machined part. A cutting length of 15 mm was fixed for each trial. Furthermore, each single cut starts 10 mm before the material side to achieve a constant flow of water and abrasive. Machining was carried out on three specimens (KT1, KT2, KT3) of the same CFRTP in order to obtain reproducibility of the results achieved.

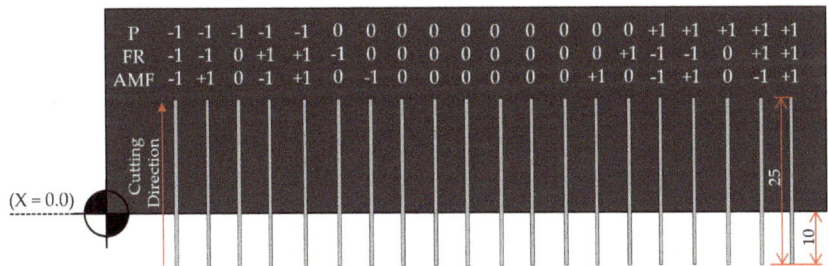

Figure 2. Experimental design scheme of CFRTP machining carried out.

On the basis of this methodology, an ANOVA analysis was developed in order to obtain the statistical influence of input parameters on output variables. Pressure, feed rate, and abrasive mass flow were changed in accordance with the fact that the experiment was conducted according to Box-Behnken design (three-level). In addition, RSM allows us to generate different contour diagrams or response surfaces from a second order polynomial Equation (1). There are several articles that have implemented this type of equation in order to develop the results obtained in the experiments carried out [31,33,34].

$$Y = C_0 + \sum_{i=1}^{k} C_i x_i + \sum_{i=1}^{k} C_{ii} x_i^2 + \sum_{i<j}^{k} C_{ij} x_i x_j + \varepsilon \tag{1}$$

Y corresponds to the expected response, in this case the kerf taper generated (KT), x_i are the parameters used in the study (P, FR, AMF), C_0, C_i, C_{ii}, C_{ij} are the regression coefficients, and ε is the random error of the model.

A stereoscopical microscope (Nikon, SMZ 800, Tokyo, Japan) was employed in order to obtain macrographies of each slot (Figure 3a). Image processing software made it possible to define the contour of the smooth cutting region (SCR) generated (Figure 3b). After machining, a first analysis was carried out to establish the range that delimits the three separate zones in section of cut. Later, the region obtained was split into 100 points. A trend line that adjust to those points was created. The intersection of this trend line with the value 0 mm (minimum thickness) and 2.08 mm (maximum thickness) was forced (Figure 3c).

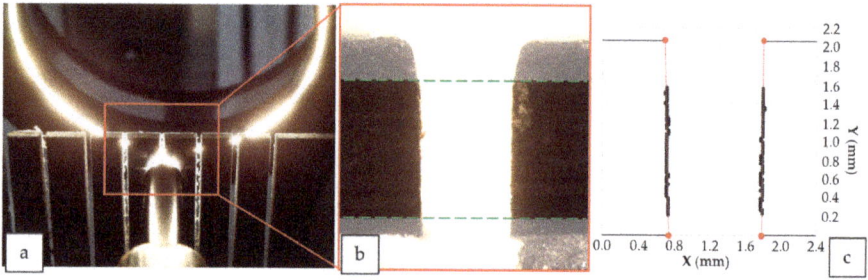

Figure 3. Macrographies obtaining procedure: (**a**) positioning in stereoscopical microscope; (**b**) image of a slot with the smooth cutting region (SCR) pointed out; (**c**) graph of the SCR points with intersection at 0.0 mm and 2.08 mm.

These intersection points are used to obtain the top kerf width (Wt, mm) and the bottom kerf width (Wb, mm). Finally, the kerf taper defect (KT) is defined as shown in Equation (2), where t is thickness of CFRTP in millimeters.

$$KT\ (°) = 2*\mathrm{atan}\left(\frac{\frac{W_t - W_b}{2}}{t}\right) \qquad (2)$$

3. Results

The kerf taper values obtained in the three specimens (KT1, KT2, KT3), as well as the average kerf taper and its standard deviation (average KT), are shown in Table 3.

All the results of the mean values given in Table 3 are positive. This means that, according to Equation (2), the upper width of the cut is always greater than the lower width. It follows that the geometric shape obtained in all machined slots has a "V" shape. This geometric effect is produced by the material thickness together with the transverse feed rate and energy amount (pressure and focusing tube diameter). These variables affect the shape distortion of the slot [35]. It should be pointed out that the kerf taper defect generated in waterjet machining is independent of the thickness of material. This defect will be greater or smaller according to this thickness, but it will always happen, even in materials of small thicknesses, such as the CFRTP used in this article.

Average kerf taper values between 2.15° and 7.79° were obtained. It is necessary to remember that there is currently no article that analyzes the taper defect in abrasive waterjet machining of thermoplastic composite materials. Some of reviewed literature [22,24,27] shows values close to the range obtained in this experiment, although some specific kerf taper values are different. It must be taken into account that the composites used in the other articles are made of a thermoset matrix. Most of them contain a higher glass transition temperature than the thermoplastic resin used in this experiment. In addition, fixed variables such as the focusing tube diameter and abrasive grain size change during the mixing process and could generate different results.

On the other hand, not all of the existing literature takes into account the three areas generated in cutting slot—The initial damage region (IDR), the smooth cutting region (SCR), and the rough cutting region (RCR). The emphasis in this article is on the independent treatment of such areas. Several authors [17,22,24,27] calculate the kerf taper defect without taking this indication into account. This could cause the kerf taper values to be altered by the rounding radius generated on the top surface of machined material. This radius occurs when the waterjet hits the surface to be machined. Abrasive particles, in a first instant of contact, meet a wall that they must pass through. Not all of them are able to do it, so they disperse along the upper surface of the material, producing a rounding at the input of cut.

Table 3. Kerf taper values obtained in all tests.

Test	Pressure [bar]	Feed Rate [mm/min]	Abrasive [g/min]	KT 1 [°]	KT 2 [°]	KT 3 [°]	Average KT [° ± (°)]
1	1200	100	170	4.71	3.23	3.92	3.95 ± 0.60
2	1200	100	340	4.09	5.25	4.67	4.67 ± 0.47
3	1200	300	225	6.14	4.92	6.25	5.77 ± 0.60
4	1200	500	170	6.51	7.18	9.58	7.76 ± 1.32
5	1200	500	340	7.53	7.80	7.73	7.69 ± 0.11
6	2500	100	225	1.39	3.34	2.91	2.55 ± 0.84
7	2500	300	170	1.96	3.62	3.41	3.00 ± 0.74
8	2500	300	225	2.43	3.43	3.13	3.00 ± 0.42
9	2500	300	225	5.12	2.16	6.39	4.55 ± 1.77
10	2500	300	225	4.22	2.03	5.98	4.07 ± 1.62
11	2500	300	225	4.77	2.26	3.25	3.43 ± 1.03
12	2500	300	225	5.12	2.85	3.16	3.71 ± 1.00
13	2500	300	225	4.54	4.40	3.26	4.07 ± 0.57
14	2500	300	340	3.39	5.59	4.70	4.56 ± 0.90
15	2500	500	225	5.88	6.10	4.94	5.64 ± 0.50
16	3400	100	170	1.61	2.11	3.08	2.27 ± 0.61
17	3400	100	340	0.75	2.91	4.84	2.83 ± 1.67
18	3400	300	225	2.16	1.48	2.81	2.15 ± 0.54
19	3400	500	170	1.52	2.69	3.31	2.51 ± 0.74
20	3400	500	340	1.68	2.16	3.68	2.51 ± 0.85

Furthermore, taking into account the RCR zone means that there may be an error when obtaining the values. In this area, the waterjet comes out of machined material, resulting in a new opening. In composite materials, carbon fibers and the matrix can become detached by creating loose yarns or cavities. Therefore, 10% of the cutting slot as was chosen as the RCR zone.

Two macrographies of the cutting section of two machined slots with different parameter combinations are shown in Figure 4. A rounded radius produced in the upper face of the cutting slot is shown in both figures. This defect is usually caused by the collision of the waterjet and abrasive particles with the top face of material. The radius obtained for Figure 4a is 0.063 mm, while the radius generated in Figure 4b is 0.037 mm.

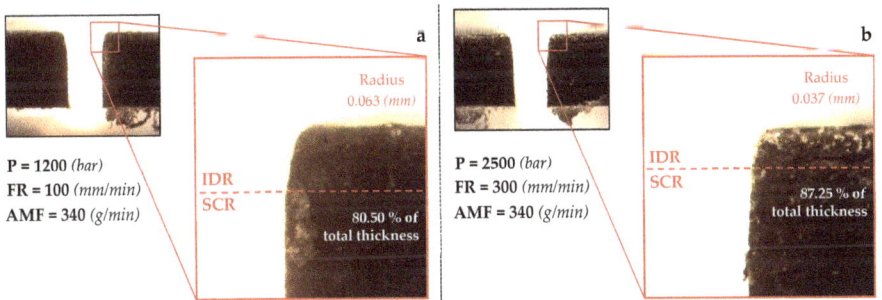

Figure 4. Cutting path macrograph for machining under conditions: (a). P 1200 bar, FR 100 mm/min, AMF 340 g/min; (b). P 2500 bar, FR 300 mm/min, AMF 340 g/min.

This confirms the randomness that can be obtained in two slots machined on the same material. In addition, the point that separates IDR zone from SCR is found at different heights. Figure 4a shows the point at a height equivalent to 80.50% of total thickness (2.08 mm), and Figure 4b shows it at 87.25% of that thickness.

To sum up, taking into account the thickness of material used in this article, cutting section was split according to 0–10% of the thickness for the rough cutting region, 10–75% for the smooth cutting region and, 75–100% for the initial damage region. This article considers the kerf taper defect in the SCR (Figure 5).

Figure 5. Detail of a real cut section where it can be seen the areas: Rough cutting region (RCR) 0–10%, smooth cutting region (SCR) 10–75%, initial damage region (IDR) 75–100%.

3.1. Statistical Analysis

An ANOVA analysis of the obtained kerf taper values is shown in the Table 4. As discussed in the methodology section, a second order quadratic model was employed. This model allows us to relate the input parameters used in experiment (pressure, feed, rate and abrasive mass flow) with the results obtained after machining. The ratio between both variables makes it possible to create greater or lesser accuracy between them. The p-value of this model is less than 0.05, which indicates that it is statistically significant. In addition, a high F-value will make variable more relevant in the analysis.

Table 4. ANOVA analysis of the kerf taper values obtained.

Source	DF	Adj SS	Adj MS	F-Value	p-Value
Model	9	48.2278	5.3586	16.95	0.00
Pressure (bar)	1	30.6617	30.6617	97.00	0.00
Feed Rate (mm/min)	1	9.8658	9.8658	31.21	0.00
Abrasive (g/min)	1	0.7668	0.7668	2.43	0.15
Error	10	3.1610	0.3161		
Lack-of-Fit	5	1.6498	0.3300	1.09	0.46
Pure Error	5	1.5112	0.3022		
Total	19	51.3888			

As can be seen in Table 4, pressure and feed rate are the most significant parameters in taper defect. This is deduced because p-value is 0, i.e., the significance is maximum. It is observed that the AMF parameter has a p-value of 0.15, which implies that this variable has less influence on the generation of the taper angle. In addition, it can be seen that the F-value for pressure variable triples to that obtained for feed rate. This allows us to deduce that, although both variables are significant, hydraulic pressure has greater weight in the generation of the kerf taper.

These statistical results are consistent with the articles related to the topic studied. Therefore, it can be concluded that, for generation of a taper defect, the order of parameters according to their influence is hydraulic pressure, feed rate, and abrasive mass flow.

Equation (3) shows mathematical model obtained for analyzed response surface methodology. This equation presents an R^2 of 93.85%, which implies a value very close to 100%. In later sections, a verification of Equation (3) will be performed from input variables shown in Table 3.

$$KT = 1.00 + (0.7P + 124.8FR + 153AMF + 0.08FR^2 - 0.17AMF^2 - 0.04P*FR - 0.13FR*AMF)*10^{-4} \quad (3)$$

3.2. Effect of Hydraulic Pressure on Kerf Taper

The effect of hydraulic pressure on variation of the kerf taper is shown in Figure 6. To discuss these data, feed rate and abrasive mass flow variables have been kept fixed and at their intermediate level (FR = 300 mm/min; AMF = 225 g/min). Three machined specimens have been classified with different colors. Thus, red color corresponds to test piece 1, yellow to test piece 2. and green to test piece 3. In addition, each figure shows the mean value of taper defect obtained for each test. This value was represented with a discontinuous line followed by mean value and resulting deviation. Pressure is one of the most influential variables in machining, therefore Figure 6 contains images that facilitate visual compression.

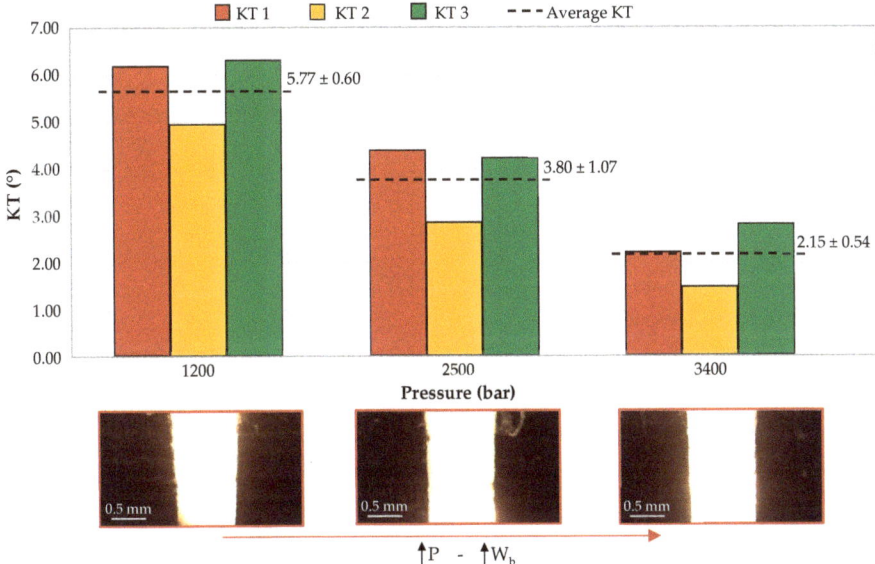

Figure 6. Variation of the kerf taper in three specimens as a function of pressure (FR = 300 mm/min; AMF = 225 g/min).

At first sight, the taper decreases as pressure increases. By taking a look at the mean values obtained and pressures involved, it can be seen how the resulting data adjust to a large extent to a linear regression. Higher pressure means an increase in kinetic energy generated by the waterjet. This increase in kinetic energy leads to a higher material removal. Consequently, the walls of the slot are subjected to a greater force in removing the material that causes a more vertical final state.

By looking at the graph with values of each specimen independently, it can be seen how specimen 2 in all cases has a taper value smaller than 1 and 3. This may have been affected by the nature of the composite. The material is composed by long carbon fibers and thermoplastic matrix sheets. The arrangement of these elements along the composite is crucial to carry out the machining. In this way, a fiber yarn displaced at the time of manufacturing or an irregular consolidation of matrix along the surface could alter its homogeneity (Figure 7). Nevertheless, the decreasing tendency of the taper defect as hydraulic pressure increases is reflected in three specimens studied. In this case, a value of

3400 bar generates the lowest taper, being 2.15°. It should be noted that the thickness of composite employed is considered thin, which could make it easier for kinetic energy created by pressure to generate a homogeneous wall in slot.

Figure 7. Macrographs of a composite cross-section with (**a**). thermoplastic matrix; (**b**). thermoset matrix.

A study carried out by Dhanawade et al. [27] shows a similar trend to that generated in this research. The material used in its study is a composite formed by carbon fibers and thermostable epoxy resin. They agree that an increase in pressure generates a decrease in taper defect. In addition, Dhanawade et al. use a hydraulic pressure that reaches 4000 bar, obtaining a taper angle that oscillates from 2.2° to 3.8°. It can be deduced that these results agree with those obtained in this article. Therefore, if the pressure parameter is analyzed in isolation, it seems that the kind of resin used in composite material does not greatly influence the kerf taper generated.

Ruiz-Garcia et al. [20] carried out a study on the analysis of the kerf taper defect in the abrasive waterjet machining of CFRP/UNS A97075 stack. In that paper, the kerf taper results obtained were found in a range of 1°. Ruiz-Garcia et al. used lower feed rates than those used in this article. This resulted in a greater homogeneity of taper defects as well as a smaller variation of them. In addition, the matrix used by Ruiz-Garcia et al. was epoxy resin, which has a higher vitreous transition temperature than the thermoplastic resin used in this experiment (TPU—polyurethane, 145 °C). The use of a thermoset resin could allow the composite to achieve greater mechanical properties than the CFRTP employed in this experiment.

3.3. Effect of Feed Rate on Kerf Taper

Figure 8 shows the effect of feed rate on the kerf taper defect. The distribution of elements in graph is similar to the one shown in Figure 6. As for previous section, this figure contains 3 images that make it easier to understand. It can be seen how, unlike for pressure, average values of kerf taper rise as feed rate increases.

A slower cut is made when a slot is made with a minimum speed of 100 mm/min, causing a more homogeneous fracture along the surface of material. In this case, waterjet is able to pass through material and remove each layer of carbon fiber and matrix in an orderly way. This effect, together with abrasive particles, reduces the influence of cohesion nature of a composite material on the generation of kerf taper. Therefore, it follows that applying a low feed rate, kerf taper achieved will be smaller, or in other words, upper and lower width of slot are values closer to each other.

On the other hand, when a high feed rate is applied, a loss of kinetic energy is generated in cutting channel. These could produce a decrease in material removal and a decrease in abrasive particles affecting the surface of material. On this occasion, the waterjet is more susceptible to the nature of the material. Different layers of carbon fiber and matrix take advantage of the reduced energy of the jet to make it difficult to pass through the composite. This can lead to the slot walls not being perpendicular, resulting in the upper width of slot being greater than the lower.

Thus, the lowest taper value will be produced for a feed rate of 100 mm/min, being 2.55°, and the highest value will be produced with a feed rate 500 mm/min, resulting in 5.64°. The lower thickness of the material employed, combined with a low feed rate, are parameters that help to create a homogeneous cut by AWJM.

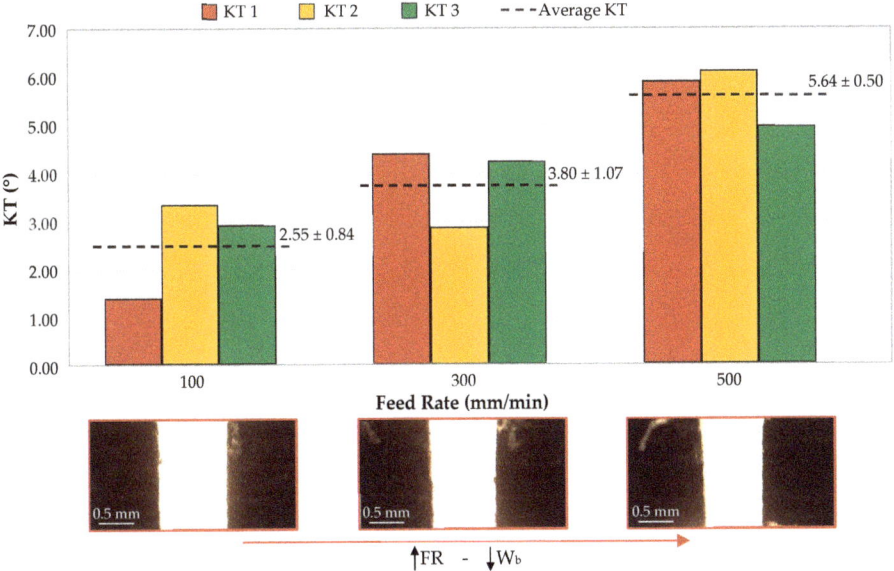

Figure 8. Variation of kerf taper in three specimens as a function of feed rate (P = 2500 bar; AMF = 225 g/min).

Everything discussed above is consistent with Wong et al. [31]. In this paper, the authors analyze the taper defects in a carbon fiber thermostable matrix (epoxy). The FR used oscillated in 1000–2500 mm/min, which is a little bit higher than that carried out in this article. However, the authors conclude that a greater feed rate implies a lower amount of abrasive particles affecting the material. The decrease in the amount of these particles caused a dirtier and more random cut.

As the cutting speed increases, the upper and lower widths of the slot decrease. However, the width of the bottom surface has a greater decreasing tendency than the top surface. This is consistent with articles that study the taper defect in other kinds of materials [36,37].

3.4. Effect of Abrasive Flow Rate on Kerf Taper

The influence of the abrasive mass flow (AMF) on generation of taper defect is shown in Figure 9. In Section 3.1, it was concluded that this parameter is the least influential of the three used in this article. However, this fact does not imply that the amount of abrasive employed does not affect the results. In fact, by looking at the trend of the mean values plotted in Figure 9, it can be seen how it ascends as the amount of abrasive increases. The upward trend caused by AMF is less than that caused by pressure and feed rate. Therefore, this parameter is the least influential of those used.

It should be noted that an excessive increase in AMF can lead to a loss of kinetic energy. In this case, abrasive particles are more likely to collide with each other, causing a more disturbed cut, which translates into greater erosion at the input of slot, a greater difference between the upper and lower widths of the slot, and greater fraying at the outlet surface of the material [38].

In this case, a smallest kerf taper will be produced with a small amount of abrasive, 170 g/min, resulting in an angle of 3.00°. For the highest amount of abrasive, an angle of 4.56° is obtained.

Ruiz-Garcia et al. [20] achieved an upward trend similar to that obtained in this experiment. The use of a higher abrasive flow means a greater difference between the upper width of the slot and the lower width. Due to that, it can be noted that the abrasive flow rate used in AWJM is not linked to the kind of resin employed in the manufacture of composite.

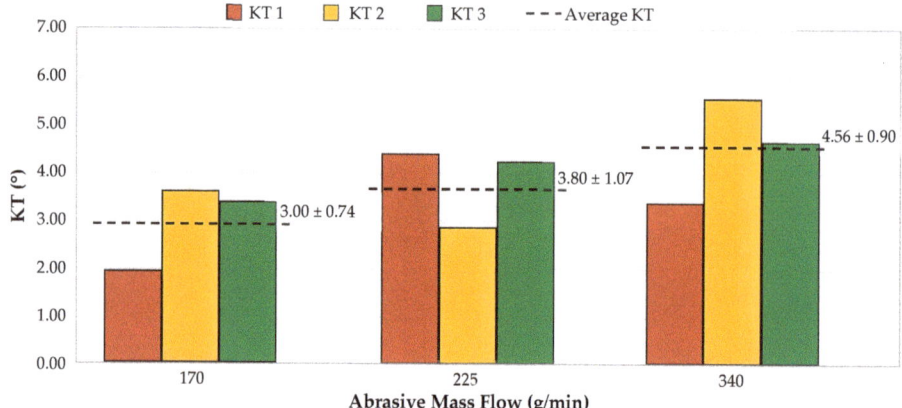

Figure 9. Variation of kerf taper in three specimens as a function of abrasive mass flow rate (P = 2500 bar; FR = 300 mm/min).

3.5. Response Surface

A response surface allows two input parameters to interact with an output variable, keeping all other parameters constant. The ANOVA analysis carried out in Section 3.1 gave hydraulic pressure and feed rate the greatest significance on the machining process. Therefore, FR and P have been represented together with the kerf taper in Figure 10, keeping AMF = 225 g/min fixed.

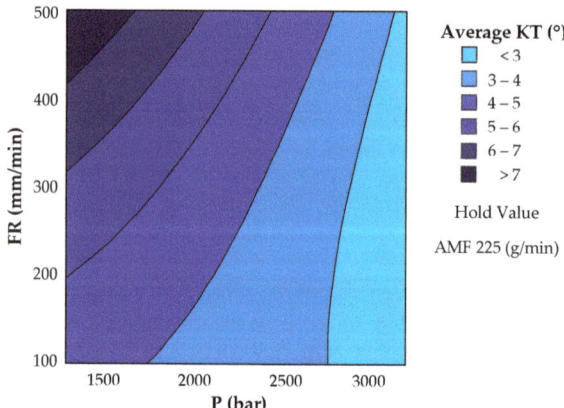

Figure 10. Contour plot of kerf taper defect as a function of most influential variables: Pressure and feed rate.

As can be seen in this figure, both parameters seem very significant. A value close to 3400 bar combined with various feed rate combinations offer the lowest kerf taper values (Figure 11). As a function of P-value, both P and FR were significant, while according to F-value, pressure tripled the influence of feed rate. This shows that a single pressure results in a range of kerf taper values (<3°) for the whole range of feed rates used in this experiment.

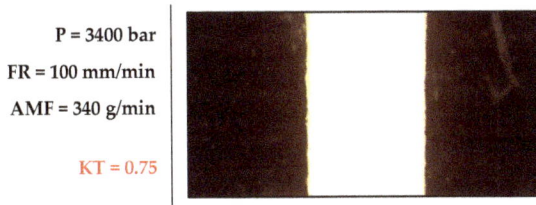

P = 3400 bar
FR = 100 mm/min
AMF = 340 g/min

KT = 0.75

Figure 11. Combination that minimizes kerf taper defect (P = 3400 bar; FR = 100 mm/min; AMF = 340 g/min).

On the other hand, a combination of 1200 bar of pressure and 500 mm/min of feed rate seems to be the most unfavorable in the study of kerf taper, and it seems to be the element that greatly increases the upward trend. As speed increases, cutting capacity decreases due to the loss of kinetic energy in the waterjet. To this effect, it would be necessary to add the reduction of the number of abrasive particles that affect the material.

In addition, the polyurethane matrix applied in this article has a melting temperature of 145 °C. Abrasive waterjet cutting is a technology that generates a lower temperature than other conventional machining technologies, such as milling or drilling [14]. This process makes it easier to machine temperature-sensitive materials. High temperatures could lead to thermoplastic matrix softening, generating defects in the final quality of the slot. The kerf taper values achieved are similar to those studied by Wong et al. [31]. This could mean that the effect of temperature on the thermoplastic matrix is not enough to alter the kerf taper.

According to our mathematical model, a combination of P = 3400 bar, FR = 172.73 mm/min, and AMF = 170 g/min should result in the lowest kerf taper value of 1.79°. The degree of desirability of the result achieved is also obtained. Individual and compound desirability evaluates how well a combination of parameters satisfies the objectives defined for output variables.

Individual desirability (d) evaluates how adjustments optimize a single response. The compound (D), on the other hand, looks at how adjustments optimize a set of responses in general. This variable has a range of 0 to 1, 1 being the ideal case, and 0 meaning that some of answers obtained are outside acceptable limits. In this case, a maximum desirability was generated, with a value of 1, which implies that the combination of cutting parameters selected offers a desirable result.

3.6. Mathematical Model Validation

Table 5 shows the taper values obtained from the complementary experimental design together with the same values predicted by Equation (3).

Table 5. Complementary DOE for validation of the mathematical model.

Test	Pressure [bar]	Feed Rate [mm/min]	Abrasive [g/min]	KT [°]	Predicted KT [°]	Error [%]
1	1200	100	225	3.73	4.13	10.81
2	1200	300	170	6.52	5.46	16.21
3	1200	300	340	5.99	5.98	0.12
4	1200	500	225	6.77	7.95	17.47
5	2500	100	170	2.23	2.75	23.19
6	2500	100	340	4.01	3.76	6.39
7	2500	500	170	4.68	4.89	4.37
8	2500	500	340	4.17	5.00	19.85
9	3400	100	225	1.60	2.28	42.33
10	3400	300	170	1.40	1.91	36.27
11	3400	300	340	1.67	2.50	49.84
12	3400	500	225	2.32	2.76	19.07

In order to evaluate the mathematical model carried out in this experiment, a comparison between combinations of parameters not used in original DOE and predicted values was carried out (Figure 12).

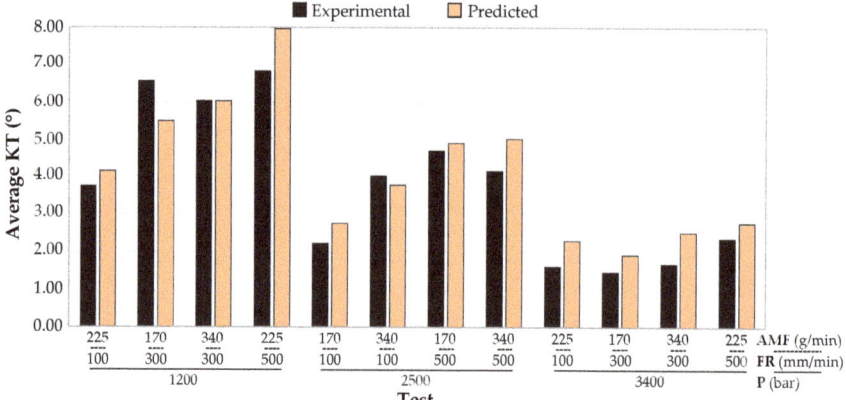

Figure 12. Kerf taper values (experimental and predicted).

It should be noted that, in most cases, experimental values are below those predicted at a similar distance. When the errors tabulated in Table 5 are observed, it can be seen that, for pressures of 1200 and 2500 bar, errors oscillate in a range of 0–20%, while highest pressure of 3400 bar generates high errors with values of up to 49.84%.

The average error obtained is 20.49%. Experimental values follow the trend of the predicted values but maintain an almost constant difference between them. This difference may be due to the anisotropy and nature of material, as seen in Figure 7. This model could be used to predict the trend that will follow the kerf taper as a function of parameters used, although the error obtained must be taken into account.

4. Conclusions

An experimental study on the influence of cutting parameters on the generation of a geometric defect, called a kerf taper, focused on the machining of carbon fiber composite materials with thermoplastic matrix, was developed. A face-centered composite experimental design (FCD) based on a response surface methodology (RSM) was development. This type of experimental design has given rise to a second order polynomial equation that relates input parameters to output variable (kerf taper).

The literature review allowed for the selection of pressure, feed rate, and abrasive mass flow as the most influential input parameters in the machining performed while keeping other parameters, such as stand-off distance or size of abrasive particles, fixed.

A second experimental design was used to verify the second order polynomial equation generated by the taper obtained, which contrasts combinations of parameters not used in the original DOE with values predicted by model. An error of 20.49% was obtained, which can be considered small if the anisotropy and nature of a composite material are taken into account.

Three zones along thickness of material have been identified—The rough cutting region at 0–10%, the smooth cutting region at 10–75%, and the initial damage region at 75–100%. In addition, only the Smooth Cutting Region was taken into account for the development of this experiment.

The kerf taper defect was studied in three specimens of the same thermoplastic composite material in order to obtain an average value and its respective deviation. Marginal graphs show how hydraulic pressure causes a decrease in taper generated, and feed rate and abrasive mas flow produce an increase in the same.

ANOVA analysis has indicated that hydraulic pressure and feed rate are the most influential parameters in abrasive waterjet machining. The slot walls become more vertical at high pressures and low feed rates. This is due to a higher concentration of energy impacting the composite to be machined, which translates into higher material removal. Also, the mathematical model obtained for analyzed response surface methodology has presented an R^2 of 93.85%.

The effect of temperature does not seem to influence the quality of the results obtained by AWJM. After a concise literature review, it seems that the results obtained in taper defect agree with those obtained by other scientific authors.

Finally, a combination of cutting parameters that minimizes kerf taper defect was found, resulting in a pressure of 3400 bar, a feed rate of 100 mm/min, and an abrasive mass flow of 340 g/min, producing an upper-lower width ratio close to 1, i.e., 0.75°. This small kerf taper defect means that for specific applications of AWJM could be considered as a high precision process.

Author Contributions: A.S. and F.B. developed machining tests. M.B. and J.S. developed data treatment. F.B., A.S., M.B., B.S., and J.S. analyzed the influence of the parameters involved. F.B. and A.S. collaborated in preparing figures and tables and F.B., A.S., M.B., B.S., and J.S. wrote the paper.

Funding: This work was developed with the support of a pre-doctoral industrial fellowship financed by NANOTURES SL, the Mechanical Engineering and Industrial Design Department and Vice-Rectorate of Transference and Technological Innovation of the University of Cadiz.

Conflicts of Interest: The authors declare no conflict of interest.

References

1. Sathish, P.; Kesavan, R.; Mahaviradhan, N. Hemp Fiber Reinforced Composites: A Review. *Int. J. Res. Appl. Sci. Eng. Technol.* **2017**, *5*, 594–595. [CrossRef]
2. Altin Karatas, M.; Gökkaya, H. A review on machinability of carbon fiber reinforced polymer (CFRP) and glass fiber reinforced polymer (GFRP) composite materials. *Def. Technol.* **2018**, *14*, 318–326. [CrossRef]
3. Pramanik, A.; Basak, A.K.; Dong, Y.; Sarker, P.K.; Uddin, M.S.; Littlefair, G.; Dixit, A.R.; Chattopadhyaya, S. Joining of carbon fibre reinforced polymer (CFRP) composites and aluminium alloys-A review. *Compos. Part A Appl. Sci. Manuf.* **2017**, *101*, 1–29. [CrossRef]
4. Witten, E.; Mathes, V.; Sauer, M.; Kühnel, M. *Composites Market Report 2018—Market Developments, Trends, Outlooks and Challenges*; AVK Federation of Reinforced Plastics: Frankfurt, Germany, 2018.
5. Dauguet, M.; Mantaux, O.; Perry, N.; Zhao, Y.F. Recycling of CFRP for high value applications: Effect of sizing removal and environmental analysis of the SuperCritical Fluid Solvolysis. *Procedia CIRP.* **2015**, *29*, 734–739. [CrossRef]
6. Masek, P.; Zeman, P.; Kolar, P. Development of a cutting tool for composites with thermoplastic matrix. *Mod. Mach. Sci. J.* **2013**, *3*, 423–427. [CrossRef]
7. Biron, M. Outline of the actual situation of plastics compared to conventional materials. In *Industrial Applications of Renewable Plastics*; William Andrew: Norwich, NY, USA, 2017; ISBN 9780323480659.
8. Goto, K.; Imai, K.; Arai, M.; Ishikawa, T. Shear and tensile joint strengths of carbon fiber-reinforced thermoplastics using ultrasonic welding. *Compos. Part A* **2019**, *116*, 126–137. [CrossRef]
9. Christmann, M.; Medina, L.; Mitschang, P. Effect of inhomogeneous temperature distribution on the impregnation process of the continuous compression molding technology. *J. Thermoplast. Compos. Mater.* **2016**, *30*, 1–18. [CrossRef]
10. Ishikawa, T.; Amaoka, K.; Masubuchi, Y.; Yamamoto, T.; Yamanaka, A.; Arai, M.; Takahashi, J. Overview of automotive structural composites technology developments in Japan. *Compos. Sci. Technol.* **2018**, *155*, 221–246. [CrossRef]
11. Biron, M. *Thermoplastics and thermoplastic composites*, 3rd ed.; William Andrew: Norwich, NY, USA, 2018; ISBN 9781856174787.
12. Olabisi, O.; Adewale, K. *Handbook of thermoplastics*, 2nd ed.; Taylor & Francis Group: New York, NY, USA, 2016; ISBN 9781466577237.
13. Masek, P.; Zeman, P.; Kolar, P.; Holesovsky, F. Edge trimming of C/PPS plates. *Int. J. Adv. Manuf. Technol.* **2019**, *101*, 157–170. [CrossRef]

14. Fu, R.; Jia, Z.; Wang, F.; Jin, Y.; Sun, D.; Yang, L.; Cheng, D. Drill-exit temperature characteristics in drilling of UD and MD CFRP composites based on infrared thermography. *Int. J. Mach. Tools Manuf.* **2018**, *135*, 24–37. [CrossRef]
15. Masek, P.; Zeman, P.; Kolar, P. Technology optimization of PPS/C composite milling using Taguchi method. In Proceedings of the 9th International Conference on Machine Tools Automation, Technology and Robotics, Prague, Czech Republic, 12–14 September 2012.
16. Kakinuma, Y.; Ishida, T.; Koike, R.; Klemme, H.; Denkena, B.; Aoyama, T. Ultrafast feed drilling of carbon fiber-reinforced thermoplastics. *Procedia CIRP.* **2015**, *35*, 91–95. [CrossRef]
17. El-Hofy, M.; Helmy, M.O.; Escobar-Palafox, G.; Kerrigan, K.; Scaife, R.; El-Hofy, H. Abrasive water jet machining of multidirectional CFRP laminates. *Procedia CIRP.* **2018**, *68*, 535–540. [CrossRef]
18. Ramulu, M.; Isvilanonda, V.; Pahuja, R.; Hashish, M. Experimental investigation of abrasive waterjet machining of titanium graphite laminates. *Int. J. Autom. Technol.* **2016**, *10*, 392–400. [CrossRef]
19. Melentiev, R.; Fang, F. Recent advances and challenges of abrasive jet machining. *CIRP J. Manuf. Sci. Technol.* **2018**, *470*, 1–20. [CrossRef]
20. Ruiz-Garcia, R.; Ares, P.F.M.; Vazquez-Martinez, J.M.; Gómez, J.S. Influence of abrasive waterjet parameters on the cutting and drilling of CFRP/UNS A97075 and UNS A97075/CFRP stacks. *Materials* **2018**, *12*, 107. [CrossRef] [PubMed]
21. Vigneshwaran, S.; Uthayakumar, M.; Arumugaprabu, V. Abrasive water jet machining of fiber-reinforced composite materials. *J. Reinf. Plast. Compos.* **2017**, *37*, 230–237. [CrossRef]
22. Pahuja, R.; Ramulu, M.; Hashish, M. Surface quality and kerf width prediction in abrasive water jet machining of metal-composite stacks. *Compos. Part B.* **2019**, *175*, 107134. [CrossRef]
23. Pahuja, R.; Ramulu, M. Abrasive water jet machining of Titanium (Ti6Al4V)–CFRP stacks – A semi-analytical modeling approach in the prediction of kerf geometry. *J. Manuf. Process.* **2019**, *39*, 327–337. [CrossRef]
24. Li, M.; Huang, M.; Chen, Y.; Gong, P.; Yang, X. Effects of processing parameters on kerf characteristics and surface integrity following abrasive waterjet slotting of Ti6Al4V/CFRP stacks. *J. Manuf. Process.* **2019**, *42*, 82–95. [CrossRef]
25. Ramulu, M.; Pahuja, R.; Hashish, M.; Isvilonanda, V. Abrasive waterjet machining effects on kerf quality in thin fiber metal laminate. In Proceedings of the WJTA-IMCA Conference and Expo, New Orleans, LA, USA, 2–4 November 2015.
26. Pahuja, R.; Ramulu, M.; Hashish, M. Abrasive waterjet profile cutting of thick Titanium/Graphite fiber metal laminate. In Proceedings of the ASME 2016 International Mechanical Engineering Congress and Exposition, Phoenix, AZ, USA, 11–17 November 2016.
27. Dhanawade, A.; Kumar, S. Experimental study of delamination and kerf geometry of carbon epoxy composite machined by abrasive water jet. *J. Compos. Mater.* **2017**, *51*, 3373–3390. [CrossRef]
28. Chen, M.; Jiang, C.; Xu, Z.; Zhan, K.; Ji, V. Experimental study on macro- and microstress state, microstructural evolution of austenitic and ferritic steel processed by shot peening. *Surf. Coat. Technol.* **2019**, *359*, 511–519. [CrossRef]
29. Mayuet, P.F.; Girot, F.; Lamíkiz, A.; Fernández-vidal, S.R.; Salguero, J.; Marcos, M. SOM/SEM based characterization of internal delaminations of CFRP samples machined by AWJM. *Procedia Eng.* **2015**, *132*, 693–700. [CrossRef]
30. Popan, I.A.; Contiu, G.; Campbell, I. Investigation on standoff distance influence on kerf characteristics in abrasive water jet cutting of composite materials. *MATEC Web Conf.* **2017**, *137*. [CrossRef]
31. Wong, M.M.I.; Azmi, A.; Lee, C.; Mansor, A. Kerf taper and delamination damage minimization of FRP hybrid composites under abrasive water-jet machining. *Int. J. Adv. Manuf. Technol.* **2018**, *94*, 1727–1744.
32. Gupta, V.; Pandey, P.M.; Garg, M.P.; Khanna, R.; Batra, N.K. Minimization of kerf taper angle and kerf width using Taguchi's method in abrasive water jet machining of marble. *Procedia Mater. Sci.* **2014**, *6*, 140–149. [CrossRef]
33. Dumbhare, P.A.; Dubey, S.; Deshpande, Y.V.; Andhare, A.B.; Barve, P.S. Modelling and multi-objective optimization of surface roughness and kerf taper angle in abrasive water jet machining of steel. *J. Brazilian Soc. Mech. Sci. Eng.* **2018**, *40*, 1–13. [CrossRef]
34. Hlaváč, L.M.; Krajcarz, D.; Hlaváčová, I.M.; Spadło, S. Precision comparison of analytical and statistical-regression models for AWJ cutting. *Precis. Eng.* **2017**, *50*, 148–159. [CrossRef]

35. Hlaváč, L.M.; Hlaváčová, I.M.; Plančár, S.; Krenický, T.; Geryk, V. Deformation of products cut on AWJ x-y tables and its suppression. *IOP Conf. Ser. Mater. Sci. Eng.* **2018**, *307*, 1–10. [CrossRef]
36. Hlaváč, L.M.; Hlaváčová, I.M.; Geryk, V.; Plančár, Š. Investigation of the taper of kerfs cut in steels by AWJ. *Int. J. Adv. Manuf. Technol.* **2015**, *77*, 1811–1818. [CrossRef]
37. Hlaváč, L.M.; Hlaváčová, I.M.; Geryk, V. Taper of kerfs made in rocks by abrasive water jet (AWJ). *Int. J. Adv. Manuf. Technol.* **2017**, *88*, 443–449. [CrossRef]
38. Azmir, M.A.; Ahsan, A.K. A study of abrasive water jet machining process on glass/epoxy composite laminate. *J. Mater. Process. Technol.* **2009**, *209*, 6168–6173. [CrossRef]

© 2019 by the authors. Licensee MDPI, Basel, Switzerland. This article is an open access article distributed under the terms and conditions of the Creative Commons Attribution (CC BY) license (http://creativecommons.org/licenses/by/4.0/).

Article

Defect Analysis and Detection of Cutting Regions in CFRP Machining Using AWJM

Pedro F. Mayuet Ares [1,*], Franck Girot Mata [2], Moisés Batista Ponce [1] and Jorge Salguero Gómez [1]

[1] Mechanical Engineering & Industrial Design Department, Faculty of Engineering, University of Cadiz, Av. Universidad de Cadiz 10, E-11519 Puerto Real-Cádiz, Spain; moises.batista@uca.es (M.B.P.); jorge.salguero@uca.es (J.S.G.)

[2] Mechanical Engineering Department, Faculty of Engineering, University of the Basque Country, Plaza Ingeniero Torres Quevedo, 1, 48013 Bilbao, Spain; franck.girot@ehu.eus

* Correspondence: pedro.mayuet@uca.es; Tel.: +34-956-483-311

Received: 31 October 2019; Accepted: 2 December 2019; Published: 5 December 2019

Abstract: The use of composite materials with a polymeric matrix, concretely carbon fiber reinforced polymer, is undergoing further development owing to the maturity reached by the forming processes and their excellent relationship in terms of specific properties. This means that they can be implemented more easily in different industrial sectors at a lower cost. However, when the components manufactured demand high dimensional and geometric requirements, they must be subjected to machining processes that cause damage to the material. As a result, alternative methods to conventional machining are increasingly being proposed. In this article, the abrasive waterjet machining process is proposed because of its advantages in terms of high production rates, absence of thermal damage and respect for the environment. In this way, it was possible to select parameters (stand-off distance, traverse feed rate, and abrasive mass flow rate) that minimize the characteristic defects of the process such as taper angle or the identification of different surface quality regions in order to eliminate striations caused by jet deviation. For this purpose, taper angle and roughness evaluations were carried out in three different zones: initial or jet inlet, intermediate, and final or jet outlet. In this way, it was possible to characterize different cutting regions with scanning electronic microscopy (SEM) and to distinguish the statistical significance of the parameters and their effects on the cut through an analysis of variance (ANOVA). This analysis has made it possible to distinguish the optimal parameters for the process.

Keywords: AWJM; waterjet; CFRP; kerf taper; surface quality

1. Introduction

Carbon fiber reinforced composite materials with a polymeric matrix (CFRP) are currently widely used in the industry mainly because they possess excellent specific and customizable properties, which allows CFRP characteristics that are impossible to achieve with other materials. In addition, owing to the maturity reached by the different forming technologies, their cost is increasingly approaching that of other structural materials, such as titanium or Inconel alloys. For this reason, these materials are frequently used in different industrial sectors, especially those that require incorporating weight reductions in their manufactured components [1,2].

Although CFRP applications are usually linked to the fields of transport, construction, and energy, in recent years, their use has begun to spread to other sectors such as the naval industry and in applications in consumer goods related to sport and entertainment [3].

However, in spite of all this, the CFRPs continue to generate problems after their conformation. In general, after lamination and curing, CFRPs require a machining process until reaching their final shape owing to the dimensional and geometric requirements demanded by different industries, especially the aerospace sector [4–6]. For this reason, it is common to perform contour milling and drilling operations on these materials with conventional machining technologies, which entails a series of issues [4,7,8].

The stiffness and abrasion of the carbon fiber together with the heat sensitivity of the matrix make the machining an industrial challenge. In general, during the machining of polymer matrix composite materials, a continuous chip is not formed, but rather a micro-chip or high hardness dust. This dust generates an abrasive environment that causes wear of the cutting tool with loss of material and initial geometry [9]. In addition, it must be removed from the cutting area to avoid damage to the cutting equipment and associated personnel. However, this is not the only problem, because the temperature generated during the process, which is increased by abrasion, causes a softening of the matrix, and even its degradation [10,11]. This can cause damage to the material and the appearance of thermal adhesion on the tool, which favors wear. This whole process is inherent to the nature of composite materials and can affect the surface integrity of the part causing damage such as delaminations, fiber fraying, degradation, or micro-cracks [12,13].

In addition, the possibility of personalization or customization of properties makes the possibilities of configuration of the material almost infinite. For the machining process, this means that the same process with the same characteristics can lead to different results.

As a consequence of the above, increasingly more alternatives to traditional machining are being proposed, where abrasive waterjet machining (AWJM) may be an alternative thanks to a relevant number of specific advantages [14–19]:

- Absence of damage due to temperature increase during cutting that could thermally damage the matrix. The water absorbs the heat generated by the impact of abrasive particles against the material.
- Tool wear is minimal compared with traditional cutting tools. In addition, this wear is independent of the material to be machined.
- Absence of high cutting forces on the workpiece and machine. Abrasive particles act as a cutting edge, generating comparatively low or negligible tangential cutting forces compared with traditional methods. Therefore, the fixation of the piece to the machine does not require complex tooling and its preparation time is reduced.
- There is no direct contact between the cutting head and the workpiece. So deformation or vibration problems are avoided.
- Although the same dimensional quality is generally not achieved as in traditional machining, the productivity rate is higher.
- Good environmental performance because no vapors or gases are generated during machining.

Although waterjet cutting is an increasingly used technology, there are difficulties in finding optimal machining parameters for CFRP that reduce the appearance of defects and improve the dimensional and geometrical quality of the machined component.

These defects produce a special negative impact on the surface integrity of the parts obtained owing to dimensional distortions of taper, a progressive increase in roughness, and striation owing mainly to the delay of the jet when it lacks kinetic energy. As a result, it is possible to find up to three possible cutting regions [20–22]:

- Initial damage region (IDR). The impact of the jet deforms the surface of the material by the successive impact of the particles.
- Smooth cutting region (SCR). The jet still possesses sufficient kinetic energy. In this region, the obtained roughness is reduced in comparison with the rest of the zones.
- Rough cutting region (RCR). Striations are detected at the outlet of the material owing to the loss of energy from the jet. The worst surface quality results are obtained in this area.

The defects mentioned are directly related to parameters such as traverse feed rate (TFR), stand-off distance (SOD), water pressure (WP), and abrasive mass flow rate (AMFR), as well as to material conditions and part thickness [23,24]. Table 1 shows the main technological parameters and their influence on the defects mentioned.

Table 1. Influence of main abrasive waterjet machining (AWJM) parameters on cutting quality. TFR, traverse feed rate; SOD, stand-off distance; WP, water pressure; AMFR, abrasive mass flow rate; RCR, rough cutting region; SCR, smooth CR.

Parameter	Influence on Taper	Influence on Roughness
TFR	An increase in traverse feed rate means that the jet remains on the cutting surface for less time. This results in less particle impact, making the machined slot narrower [25].	An increase in traverse feed rate causes an increase in roughness because there are less abrasive particles per unit area acting on the material. This effect accentuates the formation of RCR [26].
SOD	As SOD increases, the jet loses greater coherence, which translates into an increase in the diameter of the jet and therefore of T. This effect is particularly significant at the entry of the cut [27].	A larger diameter of the jet causes an initial damage region that is more deformed by the impact of particles and, therefore, produces an increase in roughness in this region. SOD loses influence as the jet penetrates the composite [28].
WP	An increase in pressure leads to an increase in the kinetic energy of the jet causing greater erosion in the material, leaving a smoother surface. The roughness worsens as the jet loses energy, affecting the formation of SCR [29].	An increase in pressure produces an increase in taper. This effect is more significant when the thickness of the material increases [30].
AMFR	There is no direct correlation between abrasive mass flow variation and T. Normally, increased abrasive flow results in a non-significant increase in taper angle [29].	An increase in abrasive mass flow produces a greater presence of abrasive particles, which avoids the appearance of SCR during cutting. As a consequence, it generally generates a good surface quality at the exit of the cut. However, it can also cause collisions between particles, resulting in a decrease in the effectiveness of the cut [31].

There are publications that use similar materials and parameters that study defects and the dimensional and geometrical quality of the tests performed. However, in the study of microgeometry or surface quality, where there are several criteria in relation to possible areas of roughness, it has been detected that there are no publications that identify the influence of parameters in different regions.

In this article, the main objective is to study the influence of the main technological parameters in AWJM in order to obtain cutting conditions that minimize the appearance of defects by analyzing each cutting region. Likewise, the detection of different cutting regions characteristic of the process will be analyzed.

For this purpose, straight cuts using AWJM technology were performed to determine the influence of the main cutting parameters on CFRP composite material in contour machining operations. Specifically, roughness was measured in three differentiated zones in order to detect the different regions (IDR, SCR, and SCR) that can be produced during cutting, trying to distinguish the border of each of them through statistical analysis and scanning electronic microscopy (SEM).

2. Materials and Methods

2.1. Experimental Development

For the experimental development, a CFRP composite plate formed with 20 layers of prepreg 0.2 ± 0.1 mm thickness (Carbures, Cádiz, Spain) was used to form a 4 mm final plate. The main characteristics of the material are shown in Table 2.

Table 2. CFRP plate features.

Type of Material	Composition	Production Method	Technical Specification
Layers of carbon fiber with epoxy resin matrix and a symmetrical stacking sequence of (0/90)	Intermediate module fiber (66%) and epoxy resin (34%)	Prepreg and autoclaved at 458° ± 5° at a pressure of 0.69 MPa	AIMS-05-01-002

The machining process consisted of the generation of 24 straight cuts, combining the parameter levels shown in Table 3. The length of each test was 180 mm in order to guarantee stable cutting conditions. For this purpose, Mecanumeric MECAJET (Mecanumeric, Marssac-sur-Tarn, France) waterjet machine was employed for cutting the composite, as shown in Figure 1. Table 4 shows the cutting parameters that were kept constant, where the pressure used in the tests is the maximum allowed by the equipment.

Table 3. Cutting parameters selected for the tests.

Parameter	Levels			
TFR (mm/min)	300	900	1500	2100
SOD (mm)	1.5	3.0	4.5	-
AMFR (g/min)	300	600	-	-

Figure 1. Mecanumeric MECAJET performing carbon fiber reinforced composite material with a polymeric matrix (CFRP) cutting.

Table 4. Constant parameters during the tests.

Orifice Diameter (mm)	Nozzle Diameter (mm)	Nozzle Lenght (mm)	Abrasive Size (μm)	Abrasive Type	Water Pressure (MPa)
0.30	0.8	94.7	120	Garnet	450

2.2. Test Evaluation

The variables evaluated were the angle of taper T and the roughness through the parameter Ra (arithmetic average of the roughness profile) distributed in three zones: Zone 1, Zone 2, and Zone 3, as shown in Figure 2. The distribution of roughness measurements was carried out as follows:

- Zone 1. At the entrance of the cut in the erosion affected zone (EAZ). This zone can be identified with IDR.
- Zone 2. Following the EAZ. Specifically, 1 mm from the entrance zone of the jet. This zone can be indentified with SCR.
- Zone 3. The measurement was made at the outlet of the jet. Specifically, to 3.5 mm of the entrance of the jet. This zone can be identified with RCR and is related to the jet delay effect.

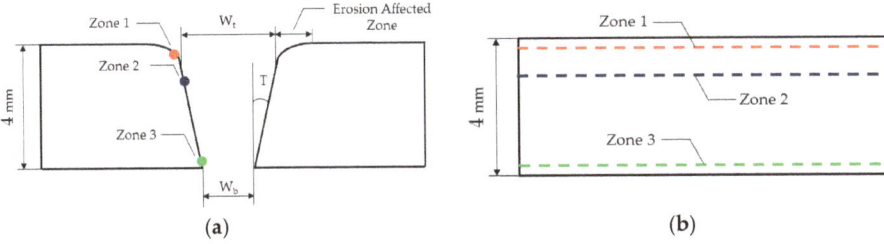

Figure 2. Defects diagram and evaluation zones of the cutting process with (**a**) parameters for measuring taper; (**b**) roughness measurement zones.

For the evaluation of straight cuts, optical evaluation of the machined material was performed by means of the scanning electronic microscope (SEM) technique. For that, a Hitachi SU 1510 (Hitachi, Tokio, Japan) microscope was used for SEM inspection. With respect to roughness measurements, a station Mahr Pertometer Concept PGK 120 (Mahr, Göttingen, Germany) was used.

For T calculation, the expression of Equation (1) was used, where T is the conical angle of the slot, W_t is the width of the slot at the jet inlet, W_b is the slot width at the jet outlet, and t is the thickness of the sample. For the measurement of W_t and W_b, the software ImageJ was used. A previous calibration was carried out to obtain a pixel/mm ratio that allows to establish a correct measurement directly on the image [32,33]. In addition, it is important to note that the area affected by erosion caused by the loss of jet coherence at the jet inlet was differentiated, as shown in Figure 3.

$$T = tan^{-1}\left(\frac{Wt - Wb}{2t}\right) \quad (1)$$

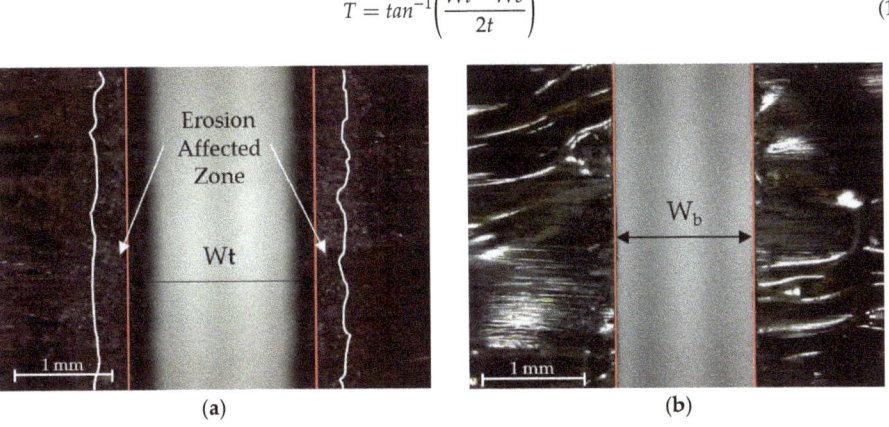

Figure 3. Difference between (**a**) cutting at material inlet cut at the entrance of the material showing the width at the top along with the erosion-affected zone; (**b**) cutting at material outlet showing the width at the bottom.

2.3. Statistical Analysis

The analysis of results was carried out by comparing combinations of parameters through interaction graphs. In this phase, an attempt was made to identify growth or decrease trends between parameters in relation to the variables studied.

In a second phase, the objective was to quantify the influence of the parameters on the cutting process, by means of an analysis of variance (ANOVA) with a 95% confidence interval. Thus, the F-value and the *p*-value were analyzed to measure the evidence against the null hypothesis and the significance (S) of each parameter.

Finally, contour graphs were represented, emphasizing the discussion of parameters whose significance was demonstrated by ANOVA. The analysis was carried using Minitab statistical software.

3. Results and Discussion

3.1. Global Analysis of Results

The results obtained after the measurement process are shown in Table 5. The interaction graphs obtained for each of the four variables are described below in order to visualize trends between the levels of parameters used.

Table 5. Results obtained during the evaluation for each combination of parameters.

Test	TFR (mm/min)	SOD (mm)	AMFR (g/min)	T (°)	Zone 1 (µm)	Zone 2 (µm)	Zone 3 (µm)
1	300	1.5	300	0.880	5.87	5.03	5.44
2	300	1.5	600	0.761	6.10	4.24	4.27
3	300	3.0	300	1.171	6.08	5.30	6.01
4	300	3.0	600	0.996	6.74	5.15	4.32
5	300	4.5	300	3.457	10.54	6.51	7.58
6	300	4.5	600	3.743	10.70	6.67	6.19
7	900	1.5	300	2.283	7.05	6.31	7.19
8	900	1.5	600	2.517	5.75	5.38	5.53
9	900	3.0	300	2.634	7.50	6.33	6.68
10	900	3.0	600	2.751	6.90	5.48	5.78
11	900	4.5	300	2.926	9.45	7.21	7.79
12	900	4.5	600	3.160	9.35	6.38	6.11
13	1500	1.5	300	3.160	9.64	6.95	9.68
14	1500	1.5	600	3.393	6.71	6.14	6.41
15	1500	3.0	300	3.277	8.61	8.26	9.64
16	1500	3.0	600	3.103	9.18	6.92	7.11
17	1500	4.5	300	4.210	11.68	9.11	11.12
18	1500	4.5	600	4.151	13.45	11.13	9.98
19	2100	1.5	300	2.692	8.71	9.46	10.49
20	2100	1.5	600	3.043	7.90	6.83	7.08
21	2100	3.0	300	3.860	9.88	10.14	10.51
22	2100	3.0	600	4.200	9.29	8.18	8.92
23	2100	4.5	300	4.908	13.98	8.87	9.65
24	2100	4.5	600	5.257	15.36	9.02	7.50

Figure 4 shows the interaction graph for the variable T, where the different marginal relationships between parameters are observed. In a first observation, it is shown how the highest results of T are obtained when the levels of the parameters SOD and TFR are 4.5 mm and 2100 mm/min, respectively. This is in good agreement with the works of [29,33]. Thus, with a high value of SOD, the jet loses coherence and increases T. An increase of the TGF means that the jet remains for less time impacting the same area of the piece with the consequent loss of its penetration capacity. The combination of these two factors with high levels favors the formation of the defect, and vice versa. AMFR does not have a defined influence for both levels. Indeed, for 300 g/min and 600 g/min, the taper does not seem to increase considerably, although it is true that, for the highest level, there seems to be a slight increase in taper [34]. This could be because of an excess of particles directed by the jet that impact each other before reaching the cutting surface, so their size varies owing to fractures between particles, as well as their kinetic energy before impact.

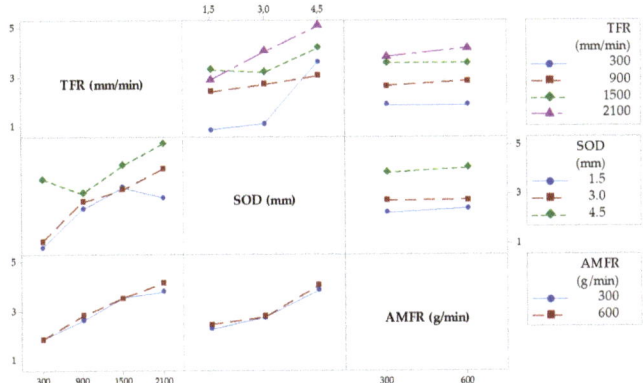

Figure 4. Interaction graph for T. TFR, traverse feed rate; SOD, stand-off distance; AMFR, abrasive mass flow rate.

Figures 5–7 shown the interaction graphs for the roughness values obtained in Zone 1, Zone 2, and Zone 3, respectively.

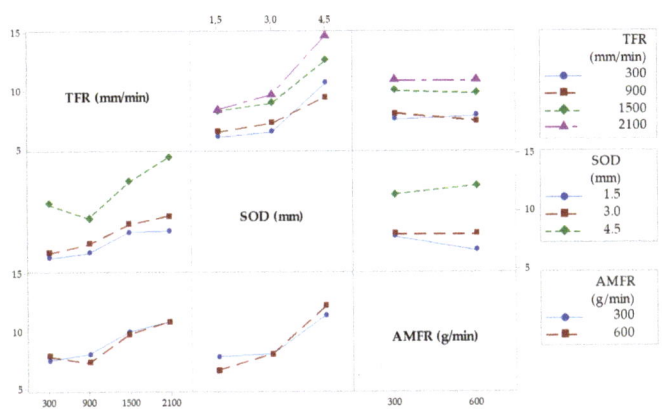

Figure 5. Interaction graph for Zone 1.

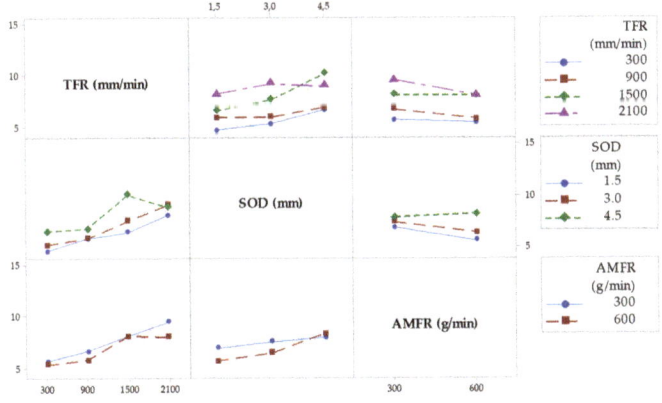

Figure 6. Interaction graph for Zone 2.

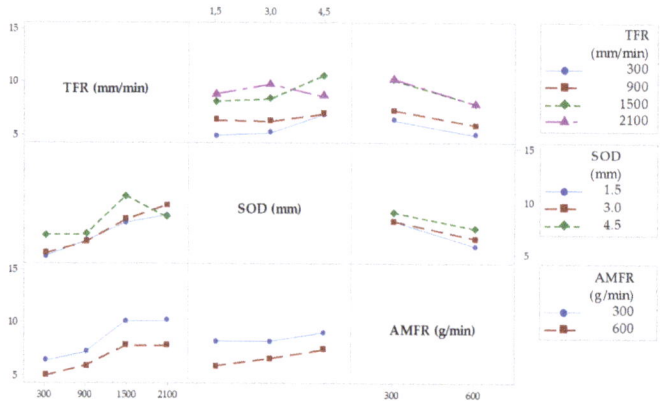

Figure 7. Interaction graph for Zone 3.

Zone 1. This zone contains the highest roughness measurements, where some Ra values reach 15 μm. This shows the existence of an initial damage region at the input of the compound produced by the random impact of particles and consequent deformation of the material at the beginning of the cut. Thus, Figure 5 shows how SOD and TFR are the parameters that influence the process, increasing the roughness as higher values are selected [22]. An example of this can be the comparison of test 1 (TFR = 300 mm/min, SOD = 1.5 mm, and AMFR = 300 g/min) and test 19 (TFR = 2100 mm/min, SOD = 1.5 mm, and AMFR = 300 g/min), where the roughness experiences an increase from 5.87 μm to 8.71 μm. Finally, the AMFR parameter does not seem to show a clear influence during the machining process, because the values in this area do not vary considerably [35]. Figure 8 shows an image taken by SEM distinguishing the transition zone from IDR to SCR. A first observation shows that the IDR zone is completely damaged by the impact of particles, while the SCR zone maintains certain integrity in terms of the structure of the material. Specifically, the particles that erode the surface of the material produce the area affected by erosion by damaging the matrix. Subsequently, the fibers without a matrix are broken in the area where the jet penetrates, generating the mechanized groove. In this process, some particles are embedded in the material in the upper zone as a result of the impact on the material or collisions of particles with each other before impacting [36,37].

Figure 8. Border between the initial damage region (IDR) and smooth cutting region (SCR) formation. Particles embedded in the surface of the material and another group fragmented as a result of the cutting process are observed. Test 12. TFR = 900 mm/min; SOD = 4.5 mm; AMFR = 600 g/min.

Zone 2. The values obtained in this area show a decrease in roughness with respect to previous values. This is because of two different factors: this zone does not show deformation caused by the random impact of particles as in the previous case and the jet still contains sufficient kinetic energy because the measurement is made in an area close to the entrance of the jet. This is in good agreement with the search for a SCR zone that shows a cut without the appearance of defects or striations [21]. Figure 8 shows how the SOD parameter reaches the maximum roughness values, around 10 μm, for the most unfavorable cutting conditions for the highest levels of SOD and TFR. It should be noted that, once again, AMFR does not seem to show a clear trend in roughness variation when values of 300 g/min or 600 g/min are selected, although it is true that, for the latter value, the roughness seems to decrease when TFR presents levels of 300 mm/min and 900 mm/min.

Zone 3. This zone presents similar values to those obtained in the previous zone, although the behavior of the parameters of roughness varies. The SOD parameter does not seem to have a clear influence on the process in this case, as the results are kept within the stability when TFR and AMFR are fixed, as shown in Figure 6. In addition, it is observed that, although TFR continues to maintain a growth in roughness for the highest feed rate values, the AMFR parameter shows an even greater change in its behavior. Thus, an increase in the abrasive level of 600 g/min shows a clear decrease in roughness compared with 300 g/min. This indicates that the jet requires more abrasiveness to maintain low roughness values in this zone. This phenomenon is shown especially at high feed rates, where roughness decreases by up to 25%. Therefore, although the definitive values show values similar to those in Zone 2, a change in the influence of parameters is observed, which will be studied in depth with the ANOVA.

Figure 9 shows an SEM image of a test carried out with the highest level of speed and the lowest level of abrasiveness. Precisely, these levels would favor the appearance of RCR, but as can be seen, the zone appears free of striations. In addition, at the outlet of the jet, there is a delamination formed by the combination of cutting parameters. Specifically, the pressure combined with the expansion of the jet in the material produces stresses that can prevent the layers of the composite from remaining together, causing their separation and the formation of cracks. As the cutting process continues, the abrasive particles lodge in the cracks and consequently expand, forming delaminations [38–40].

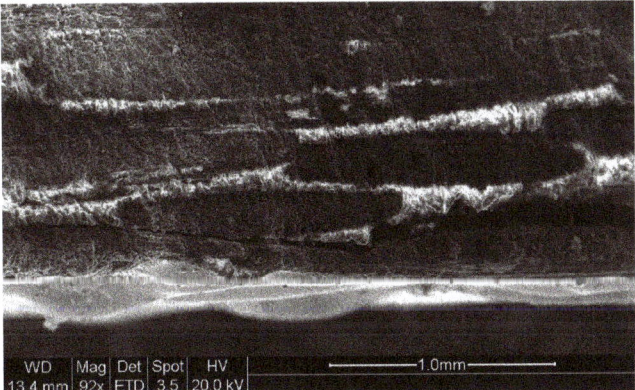

Figure 9. Absence of striation or waviness that visibly differentiates the difference between the smooth cutting region (SCR) and rough CR (RCR). Test 22. TFR = 2100 mm/min; SOD = 3.0 mm; AMFR = 300 g/min.

3.2. ANOVA

This section analyses the degree of influence of each parameter by zone. For this purpose, Table 6 shows the F-value and p-value for T, Zone 1, Zone 2, and Zone 3.

Table 6. Analysis of variance (ANOVA) of the evaluated variables.

TFR				SOD				AMFR			
Variable	F-Value	p-Value		Variable	F-Value	p-Value		Variable	F-Value	p-Value	
T	20.18	0.000	S	T	21.45	0.000	S	T	0.40	0.534	-
Zone 1	16.82	0.000	S	Zone 1	52.35	0.000	S	Zone 1	0.11	0.744	-
Zone 2	17.65	0.000	S	Zone 2	8.36	0.003	S	Zone 2	3.26	0.089	-
Zone 3	26.62	0.000	S	Zone 3	4.73	0.023	-	Zone 3	31.45	0.000	S

In Figure 10a, the main effects graph for variable T is shown. The data reflect that the parameters that have significance in the process are SOD and TFT, with F values of 21.45 and 20.18, respectively. Specifically, the SOD parameter seems to have a more definite influence on the formation of the defect when the selected level increases from 3.0 mm to 4.5 mm, as reflected in the slope of the graph, while the same increase in slope for TFR was detected in the initial levels from 300 mm/min to 900 mm/min [26]. AMRF does not seem to have a definite influence on the formation of T.

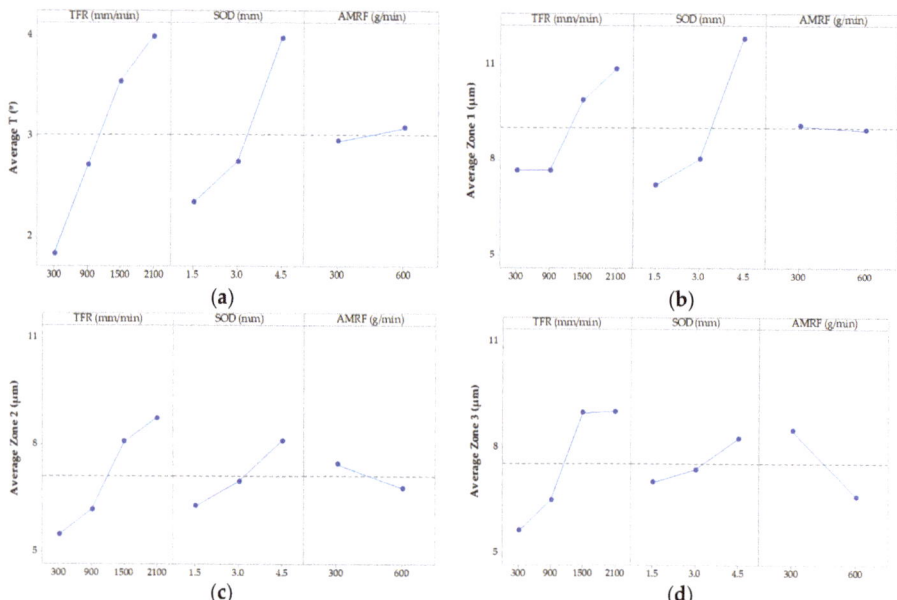

Figure 10. Parameter interaction graph for the response variable for (**a**) T; (**b**) Zone 1; (**c**) Zone 2; and (**d**) Zone 3.

As for the formation of possible roughness zones that can be detected, Figure 10b–d show the main effect graphs for Zone 1, Zone 2, and Zone 3. The degree of influence of each parameter on the three variables is described below [31,41–43]:

- Zone 1. The SOD and TFR parameters show the greatest degree of influence, although it is true that SOD shows the highest F-value, with a result of 52.35. This shows the importance of this parameter on the IDR zone owing to its effect on the coherence of the jet. This is reflected in Figure 10b with an increase in the mean value obtained as the distance increases. As for TFR, it was determined that it also has a degree of significance over the variable, but its influence is less. Thus, the figure shows that the parameter tends to increase roughness as speed increases, as reflected by the slope of Figure 10b for levels of 1500 mm/min and 2100 mm/min. The AMRF parameter has a reduced influence and has no significance on the process.

- Zone 2. In this case, the parameters that show significance with the cutting process are repeated: TFR and SOD. However, with a different level of influence, as reflected in Table 5. Thus, the F-value decreases from 52.35 in Zone 1 to 8.36 in Zone 2 for SOD, and increases from 16.82 in Zone 1 to 17.65 in Zone 2 for TFR. These data reflect that, while TFR maintains a similar degree of influence, SOD suffers a severe decrease as the jet penetrates the machined piece. These values are reflected in Figure 10c, where the highest roughness values are because of the influence of TFR for levels from 1500 mm/min and 2100 mm/min. In this case, AMFR does not seem to have a determining influence on the formation of the defect. However, the increase of the F-value from 0.11 in Zone 1 to 3.26 in Zone 2 should be highlighted. This is reflected in the slope of the figure and in the decrease of the p-value in Table 5 from 0.744 obtained in the previous zone to 0.089.
- Zone 3. In the last zone of roughness measurement, the parameters that show significance with the process are AMFR and TFR. In the first case, AMFR describes a strong relationship, showing the highest F-value with a result of 31.45, as described in Table 5. This is reflected in Figure 10d comparing the slope of the graphs obtained in the three zones, where it is observed as the slope that joins the average values of both levels of abrasive increases. This suggests that the appearance of the RCR zone is formed in machined samples with low levels of abrasiveness. As for TFR, the increase of the F-value to 26.62 illustrates what has been said so far. The degree of influence continues to increase with respect to the two previous areas and a considerable increase in roughness is observed for levels with high speed. The greatest increase in roughness occurs in the increase range of 300 mm/min to 900 mm/min. Finally, the influence of the SOD parameter decreases until it loses significance in the process, as shown in Table 6, and its reduced slope in Figure 10c with respect to the previous zones.

Therefore, everything seems to indicate that, as described, the analysis by zones reveals that, although it is possible to establish that there are two identified zones of roughness, IDR and SCR, the measurements carried out in Zone 3 reveal a change in the importance of parameters, revealing that TFR, and above all, AMFR, acquire greater significance to the detriment of SOD. Therefore, it could be considered that, with a larger plate thickness, it would have been possible to detect the presence of striations [30].

3.3. Analysis of Contour Graphs for Significant Parameters

This section shows the contour graphics as a function of the significant variables. The aim of this analysis is to evaluate the most favourable cutting conditions within the range of parameters studied.

3.3.1. Taper Analysis

Figure 11 shows the contour graph for T as a function of TFR and SOD. In this case, the AMFR parameter is irrelevant. The data obtained reflect how the cutting parameters where the taper decreases are in the range of feed rates and reduced distances. In this way, the jet can remain longer eroding the material reducing the loss of coherence of the abrasive jet. These parameters coincide with the levels of TFR = 300 mm/min and SOD = 1.5 mm, although it is reflected in the graph that distances up to 3 mm can maintain the efficiency of the cut, minimizing the growth T. In this case, the conicity can be maintained in values lower than 1°.

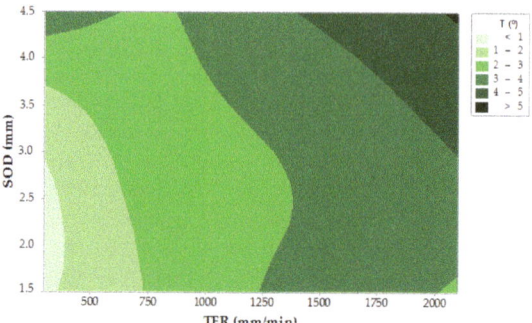

Figure 11. Contour graph of T as a function of parameters with significance: TFR and SOD.

3.3.2. Roughness Analysis

As in the previous case, Figure 12 shows the contour graphs for Zone 1, Zone 2, and Zone 3. In the first two cases, the parameters with significance in the process are TFR and SOD, while at the output of the jet, the AMRF parameter is the most relevant with TFR.

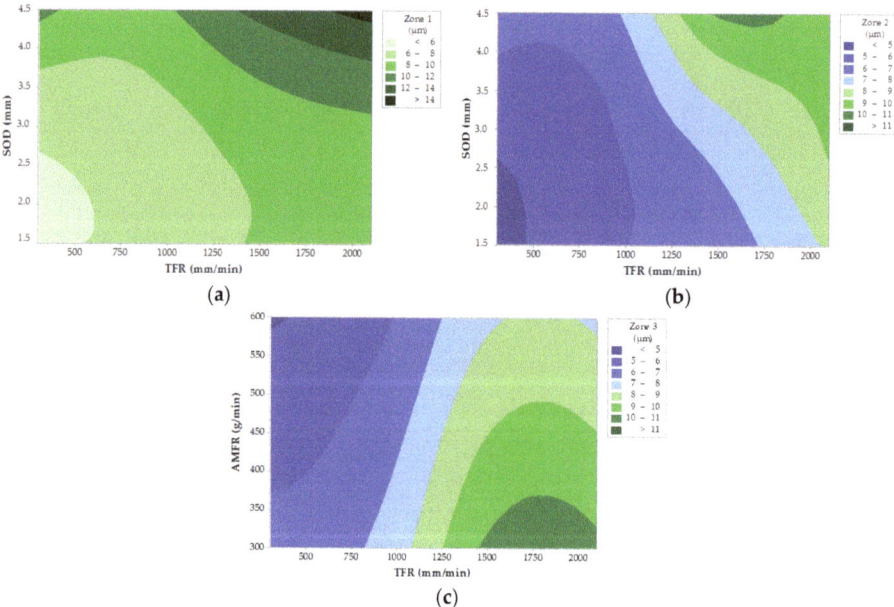

Figure 12. Roughness contour graphs as a function of significant parameters: (**a**) Zone 1: TFR and SOD; (**b**) Zone 2: TFR and SOD; and (**c**) Zone 3: TFR and AMFR.

It is shown that, globally, the results where roughness decreases occur when TFR and SOD are reduced, while the AMFR parameter increases. This is in good agreement with the works of [18,33], although it should be noted that, with this process, it has not been possible to obtain roughness levels lower than 4 µm. In addition, it is necessary to mention that Zone 1, corresponding to IDR, contains the highest roughness results, as has been revealed. For Zone 2 and Zone 3, the most favorable results are less than 5 µm. Although, it is to be expected that in Zone 3, identified as the possible RCR, the results are higher, and the comment has been made that the thickness of the specimens has a considerable influence. However, the ANOVA reveals that there is a change of trend in the significance of the variables that has allowed to reveal that SOD has no influence in Zone 3, highlighting the importance of AMFR in the cutting process.

4. Conclusions

A study was carried out on the influence of the main technological parameters on abrasive water jet involved in the formation of defects and roughness zones on CFRP material. From the study carried out in this article, the following conclusions are drawn:

- The values evaluated for T in the experiment varied between 1° and 5°. The data showed that the highest values of T are obtained when TFR and SOD reach 2100 mm/min and 4.5 mm, respectively.
- The initial assessment of surface quality clearly shows that there are at least two well-defined roughness zones: IDR and SCR. The boundary between both zones was verified using SEM techniques. Measurements taken in Zone 1 showed roughness data above 15 μm for elevated SOD and TFR, while in Zone 2, values about 10 μm were obtanined under the same conditions. This results in approximately 35% lower roughness between zones.
- Initially, the roughness analysis carried out in Zone 3 did not reveal any changes with respect to the values measured in Zone 2. However, the ANOVA of the data showed a significant change in the influence of parameters. In this sense, AMFR acquires great significance in the process, revealing its importance in the formation of RCR. Thus, in Zone 2 or SCR, the significant parameters were TGF (F-value = 17.35) and SOD (F-value = 8.36), while in Zone 3, the significant parameters were AMFR (F-value = 31.45) and TGF (F-value = 26.32).

Once the analysis of the data was completed, contour graphs with the significant parameters for each variable were represented in order to record the parameters that minimize taper and roughness defects. From these results, the following conclusions are also drawn:

- Parameters that minimize the effect of taper during cutting are achieved with SOD values between 1.5 and 3.0 mm and TFR values lower than 350 mm/min.
- The roughness data for IDR can be considerably reduced with SOD values lower than 2.0 mm and TFR values lower than 500 mm/min.
- In Zone 2 or SCR the lowest roughness data of all cutting regions are obtained with parameters similar to those recommended in the IDR.
- Roughness data in Zone 3 or RCR can be reduced when using low TFR values combined with high abrasive rates. In this experimental case, for values higher than 550 gr/min.

On the above basis, the experiment carried out shows that, in order to obtain the minimum T values and high surface quality during the cutting process of CFRP composite specimens, SOD values between 1.5 and 2.0 mm, reduced TFR values around 300 mm/min, and high AMFR values must be selected.

Finally, in order to avoid the appearance of cracks and delaminations, it is recommended to establish a compromise between the pressure used and the thickness of the composite plate. Thus, it is established as a line of future works to obtain cuts with the absence of delaminations under the considerations made throughout this article.

Author Contributions: Conceptualization, P.F.M.A.; Methodology, P.F.M.A. and F.G.M.; Software, P.F.M.A. and M.B.P.; Validation, J.S.G.; Formal Analysis, P.F.M.A. and F.G.M.; Investigation, P.F.M.A.; Resources, F.G.M. and J.S.G.; Data Curation, P.F.M.A. and F.G.M.; Writing—Original Draft Preparation, P.F.M.A.; Writing—Review & Editing, P.F.M.A. and M.B.P.; Supervision, M.B.P. and F.G.M.; Project Administration, J.S.G.

Funding: This work has received financial support from programme for the Promotion and Impulse of Research and Transfer of the University of Cadiz. Project: PR PR2017-086, Machining of composite materials of strategic use in the aeronautical industry using AWJM.

Acknowledgments: To Serge Tcherniaeff and Madalina Calamaz, for their kind help and support during the tests at ENSAM in Bordeaux. To Mariano Marcos Bárcena, for promoting the research carried out in this article.

Conflicts of Interest: The authors declare no conflict of interest.

References

1. Abrão, A.M.; Faria, P.E.; Campos Rubio, J.C.; Reis, P.; Paulo Davim, J. Drilling of fiber reinforced plastics: A review. *J. Mater. Process. Technol.* **2007**, *186*, 1–7. [CrossRef]
2. Mayuet, P.F.; Gallo, A.; Portal, A.; Arroyo, P.; Álvarez-Alcón, M.; Marcos, M. Damaged Area based Study of the Break-IN and Break-OUT Defects in the Dry Drilling of Carbon Fiber Reinforced Plastics (CFRP). *Procedia Eng.* **2013**, *63*, 743–751. [CrossRef]
3. Faraz, A.; Biermann, D.; Weinert, K. Cutting edge rounding: An innovate tool wear criterion in drilling CFRP composite laminates. *Int. J. Mach. Tools Manuf.* **2005**, *49*, 1185–1196. [CrossRef]
4. Lachaud, F.; Piquet, R.; Collombet, F.; Surcin, L. Drilling of composite structures. *J. Compos. Struct.* **2001**, *52*, 511–516. [CrossRef]
5. Langella, A.; Nele, L.; Maio, A. A torque and thrust prediction model for drilling of composite materials. *Compos. Part A Appl. Sci. Manuf.* **2005**, *36*, 83–93. [CrossRef]
6. Teti, R. Machining of Composite Materials. *CIRP Ann. Manuf. Technol.* **2002**, *51*, 611–634. [CrossRef]
7. Karpat, Y.; Değer, B.; Bahtiyar, O. Drilling thick fabric woven CFRP laminates with double point angle drill. *J. Mater. Process. Technol.* **2012**, *212*, 2117–2127. [CrossRef]
8. Shyha, I.; Soo, S.L.; Aspinwall, D.; Bradley, S. Effect of laminate configuration and feed rate on cutting performance when drilling holes in carbon fibre reinforced plastic composites. *J. Mater. Process. Technol.* **2010**, *210*, 1023–1034. [CrossRef]
9. König, W.; Wulf, C.; Graβ, P.; Willerscheid, H. Machining of fibre reinforced plastics. *Ann. CIRP* **1985**, *34*, 537–548. [CrossRef]
10. Bonnet, C.; Poulachon, G.; Rech, J.; Girard, Y.; Philippe, J. CFRP drilling: Fundamental study of local feed force and consequences on hole exit damage. *Int. J. Mach. Tools Manuf.* **2015**, *94*, 57–64. [CrossRef]
11. Lin, S.C.; Chen, I.K. Drilling carbon fiber-reinforced composite material at high speed. *Wear* **1996**, *194*, 156–162. [CrossRef]
12. Karnik, S.R.; Gaitonde, V.N.; Campos-Rubio, J.; Esteves-Correia, A.; Abrão, A.M.; Davim, J.P. Delamination analysis in high speed drilling of carbon fiber reinforced plastics (CFRP) using artificial neural network model. *Mater. Des.* **2008**, *29*, 1768–1776. [CrossRef]
13. Tsao, C.C.; Hocheng, H.; Chen, Y.C. Delamination reduction in drilling composite materials by active backup force. *CIRP Ann. Manuf. Technol.* **2012**, *61*, 91–94. [CrossRef]
14. Wang, J. A machinability study of polymer matrix composites using abrasive waterjet cutting technology. *J. Mater. Process. Technol.* **1999**, *94*, 30–35. [CrossRef]
15. El-Hofy, M.; Helmy, M.O.; Escobar-Palafox, G.; Kerrigan, K.; Scaife, R.; El-Hofy, H. Abrasive Water Jet Machining of Multidirectional CFRP Laminates. *Procedia Cirp* **2018**, *68*, 535–540. [CrossRef]
16. Hlaváč, L.M. Application of water jet description on the de-scaling process. *Int. J. Adv. Manuf. Technol.* **2015**, *80*, 721–735. [CrossRef]
17. Zeng, J. Determination of machinability and abrasive cutting properties in AWJ cutting. In Proceedings of the American Waterjet WJTA Conference, Houston, TX, USA, 19–21 August 2007.
18. Ramulu, M.; Arola, D. Water jet and abrasive water jet cutting of unidirectional graphite/epoxy composite. *Composites* **1993**, *24*, 299–308. [CrossRef]
19. Liu, H.-T. Waterjet technology for machining fine features pertaining to micromachining. *J. Manuf. Process.* **2010**, *12*, 8–18. [CrossRef]
20. Eneyew, E.D.; Ramulu, M. Experimental study of surface quality and damage when drilling unidirectional CFRP composites. *J. Mater. Res. Technol.* **2014**, *3*, 354–362. [CrossRef]
21. Begic-Hajdarevic, D.; Cekic, A.; Mehmedovic, M.; Djelmic, A. Experimental Study on Surface Roughness in Abrasive Water Jet Cutting. *Procedia Eng.* **2015**, *100*, 394–399. [CrossRef]
22. Zahavi, J.; Schmitt, G.F., Jr. Solid particle erosion of reinforced composite materials. *Wear* **1981**, *71*, 179–190. [CrossRef]
23. Nanduri, M.; Taggart, D.G.; Kim, T.J. The effects of system and geometric parameters on abrasive water jet nozzle wear. *Int. J. Mach. Tools Manuf.* **2002**, *42*, 615–623. [CrossRef]
24. Bijwe, J.; Rattan, R. Carbon fabric reinforced polyetherimide composites: Optimization of fabric content for best combination of strength and adhesive wear performance. *Wear* **2007**, *262*, 749–758. [CrossRef]

25. Arola, D.; Ramulu, M. A study of kerf characteristics in abrasive waterjet machining of graphite/epoxy composite. *J. Eng. Mater. Technol.* **1996**, *118*, 256–265. [CrossRef]
26. Palleda, M. A study of taper angles and material removal rates of drilled holes in the abrasive water jet machining process. *J. Mater. Process. Technol.* **2007**, *189*, 292–295. [CrossRef]
27. Shanmugam, D.K.; Masood, S.H. An investigation on kerf characteristics in abrasive waterjet cutting of layered composites. *J. Mater. Process. Technol.* **2009**, *209*, 3887–3893. [CrossRef]
28. Naresh, M.N. Investigation on Surface Rougness in Abrasive Water-Jet Machining by the Response Surface Method. *Mater. Manuf. Process.* **2014**, *29*, 1422–1428. [CrossRef]
29. Gupta, V.; Pandey, P.M.; Garg, M.P.; Khanna, R.; Batra, N.K. Minimization of Kerf Taper Angle and Kerf Width Using Taguchi's Method in Abrasive Water Jet Machining of Marble. *Procedia Mater. Sci.* **2014**, *6*, 140–149. [CrossRef]
30. Orbanic, H.; Junkar, M. Analysis of striation formation mechanism in abrasive water jet cutting. *Wear* **2008**, *265*, 821–830. [CrossRef]
31. Ahmed, T.M.; El Mesalamy, A.S.; Youssef, A.; El Midany, T.T. Improving surface roughness of abrasive waterjet cutting process by using statistical modeling. *CIRP J. Manuf. Sci. Technol.* **2018**, *22*, 30–36. [CrossRef]
32. Gnanavelbabu, A.; Saravanan, P.; Rajkumar, K.; Karthikeyan, S. Experimental Investigations on Multiple Responses in Abrasive Waterjet Machining of Ti-6Al-4V Alloy. *Mater. Today Proc.* **2018**, *5*, 13413–13421. [CrossRef]
33. Alberdi, A.; Suárez, A.; Artaza, T.; Escobar-Palafox, G.A.; Ridgway, K. Composite Cutting with Abrasive Water Jet. *Procedia Eng.* **2013**, *63*, 421–429. [CrossRef]
34. Shanmugam, D.K.; Wang, J.; Liu, H. Minimisation of kerf tapers in abrasive waterjet machining of alumina ceramics using a compensation technique. *Int. J. Mach. Tools Manuf.* **2008**, *48*, 1527–1534. [CrossRef]
35. Sheldon, G.L.; Finnie, I. The mechanism of material removal in the erosive cutting of brittle materials. *J. Eng. Ind.* **1966**, *88*, 393–399. [CrossRef]
36. Wang, J.; Guo, D.M. A predictive depth of penetration model for abrasive waterjet cutting of polymer matrix composites. *J. Mater. Process. Technol.* **2002**, *121*, 390–394. [CrossRef]
37. Zahavi, J.; Schmitt, G.F., Jr. Solid particle erosion of polymeric coatings. *Wear* **1981**, *71*, 191–210. [CrossRef]
38. Schwartzentruber, J.; Spelt, J.K.; Papini, M. Prediction of surface roughness in abrasive waterjet trimming of fiberreinforced polymer composites. *Int. J. Mach. Tools Manuf.* **2017**, *122*, 1–17. [CrossRef]
39. Phapale, K.; Singh, R.; Patil, S.; Singh, R.K.P. Delamination Characterization and Comparative Assessment of Delamination Control Techniques in Abrasive Water Jet Drilling of CFRP. *Procedia Manuf.* **2016**, *5*, 521–535. [CrossRef]
40. Shanmugam, D.K.; Nguyen, T.; Wang, J. A study of delamination on graphite/epoxy composites in abrasive waterjet machining. *Compos. Part A Appl. Sci. Manuf.* **2008**, *39*, 923–929. [CrossRef]
41. Schwartzentruber, J.; Spelt, J.K.; Papini, M. Modelling of delamination due to hydraulic shock when piercing anisotropic carbon-fiber laminates using an abrasive waterjet. *Int. J. Mach. Tools Manuf.* **2018**, *132*, 81–95. [CrossRef]
42. Drensky, G.; Hamed, A.; Tabakoff, W.; Abot, J. Experimental investigation of polymer matrix reinforced composite erosion characteristics. *Wear* **2011**, *270*, 146–151. [CrossRef]
43. Kantha-Babu, M.; Krishnaiah-Chetty, O.V. A study on recycling of abrasives in abrasive water jet machining. *Wear* **2003**, *254*, 763–773. [CrossRef]

© 2019 by the authors. Licensee MDPI, Basel, Switzerland. This article is an open access article distributed under the terms and conditions of the Creative Commons Attribution (CC BY) license (http://creativecommons.org/licenses/by/4.0/).

Article

On the Machinability of an Al-63%SiC Metal Matrix Composite

David Repeto *, Severo Raul Fernández-Vidal, Pedro F. Mayuet, Jorge Salguero and Moisés Batista *

Department of Mechanical Engineering & Industrial Design, Faculty of Engineering, University of Cadiz, Av. Universidad de Cádiz 10, E-11519 Puerto Real-Cádiz, Spain; raul.fernandez@uca.es (S.R.F.-V.); pedro.mayuet@uca.es (P.F.M.); jorge.salguero@uca.es (J.S.)
* Correspondence: david.repeto@uca.es (D.R.); moises.batista@uca.es (M.B.); Tel.: +34-956-483-406 (M.B)

Received: 31 October 2019; Accepted: 3 March 2020; Published: 6 March 2020

Abstract: This paper presents a preliminary study of aluminium matrix composite materials during machining, with a special focus on their behavior under conventional processes. This work will expand the knowledge of these materials, which is considered to be strategic for some industrial sectors, such as the aeronautics, electronics, and automotive sectors. Finding a machining model will allow us to define the necessary parameters when applying the materials to industry. As a previous step of the material and its machining, an experimental state-of-the-art review has been carried out, revealing a lack of studies about the composition and material properties, processes, tools, and recommended parameters. The results obtained and reflected in this paper are as follows; SiC is present in metallic matrix composite (MMC) materials in a very wide variety of sizes. A metallographic study of the material confirms the high percentage of reinforcement and very high microhardness values registered. During the machining process, tools present a very high level of wear in a very short amount of time, where chips are generated and arcs are segmented, revealing the high microhardness of the material, which is given by its high concentration of SiC. The chip shape is the same among other materials with a similar microhardness, such as Ti or its alloys. The forces registered in the machining process are quite different from conventional alloys and are more similar to the values of harder alloys, which is also the case for chip generation. The results coincide, in part, with previous studies and also give new insight into the behavior of this material, which does not conform to the assumptions for standard metallic materials, where the hypothesis of Shaffer is not directly applicable. On the other hand, here, cutting forces do not behave in accordance with the traditional model. This paper will contribute to improve the knowledge of the Al-63%SiC MMC itself and the machining behavior.

Keywords: metal matrix composite; Al-SiC; microstructure; machining

1. Introduction

Metallic matrix composite (MMC) materials are the result of the combination of a metallic alloy matrix with reinforcement [1], giving rise to a unique combination of mechanical properties, which surpasses those of the individual components [2,3]. This improvement is given by the incorporation of a ductile metal matrix of reinforcements particles with a high strength and modulus of elasticity [4,5]. When compared to monolithic alloys, MMCs offer superior strength and stiffness values, lower weight and thermic expansion coefficients, and, in some cases, the ability to operate at high temperatures [6]. Because of its exceptional properties, and despite the complexities of its behavior throughout the machining process, there exists great industrial interest in MMC's: several potential areas of application can be covered by these materials.

Each MMC is unique and its composition corresponds to a general criterion, which represents an added difficulty to its replicability of laboratory experiments, as well as subsequent industrial production.

The uneven distribution of the reinforcement accounts for the dispersion found by its behavior during the machining process, although some manufacturers that commercialize these products try to combine the matrix and straighteners in a particular manner in the lab in an attempt to control the process from the beginning.

A standard commercial material would guarantee similar results in successive experiments, although the use of a specific aluminium matrix is not currently well defined, and the choice appears to be conditioned by the matrix properties, as its behavior is improved by the addition of reinforcements. The microstructural analysis of the material revealed a distribution of reinforcements typical of an MMC, highlighting the high concentration (63%) and the differences in size, as well as the random distribution. Improving the quality and homogeneity of reinforcement particles would facilitate the MMC manufacturing process in terms of the required specifications. This is the part of the process where an exhaustive control is generally lacking in the literature.

Although the military industry has a specific interest in MMC applications (mainly for ballistics), the automotive sector has opened a broader field for these materials, motivated by the search of lighter vehicles. In the electronics sector, MMCs have direct applications in the manufacturing of microprocessors, due to their excellent heat transfer properties. Nowadays, MMCs are considered as strategic materials, forming part of the essential elements in many advanced technologies [7].

MMCs are designed according to their final purpose, based on the properties of the metal matrix (Al, Mg, Ti, Cu, etc.) and the particle reinforcement characteristics, varying by application and manufacturer [8]. Materials which are more resistant and possess a lower weight are required nowadays to replace lightweight alloys, and MMCs are in the best position to satisfy these requests. Adding ceramic reinforcements to an aluminium matrix helps to dramatically increase some properties, such as microhardness, thermal conductivity, and strength.

Al-based MMCs are usually manufactured with continuous (long fibers) or discontinuous reinforcements (short fibers, whiskers, wires, or irregular and spherical particles) [9,10].

Particulate-based metal matrix composites (PMMCs) are of particular interest, as they exhibit higher ductility and lower anisotropy than fiber-reinforced MMCs [11]. Because of this, reinforcements are distributed in a 4:1 ratio—in favor of the particles against short fibers [12]—and the selection of SiC as a reinforcement with an aluminium matrix is over 60% in this regard, because of its mechanical properties, characteristics, physical, and chemical properties, and low cost of production when compared with other reinforcements which could be used, such as Al_2O_3, with around 30% [13]. The rest of the various materials are not relevant in terms of their composition percentage, where 10%, which is quite dispersed, can be found for B_4C, TiC, or hybrids of SiC and Al_2O_3 [14–16].

In terms of reinforcement quantity, [6,11,17–23], most of applications include at least 20% reinforcement, which can be considered as the highest ratio. Regarding reinforcement sizes, these are often between 3 and 65 µm; however, three-quarters of the studied sizes are ~20 µm [24].

The number of papers found regarding the machinability of Al-based MMCs with particulate reinforcement similar to the material studied here (over 60% Al) is small, and these papers are rare [25,26]. Consequently, it was decided to investigate this kind of material, which is the main objective of this paper.

2. Machining of Metal Matrix Composites

Quigley studied factors affecting the turning [27] of MMC 5083 AlSiC with 25% Al and found that the conventional and coated tool flank wear was very high; the surface finish was better than that which was expected theoretically, due to the tool morphology and material effect (BUE); and that tool wear leads to a loss in dimensional accuracy.

El-Gallab studied tool performance and workpiece surface integrity [11] when turning Duralcan F3S.20S AlSiC (A356/20% SiC) with Al_2O_3/TiC, TiN, and PCD tools, where all of the BUE and flank wear (VB) of all forms were measured. The cutting parameters (the speed of cut, feed, and depth of cut) play a key role in determining the amount of tool flank wear, as well as the size of the built-up edge. El-Gallab also noticed [19] that no research has yet been carried out to determine the effect of the cutting parameters on the workpiece surface integrity and sub-surface damage during the machining of an AlSiC particulate-based metal matrix composite, where cutting parameters have a significant effect on the form of the chips produced. In general, at low feed rates and small cut depths, the chips tended to be continuous, and this observation contradicts Monaghan's results [28]; even though the parameters and materials were the same, there was a 5% SiC difference between them. A reduction in the surface roughness can be noticed with the increase in feed rate, and this effect is attributed to the reduction in the flank wear of the tool. This could be attributed to the stable built-up edge, which protects the tool from wear by abrasion. Manna investigated the machinability of an AlSiC MMC (A413/15% SiC) [29] and found that the lower built-up edge (BUE) is formed during the machining of AlSiC MMC at a high speed and low cut depth, while also observing a better surface finish at a high speed with a low feed rate and low cut depth.

Davim studied the optimization of cutting parameters based on orthogonal arrays [21], where turning A356/20SiCp-T6 with Poly Crystalline Diamond (PCD) tools showed that the cut speed is the main factor that affects tool wear and power consumption, whereas the feed rate is the main factor for surface roughness (Ra). Also, in 2007, Davim made a correlation between the chip compression ratio and shear plane angle or chip deformation during the turning of MMCs, showing that the shear angle decreases with the chip compression ratio. On the contrary, chip deformation has been shown to increase with the chip compression ratio [30]. The Merchant model gives, in general, an overestimation of the shear plane angle value in the cutting of Al-based MMCs.

Dabade investigated the effect of a change in size and volume fraction of reinforcement (Al/20–30 SiCp; 220–600) on the mechanism of chip formation by changing the processing conditions and tool geometries [31]. Here, the results were that at a lower cutting speed (40 m/min), thin needle-type flakes, as well as segmented chips are formed, whereas at higher cutting speeds (120 m/min), generally, semicontinuous, continuous, scrambled ribbon, and tubular helix chips are formed. The length of chip and the number of chip curls increases with the increase of feed rate at any given cutting speed and cut depth. The size and volume fraction of reinforcement significantly influences the chip formation mechanism.

One of the conclusions that can be drawn about the different manufacturing systems of MMCs is that they must evolve over time in terms of the method of incorporation of reinforcements into the matrix, so that the desired properties are established and maintained. Another concern, although to a lesser extent, is that the temperature reached during the manufacturing process should be kept as low as possible. The papers we have studied reflect the effort to manufacture a material that faithfully follows the design specifications of the experiment and does not raise concern about the scope of its cost. The objective in this study, differently to the objectives of other studies reported in the literature, is to move the manufacturing method into industry.

Although MMC production is relatively well adjusted to its final product shape, machining is unavoidable [32,33]. Nowadays, improvements of this process are the main focus of many research efforts [34]. Nevertheless, it is the combination of matrix and reinforcements that creates the enhanced properties of an MMC that hinder its machining, even for matrix-based aluminium alloys, which are considered to have a high machinability. Despite this, the machining of aluminium alloys is not exempt from difficulties, traditionally owing to the great capability of the material to adhere to tool edges and, above all, the shear angle [35].

The reinforcement itself represents an added difficulty, as the tool encounters a heterogeneous material throughout its path. If the reinforcement is harder than aluminium, as is the case, the machining becomes even more complex. Regarding MMCs with ceramic reinforcements (Al_2O_3 and SiC), the situation is further complicated by the abrasiveness of these materials, resulting in extremely damaging machining for traditional tools, shortening their service life drastically.

As the difficulty in machining MMCs increases because of the abrasion phenomenon described above, a greater force is required to pull the chips and the amount of energy required increases, which can greatly influence the total efficiency of the process [36]. The greatest obstacle encountered by MMCs in replacing monolithic alloys is the difficulties associated with their machining, as the tools required suffer from intense wear owing to the presence of the reinforcements, making it a highly inefficient process [6,37].

The shape, type, presentation, and concentration of the reinforcement greatly influence the wearing process. Similarly, a great dispersion between the responses in MMCs can be detected, as the characteristics of the particular composites integrating them can result in the manifestation of completely opposite properties.

In this paper, a study of the machinability of an Al-based MMC with a high concentration (63%) of SiC reinforcement has been performed. Machinability tests have been carried out by orthogonal machining (shaping) from some values of speed and deep of cut. The geometry of the chips obtained, as well as the characterization of cutting tool wear, have been studied by optical and scanning microscopy techniques; also, cutting forces are measured. This paper will allow to establish the basis of a parametrical analysis in regards MMC's with high content SiC reinforcement or some other materials characterized by a similar behavior.

3. Experimental Procedure

The experimental procedure has been divided into two phases: The first one develops a microstructural characterization of the MMC to be machined, and this is necessary to know the distribution of the reinforcements. The second phase is focused towards the machining of this MMC. During and after the machining, the chip behavior has also been studied via high-speed video film and metallographical methods, including Stereoscopic Optical Microscopy (SOM) and Scanning Electron Microscopy (SEM) techniques.

3.1. Material Characterization

For this study, a commercial MMC sheet (ref. AlSiC-9) composed of an aluminium UNS A03562 (A356.2) matrix with 63% SiC reinforcement was used. The composition of the A365.2 Al alloy, obtained from the webpage Matweb, is shown in Table 1.

Table 1. Composition of the metallic matrix composite (MMC) matrix alloy [38].

Alloy	Al	Cu	Fe	Mg	Mn	Si	Ti	Zn	Others
UNS Al356.2	91.30–93.20	0.10	0.12	0.30–0.45	0.05	6.5–7.5	0.20	0.05	0.15
A356.2	91.10–93.30	≤0.20	≤0.20	0.25–0.45	≤0.10	6.5–7.5	≤0.20	≤0.10	0.15

Even though this material has a commercial distribution, this does not mean that it is unnecessary to study it because it is not an alloy under international regulations. The sample was 180.17 mm wide and 136.89 mm long, with a thickness of 5.55 mm. The main usage of this plate was to support an electronic array for an insulated gate bipolar transistor (IGBT), provided by CPS Technologies Corp, Norton, MA, USA. The material properties are shown in Table 2.

Table 2. Composition and some properties of the MMC used [39].

Composition and Properties	AlSiC-9
Matrix: Aluminium alloy A 356.2	37 vol %
Reinforcement: SiC	63 vol %
Density (g/cm^3)	3.01
Thermal conductivity (W/K·m) at 298.15 K	190
Specific heat (J/g·K) at 298.15 K	0.741
Young's modulus (GPa)	188
Coulomb, shear modulus (GPa)	76

To characterize the microstructure, some samples were cut into an appropriate size and polished, prior to etching with Keller's reactant. After, optical evaluation of the etched surface was carried out, as can be seen in Figure 1, using a Nikon SMZ 800 stereoscopic microscope and a Nikon Epiphot 200 metallographic microscope (Nikon Instruments, Amsterdam, Netherlands), both of which were equipped with digital high-resolution cameras.

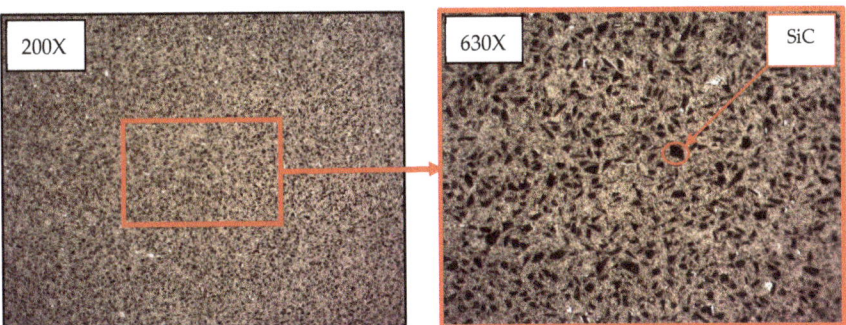

Figure 1. Metallography of the AlSiC MMC with 63% Al.

Additionally, microhardness was measured with the HMV-2 (Shimadzu, Kioto, Japan) microhardness tester to evaluate the behavior in different phases of the matrix, as well as the properties of the reinforcement particles. Forces of 980.70 mN and 1.96 N were applied, using 100 and 200 grams, respectively, during 15 seconds, which corresponded to a Vickers Hardness (HV) value between 0.1 and 0.2, respectively, according to UNE-EN ISO 6507 [40].

3.2. Machining Process

To study the machinability but avoid complex cutting geometries, it is typical to use machine tools with a rectilinear cutting movement, as this is the simplest machining model and it can be extrapolated to an oblique cutting model. Therefore, an orthogonal cutting configuration in a shaping machine (GSP 2108 R.20. GSP, Paris, France) was used, as shown in Figure 2.

Figure 2. GSP 2108 R.20 shaping machine tool.

All tests were performed in dry conditions, and a new cutting tool was used in each trial. The cutting parameters used were based on the combination of 4 cutting speeds and 2 cut depths selected from previous studies [41,42]. This is shown in Table 3.

Table 3. Machining parameters MMC AlSiC.

Cutting Speed (Vc)	20, 30, 40, 50 m/min
Depth of Cut (DoC)	0.10, 0.20 mm

Every test was carried out for at least 1 or 3 tool paths. The machining time ranged from 0.15 to 0.45 s, whereas the measured machining lengths were between 45 and 120 mm. The cutting forces were acquired by a Kistler 9257B dynamometer (Kistler Instruments, Winterthur, Switzerland), while the cutting process was filmed at high speed with a Photron APX RS (Photron, San Diego, CA, USA) camera.

A WC/6%Co "blank" cutting-tool from Sandvik (ref. H13A, without coating) was selected, with a rake angle of 0° and a clearance angle of 12°, as shown in Figure 3.

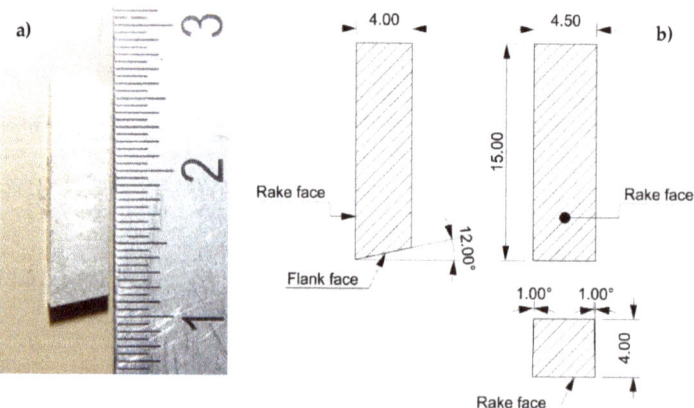

Figure 3. Blank cutting tool. (**a**) H13A from Sandvik. (**b**) Cutting geometry.

Chip morphology was analyzed according to ISO 3685:1993 (annex G). This is also a useful indicator of tool life. Angle wear can be observed by observing the tool wear via its initial profile, as shown in Figure 4.

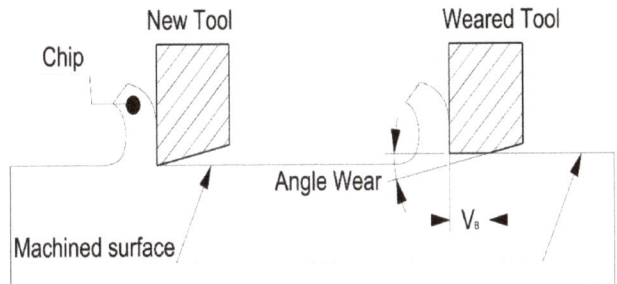

Figure 4. Orthogonal cutting configuration and evaluation zone for flank wear.

Further evaluation of the material, chips, and tools was performed with a Quanta 200 scanning electron microscope (FEI, Hillsboro, OR, USA) equipped with a Phoenix Energy dispersive X-ray spectroscopy (EDX) system (EDAX, Mahwah, NJ, USA).

4. Results

4.1. Introduction

The manufacturing process of the MMC sheet was based on the method of infiltration by pressing the aluminium alloy over a preform of SiC. Infiltration was made in a cast created to give the final shape to the sheets, and its design corresponds to the net shaping criteria and techniques. Because of the manufacturing process, a skin of aluminium is formed over the sheet surface, which can reach a thickness of 0.08 mm and would need to be removed to find the "pure" MMC.

The images captured and shown above, Figure 5; allow the observation of the irregular dispersion of the reinforcement over the MMC surface, and also, its different sizes. Reinforcement shows two singular characteristics, namely, irregular shapes and a range of different sizes. The uneven distribution of the reinforcement sizes has been reported previously [6,43].

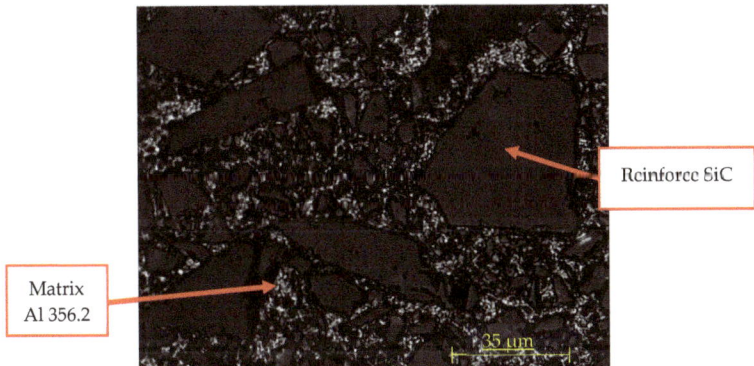

Figure 5. Stereoscopic Optical Microscopy (SOM) image at 40× magnification of the aluminium MMC with SiC reinforcements.

Using the picture processing software Image J, the areas occupied by each reinforcement were measured in four different places within the sheet. For this, a minimum-area size filter has been used for the initial pictures, so that smaller sizes are not considered, thereby eliminating the noise effect. The number of detectable size reinforcements in each of the four samples analyzed varies, almost reaching 12.50% for the those of the same material and between very similar study areas, with the lack of homogeneity being around 10%. Depending on the relative position of the reinforcement, the properties of the material can differ significantly.

The percentage of area occupied by detectable sizes of the reinforcement varies between 43.16% and 55.13%. Nonetheless, these values do not overlap with the maximum or minimum reinforcement quantities. The size and quantity of reinforcements is scattered, so that the properties of the material might have some heterogeneity.

Table 4 shows the microhardness values taken in different areas and samples of material. Relevant parameters are also included in the table. Registered values of microhardness, from highest to lowest, are shown. The highest value belongs to Sample 1 and was obtained in the reinforcement. The second value, from Sample 2, was taken in the matrix. Sample 3 had the second highest value, registered in the matrix-reinforcement interface. The last value, from Sample 4, was also taken from the matrix.

Table 4. Microhardness (HV) of the aluminium MMC.

Sample	Force (N)	Time (seconds)	HV (kg/mm^2)	Notes
1	0.98	15	1786	Measured in reinforcement
2	1.96	15	262	Measured in matrix
3	1.96	15	600	Measured in interface
4	1.96	15	186	Measured in matrix

4.2. Machining Process

The high-speed film detected the formation of burrs from the very beginning of the machining process, which obstruct the visualization of the cutting zone. Furthermore, a great number of shoots were observed, and, although the surfacing of segmented and jagged chips in all cases is significant, this is an odd behavior for the matrix material at this particular cutting depth. Both of these properties provide some information about the brittleness of the chip (Figure 6).

The literature review on chip formation established that the machining process of aluminium MMCs with SiC reinforcements generates "sawtooth" chips because of the shape of cross section used to produce them [44]. However, a significant amount of the machined material was reduced to powder. This experiment corroborates [45] that reinforcements, which have a higher hardness, cause a reduction in MMC ductility, facilitating the appearance of these kinds of jagged chips. The tool rake angle of 0°, and even negative values, noticeably influences the formation of the chips.

Regarding the shoots, these present the phenomenon described [46]. The material cannot withstand the cutting tension and it breaks and is shot away once the chip has reached its maximum size. Some of the materials studied during the literature review do not belong to the metal matrix composites with ceramic reinforcement group. These are mostly very hard steels and some alloys employed in aeronautics, such as the titanium alloy Ti6Al4V or a nickel alloy known as Inconel 718. From the results obtained, we can observe a similar behavior in the MMC in terms of the chip formation mechanism and the type of chip resulting from the machining.

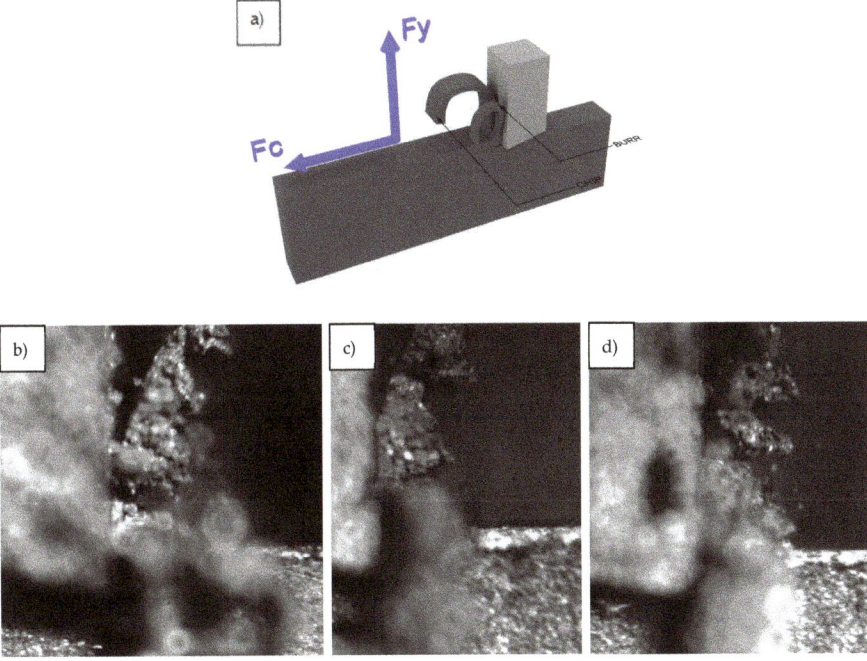

Figure 6. (**a**) Scheme of the tool position and images of the machining as obtained with high-speed film: (**b**) Vc = 20 m/min, one tool path, and d = 0.2 mm; (**c**) Vc = 30 m/min, three tool paths, and d = 0.2 mm; (**d**) Vc = 50 m/min, three tool paths, and d = 0.2 mm.

4.3. Chips Characterization

One of the characteristic aspects identified during the review of published papers was the cyclic variation of the shear angle along the machining process. This cyclic change of the shear angle in tandem with the chip formation angle can be verified [47]. In line with this, some fluctuations in the components of the cutting forces were observed while machining a nickel alloy, namely, Inconel 718 [48], although the resultant force did not change, owing to the cyclic variation of the chip segmentation (changes in the chip thickness from the shear angle). This phenomenon is still yet to be proven to occur in the case of aluminium MMC reinforced with SiC. The shootings observed during the film recording cannot allow us to confirm the adherence of the material to the cutting edge, although this is proven in later analysis.

After machining under the different experimental conditions scheduled, the morphology of the chips obtained were determined to be of the segmented type with a curved profile. Attending to International Standard Organization standard ISO 3685:1993, the chip formed during the cutting process has characteristics which are related to the work material, tool material, tool geometry, condition of the cutting edges, cutting edge position, and cutting data and conditions. According to annex G, these chips are of the 6.2 type (arched and shredded), which indicates that chip movement is produced towards the workpiece and in the direction of feed motion. However, during the cyclic chip-forming process, the shear angle is not simply a constant value, but changes cyclically as the cyclic chip formation takes place [48]. While machining ductile materials, chip formation is accompanied with very severe plastic deformation at the shear zone, where if a work material does not have enough ductility, this will result in deformation, which is limited by the crack initiation at the surface where no hydrostatic pressure exists [44]. As the reinforcement percentage is much higher in the case of the matrix, greater hardness is registered and the material is less ductile, resulting in the generation of a discontinuous chip because of high reinforcement concentration in the matrix [49].

Cutting forces in the machining of hard materials are, in spite of their hardness, not necessarily high because of the following two effects, namely, relatively small plastic deformation of the chip due to the crack formation mentioned above, and the relatively small area of tool-chip contact, which reduces the friction force [44].

The addition of SIC particle reinforcement into the aluminium matrix has caused a reduction in ductility and makes the material ideal for producing semicontinuous chips, which can be easily discarded after machining. This addition could be beneficial in terms of machinability; however, it creates some fluctuation in the force measurements [49]. Clarifying the mechanism and factors responsible for sawtooth chip formation and exploring the relationship between the formation of sawtooth chips and cutting force fluctuation is of great importance to the development of an effective cutting process monitoring strategy [48]. The face of the chip in contact with the tool is a smooth and shiny surface with crevices. This can be attributed to the lateral distortion forces induced by friction, which is a differentiating characteristic from traditional aluminium alloys [50].

The reverse or back side of the chip shows an irregular profile which is matt and discontinuous, allowing the shearing effects to show through and forming the characteristic sawtooth shape. Additionally, a burr area formed because of the fragility of the chip is observed in the high-speed film. It is possible to observe SiC particles near the chip surface, and these particles could damage the tool [51].

These effects occurred under all of the studied conditions (Figure 7). The same shine and smoothness on the exterior size of the chips was observed [48]. This phenomenon is attributed to the high tensions which were created when in contact with the tool, and the shearing tensions occurring on the tool–chip interface. In the fracture zone of the chip, a lack of homogeneity on the fracture surface can be observed, and it can be concluded that said fracture cannot be considered ductile, but its appearance resembles that of a brittle fracture. This coincides with the increasing tenancy, related to the incorporation of reinforcements to an aluminium alloy, seen with the traditional characteristics of an MMC.

4.4. Tools Characterization

In Figure 8, the rake faces of the tool are shown. The wear of the tool is significant in all cases, as can be seen by loss of tool material in the cutting edge and the material adhered in the rake face. Note that secondary adhesion is the most common mechanism related to the machining of aluminium alloys, and, in general, and it appears in this case with the MMC matrix. However, the SiC reinforcements should cause abrasion wear to the tool, and this favors material loss of the tool and provokes flank wear. Adhesion is presented almost consistently throughout the tool edge, creating an adhered layer or BUL (built-up layer). However, a clear relationship with the cutting parameters has not been revealed, although the BUL appears to increase as the machining time does, causing more damage to the tools after three tool paths.

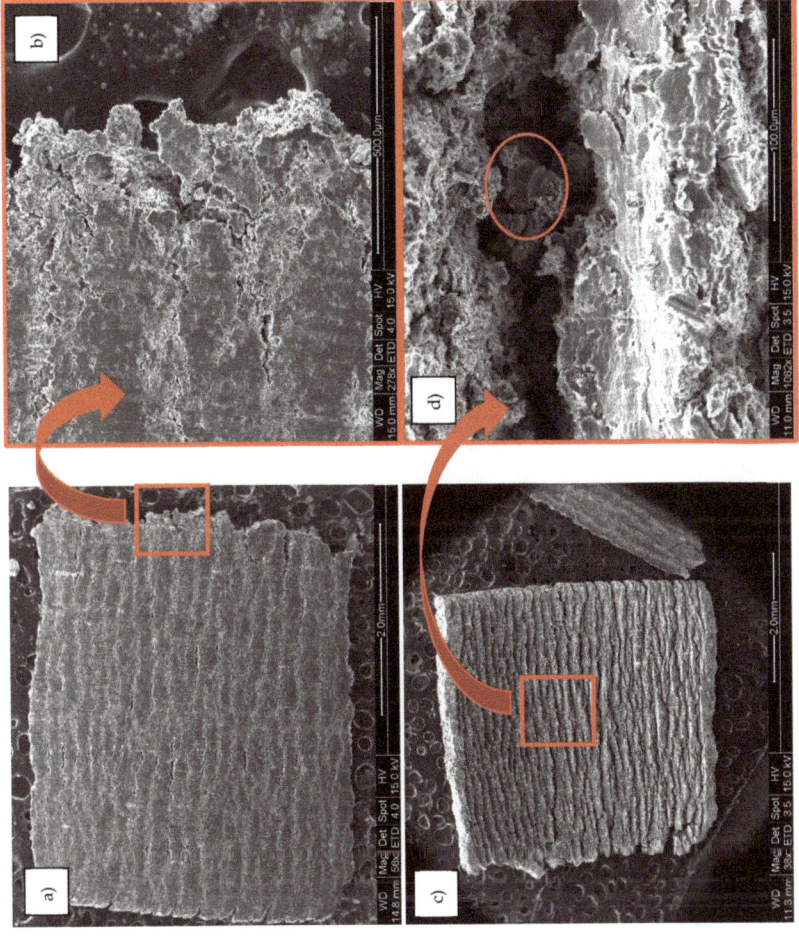

Figure 7. SEM images of MMC chip at Vc = 20 m/min, d = 0.2 mm, after three tool paths. (**a**) Front side. (**b**) Burrs at the front side. (**c**) Back side. (**d**) Reinforcement SiC particle in the back side.

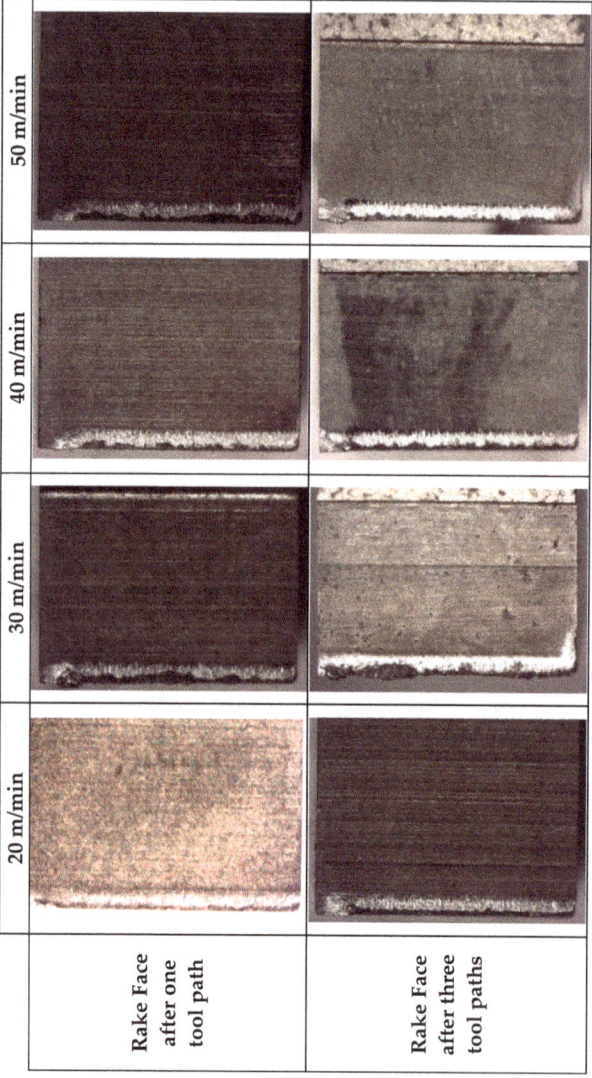

Figure 8. SOM images of tools used in machining test, with the rake face at 200X. Vc = 30 to 50 m/min, d = 0.2 mm.

The tool wear mechanism seems to be a mix between the dominant ones for each phase of the material. The secondary adhesion, distinctive for aluminium alloys (the matrix material), is boosted by high temperatures, as it is a thermomechanical mechanism. Besides, there is a loss of material produced by the abrasion that SiC reinforcements cause, which eliminates material from the tool edge and flank. This mechanism will also cause a rise in the temperature due to friction, which will favor secondary adhesion. The described mechanisms acting synergistically will result in the tool being rendered unusable after only very short machining times.

This behavior is found to be similar to those of different materials such as stacks, although it presents an added difficulty as the machining is not done in batches.

It is also remarkable that this mechanism seems to be dependent on cutting speed, as well as the machining time, and the effect increases at low velocities. This can be attributed to the change in behavior produced by friction, as an increase in friction is paired with greater abrasion, causing plasticity within the matrix, which results in an increase in adhesion, and thus the synergetic effect.

To quantify the adhesive wear, we quantified the length of contact between the tool and chip, referred to as h. After this analysis, it was noted that there is a fairly constant length, settling at 0.375 mm. This value allows quantification of the area damaged by secondary adhesion in the tool. Note that, according to the calculation made earlier, this value differs by up to 80% of the calculated value, meaning that the Shaffer model could not be applied for these materials, and that additional information is still needed to ensure the value of the shear angle. The tools were also analyzed with SEM/EDS techniques (Figure 9).

Analyzing the material adhered shows that it is material from the MMC; however, it features a different concentration of Al/SiC, depending the tool zone. In this form, zones rich in Al and others rich in SiC exist. This suggests that the reinforcement can be carried out at a high speed. This effect is in accordance with the particles shown by high-speed film and with the particles incorporated into the surface of the chip, where the effect can increase damage to the tool.

This fact could be considered secondary, because although it is of great importance from the point of view of the process, skipping particles of SiC with a high degree of abrasion and the projection itself can damage the tool, and this can also represent significant damage to equipment, meaning extreme care must be paid when machining these materials.

However, the general appearance of the tool is similar to that which is obtained during the machining of other aluminium alloys, and even the wear area, which manifests adhesion behavior that appears in the machining of other aluminium alloys in other processes [15,36]. Then, it is possible that the mechanism of predominant wear is secondary adhesion, as in most other commonly used aluminium alloys. However, some SiC particles have been deposited on the rake face. These particles can behave like the intermetallic particles in aluminium alloys, but with an increased abrasion power and concentration, and this increases abrasion and flank wear. This abrasion effect is different to all other aluminium alloys and provokes the rapid damage that has been shown previously.

The SiC particles were deposited close to the cutting edge (Figure 10), and although the rake face (blue area, tungsten rich area) was predominantly covered by aluminium (yellow area, aluminium-rich area), it appears that many areas are covered with material from the reinforcement (red area, carbon-rich area).

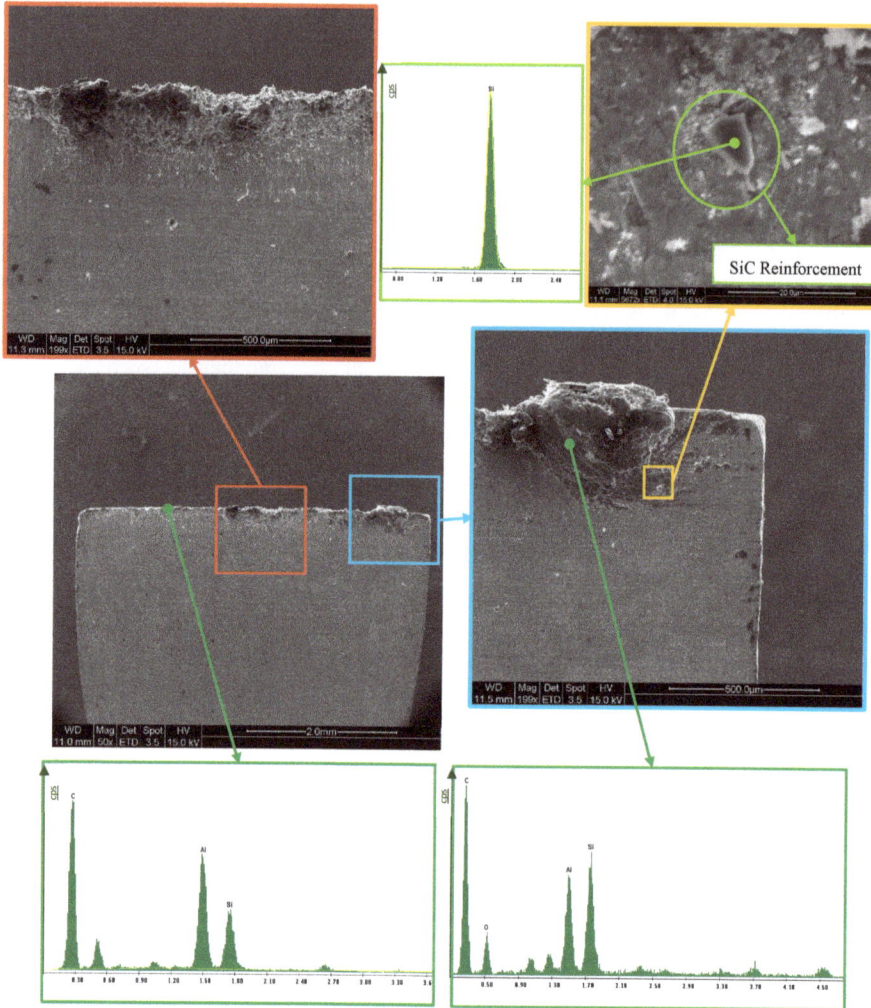

Figure 9. SEM/EDS analysis of the rake face of the cutting tool (Vc = 50 m/min, d = 0.2 mm, and one tool path). Y-axis in cps units. Two details of tool wear: figures on the **left** (red frame) are showing aluminium adhesion and figures on the **right** (blue and yellow frames) are showing SiC adhesion.

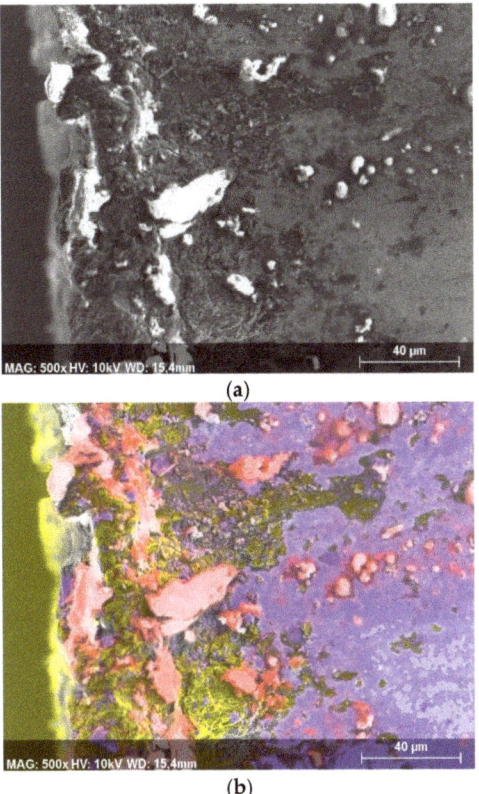

Figure 10. Scanning electron microscope analysis of the tool, where Vc = 40 m/min and d = 0.2 mm, and one tool path was used: (**a**) Original image. (**b**) Spectroscopy image with concentration by color: Al (yellow), W (blue), and Si (red).

Additionally, data taken from the characteristic zones (Figure 11) in the cutting edge show a high concentration of aluminium, which decreases when moving away from the edge (Figure 11, yellow areas). On the other hand, here, the reinforcements are distributed randomly, resulting in a non-homogeneous distribution of compounds. There are areas of very thin, electrotransparent material covered by the matrix alloy, whereas other areas only appear to be embedded with the reinforcement, as shown in the analysis earlier. It can be said that these reinforcements are firmly fixed on the tool without requiring the intervention of the aluminium (Figure 11, white zones).

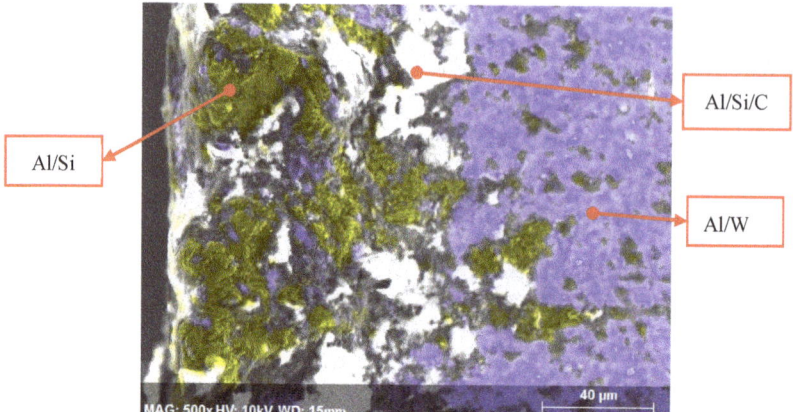

Figure 11. Spectroscopy image of compositional analysis of the cutting edge and the rake face (Vc = 20 m/min, d = 0.2 mm, one tool path).

Meanwhile, in areas close to the cutting edge, areas of both the matrix and reinforcement coexist, as has been previously seen. This suggests that although the secondary adhesion in the case of aluminium alloys is a thermomechanical process, in this material, the reinforcement induces a significant mechanical effect and, therefore, introduces the well-defined abrasion behavior.

To observe the flank wear, images of the profiles of the tools were analyzed (Figure 12). It is still possible to identify the material adhered and a high level of abrasion on the tools, denoted by the blurring observed in the high-speed film.

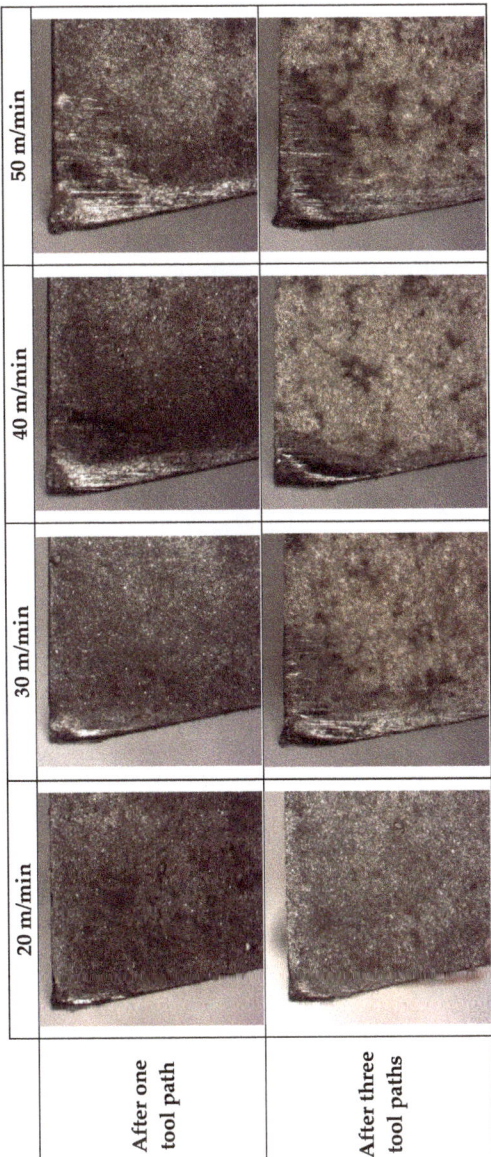

Figure 12. SOM images of the profile of the tool, where d = 0.2 mm in all of the cutting speeds studied.

It is possible to appreciate a regrown edge in all tools, where bigger sizes arise from those produced at a higher cutting speed. This occurs similarly in other aluminium alloys during analogous processes [27,51]. Furthermore, it is possible to observe the effect of material loss in the tool, as well as flank wear. This damage is increased with the increase of cutting speed and time spent machining, and it also appears after short machining times.

This might indicate that the interval of stable evolution in machining is practically nonexistent, and the loss of material from the tool is produced in such a short time that it would lead to the catastrophic breakage of the tool after only very short machining times.

After analyzing the tool with SEM (Figure 13), it is possible to see a similar behavior with abrasive particles which are stuck or embedded on different areas of the tool, and the presence of lines of abrasion marks, symptomatic of flank wear, which, in this case, differs from the previous one in intensity, since it seems to be more pronounced. This behavior is maintained almost independently of machining time, although the loss of material of the tool is very marked in the first moments of machining and progresses rapidly. Therefore, abrasion is a very important phenomenon, in addition to (or possibly even more than) the secondary adhesion, which can lead to the futility of the tool after only a very short time.

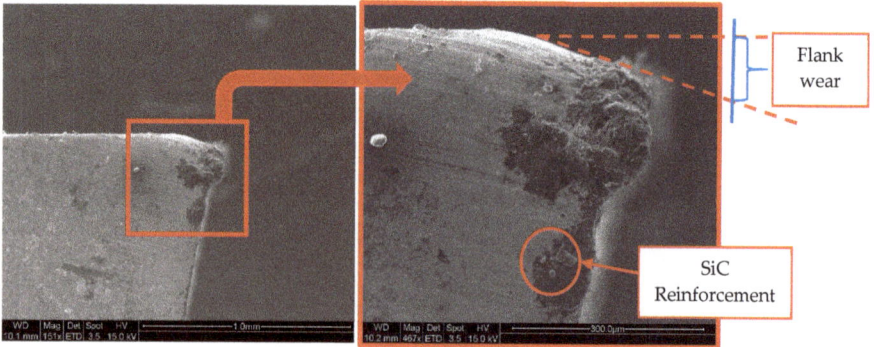

Figure 13. SEM analysis of the tool profile (Vc = 50 m/min, d = 0.2 mm, and three tool paths).

To analyze the loss of material in the flank, the profiles of the tools have been studied. The flank wear was measured using image processing techniques. The angle that forms the profile of the original rake face with the rake face after machining has been studied, in order to subsequently measure the parameter of flank wear according to ISO 3685. The data are shown in Table 5 and are represented in Figure 14.

Table 5. Values of angle wear and flank wear according to the cutting parameters.

	Machined Length	20 m/min	30 m/min	40 m/min	50 m/min
Angle wear (°)	one tool path	7.85	7.72	13.94	5.28
	three tool paths	2.56	12.12	3.95	14.39
Flank wear (μm)	one tool path	423.01	447.00	522.01	345.01
	three tool paths	435.00	636.00	597.01	489.00

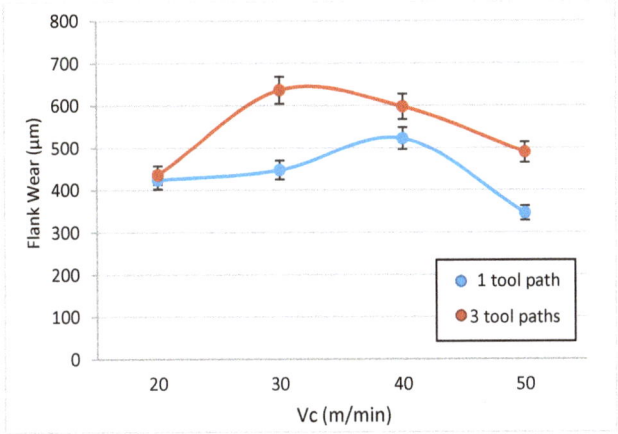

Figure 14. Flank wear evolution as a function of cutting speed and different machining times.

It can be seen that there does not seem to be a definite pattern here, and this can be related with the random distribution of the reinforcement when it is projected. However, the flank wear in all cases was greater than the flank wear accepted by ISO 3685 (300 µm), and this value is exceeded after only very short machining times.

According to this value, the tool must be replaced after only one tool path. However, after three tool paths, a striking effect takes place; however, wear increases clearly in a few cases, which would lead to an increase in the adhered material, whereas in other cases, it appears to decrease. This manifested adhesion fills the gaps that were occupied by the tool material and, therefore, this apparently causes the value to drop.

This effect is false; however, as the bonded material does not display the mechanical properties of the tool, it will tend to fall off easily, pulling particles from the tool with it.

Therefore, the general trend is an increase in the loss of material from the tool in line with the increase in cutting speed and machining time, reaching values that exceed double the value recommended for the replacement of the tool after only a very short period of machining.

This flank wear provokes a tapping effect on the tool that increases the friction and affects the cutting forces. The values of the cutting forces in the different tests are shown in Table 6.

Table 6. Average values of the cutting forces obtained in the machining tests.

Vc	Machined Length	Fc (N)	M (N)
20 m/min	one tool path	720,000	887,010
	three tool paths	832,007	1,118,255
30 m/min	one tool path	900,277	1,295,552
	three tool paths	879,159	1,351,942
40 m/min	one tool path	837,356	1,068,034
	three tool paths	859,668	1,308,143
50 m/min	one tool path	811,942	1,085,266
	three tool paths	931,908	1,434,406

4.5. Cutting Forces

The machining of MMC reinforced with SiC generate chips of a "sawtooth" type, and thus a friction phenomenon appears, which causes fluctuation cutting forces [49]. The segmentation frequency of the sawtooth chips and the fluctuation frequency of the cutting forces are highly similar. This indicates that the formation of the sawtooth chips is the most important factor influencing the periodical fluctuation of the cutting force components in the turning of Inconel 718 [48].

In the traditional cutting process, the component of force in the direction of cutting, Fc, is the most important property of the component, and it has the biggest value. The second largest component, the vertical component of the force, Fy, is greatly different. Studying the data obtained, this can be corroborated; however, Fy is much higher than the expected values, being greater than the values of Fc even after three tool paths. Increasing the cutting speed and the machining time increases the cutting forces. This is the same behavior observed during the tool wear study.

The forces registered in two axes have been combined to form resultant M; a comparative table; Table 6; and a graphic design, Figure 15.

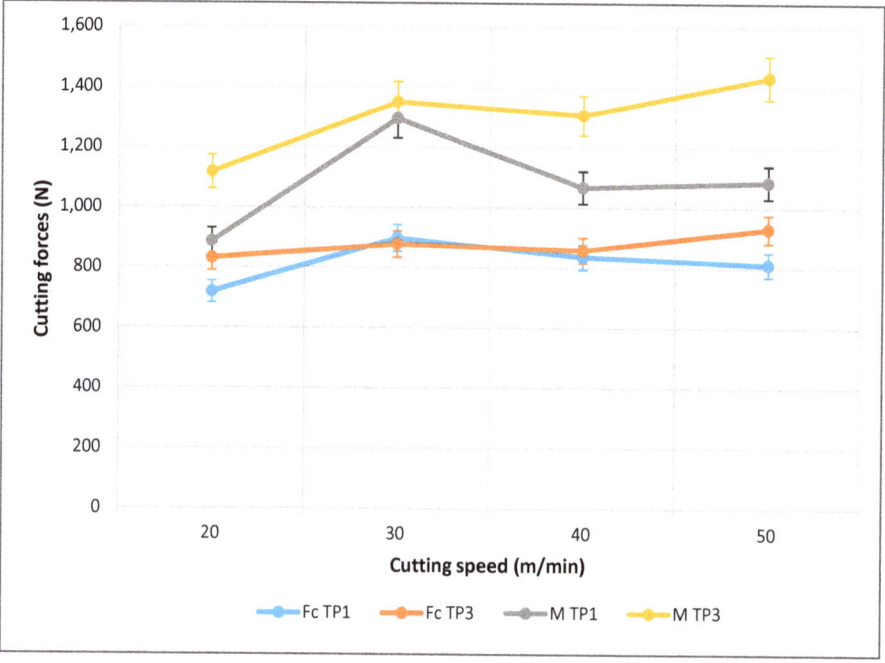

Figure 15. Cutting forces depending on the speed of cutting for several machining tool paths.

According to the composition of the resultant of the forces, the modulus has been obtained through the square root of the quadratic addition of the components, as shown in the formula below.

$$M = \sqrt{Fc^2 + Fy^2}$$

If analyzed more thoroughly (Figure 15), when considering only one tool path, it can be seen that with a Vc of 20, the values for both forces (Fc and M) are at their lowest. At this initial point, both values are increasing, where M is more significant than Fc, until reaching a peak Vc of 30, where both forces decreased until they were almost parallel up until a Vc of 40. Once this point was reached, from here to the end (Vc of 50) the values remained almost horizontal.

In the case of three tool paths, this is similar to the previous example. From Vcs of 20 to 30, the values increased; however, once reaching this peak, as in the case of one tool path, the values reduced until a Vc of 30. Between a Vc of 40 and 50, once again these values increased instead of remaining similar, as in the case of the previously registered values at a Vc of 40 for both cases, where Fc and M were both higher in the case of TP3 than TP1.

Even though the force Fc is similar in the one tool path case, the three tool path values are quite different for Fc or M, where M is always greater.

Note that although machining times are very low, this still provides evidence of the great influence of wear. This influence, as seen in the analysis of the tool, gives rise to the increase in cutting speed, which may cause increased wear and even an increase in flank wear. This would mean that the phenomenon of abrasion is increasingly more important, and, as noted, would lead to the catastrophic failure of the tool.

As seen in Figure 15, there is a dependency with the machining time that grows with the cutting speed. Slower speeds show an almost stabilized behavior, whereas the tendency is much more pronounced with increasing speeds.

On the other hand, in the case of the increased abrasion in the previously observed profile of the tool, which grows with the cutting speed, the phenomena causing this is lateral stress, and it therefore cannot be explained with standard theories about orthogonal cutting, thus an Fx force cannot exist.

When comparing the values of forces obtained with other light alloys, there are important differences to note. If compared with a Ti alloy [52], it can be seen that the forces are slightly lower when moving to depths of 0.2 mm in the range of 800 to 1500 N. However, when compared with forces that are obtained in an alloy of aluminium, it can be seen how the ranges are inferior and even with more aggressive settings and in more complex processes, they are only able to reach about 500 N [53]. This again confirms the hypothesis that increasing the concentration of reinforcement translates into increased tenacity, which causes an increase in the forces that by their inherent abrasive behavior leads to a more significant increase in damage, which is evident when considering the rapid loss of material in the cutting tools.

A brief summary of results can be found below.

- Microhardness has been studied here. The results show a wide range of values. The highest value belongs to the reinforcement material, whereas the lowest value belongs to the Al matrix. Intermediate values were obtained from matrix or interface areas with matrix reinforcement.
- Sawtooth chips were formed during machining. Some alloys employed in aeronautics, such as the titanium alloy Ti6Al4V or the nickel alloy known as Inconel 718, produce a similar behavior to that observed here, in terms of the chip formation mechanism and the type of chip resulting from the machining.
- Cyclic variation of the shear angle along the machining process can be noticed. Attending to the ISO 3685:1993 standard, the chips formed during the cutting process are of the 6.2 type. The addition of SiC particle reinforcement into the aluminium matrix caused a reduction in its ductility and makes the material ideal for producing semicontinuous chips.
- Tool wear is significant in all cases. Here, secondary adhesion appears (characteristic of Al alloys). The main factor of abrasion wear is the SiC reinforcement, which induces flank wear. A clear relationship with the cutting parameters has not been revealed. The Shaffer model could not be applied to these materials, as the values obtained differ by more than 80% of the calculated value.
- It is possible to appreciate a regrown edge in all tools, where a bigger size is found for those tools produced at higher cutting speeds. Furthermore, it is possible to observe the effect of material loss in the tool, as well as flank wear. This damage is increased with the cutting speed and the time spent machining, and it appears even after short machining times.
- The cutting force Fy showed much higher values than expected, where even after three tool paths it was above the value of Fc. Even though the force Fc is relatively similar between the one and three tool path cases, the values of Fc and M are quite different, where M is always greater than Fc.

5. Conclusions

Regarding the chip analysis, this confirmed the behavior previously described by other authors, showing a shredded arch type with a segmented or sawtooth morphology.

The tools have suffered severe wear, even in the best cutting conditions.

With regard to the cutting forces, it can be ensured that the effect of abrasion causes a predominant effect of friction on the rake face that manifests itself in a very marked increase in the force Fy.

On the other hand, the machining time is a determining factor and the increase in the wear causes a simultaneous increase in the forces, reaching values close to those found in alloys in less than one second, and surpassing those of most other aluminium alloys.

The best results have been obtained with the average values of the parameters: speed and deep of cut. The highest speed and lowest depth of cut is not always the best option.

Author Contributions: D.R., M.B., and J.S. conceived and designed the experiments; D.R., S.R.F.-V., and P.F.M. performed the experiments; D.R. and M.B. analyzed the data; D.R. and M.B. wrote the paper; S.R.F.-V., P.F.M. and J.S. revised and corrected the paper. All authors have read and agreed to the published version of the manuscript.

Funding: This work has received financial support from the Spanish Government through the Ministry of Economy, Industry, and Competitiveness, the European Union (FEDER/FSE) and the Andalusian Government (PAIDI).

Acknowledgments: The authors acknowledge CPS Technologies for providing the material used for the experimentation. A very special acknowledgement to Mariano Marcos, who initiated the development of this research (in memoriam) and to Franck Girot, who guided us in some of the procedures of MMCs.

Conflicts of Interest: The authors declare no conflicts of interest.

References

1. Cardarelli, F. *Materials Handbook: A Concise Desktop Reference*; Springer: Tucson, AZ, USA, 2008.
2. Parker, S.P. *Dictionary of Scientific & Technical Terms*, 6th ed.; McGraw-Hill: New York, NY, USA, 2003.
3. Wenzelburger, M.; Silber, M.; Gadow, R. Manufacturing of light metal matrix composites by combined thermal spray and semisolid forming processes summary of the current state of technology. *Key Eng. Mater.* **2010**, *425*, 217–244. [CrossRef]
4. Clyne, T.W. An introductory overview of MMC systems, types, and developments. *Compr. Compos. Mater.* **2000**, *3*, 1–26.
5. Cambronero, L.E.G.; Sánchez, E.; Ruiz-Roman, J.M.; Ruiz-Prieto, J.M. Mechanical characterization of AA7015 Aluminium alloy reinforced with ceramics. *J. Mater. Process. Technol.* **2003**, *143–144*, 378–383. [CrossRef]
6. Tosun, G.; Muratoglu, M. The drilling of an Al/SiCp metal-matrix composites. Part I: Microstructure. *Compos. Sci. Technol.* **2004**, *64*, 299–308. [CrossRef]
7. BCC Research. Available online: www.https://www.bccresearch.com/ (accessed on 31 December 2018).
8. Evans, A.; San Marchi, C.; Mortensen, A. *Metal Matrix Composites in Industry. An Introduction and a Survey*; Kluwer Academic Pubs: Berlin, Germany, 2003.
9. Kainer, K. *Metal Matrix Composites—Custom-Made Mtls for Automotive and Aerospace Engineering*; Wiley-VCH Verlag GmbH & Co.: Weinheim, Germany, 2006.
10. Mavhungu, S.T.; Akinlabi, E.T.; Onitiri, M.A.; Varachia, F.M. The processing techniques and behaviour of aluminum. In *Aluminium Matrix Composites for Industrial Use: Advances and Trends*; Procedia Manufacturing: Amsterdam, The Netherlands, 2017; pp. 178–182.
11. El-Gallab, M.; Sklad, M. Machining of Al-SiC particulate metal-matrix composites Part I Tool performance. *J. Mater. Process. Technol.* **1998**, *83*, 151–158. [CrossRef]
12. Bodunrin, M.O.; Alaneme, K.K.; Chown, L.H. Aluminium matrix hybrid composites: A review of reinforcement philosophies. *J. Mater. Res. Technol.* **2015**, *4*, 434–445. [CrossRef]
13. Srivyas, P.D.; Charoo, M.S. Role of Reinforcements on the Mechanical and Tribological Behavior of Aluminum Metal Matrix Composites: A Review. *Mater. Today* **2018**, *5*, 20041–20053. [CrossRef]
14. Najem, S.H. *Machinability of Al-2024 Reinforced with Al2O3 and or B4C*; Repository of College of Material Engineering, University of Babylon: Baghdad, Iraq, 2011.
15. Kishore, D.S.C.; Rao, K.P.; Mahamani, A. Investigation of cutting force, surface roughness and flank wear in turning of In-situ Al6061-TiC metal matrix composite. *Procedia Mater. Sci.* **2014**, *6*, 1040–1050. [CrossRef]
16. Cronjäger, L.; Meister, D. Machining of fibre and particle-reinforced aluminium. *Cirp Ann.* **1992**, *41*, 63–66. [CrossRef]

17. Muthukrishnan, N.; Murugan, M.; Rao, K.P. Machinability issues in turning of Al-SiC (10p) metal matrix composites. *Int. J. Adv. Manuf. Technol.* **2008**, *39*, 211–218. [CrossRef]
18. Tosun, G.; Muratoglu, M. The drilling of AlSiCp metal–matrix composites. Part II: Workpiece surface integrity. *Compos. Sci. Technol.* **2004**, *64*, 1413–1418. [CrossRef]
19. El-Gallab, M.; Sklad, M. Machining of Al-SiC particulate metal matrix composites. Part II: Workpiece surface integrity. *J. Mater. Process. Technol.* **1998**, *83*, 277–285. [CrossRef]
20. Davim, J.P.; Antonio, C.C. Optimisation of cutting conditions in machining of aluminium matrix composites using a numerical and experimental model. *J. Mater. Process. Technol.* **2001**, *112*, 78–82. [CrossRef]
21. Davim, J.P. Design of optimisation of cutting parameters for turning metal matrix composites based on the orthogonal arrays. *J. Mater. Process. Technol.* **2003**, *132*, 340–344. [CrossRef]
22. Ramulu, M.; Rao, P.N.; Kao, H. Drilling of (Al2O3) p6061 metal matrix composites. *J. Mater. Process. Technol.* **2002**, *124*, 244–254. [CrossRef]
23. Kannan, S.; Kishawy, H.A.; Deiab, I. Cutting forces and TEM analysis of the generated surface during machining metal matrix composites. *J. Mater. Process. Technol.* **2009**, *209*, 2260–2269. [CrossRef]
24. El-Kady, O.; Fathy, A. Effect of SiC particle size on the physical and mechanical properties of extruded Al matrix nanocomposites. *Mater. Des.* **2014**, *54*, 348–353. [CrossRef]
25. Wang, T.; Xie, L.J.; Wang, X.B.; Jiao, L.; Shen, J.W.; Xu, H.; Nie, F.M. Surface integrity of high speed milling of AlSiC65p aluminum matrix composites. *Procedia CIRP* **2013**, *8*, 475–480. [CrossRef]
26. Kannan, S.; Kishawy, H.A. Tribological aspects of machining aluminium metal matrix composites. *J. Mater. Process. Technol.* **2008**, *198*, 399–406. [CrossRef]
27. Quigley, O.; Monaghan, J.; O'Reilly, P. Factors affecting the machinability of an AlSiC metal-matrix composite. *J. Mater. Process. Technol.* **1994**, *43*, 21–36. [CrossRef]
28. Monaghan, J.M. The use of a quick-stop test to study the chip formation of an Si:Al metal matrix composite material and its matrix alloy. *J. Process. Adv. Mater.* **1994**, *4*, 170–179.
29. Manna, A.; Bhattacharayya, B. A study on machinability of AlSiC-MMC. *J. Mater. Process. Technol.* **2003**, *140*, 711–716. [CrossRef]
30. Davim, J.P.; Silva, J.; Baptista, A.M. Experimental cutting model of metal matrix composites (MMCs). *J. Mater. Process. Technol.* **2007**, *183*, 358–362. [CrossRef]
31. Dabade, U.A.; Joshi, S.S. Analysis of Chip Formation Mechanism in Machining of AlSicp Metal Matrix Composites. *J. Mater. Process. Technol.* **2009**, *209*, 4704–4710. [CrossRef]
32. Coelho, R.T.; Yamada, S.; Le Roux, T.; Aspinwall, D.K.; Wise, M.L.H. Conventional machining of an Aluminium based SiC Reinforced Metal Matrix Composite (MMC) alloy. In Proceedings of the 13th International Matador Conference, Manchester, UK, 31 March–1 April 1993; p. 125.
33. Chaudhary, G.; Kumar, M.; Verma, S.; Srivastav, A. Optimization of drilling parameters of hybrid metal matrix composites using response surface methodology. *Procedia Mater. Sci.* **2014**, *6*, 229–237. [CrossRef]
34. Vanarotti, M.; Shrishail, P.; Sridhar, B.R.; Venkateswarlu, K.; Kori, S.A. Study of Mechanical Properties & Residual Stresses on Post Wear Samples of A356-SiC Metal Matrix Composites. *Procedia Mater. Sci.* **2014**, *5*, 873–882.
35. Gomez-Parra, A.; Alvarez-Alcon, M.; Salguero, J.; Batista, M.; Marcos, M. Analysis of the evolution of the Built-Up Edge and Built-Up Layer formation mechanisms in the dry turning of aeronautical Aluminium alloys. *Wear* **2013**, *302*, 1209–1210. [CrossRef]
36. Ramnath, B.V.; Elanchezhian, C.; Annamalai, R.M.; Aravind, S.; Atreya, T.S.A.; Vignesh, V.; Subramanian, C. Aluminium Metal Matrix Composites: A Review. *Rev. Adv. Mater. Sci.* **2014**, *38*, 55–60.
37. Basavarajappa, S.; Chandramohan, G.; Davim, J.P.; Prabu, M.; Mukund, K.; Ashwin, M.; Prasannakumar, M. Drilling of hybrid Aluminium matrix composites. *Int. J. Adv. Manuf. Technol.* **2008**, *35*, 1244–1250. [CrossRef]
38. Available online: http://www.matweb.com/search/DataSheet.aspx?MatGUID=f38ba0e663a14183927155e5cc5d21a1&ckck=1 (accessed on 20 December 2019).
39. CPS Technologies Corp. Available online: http://www.alsic.com/data-sheets (accessed on 31 December 2018).
40. *Metallic Materials. Vickers Hardness Test. Part 1: Test Method*; AENOR: Madrid, Spain, 2006.
41. List, G.; Nouari, M.; Géhin, D.; Gomez, S.; Manaud, J.P.; Le Petitcorps, Y.; Girot, F. Wear behaviour of cemented carbide tools in dry machining of Aluminium alloy. *Wear* **2005**, *259*, 1177–1189. [CrossRef]
42. Gururaja, S.; Ramulu, M.; Pedersen, W. Machining of MMCs: A review. *Mach. Sci. Technol.* **2013**, *17*, 41–73. [CrossRef]

43. Brandes, E.A.; Brook, G.B. *Light Metals Handbook*; Elsevier Butterworth Heinemann: Oxford, UK, 1998.
44. Nakayama, K.; Arai, M.; Kanda, T. Machining Characteristics of Hard Materials. *Ann. CIRP* **1988**, *37*, 89–92. [CrossRef]
45. Pedersen, W.; Ramulu, M. Facing SiCp/Mg metal matrix composites with carbide tools. *J. Mater. Process. Technol.* **2006**, *172*, 417–423. [CrossRef]
46. Ozcatalbas, Y. Chip and built-up edge formation in the machining of in situ Al4C3-Al composite. *Mater. Des.* **2003**, *24*, 215–221. [CrossRef]
47. Komanduri, R.; Von Turkovich, B.F. New observations on the mechanism of chip formation when machining titanium alloys. *Wear* **1981**, *69*, 179–188. [CrossRef]
48. Zhang, S.; Li, J.; Zhu, X.; Lv, H. Saw-Tooth chip formation and its effect on cutting force fluctuation in turning of Inconel 718. *Int. J. Precis. Eng. Manuf.* **2013**, *14*, 957–963. [CrossRef]
49. Lin, J.T.; Bhattacharyya, D.; Ferguson, W.G. Chip formation in the machining of SiC-particle-reinforced Aluminium-matrix composites. *Compos. Sci. Technol.* **1998**, *58*, 285–291. [CrossRef]
50. Batista, M. Characterization of Secondary Adhesión Mechanisms and Influence in Tools Wear. Lightweight Alloys Dry Machining Application. Ph.D. Thesis, Cadiz University, Cadiz, Spain, 2013.
51. Álvarez, M.; Salguero, J.; Sánchez, J.A.; Huerta, M.; Marcos, M. SEM and EDS Characterisation of Layering TiOx Growth onto the Cutting Tool Surface in Hard Drilling Processes of Ti-Al-V Alloys. *Adv. Mater. Sci. Eng.* **2011**. [CrossRef]
52. Batista, M.; Calamaz, M.; Girot, F.; Salguero, J.; Marcos, M. Using Image Analysis Techniques for Single Evaluation of the Chip Shrinkage Factor in Orthogonal Cutting Process. *Key Eng. Mater.* **2012**, *504–506*, 1329–1334. [CrossRef]
53. Salguero, J.; Batista, M.; Calamaz, M.; Girot, F.; Marcos, M. Cutting Forces Parametric Model for the Dry High Speed Contour Milling of Aerospace Aluminium Alloys. *Procedia Eng.* **2013**, *63*, 735–742. [CrossRef]

© 2020 by the authors. Licensee MDPI, Basel, Switzerland. This article is an open access article distributed under the terms and conditions of the Creative Commons Attribution (CC BY) license (http://creativecommons.org/licenses/by/4.0/).

Article

Configuration Optimisation of Laser Tracker Location on Verification Process

Sergio Aguado *, Pablo Pérez, José Antonio Albajez, Jorge Santolaria and Jesús Velázquez

Design and Manufacturing Engineering Department, Universidad de Zaragoza, María Luna 3, 50018 Zaragoza, Spain; pperezm@unizar.es (P.P.); jalbajez@unizar.es (J.A.A.); jsmazo@unizar.es (J.S.); jesusve@unizar.es (J.V.)
* Correspondence: saguadoj@unizar.es

Received: 11 November 2019; Accepted: 18 December 2019; Published: 10 January 2020

Abstract: Machine tools are verified and compensated periodically to improve accuracy. The main aim of machine tool verification is to reduce the influence of quasi-static errors, especially geometric errors. As these errors show systematic behavior, their influence can be compensated. However, verification itself is influenced by random uncertainty sources that are usually not considered but affect the results. Within these uncertainty sources, laser tracker measurement noise is a random error that should not be ignored and can be reduced through adequate location of the equipment. This paper presents an algorithm able to analyse the influence of laser tracker location based on nonlinear optimisation, taking into consideration its specifications and machine tool characteristics. The developed algorithm uses the Monte Carlo method to provide a zone around the machine tool where the measurement system should be located in order to improve verification results. To achieve this aim, different parameters were defined, such as the number of tests carried out, and the number and distribution of points, and their influence on the error due to the laser tracker location analysed.

Keywords: laser tracker; machine tool; uncertainty; Monte Carlo method; verification

1. Introduction

Machine tools (MTs) are increasingly implemented in the industrial sector, which is itself increasingly competitive and seeks to increase production at a lower cost. For this, detection and reduction of MT errors is necessary.

Currently, there are two different ways to obtain MT geometric errors: direct and indirect measurement methods. Direct measurement methods consist of measuring the influence of every individual error from each axis in a particular position of the workspace of the MT [1]. Alternatively, indirect measurement methods obtain the joint influence of MT geometric errors based on multi-axis movement and its kinematic model. These are more widely used, especially in long range MTs, where direct methods require large scales, expensive dimensional measurement systems, and more time to check them [1]; so, the limitations of direct measurement cause indirect measurement to prevail in this type of machine.

Volumetric verification, using a laser tracker (LT) as a measurement system, is based on indirect measurement of geometric errors, characterising their combined effect [2]. So, the accuracy of verification results depends, among others, on errors of the MTs but also on the errors of the measurement system used. These latter errors are often ignored, and it is assumed that the performance of the measurement system is sufficiently accurate.

All measurements have a degree of uncertainty made up of systematic and random error sources. The systematic errors of LTs, such as environmental conditions or component assembly, can be estimated and compensated by software. However, random errors, such as LT measurement noise, cannot be compensated but can be reduced by the appropriate location of the measurement system, so improving verification results [3].

To find the optimal LT location, the technical specifications of the encoders, the characteristics of the MTs [4], physical restrictions such as the range of the laser tracking receiver [5], and even temperature variations [6] are required.

This paper presents a developed algorithm able to determine the influence of LT measurement noise on the verification results. The algorithm takes into consideration LT characteristics and MT workspace. In addition, the developed software uses the Monte Carlo method to provide the area where the LT should be located with its probability distribution function (PDF).

2. Materials and Methods

2.1. LT as Verification Measurement System

A LT is a portable measurement system that provides, in a spherical coordinate system, the position of a measured point. It is often composed of a laser mechanism oriented by means of angular encoders, an interferometer block, a position-sensitive device (PSD), optics responsible for the beam division, a reflector, and a control unit. Point coordinates are determined by comparing the measurement beam with the reference beam from the laser interferometer together with the combination of the azimuth and polar angle encoders of its head, which provide two rotational degrees of freedom of the LT (Figure 1).

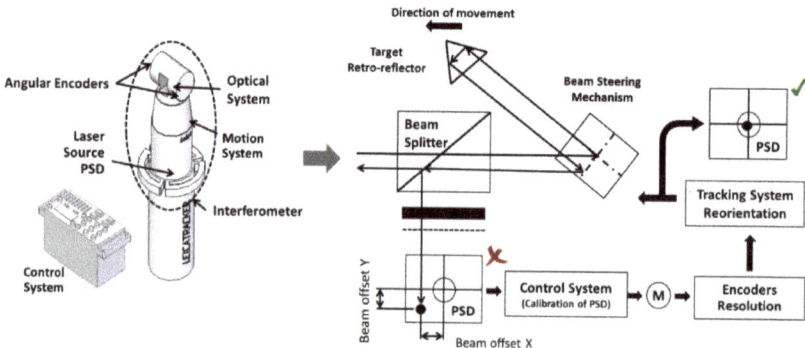

Figure 1. Errors due to encoders and sensor. Position-sensitive device (PSD).

2.1.1. Error Sources in an LT

Like any other measurement system, LTs are affected by systematic and random errors. Currently there are three standards concerning performance evaluation of LTs: ASME B.89.4.19-2005 [7], VDI/VDE 2617-10 [8], and ISO 10360-10 [9]. These three standards provide different tests to verify the performance of an LT according to the specifications of the manufacturer, reducing the influence of LT errors on the measurement.

Gallagher [10] classified error sources as: angular encoder, tracking system, and component misalignments. Knapp [11] divided sources of errors into those due to environmental factors, data capture, approximations, and simplifications.

Errors due to the interferometer and optics are the result of environmental influences and LT calibration. Atmospheric effects, variations in the speed of light, and turbulence affect the physical characteristics of the laser beam [12]. The environmental conditions, pressure, temperature, and humidity produce variations in the refractive index of the air. These variations result in errors in the laser wavelength and finally, in the measured distance [10]. In a factory workshop without a temperature-controlled environment, the temperature can significantly fluctuate through the day. In [13] authors reported an example of an aircraft assembly facility with temperature variations of 8 °C over 4 h and variation on the vertical directions of 2.2 degrees. During the aircraft assembly process, if the beginning and the ending temperatures of the measurement survey vary by more than 2.2°, then the survey is considered void and has to be repeated. Nevertheless, environmental conditions present a systematic behaviour described analytically. Therefore, the LT control unit can compensate for this influence due to its meteorological position.

Moreover, installation of LT optics introduces a series of intrinsic errors such as the Abbe error, cosine error, and depth error. If the reflector does not move parallel to the measurement axis, a cosine error will occur. In the same way, if the reflector does not move along the measurement axis of the interferometer, an Abbe error occurs. Similarly, an error of calibration between the home and reflector provides a depth error that will be transferred to all measurement points.

Additionally, the main sources of error in a PSD are its resolution and the calibration procedure that was used to determine the relationship between the sensor output and the beam offset from the centre of the target used to calculate the measured point. This is minimised by the sub-system, consisting of two stepper motors, two optical angular encoders, and a motion control card. The two motors produce the azimuthal and polar rotation of the beam tracking system, allowing the laser beam to move towards the centre of the PSD target, minimising the offset. Depending on the resolution of the encoders used, a better adjustment of the offset will be made (Figure 1).

2.1.2. LT Location in the Verification Process

The presence of LTs is increasing daily in machining and metrology companies, as tools to improve the accuracy of MTs through verification. Although LTs can be used to measure errors through geometric or pseudo-geometric verification [14], they are more frequently used in volumetric verification.

For this, the equipment should be located inside the MT kinematic chain in the same place as the workpiece [2,15]. MT kinematic chains are classified based on the movement of the workpiece and tool. The MT presented in this paper has an XFYZ configuration, where F determines the fixed part of the machine, X represents the axis that moves the part and the LT during the verification process, and Y and Z represent the axes that move with the tool [2].

The MT+LT kinematic chain mathematically links the tool centre point with the part of the machine, taking into consideration the sequence of movement and geometric errors of the MT (Equation (1)):

$$\overline{X} + \overline{R_x}\, \overline{T_{lt}} + \overline{R_x}\, \overline{R_{lt}}\, \overline{X_{lt}} = \overline{Y} + \overline{R_y}\, \overline{Z} + \overline{R_y}\, \overline{R_z}\, \overline{T} \qquad (1)$$

where $\overline{X}, \overline{Y}$, and $\overline{Z}$ represent the translational vectors of the X, Y, and Z axes, respectively, with their geometrical errors and nominal displacements. $\overline{R_x}, \overline{R_y}$, and $\overline{R_z}$ are the rotational matrices of the X, Y, and Z-axes defined by their rotational errors. $\overline{X_{lt}}$ and $\overline{T_{lt}}$ represent the translation and rotational matrices between the LT and the origin of the MT coordinate system. Finally, $\overline{T}$ describes the offset of the tool [2].

Figure 2 shows the physical space available to locate the LT. Additionally, LT angular limitations such as maximum and minimum azimuth and polar encoders, and minimum radial distance or height couplings should be taken into account when locating the LT.

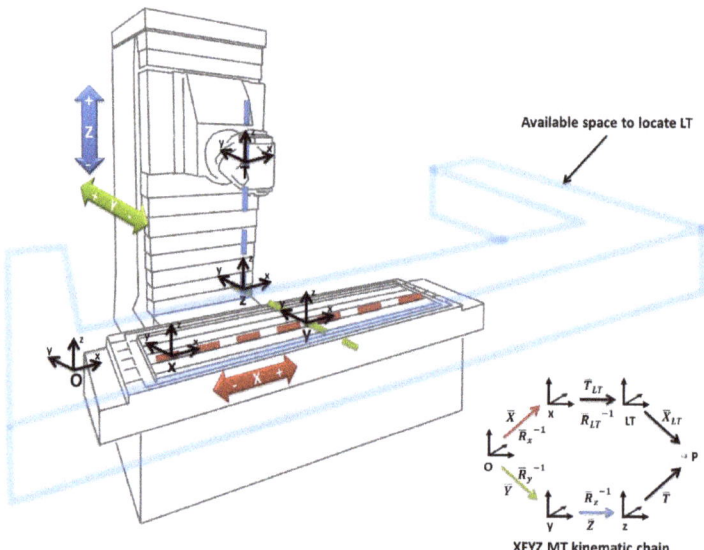

Figure 2. Machine tools (MTs) kinematic chain with XFYZ configuration, where F determines the fixed part of the machine, X represents the axis that moves the part and the LT during the verification process, and Y and Z represent the axes that move with the tool. Laser tracker (LT).

2.1.3. Influence of LT Location on MT Volumetric Verification

While systematic errors can be compensated by the LT control unit, other errors, such as the angle of incidence on the retro-reflector, or errors in the PSD sensor, due to angular encoders and the interferometer, produce a non-systematic error commonly known as measurement noise.

The influence of measurement noise on measured points is modelled with Equations (2)–(4). These Equations link data from the encoders and radial distances with their uncertainty, providing the uncertainty of a measured point in Cartesian coordinates:

$$u_x^2 = u_r^2 \cdot \sin^2\theta \cdot \cos^2\varphi + u_\theta^2 \cdot r^2 \cdot \cos^2\theta \cdot \cos^2\varphi + u_\varphi^2 \cdot r^2 \cdot \sin^2\theta \cdot \sin^2\varphi \tag{2}$$

$$u_y^2 = u_r^2 \cdot \sin^2\theta \cdot \sin^2\varphi + u_\theta^2 \cdot r^2 \cdot \cos^2\theta \cdot \sin^2\varphi + u_\varphi^2 \cdot r^2 \cdot \sin^2\theta \cdot \cos^2\varphi \tag{3}$$

$$u_z^2 = u_r^2 \cdot \cos^2\theta + u_\theta^2 \cdot r^2 \cdot \sin^2\theta \tag{4}$$

where r is the radial measured distance, u_r, the radial uncertainty, θ, the azimuth angle, u_θ, the azimuth angle uncertainty, φ, the polar angle, and u_φ, the polar angle uncertainty.

As LTs work with an absolute coordinate system and MT to verify, nominal MT points are not in the same coordinate system when are measured. So, their real uncertainty depends on verification of the LT location in the MT workspace.

2.2. LT Location Algorithm

2.2.1. Working Principles

The main aim of the developed algorithm was to provide the location of the area where the influence of the measurement system noise is smaller than the admissible error. The Guide to the Expression of the Uncertainty in Measurement (GUM) provides a framework for evaluating and expressing measurement uncertainty evaluating type A, type B, and combined uncertainties. Type A uncertainty is evaluated using statistical means, while type B is only evaluated based on experience or other information. However, the estimation of uncertainties using GUM relies on assumptions, such as non-linearity of the mathematical model, that are not always fulfilled [16]. In these cases, supplement 1 to the GUM describes the problem of uncertainty evaluation in terms of probability density functions to obtain the best estimate thorough the Monte Carlo method.

In this case, the influence of measurement noise is obtained through optimisation based on the Levenberg–Marquardt method, taking into consideration the following information:

1. Nominal MT verification points.
2. LT characteristics and limitations.
3. Limits of LT location.
4. Optimisation criteria to minimise the influence of uncertainty.
5. Number of Monte Carlo tests used to determine the location area.

The working principle of the developed algorithm is presented in Figure 3. First, the user introduces configuration parameters: LT characteristics, angular and radial uncertainties, limits of available workspace, maximum admissible error, mesh of measurement points, number of tests to simulate, and convergence criteria.

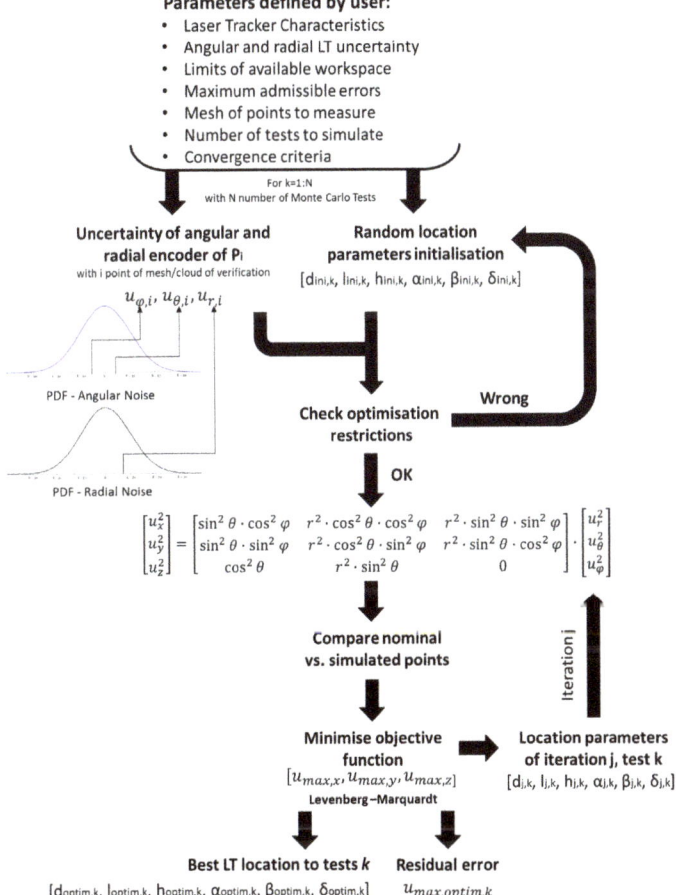

Figure 3. Working principle of the location algorithm.

Then, the algorithm begins to perform a test loop for k = 1 to k = n, with n being the number of tests defined by the user. Next, the algorithm randomly takes a value from the angular and radial PDF for each point. These values will be fixed throughout the test k, changing from one test to the next. Simultaneously, the algorithm looks for a random position within the available space for the initial location parameters. These parameters are defined by a 1 × 6 vector (d, l, h, α, β, δ), which transforms coordinates from the MT coordinate system to the LT coordinate system. Parameters d, l, and h represent a translation between the MT coordinate system and the LT coordinate system on the x, y, and z-axes, respectively, and α, β, and δ are the Euler angles that relate the orientation of the LT coordinate system to that of the machine tool, rotating first around the x-axis, then the y-axis, and finally around the z-axis.

If the restrictions are met, then the uncertainty of each point is calculated using Equations (2)–(4). If not, the algorithm looks for others. Afterward, the objective function (5) is calculated:

$$u_{max} = \left(u_{max,x}^2 + u_{max,y}^2 + u_{max,z}^2\right)^{\frac{1}{2}}. \tag{5}$$

This function is defined from a 1 × 3 vector made up from ($u_{max,x}$, $u_{max,y}$, $u_{max,z}$), considering the most restrictive criteria as admissible errors (all maximum uncertainties are at the same point). In this way, the influence of measurement uncertainty in the verification points will always be equal to or less than the residual optimisation result.

During optimisation, the algorithm modifies in each iteration j, the location parameters d, l, h, α, β, and δ, changing the spherical coordinates r, θ, and φ of each point to minimise the uncertainty influence.

When the optimisation is finished, the algorithm returns the optimisation parameters with the residual error. If the residual error is less than the admissible error introduced by the user, the algorithm stops. If not, the software divides the MT workspace into two areas and repeats the process. Moreover, the algorithm provides the PDF that defines the uncertainty behaviour depending on the location of the LT.

2.2.2. Case Study

All tests carried out had common simulation conditions: (a) the workspace to verify, defined by its limits of movement: 0 mm ≤ x ≤ 1500 mm, 0 mm ≤ y ≤ 600 mm, and 0 mm ≤ z ≤ 400 mm. (b) The available workspace around the MT where the LT could be located. This space was divided into two areas: narrow and wide. The narrow area had as available location parameters: 350 mm ≤ h ≤ 2000 mm, −500 mm ≤ d ≤ 2000 mm, and −2000 mm ≤ l ≤ −500 mm. The wide area had as available location parameters: 350 mm ≤ h ≤ 2000 mm, −2000 mm ≤ d ≤ −500 mm, and −500 mm ≤ l ≤ 2000 mm. As an additional restriction, the algorithm did not allow location of the LT inside the verification workspace (Figure 4). (c) The LT limits introduced in the algorithm were: azimuth angle θ −235° ≤ θ ≤ 235° and polar angle φ −60° ≤ φ ≤ 77°. (d) The PDF that defined the angular and radial uncertainties were normal distributions with μ = 20 μrad and σ = 1.5 μrad for the angular encoder and 4 μm ± 0.8 μm/m for the radial. Finally, the optimisation criteria limits were the same for all tests. These limits were: maximum iterations set at 1000, the minimum parameter variation set as 1×10^{-12} and the minimum objective function variation set as 1×10^{-5}.

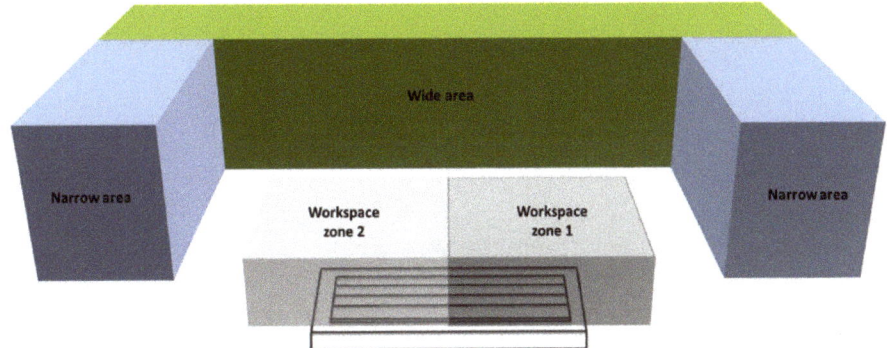

Figure 4. Admissible LT locations areas and workspace zones.

This paper studied the influence of the spatial distribution of MT workspace points, the number of points used to determine the LT location and the number of Monte Carlo tests used. Point distributions can be a mesh or a cloud. The number of points studied were 48 or 175 (Figure 5), and the number of Monte Carlo tests carried out were 100, 1000, and 10,000.

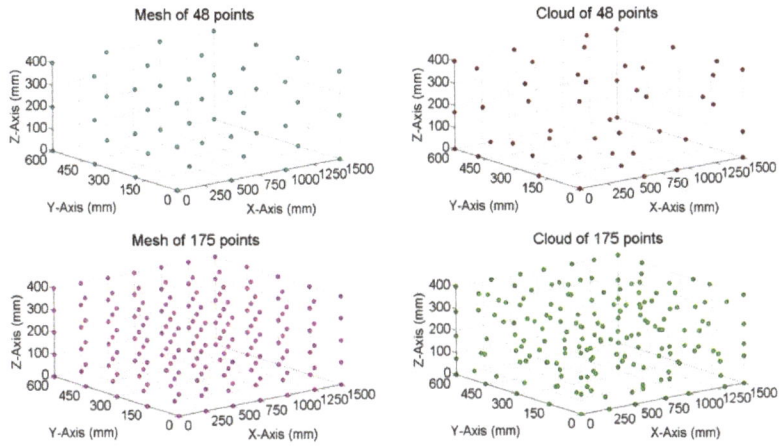

Figure 5. Distribution and number of points studied.

3. Results

3.1. Uncertainty Due to LT Location

The first tests carried out to study the uncertainty of locating an LT in the narrow and wide areas used a mesh distribution of points, with 175 points and 10,000 Monte Carlo tests to obtain optimal values of d, l, h, α, β, and δ.

As the colourmap of Figure 6 shows, when the LT is located in the wide area the error range was from 27.1 to 72.0 μm. That is to say, the test with the least influence of LT noise with specific values of $u_{r,i}$, $u_{\theta,i}$, and $u_{\varphi,i}$ with $i = 1..175$, provides a maximum uncertainty value of 27.1 μm, while the optimal parameters d, l, h, α, β, and δ in the test with the maximum uncertainty produce a value of 72.0 μm, taking into account that each test had different initial location parameters in the available workspace of Figure 4, with an initial error higher than final one presented in Figures 6 and 7 (residual error).

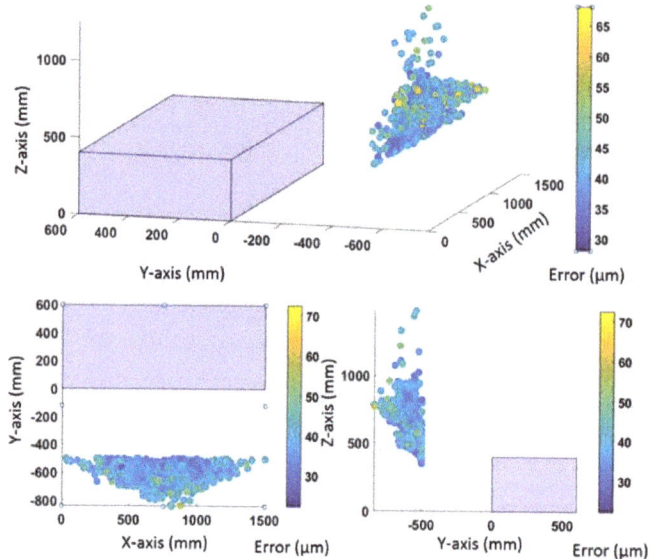

Figure 6. Error and LT location area in the wide zone using a laser–residual error.

When the LT is located in the wide area (Figure 6) there is a zone of conical shape where the tests present a high concentration of optimal locations with uncertainty values between 27.1 and 72.0 µm. So, the LT should be located in the wide area between −830 mm ≤ d ≤ −500 mm, 500 mm ≤ l ≤ 1000 mm, and 700 mm ≤ h ≤ 850 mm where the cone is registered.

When the LT is located in the narrow area, as shown in Figure 7a, noise uncertainty due to LT location increases from 56.2 µm to 170.5 µm. However, when the LT is located in the narrow zone there is an area to locate the LT of rectangular shape with l = −500 mm, 350 mm ≤ h ≤ 600 mm, and 0 mm ≤ d ≤ 600 mm, where the uncertainty is less than 115 µm. Figure 7b shows the histogram of residual errors, which allows study of the PDF that defines the behaviour of LT location influence. These errors are similar in the narrow and wide areas.

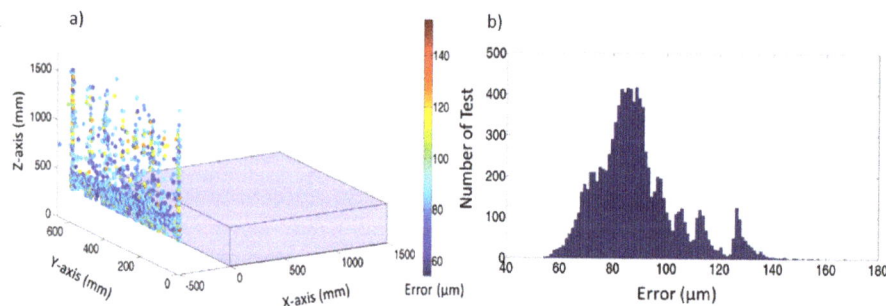

Figure 7. (**a**) Error and LT location area in the narrow zone using an LT. (**b**) Histogram of residual error.

When the residual error is higher than the introduced admissible error, the algorithm divides the verification area so that x = 750 mm (named as workspace 1 and workspace 2 in Figure 4). Then, the software analyses the influence of LT on these areas as independent workspaces, maintaining the location conditions. Table 1 compares the maximum and minimum errors when the MT workspace is divided.

Table 1. Influence of LT uncertainty depending on location and number of devices.

Zones	Workspace Divided Into Two Zones; Two LTs				Workspace 1 Zone, 1 LT	
	Min. Error (μm)		Max. Error (μm)		Min. Error (μm)	Max. Error (μm)
	Space 1	Space 2	Space 1	Space 2	Unique Workspace	Unique Workspace
Narrow	20.6	54.0	47.1	167.9	56.2	170.5
Wide	22.1	23.7	54.1	49.9	27.1	72.0

This shows that there is a relevant reduction inside the new workspace near to the LT in the narrow zone, where the minimum influence is reduced from 56.2 μm to 20.6 μm and the maximum is from 170.5 μm to 47.1 μm in Workspace 1, a reduction of approximately 70%. In the wide zone, Workspace 2, the reduction is not meaningful, around 3%. If two LTs are located on wide zones, the influence of LT uncertainty, minimum and maximum, is around 20% and 25% smaller, respectively.

3.2. Influence of Design Conditions on Location Results

Tests carried out using a unique LT in narrow or wide areas, as presented in the previous section, required a computational cost of around 6 h using a commercial PC. This was increased when the MT workspace was divided.

This value is too high for in situ machine verification. To reduce it, several configurations were tested, to study the influence of:

- Number of tests (100, 1000, or 10,000).
- Number of points (48 or 175).
- Distribution of points (mesh or cloud).
- Available workspace (narrow or wide).
- Number of LTs (1 or 2).

Due to the very large number of tests carried out, only the more relevant ones are presented here. Table 2 presents the computational cost of different design configurations depending on the number of points and tests. The distribution of points did not have a significant influence on computational cost.

Table 2. Computational time for different test configurations.

Points	Tests	Time	Points	Tests	Time
48	100	1′35″	175	100	4′20″
48	1000	14′32″	175	1000	42′31″
48	10,000	2 h 3′11″	175	10,000	6 h 37′56″

To study whether errors are significantly affected by different design configurations is necessary. Figure 8 shows the errors introduced by a unique LT in the wide zone, depending on the test configuration. The blue column represents the mean error of tests performed for each configuration. The red vertical lines show the range of the error produced for each configuration. The upper end is the maximum error chosen for a test and the lower end the minimum. For example, the design of a configuration consisting of a mesh of 175 points and 10,000 tests has an average error of 40.2 μm, a maximum error of 72.0 μm, and a minimum error of 27.1 μm, with a total range of 44.9 μm. The ninth column of the graph in Figure 8 is equivalent to Figure 6 and the results of the lower left configuration in Table 1.

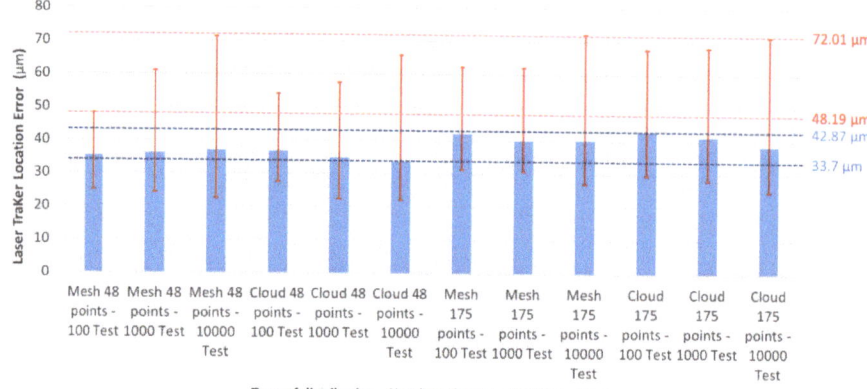

Figure 8. Error of LT location in the wide zone using one LT, depending on the type and number of tests, and the distribution and number of points.

In Figure 8 it can be seen that the mean error due to LT location is similar in all configurations, with a range of 33.7–42.87 µm. However, the maximum error range is 48.19–72.01 µm. Also, there seems to be a lower limit around the zone of 20 µm.

The configuration parameters that have more influence on measurement error due to LT location were studied based on statistical design of experiment (DOE) [17]. Three input parameters were studied: parameter A = points distribution: mesh or cloud; parameter B = number of points: 48 or 175; and parameter C = number of tests: 100 or 10,000.

Figure 9 shows the results of the DOE applied to the maximum and mean error, representing the effect and the basic contribution of each parameter. The most relevant is the number of points in both cases. The greater the number of points, the greater the error. The number of tests has a large influence on maximum error but not on the average, causing different effects. The type of point distribution is the least relevant parameter in error due to LT location. The combination has the opposite effect on maximum error, compared to individual (Figure 9a). Similar behaviour is observed in the averages (Figure 9b).

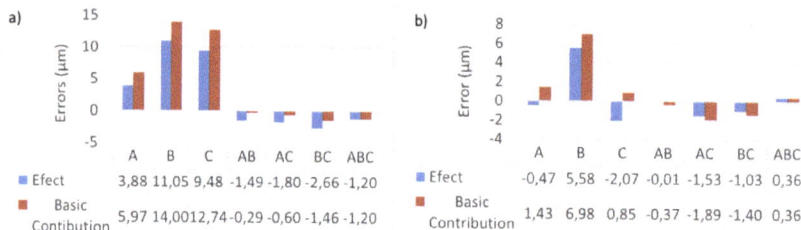

Figure 9. Effect and basic contribution of point distribution, number of points and number of tests on maximum error (**a**) and average error (**b**).

To study whether the influence of these parameters can be modelled as lineal regressions, the Scheffler regression function was used [18] to give the maximum and average errors (Equations (6) and (7)):

$$\text{max error } (\mu m) = 62.77 + 1.94A + 5.225B + 4.740C - 0.745AB - 0.900AC - 1.330BC - 0.600ABC \quad (6)$$

$$\text{average error } (\mu m) = 38.187 - 0.234A + 2.789B - 1.036C - 0.004AB - 0.764AC - 0.516BC + 0.181ABC. \quad (7)$$

To validate the adequacy of Equations (6) and (7), tests of different configurations, with 1000 tests, 48 and 175 points, and mesh and cloud distributions were used. In Table 3 we can see that these parameters do not have a linear behaviour, as might be anticipated from Figure 8.

Table 3. Adequacy of Scheffler regression functions for the average and maximum error values.

Configuration			Real Value		Estimated Value	
Type	Points	Tests	Mean (μm)	Max. (μm)	Mean (μm)	Max. (μm)
Mesh	48	1000	36.1	60.9	35.3	49.4
Cloud	48	1000	34.8	57.4	36.4	55.2
Mesh	175	1000	40.2	72.0	42.2	63.1
Cloud	175	1000	41.3	68.2	42.5	67.9

Table 4 shows the influence, in the wide zone, of distribution, number of points, and number of tests if the MT workspace is divided. Conclusions drawn from the results obtained are the same as those of the wide zone using one LT. The same is seen with the DOE tests.

Table 4. Influence of design parameters on wide zone LT location area and after division of the MT workspace.

Configuration			Wide Zone-Workspace Zone 1			Wide Zone-Workspace Zone 2		
Type	Points	Tests	Mean (μm)	Max. (μm)	Min. (μm)	Mean (μm)	Max. (μm)	Min. (μm)
Mesh	48	100	25.2	34.3	20.3	25.2	34.3	20.3
Mesh	48	1000	24.6	35.1	18.9	24.6	35.1	18.9
Mesh	48	10,000	24.4	35.1	18.6	21.3	35.1	18.6
Cloud	48	100	25.9	41.2	20.8	24.8	41.2	19.4
Cloud	48	1000	27.6	45.3	18.5	27.1	45.4	18.7
Cloud	48	10,000	25.7	48.1	18.7	25.3	48.2	18.90
Mesh	175	100	28.1	43.3	23.0	31.9	44.3	25.3
Mesh	175	1000	29.5	45.1	22.3	30.4	46.2	23.3
Mesh	175	10,000	29.2	54.1	22.1	29.5	49.8	23.7
Cloud	175	100	29.9	43.4	25.4	29.5	44.6	25.5
Cloud	175	1000	28.9	50.7	21.6	29.7	51.4	22.5
Cloud	175	10,000	30.5	48.1	18.7	30.1	48.2	19.3

Similar results were obtained when dividing the workspace volume into two areas: regardless of the design, there is a reduction in the maximum and average values of the error introduced. The error in Zone 1, compared to using only one LT, reduced from 34% to 21% in average values and from 50% to 21% in maximum error. In Zone 2, the average error was reduced from 42% to 22% and the maximum error from 50% to 21% (Table 4, Figure 10). As shown in Figure 10, when the MT workspace is divided, the range of error is reduced from 50.2 μm to 35.6 μm, and a minimum error support zone is also found at around 20 μm.

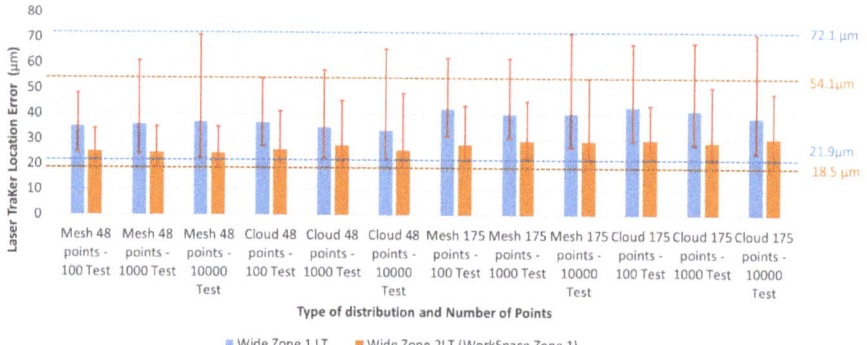

Figure 10. Error of the LT location in the wide zone. One zone vs. two zones with different configurations.

4. Discussion

Tests carried out show that there is no unique optimal position to locate the LT. As its uncertainty is defined by a PDF, each verification point will be affected by different values in each test. Therefore, there is one area where the PDF of LT influence is optimum. This depends on the measurement systems characteristic, therefore, the first step is to provide an adequate equipment characterization.

When only one LT is used to verify the whole MT workspace, the verification results can be improved by locating the LT in the wide area, inside the estimated zone with a cone shape. The Monte Carlo analysis provides an uncertainty range from 27.1 to 72 µm, providing a maximum error around 60% smaller than that obtained when the LT is located in the narrow area.

If the residual error obtained is too high, the division of the workspace into two zones provides an improvement in uncertainty due to LT location. This is especially relevant in the narrow zone, where the maximum error in workspace 1 is reduced by around 70%. If two LTs are used and located in the wide zone, their influence compared to the use of just one LT is improved by around 25%. Thus, the use of two workspaces and two LTs reduces their location influence. These results show that, in these cases, the greater the distance of measurement, the greater the mistake is committed. Therefore, the LT might be placed near the workspace to verify. One should recall that there is a minimum distance allowable for each equipment.

Tests carried out to study the influence of design configuration show that the number of points is the most relevant parameter, followed by the number of tests, and the points distribution. Moreover, we demonstrate that their relationship cannot be modelled as linear regression functions. Therefore, users should assess the computation costs against the accuracy of the method to determine the configuration parameters.

Author Contributions: Conceptualisation: S.A., J.A.A. and J.S.; Methodology: S.A., P.P. and J.A.A.; Formal Analysis: S.A., J.S., P.P. and J.V.; Resources: J.A.A., P.P., J.V. and S.A.; Writing—original draft preparation: S.A., P.P. and J.A.A.; Writing—reviewing editing: J.A.A., J.S., P.P. and S.A.; Software: S.A. and J.S. All authors have read and agreed to the published version of the manuscript.

Funding: This work was supported by the Ministerio de Economía, Industria y Competitividad de España with project number Reto 2017-DPI2017-90106-R., and by the Aragon Government (Department of Industry and Innovation) through the Research Activity Grant for research groups recognised by the Aragon Government (T56_17R Manufacturing Engineering and Advanced Metrology Group).

Conflicts of Interest: The authors declare no conflict of interest.

References

1. Schwenke, H.; Knapp, W.; Haitjema, H.; Weckenmann, A.; Schmitt, R.; Delbressine, F. Geometric error measurement and compensation of machines-An update. *CIRP Ann.-Manuf. Technol.* **2008**, *57*, 660–675. [CrossRef]
2. Aguado, S.; Santolaria, J.; Samper, D.; Aguilar, J.J.; Velázquez, J. Improving a real milling machine accuracy through an indirect measurement of its geometric errors. *J. Manuf. Syst.* **2016**, *40*, 26–36. [CrossRef]
3. Aguado, S.; Velazquez, J.; Samper, D.; Santolaria, J. Modelling of computer-assisted machine tool volumetric verification process. *Int. J. Simul. Model.* **2016**, *15*, 497–510. [CrossRef]
4. Aguado, S.; Pérez, P.; Albajez, J.A.; Santolaria, J.; Velázquez, J. Study on Machine Tool Positioning Uncertainty Due to Volumetric Verification. *Sensors* **2019**, *19*, 2847. [CrossRef] [PubMed]
5. Wang, H.; Shao, Z.; Fan, Z.; Han, Z. Optimization of laser trackers locations for position measurement. In Proceedings of the 2018 IEEE International Instrumentation and Measurement Technology Conference (I2MTC), Houston, TX, USA, 14–17 May 2018; pp. 1–6. [CrossRef]
6. Zhu, X.; Zheng, L.; Tang, X. Configuration Optimization of Laser Tracker Stations for Large-scale Components in Non-uniform Temperature Field Using Monte-carlo Method. *Procedia CIRP* **2016**, *56*, 261–266. [CrossRef]
7. *ASME B89.4.19-2005 Performance Evaluation of Laser Based Spherical Coordinate Measurement Systems*; AmericanSociety of Mechanical Engineer: New York, NY, USA, 2005.
8. VDI/VDE 2617 Part 10. *Accuracy of Coordinate Measuring Machines–Characteristics and their Checking–Acceptance and Reverification Tests of Laser Trackers*; VDI/VDE Innovation + Technik GmbH: Berlin, Germany, 2008.
9. ISO 10360-10:2016. *Geometrical Product Specifications (GPS)–Acceptance and Reverification Tests for Coordinate Measuring Systems (CMS)–Part 10: Laser Trackers for Measuring Point-to-Point Distances*; ISO: Geneva, Switzerland, 2009.
10. Gallagher, B.B. *Optical Shop Applications for Laser Tracker Metrology Systems*; The University of Arizona Press: Tucson, AZ, USA, 2003; p. 205.
11. Knapp, W. Measurement Uncertainty and Machine Tool Testing. *CIRP Ann.* **2002**, *51*, 459–462. [CrossRef]
12. Pérez Muñoz, P.; Albajez García, J.A.; Santolaria Mazo, J. Analysis of the initial thermal stabilization and air turbulences effects on Laser Tracker measurements. *J. Manuf. Syst.* **2016**, *41*, 277–286. [CrossRef]
13. Muske, S.; Salisbury, D.; Salerno, R.; Calkins, J. 747 Data Management SystemDevelopment and Implementation. In Proceedings of the CMSC Conference and the 2000 Boeing Large Scale Metrology Conference, Long Beach, CA, USA, 23–24 February 2000.
14. Costa, D.; Albajez, J.A.; Yagüe-Fabra, J.A.; Velázquez, J. Verification of Machine Tools Using Multilateration and Geometrical Approach. *Nanomanuf. Metrol.* **2018**. [CrossRef]
15. Aguado, S.; Samper, D.; Santolaria, J.; Aguilar, J.J. Volumetric verification of multiaxis machine tool using laser tracker. *Sci. World J.* **2014**. [CrossRef] [PubMed]
16. *ISO/IEC Guide 98-3:2008—Uncertainty of Measurement—Part. 3: Guide to the Expression of Uncertainty in Measurement (GUM:1995)*; ISO: Geneva, Switzerland, 2008.
17. Mason, R.L.; Gunst, R.F.; Hess, J.L. *Statical Design and Analysis of Experiments with Applications to Engineering and Sicicence*, 2nd ed.; John Wiley & Sons: New York, NY, USA, 2003; ISBN 0-471-37216-1.
18. Weisberg, S. *Applied Linear Regressions*; John Wiley & Sons: New York, NY, USA, 2005; ISBN 9-780-471-70409-6.

© 2020 by the authors. Licensee MDPI, Basel, Switzerland. This article is an open access article distributed under the terms and conditions of the Creative Commons Attribution (CC BY) license (http://creativecommons.org/licenses/by/4.0/).

Article

Estimation of an Upper Bound to the Value of the Step Potentials in Two-Layered Soils from Grounding Resistance Measurements

Jorge Moreno [1], Pascual Simon [2], Eduardo Faleiro [1,*], Gabriel Asensio [1] and Jose Antonio Fernandez [3]

1. Polytechnic University of Madrid (UPM), Escuela Técnica Superior de Ingeniería y Diseño Industrial (ETSIDI), Ronda de Valencia 3, 28012 Madrid, Spain; jorge.moreno@upm.es (J.M.); gabriel.asensio@upm.es (G.A.)
2. LCOE Laboratory, Calle Diesel 13. P.I. El Lomo, 28906 Madrid, Spain; psimon@ffii.es
3. Union Fenosa Distribución (UFD). Avenida San Luis 77, 28033 Madrid, Spain; jafernandez@ufd.es
* Correspondence: eduardo.faleiro@upm.es

Received: 19 October 2019; Accepted: 6 January 2020; Published: 8 January 2020

Abstract: Due to the constant updating of regulatory standards on safety issues in electrical installations, limits are established for the maximum step potential that an installation can hold in a ground fault situation. In this paper, an upper bound to the maximum value of the step potentials arising in the soil surface when a fault takes place in a grounded electrical installation is estimated by means of a simple procedure. The direct measurement of the grounding electrode resistance together with some information about the soil resistivity and the knowledge of characteristic parameters of the electrode are used for the calculation of that upper bound. The procedure is tested at numerical simulation level by using different electrodes in several different scenarios corresponding to two-layered soils with different resistivity ratios. The dependency of the calculated upper bound with the electrode burial depth is also studied. Finally, a real case study is presented, and the results of the field measurements are shown as an example of the validity of the procedure.

Keywords: grounding electrodes in two-layered soils; step and touch potentials; step potentials upper bound

---

1. Introduction

All electrical facilities such as transformation centers substations or transmission towers need to have a ground protection system installed that guarantees the evacuation to the ground of fault currents that could otherwise seriously damage people and the facility itself. For this purpose, a metallic electrode is buried into the ground which receives and disperses the fault currents by raising its own electric potential and that of its surroundings. This is the so-called grounding electrode.

When a fault takes place in a facility equipped with a grounding electrode, the surface of the ground experiences a rise in electrical potential that can put people and equipment at risk in its area of influence [1]. Among the magnitudes most frequently used to quantify the effect of a ground fault in an installation are the step potentials and the touch potentials. The touch potential is the potential difference between the grounding electrode and some point on the ground surface. The step potential is the potential difference between two points of the ground separated by a distance of one meter, and it is a measure of the gradient of the absolute potentials generated in the soil. The step potential reaches maximum values at the ground points near the buried electrode, that is, areas in which technical staff may be working.

The authorities responsible for ensuring the safety of electrical installations establish maximum values for the step and touch potentials that must be met in case of fault. The European Union, for example, establishes general regulations that each member country adapts to its own standards [2,3]. These potentials should be measured directly in the field unless there is an indirect procedure that ensures compliance with the maximum values officially established without the need to be directly measured. Having such a procedure not only means a considerable economic saving given the cost of direct measurement on the ground but also a considerable simplification when the step potentials must be estimated in urban areas where direct measurement is often difficult. From decades, many authors have spent great efforts to propose calculation methods for step potentials, from the exclusive use of approximate analytical formulas [4,5] to using complex numerical techniques [6], while they were measured in the field by using various strategies [7,8]. However, none of the methods known to the authors of this paper is as simple and quick to apply as the one proposed here [9].

The motivation of the present work is in part of an economic nature. Regulations regarding safety in electrical installations are constantly being updated. The tolerance with respect to step and touch potentials is frequently reviewed, so it is necessary to check if the facilities comply with the updated regulatory framework. As stated before, direct measurements in the field are usually expensive, so an estimation procedure that can avoid such measurements will always be welcomed by electric companies.

In the present paper, a procedure for estimating an upper limit to the value of the step potentials generated at the ground surface by a grounding electrode driving a fault current to ground is proposed. The comparison of such an upper limit with the maximum regulatory values for the installation will allow discarding the direct measurement if these values are above the obtained limit. Otherwise, a direct measurement will be necessary to decide whether the grounding system complies with the regulation [10].

The procedure is based on the maximization of the quotient of two characteristic parameters of the grounding electrode involving the resistance of the electrode and the normalized step potentials that it generates in the soil surface when a fault takes place.

To make the procedure known in detail, the paper is organized as follows. After the present introduction, in Section 2 of the paper, the fundamentals of the procedure are detailed, and in Section 3, the procedure is tested in some numerical experiments by using two-layered soils. In Section 4, we present a real case of study to which the procedure is applied and finally, in Section 5, the conclusions of the paper are collected.

2. Basics of the Procedure

The procedure is based on establishing a relationship between the step potentials generated by an energized grounding electrode and the potential acquired by the electrode itself in that situation. Both potentials are expressed as a function of the coefficients K_r and K_p, which are defined as

$$K_r = \frac{U_r}{\rho_r \cdot I} K_p = \frac{U_p}{\rho_p \cdot I} \qquad (1)$$

where U_r and U_p are the potential of the electrode and the maximum step potential generated in the soil surface when a current I is leaked to the ground, respectively. The resistivities ρ_r and ρ_p correspond to the equivalent resistivity of the soil at the depth of the electrode and the equivalent resistivity at surface level, which is where the step potentials are measured, respectively. Equivalent resistivities are not true resistivities in general but represent the resistivity that a homogeneous soil would have, so that the potential generated by the electrode in such soil coincides with the value of the potential in the true soil. Regarding the step potential, for each electrode, the procedure to calculate this potential is usually given as a standard rule and generally corresponds to the highest possible value among those evaluated at locations accessible by the company staff who may be near the electrode when a fault event takes place. According to the type of electrode, it is frequent that the procedure to calculate the maximum step potential is prefixed by the manufacturer in the specifications of the electrode itself. The calculations of K_r and K_p are made by calculating the potential acquired by the electrode and the maximum step potential created in the soil surface according to the electrode specifications, when both the resistivity and the electric current are of unit value. This calculation is usually carried out by means of a computer simulation in a homogeneous semi-infinite soil where the electrode is placed in the position and at the depth that it will really have. The simulation method is the one commonly used in this type of problems, that is, boundary elements together with the thin wire approach for the electrodes and the method of moments for the numerical solution of Maxwell's equations with boundary conditions only at the ground surface [9]. The main Spanish Electric Companies, for example, have an extensive database with the specifications of the grounding electrodes used in their facilities, among which are the values of K_r and K_p. These parameters may have been supplied by the manufacturers or are instead theoretically calculated from the knowledge of the shape and size of the electrodes as has been done in this paper.

From the expressions (1) it is obtained

$$\frac{U_p}{U_r} = \frac{K_p}{K_r} \cdot \frac{\rho_p}{\rho_r} \tag{2}$$

where only remains to determine what value to assign to the quotient of resistivities ρ_p/ρ_r keeping in mind that they represent equivalent resistivities. For a truly homogeneous soil, ρ_r and ρ_p correspond to the actual resistivity and thus its ratio is the unit. For a two-layer soil of parameters ρ_1, ρ_2 and h, we must distinguish between two possibilities. If $\rho_1 > \rho_2$, then $1 \leq \rho_p/\rho_r \leq \rho_1/\rho_2$, therefore an upper bound is obtained when the quotient takes the value ρ_1/ρ_2. Nevertheless, if $\rho_1 < \rho_2$, then $\rho_p/\rho_r \leq 1$ and the upper bound is obtained when ρ_p/ρ_r takes the unit value. Therefore, once the type of two-layer soil is established, the procedure for estimating the maximum step potential as a function of the electrode potential is established as follows:

$$U_p = U_r \cdot \frac{K_p}{K_r}, \ \rho_1 < \rho_2 \quad U_p = U_r \cdot \frac{K_p}{K_r} \cdot \frac{\rho_1}{\rho_2}, \ \rho_1 > \rho_2 \tag{3}$$

The determination of the type of two-layer soil is important only in the case that $\rho_1 > \rho_2$. If as is often the case, the soil has initially been assumed to be homogeneous with resistivity ρ_s, a measurement of the resistance R of the electrode by some indirect procedure will provide us not only $U_r = I \cdot R$ but also will give us the value of the equivalent resistivity $\rho_r = R/K_r$. Thus, from the expressions (3) results

$$U_p = I \cdot R \cdot \frac{K_p}{K_r}, \ \rho_1 < \rho_2 \quad U_p = I \cdot R \cdot \frac{K_p}{K_r} \cdot \frac{\rho_s}{\rho_r}, \ \rho_1 > \rho_2 \tag{4}$$

where the surface resistivity ρ_p in (1) or (2) has been replaced by the equivalent soil resistivity ρ_s estimated in the design and construction stage of the grounding system.

3. Preliminary Validation of the Procedure

In this section, the step potentials generated by two types of grounding electrode will be calculated by means of numerical experiments. As input data, the values of the K_r and K_p coefficients will be assumed known, since it will be admitted that they form part of the specifications of the electrodes. The two-layer structure of the soil will also be known, although the expression (4) will also be applied, evaluating ρ_r and ρ_s from the data when necessary. For the validation purposes, the response of the electrode to the injection of a current I will be simulated, obtaining the potentials $U_{r,sim}$ and $U_{p,sim}$, according to the specifications of the electrode. By applying the proposed method using the expressions (3) or (4), the calculated step potential $U_{p,calc}$ will be obtained and is compared with the simulated $U_{p,sim}$ which would correspond to the really measured potential. For the simulation purposes, a specific software developed by the authors has been used here in which a generic electrode buried in a multilayered soil is excited by a fault current. The grounding resistance and electric potentials are obtained from the currents leaked from the electrode to the ground, after numerically solving the

Maxwell equations [11]. The results are summarized in Table 1, while the details of the calculations are shown below.

This section may be divided according to the type of electrode and the type of soil it is buried into. It should provide a concise and precise description of the numerical experiments results, their interpretation as well as the conclusions that can be drawn.

3.1. Vertical Rod

We consider first a vertical rod of longitude $L = 1$ m and radius $r = 0.005$ m, buried at 0.5 m. The maximum step potential is specified as the electrical potential difference between two points that are 1 m apart, the point closest to the electrode being 1 m from the vertical that contains the electrode itself. The coefficients K_r and K_p can be easily calculated resulting in values $K_r = 0.828$ m^{-1} and $K_p = 0.043$ m^{-1}, where $K_p/K_r = 0.052$. We will assume for simplicity that the injected current is $I = 1$ A. Four different types of soil are going to be considered.

3.1.1. Type 1

Two-layer soil $\rho_1 = 200$ Ωm, $\rho_2 = 100$ Ωm and $h = 3$ m. With this data, $U_{r,sim} = 162.27$ V and $U_{p,sim} = 8.37$ V are obtained. By taking $\rho_s = 200$ Ωm and $\rho_r = U_{r,sim}/K_r = 196.03$ Ωm, the step potential calculated from (4) results $U_{p,calc} = 8.61$ V.

3.1.2. Type 2

Two-layer soil $\rho_1 = 100$ Ωm, $\rho_2 = 200$ Ωm and $h = 3$ m. With this data, $U_{r,sim} = 85.06$ V and $U_{p,sim} = 4.35$ V are obtained. By applying (3), it results $U_{p,calc} = 4.42$ V.

3.1.3. Type 3

Two-layer soil $\rho_1 = 2000$ Ωm, $\rho_2 = 100$ Ωm and $h = 3$ m. With this data, $U_{r,sim} = 1581.10$ V and $U_{p,sim} = 81.34$ V are obtained. By taking $\rho_s = 2000$ Ωm and $\rho_r = U_{r,sim}/K_r = 1910.01$ Ωm, the step potential calculated from (4) results $U_{p,calc} = 86.09$ V.

3.1.4. Type 4

Two-layer soil $\rho_1 = 100$ Ωm, $\rho_2 = 2000$ Ωm and $h = 3$ m. With this data, $U_{r,sim} = 95.65$V and $U_{p,sim} = 4.51$ V are obtained. By applying (3), it results $U_{p,calc} = 4.97$ V.

3.2. Complex Electrode

Horizontal square frame type electrode of 2.6 m side with vertical rods of 1.5 m length in the vertices and conductors of radius $r = 0.005$ m form the whole set buried 0.5 m from the surface, as shown in Figure 1. Such an electrode is part of the grounding system of a transmission tower, as shown in Figure 2. This figure is a schematic representation of a transmission tower with the grounding electrode of Figure 1. The step potential measurement system requires a device that injects current into the grounding electrode, which returns to through another remote electrode. The potential difference is measured with a high impedance voltmeter. The maximum step potential is specified as the existing electric potential difference between two points of the horizontal frame diagonal separated by a distance of 1 m, the point closest to the electrode being located on the vertical of any of its vertices as is also shown in Figure 2.

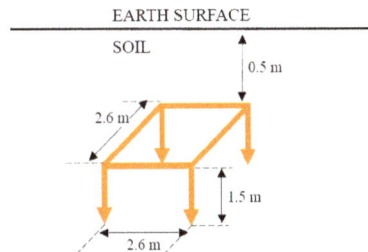

Figure 1. The complex electrode considered as a second example for validation.

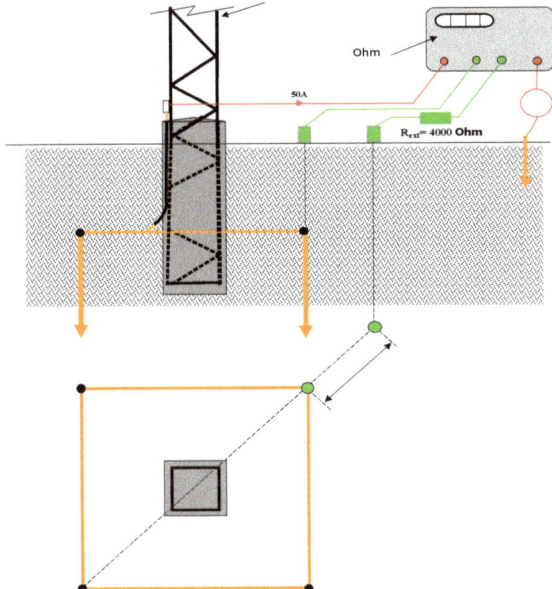

Figure 2. Step potential measurement for the grounding electrode of a power transmission line tower.

The coefficients K_r and K_p are calculated, resulting in the values, $K_r = 0.126$ m^{-1} and $K_p = 0.028$ m^{-1} from which $K_p/K_r = 0.226$ is obtained.

3.2.1. Type 1

Two-layer soil $\rho_1 = 200$ Ωm, $\rho_2 = 100$ Ωm and $h = 3$ m. With this data, $U_{r,sim} = 22.14$ V and $U_{p,sim} = 5.50$ V are obtained. By taking $\rho_s = 200$ Ωm and $\rho_r = U_{r,sim}/K_r = 176.31$ Ωm, the step potential calculated from (4) results $U_{p,calc} = 5.68$ V.

3.2.2. Type 2

Two-layer soil $\rho_1 = 100$ Ωm, $\rho_2 = 200$ Ωm and $h = 3$ m. With this data, $U_{r,sim} = 14.67$ V and $U_{p,sim} = 2.94$ V are obtained. By applying (3), it results $U_{p,calc} = 3.32$ V.

3.2.3. Type 3

Two-layer soil $\rho_1 = 2000$ Ωm, $\rho_2 = 100$ Ωm and $h = 3$ m. With this data, $U_{r,sim} = 184.87$ V and $U_{p,sim} = 52.30$ V are obtained. By taking $\rho_s = 2000$ Ωm and $\rho_r = U_{r,sim}/K_r = 1471.90$ Ωm, the step potential calculated from (4) results $U_{p,calc} = 56.80$ V.

3.2.4. Type 4

Two-layer soil $\rho_1 = 100$ Ωm, $\rho_2 = 2000$ Ωm and $h = 3$ m. With this data, $U_{r,sim} = 24.90$ V and $U_{p,sim} = 3.15$ V are obtained. By applying (3), it results $U_{p,calc} = 5.63$ V.

3.3. The Grounding Electrode of the Balaidos High-Voltage Substation

As the last example of application of the proposed method, the grounding electrode of the Balaidos high-voltage substation, belonging to the Spanish Electric Company *Unión Fenosa* near the city of Vigo (northwest of Spain). The grounding electrode is a mesh of 188 cylindrical conductors (diameter: 11.28 mm) buried at a depth of 80 cm, covering an area of about 200 m². Figure 3 shows the electrode profile in the XY plane. Although the resistivities used in the simulation are fictitious, the electrode is completely real and is buried in a soil of resistivity close to 60 Ωm. For this example, only extreme two-layer soil type 3 and type 4 are considered. The maximum step potential is defined in a similar way to the previous example.

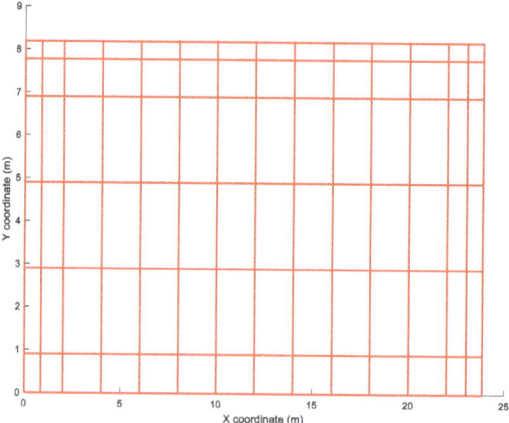

Figure 3. Representation in the XY plane of the grounding electrode of the Balaidos high-voltage substation (Spain).

The coefficients K_r and K_p are calculated, resulting in the values, $K_r = 0.028$ m^{-1} and $K_p = 0.004$ m^{-1} from which it is obtained $K_p/K_r = 0.142$.

3.3.1. Type 1

Two-layer soil $\rho_1 = 200$ Ωm, $\rho_2 = 100$ Ωm and $h = 3$ m. With this data, $U_{r,sim} = 3.84$ V and $U_{p,sim} = 0.62$ V are obtained. By taking $\rho_s = 200$ Ωm and $\rho_r = U_{r,sim}/K_r = 137.14$ Ωm, the step potential calculated from (4) results $U_{p,calc} = 0.80$ V.

3.3.2. Type 4

Two-layer soil $\rho_1 = 100$ Ωm, $\rho_2 = 200$ Ωm and $h = 3$ m. With this data, $U_{r,sim} = 4.04$ V and $U_{p,sim} = 0.45$ V are obtained. By applying (3), it results $U_{p,calc} = 0.57$ V.

3.3.3. Type 3

Two-layer soil $\rho_1 = 2000$ Ωm, $\rho_2 = 100$ Ωm and $h = 3$ m. With this data, $U_{r,sim} = 20.31$ V and $U_{p,sim} = 4.79$ V are obtained. By taking $\rho_s = 2000$ Ωm and $\rho_r = U_{r,sim}/K_r = 739.45$ Ωm, the step potential calculated from (4) results $U_{p,calc} = 7.81$ V.

3.3.4. Type 4

Two-layer soil $\rho_1 = 100$ Ωm, $\rho_2 = 2000$ Ωm and $h = 3$ m. With this data, $U_{r,sim} = 12.14$ V and $U_{p,sim} = 0.70$ V are obtained. By applying (3), it results $U_{p,calc} = 1.73$ V.

Table 1 shows the potential value of the, $U_{r,sim}$, obtained by simulation, which in this calculation coincides with its resistance since the injected current is 1A. The table also shows the simulated step potentials $U_{p,sim}$, which can be taken as those actually existing in the ground, and their upper levels calculated by (3) and (4) $U_{p,calc}$. The table shows that the best bounds to step potential are calculated for soils where $\rho_1 > \rho_2$. Otherwise, the bounds are not so good, being able to reach a large difference rate between the true value and the bound, although it can be affirmed that, far from the real value, these bounds represent an absolute limit to the value of the step potential.

Table 1. Results of the numerical tests for validating the proposed procedure.

Electrode	Two-Layer Soil	$U_{r,sim}$ (V)	$U_{p,sim}$ (V)	$U_{p,calc}$ (V)	Diff (%)
Vertical rod	Type 1	162.27	8.37	8.61	2.8
	Type 2	85.06	4.35	4.42	1.8
	Type 3	1581.10	81.34	86.09	5.8
	Type 4	95.65	4.51	4.97	10.2
Complex	Type 1	22.15	5.50	5.68	3.2
	Type 2	14.67	2.94	3.32	12.8
	Type 3	184.87	52.30	56.80	8.6
	Type 4	24.90	3.15	5.63	78.9
Balaidos	Type 1	3.84	0.62	0.80	22.5
	Type 2	4.04	0.45	0.57	21.1
	Type 3	20.31	4.79	7.81	38.6
	Type 4	12.14	0.70	1.73	59.6

For other burial depths of the electrodes, it is necessary to recalculate the values of K_r and K_p and determine the rest of the parameters in order to apply the expressions (4). Figure 4 shows all cases studied for the so-called complex electrode when its burial depth varies from $d = 0$ to 1.5 m, before the electrode crosses the interface between the two layers of the soil. The subfigures show the difference rate between the true value of the step potential and the bound as the electrode varies its burial depth. In Figure 4 it is also observed that the upper bound is very close to the real value of the step potential in when $\rho_1 > \rho_2$, both values being not very far apart. On the other hand, the worst result is obtained when the resistivities involved are very different, and also $\rho_1 < \rho_2$. Nonetheless, it can be verified that the value of the bound for the step potential is always above the true value of this magnitude.

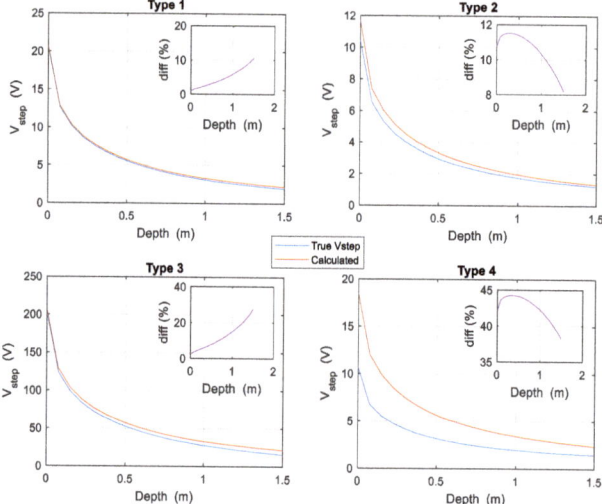

Figure 4. Step potentials created by the complex electrode of Figure 1 for all types of two-layered soils considered, as a function of the burial depth. The upper bound is shown in continuous red line while the real value is in continuous blue line.

Figure 5 shows all cases studied for the Balaidos grounding grid, where the burial depth of the electrode varies also from $d = 0$ to 1.5 m, before the electrode crosses the interface between the two layers of the soil. The subfigures also show the difference rate between the true value of the step potential and the upper bound as a function of the burial depth. In the same way as described in Figure 4, the upper bound is close to the actual value of the step potential when $\rho_1 > \rho_2$ while the opposite occurs when $\rho_1 < \rho_2$ although it is always above the real value of this magnitude.

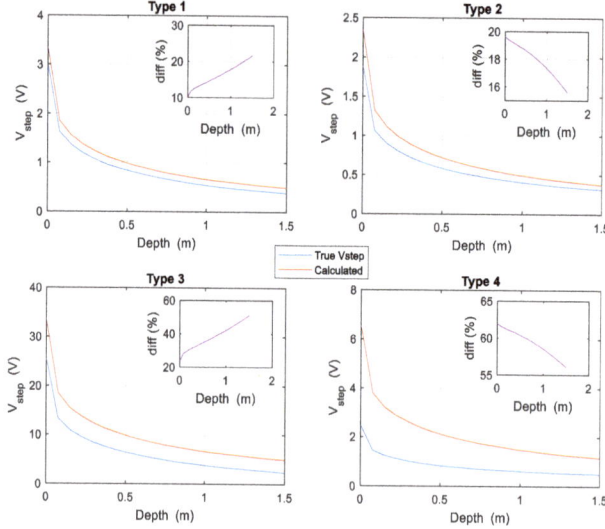

Figure 5. Step potentials generated by the Balaidos grounding electrode for all types of the two-layered soils considered, as a function of the burial depth. The upper bound is shown in continuous red line while the real value is in continuous blue line.

Finally, it is only necessary to comment on the methods to measure the resistance of the electrode, in order to apply the expressions (4). Under general conditions, the measurement is made with a tellurometer using the fall-off-potential method, bearing in mind that the soil is modeled as multi-layered. This method requires that the grounding electrode to be disconnected from the installation. For grounding electrodes that are part of an interconnected system, the clamp-on or stakeless method [12] can be used. It is a fast and reliable non-invasive method. Figure 6 graphically shows a scheme of the clamp-on method. By means of the clamp, an electric current is induced in the grounding system at the same time that the clamp itself measures the resistance of the grounding electrode.

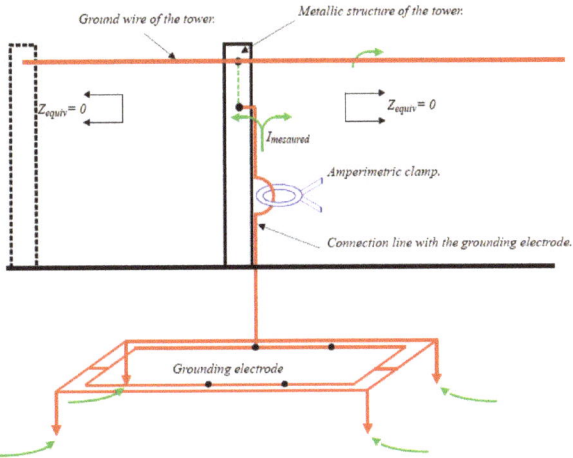

Figure 6. Measurement of the grounding resistance by means of the clamp-on method.

As a final comment, the more information about the electrode and the electrical structure of the ground, the better the estimate of the upper bound. In cases where the information is incomplete, the bound will be overestimated although it will still be valid for comparison purposes. In these situations, it is possible that this overestimated value of the upper bound exceeds the limits established by the regulatory frameworks, and it is necessary to make direct measurements of the step potentials.

4. A Case of Study

As an example of application, a transmission tower, which can be accessed by qualified staff, equipped with a lightning conductor and a non-interconnected grounding system is considered. The tower belongs to the company Union Fenosa Distribucion S.A. and is located in the Spanish region of Consuegra (Toledo). In the construction stage, a homogeneous ground resistivity of 94.95 Ωm was estimated. There is no data available on the multilayer structure of the soil nor on the exact type of electrode used in the installation, but a measurement of the grounding resistance by injecting a current of 5 A with an amperimetric clamp gives a value of $R = 5.01$ Ω and a step potential value of 0.62 V. To compensate for the lack of information on the type of electrode used, the value of the quotient K_p/K_r will be taken as the largest of the values of the electrode database used by the company in similar facilities, which is estimated at $K_p/K_r = 0.3$. Assuming, for simplicity that $\rho_1 = \rho_2$, the application of (4) gives a value for the upper bound of the step potential of $U_{p,calc} = 373$ V. In order to compare this calculated potential, considered as an upper bound, with what would be obtained in a real situation, it is necessary to know that the installation could generate a fault current of 247.90 A, which taking into account the step potential generated by the current injected could produce a real step potential of $U_{p,meas} = 30.90$ V, far below $U_{p,calc}$, the upper limit estimated by (4).

5. Conclusions

Throughout this paper a procedure to obtain an upper bound to the maximum step potential generated in the ground surface by a grounding electrode excited by a fault current has been presented. The procedure does not replace in any case an accurate assessment of the step potentials but serves to discard the direct measures of these potentials when regulatory standards related to safety are met. The procedure requires knowledge of some characteristic parameters of the electrode K_r and K_p, generally supplied by the manufacturer, although they can be theoretically calculated from the shape and size of the conductors, and some data on the resistivity of the ground, being the most convenient knowledge of the resistivities associated with the two-layer model. Besides this, a direct measurement of the grounding resistance of the electrode is necessary, which can be done by some non-invasive technique such as the clamp-on method. The procedure has been tested on three electrodes in four types of two-layer soils with different resistivity ratios but always assuming that the electrode is located in the upper layer.

The highest percentage differences between the upper bound and the true value of the maximum step potential are found for high values of the resistivity ratios, especially when $\rho_1 < \rho_2$. In these situations, the procedure may not serve to discard direct measures of the step potentials. In other situations studied, these differences are relatively small, which allows a greater capacity to comply with regulatory standards, being able to avoid direct measurements.

In order to study the variation with the burial depth of the electrode of the difference between the simulated step potentials and those calculated with the proposed method, a calculation is made by varying the burial depth up to 1.5 m, and the results are presented in Figures 4 and 5. The Figures also show the variation of the percentage difference between both step potentials.

Finally, a case study has been presented in which, due to the lack of information on the electrode, the upper bound obtained has been greatly overestimated with respect to the true value of the step potential. Although it is unfortunately a common situation, the method always gives a value of the step potential that is always above its real value and thus valid for comparison with regulatory standards.

Author Contributions: All authors were involved to some extent in the different parts of the paper. In particular, J.M., P.S. and J.A.F. have contributed in a special way in the case study. All authors have read and agreed to the published version of the manuscript.

Funding: This research received no external funding.

Acknowledgments: The authors would like to thank both the Department of Applied Mathematics and the IEEF Department of the *Escuela Técnica Superior de Ingeniería y Diseño Industrial* (ETSIDI) at the Polytechnic University of Madrid (UPM) for their support to the undertaking of the research summarized here. We also would like to thank for the support provided by *Unión Fenosa Distribución S.A.* to perform this work.

Conflicts of Interest: The authors declare no conflict of interest.

References

1. Energy Networks Association. *A Guide for Assessing the Rise of Earth Potential at Electrical Installations*; Energy Networks Association: London, UK, 2017.
2. NEN EN 50522:2010. Available online: https://infostore.saiglobal.com/en-us/Standards/NEN-EN-50522-2010-799499_SAIG_NEN_NEN_1915943/ (accessed on 5 July 2019).
3. UNE EN 50522:2012. Available online: https://infostore.saiglobal.com/en-us/standards/une-en-50522-2012-26163_SAIG_AENOR_AENOR_57403/ (accessed on 5 July 2019).
4. Chen, C.X.; Xie, G.R. A new formula for calculation of the maximum step voltage in a substation grounding grid. *Electr. Power Syst. Res.* **1988**, *14*, 41–49. [CrossRef]
5. Chen, L.H.; Chen, J.F.; Liang, T.J.; Wang, W.I. Calculations of ground resistance and step voltage for buried ground rod with insulation lead. *Electr. Power Syst. Res.* **2008**, *78*, 995–1007. [CrossRef]
6. Ayodele, T.R.; Ogunjuyigbe, A.S.O.; Oyewole, O.E. Comparative assessment of the effect of earthing grid configurations on the earthing system using IEEE and Finite Element Methods. *Eng. Sci. Technol. Int. J.* **2018**, *21*, 970–983. [CrossRef]

7. Meliopoulos, A.P.S.; Patel, S.; Cokkinides, G.J. A new method and instrument for touch and step voltage measurements. *IEEE Trans. Power Deliv.* **1994**, *9*, 1850–1860. [CrossRef]
8. Nikolovski, S.; Knezevic, G.; Baus, Z. Assessment of step and touch voltages for different multilayer soil models of complex grounding grid. *Int. J. Electr. Comput. Eng.* **2016**, *6*, 1441.
9. Kostic, V.I.; Raicevic, N.B. An alternative approach for touch and step voltages measurement in high-voltage substations. *Electr. Power Syst. Res.* **2016**, *130*, 59–66. [CrossRef]
10. He, J.; Zhang, B.; Zeng, R. Maximum Limit of Allowable Ground Potential Rise of Substation Grounding System. *IEEE Trans. Ind. Appl.* **2015**, *51*, 5010–5016. [CrossRef]
11. Faleiro, E.; Asensio, G.; Moreno, J.; Simón, P.; Denche, G.; García, D. Modelling and simulation of the grounding system of a class of power transmission line towers involving inhomogeneous conductive media. *Electr. Power Syst. Res.* **2016**, *136*, 154–162. [CrossRef]
12. IEEE Guide for Measuring Earth Resistivity. *Ground Impedance and Earth Surface Potentials of a Grounding System*; IEEE Power and Energy Society Publications: New York, NY, USA, 2012.

© 2020 by the authors. Licensee MDPI, Basel, Switzerland. This article is an open access article distributed under the terms and conditions of the Creative Commons Attribution (CC BY) license (http://creativecommons.org/licenses/by/4.0/).

Article

Enhanced Positioning Algorithm Using a Single Image in an LCD-Camera System by Mesh Elements' Recalculation and Angle Error Orientation

Óscar de Francisco Ortiz [1,*], Manuel Estrems Amestoy [2], Horacio T. Sánchez Reinoso [2] and Julio Carrero-Blanco Martínez-Hombre [2]

1. Department of Engineering and Applied Technologies, University Center of Defense, San Javier Air Force Base, MDE-UPCT, 30720 Santiago de la Ribera, Spain
2. Mechanics, Materials and iManufacturing Engineering department, Technical University of Cartagena, 30202 Cartagena, Spain; manuel.estrems@upct.es (M.E.A.); horacio.sanchez@upct.es (H.T.S.R.); julio.carrero@upct.es (J.C.-B.M.-H.)
* Correspondence: oscar.defrancisco@cud.upct.es; Tel.: +34-968-189918

Received: 31 October 2019; Accepted: 11 December 2019; Published: 16 December 2019

Abstract: In this article, we present a method to position the tool in a micromachine system based on a camera-LCD screen positioning system that also provides information about angular deviations of the tool axis during its running. Both position and angular deviations are obtained by reducing a matrix of LEDs in the image to a single rectangle in the conical perspective that is treated by a photogrammetry method. This method computes the coordinates and orientation of the camera with respect to the fixed screen coordinate system. The used image consists of 5 × 5 lit LEDs, which are analyzed by the algorithm to determine a rectangle with known dimensions. The coordinates of the vertices of the rectangle in space are obtained by an inverse perspective computation from the image. The method presents a good approximation of the central point of the rectangle and provides the inclination of the workpiece with respect to the LCD screen reference system of coordinates. A test of the method is designed with the assistance of a Coordinate Measurement Machine (CMM) to check the accuracy of the positioning method. The performed test delivers a good accuracy in the position measurement of the designed method. A high dispersion in the angular deviation is detected, although the orientation of the inclination is appropriate in almost every case. This is due to the small values of the angles that makes the trigonometric function approximations very erratic. This method is a good starting point for the compensation of angular deviation in vision based micromachine tools, which is the principal source of errors in these operations and represents the main volume in the cost of machine elements' parts.

Keywords: image processing; position control; accuracy; micromachines; position compensation; inverse conical perspective; micromanufacturing; manufacturing systems; mechatronics

1. Introduction

Positioning systems are increasingly present in all industrial processes. Furthermore, technology requires progressively more precise systems capable of positioning rapidly and robustly. The cost of those is one of the key factors to integrate high precision systems.

Thanks to the advances in screen and camera technology, positioning algorithms that analyze a pattern shown in a photographic image have been developed [1,2]. More recently, camera-screen positioning systems with dedicated artificial vision algorithms [3–5] have provided high precision at a very interesting cost compared to other positioning technologies such as encoders or resolvers.

Vision positioning systems are increasingly common in process automation [6–10], autonomous driving [11–15], or augmented reality assistants [16–20]. Indeed, this is one of the most promising elements in the Industry 4.0 revolution. However, the current positioning systems used in the machine tool industry based on high precision encoders and sensors are limited by their cost. Therefore, machine tools used for micro-manufacturing have very high prices and require large floor space. Due to this, in micro-manufacturing, the methods that use vision can be competitive by including high performance commercial elements and reducing space such as cameras and mobile phones' LCD screens. In addition, such devices are increasing in definition and resolution, providing vision with much better accuracy.

The methodology used in this article to calculate the position and orientation of the camera in relation to the screen is based on pose determination [21–24], which is used to estimate the position and orientation of one calibrated camera. Several similar methods for calculating the position and orientation of a camera in space using a single image have been described and presented [22,25,26]. Nevertheless, pose estimation and marker detection are widely used tasks for many other technological applications such as autonomous robots [27–29], unmanned vehicles [30–37], and virtual assistants [38–41], among others.

Consequently, this article presents an enhanced method of recalculating the center of the image used by the positioning algorithm in an LCD-camera system, similar to that developed by de Francisco [4] and improved in subsequent studies [42], but being completely different from such previous studies regarding the procedure to calculate the positioning of the part with respect to the reference system of the screen. In previous works, the positioning was obtained through the global center of gravity of the 25 selected LEDs in the image. In this work, the position of the piece is calculated by previously determining an equivalent square obtained by means of regressions of the different lines that form the grid of the 25 LEDs.

In addition, this manuscript also presents the calculation and correction of the orientation angle, which, although very small, always influences the precision positioning due to the large distance between the location of the cutter and the screen. The new method is based on the calculation of the equivalent quadrangle that allows not only the positioning of the center of the image, but also the inclination. The method uses the treatment of an image to obtain the pixel coordinates of a 5×5 dot matrix that serves to locate the focus and orientation of the camera, where the error is due to the distance between the focus and the screen and can be assumed as sine error.

2. Materials and Methods

2.1. Experimental Setup and Measurements

The experimental study was applied to a two-dimensional control system (X and Y). Figure 1 shows the model of the Micromachine Tool (MMT) demonstrator developed for this research. Two stepper motors (ST28, 12, 280 mA) controlled and moved two precision guides (IKO BSR2080 50 mm stroke), which were connected to a M3 ball screw/nut. The LCD screen used provided a 1136×640 pixel resolution, 326 ppi, and 0.078 mm dot pitch. The screen size was 88.5×49.9 mm. Both stepper motors were controlled by the digital output signals provided by an NI 6001-USB data acquisition card connected to the USB port of a laptop computer. The output signal of the acquisition card was treated by a pre-amplification power station composed of two L293 H-bridges. The control was programmed in LabVIEW. It received the image captured by the camera and processed it according to an image enhancing process. It consisted of an image mask application with color plane extraction, fuzzy pixel removal, small object removal, and particle analysis of the mass center of each evaluated pixel. Once it was processed using the developed artificial vision algorithm, it provided the positioning feedback signals needed to move the X and Y axes.

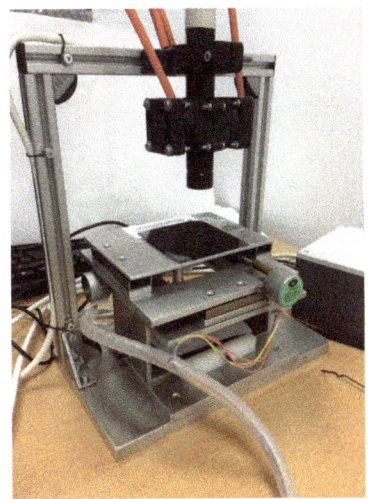

Figure 1. Model of the micromachine tool demonstrator used during the experimental test.

The images were taken by the camera included in the MMT, a Model MITSAI 1.3M digital camera with a resolution of 1280 × 1024 pixels (1.3 MPixels). To analyze the position, a Coordinate Measuring Machine (CMM) Pioneer DEA 03.10.06 with measuring strokes 600 × 1000 × 600 mm was used (Figure 2). The maximum permissible error of the DEA in the measurements was 2.8 + 4.0 L/1000 μm. The software used for the measurements was PC-DMIS.

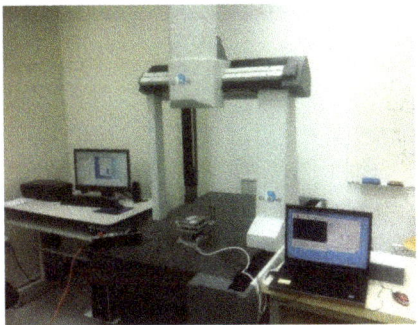

Figure 2. Setup used during the experimental test for the measurement with the Micromachine Tool (MMT) and the Coordinate Measuring Machine (CMM).

Several tests were performed over a 2 × 2 gap pattern using the camera-LCD algorithm. The simulation consisted of testing a 5 mm X axis movement using 10 steps of 0.5 mm. Each travel was repeated 3 times in both the forward and backward direction, according to the the VDI/DGQ 3441 standard: Statistical Testing of the Operational and Positional Accuracy of Machine Tools - Basis.

2.2. Image Acquisition

Image acquisition was done using a procedure developed by the authors in VBA similar to that performed by software such as ImageJ© in its tool "Analyze Particles ...".

The image may not be focused, although many webcams have autofocus mechanisms that make the focal length variable. In our case, it was unimportant because what matters was the bulk and its center of gravity. It should also be noted that if the extraction was from the complete image, the image usually contained the spherical errors of the lenses that focused the image onto the sensor.

In our case, to speed up the process and calculations, only the central area from the BMP image file that included all 25 LEDs was extracted. Only the red layer was analyzed because it was proven to be the most efficient and the only one used to generate the image. Given the size of the LEDs, an image size of 600 × 600 was sufficient to ensure the presence of at least 25 LEDs in the image.

3. Obtaining the Equivalent Quadrilateral

Once the 25 coordinates of the centers of the LEDs were obtained, as seen in Figure 3, these data had to be statistically treated to obtain four vertices of a quadrilateral that collected information about the coordinates of the 25 points. With this quadrilateral and knowing the real side dimensions given by the size of the pixels, the position and orientation of the camera with respect to this square were obtained.

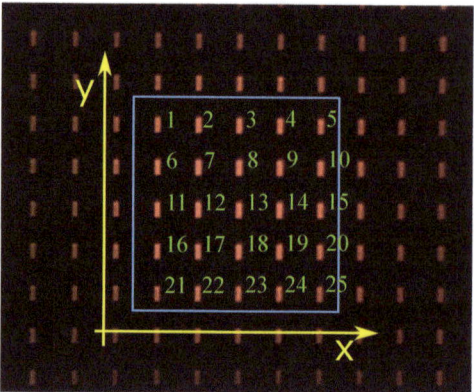

Figure 3. The 5 × 5 mesh captured by the camera with numbered elements.

3.1. Regression of Lines

From the analysis of the 5 × 5 grid, different horizontal lines could be segregated, rearranging the table of coordinates by values in y, obtaining 5 groups of 5 values corresponding to the horizontal lines. Reordering by the values in x, the vertical lines were obtained in the same way. The 5 horizontal lines must be translated into 2 lines, the same with the vertical lines, so that the intersection of the four lines gave rise to the 4 vertices of the quadrilateral that represented a square in conical perspective. The two vanishing points were obtained by the intersection of opposite sides.

Figure 4 shows the regression lines, vertical and horizontal, that represented the different groups of points. The slope and interception terms of the lines followed a tendency that could be anyway also found as shown in Figure 5. These tendencies allowed the calculation of the different slopes in the extreme lines of the rectangle that represented adequately the 25 points, as the border of a chessboard included the dimension and position of the interior squares. The correlated lines and the rectangle used to determine the position and inclination of the axis of the camera are represented in Figure 6.

Since the angles of the slopes had very small variation, the line equations had the form $y = m_i x + n_i$. The intersection of a horizontal line with another vertical line is given by Equation (1):

$$x = \frac{n_j + m_j n_i}{1 - m_i m_j} \qquad (1)$$

where subindex i corresponds to horizontal lines, while subindex j corresponds to vertical lines.

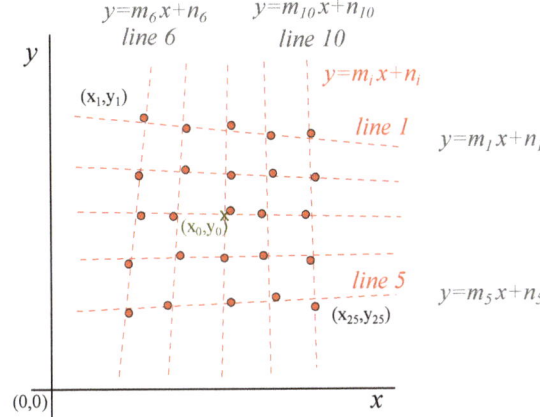

Figure 4. Regression lines in the 5 × 5 elements used in the image analysis.

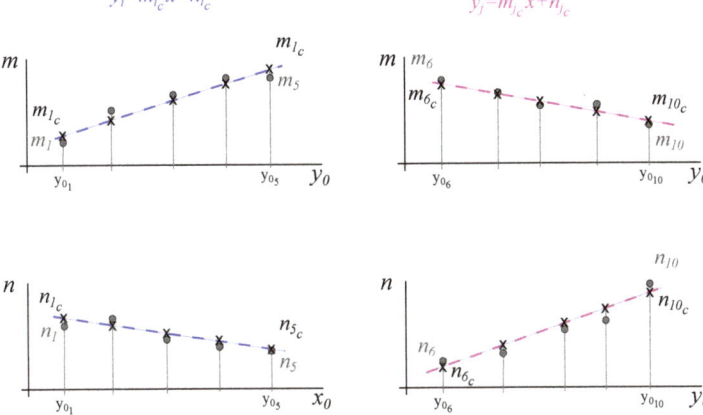

Figure 5. Regression lines (compensation) to optimize the position of the lines.

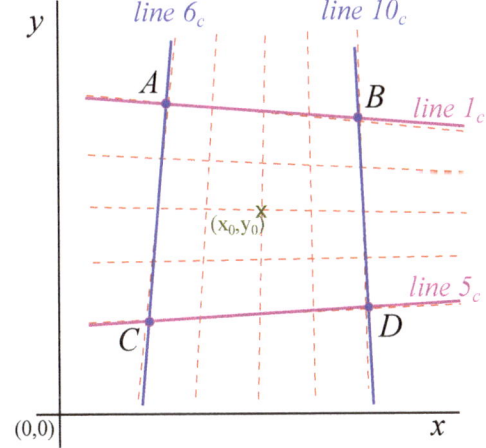

Figure 6. Final distribution of the rectangle used for the position and angle correction.

The steps to obtain the two horizontal lines were the following:

1. Sort by coordinate the data of the grid table obtained.
2. Separate this into groups of 5 points as they belong to the same line by similarity in coordinate y.
3. Perform regression of the five groups obtaining the equations of the five lines $y = m_i x + n_i$, $i = 1\ldots5$.
4. In a similar way, the data were sorted by x coordinates, then separated into 5 groups of 5 points, and the regression was performed obtaining the equations $y = m_j x + n_j$, $j = 6\ldots10$.
5. Obtain intersection points (x_i, y_i) from the central vertical line $y = m_8 y + n_8$ with each of the horizontal lines.
6. Accomplish the regression of the slopes of the horizontal lines m_i based on the vertical intersection coordinates y_i; thereby, the slope was obtained based on the vertical intersection.
7. In such a manner, we proceeded to select the slopes and the points through which the two horizontal lines indicated would be selected. The points of the extreme horizontal lines 1 and 5 were chosen. The two slopes of these two lines were calculated by means of the regression of the 5 slopes. The line was forced to pass through the intersection points of these extreme points, with the following remaining equations of the lines (Equations (2) and (3)):

$$y = m(y_1)x + (y_1 - m(y_1)x_1) \qquad (2)$$

$$y = m(y_5)x + (y_5 - m(y_1)x_5) \qquad (3)$$

A similar method was used for the calculation of the two vertical lines.

The intersection of the two almost parallel lines provided the 4 extreme vertices A, B, C, and D (Figure 6) that were introduced to the program of the inverse conical perspective to obtain the position and orientation of the camera in relation to the fixed coordinates located and oriented with the square that represented the grid of departure. The position and orientation of the contact point with respect to the screen reference system were obtained using an improvement of the method developed by Haralik for the rectangle reconstruction [22].

3.2. Example of the Calculation of Vertices

To obtain the straight lines, the slopes of the linear regression lines through the data point in the x and y arrays were calculated. In addition, the point at which a line intersected the y axis by using existing x and y values was also calculated for each vertex. The interception point was based on a best fit regression line drawn through the known x and y values, using an internal algorithm for the least squares regression procedure.

The starting point was the table of the centers of each of the zones sorted by the number of pixels comprising the area called mass (Table 1) as the number of pixels that included each zone.

Next, they were sorted by the y coordinates and were classified into groups that corresponded to the horizontal coordinates (Table 2).

As a result, the 5 horizontal lines were obtained $y = m_i x + n_i$, $j = 1\ldots5$ (Table 3).

In the same way, we proceeded to obtain the vertical lines $y = m_j + n_j$, $j = 6\ldots10$ (Table 4).

It is noted in Tables 3 and 4 that all coefficients m and n had a tendency that could be also the object of a regression. This indicated that the lines were not parallel as they had a vanishing point, and the plane that contained all the lines was not perpendicular to the focal line of the camera.

To find the two horizontal lines that represented the 5 lines, we proceeded to find the intersection points of the vertical center line with the 5 horizontal lines, obtaining the intersection points (x_i, y_i). The slope m_i was correlated with the vertical coordinates y_i, obtaining the two slopes of the representative lines as lines that had a slope $m(y_1)$ and $m(y_5)$ and passing them through Points 1 and 5, respectively (Table 5).

Table 1. Table of the center of gravity example for Image 1.

Element #	x	y	Mass
1	112.591	195.242	1024
2	113.366	82.265	1015
3	226.495	196.041	990
4	112.105	309.020	982
5	227.189	83.003	968
6	226.096	309.625	961
7	340.272	197.033	947
8	340.892	83.472	943
9	112.080	423.237	938
10	567.863	198.706	925
11	453.996	197.528	917
12	454.825	84.746	897
13	339.734	310.306	890
14	568.349	86.025	873
15	225.773	423.599	872
16	453.313	311.012	857
17	567.139	311.749	854
18	111.869	536.665	833
19	339.310	423.985	813
20	566.355	425.027	786
21	225.561	537.419	776
22	452.760	424.705	757
23	339.075	537.371	717
24	452.296	537.806	682
25	565.838	538.361	673

Table 2. Groups of points corresponding to horizontal lines.

x	y
565.838	538.361
452.296	537.806
225.561	537.419
339.075	537.371
111.869	536.665
566.355	425.027
452.760	424.705
339.310	423.985
225.773	423.599
112.080	423.237
567.139	311.749
453.313	311.012
339.734	310.306
226.096	309.625
112.105	309.020
567.863	198.706
453.996	197.528
340.272	197.033
226.495	196.041
112.591	195.242
568.349	86.025
454.825	84.746
340.892	83.472
227.189	83.003
113.366	82.265

Once having performed the regression of m with respect to y_i, a function of the slope that varied regularly across the different heights ($m = -1.136 \times 10_i^{-3} + 9.331 \times 10^{-3}$) was obtained. As a result, the equations of the horizontal lines that passed through Points 1 and 5 could be calculated.

$$y = -6108 \times 10^{-3}x + 540.807 \tag{4}$$

$$y = -9533 \times 10^{-4} x + 83.981 \qquad (5)$$

Table 3. Coefficients of horizontal lines.

m	n
3.331×10^{-3}	536.395
4.127×10^{-3}	422.710
6.018×10^{-3}	308.298
7.393×10^{-3}	194.394
8.142×10^{-3}	81.126

Table 4. Coefficients vertical lines.

m	n
-5.774×10^{-3}	568.910
-5.553×10^{-3}	455.166
-4.050×10^{-3}	341.114
-3.500×10^{-3}	227.307
-3.080×10^{-3}	113.355

Table 5. Intersection points of the central vertical line with horizontal lines.

Point	x_i	y_i
1	338.937	537.525
2	339.396	424.111
3	339.857	310.344
4	340.316	196.911
5	340.774	83.901

In the same way, we proceeded to obtain the 2 representative vertical lines:

$$x = -6444 \times 10^{-3} y + 570.765 \qquad (6)$$

$$x = -1277 \times 10^{-3} y + 112.547 \qquad (7)$$

The intersection of the opposite lines of this square provided the coordinates of the four vertices that represented the 25 LEDs (Table 6).

Table 6. Quadrilateral points to deal with the reverse perspective program.

Point	x	y
1	111.857	540.124
2	567.302	537.343
3	570.227	83.437
4	112.439	83.874

The vertices represented in Table 6 were treated using the photogrammetry method of the reconstruction of a rectangle described in Estrems [24], and the coordinates of the camera with respected to the square coordinate system were obtained, as well as the cosine direction of the focal line in this system.

$$a = \sqrt{d_f^2 - \left(d_f \cdot \cos b\right)^2} = d_f \sqrt{1 - \cos^2 b} \qquad (8)$$

$$a_2 - a_1 = d_f \left(\sqrt{1 - \cos^2 b_2} - \sqrt{1 - \cos^2 b_1}\right) \qquad (9)$$

In Figure 7, the Abbe error a is represented and calculated by the focal distance d_f and the cosine in the z direction cos b. The Abbe error is calculated in Equation (8), and the step error due to the variation of angle b during the movement is compensated at each point according to Equation (9).

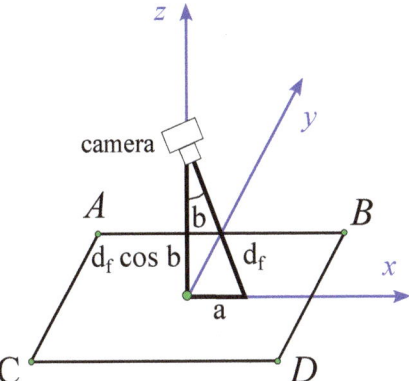

Figure 7. Position error in the vision system due to camera inclination for axis direction movement.

4. Experimental Results and Discussion

The data obtained in the experimental test are described in Tables 7–12, where CMM is the distance measure by the Coordinate Measuring Machine in the movements done by the MMT in every step during the test; Image is the distance moved in every step analyzed by the vision system; Error Image is the error provided by the vision system in every step after a comparison with the distance moved provided by the CMM; Compensation is the distance compensated due to angle error calculated in every step; Image compensated is the distance measured by the CMM after the application of the compensation calculated; Error compensation is the error after the compensation is applied; and Coincidence provides information about the coincidence in the orientation calculated for the angle compensation (YES means orientation coincidence, and NO means the angle orientation calculated is opposite the compensation required to minimize the error).

Table 7. Data for Run #1 forward with a mesh of 5 × 5 LEDs (values in µm).

CMM	Image	Error Image	Compensation	Image Compensated	Error Compensation	Coincidence
501	−501.897	−0.897	0.603	−501.294	−0.294	YES
1000	−995.031	4969	−4.432	−999.463	0.537	YES
1487	−1483.389	3611	−0.714	−1484.102	2.898	YES
1998	−1984.546	13.454	−2.164	−1986.710	11.290	YES
2489	−2489.081	−0.081	0.471	−2488.610	0.390	YES
2988	−2986.138	1.862	1.062	−2985.076	2.924	NO
3490	−3487.323	2.677	0.802	−3486.521	3.479	NO

Table 8. Data for Run #1 backward with a mesh of 5 × 5 LEDs (values in µm).

CMM	Image	Error Image	Compensation	Image Compensated	Error Compensation	Coincidence
496	−496.099	−0.099	2.394	−493.705	2.295	YES
992	−992.400	−0.400	−1.744	−994.144	−2.144	NO
1482	−1493.937	−11.937	0.505	−1493.432	−11.432	YES
1991	−1993.130	−2.130	0.632	−1992.498	−1.498	YES
2494	−2495.707	−1.707	0.912	−2494.795	−0.795	YES
2993	−2994.327	−1.327	2.214	−2992.113	0.887	YES
3494	−3493.969	0.031	−1.950	−3495.918	−1.918	YES

Table 9. Data for Run #2 forward with a mesh of 5 × 5 LEDs (values in μm).

CMM	Image	Error Image	Compensation	Image Compensated	Error Compensation	Coincidence
490	−498.217	−8.217	1.671	−496.545	−6.545	YES
985	−989.868	−4.868	2.260	−987.608	−2.608	YES
1481	−1489.347	−8.347	−0.803	−1490.150	−9.150	NO
1976	−1993.442	−17.442	1.153	−1992.290	−16.290	YES
2493	−2499.212	−6.212	0.604	−2498.608	−5.608	YES
2991	−2996.982	−5.982	2.728	−2994.254	−3.254	YES
3485	−3494.890	−9.890	1.978	−3492.912	−7.912	YES

Table 10. Data for Run #2 backward with a mesh of 5 × 5 LEDs (values in μm).

CMM	Image	Error Image	Compensation	Image Compensated	Error Compensation	Coincidence
495	−498.695	−3.695	1.631	−497.065	−2.065	YES
991	−993.169	−2.169	2.490	−990.679	0.321	YES
1486	−1497.023	−11.023	4.948	−1492.075	−6.075	YES
2003	−1994.767	8.233	−0.105	−1994.872	8.128	YES
2501	−2493.756	7.244	−0.142	−2493.898	7.102	YES
2995	−2995.898	−0.898	1.446	−2994.452	0.548	YES
3494	−3494.915	−0.915	2.048	−3492.867	1.133	YES

Table 11. Data for Run #3 forward with a mesh of 5 × 5 LEDs (values in μm).

CMM	Image	Error Image	Compensation	Image Compensated	Error Compensation	Coincidence
496	−500.617	−4.617	1.720	−498.897	−2.897	YES
994	−998.341	−4.341	1.814	−996.527	−2.527	YES
1497	−1496.865	0.135	−0.990	−1497.855	−0.855	YES
1997	−1997.961	−0.961	−0.007	−1997.968	−0.968	NO
2500	−2502.260	−2.260	0.168	−2502.092	−2.092	YES
2996	−3007.135	−11.135	2.567	−3004.567	−8.567	YES
3498	−3503.156	−5.156	0.660	−3502.496	−4.496	YES

Table 12. Data for Run #3 backward with a mesh of 5 × 5 LEDs (values in μm).

CMM	Image	Error Image	Compensation	Image Compensated	Error Compensation	Coincidence
498	−497.074	0.926	−0.532	−497.606	0.394	YES
999	−991.402	7.598	−4.628	−996.030	2.970	YES
1492	−1493.747	−1.747	2.333	−1491.414	0.586	YES
1992	−1992.997	−0.997	1.684	−1991.313	0.687	YES
2487	−2491.310	−4.310	1.007	−2490.303	−3.303	YES
2985	−2993.930	−8.930	0.791	−2993.139	−8.139	YES
3486	−3493.567	−7.567	−3.367	−3496.934	−10.934	NO

Table 13 summarizes the mean errors ($\bar{e}$) and the standard deviation errors (σ) calculated in each run of the experimental tests. The global mean (4.786 μm) and standard deviation (5.698 μm) were also calculated.

Table 13. Summary of the errors, in absolute value, provided by the proposed vision positioning algorithm.

	#1 Forward	#1 Backward	#2 Forward	#2 Backward	#3 Forward	#3 Backward	Global
Mean (μm)	3.936	2.519	8.708	4.883	4.086	4.582	4.786
Mean (%)	0.79	0.50	1.74	0.98	0.8	0.92	0.96
σ (μm)	4.771	4.238	4.212	6.586	3.699	5.567	5.698
σ (%)	0.95	0.85	0.84	1.32	0.74	1.11	1.19

As is seen in the graphs of Figure 8, the precision of positioning depended strongly on the initial error function, so the variation of the error was less than ±2μm, except several discrete points that were measured in a transition between columns of LEDs that were not so homogeneous in the LCD.

One remarkable result was that the orientation of the compensation error coincided with the sign of the error checked at almost all points. This indicated that there was variation of the orientation of the focal line with respect to the screen coordinate system, although this was not applied efficiently to improve the precision of measurement. This was probably due to the problems evaluating the trigonometric functions in angles with values less than 10^{-3} radians.

Therefore, due to the distance from the vanishing points of the lines, the estimation of the angular error did not turn out to be very precise in its quantification, although it provided qualitative descriptors on the direction and magnitude of the variation in the inclination of the camera.

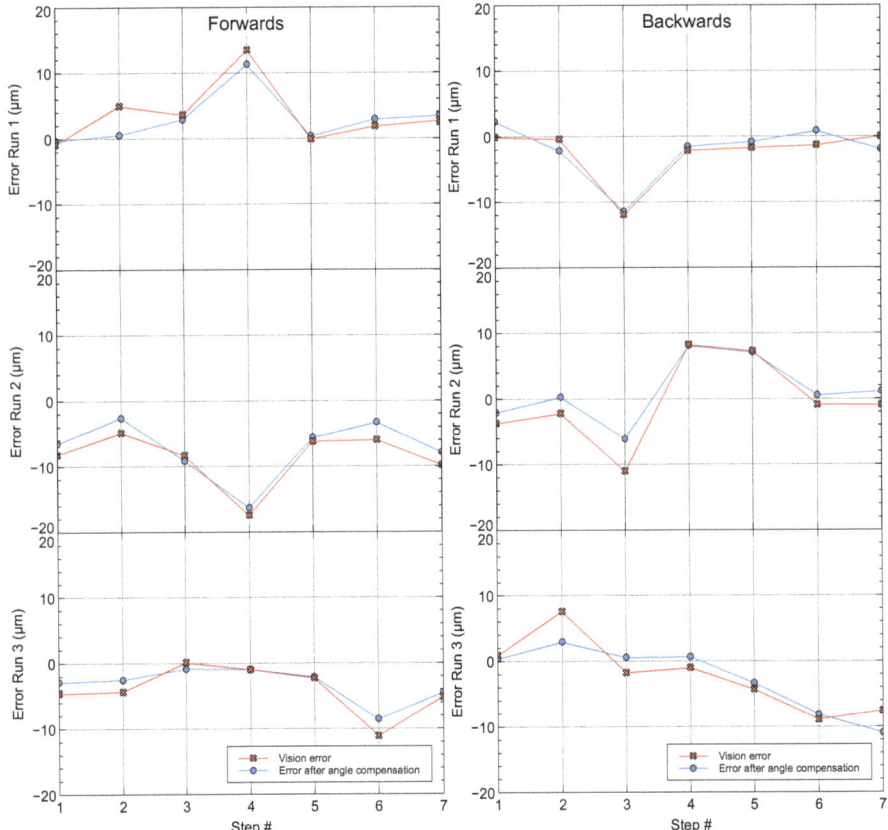

Figure 8. Error due to the vision system before and after angle compensation was applied.

5. Conclusions

A new method was developed to position the tool in a micromachine system based on a camera-LCD screen positioning system that also provided information on the angular deviations of the tool axis during operation.

The method gave a good approximation of the center point of the rectangle with a mean error of 0.96%, considering not only the vision algorithm, but also the mechanical test device, and provided the inclination of the workpiece with respect to the LCD-screen reference coordinate system.

The equivalent square was calculated as regression of the lines that could be drawn through the centers of gravity of each of the LEDs. The lack of parallelism between the sides of the square indicated an inclination of the camera axis relative to perpendicular to the screen. The variation of this inclination introduced errors in the displacements that were added to the simple displacement of the center of gravity and whose compensation was also calculated in this article.

A test of the method was designed with the assistance of a Coordinate Measurement Machine (CMM) to verify the accuracy of the positioning method. The test performed provided good accuracy in measuring the position of the designed method, but a high dispersion in the angular deviation was detected, although the orientation of the inclination was appropriate in almost all cases (85.7%). This was due to the small value of the angles that made the approximations of the trigonometric functions very erratic. With accurate formulas to approximate trigonometric functions for small angles, the method could help in obtaining more accurate measurements.

Author Contributions: Conceptualization, Ó.d.F.O. and M.E.A.; methodology, Ó.d.F.O.; software, Ó.d.F.O. and M.E.A.; validation, Ó.d.F.O., H.T.S.R., and J.C.-B.M.-H.; formal analysis, Ó.d.F.O., H.T.S.R., and J.C.-B.M.-H.; writing, original draft preparation, Ó.d.F.O.; writing, review and editing, M.E.A., H.T.S.R., and J.C.-B.M.-H.; visualization, Ó.d.F.O., M.E.A., H.T.S.R., and J.C.-B.M.-H.

Funding: This research was funded by Ingeniería Murciana SL by a private contract.

Acknowledgments: The authors want to thank the University Center of Defense at the Spanish Air Force Academy, MDE-UPCT, for financial support.

Conflicts of Interest: The authors declare no conflict of interest.

Abbreviations

The following abbreviations are used in this manuscript:

LCD   Liquid Crystal Display
MMT   Micromachine Tool
ppi   points per inch
CMM   Coordinates Measuring Machine

References

1. Leviton, D.B.; Kirk, J.; Lobsinger, L. Ultra-high resolution Cartesian absolute optical encoder. In *Recent Developments in Traceable Dimensional Measurements II, Proceedings of the SPIE's 48th Annual Meeting, San Diego, CA, USA, 2–4 August 2003*; SPIE: Bellingham, WA, USA, 2003; pp. 111–121,doi:10.1117/12.518376. [CrossRef]
2. Leviton, D.B. Method and Apparatus for Two-Dimensional Absolute Optical Encoding. U.S. Patent 6765195 B1, 20 July 2004.
3. Montes, C.A.; Ziegert, J.C.; Wong, C.; Mears, L.; Tucker, T. 2-D absolute positioning system for real time control applications. In Proceedings of the Twenty-Fourth Annual Meeting of the American Society for Precision Engineering, Rosemont, IL, USA, 13 September 2010.
4. De Francisco-Ortiz, O.; Sánchez-Reinoso, H.; Estrems-Amestoy, M. Development of a Robust and Accurate Positioning System in Micromachining Based on CAMERA and LCD Screen. *Procedia Eng.* **2015**, *132*, 8, doi:10.1016/j.proeng.2015.12.575. [CrossRef]
5. De Francisco Ortiz, O.; Sánchez Reinoso, H.; Estrems Amestoy, M.; Carrero-Blanco Martinez-Hombre, J. Position precision improvement throughout controlled led paths by artificial vision in micromachining processes. *Procedia Manuf.* **2017**, *13*, 197–204, doi:10.1016/j.promfg.2017.09.044. [CrossRef]
6. Byun, S.; Kim, M. Real-Time Positioning and Orienting of Pallets Based on Monocular Vision. In Proceedings of the 2008 20th IEEE International Conference on Tools with Artificial Intelligence, Dayton, OH, USA, 3–5 November 2008; Volume 2, pp. 505–508, doi:10.1109/ICTAI.2008.124. [CrossRef]
7. Zhang, B.; Wang, J.; Rossano, G.; Martinez, C.; Kock, S. Vision-guided robot alignment for scalable, flexible assembly automation. In Proceedings of the 2011 IEEE International Conference on Robotics and Biomimetics, Karon Beach, Phuket, Thailand, 7–11 December 2011; pp. 944–951, doi:10.1109/ROBIO.2011.6181409. [CrossRef]
8. Zhou, K.; Wang, X.J.; Wang, Z.; Wei, H.; Yin, L. Complete Initial Solutions for Iterative Pose Estimation from Planar Objects. *IEEE Access* **2018**, *6*, 22257–22266, doi:10.1109/access.2018.2827565. [CrossRef]

9. Lyu, D.; Xia, H.; Wang, C. Research on the effect of image size on real-time performance of robot vision positioning. *EURASIP J. Image Video Process.* **2018**, *2018*, 112, doi:10.1186/s13640-018-0328-0. [CrossRef]
10. Montijano, E.; Cristofalo, E.; Zhou, D.; Schwager, M.; Sagüés, C. Vision-Based Distributed Formation Control Without an External Positioning System. *IEEE Trans. Robot.* **2016**, *32*, 339–351, doi:10.1109/TRO.2016.2523542. [CrossRef]
11. Yang, S.; Jiang, R.; Wang, H.; Ge, S.S. Road Constrained Monocular Visual Localization Using Gaussian-Gaussian Cloud Model. *IEEE Trans. Intell. Transp. Syst.* **2017**, *18*, 3449–3456, doi:10.1109/TITS.2017.2685436. [CrossRef]
12. Guo, D.; Wang, H.; Leang, K.K. Nonlinear vision-based observer for visual servo control of an aerial robot in global positioning system denied environments. *J. Mech. Robot.* **2018**, *10*, 061018. [CrossRef]
13. Vivacqua, R.P.D.; Bertozzi, M.; Cerri, P.; Martins, F.N.; Vassallo, R.F. Self-Localization Based on Visual Lane Marking Maps: An Accurate Low-Cost Approach for Autonomous Driving. *IEEE Trans. Intell. Transp. Syst.* **2018**, *19*, 582–597. doi:10.1109/TITS.2017.2752461. [CrossRef]
14. Fang, J.; Wang, Z.; Zhang, H.; Zong, W. Self-localization of Intelligent Vehicles Based on Environmental Contours. In Proceedings of the 2018 3rd International Conference on Advanced Robotics and Mechatronics (ICARM), Singapore, 18–20 July 2018; pp. 624–629. doi:10.1109/ICARM.2018.8610687. [CrossRef]
15. Islam, K.T.; Wijewickrema, S.; Pervez, M.; O'Leary, S. Road Trail Classification using Color Images for Autonomous Vehicle Navigation. In Proceedings of the 2018 Digital Image Computing: Techniques and Applications (DICTA), Canberra, Australia, 10–13 December 2018; pp. 1–5, doi:10.1109/DICTA.2018.8615834. [CrossRef]
16. You, S.; Neumann, U.; Azuma, R. Hybrid inertial and vision tracking for augmented reality registration. In Proceedings of the IEEE Virtual Reality (Cat. No. 99CB36316), Houston, TX, USA, 13–17 March 1999; pp. 260–267, doi:10.1109/VR.1999.756960. [CrossRef]
17. Kim, J.; Jun, H. Vision-based location positioning using augmented reality for indoor navigation. *IEEE Trans. Consum. Electron.* **2008**, *54*, 954–962, doi:10.1109/TCE.2008.4637573. [CrossRef]
18. Samarasekera, S.; Oskiper, T.; Kumar, R.; Sizintsev, M.; Branzoi, V. Augmented Reality Vision System for Tracking and Geolocating Objects of Interest. U.S. Patent 9,495,783, 15 November 2016.
19. Suenaga, H.; Tran, H.H.; Liao, H.; Masamune, K.; Dohi, T.; Hoshi, K.; Takato, T. Vision-based markerless registration using stereo vision and an augmented reality surgical navigation system: A pilot study. *BMC Med. Imaging* **2015**, *15*, 51. [CrossRef]
20. Rajeev, S.; Wan, Q.; Yau, K.; Panetta, K.; Agaian, S.S. Augmented reality-based vision-aid indoor navigation system in GPS denied environment. In *Proceedings of the SPIE 10993, Mobile Multimedia/Image Processing, Security, and Applications*; SPIE: Baltimore, MA, USA, 2019, doi:10.1117/12.2519224. [CrossRef]
21. Wefelscheid, C.; Wekel, T.; Hellwich, O. Monocular Rectangle Reconstruction Based on Direct Linear Transformation. In Proceedings of the International Conference on Computer Vision Theory and Applications (VISAPP 2011), Vilamoura, Portugal, 5–7 March 2011; pp. 271–276.
22. Haralick, R.M. Determining camera parameters from the perspective projection of a rectangle. *Pattern Recognit.* **1989**, *22*, 225–230, doi:10.1016/0031-3203(89)90071-x. [CrossRef]
23. Quan, L.; Lan, Z. Linear N-point camera pose determination. *IEEE Trans. Pattern Anal. Mach. Intell.* **1999**, *21*, 774–780, doi:10.1109/34.784291. [CrossRef]
24. Estrems Amestoy, M.; de Francisco Ortiz, O. Global Positioning from a Single Image of a Rectangle in Conical Perspective. *Sensors* **2019**, *19*, 5432, doi:10.3390/s19245432. [CrossRef] [PubMed]
25. Abidi, M.A.; Chandra, T. A new efficient and direct solution for pose estimation using quadrangular targets—Algorithm and evaluation. *IEEE Trans. Pattern Anal. Mach. Intell.* **1995**, *17*, 534–538, doi:10.1109/34.391388. [CrossRef]
26. Haralick, R.M. Using perspective transformations in scene analysis. *Comput. Graph. Image Process.* **1980**, *13*, 191–221, doi:10.1016/0146-664x(80)90046-5. [CrossRef]
27. Sim, R.; Little, J. Autonomous vision-based robotic exploration and mapping using hybrid maps and particle filters. *Image Vis. Comput.* **2009**, *27*, 167–177, doi:10.1016/j.imavis.2008.04.003. [CrossRef]
28. Valencia-Garcia, R.; Martinez-Béjar, R.; Gasparetto, A. An intelligent framework for simulating robot-assisted surgical operations. *Expert Syst. Appl.* **2005**, *28*, 425–433, doi:10.1016/j.eswa.2004.12.003. [CrossRef]

29. Pichler, A.; Akkaladevi, S.; Ikeda, M.; Hofmann, M.; Plasch, M.; Wögerer, C.; Fritz, G. Towards Shared Autonomy for Robotic Tasks in Manufacturing. *Procedia Manuf.* **2017**, *11*, 72–82, doi:10.1016/j.promfg.2017.07.139. [CrossRef]
30. Broggi, A.; Dickmanns, E. Applications of computer vision to intelligent vehicles. *Image Vis. Comput.* **2000**, *18*, 365–366. [CrossRef]
31. Patterson, T.; McClean, S.; Morrow, P.; Parr, G.; Luo, C. Timely autonomous identification of UAV safe landing zones. *Image Vis. Comput.* **2014**, *32*, 568–578, doi:10.1016/j.imavis.2014.06.006. [CrossRef]
32. González, D.; Pérez, J.; Milanés, V. Parametric-based path generation for automated vehicles at roundabouts. *Expert Syst. Appl.* **2017**, *71*, 332–341, doi:10.1016/j.eswa.2016.11.023. [CrossRef]
33. Sanchez-Lopez, J.; Pestana, J.; De La Puente, P.; Campoy, P. A reliable open-source system architecture for the fast designing and prototyping of autonomous multi-UAV systems: Simulation and experimentation. *J. Intell. Robot. Syst.* **2015**, *84*, 779–797. [CrossRef]
34. Olivares-Mendez, M.; Kannan, S.; Voos, H. Vision based fuzzy control autonomous landing with UAVs: From V-REP to real experiments. In Proceedings of the 2015 23rd Mediterranean Conference on Control and Automation (MED), Torremolinos, Spain, 16–19 June 2015; pp. 14–21, doi:10.1109/MED.2015.7158723. [CrossRef]
35. Romero-Ramirez, F.J.; Muñoz-Salinas, R.; Medina-Carnicer, R. Speeded up detection of squared fiducial markers. *Image Vis. Comput.* **2018**, *76*, 38–47, doi:10.1016/j.imavis.2018.05.004. [CrossRef]
36. Germanese, D.; Leone, G.R.; Moroni, D.; Pascali, M.A.; Tampucci, M. Long-Term Monitoring of Crack Patterns in Historic Structures Using UAVs and Planar Markers: A Preliminary Study. *J. Imaging* **2018**, *4*, 99, doi:10.3390/jimaging4080099. [CrossRef]
37. An, G.H.; Lee, S.; Seo, M.W.; Yun, K.; Cheong, W.S.; Kang, S.J. Charuco Board-Based Omnidirectional Camera Calibration Method. *Electronics* **2018**, *7*, 421. doi:10.3390/electronics7120421. [CrossRef]
38. Pflugi, S.; Vasireddy, R.; Lerch, T.; Ecker, T.; Tannast, M.; Boemke, N.; Siebenrock, K.; Zheng, G. Augmented marker tracking for peri-acetabular osteotomy surgery. *Conf. Proc. IEEE Eng. Med. Biol. Soc.* **2017**, *2017*, 937–941. doi:10.1109/EMBC.2017.8036979. [CrossRef]
39. Lima, J.P.; Roberto, R.; Simões, F.; Almeida, M.; Figueiredo, L.; Teixeira, J.M.; Teichrieb, V. Markerless tracking system for augmented reality in the automotive industry. *Expert Syst. Appl.* **2017**, *82*, 100–114. [CrossRef]
40. Chen, P.; Peng, Z.; Li, D.; Yang, L. An improved augmented reality system based on AndAR. *J. Vis. Commun. Image Represent.* **2016**, *37*, 63–69, doi:10.1016/j.jvcir.2015.06.016. [CrossRef]
41. Khattak, S.; Cowan, B.; Chepurna, I.; Hogue, A. A real-time reconstructed 3D environment augmented with virtual objects rendered with correct occlusion. In Proceedings of the 2014 IEEE Games Media Entertainment, Toronto, ON, Canada, 22–24 October 2014; pp. 1–8.
42. De Francisco Ortiz, O.; Sánchez Reinoso, H.; Estrems Amestoy, M.; Carrero-Blanco, J. Improved Artificial Vision Algorithm in a 2-DOF Positioning System operated under feedback control in micromachining. In *Euspen's 17th International Conference & Exhibition Proceedings*; Billington, D., Phillips, D., Eds.; Euspen: Northampton, UK, 2017; pp. 460–461.

© 2019 by the authors. Licensee MDPI, Basel, Switzerland. This article is an open access article distributed under the terms and conditions of the Creative Commons Attribution (CC BY) license (http://creativecommons.org/licenses/by/4.0/).

Article

Industrial Calibration Procedure for Confocal Microscopes

Alberto Mínguez Martínez [1,2,*] and Jesús de Vicente y Oliva [1,2,*]

1. Laboratory of Metrology and Metrotechnics (LMM), School of Industrial Engineering, Technical University of Madrid; c./José Gutiérrez Abascal, 2, 28006 Madrid, Spain
2. Laser Center, Technical University of Madrid, Campus Sur, Edificio "La Arboleda"; c./Alan Turing, 1, 28031 Madrid, Spain
* Correspondence: a.minguezm@upm.es (A.M.M.); jvo@etsii.upm.es (J.d.V.y.O.)

Received: 10 November 2019; Accepted: 9 December 2019; Published: 10 December 2019

Abstract: Coordinate metrology techniques are widely used in industry to carry out dimensional measurements. For applications involving measurements in the submillimeter range, the use of optical, non-contact instruments with suitable traceability is usually advisable. One of the most used instruments to perform measurements of this type is the confocal microscope. In this paper, the authors present a complete calibration procedure for confocal microscopes designed to be implemented preferably in workshops or industrial environments rather than in research and development departments. Therefore, it has been designed to be as simple as possible. The procedure was designed without forgetting any of the key aspects that need to be taken into account and is based on classical reference material standards. These standards can be easily found in industrial dimensional laboratories and easily calibrated in accredited calibration laboratories. The procedure described in this paper can be easily adapted to calibrate other optical instruments (e.g., focus variation microscopes) that perform 3D dimensional measurements in the submillimeter range.

Keywords: coordinate metrology; confocal microscopy; measurement; calibration; traceability; uncertainty; quality assessment

1. Introduction

In industry, coordinate measuring machines (CMMs) are widely used to carry out dimensional measurements, because these kinds of machines are capable of measuring many different types of geometries with great flexibility and sufficient accuracy. For this reason, CMMs can be considered as universal measuring devices [1]. In addition, modern advanced manufacturing processes demand a deep study of surface textures. Probably the most common method for surface texture verification up to recent years was to perform roughness measurements with a roughness measuring machine—usually a 2D stylus instrument [2]. In many cases, manufacturers prefer to verify texture and geometry without mechanical contact between the instrument and surface [3]. Due to this, optical instruments for coordinate metrology have been developed. ISO 25178-6 lists optical methods for measuring surface texture [4], and among them, it includes confocal microscopy, which permits both dimensional and 2D/3D roughness measurements [5] without mechanical contact.

The confocal microscope was developed in 1955 by Minsky [6,7] and allows images of optical sections of samples to be obtained, from which the full 3D object geometry can be reconstructed. Confocal microscopy is also important, because it is a powerful tool for observation and measurement at both scientific research and workshop levels. It presents the following advantages [8]:

- It adds the Z-axis to traditional measuring optical microscopes, which only work in the XY plane.

- It allows analysis of the 3D geometry of the object surface and characterization of its quality from data points acquired while scanning it.
- Its lateral resolution is better than in traditional optical microscopy.
- It permits more precise 3D images of the objects being measured to be obtained that are of higher quality and in less time compared to other methods. This allows many useful measurements to be carried out in short intervals of time.
- Transparent specimens can be observed, as can sections with a certain thickness, without the need to section the object under study.

Confocal microscopy has applications in many fields, both in research and industrial applications. This type of microscope is widely used in biomedical science, material science, and surface quality metrology at micro and macro scales [9]. However, there is no standardized procedure to calibrate and provide traceability to these instruments in the fields of coordinate metrology and surface texture metrology. Intense work is being done around ISO standard 25178-700, but it is still in the draft stage. The calibration of measurement instruments is crucial to maintain the traceability of measurement results [10]. As is widely known, traceability can be defined as the property of a measurement result by which it can be related to a reference through an uninterrupted and documented chain of calibrations, each of which contributes to measurement uncertainty [11].

The purpose of this paper is to describe a way to provide suitable traceability to a confocal microscope when performing metrological activities using single topography measurements, in the fields of both coordinate metrology and roughness metrology. Please note that when image stitching is not used (single topography), there is no movement of the XY stage, and there is therefore no need to calibrate the displacements of this stage. The calibration procedure presented by the authors is intended to be simple and is based on classical mechanical standards. Note that the objective was not to perform a state-of-the-art calibration of a confocal microscope [12–15], neither was it to achieve very low uncertainties; the objective was to ensure adequate traceability with adequate uncertainty estimation in the field of dimensional metrology in the submillimeter range.

Prior to carrying out the calibration, an analysis of the operating principle of the confocal microscope is necessary for a better understanding of the device. This kind of microscope usually uses a low-power, high-intensity, monochromatic laser system for illumination [16–19]. A laser beam passes through a beam splitter, and one of the beams is then redirected to the sample, passing through complex optics [7]. Once the scanning surface is illuminated, the reflected beam travels back along the same path. If the illumination is properly focused on the surface, the reflected beam will go to the detector without losing intensity, but if the surface is out of focus, the intensity will be lower. The filtered beam arrives at the detector and a computer system processes the signal, making a 3D reconstruction of the surface [9,16,18].

Several factors affect the quality of these measurements [5,20]:

- Metrological characteristics of the instrument: measurement noise, flatness deviation, non-linearity errors, amplification coefficients, and perpendicularity errors between axes.
- Instrument geometry: alignment of components and the XY stage and rotary stage error motions.
- Source characteristics: focal spot size and drift.
- Detector characteristics: pixel response, uniformity and linearity, detector offset, and bad pixels.
- Reconstruction and data processing: surface determinations, data representation, and calculation approaches.
- Environmental conditions: temperature, humidity, and vibration.

As can be seen in Figure 1, the confocal microscope projects, through a complex optical system, illumination patterns over the surface that is being explored and captures the returned beam through the same pattern of illumination. As a result, it is possible to discriminate if the returned beams are out of focus and filter them [5,7,16,17,21]. In Figure 2, this property is shown in detail.

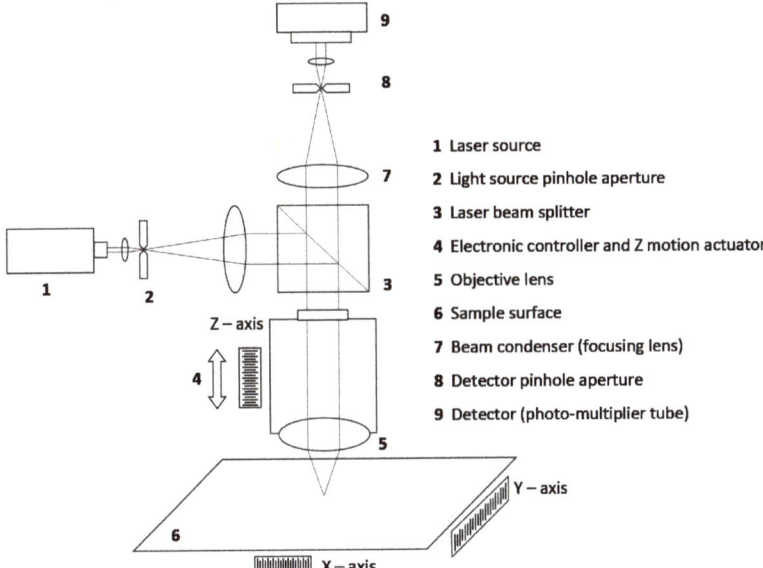

Figure 1. Scheme of a confocal microscope [5,18,22].

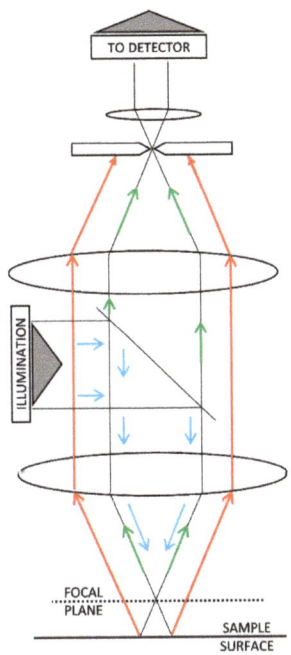

Figure 2. Filtration of out-of-focus signal (red) in confocal microscopy [5,7,16,17,21] in comparison with in-focus signal (green).

Once the in-focus image goes to the detector, the computational treatment starts. The electronic controller moves the objective along the Z-axis in order to permit the confocal microscope to capture

2D images at different z coordinates. As 2D images are composed of pixels, 3D images obtained with confocal microscopes are composed of voxels, as shown in Figure 3. If an interpolation if made between consecutive 2D images, it is possible to create a 3D model of the scanned surface [23].

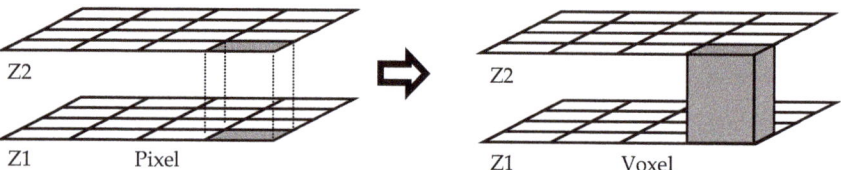

Figure 3. Transformation from pixel to voxel.

In order to achieve dimensional traceability in dimensional measurements carried out with confocal microscopes, it is necessary to perform a 3D calibration of the instrument. This calibration should provide estimations of voxel sizes along X, Y, and Z axes, but it would also be advisable to provide estimations of perpendicularity errors between axes. Additionally, as the confocal microscope can be used to perform roughness measurements, a specific calibration of the instrument for roughness measurements is advisable.

2. Materials and Methods

In order to ease the understanding of the calibration procedures described later on, a calibration example was carried with the following confocal microscope and software:

- Leica DCM3D confocal microscope (Wetzlar, Germany) with a 10× objective (EPI-L, NA = 0,30). Field of view 1270 µm × 952 µm (768 × 576 pixels); 1.65 µm nominal voxel width. The overall range of the Z-axis is 944 µm using 2 µm axial steps (voxel height), but the instrument is used in a reduced working range of only 100 µm.
- SensoSCAN—LeicaSCAN DCM3D 3.41.0 software developed by Sensofar Tech Ltd. (Terrassa, Spain).

In this paper, we propose a calibration procedure that is only valid for single topography measurements—that is, without using image stitching. In single topography measurements, the XY stage is not moved during measurement, and its errors do not contribute to uncertainty. When using image stitching (extended topography measurements), the XY stage does contribute to uncertainty, and the calibration procedure described in this paper should be updated using techniques such as those described in [14,15]. The complete calibration procedure includes the following:

- Calibration of the X and Y scales, using a stage micrometer as a reference measurement standard.
- Estimation of perpendicularity error between X and Y axes.
- Estimation of the flatness deviation of the focal plane using an optical flat.
- Calibration of Z scale using a calibrated steel sphere.
- Calibration of the confocal microscope for the measurement of 2D roughness using periodic and aperiodic 2D roughness measurement standards.
- All uncertainties are estimated following the mainstream GUM method (Guide to the Expression of Uncertainty in Measurement [24]) or EA-04/02 M:2013 document [25], as they are standard procedures in calibration laboratories accredited under ISO 17025 [26].

All reference measurement standards used were chosen to be:

- Easy to find.
- Easy to calibrate with low enough uncertainties in National Measurement Institutes (NMIs) or preferably in accredited calibration laboratories (ACLs).
- Stable mechanical artifacts that could guarantee long recalibration intervals.

- Common in the field of dimensional metrology in order to facilitate their acquisition, calibration, and correct use.

2.1. Flatness Verification

Before calibrating the X and Y axes, a flatness verification must be performed. This is necessary because it is often observed that the XY plane of confocal microscopes is slightly curved. This is evident when exploring a flat surface such as an optical flat whose total flatness error is usually lower than 50 nm. In these cases, the reference flat surface when observed by the confocal microscope appears curved, as if it was a cap of a sphere or an ellipsoid. According to manufacturers, this error is usually small enough, but it is impossible to carry out an accurate measurement without taking this component of uncertainty into account [27].

For this verification, the authors propose following a procedure based on [28], but using a confocal microscope instead of an interferometer. The software of the confocal microscope provides a topographic map of the explored surface from which the total flatness error (peak to peak) or the RMS (root mean square error) flatness can be estimated.

The calibration is done in two positions (0° and 90°) and therefore, two measurements are obtained (Figure 4):

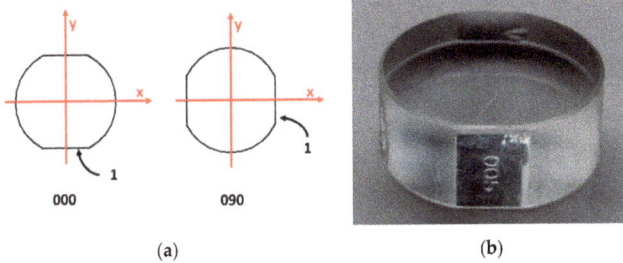

Figure 4. (a) Positions of the optical flat during calibration; (b) optical flat.

For this calibration, the authors recommend using the RMS flatness deviation, because it is more statistically stable than the total flatness deviation. As discussed in [29] (Section 1.3.5.6 "Measures of Scale"), the total flatness deviation—which is equivalent to the range—is very sensible to the presence of outliers because it is determined as the difference between the two most extreme points. The problem increases with the number N of points from which the range is determined. In our case, the number of points was very large ($N = 768 \times 576 = 442,368$). For these reasons, other parameters such as the standard deviation (equivalent to the root mean square error), or even better, the median absolute deviation (MAD) should be used. Since confocal software always includes the RMS flatness deviation and it is not very easy to include MAD, we recommend the RMS flatness deviation.

2.2. XY Plane Calibration

In the literature, it is possible to find several procedures for this calibration. Following the studies of de Vicente et al. [30] and Guarneros et al. [31], it is possible to calibrate X and Y scales and estimate their perpendicularity error by making measurements of a stage micrometer in four positions (Figure 5).

A stage micrometer is easy to calibrate in a National Measurement Institute (NMI) or in an accredited calibration laboratory (ACL) with sufficiently small uncertainty (equal to or lower than 1 µm) for the calibration of a confocal microscope.

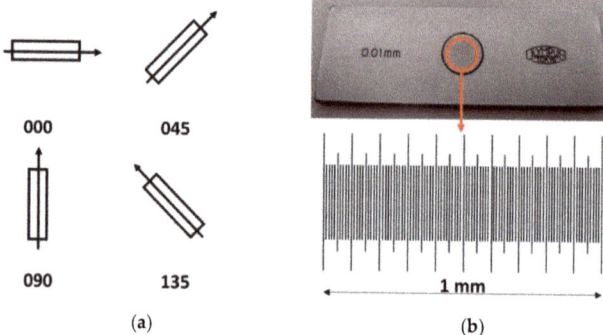

Figure 5. (a) The different positions of scanning for the stage micrometer; (b) the stage micrometer used during calibration.

It is strongly recommended that the stage micrometer should be metallic and have the marks engraved, not painted, such as those used to calibrate metallographic microscopes. Marks painted over glass are difficult to detect with a confocal instrument.

The matrix model proposed for calibration by de Vicente et al. [32] is as follows:

$$\begin{bmatrix} x \\ y \end{bmatrix} = \begin{bmatrix} p \\ q \end{bmatrix} + \begin{bmatrix} c_{xy} + a & \theta/2 \\ \theta/2 & c_{xy} - a \end{bmatrix} \cdot \begin{bmatrix} p \\ q \end{bmatrix}, \qquad (1)$$

where (p, q) are the readings directly provided by the confocal microscope for the Cartesian coordinates in the XY plane. (x, y) are the corrected Cartesian coordinates once the calibration parameters c_{xy}, a, and θ have been applied using the previous matrix model.

The meanings of these three parameters are as follows:

c_{xy} represents the deviation of actual pixel width w_{xy} from the nominal pixel width $w_{xy,nom}$:

$$w_{xy} = w_{xy,nom} \cdot (1 + c_{xy}). \qquad (2)$$

a represents the difference between pixel widths along X-axis (w_x) and Y-axis (w_y):

$$w_x = w_{xy,nom} \cdot (1 + c_{xy} + a), \qquad (3)$$

$$w_y = w_{xy,nom} \cdot (1 + c_{xy} - a), \qquad (4)$$

$$w_{xy} = \frac{(w_x + w_x)}{2}. \qquad (5)$$

θ represents the perpendicularity error between the X-axis and Y-axis. The actual angle between these axes is $\pi/2 - \theta$.

The amplification coefficients α_x, α_y, and α_z of the axes (according to ISO 25178-70 [33]) are:

$$\alpha_x = 1 + c_{xy} + a, \qquad (6)$$

$$\alpha_y = 1 + c_{xy} - a, \qquad (7)$$

$$\alpha_z = 1 + c_z. \qquad (8)$$

We recommend using the average pitch ℓ of the stage micrometers. ℓ is the average of all individual pitches (distances between two consecutive marks) observed in the images provided by the confocal microscope. Figure 6 summarizes the measurement of the stage micrometer in one position. Using special software for this task, written in Matlab®R2019a and developed at the Laboratorio de Metrología y Metrotecnia (LMM), it is possible to automatically detect and estimate the distance d_i of each mark from the zero mark. Using this software, all the distances (pitches) between two consecutive marks in the stage micrometers were estimated (Figure 6a). Moreover, pitches can be measured in different positions—in the middle and in higher and lower positions—which permits the estimation of the repeatability during the pitch measurements. In Figure 6a, for each pitch, the average value is represented by a circle, and the measurement variability around this value is represented with a vertical line. In order to estimate the non-linearity errors e_i (Figure 6b), a straight line $d_i \cong m + \ell \cdot i$ was fitted, and the errors were estimated as $e_i = d_i - (m + \ell \cdot i)$. The coefficient m represents the deviation of the zero mark from its estimated position and ℓ represents the average pitch.

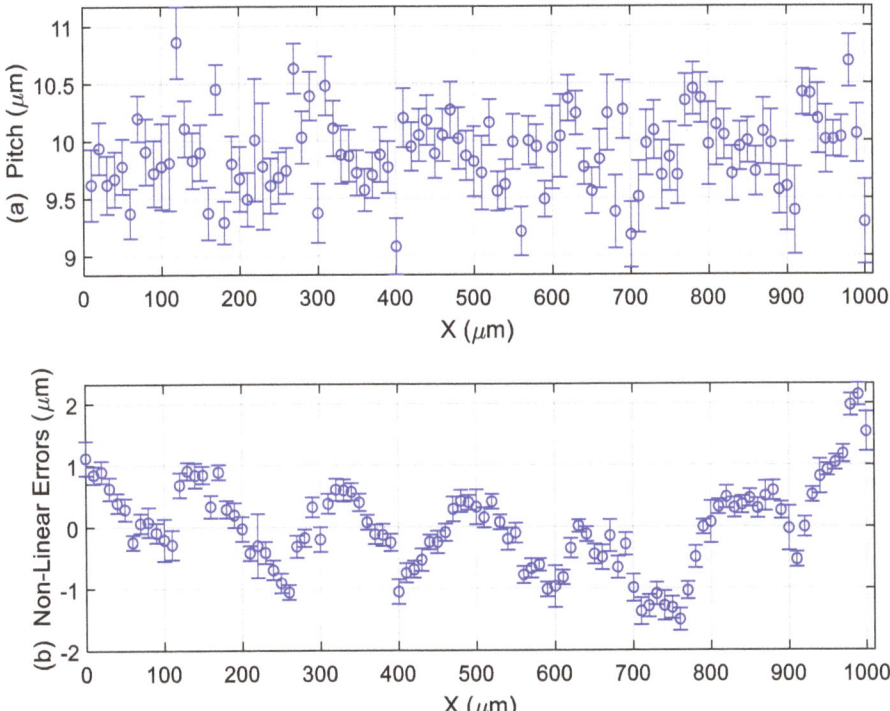

Figure 6. Measurement of stage micrometer in position $0°$: (**a**) pitch measurements results in μm; (**b**) non-linear errors in μm.

If the previously mentioned special software is not available, measurements of distances between marks must be done by hand. However, although it is more laborious, the entire procedure described above for the estimation of the average pitch ℓ and non-linearity errors e_i can be carried out without any problem.

Let ℓ_0 be the average pitch of the stage micrometer certified by a suitable laboratory with a standard uncertainty $u(\ell_0)$. ℓ_1, ℓ_2, ℓ_3, and ℓ_4 are the average pitches measured with the confocal microscope in positions $0°$, $90°$, $45°$, and $135°$, respectively. Their corresponding standard uncertainties are $u(\ell_1)$, $u(\ell_2)$, $u(\ell_3)$, and $u(\ell_4)$, where only the variability observed in Figure 6 (or equivalent ones) was taken into account.

When the matrix model is applied to positions 0°, 90°, 45°, and 135°, we obtain the following expressions that permit simple estimations of calibration parameters c_{xy}, a, and θ:

$$\text{Position } 0°: \ell_1 \cdot (1 + c_{xy} + a) \cong \ell_0, \tag{9}$$

$$\text{Position } 90°: \ell_2 \cdot (1 + c_{xy} - a) \cong \ell_0, \tag{10}$$

$$\text{Position } 45°: \ell_3 \cdot \left(1 + c_{xy} + \frac{\theta}{2}\right) \cong \ell_0, \tag{11}$$

$$\text{Position } 135°: \ell_4 \cdot \left(1 + c_{xy} - \frac{\theta}{2}\right) \cong \ell_0. \tag{12}$$

From these expressions, it is easy to conclude that possible estimations of c_{xy}, a, and θ are:

$$c_{xy} = \frac{\ell_0}{4} \cdot \left(\frac{1}{\ell_1} + \frac{1}{\ell_2} + \frac{1}{\ell_3} + \frac{1}{\ell_4}\right) - 1, \tag{13}$$

$$\text{with } u(c_{xy}) = \frac{\sqrt{u^2(\ell_0) + [u^2(\ell_1) + u^2(\ell_2) + u^2(\ell_3) + u^2(\ell_4)]/16}}{\ell_0}, \tag{14}$$

$$a = \frac{\ell_0}{2} \cdot \left(\frac{1}{\ell_1} - \frac{1}{\ell_2}\right), \tag{15}$$

$$\text{with } u(a) = \frac{\sqrt{u^2(\ell_1) + u^2(\ell_2)}}{2\ell_0}. \tag{16}$$

$$\theta = \ell_0 \cdot \left(\frac{1}{\ell_3} - \frac{1}{\ell_4}\right), \tag{17}$$

$$\text{with } u(\theta) = \frac{\sqrt{u^2(\ell_3) + u^2(\ell_4)}}{\ell_0}. \tag{18}$$

Correlations between these parameters (c_{xy}, a, and θ) are usually very small (lower than 0.01). Therefore, these correlations can be neglected.

2.3. Z-Axis Calibration

Document [34] proposes calibrating the Z-axis using a step gauge built with gauge blocks over an optical flat (Figure 7). However, the short field of view of confocal microscopes makes it difficult to carry out the calibration with this type of measurement standard.

To solve this problem, several authors [5,35] have proposed the use of step height standards (Figure 8). Wang et al. [35] used them with the nominal values 24, 7, 2, and 0.7 µm. This kind of measurement standard has several grooves whose nominal depths cover the range of use of the confocal microscope on the Z-axis. Following their procedures, every groove has to be measured 10 times, changing the position of the standard on the objective.

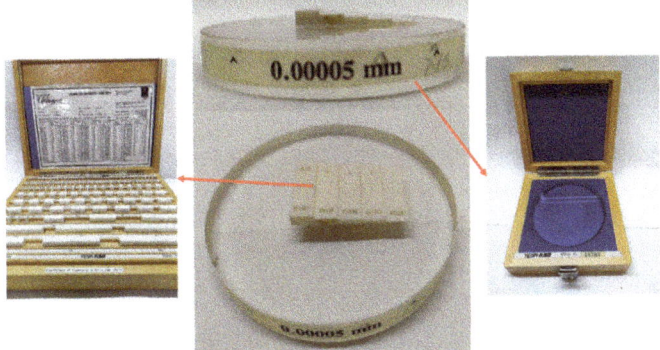

Figure 7. A step gauge built on an optical flat (with total flatness error 0.00005 mm) and a set of gauge blocks.

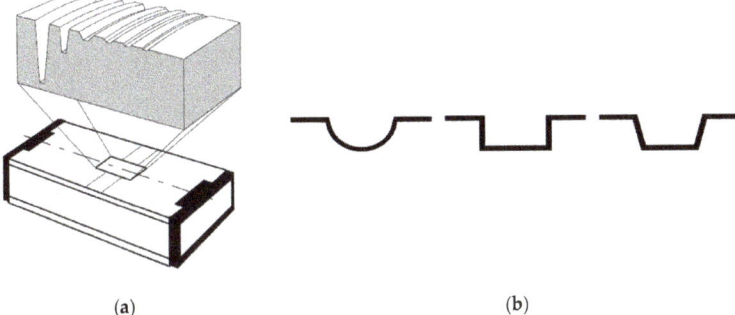

Figure 8. This figure shows (**a**) a typical model of a step height standard; and (**b**) different models of step height standards' grooves (ISO 5436-1 types A and B) [30,35–37].

This kind of standard is typically used for roughness calibration. If the purpose is to make a calibration on the Z-axis, these standards have the limitation of the groove depth, which usually is small to cover the range of the Z-axis.

In order to solve this problem, we propose using a small metallic sphere, as shown in Figure 9, with a nominal diameter between 1 and 10 mm, similar to the one used in [38]. This kind of measurement standard is easy to find and easy to calibrate in both NMIs and ACLs with uncertainties equal to or lower than 0.5 μm. The software of confocal microscopes usually permits a spherical surface to be fit to the points detected over the surface of the spherical measurement standard. If not, coordinates (p, q, r) obtained with the confocal microscope can be exported to a text file and processed with a routine similar to that described in Appendix A. Therefore, it is possible to compare the certified diameter D_0 of the sphere against the diameter D_m of the spherical surface fitted by the confocal microscope. We propose the use of an extended matrix model to take into account the calibration of the Z-axis:

$$\begin{bmatrix} x \\ y \\ z \end{bmatrix} = \begin{bmatrix} 1+c_{xy}+a & \theta/2 & 0 \\ \theta/2 & 1+c_{xy}-a & 0 \\ 0 & 0 & 1+c_z \end{bmatrix} \cdot \begin{bmatrix} p \\ q \\ r \end{bmatrix} \qquad (19)$$

where p, q, and r are readings provided by the confocal microscope for the Cartesian coordinates x, y, and z. The calibration parameters are those described in Section 2.2 (c_{xy}, a, θ),

and the new parameter c_z is introduced to permit the calibration in the Z-axis. The corrected z coordinate is:
$$z = (1+c_z)\cdot r. \tag{20}$$

This simple matrix model supposes that there is no (or negligible) perpendicular error between the Z-axis and the XY plane. This hypothesis is very close to reality when the Z-axis range is clearly lower than ranges of the X and Y axes. When the Z-axis range is equal to or greater than X and Y ranges, a more complex model must be used (zero terms in the matrix of the model are no longer zero—see, for example, [39]). It is easy to demonstrate that, using the matrix model, the corrected diameter D of the spherical surface fitted by the confocal microscope software is:

$$D = D_m \cdot \frac{1+2c_{xy}}{1+c_z} \tag{21}$$

where D_m is the diameter provided by the confocal microscope prior to applying any calibration parameter. Therefore, an estimation of c_z is

$$c_z = \frac{D_m}{D_0}\cdot(1+2c_{xy}) - 1 \tag{22}$$

where D_0 is the certified diameter of the sphere by the ACL. The standard uncertainty of c_z is

$$u(c_z) = \sqrt{\frac{u^2(D_0) + u^2(D_m)}{D_0^2} + 4u^2(c_{xy})}. \tag{23}$$

Equation (22) of c_z shows a clear positive dependency with c_{xy}. Therefore, the correlation coefficient $r(c_z, c_{xy})$ should be estimated, and it can be done using the following expression:

$$r(c_z, c_{xy}) = 2\cdot\frac{u(c_{xy})}{u(c_z)}. \tag{24}$$

Note that the correlation coefficient is denoted as $r(c_z, c_{xy})$. Do not confuse it with the reading of the confocal microscope for the Z-axis, which is denoted as r.

Figure 9. Steel sphere used in calibration.

2.4. Calibration for Roughness Measurements

The calibration of the Z-axis against the reference sphere (previous section) guarantees the traceability of the vertical measurements performed with the confocal microscope to the SI unit of length (the meter). Therefore, any vertical roughness parameter will have an adequate traceability once the instrument has been calibrated along its Z-axis. Notwithstanding, we followed the recommendation included in documents DKD-R 4-2 [40–42], which propose performing an additional calibration against roughness standards to validate the Z-axis calibration for roughness measurements.

Many parameters are used to characterize surface texture. Among the 2D roughness parameters, one of the most widely used is the R_a parameter, which is the arithmetic mean of the absolute values of the profile deviations from the mean line of the roughness profile [43]. We only consider the R_a parameter during calibration, but readers interested in other 2D roughness vertical parameters (R_q, R_p, R_v, R_z, ...) can use the same calibration procedure described in this paper but with minor variations. Calibration was performed in the range $0.1 < R_a \leq 2$ µm. For this range, according to ISO 4288 [44], the sampling length should be $l_r = 0.8$ mm, which is possible to carry out with a field of view of 1270 µm × 952 µm. For $R_a > 2$ µm, the sampling length should be $l_r = 2.5$ mm or higher, and it is impossible to achieve this with a field of view of 1270 µm × 952 µm (10× objective). It makes no sense to measure roughness lower than $R_a = 0.1$ µm with an instrument with repeatability in the Z-axis of around 0.5 µm. Therefore, calibration for $R_a < 0.1$ µm and $R_a > 2$ µm was discarded.

Figure 10a shows three metallic, aperiodic 2D roughness standards. Figure 10b shows three glass, periodic 2D roughness standards. Other types of 2D roughness profile types are described in Section 7 of ISO 25178-70 [33]. We recommend the use of aperiodic standards because they cover a wide range of wavelengths, in contrast to periodic standards that only cover a single wavelength. However, periodic standards were used in this case in order to complete the range of measurements between 0.1 µm and 2 µm for different calibration points and for different materials (glass instead of metallic items).

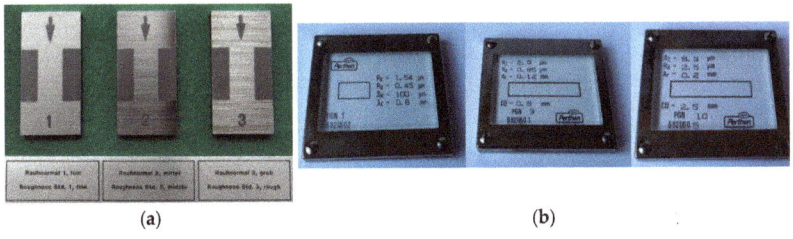

Figure 10. Step height standards used during calibration: (**a**) aperiodic, metallic standards and (**b**) periodic, glass standards.

These standards were measured over five different zones in two different orientations (Figure 11a,b). In each zone, the measurement was carried out along a line perpendicular to the roughness lines (Figure 11c) and located at the center of the zone. Therefore, a total of $2 \times 5 = 10$ roughness measurements were obtained from each standard.

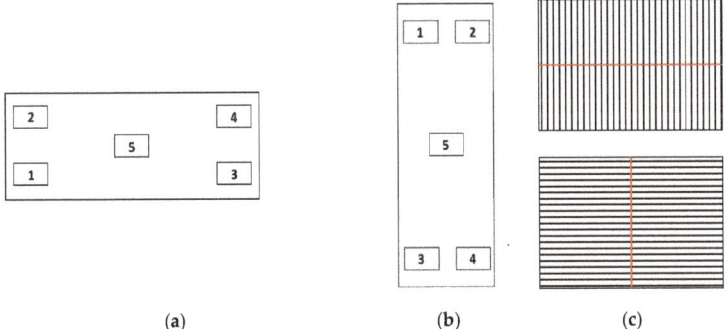

Figure 11. Location of the five scanning positions for roughness calibration: (**a**) horizontal orientation; (**b**) vertical orientation; and (**c**) location of measurement lines.

We recommend using at least three different roughness standards with nominal values of R_a uniformly distributed along the range where the instrument must be calibrated. However, it is advisable to use five or more standards and, if possible, standards made of different materials (e.g., metallic and glass).

It is important to note that there will be differences between measurements obtained with a confocal microscope and measurements obtained with a stylus instrument [4,45].

The main reasons for this are as follows.

- The way the surface is detected is totally different: microscopes use light, and stylus instruments use a mechanical tip. Usually, optical instruments present higher instrument noise than stylus instruments. Possible reasons are the effects of multiple scattering and discontinuities [45]. As a consequence, optical instruments tend to overestimate surface roughness.
- Stylus instruments permit evaluation lengths l_n that are as long as necessary (see ISO 4288 [44]). Microscopes usually have small fields of view that limit the maximum length of the profile that can be scanned. For example, for samples with 0.1 μm < R_a ≤ 2 μm, ISO 4288 recommends using five sampling lengths $l_r = 0.8$ mm for a total evaluation length $l_n = 4$ mm. This is not a problem for stylus instruments, which can cope with longer evaluation lengths (up to 100 mm in some cases). However, the confocal microscope described at the beginning of Section 2 has a maximum evaluation length of 1.27 mm. Therefore, only one sampling length $l_r = 0.8$ mm could be used. Using only one sampling length instead of five usually causes a bias toward lower R_a values, which are accompanied by an increase in variability. The effect is considerably higher when even the sampling length l_r has to be reduced.

In order to ensure a good match between roughness measurements performed with stylus instruments and optical instruments, the concept of "bandwidth matching" should be correctly applied. This term refers to the good correspondence between the spectral bandwidths of two different instruments used for roughness measurements [45]. The effective spectral bandwidth of a roughness-measuring instrument is limited by the two cut-off wavelengths, λ_S (a high-pass filter), and λ_C (a low-pass filter), and it is influenced by the X-axis resolution and tip radius in stylus instruments or lateral resolution and pixel size in optical instruments.

2.5. Summary of Characteristics of Measurement Standards Used during Calibration

In this section, the nominal values and the uncertainties of the different reference measurement standards used during calibration are summarized. All of them were calibrated in ACLs.

We include the calibration of the confocal microscope against roughness standard #6 only for informative purposes. Its measurements were made using a sampling length of $l_r = 0.8$ mm because of the reduced field of view of the instrument (with an 10× objective). However, ISO 4288 [44] recommends the use of a sampling length $l_r = 2.5$ mm, which is a measurement that is impossible to achieve with a 10× objective.

3. Results

3.1. Flatness Verification

The following figure shows a topographic image of the optical flat of Figure 4 that was used as a flatness calibration surface. This optical flat was previously calibrated in an accredited laboratory. The total flatness error was 118 nm with a standard uncertainty of 25 nm ($k = 1$), and its RMS flatness was 28 nm with a standard uncertainty of 7 nm ($k = 1$); see Table 1.

Table 1. Nominal values and the uncertainties of the material reference standards used during calibration.[1] S_m is a spacing parameter defined as the mean spacing between peaks. S_m values included in this table are only informative. RMS: root mean square error.

Reference Measurement Std.	Parameter	Certified Value (μm)	Std. Uncertainty (k=1)(μm)
Optical flat	Total flatness error	0.118	0.025
	RMS flatness	0.028	0.007
Stage Micrometer	Average pitch ℓ_0	9.980	0.005
Sphere	Diameter D_o	4 001.08	0.25
Roughness std. #1 metallic, aperiodic	R_a (R_0)	0.183	0.039
	S_m[1]	48	
Roughness std. #2 metallic, aperiodic	R_a (R_0)	0.512	0.041
	S_m[1]	185	
Roughness std. #3 metallic, aperiodic	R_a (R_0)	1.677	0.057
	S_m[1]	176	
Roughness std. #4 glass, periodic	R_a (R_0)	0.460	0.030
	S_m[1]	100	
Roughness std. #5 metallic, aperiodic	R_a (R_0)	0.850	0.030
	S_m[1]	120	
Roughness std. #6 glass, periodic	R_a (R_0)	2.440	0.080
	S_m[1]	200	

Figure 12 shows the absence of significant curvature in the XY plane. A slight uncorrected curvature of about 0.6 μm (peak to peak) was observed, which is small and could be neglected when compared with the Z-axis axial step (2.0 μm) and observed instrument noise (about 1.0 μm peak to peak). This could be an empirical demonstration of a good adjustment and/or correction of the microscope by the manufacturer. In a situation such as this, there is no need to apply any further correction to compensate the curvature of the XY plane.

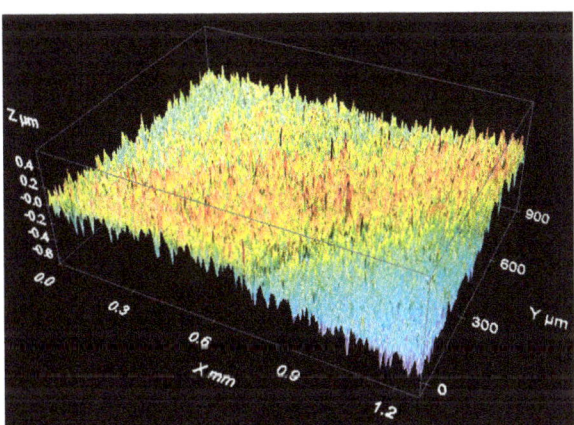

Figure 12. Result of the flatness measurement performed over the optical flat.

Table 2 shows the results of the measurements performed with the confocal microscope (in both positions 0° and 90°).

Table 2. RMS flatness measured with the confocal microscope in positions 0° and 90°.

Position	RMS Flatness (μm)
0°	0.48
90°	0.59

The RMS values of Table 2 are small when compared with the Z-axis axial step of 2.0 μm. Therefore, they were probably caused by the lack of repeatability of the instrument. In any case, the most conservative option is to estimate a component of the uncertainty associated with the possible curvature of the XY plane equal to the average value of both RMS values of Table 2:

$$u_{\text{FLT}} = 0.54 \text{ μm} \tag{25}$$

A better estimation for u_{FLT} would likely be to quadratically subtract the RMS flatness of the optical flat (0.28 μm):

$$u_{\text{FLT}} = \sqrt{(0.54 \text{ μm})^2 - (0.028 \text{ μm})^2} = 0.539 \text{ μm} \tag{26}$$

Regardless, we considered that the first estimation ($u_{\text{FLT}} = 0.54$ μm) is slightly more conservative and clearly simpler.

3.2. XY Plane Calibration

Figure 13 shows the four positions (0°, 45°, 90°, 135°) in which the stage micrometer was measured in the confocal microscope during the XY plane calibration.

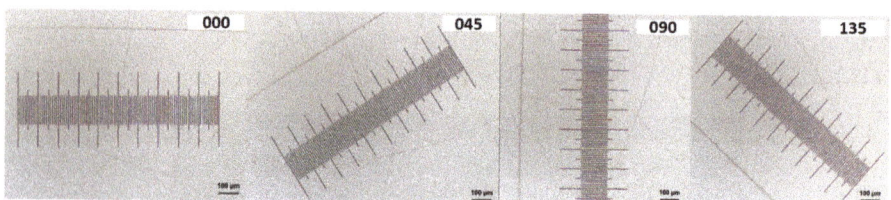

Figure 13. Different measurement positions (0°, 45°, 90°, 135°) of the stage micrometer.

In each position, the average pitch ℓ_i was determined from readings provided by the confocal microscope. The results are shown in Table 3.

Table 3. Measurements of the average pitch ℓ_i in different orientations.

Position		Average Pitch ℓ_i (μm)	Uncertainty $u(\ell_i)$ (μm)	Repeatability s (μm)	Non-Linearity RMS (μm)
1	0°	9.892 34	0.000 53	0.34	0.71
2	45°	9.897 33	0.000 57	0.38	0.60
3	90°	9.891 56	0.000 49	0.42	0.69
4	135°	9.889 50	0.000 47	0.34	0.61

The average value for repeatability in the XY plane was $s_r(x) = s_r(y) = 0.4$ μm. This is a reasonable value when compared with the 1.65 μm lateral resolution (nominal voxel width).

The stage micrometer had a certified average pitch $\ell_0 = 9.980$ μm with a standard uncertainty $u(\ell_0) = 0.005$ μm.

Using the expression for Section 2.2, we obtained the following estimations for calibration parameters c_{xy}, a, and θ:

$$c_{xy} = \frac{\ell_0}{4} \cdot \left(\frac{1}{\ell_1} + \frac{1}{\ell_2} + \frac{1}{\ell_3} + \frac{1}{\ell_4}\right) - 1 = 0.00883 \tag{27}$$

$$\text{with } u(c_{xy}) = \frac{\sqrt{u^2(\ell_0) + [u^2(\ell_1) + u^2(\ell_2) + u^2(\ell_3) + u^2(\ell_4)]/16}}{\ell_0} = 0.00050 \tag{28}$$

$$a = \frac{\ell_0}{2} \cdot \left(\frac{1}{\ell_1} - \frac{1}{\ell_2}\right) = -0.000040 \tag{29}$$

$$\text{with } u(a) = \frac{\sqrt{u^2(\ell_1) + u^2(\ell_2)}}{2\ell_0} = 0.000036 \tag{30}$$

$$\theta = \ell_0 \cdot \left(\frac{1}{\ell_3} - \frac{1}{\ell_4}\right) = -0.000798 \tag{31}$$

$$\text{with } u(\theta) = \frac{\sqrt{u^2(\ell_3) + u^2(\ell_4)}}{\ell_0} = 0.000074 \tag{32}$$

All three parameters are dimensionless.

Observing the non-linearity RMS values in Table 3, the overall standard uncertainty estimation for non-linearity in the XY-plane was $u_{NL,xy} = 0.7$ µm.

3.3. Z-Axis Calibration

Figure 14 shows an example of a measurement of the spherical cap of a stainless steel reference sphere with a 4-mm nominal diameter (see Figure 9). It is a three-dimensional reconstruction of the sphere surface.

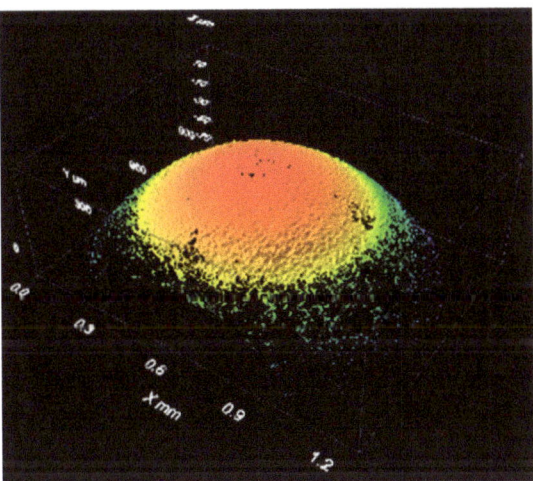

Figure 14. Results of the measurement of a bearing sphere with white light.

Using this information, the confocal microscope software can perform a least-square fitting to a spherical surface from which we could estimate the diameter of the sphere and RMS error of the fit.

If the confocal software does not permit fitting a spherical surface, a code similar to the one described and listed in Appendix A can be used.

In this calibration, two different types of illumination were used (white and blue light), and measurements were taken in three orientations: 0°, 45°, and 90°. Finally, there were $n = 6$ measurements. Results obtained during the measurement of the bearing steel sphere of Figure 9 are presented in Table 4.

Table 4. Root mean square error and diameter D_m of the spherical caps fitted using least squares.

Position	Illumination	RMS Error (µm)	Diameter D_m (mm)
0°	Blue	0.86	3.9740
45°	Blue	1.08	3.9562
90°	Blue	1.08	3.9638
0°	White	0.86	3.9740
45°	White	0.89	3.9828
90°	White	0.87	3.9766

The average value $\overline{D}_m$ of the six diameters D_m was $\overline{D}_m = 3.9712$ mm, and the standard deviation $s(D_m)$ was 0.0096 mm. We estimated $u(\overline{D}_m)$ as:

$$u(\overline{D}_m) = \frac{s(D_m)}{\sqrt{n}} = 0.0039 \text{ mm}. \tag{33}$$

The RMS error is an estimation of the repeatability in the Z-axis, which probably includes the non-linearity in the Z-axis. The mean value for this Z-axis repeatability was $s_r(z) = 0.8$ µm, which seems to be a reasonable value when compared with the Z-axis axial step of 2 µm.

The certified diameter D_0 of the reference sphere was $D_0 = 4.0011$ mm with a standard uncertainty $u(D_0) = 0.25$ µm.

Using the expression of Section 2.3, the Z-axis calibration parameter c_z can be estimated as follows:

$$c_z = \frac{D_m}{D_0} \cdot (1 + 2c_{xy}) - 1 = 0.0101 \tag{34}$$

$$\text{with } u(c_z) = \sqrt{\frac{u^2(D_0) + u^2(\overline{D}_m)}{D_0^2} + 4u^2(c_{xy})} = 0.0014 \tag{35}$$

The correlation coefficient $r(c_z, c_{xy})$ was

$$r(c_z, c_{xy}) = 2 \cdot \frac{u(c_{xy})}{u(c_z)} = 0.72 \tag{36}$$

This correlation coefficient is clearly higher than zero, showing a strong positive correlation between c_z and c_{xy} that should be taken into account after calibration when needed. Correlation coefficients $r(c_z, a)$ and $r(c_z, \theta)$ are usually very small (lower than 0.01); therefore, correlation between c_z and parameters a and θ could be neglected.

3.4. Calibration for Roughness Measurements

As an example of data acquisition results when measuring a material roughness standard, the following figures show three-dimensional reconstructions of the surface of an aperiodic, metallic roughness standard (Figure 15) and a periodic, glass roughness standard (Figure 16).

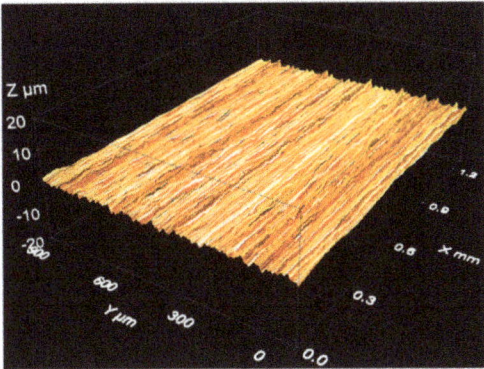

Figure 15. Measurement of an aperiodic roughness standard with a confocal microscope: 3D view of the measurement results.

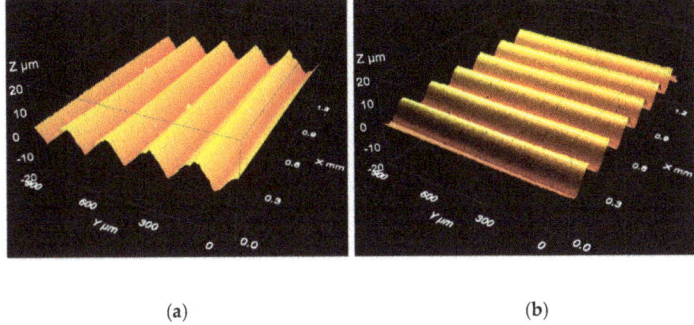

(a)   (b)

Figure 16. Measurement of a periodic roughness standard with a confocal microscope: 3D view of the measurement results with roughness lines (**a**) parallel to the X-axis and (**b**) parallel to the Y-axis.

Calibration was performed by repeating the measurements of six roughness standards 10 times (five zones, two orientations; see Table 1). The results are summarized in Table 5, showing average results $\overline{R}$ of the 10 repeated measurements and corresponding standard deviations $s(R)$. Direct readings R provided by the confocal microscope were obtained prior to introducing the calibration parameter c_z using only one sampling length $l_r = 0.8$ mm.

Therefore, these readings should be corrected by applying the following expression to take into account the Z-axis calibration:

$$R_{corrected} = \overline{R} \cdot (1 + c_z) \tag{37}$$

The authors followed the recommendations of ISO 4288 [44] that, for $0.1 < R_a \leq 2$ mm, recommend five sampling lengths $l_r = 0.8$ mm for a total evaluation length of $l_n = 4$ mm. Due to the limitations of the instrument's field of view (see Section 2), only one sampling length $l_r = 0.8$ mm could be used. This reduction in the number of sampling lengths from five to one caused slightly lower values for R_a and higher variabilities [45].

It can be concluded from Table 5 that a typical value for $s(R)$ was $s(R) = 0.07$ µm, which was the quadratic average of repeatabilities of the first five standards.

Note that corrected values $\overline{R} \cdot (1 + c_z)$ were always higher than the certified values R_0 (compare the results from Table 5 to those from Table 1). It seems that for surface roughness similar to the nominal voxel height ($w_z = 2$ µm), readings provided by the confocal microscope presented a positive bias caused by noise observed, for example, when measuring an optical flat (see Section 3.1, Figure 12).

The RMS flatness observed when measuring the optical flat (0.54 µm) was slightly higher than the R_a values that were observed when measuring roughness standard #1, which is a quasi-flat surface (certified value $R_a = 0.183$ µm) for an instrument with a voxel height of 2 µm. The definition of R_a is similar but not equal to the definition of RMS flatness, but most importantly, R_a was evaluated after filtering the readings using a low-pass filter (defined through the sampling length l_r).

Table 5. Results obtained when calibrating the confocal microscope described in Section 2 using six roughness standards (Table 1).

Reference Meas. Std.	Average R_a $\overline{R}$ (µm)	Repeatability $s(R)$ (µm)	Corrected R_a $R \cdot (1+c_z)$ (µm)	Bias Estimation b (µm)	Standard Uncertainty $u(b)$ (µm)
Roughness std. #1	0.43	0.06	0.43	0.25	0.04
Roughness std. #2	0.59	0.06	0.60	0.08	0.05
Roughness std. #3	1.70	0.11	1.71	0.04	0.07
Roughness std. #4	0.51	0.04	0.52	0.06	0.03
Roughness std. #5	0.95	0.05	0.96	0.11	0.03
Roughness std. #6 [1]	2.50	0.06	2.53	0.09	0.08

[1] Values obtained when measuring roughness standard #6 are included in this table only for informative reasons. Measurements of this standard were made using a sampling length $l_r = 0.8$ mm, because of the reduced field of view of the instrument, instead of a sampling length $l_r = 2.5$ mm, as recommended by ISO 4288 [44].

We suggest estimating the positive bias at each calibration point using the following expression, where R_0 is the R_a certified value for the standard used at each calibration point:

$$b = \overline{R} \cdot (1 + c_z) - R_0. \tag{38}$$

Its corresponding standard uncertainty $u(b)$ is:

$$u(b) = \sqrt{u^2(R_0) + \overline{R}^2 \cdot u^2(c_z) + \frac{s^2(R)}{n}}. \tag{39}$$

Using this approach, the calibration results are those values, b_i and $u(b_i)$, which are presented in the two columns on the right side of Table 5. Index i refers to the roughness standard used. These results are represented graphically in Figure 17 in order to analyze their metrological compatibility. Red lines represent values corresponding to metallic, aperiodic standards #1, #2, and #3. Green lines represent values corresponding to standards #4 and #5. The blue line is the result from standard #6 that will not be taken into account. Vertical lines represents uncertainty intervals $b_i \pm U(b_i)$ where the expanded uncertainties $U(b_i) = k \cdot u(\overline{b})$ were evaluated for a coverage factor $k = 2$ (see Section 6.2.1 in [24] for definitions of coverage factor and expanded uncertainty). When analyzing the compatibility between measurement results, it is very common to use a coverage factor of $k = 2$ to estimate the expanded uncertainties. The horizontal black solid line in Figure 17 corresponds to the average value $\overline{b}$ of the first $N = 5$ roughness standards:

$$\overline{b} = \frac{\sum_{i=1}^{N} b_i}{N} = \frac{b_1 + b_2 + b_3 + b_4 + b_5}{5} = 0.11 \text{ µm} \tag{40}$$

In order to make a correct estimation of average bias $\overline{b}$, the correlation between bias b_i, b_j at each calibration points should be taken into account for the following reasons:

- Dominant contributions to uncertainties $u(b_i)$ are the calibration uncertainties $u(R_0)$ of the roughness standards.
- There is a high probability that all roughness standards were calibrated in the same calibration laboratory. Therefore, there will be strong correlation between them.

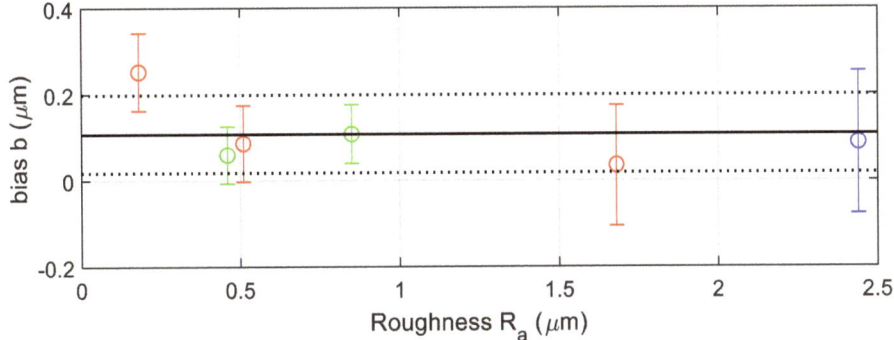

Figure 17. Bias observed at each calibration point (roughness calibration).

There will be a high correlation between bias b_i. We performed estimations in different situations, and it is possible to see correlation coefficients $r(b_i, b_j)$ as high as +0.8.

In order to simplify calculations, we suggest assuming $r(b_i, b_j) = +1$, which leads to higher estimations for the uncertainty $u(\bar{b})$ of $\bar{b}$. Then, it can be demonstrated that $u(\bar{b})$ was:

$$u(\bar{b}) = \frac{1}{N}\sum_{i=1}^{N} u(b_i) = \frac{u(b_1) + u(b_2) + u(b_3) + u(b_4) + u(b_5)}{5} = 0.05 \text{ μm} \qquad (41)$$

In Figure 17, the uncertainty interval $\bar{b} \pm U(\bar{b})$ is represented by the space between the higher and lower black dotted lines. $U(\bar{b}) = k \cdot u(\bar{b})$ is the expanded uncertainty of $\bar{b}$ evaluated with a coverage factor $k = 2$. Please note that all the uncertainty intervals $b_i \pm U(b_i)$ overlap the interval $\bar{b} \pm U(\bar{b})$. Notwithstanding, point b_1 is outside the interval $\bar{b} \pm U(\bar{b})$. This could indicate that some variability of the bias b was not taken into account in $u(\bar{b})$. Therefore, a conservative approach would be to assume that there is a variability represented by δb that should be added to $u(\bar{b})$. Suppose that δb is a uniform random variable of null mean and a full range $b_{max} - b_{min}$. Then, its standard uncertainty would be:

$$u(\delta b) = \frac{b_{max} - b_{min}}{\sqrt{12}} = 0.06 \text{ μm} \qquad (42)$$

In order to estimate the noise of the instrument, according to [40–42], we repeated 10 measurements over an optical flat (that of Figure 4) in two orientations: 0° and 90°. For a confocal microscope, an optical flat is a specimen with null roughness (very small in comparison with its noise). Therefore, values of R_a obtained over an optical flat are a very good estimation of the instrument noise. The average value and the standard deviation of the 10 R_a values were

$$\bar{R}_a = 0.09 \text{ μm} \qquad (43)$$

$$s(R_a) = 0.003 \text{ μm} \qquad (44)$$

Therefore, a good estimation for the uncertainty component associated with noise instrument is

$$u_{noise} = \bar{R}_a = 0.09 \text{ μm} \qquad (45)$$

4. Discussion

Table 6 summarizes the results obtained during the confocal microscope calibration (Section 2).

Table 6. Results of calibration.

Parameter	Value	Units	Standard Uncertainty
c_{xy}	0.008 83	-	0.00050
a	−0.000040	-	0.000036
θ	−0.000798	-	0.000074
c_z	0.0101	-	0.0014
$r(c_{xy}, c_z)$	0.72	-	-
u_{FLT}	0.54	μm	-
$u_{NL,xy}$	0.70	μm	-
$s_r(x) = s_r(y)$	0.40	μm	-
$s_r(z)$	0.80 [1]	μm	-
$\bar{b}$	0.11	μm	0.05
δb	0	μm	0.06
$s(R)$	0.07	μm	-
u_{noise}	0.09	μm	-

[1] Non-linearity in Z-axis is included in $s_r(z)$.

Note that these results are only valid for measurements made with the same objective (10×). If other objectives are used, a whole recalibration is needed for each new objective.

The effects and their uncertainties were the highest for parameters c_{xy} and c_z. If their effects were not corrected, their contribution to the relative expanded uncertainty would be around 1%.

Fortunately, the software of confocal microscopes usually permits users to introduce their value in order to compensate their effects. If this compensation is done, their contribution to the relative expanded uncertainty is reduced to 0.3%.

The effect of parameter a (difference between pixel lengths along X and Y axes) was negligible. Its absolute value was lower than its expanded uncertainty $U(a) = k \cdot u(a)$ (for $k = 2$); therefore, the null hypothesis $a = 0$ could not be rejected. Its contribution to the relative expanded uncertainty was very low (around 0.01%).

The effect of parameter θ (perpendicularity error between X and Y axes) seemed to be significant (its absolute value was clearly higher than its expanded uncertainty), but its contribution to the relative expanded uncertainty (around 0.1%) was clearly negligible in comparison with c_{xy} and c_z.

The contributions of XY plane RMS flatness (u_{FLT}) and the non-linearity in X and Y axes were clearly lower than the voxel dimensions ($w_{xy} = 1.65$ μm and $w_z = 2$ μm). Therefore, the instrument adjustment performed by the manufacturer seems to have been good.

Repeatabilities in the XY plane and in the Z-axis, in comparison with the voxels dimensions, were low. Again, this can be used to conclude that the instrument was working well.

In roughness measurements (which only apply when using the R_a parameter), the repeatability $s(R)$, average bias $\bar{b}$, bias variability $u(\delta b)$, and instrument noise u_{noise} were very small in comparison with voxel height $w_z = 2$ μm.

4.1. Expanded Uncertainty Estimation for Length Measurements in the XY Plane

As was pointed out, the instrument software usually permits users to introduce parameters c_{xy} and c_z in order to apply the corresponding corrections. On the contrary, parameters a and θ cannot be introduced. Therefore, the effect of uncorrected, non-null parameters a and θ would be taken into account as a systematic effect whose equivalent standard uncertainties would respectively be $|a|/\sqrt{3}$ and $|\theta|/\sqrt{3}$. We supposed that it was equivalent to the introduction of two components, δa and $\delta \theta$, uniformly distributed along $[-a. + a]$ and $[-\theta. + \theta a]$, respectively.

For length measurements performed in the XY plane, the following expression could be a good estimate of its expanded uncertainty, where the pixel width component was estimated as $w_{xy}/\sqrt{12}$ (uniformly distributed between $\pm w_{xy}/2$):

$$U(L_{xy}) = k \cdot \sqrt{L_{xy}^2 \cdot \{u^2(c_{xy}) + \frac{a_r^2}{3} + u^2(a) + \frac{1}{2}[\frac{\theta^2}{3} + u^2(\theta)]\} + u_{NL,xy}^2 + s_r^2(x) + \frac{w_{xy}^2}{12}} \leq \quad (46)$$
$$\leq 1.9 \text{ μm} + \frac{L}{1600}.$$

Uncertainty components for which no distribution was described were supposed to have normal distributions. In such situations, where there are many uncertainty components (eight components) and most of them are normally distributed and they contribute similarly to the total combined uncertainty, it can be supposed that the output variable L_{xy} is normally distributed [25]. Therefore, a coverage factor $k = 2$ can be used when computing the expanded uncertainty, assuming a coverage probability of approximately 95%.

4.2. Expanded Uncertainty Estimation for Height Measurements along the Z-Axis

For height measurement ($0 \leq h \leq 100$ μm—the Z range approximately covered by the sphere cap measured), the following expression gives us a reasonable estimation of its expanded uncertainty $U(h)$ for a coverage factor $k = 2$:

$$U(h) = k \cdot \sqrt{h^2 \cdot u^2(c_z) + u_{FLT}^2 + s_r^2(z) + \frac{w_z^2}{12}} \leq 2.2 \text{ μm} + \frac{h}{120} \quad (47)$$

Now there are four uncertainty components, where three are distributed normally, and only one w_z is distributed uniformly, but w_z is never the dominant contribution. Again, in a situation such as this, a coverage factor $k = 2$ can be used when computing the expanded uncertainty, assuming a coverage probability of approximately 95% [25].

4.3. Expanded Uncertainty for Roughness Measurements

Following the recommendations of DKD-R 4-2 [40–42], a model for a corrected R_a roughness measurement performed after instrument calibration would be:

$$R_a = \overline{R} \cdot (1 + c_z) - (\overline{b} + \delta b) + \delta R_{noise} \quad (48)$$

where now $\overline{R}$ is the average of m repeated measurements made over the specimen being measured, and δR_{noise} is a random variable of null mean distributed normally with standard deviation u_{noise}. The standard uncertainty of R_a is

$$u(R_a) = \sqrt{\frac{s^2(R)}{m} + \overline{R}^2 \cdot u^2(c_z) + u^2(\overline{b}) + u^2(\delta b) + u_{noise}^2} \quad (49)$$

The expanded uncertainty $U(R_a)$, using a coverage factor k is

$$U(R_a) = k \cdot \sqrt{\frac{s^2(R)}{m} + \overline{R}^2 \cdot u^2(c_z) + u^2(\overline{b}) + u^2(\delta b) + u_{noise}^2} \quad (50)$$

In this case, there are five uncertainty components, δb is distributed uniformly, and $\overline{R}$ follows a t-Student distribution with $\nu = m - 1$ degrees of freedom. If we compute the degrees of freedom of the output variable R_a, the result is approximately $\nu(R_a) = 30$. With $\nu(R_a) > 10$, it is possible to use a coverage factor $k = 2$ corresponding to a coverage probability of approximately 95% [25]. Then, assuming that measurements will be repeated $m = 3$ times, the expanded uncertainty would be

$$U(R_a) < 0.25 \text{ μm} \quad (51)$$

This value is very good for an instrument with a voxel height of $w_z = 2$ µm.

4.4. Propagation of Uncertainty When Measuring the Radius of a Cylindrical Surface

As an example of uncertainty propagation in dimensional measurements not directly covered in Sections 4.2 and 4.3, we present the case of a measurement of the radius of a cylindrical surface (see Figure 18) of a steel bar with a nominal value of 2.75 mm.

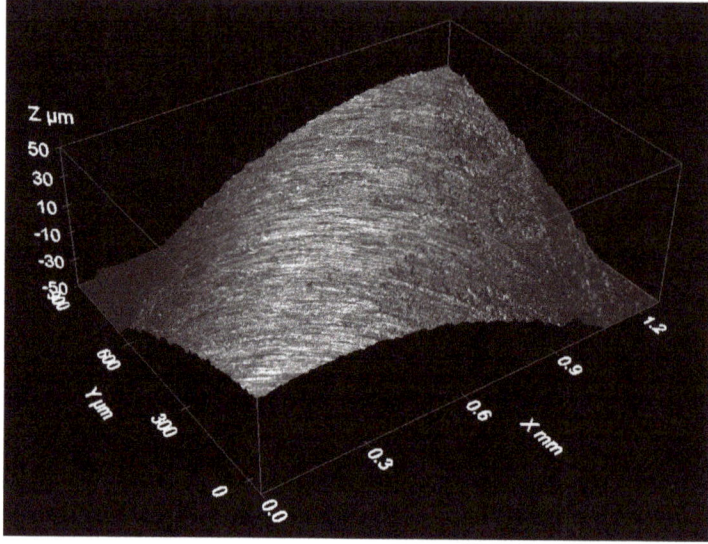

Figure 18. Measurement of a cylindrical surface.

The uncertainty propagation was done using Monte Carlo simulation [46]. The model function used during the simulation of the coordinates of the surface's points was the following:

$$\begin{bmatrix} x \\ y \\ z \end{bmatrix} = \begin{bmatrix} 1 + c_{xy} + a + \delta a & (\theta + \delta\theta)/2 & 0 \\ (\theta + \delta\theta)/2 & 1 + c_{xy} - a - \delta a & 0 \\ 0 & 0 & 1 + c_z \end{bmatrix} \cdot \begin{bmatrix} p \\ q \\ r \end{bmatrix} + \begin{bmatrix} \delta x \\ \delta y \\ \delta z \end{bmatrix} \quad (52)$$

where (p, q, r) are the coordinates provided by the confocal microscope during the measurement. They were not simulated. Table 7 enumerates variables that were simulated and how the simulation was performed. Corrections a and t were not applied, because the instrument software cannot take them into account. In order to take into account the effect of not applying these corrections, δa and δt were introduced, and a and t are supposed to be normal distributions with zero mean (not applying corrections) and typical uncertainty $u(a)$ and $u(t)$. δa and δt have uniform distribution with zero mean and typical uncertainties $|a|/\sqrt{3}$ and $|\theta|/\sqrt{3}$. δp, δq, and δr represent the repeatability effects over coordinates $x, y,$ and z. We supposed that they had normal distributions with zero mean and typical uncertainties $u(\delta p) = (\delta q) = s_r(x) = s_r(y)$ and $u(\delta r) = s_{LS} = 1.0$ µm, where s_{LS} is the root mean squared error observed when fitting a cylindrical surface to measured points (x, y, z) using a least-squares fit.

Table 7. Variables simulation.[1] Corrections a and t were not applied.

Variable	Mean Value	Units	Standard Uncertainty	Distribution Type
c_{xy}	0.00883	-	0.00050	Normal
a	0 [1]	-	0.000036	Normal
δa	0		$u(\delta a) = \frac{a}{\sqrt{3}} = 0.000023$	Uniform
θ	0 [1]	-	0.000074	Normal
δa	0		$u(\delta a) = \frac{a}{\sqrt{3}} = 0.00046$	Uniform
c_z	0.0101	-	0.0014	Normal
$r(c_{xy}, c_z)$	0.72	-	-	-
$\delta p, \delta q$	0	μm	$s_r(x) = s_r(y) = 0.40$	Normal
δr	0	μm	$s_{LS} = 1.0$	Normal

A total of $N = 10^4$ simulations were generated. For each simulation, a value R_i was obtained for the radius of the cylindrical surface. Figure 19 shows the histogram of the simulated radius of the cylindrical surface. The red smooth line represents the best approximation of the histogram through a normal distribution. Differences between the red line and the histogram were small enough to be negligible. It is likely that upon increasing the number of simulations (the advisable value for N when there is no time limit during the execution to the simulation process is $N = 10^6$ [46]), these differences would be smaller. For this reason, a coverage factor $k = 2$ (corresponding to an approximate coverage probability of 95%) was chosen in previous sections: final distributions are usually very close to normal distribution where $k = 2$ corresponds to a coverage probability of 95.45%.

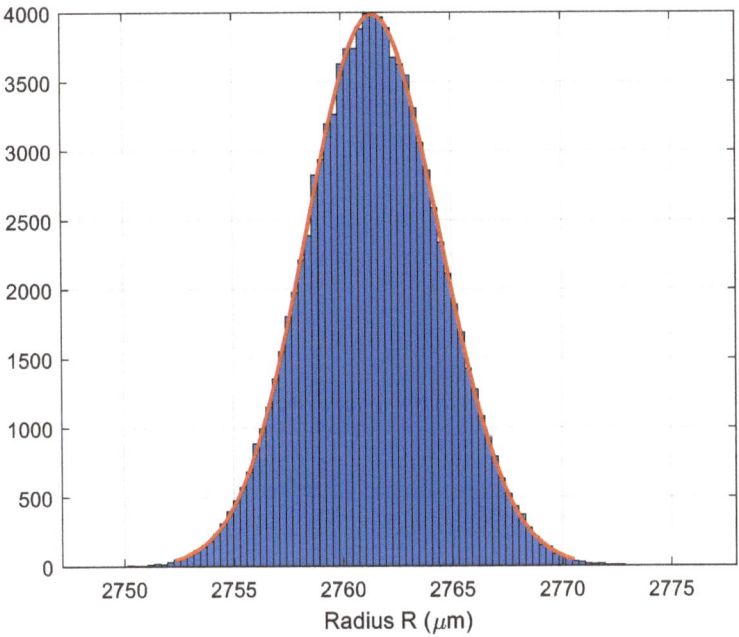

Figure 19. Histogram of the radius R of the cylindrical surface.

The final result was $R = (2.7907 \pm 0.0061)$ mm, where the expanded uncertainty $U(R) = k \cdot s(R)$, where now $s(R)$ is the standard deviation of N simulated values R_i. A coverage factor $k = 2$ for a coverage probability of approximately 95% was used.

A similar approach could be used to propagate uncertainties when using the confocal microscope for other types of dimensional or angular measurements.

5. Conclusions

A complete calibration procedure that provides adequate traceability to confocal microscopes used in submillimeter coordinate metrology was presented. This procedure provides adequate traceability for length and roughness measurements performed with confocal microscopes and can be easily adapted to calibrate other 3D optical instruments (e.g., focus variation microscopes). The calibration procedure is as simple as possible, as it was designed to be implemented in industrial environments. Reference material standards were chosen to be easy to find and easy to calibrate again in industrial environments. The calibration procedure covers all the key points of operation of a confocal microscope. It permits the estimation of:

- Amplification coefficients $\alpha_x = 1 + c_{xy} + a$, $\alpha_y = 1 + c_{xy} - a$, and $\alpha_z = 1 + c_z$.
- Non-linearity errors.
- Perpendicularity error θ between X and Y axes.
- Relative difference $2a$ in pixel dimensions along X and Y axes.
- Repeatabilities when measuring lengths or heights.
- Flatness deviations in the XY plane.
- Bias deviation b when measuring roughness.
- Instrument noise when measuring roughness.
- Repeatability when measuring roughness.

Some of these parameters (amplification coefficients, flatness deviation in XY plane) can usually be introduced in the instrument software to compensate for their effects. Others cannot be compensated (i.e., θ, a) but if high values are detected, the user can ask the instrument manufacturer to adjust and/or repair the instrument to reduce their effects. Even if they are not introduced, an alternative approach is presented here to account for the fact that these corrections were not applied.

Uncertainty estimations were carried out for all parameters following the mainstream GUM method. In addition, for measurements of lengths and roughness, expressions for expanded uncertainties of measurement carried out by the instrument were provided. There are other types of measurements, such as angular measurements, that were not addressed in this paper due to limitations in the extent of the text. Notwithstanding, all the information needed to propagate uncertainties to these other types of measurements is provided in the paper, as can be demonstrated through an example solved using Monte Carlo simulation.

The procedure described in this paper can be easily adapted to calibrate other optical instruments in the submillimeter range, which are capable of providing 3D information of surfaces being observed by them (e.g., focus variation microscopes). For example, the material of some of the reference material standards would have to be changed, but the core of the procedure would remain the same.

Author Contributions: Conceptualization, A.M.M. and J.d.V.y.O.; Investigation, A.M.M.; Methodology, A.M.M. and J.d.V.y.O.; Software, A.M.M. and J.d.V.y.O.; Supervision, J.d.V.y.O.; Validation, J.d.V.y.O.; Writing—Original draft, A.M.M.; Writing—Review and editing, J.d.V.y.O.

Funding: This research received no external funding.

Conflicts of Interest: The authors declare no conflicts of interest.

Appendix A Algorithm for Spherical Cap Fitting

At the end of this appendix, we include the source code of a routine written to be executed in Matlab®(www.mathworks.com) or GNU Octave (www.gnu.org/software/octave/) that performs a least-squares fitting of n points with Cartesian coordinates (x_i, y_i, z_i) to a spherical surface of radius R and center at (x_C, y_C, z_C). The following equation is used to describe the surface:

$$z(x,y) = f(\mathbf{p}, x, y) = z_C + \sqrt{R^2 - (x - x_C)^2 - (y - y_C)^2}. \tag{A1}$$

Vector $\mathbf{p} = [R\ x_C\ y_C\ z_C]$ stores the four parameters that define the spherical surface.

This expression works well when performing a least-square fitting when points (x_i, y_i, z_i) are over a spherical cap near $z_{MAX} = z_C + R$ (sphere north pole) or $z_{MIN} = z_C - R$ (sphere south pole).

The algorithm needs an initial solution $\mathbf{p}_0 = \left[R^{(0)}\ x_C^{(0)}\ y_C^{(0)}\ z_C^{(0)}\right]$ from which starts an iterative process. In each iteration k, the following system of equations is solved by least squares:

$$\frac{R^{(k-1)}}{h_i} \cdot \Delta R + \frac{x_i - x_C^{(k-1)}}{h_i} \cdot \Delta x_C + \frac{y_i - y_C^{(k-1)}}{h_i} \cdot \Delta y_C + \Delta z_C = z_i - f(\mathbf{p}_{k-1}, x_i, y_i), \tag{A2}$$

where $i = 1, 2, \cdots, n$ and $h_i = \sqrt{\left(R^{(k-1)}\right)^2 - \left(x_i - x_C^{(k-1)}\right)^2 - \left(y_i - y_C^{(k-1)}\right)^2}$.

In each iteration, a detection of outliers is made following the method described in Section 1.3.5.17 of [29]. The z coordinates obtained with a confocal microscope are usually noisy enough to make the adjustment by least squares of the surface very difficult. Therefore, a good outlier detection algorithm must be used. The iterative process finishes when no more outliers are detected and no decrease in RMS residual is observed.

```
function [R,p,s,e,Cp,xn,yn,zn]=SphericalCapFit(x,y,z,xo,yo,zo,Ro);
    % Fit a spherical surface to probing points (x,y,z)
    % with outliers detection.
    %
    % Equation of the spherical surface
    %
    % z(x,y) = zo + sqrt( R^2 - (x-xo)^2 - (y-yo)^2 )
    %
    % It is needed an initial solution:
    %    Sphere Center: (xo,yo,zo)
    %    Sphere Radius: Ro
    %
    % Output variables
    %    R :      Sphere Radius
    %    p :      Shpere Parameters Vector, p=[R xo yo zo]
    %    s :      RMS Residual
    %    e :      Vector of residuals
    %    Cp :     Covariance matrix of p
    %    xn,yn,zn: Coordinates of points not discarded

    % Initial parameters
    p =[ Ro xo yo zo ]';

    while true
        R    = p(1); xo=p(2); yo=p(3); zo=p(4);
        h    = sqrt( R^2 - (x-xo).^2 - (y-yo).^2 );
        zest = zo + h ;

        A    = [ R./h  (x-xo)./h  (y-yo)./h  ones(size(x)) ];

        dp   = A\(z-zest) ;
        e    = (z-zest)-A*dp ;      % Residuals

        % Outliers detection following
        % section 1.3.5.17. Detection of Outliers
        % NIST - Engineering Statistical Handbook
        % https://www.itl.nist.gov/div898/handbook/eda/section3/eda35h.htm
        se   = 1/0.6745*median(abs(e));  % Robust estimation of standard deviation
        ii   = find(abs(e)<3.5*se);      % Outliers detection

        if (length(ii)==length(e))&&(se/se_ans>=1)
            break                        % End of the iterations
        else
            se_ans=se;
        end

        x=x(ii); y=y(ii); z=z(ii); e=e(ii);  % Outliers elimination
        p = p + dp;
    end

    % Uncertainty Estimation
    se = sqrt(sum(e.^2)/(length(e)-length(p))) ;
    Cp = se^2*inv(A'*A); % Covariance matrix of parameters p

    R = p(1);             % Estimate of R
    xn=x; yn=y; zn=z;    % Not discarded points coordinates

    % Residuals
    zest=p(4)+sqrt(p(1)^2-(x-p(2)).^2-(y-p(3)).^2);
    e = z - zest ;

    % RMS residual
    s = sqrt(sum(e.^2)/(length(e)-length(p))) ;
end
```

References

1. Carmignato, S.; Voltan, A.; Savio, E. Metrological performance of optical coordinate measuring machines under industrial conditions. *CIRP Ann. Manuf. Technol.* **2010**, *59*, 497–500. [CrossRef]
2. European Committee for Standardization (CEN). *ISO 3274:1996 Geometrical Product Specifications (GPS) Surface Texture: Profile Method. Nominal Characteristics of Contact (stylus) Instruments*; CEN: Bruxelles, Belgium, 1996.
3. Leach, R.K. Towards a complete framework for calibration of optical surface and coordinate measuring instruments. In Proceedings of the SPIE Optical Metrology Plenary Session, ICM, Munich, Germany, 26 June 2019.
4. European Committee for Standardization (CEN). *ISO 25178-6:2010 Geometrical Product Specification (GPS) Surface Texture: Areal—Part 6: Classification Methods for Measuring Surface Texture*; CEN: Bruxelles, Belgium, 2010.

5. Giusca, C.; Leach, R.K. *Measurement Good Practice Guide (No. 128): Calibration of the Metrological Characteristics of Imaging Confocal Microscopes (ICMs)*; National Physical Laboratory (NPL): Teddington, UK, 2012.
6. Minsky, M. Memoir on Inventing the Confocal Scanning Microscope. *Scanning* **1988**, *10*. [CrossRef]
7. Claxton, N.S.; Fellers, T.J.; Davidson, M.W. Laser scanning confocal microscopy. In *Encyclopedia of Medical Devices and Instrumentation*, 2nd ed.; Webster, J.G., Ed.; John Wiley & Sons, Inc.: Hoboken, NJ, USA, 2006; pp. 1–37. [CrossRef]
8. Watson, T. Fact and Artefact in Confocal Microscopy. *Adv. Dent. Res.* **1997**, *11*, 128–138. [CrossRef]
9. Cheng, C.; Wang, J.; Leach, R.K.; Lu, W.; Liu, X.; Jiang, X. Corrected parabolic fitting for height extraction in confocal microscopy. *Opt. Express* **2019**, *27*, 3682–3697. [CrossRef]
10. Alburayt, A.; Syam, W.P.; Leach, R.K. Lateral scale calibration for focus variation microscopy. *Meas. Sci. Technol.* **2019**, *29*. [CrossRef]
11. Joint Committee for Guides in Metrology (JCGM). *International Vocabulary of Metrology (VIM)—Basic and General Concepts and Associated Terms*, 3rd ed.; JCGM: Paris, France, 2012.
12. Leach, R.K.; Giusca, C.; Haitjema, H.; Evans, C.; Jiang, X. Calibration and verification of areal surface texture measuring instruments. *CIRP Ann. Manuf. Technol.* **2015**, *64*, 797–813. [CrossRef]
13. Caja, J.; Sanz, A.; Maresca, P.; Fernández, T.; Wang, C. Some Considerations about the Use of Contact and Confocal Microscopy Methods in Surface Texture Measurement. *Materials* **2018**, *11*, 1484. [CrossRef]
14. Wang, C.; Gómez, E.; Yu, Y. Characterization and correction of the geometric errors in using confocal microscope for extended topography measurement. Part I: Models, Algorithms Development and Validation. *Electronics* **2019**, *8*, 733. [CrossRef]
15. Wang, C.; Gómez, E.; Yu, Y. Characterization and Correction of the Geometric Errors Using a Confocal Microscope for Extended Topography Measurement, Part II: Experimental Study and Uncertainty Evaluation. *Electronics* **2019**, *8*, 1217. [CrossRef]
16. Webb, R.H. Confocal optical microscopy. *Rep. Prog. Phys.* **1996**, *59*, 427–471. [CrossRef]
17. Webb, R.H. Theoretical Basis of Confocal Microscopy. *Meth. Enzymol.* **1999**, *307*, 3–20. [CrossRef] [PubMed]
18. Salerni, G. Uso de la microscopía confocal de reflectancia en dermatología. *Dermatol. Argent.* **2011**, *17*, 230–235.
19. Confocal Microscopy, Molecular Expressions. Available online: https://micro.magnet.fsu.edu/primer/techniques/confocal/index.html (accessed on 26 October 2019).
20. Leach, R.K.; Bourell, D.; Carmignato, S.; Donmez, A.; Senin, N.; Dewulf, W. Geometrical metrology for metal additive manufacturing. *CIRP Ann. Manuf. Technol.* **2019**, *68*, 677–700. [CrossRef]
21. Introduction to Confocal Microscopy. Available online: https://www.olympus-lifescience.com/en/microscope-resource/primer/techniques/confocal/confocalintro/ (accessed on 26 October 2019).
22. Tata, B.V.R.; Raj, B. Confocal laser scanning microscopy: Applications in material science and technology. *Bull. Mater. Sci.* **1998**, *21*, 263–278. [CrossRef]
23. Cohen-Or, D.; Kaufmann, A. Fundamentals of Surface Voxelization. *CVGIP: Graph. Model. Image Process.* **1995**, *57*, 453–461. [CrossRef]
24. Joint Committee for Guides in Metrology, Working Group 1 (JCGM/WG 1). *JCGM 100:2008 Evaluation of Measurement Data—Guide to the Expression of Uncertainty in Measurement (GUM)*, 1st ed.; JCGM: Paris, France, 2008.
25. European Accreditation (EA). *EA-4/02—Evaluation of the Uncertainty of Measurement in Calibration*; EA: London, UK, 2013.
26. European Committee for Standardization (CEN). *ISO/IEC 17025:2017 General Requirements for the Competence of Testing and Calibration*; CEN: Bruxelles, Belgium, 2017.
27. SENSOFAR-TECH LTD. *Application Note: Flatness Error on Imaging Confocal Microscopes*; Sensofar: Tarrasa, Spain, 2009.
28. Centro Español de Metrología (CEM). *Procedimiento DI-035 para la Calibración de Patrones de Planitud de Vidrio*; CEM: Madrid, Spain, 2004.
29. NIST/SEMATECH e-Handbook of Statistical Methods. Available online: http://www.itl.nist.gov/div898/handbook/ (accessed on 27 November 2019).
30. De Vicente, J.; Molpeceres, C.; Guarneros, O.; García-Ballesteros, J. *Calibración de Microscopios Confocales*; XVII National Congress of Mechanical Engineering: Gijón, Spain, 2008.
31. Guarneros, O.; De Vicente, J.; Maya, M.; Ocaña, J.L.; Molpeceres, C.; García-Ballesteros, J.; Rodríguez, S.; Duran, H. Uncertainty Estimation for Performance Evaluation of a Confocal Microscope as Metrology Equipment. *MAPAN—J. Metrol. Soc. I* **2014**, *29*, 29–42. [CrossRef]

32. De Vicente, J.; Sánchez-Pérez, A.M.; Maresca, P.; Caja, J.; Gómez, E. A model to transform a commercial flatbed scanner into a two-coordinates measuring machine. *Measurement* **2015**, *73*, 304–312. [CrossRef]
33. European Committee for Standardization (CEN). *ISO 25178-70 Geometrical Product Specification (GPS) Surface Texture: Areal—Part 70: Material Measures*; CEN: Bruxelles, Belgium, 2014.
34. Centro Español de Metrología (CEM). *Procedimiento DI-004 para la Calibración de Medidoras de una Coordenada Vertical*; CEM: Madrid, Spain, 2013.
35. Wang, C.; Caja, J.; Gómez, E.; Maresca, P. Procedure for Calibrating the Z-axis of a Confocal Microscope: Application for the Evaluation of Structured Surfaces. *Sensors* **2019**, *19*, 527. [CrossRef] [PubMed]
36. European Committee for Standardization (CEN). *ISO 5436-1:2000 Geometrical Product Specipications (GPS) Surface Texture: Profile Method. Measurement Standards—Part 1: Material Measures*; CEN: Bruxelles, Belgium, 2000.
37. Calibration standards, HALLE Präzisions-Kalibriernormale GmbH. Available online: http://halle-normale.de/framesets/englisch/products/products.html (accessed on 29 October 2019).
38. Balcon, M.; Carmignato, S.; Savio, E. Performance verification of a confocal microscope for 3D metrology tasks. *Qual.—Access Success* **2012**, *13*, 63–66.
39. De Vicente, J.; Raya, F. Simplified Statistical Method for Uncertainty Estimation in Coordinate Metrology. In Proceedings of the 9th International Metrology Congress, Bordeaux, France, 18–21 October 1999.
40. Deutscher Kalibrierdienst (DKD). *Guideline DKD 4-2 Calibration of Measuring Instruments and Standards for Roughness Measuring Technique (Sheet 1: Calibration of Standards for Roughness Measuring Technique)*; DKD: Braunschweig, Germany, 2011.
41. Deutscher Kalibrierdienst (DKD). *Guideline DKD-R 4-2 Calibration of Devices and Standards for Roughness Metrology (Sheet 2: Calibration of the Vertical Measuring System of Stylus Instrument)*; DKD: Braunschweig, Germany, 2011.
42. Deutscher Kalibrierdienst (DKD). *Guideline DKD-R 4-2 Calibration of Devices and Standards for Roughness Metrology (Sheet 3: Calibration of Standards with Periodic Profiles in Horizontal Direction by Means of Stylus Instrument)*; DKD: Braunschweig, Germany, 2011.
43. European Committee for Standardization (CEN). *ISO 4287:1999 Geometrical Product Specifications (GPS) Surface Texture: Profile Method. Terms, Definitions and Surface Texture Parameters*; CEN: Bruxelles, Belgium, 1999.
44. European Committee for Standardization (CEN). *ISO 4288:1996 Geometrical Product Specifications (GPS) Surface Texture: Profile Method. Rules and Procedures for the Assessment of Surface Texture*; CEN: Bruxelles, Belgium, 1996.
45. Leach, R.K.; Haitjema, H. Bandwidth characteristics and comparisons of surface texture measuring instruments. *Meas. Sci. Technol.* **2010**, *21*. [CrossRef]
46. Joint Committee for Guides in Metrology, Working Group 1 (JCGM/WG 1). *JCGM 101:2008 Evaluation of Measurement Data—Supplement 1 to the "Guide to the Expression of Uncertainty in Measurement"—Propagation of Distributions using a Monte Carlo Method (GUM-S1)*, 1st ed.; JCGM: Paris, France, 2008.

© 2019 by the authors. Licensee MDPI, Basel, Switzerland. This article is an open access article distributed under the terms and conditions of the Creative Commons Attribution (CC BY) license (http://creativecommons.org/licenses/by/4.0/).

Article

Development and Validation of a Calibration Gauge for Length Measurement Systems

Francisco Javier Brosed *, Raquel Acero Cacho, Sergio Aguado, Marta Herrer, Juan José Aguilar and Jorge Santolaria Mazo

Design and Manufacturing Engineering Department, Universidad de Zaragoza, María Luna 3, 50018 Zaragoza, Spain; racero@unizar.es (R.A.C.); saguadoj@unizar.es (S.A.); 680722@unizar.es (M.H.); jaguilar@unizar.es (J.J.A.); jsmazo@unizar.es (J.S.M.)
* Correspondence: fjbrosed@unizar.es

Received: 31 October 2019; Accepted: 27 November 2019; Published: 29 November 2019

Abstract: Due to accuracy requirements, robots and machine-tools need to be periodically verified and calibrated through associated verification systems that sometimes use extensible guidance systems. This work presents the development of a reference artefact to evaluate the performance characteristics of different extensible precision guidance systems applicable to robot and machine tool verification. To this end, we present the design, modeling, manufacture and experimental validation of a reference artefact to evaluate the behavior of these extensible guidance systems. The system should be compatible with customized designed guides, as well as with commercial and existing telescopic guidance systems. Different design proposals are evaluated with finite element analysis, and two final prototypes are experimentally tested assuring that the design performs the expected function. An estimation of the uncertainty of the reference artefact is evaluated with a Monte Carlo simulation.

Keywords: calibration artifact; kinematic support; dimensional metrology; machine tool; length measurement

1. Introduction

Volumetric verification is a verification technique to improve the accuracy of machine tools (MTs) and robots based on indirect measurement [1]. It uses the combined effect of all geometric errors through a parameter identification process [2]. Many studies have been carried out for its application to coordinate measurement machines (CMMs) and machine tools (MTs) [3,4]. The increasing implementation of this verification technique in the field of machine tool verification has led to the development of verification procedures that depend on different factors such as the type of machine, the non-geometric errors of the machine, the system and measurement technique applied, etc. [5]. The result of the equipment's verification is linked to the calibration of the measurement system used, procedure which is normally carried out in accordance with the applicable standards. This applies to measuring instruments commonly used in volumetric verification such as laser trackers [6]. However, in some cases, the lack of guidelines or standards makes it necessary to develop internal calibration procedures and to use specific reference gauges [7,8].

Therefore, this work presents the development of a reference artefact to calibrate extensible guidance systems used in machine tool and robot verification procedures. The reference artefact materializes several working positions and lengths with a fixed reference origin. The reference origin consists of a nest for a precision sphere, and the working positions include different nests with precision spheres and kinematic couplings. The mechanical repeatability of the reference artefact for the nests' positioning in the different working locations is achieved with kinematic couplings configuration of spheres and cylinders. The design of the artefact will also compensate the errors associated with its deflections [9,10]. In [9], an analysis of the measured length of long artifacts is showed, and the use of tubes instead of solid bars is recommended to reduce the elongation due to self-loading. In [10], the author gives an estimation of the location of the supports to obtain parallel surfaces at the end of a bar (Airy points) and another option (Bessel points) to obtain the minimum change in length.

The paper is structured as follows. Firstly, the authors analyze the requirements of the design and the structure of the reference artefact. Secondly, it is performed an evaluation of the different gauge design proposals by means of a finite element simulation in Solid Edge. In this analysis, the displacement generated in the gauge due to the load application is measured for each case. Then, the design proposals selected are manufactured by 3D printing, and these prototypes are used in the experimental testing and measured with a CMM (coordinate measurement machine). Finally, after optimizing the design with the feedback of simulation and experimental testing, the paper presents an uncertainty estimation of the designed calibration system.

2. Materials and Methods

The calibration artefact has to materialize the calibration positions for a length measurement instrument. The instrument consists of a system that measures the distance between two spheres. One of the spheres is fixed to the instrument and the other is fixed to the machine tool, robot or coordinate measuring machine under verification. As it can be seen in Figure 1, the gauge is composed of a sphere (1) and a support (5) to hold the sphere fixed to the machine tool under verification (6), being both located at the edges of the artefact. Between both sides, there is an interferometer (2) to measure the different distances that will be materialized in the gauge. These different lengths are achieved with a telescopic system (3) that also assures the alignment of the interferometer and the retroreflector (6). The interferometer is located on the left side of the gauge closed to the fix sphere (1). The retroreflector will be located on the other side of the system close to the sphere fixed to the machine tool. The calibration artefact should be able to calibrate measurement instruments with a measurement range from 400 mm to 1600 mm (max. and min. in Figure 1). Once calibrated, the instrument will give the distance between the centers of the two spheres.

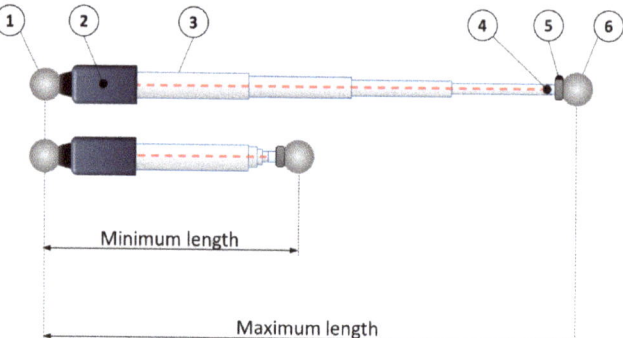

Figure 1. Scheme and components of the measurement instrument. (1) sphere fixed to the instrument; (2) interferometer; (3) telescopic system; (4) retroreflector; (5) magnetic holder; (6) sphere fixed to the machine tool.

The calibration artefact has a fixed magnetic sphere-holder to lock the position of the sphere fixed to the instrument (1) and several kinematic supports to obtain a repeatable positioning of a sphere. When the sphere of the instrument (1) is locked in the magnetic sphere holder and the other side of the instrument (5) reaches the sphere fixed to the machine tool, a calibrated length materializes in the gauge. The defined nominal lengths of the calibration artefact range from 400 to 1600 mm.

During the calibration of the measurement instrument, the calibration artefact rests in a flat surface. Therefore, the artefact incorporates three support legs on its base to assure its stability in the calibration process.

The main components under analysis in the design of the gauge are the following: the position of the support points to minimize the deformation in the length measurement, the kinematic couplings support that allows the movable sphere positioning with high repeatability and the mechanical structure to materialize the calibration lengths.

During the calibration of the measurement instrument, we need a repeatable positioning of the movable sphere to materialize the calibration positions. For this purpose, a kinematic base has been designed with calibrated spheres and cylinders (6-points 3-cylinders). The kinematic contact has two parts (upper and lower). In the lower part, six spheres are fixed in three pairs located at 120° meanwhile in the upper part, three cylinders are fixed with its axis located at 120° and pointing to the center of the geometrical distribution (Figure 2). Each interface provides two constraints, totaling six constraints for the system. The best stability is achieved when the axes of the contact planes bisect the coupling triangle with each interface as a vertex of this triangle. Four spheres secure the position of the cylinders in the upper part. The upper and lower parts are fixed with magnets located in the center of the geometrical distribution (in the upper and lower part respectively).

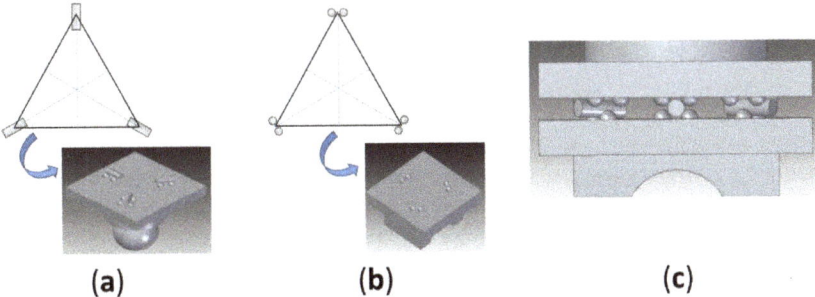

Figure 2. Kinematic support for the mobile sphere. (**a**) Distribution and orientation of the cylinders; (**b**) distribution and orientation of the spheres; (**c**) model of the kinematic support mounted, contact between spheres and cylinders.

The main element of the artefact is a tube that goes through the other parts of the assembly. The junction between the tube and the other parts (magnetic holder or kinematic support for the magnetic holder) is materialized with a flange.

Two design proposals for the artefact structure are evaluated. The first prototype is a single tube structure in which the line of the measurement points is parallel to the bar and is located beyond the structure (Figure 3a). Each flange has been designed to hold the bar and locate the magnetic holder, in one case, and the kinematic support for the magnetic holder, in the rest of the cases, defining each measurement position (Figure 3b,c).

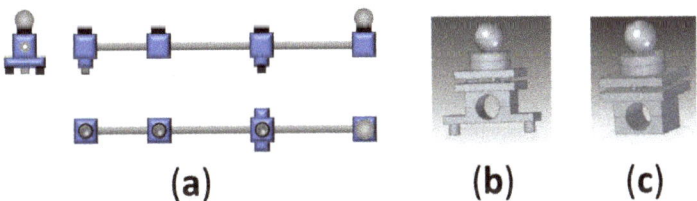

Figure 3. (**a**) Single tube structure for the calibration artefact; (**b**) flange of the single bar artefact with standing legs; (**c**) flange of the single bar artefact without standing legs.

The second prototype is a double tube structure that locates the line of the measurement points between both bars (Figure 4a). The flanges hold the bars and support a base where the magnetic holder is located in the first point and the kinematic supports in the other cases (Figure 4b).

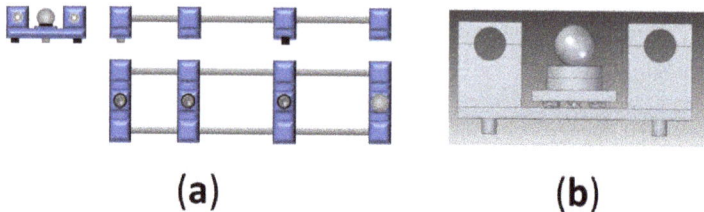

Figure 4. (a) Double tube structure for the calibration artefact; (b) flange for the double bar structure with the base and he kinematic support.

A horizontal bar of great length requires two points of support in the direction of its length to be stable. The position of these standing legs determines the action of the gravity of this bar; depending on how it is supported, measurement errors can be caused [11]. Therefore, if the supports are positioned at the ends, it will warp in the center causing the ends to come closer and tilt upwards. On the contrary, if the two supports are positioned in the middle, the bar will be bent at the ends [9]. From [9] the use of tubes instead of bars to reduce the elongation due to self-loading is taken as well as the location of the spheres in the neutral bending surface.

The distances between the supports of the bar have been defined using Airy and Bessel methodologies and comparing the results of the deformation. A bar supported at its Airy points has parallel ends and supported at its Bessel points has maximum length due to deflection reduction. The value of the Airy and Bessel points has been taken from [10]. The distance between supports (a) and the position of each support (Lmin and Lmid), for a simple bar of 1600 mm length (L), appears in Table 1, and the deformation obtained in the bar appears in Figure 5.

Table 1. Values of the Airy and Bessel point.

Parameter	Airy	Bessel
L (mm)	1600	1600
Factor	0.57735	0.55940
a (mm)	923.76	895.04
Lmin (mm)	338.12	353.5
Lmid (mm)	1261.88	1247.88

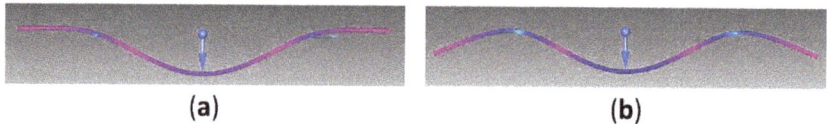

Figure 5. Results of the simulation of a simple bar (aluminum) with two supports following each two methods. (a) Simple bar supported at the Airy points; (b) simple bar supported at the Bessel points.

Four different positions of the supports are proposed; two of them following the Airy and Bessel methodologies (Figure 6a). The other two configurations locate the supports in the reference flange (point 0, Figure 6b) and in the flange that materializes Lmid (point 2, Figure 6b) in the third case and in the reference flange (point 0, Figure 6c) and in the flange that materializes Lmax (point 3, Figure 6c) in the fourth case.

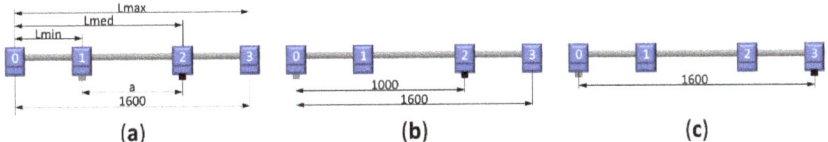

Figure 6. Scheme of the four locations of the supports and the location of the loads in each simulation; there is a load of 1N in point 0 for every simulation and another load of 1N in point 1 for the simulation of Lmin, in point 2 for the simulation of Lmid, and in point 3 for the simulation of Lmax. (**a**) Airy and Bessel points; (**b**) supports located in point 0 (reference) and in Lmid; (**c**) supports located in point 0 (reference) and in Lmax.

3. Results

This section provides the description of the main results of the components analysis and the development of the artefact. As the previous section indicates, the main components under analysis are the following: the position of the support points to minimize the deformation in the length measurement, the kinematic couplings support that will allow the movable sphere positioning with high repeatability, and the mechanical structure to materialize the calibration lengths.

3.1. Design Selection

In order to define the position of the standing legs, a finite element analysis of the deformation of the structure has been carried out (twenty-four simulations, twelve for each prototype, were performed). The study analyses four different positions of the supports, and the deformation occurred when the measurement system was placed in the three different measurement positions (Lmin, Lmid, and Lmax). The measurement system will rest in the reference position and in the position under verification (Lmin, Lmid, or Lmax). Therefore, in the analysis there is a load of 1N in the reference position (position of the magnetic holder, point 0, Figure 6) and another load of the same value in the measurement position for each case. The four positions of the supports are the Airy point (a = 923.76 mm), the Bessel points (a = 895.04 mm, Figure 6a), the supports located in the reference position and in the Lmid (Figure 6b), and finally, the supports located in the reference position and in the Lmax (Figure 6c). All the combinations make twenty-four simulations (2 prototypes, 4 supporting leg configuration and 3 load configurations).

The material properties taken into account for the structural analysis are shown in Table 2.

Table 2. Material properties for prototypes 1 and 2 (aluminum 6061) and prototype 3 (carbon fiber).

Property	Al 6061	Carbon Fiber
Density (T/m^3)	2.7	1.6
Young Module (GPa)	68.9	393.3
Poisson Coeficient	0.330	0.100

The increment of the measurement distance or the measurement error ΔL_{MEAS} in Equation (1) characterizes the deformation of the structure.

$$\Delta L_{MEAS} = \sqrt{(L_n + \Delta x_n - \Delta x_0)^2 + (\Delta y_n - \Delta y_0)^2 + (\Delta z_n - \Delta z_0)^2}, \quad (1)$$

where L_n is the nominal distance of the measurement point (n = Lmin, Lmid and Lmax); (Δx_n, Δy_n, Δz_n) are the displacements of the measurement point due to the deformation of the structure, and (Δx_0, Δy_0, Δz_0) are the displacements of the reference point due to the deformation of the structure.

Combining the four proposed positions of the supports and the three different pairs of loads, twelve values of simulated measurement error have been obtained for each prototype after the analysis of the prototypes using finite elements software (Solid Edge ST8, Siemens PLM Software, Plano, TX, USA) (Figure 7).

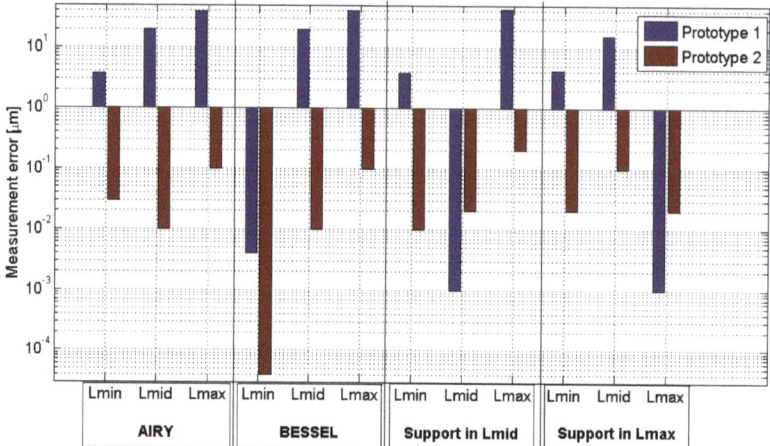

Figure 7. Measurement error in μm for the twelve cases for each two prototypes. These results have been obtained after the analysis of the prototypes using finite elements software.

The localization of the spheres beyond the structure line amplifies the measurement error due to the deformation in prototype 1. Based on that, prototype 1 was discarded and the results shown in the paper correspond to prototype 2. The measurement errors in prototype 2 are minimum using the Bessel points and do not exceed from 0.1 μm, value obtained when the system is loaded in Lmax position (points 0 and 3, Figure 6) and lower in the other cases (Lmin and Lmid) (according with the simulation results using finite elements software (Solid Edge ST8, Siemens PLM Software, Plano, TX, USA)).

After the simulation with the 3D models of prototypes 1 and 2, a kinematic support prototype was manufactured by additive manufacturing (Figure 8) and tested using a CMM. The kinematic supports have been tested measuring the repeatability of the manufactured kinematic supports with the spheres and the cylinders. The position measurement repeatability of each location obtained after ten iterations was 8 μm (measured with a CMM).

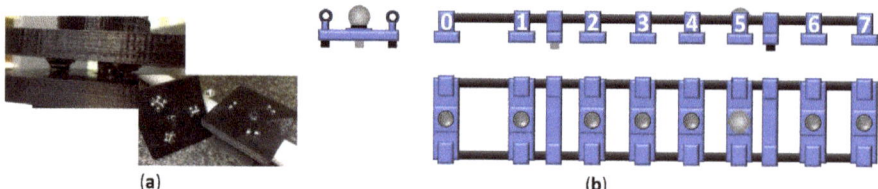

Figure 8. (a) Kinematic supports samples manufactured by additive manufacturing; (b) prototype 3.

Once the kinematic supports have been tested (repeatability measured with a CMM), and the adequacy of the Bessel points for this application has been proved (simulation using finite elements software (Solid Edge ST8, Siemens PLM Software, Plano, TX, USA), we manufactured a new design with aluminum flanges and carbon fiber structure tubes (27″ diameter, 1830 mm length). The number of measurement points increments to seven from three in the previous prototypes (Table 3). In this case, the values of the measurement errors obtained for each measurement position are under 0.1 μm (measured with CMM).

Table 3. Bessel and measurement points for prototype 3 (proposed nominal values).

A: $n = 0$	B: $n = 1$	Bessel 1	C: $n = 2$	D: $n = 3$	E: $n = 4$	F: $n = 5$	Bessel 2	G: $n = 6$	H: $n = 7$
0.00	280.00	396.54	535.00	785.00	1040.00	1295.00	1403.46	1545.00	1800.00
					(mm)				

3.2. Manufacture, Assembly, and Performance

The flanges, the bases where the kinematic supports are located, and the standing legs of the artefact were made of Aluminum 6061 for the prototype 3. The flanges join the carbon fiber tubes with the base that contains the magnetic holder for the position A (n = 0, Figure 8b) and with the base that contains the kinematic supports for the rest of the cases, positions B to H (n from 1 to 7, Figure 8b).

The flange geometry has been redesigned to adequate it to a wire EDM manufacturing process (Figure 9a,b).

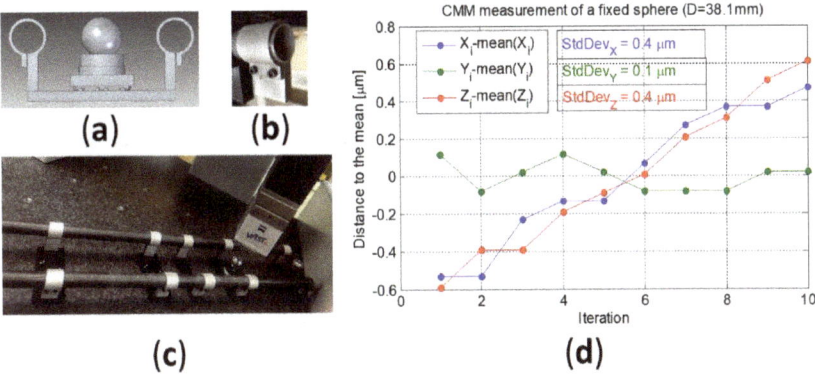

Figure 9. (a) Model of the flanges in prototype 3; (b) detail of the flange embracing a tube; (c) prototype 3 with the new flanges and carbon fiber tubes during performance test in coordinate measurement machine (CMM); (d) CMM measuring a fixed sphere (Diameter 38.1 mm), distance to the mean of the ten iterations in mm. Standard deviation values of each coordinate are included next to the legend.

The position of the spheres has been measured with a coordinate measurement machine (CMM) to obtain the uncertainty of the calibration artefact. First, the repeatability of the CMM measuring a sphere with a diameter of 38.1 mm ($1\frac{1}{2}''$) has been estimated in 0.4 µm. The measurement position number 3 (Figure 8b) was measured ten times with the CMM without removing the sphere from the kinematic support (Figure 9d).

A second measurement with ten iterations was carried out assembling and disassembling the kinematic support with the sphere located in position number 3 (not fixed sphere, Figure 10). The results are compared with those obtained without removing the kinematic support (fixed sphere, Figure 10).

The standard deviation values of the sample in X, Y, and Z coordinates are 0.4, 0.1, and 0.4 µm, respectively, without removing the kinematic support (fixed sphere). Disassembling the kinematic support with the sphere located in position number 3 (not fixed sphere) are 0.4, 0.2, and 0.5 µm.

Z coordinate is more sensible to the movements when mounting and demounting the kinematic support but, in any case, the standard deviation of the sample is low enough for the application.

The next step, after measuring the repeatability of the kinematic supports, is to check the effect of the deformation in the measurement length corresponding to each sphere position. To evaluate this effect, the measurement of the reference position (A, number 0 in Figure 8b) and the other positions has been carried out moving the sphere with the kinematic support from B (position number 1 in Figure 8b) to H (position number 7 in Figure 8b). The procedure is repeated ten times. The measurement results allow estimating the repeatability of the measurement length in each position as the standard deviation of ten repetitions in each position (Figure 11).

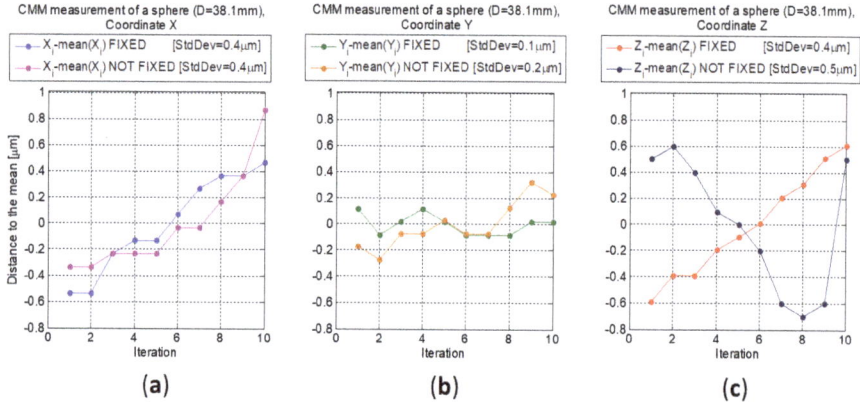

Figure 10. CMM measuring a sphere (Diameter 38.1 mm), distance to the mean of the ten iterations in mm in each coordinate. Standard deviation values of each coordinate (fixed and not fixed) are included in the legend. The results are compared with those obtained measuring a fixed sphere. (**a**) Distance to the mean of the ten iterations in mm in X coordinate; (**b**) Distance to the mean of the ten iterations in mm in Y coordinate; (**c**) Distance to the mean of the ten iterations in mm in Z coordinate.

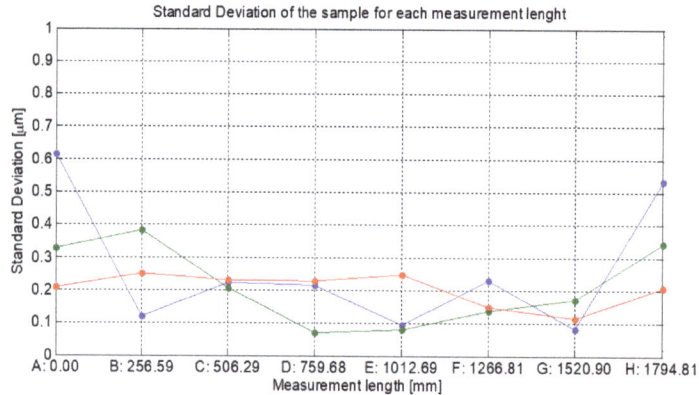

Figure 11. CMM measuring a sphere (Diameter 38.1 mm) in each position of the artefact. The graph shows the standard deviation of the sample (ten iterations) for each measurement length (X, Y, and Z coordinates). The abscissa identifies the mean value of the measurement length of each position (from A to H).

3.3. Calibration Artefact Uncertainty Estimation with Monte Carlo Simulation

Monte Carlo (MC) method is widely used in measurement uncertainty estimation procedures [12,13], for example applied to coordinate measuring machines (CMMs) uncertainty analysis [14], [15–17]. In this work, we used the MC simulation to estimate the uncertainty of the reference artefact in the calibration of length measurement systems. The uncertainty values have been calculated according with the Guide to the expression of uncertainty in measurement GUM [18,19] using a confidence level of 95% (k = 2).

The input data for the MC simulation method are the probability distributions of the variability of the different error sources. In this case, the main error source is the variability of the positioning of each calibration point.

The nominal value of each position coordinate is the mean value obtained from the CMM measurement. The standard deviation of the distribution of each position is also the standard deviation of the CMM measurements for each position (Figure 11). Then, each position (from A to H) is measured with a CMM and the repeatability of the X, Y, and Z coordinates is evaluated modelling its distribution as a normal distribution. When the effect of the distribution of each variable that influences the measurement length materialized by the calibration artefact is considered, the distribution of the measurement result can be estimated (Figure 12).

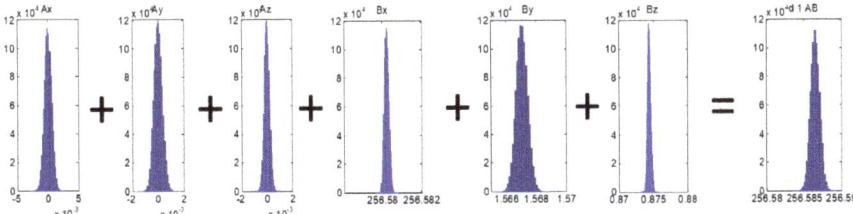

Figure 12. Distribution of the values for each set of coordinates of A and B positions. The distance between this two positions materializes the measurement length number 1 (d1, AB). The distribution of the measurement length can be obtained by simulation with the Monte Carlo method.

The results of the Monte Carlo simulation also depend on the number of iterations carried out. When the number of iterations is low, the results are not representative but as the number of results increases, the values obtained for the uncertainty converge (Figure 13).

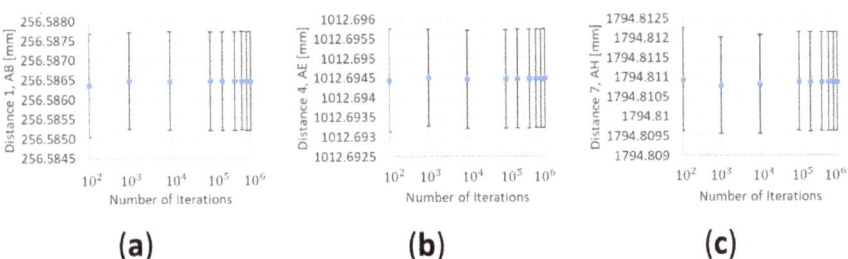

Figure 13. Evolution of the results as the number of iterations increase from 10^2 to 10^6. (**a**) Results for the first measurement length materialized between point A and point B of the calibration artefact; (**b**) results for the fourth measurement length materialized between point A and point E of the calibration artefact; (**c**) results for the seventh measurement length materialized between point A and point H of the calibration artefact.

The results of the MC simulation are shown in Table 4 including the uncertainty value for each length measurement. The uncertainty distribution for each measurement length (from distance between n=0 to n=1 in Figure 8b, distance 1 AB, to distance between n=0 to n=7 in Figure 8b, distance 1 AH) is shown in Figure 14 obtaining values below ±1.60 μm for all the positions in the study.

Table 4. Bessel and measurement points for prototype 3 (proposed nominal values).

Position	Number, n	Measurement length (mm)	Uncertainty (k = 2) (μm)
B	1	256.586	±1.25
C	2	506.29	±1.30
D	3	759.684	±1.30
E	4	1012.694	±1.20
F	5	1266.808	±1.30
G	6	1520.905	±1.20
H	7	1794.811	±1.60

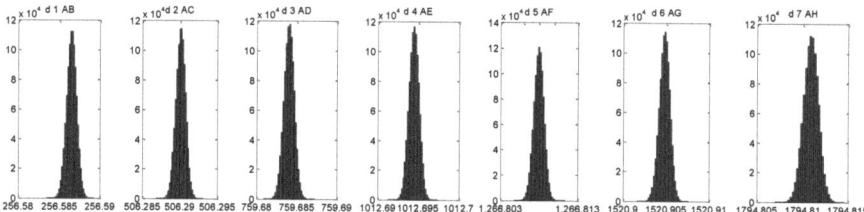

Figure 14. The uncertainty distribution for each measurement length (from distance d1, AB to distance d7, AH).

4. Discussion

This work presented the design, development, manufacturing, and experimental validation of a reference artefact to calibrate extensible guidance systems used in machine tool and robot verification procedures. The artefact uses spheres and spherical nests with kinematic supports that assure the high repeatability of the system. Different design proposals were evaluated with finite element analysis, and two final prototypes were experimentally tested assuring that the design of kinematic couplings performs the expected function. The paper finally presents the uncertainty estimation of the calibration artifact using a Monte Carlo simulation (MC).

We could conclude from the results of the Monte Carlo simulation that the calibration uncertainty of the artefact designed for length measurement systems could be adequate for the application, considering tests carried out in a horizontal position.

The calibration artefact presented in this work can be used to test the telescopic system not only in a horizontal position but also by varying the angle and reaching an upright position. Therefore, simulation and experimental validation would be necessary in these conditions in the future, although it is expected that the configuration of the most precise calibration artefact would be the same as the one presented in this paper for horizontal tests.

Author Contributions: Conceptualization, J.J.A., M.H. and J.S.M.; methodology, J.J.A., M.H. and R.A.C.; software and validation, J.J.A., M.H. and S.A.; formal analysis, S.A., J.S.M. and F.J.B.; investigation, J.J.A. and M.H., S.A.; resources, R.A.C., J.S.M., F.J.B.; writing—original draft preparation, R.A.C., S.A. and F.J.B.; writing—review and editing, R.A.C., F.J.B. and J.S.M.

Funding: This research was funded by the Ministerio de Economía, Industria y Competitividad with project number Reto 2017-DPI2017-90106-R, by Aragon Government (Department of Industry and Innovation) through the Research Activity Grant for research groups recognized by the Aragon Government (T56_17R Manufacturing Engineering and Advanced Metrology Group) co-funded with European Union ERDF funds (European Regional Development Fund 2014-2020, "Construyendo Europa desde Aragón") and by the Universidad de Zaragoza and the Centro Universitario de la Defensa, project number UZCUD2018-TEC-04.

Conflicts of Interest: The authors declare no conflict of interest. The funders had no role in the design of the study; in the collection, analyses, or interpretation of data; in the writing of the manuscript; or in the decision to publish the results.

References

1. Longstaff, A.P.; Fletcher, S.; Parkinson, S.; Myers, A. The role of measurement and modelling of machine tools in improving product quality. *Int. J. Metrol. Qual.* **2013**, *4*, 177–184. [CrossRef]
2. Jung, J.H.; Choi, J.P.; Lee, S.J. Machining accuracy enhancement by compensating for volumetric errors of a machine tool and on-machine measurement. *J. Mater. Process. Technol.* **2006**, *174*, 56–66. [CrossRef]
3. Aguado, S.; Santolaria, J.; Samper, D.; Velazquez, J.; Aguilar, J.J. Empirical analysis of the efficient use of geometric error identification in a machine tool by tracking measurement techniques. *Meas. Sci. Technol.* **2016**, *27*, 12. [CrossRef]
4. He, Z.; Fu, J.; Zhang, X.; Shen, H. A uniform expression model for volumetric errors of machine tools. *Int. J. Mach. Tools Manuf.* **2016**, *100*, 93–104. [CrossRef]
5. Aguado, S.; Santolaria, J.; Samper, J.; Aguilar, J.J. Protocol for machine tool volumetric verification using commercial laser tracker. *Int. J. Adv. Manuf. Technol.* **2014**, *75*, 425–444. [CrossRef]
6. Wang, J.; Guo, J. Research on volumetric error compensation for NC machine tool based on laser tracker measurement. *Technol. Sci.* **2012**, *55*, 3000–3009. [CrossRef]
7. Hudlemeyer, A.; Sawyer, D.; Blackburn, C.J.; Lee, V.D.; Meuret, M.; Shakarji, C.M. Considerations for Design and In-Situ Calibration of High Accuracy Length Artifacts for Field Testing of Laser Trackers. National Institute of Standards and Technology. March 01th 2015. Available online: https://www.nist.gov/publications/considerations-design-and-situ-calibration-high-accuracy-length-artifacts-field-testing (accessed on 25 October 2019).
8. Zhao, H.; Yu, L.; Xia, H.; Li, W.; Jiang, Y.; Jia, H. 3D artifact for calibrating kinematic parameters of articulated arm coordinate measuring machines. *Meas. Sci. Technol.* **2018**, *29*, 8. [CrossRef]
9. Sawyer, D.; Parry, B.; Phillips, S.; Blackburn, C.; Muralikrishnan, B. A model for geometry-dependent errors in length artefacts. *J. Res. Natl. Inst. Stand. Technol.* **2012**, *117*, 216–230. [CrossRef] [PubMed]
10. Verdirame, J. Airy Points, Bessel Points, Minimum Gravity Sag and Vibration Nodal Points of Uniform Beams. Mechanics and Machines 2016. Available online: http://www.mechanicsandmachines.com/?p=330 (accessed on 25 October 2019).
11. Köning, R.; Przebierala, B.; Weichert, C.; Flügge, J.; Bosse, H. A revised treatment of the influence of the sample support on the measurement of line scales and the consequences for its use to disseminate the unit of length. *Metrologia* **2009**, *46*, 187–195.
12. Cox, M.G.; Siebert, B.R.L. The use of a Monte Carlo method for evaluating uncertainty and expanded uncertainty. *Metrologia* **2006**, *43*, S178–S188. [CrossRef]
13. Schwenke, H.; Siebert, B.R.L.; Wäldele, F.; Kunzmann, H. Assessment of Uncertainties in Dimensional Metrology by Monte Carlo Simulation: Proposal of a Modular and Visual Software. *CIRP Ann. Manuf. Technol.* **2000**, *49*, 395–398. [CrossRef]
14. Balsamo, A.; Di Ciommo, M.; Mugno, R.; Rebaglia, B.I.; Ricci, E.; Grella, R. Evaluation of CMM Uncertainty Through Monte Carlo Simulations. *CIRP Ann. Manuf. Technol.* **1999**, *48*, 425–428. [CrossRef]
15. Romdhani, F.; Hennebelle, F.; Ge, M.; Juillion, P.; Coquet, R.; Fontaine, J.F. Methodology for the assessment of measuring uncertainties of articulated arm coordinate measuring machines. *Meas. Sci. Technol.* **2014**, *25*, 125008. [CrossRef]
16. Sładek, J.; Gąska, A. Evaluation of coordinate measurement uncertainty with use of virtual machine model based on Monte Carlo method. *Measurement* **2012**, *45*, 1564–1575. [CrossRef]
17. Pérez, P.; Aguado, S.; Albajez, J.A.; Santolaria, J. Influence of laser tracker noise on the uncertainty of machine tool volumetric verification using the Monte Carlo method. *Measurement* **2019**, *133*, 81–90. [CrossRef]
18. BIPM; IEC; IFCC; ILAC; ISO; IUPAC; IUPAP; OIML. *Evaluation of Measurement Data—Guide to the Expression of Uncertainty in Measurement*; JCGM 100 Bureau International des Poids et Mesures (BIPM): Sèvres, France, 2008.
19. BIPM; IEC; IFCC; ILAC; ISO; IUPAC; IUPAP; OIML. *Evaluation of Measurement Data—Supplement 1 to the Guide to the Expression of Uncertainty in Measurement—Propagation of Distributions Using a Monte Carlo Method*; JCGM 101 Bureau International des Poids et Mesures (BIPM): Sèvres, France, 2008.

© 2019 by the authors. Licensee MDPI, Basel, Switzerland. This article is an open access article distributed under the terms and conditions of the Creative Commons Attribution (CC BY) license (http://creativecommons.org/licenses/by/4.0/).

Article

2D–3D Digital Image Correlation Comparative Analysis for Indentation Process

Carolina Bermudo Gamboa, Sergio Martín-Béjar *, F. Javier Trujillo Vilches, G. Castillo López and Lorenzo Sevilla Hurtado

Department of Civil, Material and Manufacturing Engineering, EII University of Malaga C/Dr, Ortiz Ramos s/n, E-29071 Malaga, Spain; bgamboa@uma.es (C.B.G.); trujillov@uma.es (F.J.T.V.); gcastillo@uma.es (G.C.L.); lsevilla@uma.es (L.S.H.)
* Correspondence: smartinb@uma.es

Received: 28 October 2019; Accepted: 9 December 2019; Published: 11 December 2019

Abstract: Nowadays, localized forming operations, such as incremental forming processes, are being developed as an alternative to conventional machining or forming techniques. An indentation process is the main action that takes places in these forming activities, allowing small, localized deformations. It is essential to have the knowledge of the material behavior under the punch and the transmitted forces to achieve correct control of the entire procedure. This paper presents the work carried out with the digital image correlation (DIC) technique applied to the study of the material flow that takes place under an indentation process. The material flow analysis is performed under 2D and 3D conditions, establishing the methodology for the calibration and implementation for each alternative. Two-dimensional DIC has been proven to be a satisfactory technique compared with the 3D method, showing results in good agreement with experimental tests and models developed by the finite element method. Notwithstanding, part of the indented material flows under the punch, emerging on the front surface and generating a dead zone that can only be addressed with a 3D technique. So, the main objective is to carry out a comparison between the 2D and 3D techniques to identify if the 3D application could be mandatory for this type of process. Also, a 2D–3D mix analysis is proposed for study cases in which it is necessary to know the material flow in that specific area of the workpiece.

Keywords: digital image correlation; indentation process; incremental forming; finite element method; material flow

1. Introduction

Plastic deformation is one of the main processes in the current manufacturing industry. The material is progressively shaped from a simpler geometry to a final complex shape. Therefore, the knowledge and control of the material flow is essential to achieve a correct product and improve the entire sequence. The indentation process is still one of the most commonly applied processes needed to investigate the material characteristics in general. Recent works show how the technique is still evolving and can offer essential information [1–4].

From a manufacturing perspective, the indentation process is applied to innovative approaches, as part of localized forming operations, such as incremental forming processes (IFPs). IFPs are relatively new processes that are mainly applied on sheet forming processes [5]. Complex shapes are obtained according to the predetermined trajectory that the forming tool follows, generally controlled by Computer Numerical Control (CNC) machines. Manufacturing small batches, formability limits, forming process, shape accuracy, springback problems, and so on can be analyzed and solved. However, the principal advantage is that a specific tooling is not needed, thus giving the process great flexibility as well as great formability compared with other forming processes. These processes are presented as interesting alternatives in important industries such as the medical product manufacturing, aeronautics, and transport industries, among other applications [6–10].

From the bulk forming perspective, incremental forming is an old and well-known technique within the metal working field. Nowadays, these kind of processes are employed to obtain a wide range of products and are continuously evolving and adapting with the newest technologies [11]. Incremental sheet-bulk metal forming technology can be found in the automotive industry for the manufacturing of geared components. An indenter with the final gear form is pressed to the semi-finished products. This process allows avoiding the conventional hardening and heat treatment [12–14].

In order to understand the process and establish the final workpiece shape, a deep analysis of the stresses and strains is essential. Currently, different numerical (analytical methodologies, finite element method (FEM), and so on) and experimental techniques such as digital image correlation (DIC) can be applied.

Analytical approaches, like the upper bound technique or the slip line theory, have been proven to provide good results for a wide variety of processes. Different works can be found where the material flow field and the forces needed to achieve certain deformation are obtained. Through the comparison with other techniques like FEM or experimental tests, it has been proven that the analytical methods can be accurate [15–18]. However, the application is complex, owing to the large number of variables involved. Therefore, the accuracy of the results strongly depends on the analysis optimization. Nevertheless, the results obtained can be considered to be a good approximation for facilitating the decision-making. When a great number of variables must be studied and controlled, the analytical methods can be excessively complex and, therefore, resources like FEM, DIC, and so on are well considered.

FEM and other general numerical simulation methods reduce the number of required experiments, as well as replace the more expensive and complex experiments, thus being a great tool to optimize the final shape of the workpiece [19]. An extensive range of materials, processes, and variables can be simulated. However, the results obtained always rely on the mathematical model behind this technology, and it is primordially to dispose the adequate material law, being difficult to obtain under extreme conditions. Also, remeshing techniques optimization is needed to avoid element distortion, inducing numerical errors and computational time increment [20,21]. Experimental analysis to obtain new material models and validations is commonly required. So, the FEM approach can be an expensive resource owing to the time consumption and experimental support in those case studies.

Focusing on the material flow analysis and material behavior, DIC is growing as a good alternative, showing certain advantages over other methods that make it more suitable. One of the main advantages is the non-contact optical application, which reduces the specimens' interaction, providing high resolution results in wide measurement ranges [22,23]. Measuring displacements and strains in real time is also possible. After the image capture of the sample deformation process, a mathematical correlation analysis is carried out, displaying bulk deformation by comparing deformed and un-deformed digital images, determining the displacements field and mapping their distribution [24–27]. Another important advantage is that this technique can be applied to 2D or 3D case studies without increasing much the difficulty of the analysis.

The 2D DIC technology offers higher accuracy and reduced computational complexity. Various studies show the application breadth of it. For instance, S. Roux et al. applies the DIC technique to detect cracks and estimate the stress intensity factor on SiC samples [28], as several works show how suitable this technique is when applied to fracture case studies [29–32]. On the study of material characterization, mechanical properties, such as the Young's modulus, Poisson's ratio, anisotropic plastic ratio parameters, and flow curves, are easily obtained with 2D DIC [33]. In general, DIC is a technology widely applied today, presenting a fair comparison with the results obtained through classical measurement methods [34–37].

For more complex processes and to avoid limiting the study to planar specimens and no-out-of-plane motion, a 3D analysis can be achieved. The number of cameras has to be increased and positioned to take the images from different angles. The same analysis methodology is applied and the computational time does not increase as much as in other analysis techniques, such as 3D FEM [38,39]. The methodology is not only applicable to 3D objects, but also to objects in motion [24,40] and objects subjected to large deformations [41,42].

Analyzed samples must present an irregular pattern on the surface in order to be studied by DIC technology. If the workpiece does not present an irregular pattern on the surface on its own, this must be created. This pattern consists of a random mottling or speckle that allows recognizing the position of every single point before and after deformation (Figure 1). To be able to analyze the images obtained through the deformation process, it is necessary to select a subset. The subset is recognized as a portion of the pattern selected for tracking, being placed in the reference image from which the displacements are calculated. After the images have been captured during the deformation process, a successive comparison is made between them. The subset in the deformed image is matched to the subset from the reference image to evaluate the selected subset displacement. Correlation algorithms (Equation (1)) [23,43–45] work by locating the subset from the reference image within the new image. The pixels of the reference subset are associated with a number depending on the grey level. These values depend on the number of bits of the CCD camera. Figure 2 shows an example where a value of 100 is assigned for white and 0 for black. The algorithm applied use a grey value interpolation, being able to choose between 4, 6, or 8 tap splines in this case. For a greater accuracy, the highest order spline must be selected.

$$C(x,y,u,v) = \sum_{i,j=\frac{-n}{2}}^{\frac{n}{2}} (I(x+i,y+j) - \Gamma(x+u+i,y+v+j))^2, \qquad (1)$$

where

$C(x,y,u,v)$: correlation function.
x, y: pixel coordinates values in the reference image.
u, v: displacement in x and y coordinates.
$n/2$: should be taken as an integer.
$I(x+i, y+j)$: value associated to pixel in position $(x+i, y+j)$ (reference image).
$\Gamma(x+u+i, y+v+j)$: value associated to pixel in position $(x+u+i, y+v+j)$ (deformed image).
$(x+i)$, $(y+j)$: pixel values of image before deformation.
$(x+u+i)$, $(y+v+j)$: pixel values of image after deformation.

Figure 1. Example of random subset painted in sample.

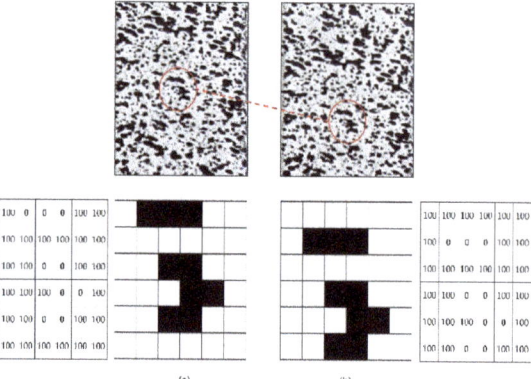

Figure 2. Subset deformation (**a**) before and (**b**) after deformation.

The lowest C value gives the best correlation possible, giving the pixel new position (x, y) after deformation as well as the horizontal and vertical displacements (u, v) [45,46].

The coordinates of a Q point around the subset centred P in the reference subset can be mapped to Q' point in the target subset using displacement mapping first order shape functions [22,47]. Thus, the first order shape function that allows translation, rotation, shear, normal strains, and their combinations of the subset Equation (2) and Equation (3):

$$x'_i = x_i + u_x \Delta x + u_y \Delta y, \qquad (2)$$

$$y'_i = y_i + v_x \Delta x + v_y \Delta y, \qquad (3)$$

where

x'_i, y'_i is the mapped position of Q point.
x_i, y_i is the position of Q point in the reference image.
u_x, u_y, v_x, v_y are the first-order displacement gradients of the reference subset.
Δx is the x distance between Q and P points in the reference subset.
Δy is the y distance between Q and P points in the reference subset.

Between two consecutives frames, the hypothesis of small strains is applicable. So, the Cauchy–Almansi tensor can be applied to obtain the strain field (Equation (4)):

$$\varepsilon = \begin{pmatrix} \varepsilon_{xx} & \varepsilon_{yx} \\ \varepsilon_{xy} & \varepsilon_{yy} \end{pmatrix} = \begin{pmatrix} u_x & \frac{1}{2}(u_y + v_x) \\ \frac{1}{2}(u_y + v_x) & v_y \end{pmatrix}, \qquad (4)$$

where

ε_{xx} and ε_{yy} are the longitudinal strains in x and y directions, respectively.
ε_{xy} is the angular strain.

Previous study focused on the 2D DIC technology for the study of the material flow in an indentation process, comparing the displacement load curve and the strain maps obtained from the material surface with FEM [23]. Even though the tests specimens meet the necessary characteristics to be mostly assimilated to plane strain condition, part of the material was projected outwards from the plane (Z axis), and so it was not possible to analyze part of the studied area, requiring a 3D approach. This work presents a comparative study with 2D and 3D DIC technology, showing the methodology implemented and the results achieved in each case study. A 2D–3D mix analysis is also proposed for studying cases like the one addressed, where a protuberance is projected on the vertical surface of the workpiece during the punch penetration in the indentation process, corresponding to the dead material added to the punch. Being an out-of-plane situation, the 2D analysis does not reach the analysis in that particular area of the workpiece. Working with 3D DIC, the Z axis towards the exterior of the tested specimen can be taken into account, carrying out a complete and more accurate study of the deformation during indentation if necessary.

2. Materials and Methods

The present work considers different approaches to address the case study presented. Experimental tests analyzed by the 2D and 3D DIC technique and FEM. Two different experimental tests were carried out. Compression tests to obtain the material behavior for the FEM model and the indentation tests for the flow analysis. Once the material was characterized, the indentation tests applying the 2D and 3D DIC analysis were performed, adapting the methodology for each case study. The results obtained from both methodologies (FEM and DIC) were compared.

In the following sections, the equipment used is described, as well as the material, test specimens, and the performance of the tests.

2.1. Materials and Specimens

To achieve a deep indentation, the specimens were manufactured with 99% tin (NB1101003). Tin bars were melted in order to obtain sand casted bulk ingots from which the specimens (40 × 30 × 30 mm) are obtained (Figure 3a) machined from the previous ingots (Figure 3b,c). Knowing that the tests are performed under plane strain conditions and the indenter is 4 mm wide, each specimen depth must be between 6 and 10 times the width. So, most of the specimen behaves as in plane strain conditions, except for the vicinity of the free surfaces, where a plane stress behaviour can be considerate. A steel AISI 304 punch was used. Also, a restraining tool was needed to avoid punch inclination during the indentation process. Figure 3d shows the restraining tool designed. It provides embedding to the upper area of the punch. A lateral compression force, provided by two fixing screws, stabilizes the punch. The fixing screws push a bar that homogeneously distributes the force over the whole embedded punch surface, making sure that it does not affect the indentation results.

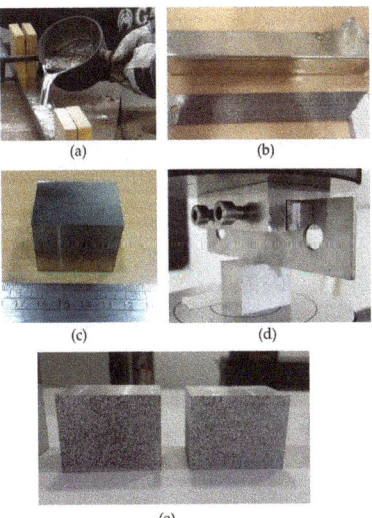

Figure 3. (**a**) Casting process, (**b**) tin ingot, (**c**) machined sample, (**d**) fixing tool for the stabilization of the indentation process, and (**e**) mottling pattern.

Owing to the lack of a natural pattern on the surface of the samples (Figure 3) and being essential to avoid specular reflections that can saturate camera sensors, an artificial pattern was conferred to each specimen. The samples were painted with spray paint in order to generate a random mottling. A white cover was sprayed, making sure the light would not reflect in the metal surface and saturates the image. After the white coat dried, a black mottling could be also sprayed from a greater distance to avoid thick droplets (Figure 3e). This pattern allows points recognition before and after the deformation process. The speckles should be neither too small nor too large. If the pattern is too large, it will be necessary to increase the subset size, but at the cost of spatial resolution. If the pattern is too small, the image will be very sensitive to defocus and also the resolution of the camera may not be enough accurately and an aliasing problem will appear. In this work, the size of the speckles was selected in order to have at least five speckles in the subset [48].

2.2. Equipment

Although the equipment used for the image capture is similar for 2D and 3D DIC methodology, the after treatment differs. The main differences are in the calibration process before the images capture and the images analysis after the images capture.

For the 2D analysis, an Allied digital camera Stingray F-504 (Allied Vision, Stadtroda, Germany) of 5 megapixels was used, with a cell size of 3.45 µm × 3.45 µm. The camera was equipped with a Pentax C7528-M lens (Ricoh, Barcenola, España). This lens is specially designed for image processing applications. It is purposely designed to maximize the picture performance at short distances with a 75 mm focal length. The field of view (fov) for a distance of 0.6 m was 62.65 mm with a pixel size of 30 µm. A 2 Hz frame acquisition frequency was established for the images capture, treating them with the software VIC SNAP [49] and VIC 2D [50] after the test were conducted.

For the 3D analysis, a binocular stereovision was needed. So, two cameras Grasshopper3 GS3-U3-123S6M-C (Flir, Wilsonville, OR, USA) of 12.3 megapixels were used, with a cell size of 3.45 µm × 3.45 µm. The cameras were equipped with a Fujinon HF50HA-1B lens (Fujifilm, Japan) with a 50 mm focal length. The field of view (fov) for a distance of 0.6 m was 80 mm with a pixel size of 30 µm. The cameras were placed around the specimens, providing enough information for the 3D study. For the test area illumination, a Hedler spotlight DX 15 (metal 150 W Halide lamp) (Hendler Systemlicht, Runkel, Germany) was used in both 2D and 3D analysis. Figure 4 shows the sets for both analyses.

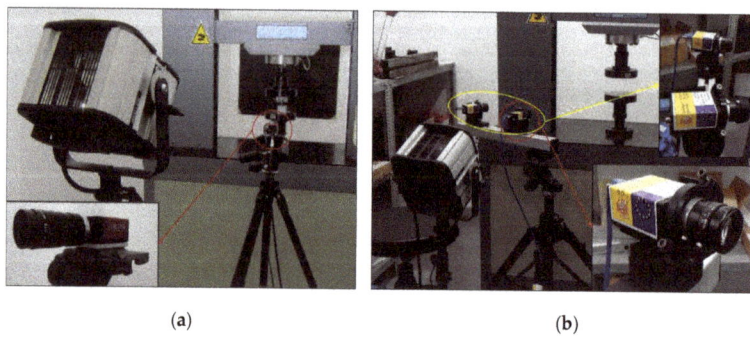

(a) (b)

Figure 4. Test set for the (**a**) 2D and (**b**) 3D digital image correlation (DIC) analysis.

The calibration procedure for 2D and 3D DIC differs from each other. During the 2D image capture, it is only necessary to capture an image with the sample and a reference element near it, as Figure 5a shows. Once the indentation process is completed and all the images are captured and stored, the calibration process can take place with VIC 2D. The first image is the one taken where the specimen appears near a ruler in mm as the reference element (Figure 5a). This image is analysed by the software. In this case, having the ruler near the specimen, a line of a certain length is drawn, scaling the length of the line with the ruler. Having that reference, the software can calibrate itself and process the rest of the images.

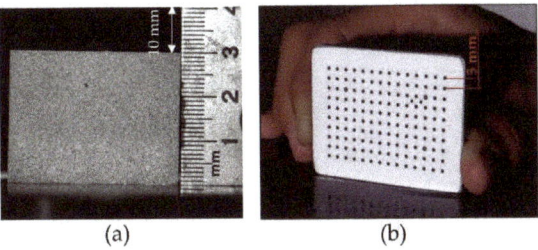

Figure 5. (**a**) Calibration for 2D and (**b**) 3D with the 3D pattern.

The 3D calibration requires more stages because it needs to calibrate the intersection of two optical rays formulated in a common coordinate system, being a stereo-triangulation process. A calibration target from Correlated Solutions Inc. was selected in order to cover the fov, with a spaced hole array every 5 mm (Figure 5b). This calibration target is placed where the specimens are going to proceed the experimental tests, undergoing through arbitrary motions along the three axis, capturing at least five images per axis, rotating the pattern left/right and up/down, and taking four or five more aleatory position images. The calibration system has a recognition software that determines the correspondences between the target points from the images captured, previously knowing the shape and scale of the target used, as seen in Figure 5b. The recognition program determines the situation of the cameras by the correspondence between the images captured with both cameras. After the calibration process, the specimens need to be place in the same position as the special target was situated, so it is essential to maintain the cameras position, focus, zoom, and illumination, marking the test area for a good positioning between tests.

A universal tension-compression machine Servosis ME 405 (Servosis Teaching Machines, Madrid, Spain), equipped with a 5 kN load cell, was used for the indentation tests. To improve the image capturing and displacement precision, the tests were set at 1 mm/min speed, synchronizing test start and ending with the image capture. The tension-compression machine allows obtaining the load forces and displacements of the tool. With previous synchronization of the digital image acquisition, it is possible to know the load-time and displacement-time evolution for each test performed.

For the image analysis, it is necessary to identify the window or area of interest in which the analysis will be performed. Figure 6 shows this window as a red area. After a subset is selected, it is necessary to find the new position of this subset in the next image. It must move the subset around the area of interest by means of a selected distance defined as a number of pixels (step) and calculate the correlation, according to Equation (1). A low value of the step could give a higher accuracy of results, but the computational cost increase.

In previous studies [23], a parametric step and subset study was carried out to obtain its optimum values, taking into account the computation time and the precision obtained. The step and subset are established in 2 and 45, respectively, to obtain a confidence below 0.001 pixels for the match location, which offers good quality results (Figure 6a). The bigger the confidence gets, the more information that can be lost during the process. The step size controls the analyzed data density. Having the step set at 2, the software carries out the correlation at every other pixel in both the horizontal and vertical direction. A low step number leads to more accurate analysis, but it increases the resolution time, varying inversely with the square of the step size.

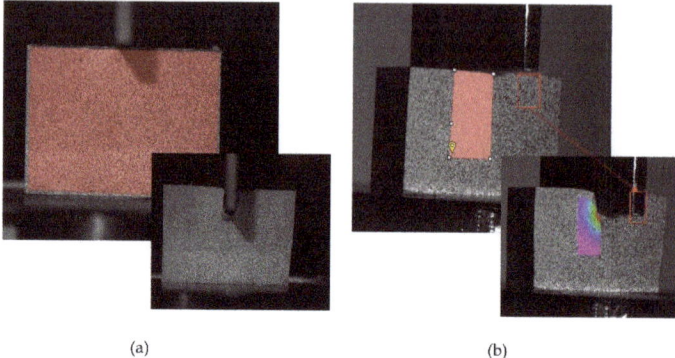

Figure 6. Subset placement at the beginning of the tests and test completed for (**a**) 2D and (**b**) 3D DIC analysis.

For the 3D analysis, the area of interest is placed only on the nearest half of the specimen to avoid interferences when the punch starts penetrating (Figure 6b). The 3D software, 3D VIC (version, company, city, country) [51], suggests a subset size depending on the image supplied, adapting the size for each sample. The step size is established in 3. Once a first correlation is launched, a von Mises analysis can be performed in order to get the displacement and tensions from each test. So, the analysis time increments increase considerably in this case compared with the 2D study.

On the basis of previous studies [52–54], the optimal mesh is established with 1000 elements, two mesh windows with a 1/10 relation, and remeshing every two steps to avoid element distortion. A plane strain conditions is considered for the 2D analysis taking into account the dimensions of the specimen and the punch. Vertical displacements are fixed at the workpiece base without friction. The elements used are two-dimensional plane strain elements of four nodes. For the punch-workpiece contact, a 0.12 shear type friction was considered [50]. For the material behaviour, a tabular data format $(\bar{\sigma} = \bar{\sigma}[\bar{\varepsilon}, \dot{\varepsilon}, T]$, where $\bar{\sigma}$ is the flow stress, $\bar{\varepsilon}$ is the effective plastic strain, $\dot{\varepsilon}$ is the effective strain rate, and T is the temperature) was selected [50] in order to introduce data obtained previously from the compression tests performed to this aim.

3. Results

Figure 7 shows the compression tests implemented to obtain the material behaviour necessary for the FEM model. Figure 7a shows the compression values obtained from the experimental tests and FEM simulation, in order to validate the material model introduced and adjust it for the indentation models.

Regarding the indentation tests, five specimens were tested for both the 2D and 3D DIC analysis, being 10 indentations tests in total. Figure 8 shows the comparison between the results mean obtained from the indentation tests, as well as the 3D and 2D FEM analysis. It can be seen that numerical and experimental results are in good agreement. For a deep indentation, starting from 5 mm, FEM forces are higher than those obtained experimentally. This can be because of a greater element deformation and a coarse mesh at that stage of the process. Nevertheless, 2D and 3D FEM models offer results according to the test performed, being a good approximation for a first analysis of the indentation process.

Figure 9 presents the main differences between the 2D and 3D FEM models. Because of the necessity of a larger mesh to cover the whole 3D model, the 2D analysis takes much less computing time. Nevertheless, with the 2D analysis, is not possible to simulate the material nose or dead zone that is developed under the punch while the indentation is being performed (Figure 10). This dead zone can be observed during the 3D simulation and the material flow can be examined. However, knowing that the main material flow occurs on the surroundings of the material nose and that strain and stress results obtained from both models are similar, the 2D model can be considered as an adequate solution for this case study.

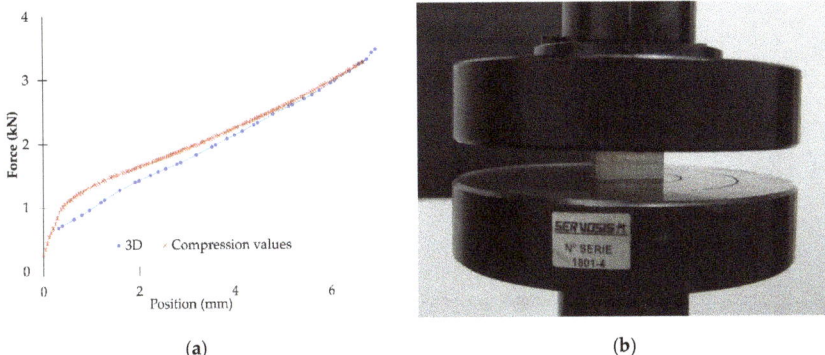

Figure 7. Results obtained from the compression tests, (**a**) experimental and 3D FEM simulation and (**b**) specimen subjected to the compression test.

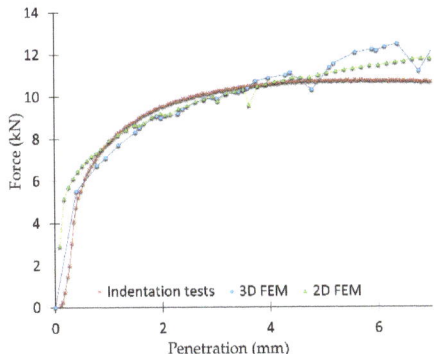

Figure 8. Comparison between results obtained from the indentation test and the 2D and 3D models developed.

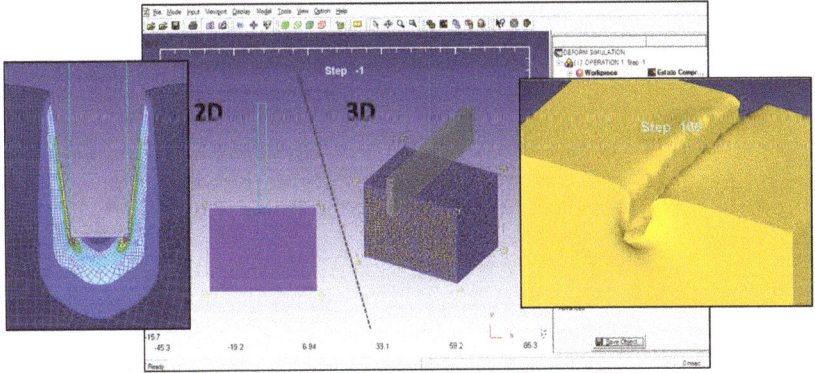

Figure 9. Comparison between results obtained from the indentation process and the 2D and 3D models developed.

Figure 10. Specimen after an indentation process and detail of the material nose developed under the punch during its penetration.

Figure 11 presents the main differences between the 2D and 3D DIC analysis performed with 2D and 3D VIC software. On the one hand, for the 2D study, the whole front surface is selected for the image analysis, being the main strains concentrated under the punch. For the 3D study, only half of the specimen can be selected. Owing to the cameras' positioning, it is not possible to have a frontal capture. As the punch progresses, it hides part of the specimen. However, this case study considers a symmetric specimen, so it is possible to reduce the analysis to half. Also, even selecting the entire half of the sample, the selection is reduced to the area where the main material flow takes place during the correlation analysis.

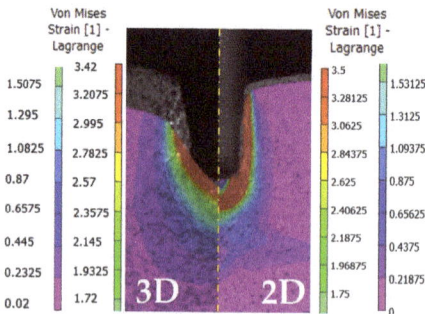

Figure 11. Results obtained with 3D and 2D DIC analysis.

On the other hand, the 3D DIC analysis takes more computational time, and achieving results with a good confidence value is a greater challenge, owing to the combination of two cameras and the random pattern. Figure 12 shows part of the obtained results for the 3D analysis, being the von Mises strain values represented in Figure 12a, with its virtual representation in Figure 12b.

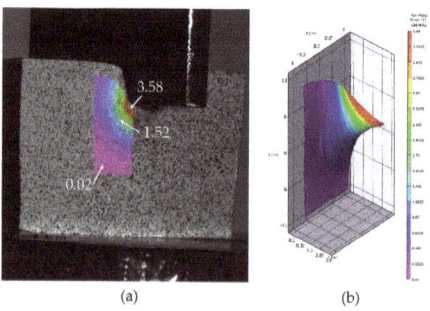

Figure 12. Von Mises strains obtained with 3D VIC, (**a**) projected on the specimen 1, (**b**) 3D representation for specimen 1.

Focusing on the von Mises strain results under the punch, the maximum strain obtained for the 2D analysis is over 3.5, while the 3D analysis shows a maximum of 3.58 over the bulge that emerges under the punch and the FEM results are over 3.1 (Figure 13). For the Z displacement (Figure 10), the whole specimen set was measured after the indentation test, presenting an average displacement of 3.07 mm. So, the error with the correlation displacement results is set on 2.67% (average).

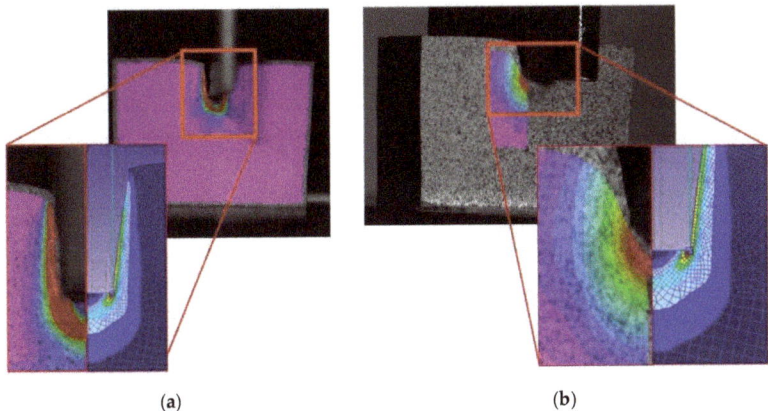

(a)   (b)

Figure 13. Von Mises comparison between (**a**) 2D DIC and 2D finite element method (FEM) analysis and (**b**) 3D DIC and 2D FEM analysis.

4. Conclusions

The aim of the present work is to compare the application and results of the 2D and 3D DIC technique analyzing the material flow that takes place in a deep indentation process. Also, a validation between DIC perspectives and FEM is made in order to establish the methodology and select the optimal analysis method, showing the advantages and disadvantages of each one.

Both DIC methods can be presented as efficient for the analysis carried out, adequately identifying the material flow field and von Mises strains on the specimens studied. The von Mises strain results obtained are in good agreement with each other and with the results obtained from the FEM analysis, 3.5 (2D), 3.58 (3D), and 3.1 (FEM). Nevertheless, depending on the analysis purpose, the 3D or 2D technique can be adjusted better or worse to the study. If the specific area of material generated under the punch needs to be examined, in order to know how much it grows towards the Z axis or the material flow that occurs just at that precise point, 2D technology is not able to provide such information. Focusing on the dead material under the punch, the 3D DIC analysis provides Z displacement of 2.99 mm versus an average displacement of 3.07 mm measured on the specimens tested and 2.90 mm provided by FEM results.

For a general analysis of the process, the more adequate method proposed is the 2D DIC analysis versus its 3D variant. The time for the set implementation is considerably reduced (about a 60% reduction) because there is no need to calibrate the camera before the image capture. During the 2D analysis, it is possible to integrate the calibration, having one of the capture images with a reference measurement element. The 3D analysis needs a proper calibration of the axis before placing the specimens and starting the indentation tests. Furthermore, the computational time for the images processing for the 2D technique versus the 3D option is also reduced. With the results obtained being similar, the 3D DIC is not the ideal solution for a general analysis in this case study.

Therefore, a mixed 2D–3D analysis is proposed, with it being possible to place both cameras so that the captured images can be used for a 2D analysis. In the case in which the study of a specific point of the material that evolves along the Z axis is needed, it is possible to implement the 3D analysis only integrating the images of the second camera with the 3D software.

Author Contributions: Conceptualization, C.B.G.; methodology, C.B.G., G.C.L., and S.M.-B.; software, C.B.G., G.C.L., and F.J.T.V.; validation, C.B.G., G.C.L., and S.M.-B.; formal analysis, C.B.G., F.J.T.V., and L.S.H.; writing—original draft preparation, C.B.G.; writing—review and editing, L.S.H., G.C.L., and F.J.T.V.; supervision, L.S.H.

Funding: This research received no external funding.

Acknowledgments: The authors want to thank the University of Malaga—Andalucía Tech, International Campus of Excellence and the state subprogram of scientific and technical infrastructures and equipment, co-financed by the Ministry of Economy and Competitiveness and with European Union Structural Funds (Ref: UNMA 15-CE-3571).

Conflicts of Interest: The authors declare no conflict of interest.

References

1. Wagih, A.; Attia, M.A.; AbdelRahman, A.A.; Bendine, K.; Sebaey, T.A. On the indentation of elastoplastic functionally graded materials. *Mech. Mater.* **2019**, *129*, 169–188. [CrossRef]
2. Huang, L.Y.; Guan, K.S.; Xu, T.; Zhang, J.M.; Wang, Q.Q. Investigation of the mechanical properties of steel using instrumented indentation test with simulated annealing particle swarm optimization. *Theor. Appl. Fract. Mech.* **2019**, *102*, 116–121. [CrossRef]
3. Wu, S.B.; Guan, K.S. Evaluation of tensile properties of austenitic stainless steel 316L with linear hardening by modified indentation method. *Mater. Sci. Technol.* **2014**, *30*, 1404–1409. [CrossRef]
4. Udalov, A.; Parshin, S.; Udalov, A. Indentation size effect during measuring the hardness of materials by pyramidal indenter. *Mater. Today Proc.* **2019**. [CrossRef]
5. Suresh, K.; Kumar, P.G.; Priyadarshini, A.; Kotkunde, N. Analysis of formability in incremental forming processes. *Mater. Today Proc.* **2018**, *5*, 18905–18910. [CrossRef]
6. Bouhamed, A.; Jrad, H.; Mars, J.; Wali, M.; Gamaoun, F.; Dammak, F. Homogenization of elasto-plastic functionally graded material based on representative volume element: Application to incremental forming process. *Int. J. Mech. Sci.* **2019**, *160*, 412–420. [CrossRef]
7. Ambrogio, G.; Palumbo, G.; Sgambitterra, E.; Guglielmi, P.; Piccininni, A.; De Napoli, L.; Villa, T.; Fragomeni, G. Experimental investigation of the mechanical performances of titanium cranial prostheses manufactured by super plastic forming and single-point incremental forming. *Int. J. Adv. Manuf. Technol.* **2018**, *98*, 1489–1503. [CrossRef]
8. Palumbo, G.; Brandizzi, M. Experimental investigations on the single point incremental forming of a titanium alloy component combining static heating with high tool rotation speed. *Mater. Des.* **2012**, *40*, 43–51. [CrossRef]
9. Fragapane, S.; Giallanza, A.; Cannizzaro, L.; Pasta, A.; Marannano, G. Experimental and numerical analysis of aluminum-aluminum bolted joints subject to an indentation process. *Int. J. Fatigue* **2015**, *80*, 332–340. [CrossRef]
10. Marannano, G.; Virzì Mariotti, G.; D'Acquisto, L.; Restivo, G.; Gianaris, N. Effect of Cold Working and Ring Indentation on Fatigue Life of Aluminum Alloy Specimens. *Exp. Tech.* **2015**, *39*, 19–27. [CrossRef]
11. Groche, P.; Fritsche, D.; Tekkaya, E.A.; Allwood, J.M.; Hirt, G.; Neugebauer, R. Incremental Bulk Metal Forming. *CIRP Ann. Manuf. Technol.* **2007**, *56*, 635–656. [CrossRef]
12. Wernicke, S.; Gies, S.; Tekkaya, A.E. Manufacturing of hybrid gears by incremental sheet-bulk metal forming. *Procedia Manuf.* **2019**, *27*, 152–157. [CrossRef]
13. Sieczkarek, P.; Wernicke, S.; Gies, S.; Tekkaya, A.E.; Krebs, E.; Wiederkehr, P.; Biermann, D.; Tillmann, W.; Stangier, D. Improvement strategies for the formfilling in incremental gear forming processes. *Prod. Eng.* **2017**, *11*, 623–631. [CrossRef]
14. Pilz, F.; Merklein, M. Comparison of extrusion processes in sheet-bulk metal forming for production of filigree functional elements. *CIRP J. Manuf. Sci. Technol.* **2019**. [CrossRef]
15. Bermudo, C.; Sevilla, L.; Martín, F.; Trujillo, F.J. Hardening effect analysis by modular upper bound and finite element methods in indentation of aluminum, steel, titanium and superalloys. *Materials* **2017**, *10*, 556. [CrossRef] [PubMed]
16. Moncada, A.; Martín, F.; Sevilla, L.; Camacho, A.M.; Sebastián, M.A. Analysis of Ring Compression Test by Upper Bound Theorem as Special Case of Non-symmetric Part. *Procedia Eng.* **2015**, *132*, 334–341. [CrossRef]
17. Wu, Y.; Dong, X.; Yu, Q. An upper bound solution of axial metal flow in cold radial forging process of rods. *Int. J. Mech. Sci.* **2014**, *85*, 120–129. [CrossRef]
18. Wu, Y.; Dong, X. An upper bound model with continuous velocity field for strain inhomogeneity analysis in radial forging process. *Int. J. Mech. Sci.* **2016**, *115*, 385–391. [CrossRef]

19. López-Chipres, E.; García-Sanchez, E.; Ortiz-Cuellar, E.; Hernandez-Rodriguez, M.A.L.; Colás, R. Optimization of the Severe Plastic Deformation Processes for the Grain Refinement of Al6060 Alloy Using 3D FEM Analysis. *J. Mater. Eng. Perform.* **2010**. [CrossRef]
20. Hofmeister, B.; Bruns, M.; Rolfes, R. Finite element model updating using deterministic optimisation: A global pattern search approach. *Eng. Struct.* **2019**, *195*, 373–381. [CrossRef]
21. Skozrit, I.; Frančeski, J.; Tonković, Z.; Surjak, M.; Krstulović-Opara, L.; Vesenjak, M.; Kodvanj, J.; Gunjević, B.; Lončarić, D. Validation of Numerical Model by Means of Digital Image Correlation and Thermography. *Procedia Eng.* **2015**, *101*, 450–458. [CrossRef]
22. Pan, B.; Qian, K.; Xie, H.; Asundi, A. Two-dimensional digital image correlation for in-plane displacement and strain measurement: A review. *Meas. Sci. Technol.* **2009**, *20*, 062001. [CrossRef]
23. Bermudo, C.; Sevilla, L.; Castillo, G. Material flow analysis in indentation by two-dimensional digital image correlation and finite elements method. *Materials* **2017**, *10*, 674. [CrossRef] [PubMed]
24. Sousa, P.J.; Barros, F.; Tavares, P.J. Displacement measurement and shape acquisition of an RC helicopter blade using Digital Image Correlation. *Procedia Struct. Integr.* **2017**, *5*, 1253–1259. [CrossRef]
25. Ho, C.-C.; Chang, Y.-J.; Hsu, J.-C.; Kuo, C.-L.; Kuo, S.-K.; Lee, G.-H. Residual Strain Measurement Using Wire EDM and DIC in Aluminum. *Inventions* **2016**, *1*, 4. [CrossRef]
26. Bruck, H.A.; McNeill, S.R.; Sutton, M.A.; Peters, W.H. Digital image correlation using Newton-Raphson method of partial differential correction. *Exp. Mech.* **1989**, *29*, 261–267. [CrossRef]
27. Park, J.; Yoon, S.; Kwon, T.-H.; Park, K. Assessment of speckle-pattern quality in digital image correlation based on gray intensity and speckle morphology. *Opt. Lasers Eng.* **2017**, *91*, 62–72. [CrossRef]
28. Roux, S.; Réthoré, J.; Hild, F. Digital image correlation and fracture: An advanced technique for estimating stress intensity factors of 2D and 3D cracks. *J. Phys. D. Appl. Phys.* **2009**, *42*, 214004. [CrossRef]
29. Lin, Q.; Wan, B.; Wang, Y.; Lu, Y.; Labuz, J.F. Unifying acoustic emission and digital imaging observations of quasi-brittle fracture. *Theor. Appl. Fract. Mech.* **2019**, *103*, 102301. [CrossRef]
30. Hao Kan, W.; Albino, C.; Dias-da-Costa, D.; Dolman, K.; Lucey, T.; Tang, X.; Cairney, J.; Proust, G. Fracture toughness testing using photogrammetry and digital image correlation. *MethodsX* **2018**, *5*, 1166–1177. [CrossRef]
31. Bertelsen, I.M.G.; Kragh, C.; Cardinaud, G.; Ottosen, L.M.; Fischer, G. Quantification of plastic shrinkage cracking in mortars using digital image correlation. *Cem. Concr. Res.* **2019**, *123*. [CrossRef]
32. Fayyad, T.M.; Lees, J.M. Application of Digital Image Correlation to reinforced concrete fracture. *Procedia Mater. Sci.* **2014**, *3*, 1585–1590. [CrossRef]
33. Nguyen, V.-T.; Kwon, S.-J.; Kwon, O.-H.; Kim, Y.-S. Mechanical Properties Identification of Sheet Metals by 2D-Digital Image Correlation Method. *Procedia Eng.* **2017**, *184*, 381–389. [CrossRef]
34. del Rey Castillo, E.; Allen, T.; Henry, R.; Griffith, M.; Ingham, J. Digital image correlation (DIC) for measurement of strains and displacements in coarse, low volume-fraction FRP composites used in civil infrastructure. *Compos. Struct.* **2019**, *212*, 43–57. [CrossRef]
35. Kumar, S.L.; Aravind, H.B.; Hossiney, N. Digital Image Correlation (DIC) for measuring strain in brick masonry specimen using Ncorr open source 2D MATLAB program. *Results Eng.* **2019**, 100061. [CrossRef]
36. Hensley, S.; Christensen, M.; Small, S.; Archer, D.; Lakes, E.; Rogge, R. Digital image correlation techniques for strain measurement in a variety of biomechanical test models. *Acta Bioeng. Biomech.* **2017**, *19*, 187–195.
37. Yang, L.; Smith, L.; Gotherkar, A.; Chen, X. *Measure Strain Distribution Using Digital Image Correlation (DIC) for Tensile Tests*; Oakland University: Oakland, Rochester, ML, USA, 2010.
38. Chen, Y.; Wei, J.; Huang, H.; Jin, W.; Yu, Q. Application of 3D-DIC to characterize the effect of aggregate size and volume on non-uniform shrinkage strain distribution in concrete. *Cem. Concr. Compos.* **2018**, *86*, 178–189. [CrossRef]
39. Aydin, M.; Wu, X.; Cetinkaya, K.; Yasar, M.; Kadi, I. Application of Digital Image Correlation technique to Erichsen Cupping Test. *Eng. Sci. Technol. Int. J.* **2018**, *21*, 760–768. [CrossRef]
40. Helfrick, M.N.; Niezrecki, C.; Avitabile, P.; Schmidt, T. 3D digital image correlation methods for full-field vibration measurement. *Mech. Syst. Signal Process.* **2011**, *25*, 917–927. [CrossRef]
41. Molina-Viedma, A.J.; Felipe-Sesé, L.; López-Alba, E.; Díaz, F. High frequency mode shapes characterisation using Digital Image Correlation and phase-based motion magnification. *Mech. Syst. Signal Process.* **2018**, *102*, 245–261. [CrossRef]
42. Guo, X.; Liang, J.; Xiao, Z.; Cao, B. Digital image correlation for large deformation applied in Ti alloy compression and tension test. *Opt. Int. J. Light Electron Opt.* **2014**, *125*, 5316–5322. [CrossRef]

43. Castillo López, G.; Carrasco Vela, G. Correlación numérico-experimental del ensayo de tracción. In *XX Congreso Nacional de Ingeniería Mecánica*; Universidad de Málaga: Málaga, Spain, 2014.
44. Correlated Solutions Inc. *Digital Image Correlation: Principles and Software*; University of South Carolina: Columbia, SC, USA, 2009.
45. Lakshmi Aparna, M.; Chaitanya, G.; Srinivas, K.; Rao, J.A. Fatigue Testing of Continuous GFRP Composites Using Digital Image Correlation (DIC) Technique a Review. *Mater. Today Proc.* **2015**, *2*, 3125–3131. [CrossRef]
46. Schreier, H.; Orteu, J.-J.; Sutton, M.A. *Image Correlation for Shape, Motion and Deformation Measurements*; Springer US: Boston, MA, USA, 2009; ISBN 978-0-387-78746-6.
47. Schreier, H.W.; Sutton, M.A. Systematic errors in digital image correlation due to undermatched subset shape functions. *Exp. Mech.* **2002**, *42*, 303–310. [CrossRef]
48. Correlated Solutions Inc. *CSI Application Note AN-525. Speckle Pattern Fundamentals*; Correlated Solutions: Irmo, SC, USA, 2009.
49. Correlated Solutions VIC SNAP. Available online: https://www.correlatedsolutions.com/vic-snap-remote/ (accessed on 10 December 2009).
50. Correlated Solutions. *VIC-2D Reference Manual*; Correlated Solutions: Irmo, SC, USA, 2009; p. 59.
51. Correlated Solutions. *VIC-3D Reference Manual*; Correlated Solutions: Irmo, SC, USA, 2018.
52. Fluhrer, J. *Deform. Design Environment for Forging. User's Manual*; Scientific Forming Technologies Corporation: Columbus, OH, USA, 2010.
53. Bermudo, C.; Sevilla, L.; Martín, F.; Trujillo, F.J. Study of the Tool Geometry Influence in Indentation for the Analysis and Validation of the New Modular Upper Bound Technique. *Appl. Sci.* **2016**, *6*, 203. [CrossRef]
54. Bermudo Gamboa, C. *Análisis, Desarrollo y Validación del Método del Límite Superior en Procesos de Conformado por Indentación*; Servicio de Publicaciones y Divulgación Científica: Málaga, Spain, 2015.

© 2019 by the authors. Licensee MDPI, Basel, Switzerland. This article is an open access article distributed under the terms and conditions of the Creative Commons Attribution (CC BY) license (http://creativecommons.org/licenses/by/4.0/).

Article

Parametric Analysis of the Mandrel Geometrical Data in a Cold Expansion Process of Small Holes Drilled in Thick Plates

Jose Calaf-Chica [1,*], Marta María Marín [2], Eva María Rubio [2], Roberto Teti [3] and Tiziana Segreto [3]

1. Department of Civil Engineering, University of Burgos, Av. Cantabria s/n, 09007 Burgos, Spain
2. Department of Manufacturing Engineering, Universidad Nacional de Educación a Distancia (UNED), St. Juan del Rosal 12, E28040 Madrid, Spain; mmarin@ind.uned.es (M.M.M.); erubio@ind.uned.es (E.M.R.)
3. Department of Chemical, Materials and Industrial Production Engineering, University of Naples Federico II, Piazzale Tecchio, 80, 80125 Naples, Italy; roberto.teti@unina.it (R.T.); tsegreto@unina.it (T.S.)
* Correspondence: jcalaf@ubu.es; Tel.: +34-660-040-883

Received: 31 October 2019; Accepted: 3 December 2019; Published: 8 December 2019

Abstract: Cold expansion technology is a cold-forming process widely used in aeronautics to extend the fatigue life of riveted and bolted holes. During this process, an oversized mandrel is pushed through the hole in order to yield it and generate compressive residual stresses contributing to the fatigue life extension of the hole. In this paper, a parametric analysis of the mandrel geometrical data (inlet angle straight zone length and diametric interference) and their influence on the residual stresses was carried out using a finite element method (FEM). The obtained results were compared with the conclusions presented in a previous parametric FEM analysis on the influence of the swage geometry in a swaging cold-forming process of gun barrels. This process could be considered, in a simplified way, as a scale-up of the cold expansion process of small holes, and this investigation demonstrated the influence of the diameter ratio (K) on the relation between the mandrel or swage geometry and the residual stresses obtained after the cold-forming process.

Keywords: cold expansion; mandrel; cold-forming; swaging

1. Introduction

The autofrettage cold-forming process for thick-walled cylinders has its origin in artillery technology of the 19th century. Gun barrels showed a significant development in range, power and reliability because of the application of an innovative manufacturing process: the generation of internal compressions at the bore with a hoop installed on the barrel with enough interference. In the 1920s, a new cold-forming process was developed to obtain similar residual compressive stresses using a hydraulic pressure into the barrel. It was called hydraulic "autofrettage" using a French term which means "self-hooping" due to the absence of hoops to obtain the residual stresses [1]. Development of new steel alloys during the first half of the 20th century with higher yield and ultimate tensile strengths generated an increment of the necessary autofrettage pressure to reach the requirements of residual stresses. The inherent process difficulties to reach these pressure levels motivated the development of alternative processes (explosive autofrettage [2,3], double-layer autofrettage [4], thermal autofrettage [5], rotational autofrettage [6], etc.). In this context, around the middle of the last century, an innovative cold-forming process in gun barrels was presented [7]: the mechanical autofrettage or swaging. In this process, a swage with a diametrical interference with the barrel is forced to pass through bore, generating compressive residual stresses. The necessary pressure to move the swage was significantly lower than the equivalent hydraulic pressure in a hydraulic autofrettage.

The cold-forming process of swages showed alternative applications in aeronautics in the 1970s to extend the fatigue life of fastened parts on aircrafts. In 1974, Boeing's materials research and development developed a swaging process called "coldworking" which was applied to the fastened holes to extend their fatigue life [8,9]. An oversized swage or mandrel was pushed through the hole to yield the inner diameter of the hole. The remaining elastic zone after the swaging process generated a residual compressive hoop stress which was the origin of the delay in the crack initiation. Although there are many investigations related to the effect of the cold expansion on holes [10,11] and failure mode analyses of the riveted assemblies [12], there are less studies centred on the geometrical parameters of the cold expansion process. An optimization of the mandrel geometry would derive on an increase of the compressive residual stresses obtained after the cold forming process and an increase in the fatigue life of the fastened assemblies. The cost of maintenance in the aeronautical industry is directly related to the frequency of the periodic inspections. This frequency is calculated with analytical and numerical analyses of crack propagation (fatigue and damage tolerance analyses). Most of the crack initiation locations in aeronautics are considered in fastened holes, and an aircraft is plenty of holes. The cold expansion process is a technique to expand the fatigue life of holes, retarding the initiation of a crack and slowing down the crack propagation. Therefore, this extension in fatigue life would reduce the necessary frequency of some periodic inspections, reducing the cost of maintenance, and expanding the economic life of the airplanes.

The main difference between both applications of this cold-forming process, swaging of thick-walled cylinders and cold expansion of fastened holes in aircrafts, would seem to be only in the size of the hole: big holes in thick-walled cylinders and small holes in aeronautics. Swage geometry and how it affects to the residual stresses could be similar in both applications, because it is a scale-up of a similar cold-forming process.

Figure 1 shows a schematic representation of the most important parts in a cold-forming process of swaging:

1. The swage or mandrel. A part manufactured with tungsten carbide alloy, high-strength steel or a combination of different materials, and it is pushed through the hole during the cold expansion process.
2. The cylinder or hole. The part that is cold-formed. It could be a cylinder or a hole.

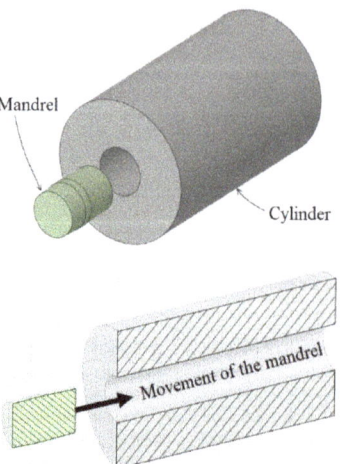

Figure 1. Schematic representation of the swaging process.

Figure 2 shows the application of the swaging process in a hole drilled in a thick plate representing a sketch of the application of the cold expansion process for fastened holes in aeronautics. When the ratio between the hole diameter and the thickness of the plate is high enough (thick plates and plane strain condition), the main difference between thick-walled cylinders and holes in plates is in the K ratio or diameter ratio of the cylinder ($K = D_{ext}/D_{int}$). In the thick-walled cylinders, K is limited to small values, and it takes very high values for holes in plates.

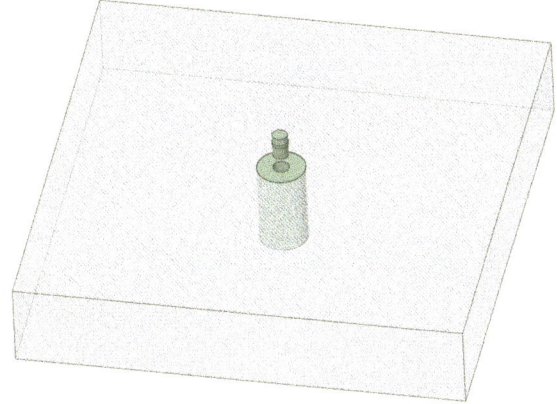

Figure 2. Sketch of a cold expansion process in a small hole.

In both cases (Figures 1 and 2), the residual stress field obtained after the cold-forming swaging process is typically represented in a polar coordinate system. The principal stresses (σ_1 = hoop, σ_2 = longitudinal, and σ_3 = radial) are shown in Figure 3. In Figure 4 an example of the residual stress field is reported along the normalized radial distance $r = (D - D_{int})/(D_{ext} - D_{int})$ at the middle longitudinal distance of a hole. After the cold expansion forming process, a compressive residual hoop stress (Figure 4a) is generated in the inner radius of the hole preventing and retarding any crack initiation.

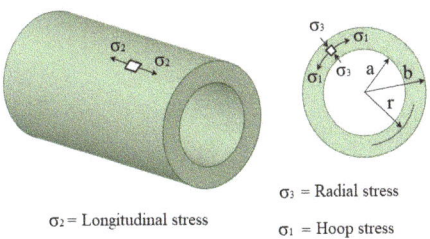

Figure 3. Sketch of stress field nomenclature.

The main geometrical parameters of the mandrel are: the inlet angle (α), the outlet angle (β), the length of the straight zone (SZ) and the major diameter (d_s) (see Figure 5). The swage major diameter is important just related to the hole diameter and the interference between them (int).

The most important reference for the investigation of the influence of the swage geometrical data for thick-walled cylinders was the research published by Gibson et al. [13,14]. A systematic finite element method (FEM) analysis of a swaging process in a cylinder with $K = 2.257$ was performed to obtain the residual stress field and quantify the influence of the geometrical parameters of the swage in the stress field. Gibson achieved that an increment of the inlet and outlet angles and a reduction of the length of the straight zone generated an increase of the residual hoop stresses at the bore of the hole. Moreover, these increments of the inlet and outlet angles and the reduction of the length of the straight zone generated a reduction of the yielded area after the cold expansion process.

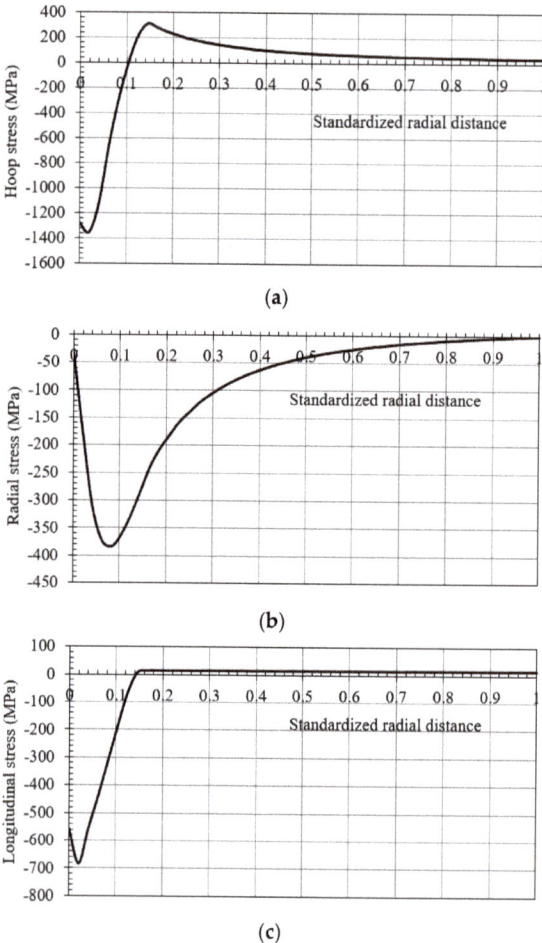

Figure 4. Residual (**a**) hoop, (**b**) radial and (**c**) longitudinal stresses after a cold expansion forming process.

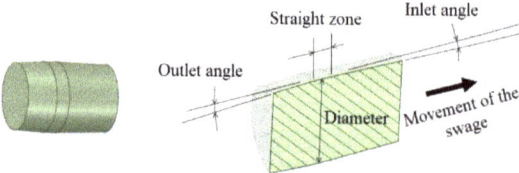

Figure 5. Sketch of a mandrel or swage with the geometrical parameters.

In the case of the cold expansion of riveted holes in aeronautics, there are less research related to the influence of the mandrel geometrical parameters on the residual stresses. There are investigations using steel balls as mandrels [15] or swages, as previously shown in Figure 5 [16], but there is no systematic FEM analysis such as the Gibson's study applied to small holes.

Starting from Gibson's study on a parametric analysis of the mandrel geometry and its influence on in the residual stresses [13,14], this paper is focused on the specific cold-forming application for small holes drilled in thick plates. Similarities and differences in the cold-forming process applied to small holes and thick-walled cylinders were performed to verify or refute the applicability of the conclusions published by Gibson et al. for both scenarios.

2. Materials and Methods

This investigation was centred on a parametric numerical analysis through a finite element method (FEM) of the mandrel geometrical data during a cold expansion process for a small hole. An ANSYS v19 software was used to simulate the cold-forming process. The evaluated mandrel geometrical parameters were: the inlet angle, the length of the straight zone and the interference between the swage and the hole. The outlet angle was fixed at $\beta = 3°$. A full factorial DOE (Design of Experiments) was designed to analyse every possible combination of the geometrical parameters (inlet angle, length of the straight zone and diametric interference).

Table 1 summarized the geometrical data included in the performed FEM analysis: four levels for the inlet angle (α), three levels for the straight zone length (SZ), and three levels for the diametric interference (int). A total number of 36 simulations for the cold expansion forming process were done (full factorial DOE). The selected ranges for each geometrical parameter were based on the optimum values deduced in previous investigations by Gibson [13] and O'Hara [17].

Table 1. Mandrel geometry analysed in the parametric FEM simulations.

Geometrical Parameters	Value
Inlet angle (α)	0.75, 1.00, 1.50, 2.00°
Length of the straight zone (SZ)	0.5, 1.0, 1.5 mm
Diametric interference (int)	0.10, 0.15, 0.20 mm

For the plate geometry, a high ratio $L = t/D_{hole} = 22$ was selected to consider a thick plate (where $t = 75$ mm was the plate thickness and $D_{hole} = 3.4$ mm was the hole diameter; see Figure 6). As mentioned previously, the aim of the L ratio was to avoid the effect of the hole ends in order to evaluate and compare the results with the thick-walled cylinders.

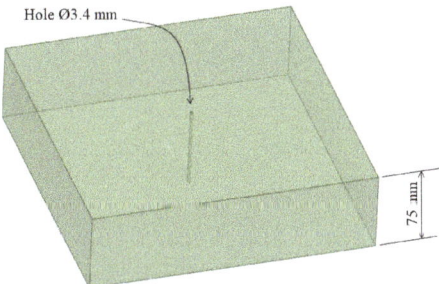

Figure 6. Sketch of the drilled plate.

In the specific case of the small holes, mandrels are usually manufactured with high strength steels, and considering the advances of powder metallurgy for tungsten carbide parts development, in this paper, a CTS12L alloy manufactured by Ceratizit [18] was selected for the high-strength and high-stiffness values (ultimate tensile strength of $\sigma u = 3500$ MPa and Young's modulus of $E = 630$ GPa).

Moreover, the material selected for the plate was the precipitation hardening stainless steel 15-5 PH H1025. Whose mechanical properties were listed in Table 2 as specified in the AMS5659 standard [19] As a note, it is important to point out that the alteration of plate material with lower plastic mechanical properties would derive in a reduction of the maximum residual stresses that cold expansion process could obtain, but the influence of each geometrical parameter evaluated in this investigation would have similar tendencies in the residual stresses.

Table 2. Mechanical properties of stainless steel 15-5PH H1025 (AMS5659).

Young's modulus (GPa)	190
Yield strength (MPa)	1000
Ultimate tensile strength (MPa)	1069
Elongation at fracture (%)	12

The stress-strain curve of the considered steel was approximated with the Ramberg–Osgood hardening law [20] (see Equation (1)). Equation (2) deduced by Kamaya et al. [21] was used to calculate the hardening coefficient n, obtaining $n = 30.3$. The plastic hardening was simulated with a kinematic hardening model to introduce the Bauschinger effect in the cold forming process.

$$\varepsilon = \sigma/E + \varepsilon_{offset}\,(\sigma/\sigma_y)^n \tag{1}$$

$$n = 3.93\{ln(\sigma_{u\_eng}/\sigma_y)\}^{-0.754} \tag{2}$$

where:

E: Young's modulus,

ε: true strain,

ε_{offset}: offset strain (0.002) to obtain the yield strength,

σ: true stress,

n: hardening coefficient,

σ_{u_eng}: engineering ultimate tensile strength, and

σ_y: yield strength.

To reduce the computational requirement of the numerical analyses, an axisymmetric model was considered. In the case of a thick plate, the outer diameter of the part needs to be defined (see Figure 7). The Lamé's equation for the hoop stress on the inner diameter of a thick-walled cylinder with an inner pressure of 1000 MPa (see Equation (3)) was used to fix a specific value for the outer diameter.

$$\sigma_{hoop} = (D_{ext}^2 + D_{int}^2)/(D_{ext}^2 - D_{int}^2) \cdot P_{int} \tag{3}$$

where:

D_{ext}: outer diameter,

D_{int}: inner diameter,

P_{int}: internal pressure, and

σ_{hoop}: hoop stress.

Fixing the inner diameter at $D_{int} = 3.4$ mm and the internal pressure at $P_{int} = 1000$ MPa, the influence of the outer diameter (D_{ext}) on the hoop stress at the hole's bore tended to a stabilized value equal to the internal pressure (see Figure 8). Therefore, an outer diameter of 30 mm could be considered equivalent to a distant hole from any discontinuity (other holes or plate ends). Figure 9 shows the considered geometry for the axisymmetric model of the plate.

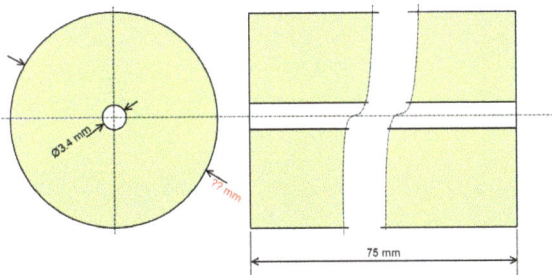

Figure 7. Geometry of the thick plate as an axisymmetric model.

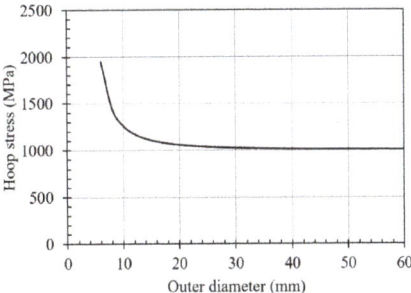

Figure 8. Hoop stress (σ_{hoop}) in the inner diameter vs. the outer diameter.

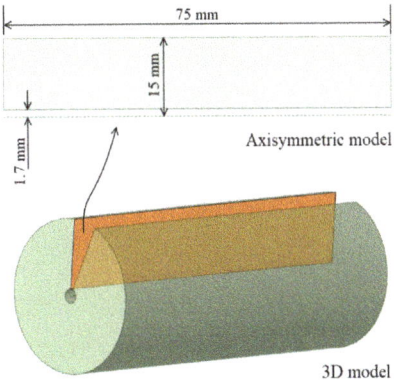

Figure 9. Geometry of the axisymmetric model.

The FEM simulations were identified with the ID AeExxx-Ixxx-Lrxx, where: "Ae" represents the inlet angle of the mandrel followed by a numerical value in tenths of a degree; "I", the diametric interference in hundredths of a millimeter; and "Lr", the length of the straight zone in tenths of a millimeter. For example, the ID Ae200-I020-Lr15 matches with the mandrel of: inlet angle equal to $\alpha = 2°$, diametric interference of $int = 0.2$ mm and length of the straight zone of $SZ = 1.5$ mm.

For the contact between the mandrel and the hole, a friction coefficient $\mu = 0.05$ (typical for a lubricated contact) was fixed for the simulations. A sensitivity mesh analysis was performed with fine mesh sizes of $ms = 0.4, 0.3$ and 0.2 mm established in the edge of the inner diameter of the plate and the outer face of the mandrel with Quad8 and Tria6 elements (see Figure 10). Figure 11 shows the residual hoop stresses obtained after the cold expansion process for the different ms values. These results showed no significant alterations and the mesh size of 0.2 mm was established as the standard for this

FEM analysis, showing an orthogonal quality of $OQ = \{0.602, 1.000\}$ and a skewness of $S = \{1.114 \times 10^{-3}, 0.591\}$. These mesh metrics are enough good to guarantee no convergence problems related with mesh quality. The boundary conditions of the model were a restricted longitudinal movement of the inlet edge of the plate (B location of Figure 12), and an imposed longitudinal movement of the outlet edge of the mandrel equal to 86 mm (A location of Figure 12).

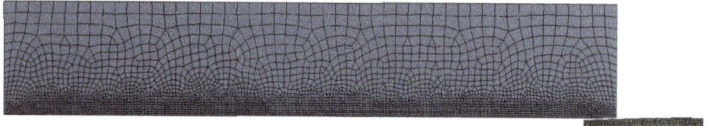

Figure 10. Mesh of the cold expansion simulation ($ms = 0.2$ mm).

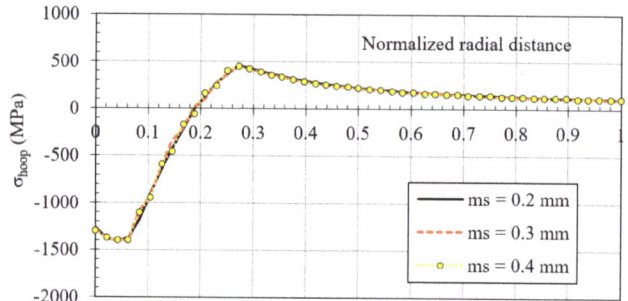

Figure 11. Sensitivity mesh analysis.

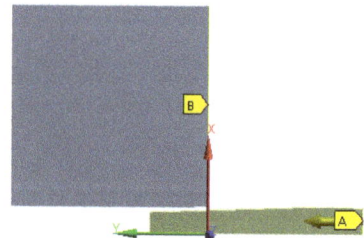

Figure 12. Boundary conditions of the FEM model.

This FEM model was verified with a comparison with the Gibson FEM results [13], just changing the geometrical parameters and material properties of the previously shown model adapting it to the Gibson cylinder geometry. Table 3 shows the corresponding geometrical data and the mechanical properties and Figure 13 shows the FEM model used for this comparison. Figure 14 represents the comparison of the residual hoop stresses (normalized by the yield strength) versus the normalized radial position (from 0: inner radius to 1: outer radius). Similar results ensured the feasibility of the FEM model designed for this investigation.

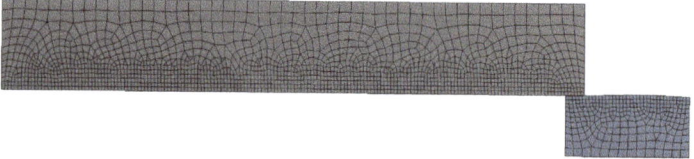

Figure 13. FEM model for the comparison with the Gibson FEM model.

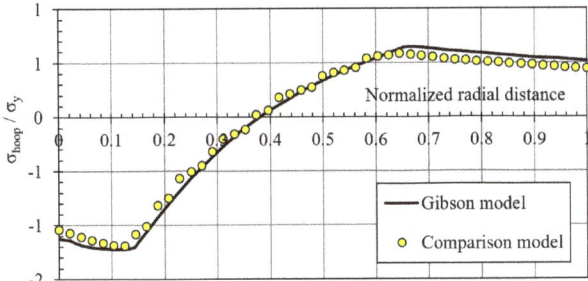

Figure 14. Comparison between residual hoop stresses obtained with the FEM model and the Gibson FEM model.

Table 3. Geometry and mechanical properties of the Gibson FEM model [13].

Cylinder	
Young's modulus (GPa)	209
Poisson's ratio	0.3
Yield strength (MPa)	1195
Plastic modulus (GPa)	3.723
Mandrel	
Young's modulus (GPa)	500
Poisson's ratio	0.24
Geometry	
Cylinder inner radius (mm)	52.5
Cylinder outer radius (mm)	131.25
Cylinder length (mm)	525
Mandrel outer radius (mm)	53.81
Inlet angle (°)	1.5
Outlet angle (°)	3.0
Length of the straight zone(mm)	6.3

3. Results

The results obtained from the FEM simulations of the mandrel geometrical data during the cold expansion process were reported in Figures 15–19. Figure 15 shows the Von Mises stress during the cold expansion process simulation, where the mandrel has been pushed up to the half of the plate thickness. This figure represents the simulation ID Ae200-I020-Lr15 (inlet angle: 2.0°; diametric interference: 0.2 mm; and straight zone length: 1.5 mm). The most critical Von Mises stress is in the mandrel with values close to 3000 MPa. Therefore, this part needs to have a very high yield strength to guarantee no plastic behaviour of the mandrel during the process.

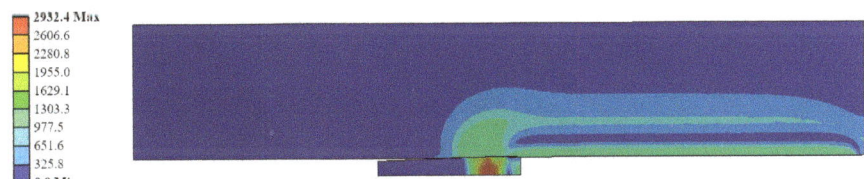

Figure 15. Von Mises stress (σ_{VM}) during the cold expansion process.

Radial stress field is represented for the same ID Ae200-I020-Lr15 in Figure 16. The most stressed location matched with the contact between the mandrel and the hole in the straight zone of the mandrel. The contact pressure reached up to −3600 MPa. This is feasible due to the high hydrostatic component in the stress field: the hoop stress reached −3600 MPa (see Figure 17) and the longitudinal stress showed values of −2600 MPa (see Figure 18).

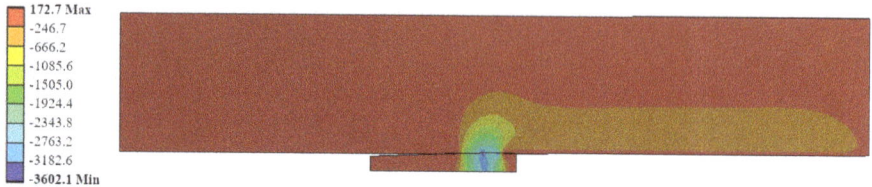

Figure 16. Radial stress (σ_{rad}) during the cold expansion process.

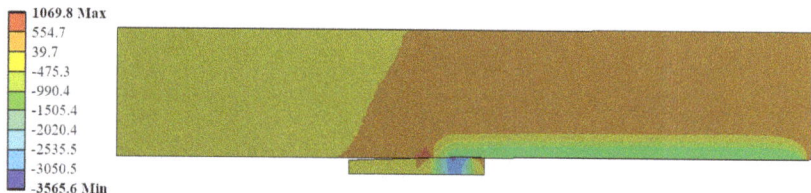

Figure 17. Hoop stress (σ_{hoop}) during the cold expansion process.

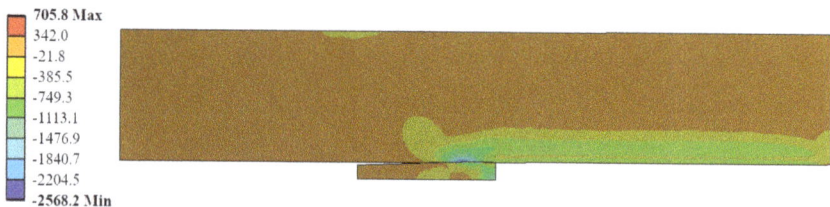

Figure 18. Longitudinal stress (σ_{long}) during the cold expansion process.

Figure 19 shows the triaxiality for the same ID Ae200-I020-Lr15. Focusing the analysis in the region where the most critical stresses shown in the previous figures (i.e., the contact region between the mandrel and the hole), the triaxiality reached $FT = -2.29$ in the most critical location for the plate. This fact verified the previous hypothesis: with the high compressive hydrostatic component of the stress field (equivalent to a very negative triaxiality) was feasible to reach a very high contact pressure during the cold-forming process.

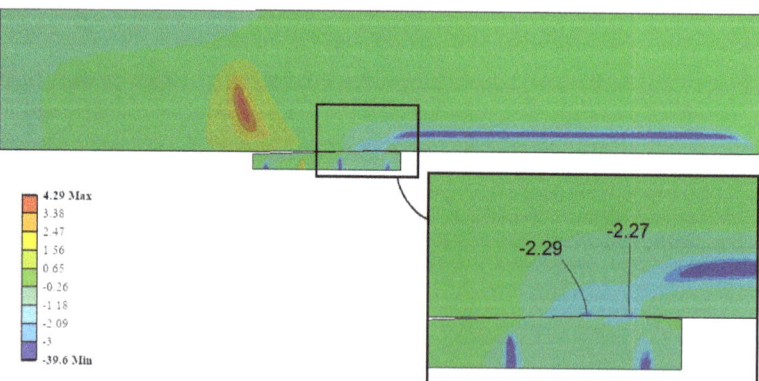

Figure 19. Triaxiality during the cold expansion process.

Figure 20a–c displays the residual hoop, radial and longitudinal stresses along the normalized radial distance in the plate (0, the hole radius; 1, the outer radius) after the cold expansion process for the IDs Ae075-I010-Lr05 /-Lr10 and /-Lr15. Changes in the length of the mandrel straight zone did not generate any significant variation in the residual stresses after the process. Figure 20d shows the maximum load needed to push the mandrel through the hole. This process parameter showed an increase with the increase of the length of the straight zone.

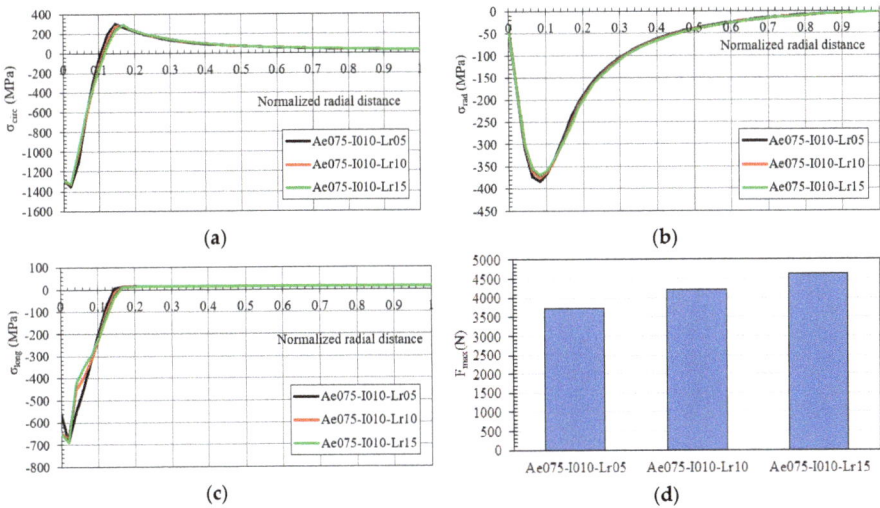

Figure 20. (a) Hoop; (b) radial; (c) longitudinal residual stresses along the normalized radial distance after the cold expansion process and (d) maximum pushing load in the mandrel (IDs Ae075-I010).

Figure 21 shows the hoop residual stress and the maximum pushing load of the mandrel for the IDs Ae075-I015 and Ae075-I020. Similar results were obtained: the modification of the straight zone length did not generate any significant variation of the residual stresses on mandrels with an inlet angle of 0.75°, and diametric interferences of 0.15 and 0.20 mm with the hole. Interference growth generated:

1. An increase of the maximum compressive residual hoop stress: Ae075-I010: $\sigma_{hoop} = -1350$ MPa, Ae075-I015: $\sigma_{hoop} = -1400$ MPa and Ae075-I020: $\sigma_{hoop} = -1500$ MPa.
2. An increase of the maximum pushing load of the mandrel.

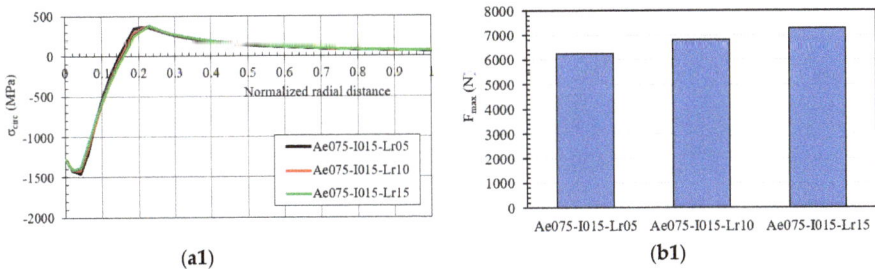

Figure 21. Cont.

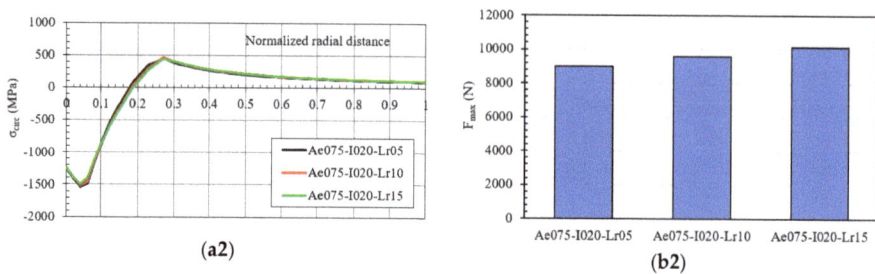

Figure 21. (**a**) Hoop residual stress along the normalized radial distance after the cold expansion process and (**b**) maximum pushing load in the mandrel [(**1**) IDs Ae075-I015; (**2**) IDs Ae075-I020].

In Figure 22, the residual hoop stress and the maximum pushing load of the mandrel were reported for the IDs with an inlet angle of 1.0°: Ae100-I010, Ae100-I015 and Ae100-I020 with /-Lr05, /-Lr10 and /-Lr15. The variation of the straight zone length did not generate any change in the residual stresses, increasing only the maximum pushing load of the mandrel. Comparing these results with the inlet angle previously considered equal to 0.75°, the residual hoop stress did not change significantly, but the maximum pushing load of the mandrel diminished with the increase of the inlet angle.

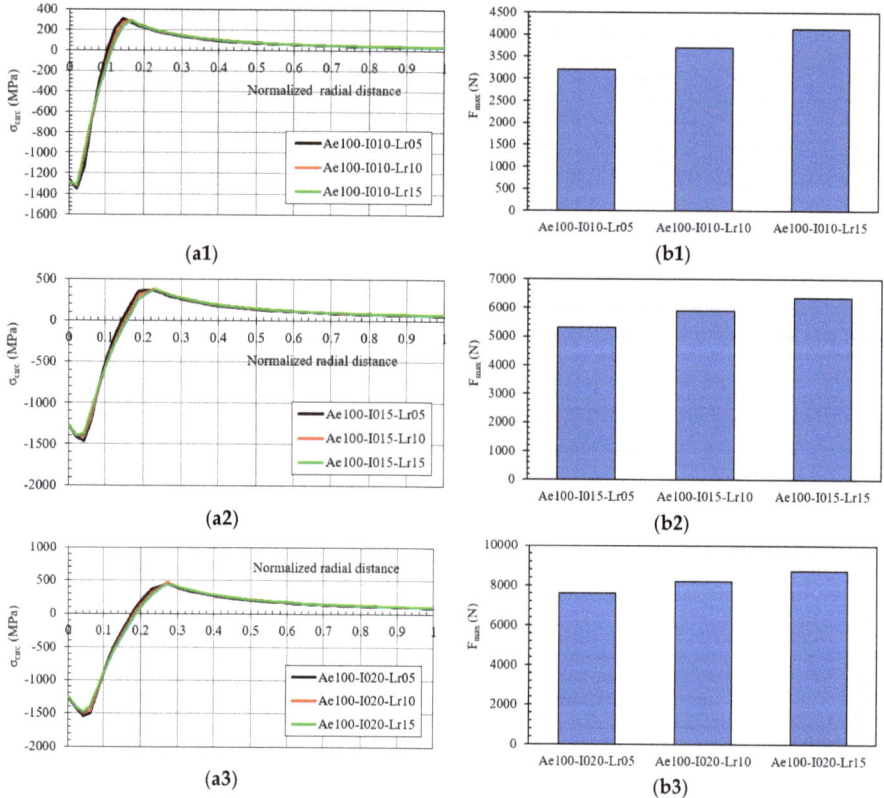

Figure 22. (**a**) Hoop residual stress along the normalized radial distance after the cold expansion process, and (**b**) maximum pushing load in the mandrel ((**1**) IDs Ae100-I010; (**2**) IDs Ae100-I015; (**3**) IDs Ae100-I020).

The results obtained for inlet angles equal to 1.5° and 2.0° are reported in Figures 23 and 24, respectively.

1. Increase of the inlet angle generated a reduction of the maximum pushing load of the mandrel.
2. Increase of the straight zone length generated an increase of the maximum pushing load of the mandrel.
3. Variation of the inlet angle and the straight zone length did not show any significant influence in the residual stresses in the hole.

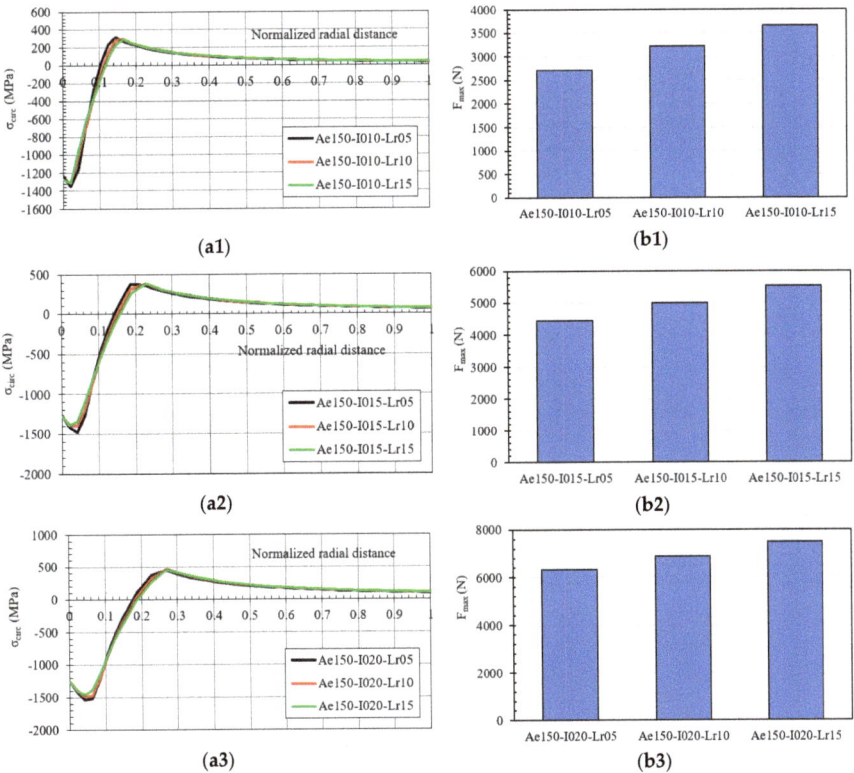

Figure 23. (a) Hoop residual stress along the normalized radial distance after the cold expansion process and (b) maximum pushing load in the mandrel ((**1**) IDs Ae150-I010; (**2**) IDs Ae150-I015; (**3**) IDs Ae150-I020).

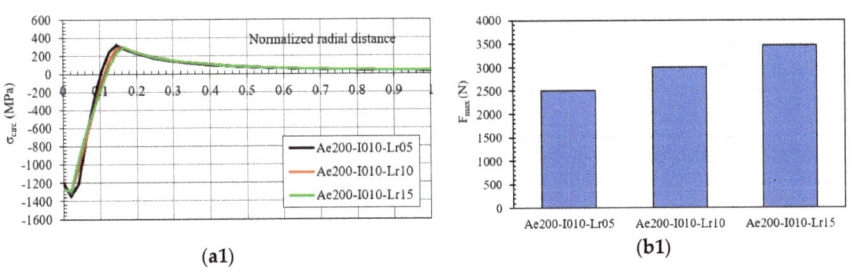

Figure 24. *Cont.*

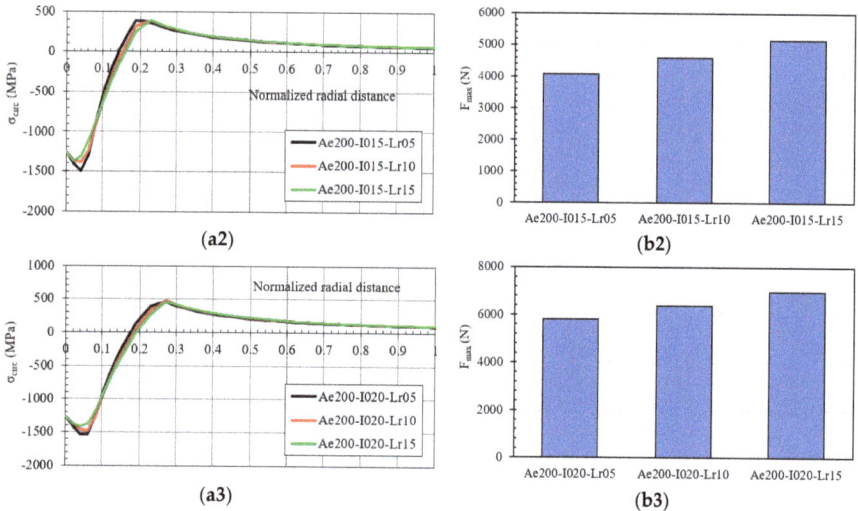

Figure 24. (**a**) Hoop residual stress along the normalized radial distance after the cold expansion process, and (**b**) maximum pushing load in the mandrel ((**1**) IDs Ae200-I010; (**2**) IDs Ae200-I015; (**3**) IDs Ae200-I020).

4. Discussion

The geometrical parameters of the mandrel evaluated in this study, inlet angle and the straight zone length, showed no significant influence in the residual stresses after the cold expansion process. This behaviour did not match the results showed by the numerical analysis performed by Gibson et al. [5,6] where there was significant variation of the hoop stress of thick-walled cylinders when the geometrical parameters of the swage were modified. Figure 25 shows the residual hoop stress obtained from the analysis performed by Gibson [5], where the influence of the straight zone length was analysed. An increase of this mandrel length did not significantly change the residual hoop stress, but from the length 1.0 up to 4.50, the compressive hoop stress at the smaller diameters of the cylinder showed an appreciable reduction. Therefore, the length value of 1.0 appeared as a limit.

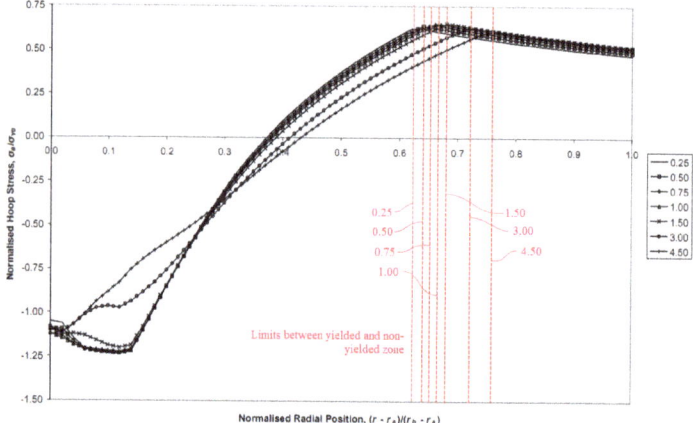

Figure 25. Normalized hoop residual stress along the normalized radial distance for different lengths of the straight zone in the Gibson et al. analysis [13].

This contradiction between the swaging of thick-walled cylinders and the cold expansion of small holes could have the explanation in the difference between the diameter ratio K of both geometries. The hole analysed in this investigation had a $K = 30/3.4 = 8.8$, whereas the Gibson's study analysed a cylinder with $K = 2.257$. Other investigations, focused on the analysis of the geometrical parameters of the swage, followed similar K ratios: O'Hara ($K = 2.257$ [17]), Bihamta et al. ($K = 3.18$ [22]) and Alinezhad et al. ($K = 3.18$ [23]). Therefore, the K value used in thick-walled cylinders ranges from 2 to 4, a much smaller interval than the K's value typically used in cold expansion processes. Could the K ratio be the origin of the previously mentioned contradiction? It could be answered with another question: what is the origin of the residual stresses after a swaging or cold expansion process? The remaining part of the cylinder or plate which has not been yielded during the process is the main part that is trying to return to its original position. However, the yielded zone does not want to return to that original position. Therefore, the non-yielded zone presses the yielded zone. If there is not a sufficient non-yielded zone, the residual compressive hoop stresses in the yielded zone will be significantly reduced. The vertical red dashed lines in Figure 25 shows the limit between the yielded and the non-yielded zones for the Gibson's model. The increase of the straight zone length of the swage generates an increase of the yielded zone. Thus, the region of the cylinder which must press the yielded zone to generate the residual compressive hoop stress is smaller and smaller. When there is not a sufficient non-yielded zone, the residual compressive hoop stress ends diminish. This means that an increase of the straight zone length of the swage does not reduces the capability of the swage to yield the inner diameter of the hole. This capability is increased but if there is not sufficient outer diameter to press the hole after the cold-forming process, an increase of the yielded zone reduces the residual stresses. This could explain the lack of variation of the residual stresses of the small hole evaluated in this investigation. There is much more non-yielded zone in the plate, and the modification of the straight zone length of the mandrel did not reduce enough this zone to show consequences in the residual stresses.

In order to verify this approach, a FEM simulation of the cold expansion process evaluated in this investigation was repeated, but modifying the outer diameter from 30 mm to the obtained value with a $K = 2.257$ (the same K value used in the Gibson's study): $D_{ext} = 3.4 \times 2.257 = 7.67$ mm. Two lengths of the straight zone were evaluated: 0.5 and 1.5 mm. The remaining geometrical parameters were:

- Diametric interference: 0.1 mm
- Inlet angle: 2°
- Outlet angle: 3°

Figure 26 shows the residual hoop stress for straight zone lengths equal to 0.5 and 1.5 mm) along the normalized radial distance for the cold expansion process with a diameter ratio of $K = 2.257$ instead of $K = 8.8$ used in this investigation. The reduction of the compressive residual hoop stress for the inner diameters for $SZ = 1.5$ mm is similar to the behaviour shown by Gibson's model, but the reason for this reduction is not justified only by the increase of the straight zone length. This increment generated an extension in the diameter of the yielded zone and the absence of enough non-yielded material to press the rest of the cylinder was the main cause for this reduction of the compressive residual hoop stresses. Consequently, the dependency of the residual hoop stresses with the straight zone length also depends on the K ratio.

In a cold expansion cold-forming process, the K ratio could show small values if the formed hole is close to any discontinuity (such as another hole or the end of the plate). It means that the influence of the straight zone length of the mandrel should be considered in those scenarios.

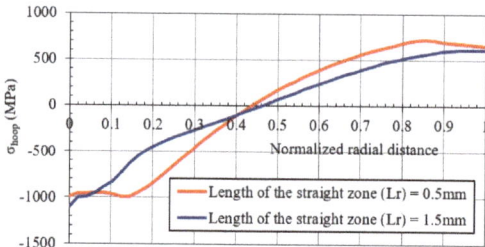

Figure 26. Hoop residual stress along the normalized radial distance after the cold expansion process with $K = 2.257$.

5. Conclusions

The main objective of this investigation was focused on a parametric analysis of the influence of the mandrel geometry in the residual stresses obtained after a cold expansion process of small holes with high ratio of plate thickness vs. hole diameter. Related to this general objective, this research has concluded the following:

1. K ratio has a considerable weight on the influence of the mandrel geometrical parameters for the residual stresses. High values of K ratio promote non-significant influence on the inlet angle variation and the straight zone length of the mandrel for the residual stresses of the cold-forming process.
2. Smaller inlet angles and higher the straight zone lengths generate an increase of the boundary diameter between the yielded and non-yielded zones of the plate. If K ratio is small enough, the non-yielded zone has not the capability to press the yielded zone and, consequently, the residual hoop stresses of the inner diameters diminishes.
3. The non-significant influence of the inlet angle and the straight zone length for the residual stresses of holes with high K ratios could mean that these values are not critical for the process, but the influence of these parameters on the pushing load of the mandrel is enough significant to consider them during any design. This numerical investigation showed that reducing the straight zone length and increasing the inlet angle, the pushing load is significantly reduced.

Author Contributions: Conceptualization: J.C.-C.; methodology, J.C.-C. and M.M.M.; validation: J.C.-C. and M.M.M.; formal analysis: J.C.-C. and M.M.M.; investigation: J.C.-C., M.M.M., E.M.R. and R.T.; resources: M.M.M. and E.M.R.; data curation: J.C.C. and M.M.M.; writing—original draft preparation: J.C.-C.; writing—review and editing: J.C.-C., M.M.M., E.M.R., R.T and T.S.; visualization: J.C.-C.; supervision: M.M.M., E.M.R. and R.T.; project administration: M.M.M., E.M.R. and R.T.

Funding: This work has been funding, in part, by grants from Industrial Engineering School-UNED (REF2019-ICF05 and REF2019-ICF08).

Acknowledgments: The authors thank to the Research Group "Industrial Production and Manufacturing Engineering (IPME)" the given support during the development of the present work.

Conflicts of Interest: The authors declare no conflict of interest.

References

1. Thomson, H.C.; Mayo, L. *The Ordnance Department: Procurement and Supply*; Office of the Chief of Military History: Washington, DC, USA, 1960.
2. Mote, J.D. *Explosive Autofrettage of Cannon Barrels*; Army Materials and Mechanics Research Center: Watertown, DC, USA, 1971.
3. Shufen, R.; Dixit, U.S. A review of theoretical and experimental research on various autofrettage processes. *J. Press. Vessel Technol.* **2018**, *140*, 050802. [CrossRef]
4. Hu, Z. Design and modelling of internally pressurized thick-walled cylinder. In Proceedings of the NDIA Conference, Dallas, TX, USA, 17–20 May 2010.

5. Shufen, R.; Dixit, U.S. An analysis of thermal autofrettage process with heat treatment. *Int. J. Mech. Sci.* **2018**, *144*, 134–145. [CrossRef]
6. Zare, H.R.; Darijani, H. Strengthening and design of the linear hardening thick-walled cylinders using the new method of rotational autofrettage. *Int. J. Mech. Sci.* **2017**, *124–125*, 1–8. [CrossRef]
7. Davidson, T.E.; Barton, C.S.; Reiner, A.N.; Kendall, D.P. New approach to the autofrettage of high-strength cylinders. *Exp. Mech.* **1962**, *2*, 33–40. [CrossRef]
8. Phillips, J.L. *Sleeve Coldworking of Fastener Holes*; AFML-TR-74-10; Air Force Materials Laboratory: Wright-Patterson Air Force Base, OH, USA, 1974.
9. Boeing Aircraft Co. *Sleeve Cold Working of Holes in Aluminum Structure*; BAC 5973; Boeing Aircraft Co., Process Specification: Chicago, IL, USA, 1975.
10. Fallahnezhad, K.; Steele, A.; Oskouei, R.H. Failure mode analysis of aluminium alloy 2024-T3 in double-lap bolted joints with single and double fasteners; A numerical and experimental study. *Materials* **2015**, *8*, 3195–3209. [CrossRef]
11. Hou, S.; Zhu, Y.; Cai, Z.; Wang, Y.; Ni, Y.; Du, X. Effect of hole cold expansion on fatigue performance of corroded 7B04-T6 aluminium alloy. *Int. J. Fatigue* **2019**, *126*, 210–220.
12. Anil Kumar, S.; Mahendra Babu, C. Effect of proximity hole on induced residual stresses during cold expansion of adjacent holes. *Mater. Today* **2018**, *5*, 5709–5715. [CrossRef]
13. Gibson, M.C. Determination of Residual Stress Distributions in Autofrettaged Thick-Walled Cylinders. Ph.D. Thesis, Cranfield University, Cranfield, UK, 2008.
14. Gibson, M.C.; Hameed, A.; Hetherington, J.G. Investigation of residual stress development during swage autofrettage, using finite element analysis. *J. Press. Vessel Technol.* **2014**, *136*, 43–50. [CrossRef]
15. Su, M.; Amrouche, A.; Mesmacque, G.; Benseddiq, N. Numerical study of double cold expansion of the hole at crack tip and the influence on the residual stresses field. *Comput. Mater. Sci.* **2008**, *41*, 350–355. [CrossRef]
16. Yuan, X.; Yue, Z.F.; Wen, S.F.; Li, L.; Feng, T. Numerical and experimental investigation of the cold expansion process with split sleeve in titanium alloy TC4. *Int. J. Fatigue* **2015**, *77*, 78–85. [CrossRef]
17. O'Hara, G.P. *Analysis of the Swage Autofrettage Process*; ARCCB-TR-92016; Benét Laboratories (Watervliet Arsenal): Watervliet, NY, USA, 1992.
18. Ceratizit. *Wear Parts, Main Catalogue*; Ceratizit: Mamer, Luxembourg, 2019.
19. SAE International. *Steel, Corrosion-Resistant, Bars, Wire, Forgings, Rings, and Extrusions 15Cr-4.5Ni-0.30Cb(Nb)-3.5Cu*; SAE Standard AMS5659; SAE International: Warrendale, PA, USA, 2016.
20. Ramberg, W.; Osgood, W.R. *Description of Stress-Strain Curves by Three Parameters*; Technical Note 902; NACA: Washington, DC, USA, 1943.
21. Kamaya, M. Ramberg-Osgood type stress-strain curve estimation using yield and ultimate strengths for failure assessments. *Int. J. Press. Vessels Pip.* **2016**, *137*, 1–12. [CrossRef]
22. Bihamta, R.; Movahhedy, M.R.; Mashreghi, A.R. A numerical study of swage autofrettage of thick-walled tubes. *Mater. Des.* **2007**, *28*, 804–815. [CrossRef]
23. Alinezhad, P.; Bihamta, R. A study on the tool geometry effects in the swage autofrettage process. *Adv. Mater. Res.* **2012**, *433–440*, 2206–2211. [CrossRef]

© 2019 by the authors. Licensee MDPI, Basel, Switzerland. This article is an open access article distributed under the terms and conditions of the Creative Commons Attribution (CC BY) license (http://creativecommons.org/licenses/by/4.0/).

Article

Shipbuilding 4.0 Index Approaching Supply Chain

Magdalena Ramirez-Peña [1,*], Francisco J. Abad Fraga [2], Alejandro J. Sánchez Sotano [1] and Moises Batista [1,*]

[1] Department of Mechanical Engineering and Industrial Design, School of Engineering, University of Cádiz, Av. Universidad de Cádiz, 10, E-11519 Puerto Real (Cádiz), Spain; alejandrojavier.sanchez@uca.es

[2] Navantia S.A., SME Astillero Bahía de Cádiz, Polígono Astilleros s/n, E-11519 Puerto Real (Cádiz), Spain; fabad@navantia.es

* Correspondence: magdalena.ramirez@uca.es (M.R.-P.); moises.batista@uca.es (M.B.); Tel.: +34-956483200 (M.B.)

Received: 31 October 2019; Accepted: 6 December 2019; Published: 10 December 2019

Abstract: The shipbuilding industry shows a special interest in adapting to the changes proposed by the industry 4.0. This article bets on the development of an index that indicates the current situation considering that supply chain is a key factor in any type of change, and at the same time it serves as a control tool in the implementation of improvements. The proposed indices provide a first definition of the paradigm or paradigms that best fit the supply chain in order to improve its sustainability and a second definition, regarding the key enabling technologies for Industry 4.0. The values obtained put shipbuilding on the road to industry 4.0 while suggesting categorized planning of technologies.

Keywords: shipbuilding; LARG paradigm; supply chain

1. Introduction

Every industrial revolution has brought important improvements in terms of manufacturing. Since 2015, the industry is working on the so-called fourth industrial revolution or Industry 4.0. This fourth industrial revolution introduces new advanced production models with new technologies that allow the digitalization of processes, services, and even business models [1]. Among other innovations, Industry 4.0 (I4.0) gives rise to the inclusion of the social aspect in the definition of the performance model of manufacturing processes, thus completing the economic, energy, environmental, and functional aspects considered until now [2].

It is also very common nowadays to use the distributed manufacturing systems which consist of manufacturing components in different physical locations and then going through supply chain management, bringing them together for the final assembly of a complex product [3]. Within the shipbuilding industry, there are two distinct fields of work, one dedicated to the repair, maintenance, or improvement of ships already built and the second dedicated to new ships. Focusing on the new construction and referring to it as shipbuilding could be consider as a case of distributed manufacturing, where the different blocks that constitute the ship built in different workshops belonging to the same manufacturing center are assembled afterwards in the dock. Therefore, shipbuilding is a complex manufacturing process that must adapt to I4.0 in order to progress. In this case, shipbuilding is a complex industry, with a complex structure composed of a large number of suppliers belonging to different locations, sizes, and typologies [4]. In addition, any small change made by each part of this structure not only affects the rest of the members but can also have enormous consequences. In this type of complex manufacturing [5], supply chain (SC) is a key factor to improve the efficiency of the shipyard in adapting to I4.0 [6]. Thus, the digitization—objective of the I4.0—of the supply chain will provide it with the agility and efficiency that shipbuilding needs to be more profitable [7,8].

Supply chain is the set of the flows of materials and information that take place within a company from the suppliers of raw materials to the consumer of the final product [9]. It is the concept that connects companies to their suppliers [10], as well as having among its activities the control of logistic activities [11] and the responsibility of analyzing purchases [12]. Supply chain represents one of the areas with the greatest investment in successful companies as it has become a strategic tool with a multidisciplinary and transversal character that affects all strategic levels of the company. It affects the sector and the market where the company is going to compete, defined by the corporate strategy, how it is going to compete, defined by the competitive strategy and of course each of the affected areas within the company, defined by the functional strategy [13].

Within the objectives of the supply chain are a rapid response to demand, flexible manufacturing, cost reduction, and inventory reduction. In addition, through the achievement of these objectives, SC aims to achieve improved competitiveness and sustainability of the company [14]. In the framework of Industry 4.0, the main objective of the supply chain must be total visibility of all product movements for each member of the chain as well as a total integration [15]. The most important paradigms on the supply chain found in the literature, under the perspective of sustainability shows the paradigm LARG. It is the paradigm defined under the acronym LARG: Lean, Agile, Resilient, and Green [14].

Lean Paradigm: The principles on which the Lean philosophy is based and its practices, make them ideal for the supply chain consisting of a network of business units or even independent companies, becoming challenging because of the complexity of management [16]. Among its contributions are:

- A collaborative relationship between its members of mutual trust and long-lasting commitment.
- Few and closer suppliers with low vertical integration are preferred.
- Multifaceted criteria approach is recommended on the capacity and benefit of suppliers and on the previous relationship.
- Software development for suppliers.
- The involvement of suppliers from the early stages of new product design and development processes.
- Frequent feedback allows risks, benefits, and solutions to be shared [17].

Agile Paradigm: The agile supply chain must know what is happening in the market in order to be able to respond as quickly and close as possible to reality [18]. It is through the integration of partners where the acquisitions of new skills allow them to respond quickly to the constant changes in the market [19]. The key elements are their dynamic structure and the visibility of information configured from beginning to end of an event-based management, such as relationships. For some authors, the supply chain should be adjusted when there is a question of a production of a considered volume with little variety, in a predictable, controllable business environment, whereas, if it is a question of unpredictable changes in the market, a small volume and a great variety are required, in this case an agile paradigm is required. Other authors such as Naylor et al. [20] introduce the term "Leagile" for the supply chain whenever demand is variable and there is a wide variety of products.

Resilient Paradigm: Resilience is the ability to overcome the disturbances suffered and recover the state in which it was before the disturbance. Based on this definition, the supply chain must have this characteristic and understand resilience as the capacity that the organization must have to continuously adjust the supply chain of events that may alter the balance of its activities [21]. One of the objectives of the resilient supply chain is to avoid a change to an undesirable state [14]. One way to achieve this is to design strategies to restore the previous state of the system [22]. Among the must have characteristics resilient supply chain, most authors agree on the total visibility of it; characteristic shared with the Agile paradigm and with Industry 4.0 [23,24]. However, several authors consider this paradigm very costly and complicated to implement. Therefore, they consider Lean Production and/or Six Sigma option as an alternative that provides flexibility and a corporate culture that could also provide resilient to supply chain [25].

Green Paradigm: This paradigm offers different approaches, from the perspective of supplier management in terms of risks and returns, in terms of supply chain management for sustainable products or both at the same time [26]. This ecological supply chain term is the consideration of environmental extension within supply chain management from the stages of product design, to the manufacturing process itself, until the delivery to the final consumer and even to the end-of-life management of the manufactured product [27]. This paradigm even lead the determination that through the greening of the different stages of the supply chain an integrated green supply chain can be achieved, which would lead to an increase in competitiveness and economic performance [28,29]. Sustainable supply chain also consider coordination of economic, environmental, and social considerations [30,31].

Looking for quantifiers on the supply chain in the literature, different parameters analysis are detected. Of those who seek the measure of their performance, some do so quantitatively, others qualitatively and there are those who analyze from both perspectives, identifying parameters such as visibility as Lia et al. did [6]. There are approaches to improving supply chain performance at different stages of the product life cycle by applying different linguistic scales to assess uncertain supply behavior or as in the case of Chang et al. [32] studying the selection of suppliers using fuzzy logic. This approach allows companies the assessment with no limitation on categories of scale and data [33]. However, there are other approaches such as improving the decision-making process. Wang et al. researched to provide decision-makers with rapid access to the practical performance of suppliers supply [34].

Alternatively, and as decision support process for incomplete hesitant fuzzy preference relations, the qualification of the supply chain is evaluated [35]. Other studies show focus on relieving the complexity of the aggregation and evaluation process by showing the connection between product strategies and supply chain performance [36]. In addition, there are benchmarking tools that develop indices to measure the agility of the supply chain such as Lin et al. [37]. Also to evaluate parameters of the supply chain itself as green or resilient [38] or even several at once as is the case of the LARG index, which evaluates the supply chain from the perspective of Lean, Agile, Resilient, and Green researched by Azevedo et al. [39].

In this case, the aim is to evaluate a shipyard in the process of adapting to industry 4.0 through its supply chain. There are already studies in which a conceptual model is being developed that separately confronts the different paradigms that make up the LARG paradigm and confronts them with the enabling technologies of Industry 4.0 in the field of shipbuilding [40]. In the case at hand and based on the previous studies of Azevedo [38,39], the Delphi method is used as a strategic information-gathering tool considered appropriate for supply chain [39]. The Delphi method consists of the technique that allows information gathered through consultation with experts [39]. It is an iterative process based on the anonymity of the answers, which allows the analysis of the answers. It is composed of several phases: 1. Definition of the research problem, 2. Determination of the participants to take part, 3. Elaboration of the questionnaire establishing the number of necessary rounds, and 4. Results [41]. A small number of participants (6–30) makes this technique best suited to scientific evidence and social values [42], coinciding with the new aspect that introduces the I4.0 in the economic performance of companies [43].

There is no previous experience of quantifying the contribution of the LARG paradigm in the shipbuilding industry, only the experience of Azevedo in the automotive industry [39], or even of the 4.0 industry in general. However, knowing that SC determine KETs is possible to look for the most important practices associated with each technology for each paradigm of the supply chain. Analyzing these practices, the relationship between the technologies and the supply chain is analyzed at the same time. Then the method developed allows evaluating how LARG is the shipbuilding industry related to its practices. In the second phase, in order to know how 4.0 SC is through the evaluation of the implementation of KETs according to each of them to the supply chain paradigms. In this case, in addition to carrying out both quantifications, a special index is created to know how advanced is the adaptation to the 4.0 industry of the shipbuilding supply chain, this being the main objective of this paper. Based on all the above, the importance of the supply chain to a company remains latent, especially for companies as complex as those dedicated to shipbuilding, specifically to the shipyard. The purpose of the article is to define an index that shows the situation in which a shipbuilding company has in relation to its adaptation to Industry 4.0, addressing its supply chain. The proposal is, on first place to define the paradigms that best suit the achievement of sustainability in the company. Second, analyze how each of the enabling technologies of Industry 4.0 affects the supply chain through the evaluation of the results obtained with the Delphi method. This evaluation of results will allow the shipyard to establish under which paradigms of the supply chain to work as well as to know which technologies will allow it to fully adapt to industry 4.0.

2. Experimental Methodology

As already mentioned, the experimental methodology follows the Delphi method. Previously, the Delphi method has already been used in issues related to supply chain. A collaboration index between retailers and manufacturers studied by Anbanandam et al. [44], Supply Chain Fragility Index by Stonebraker et al. [45], performance SC index proposed by Nunlee et al. [46], or risk assessment index studied by Rao and Schoenherr [47]. In this communication, the proposal is to develop a shipbuilding 4.0 index. Figure 1 shows a general diagram developed below:

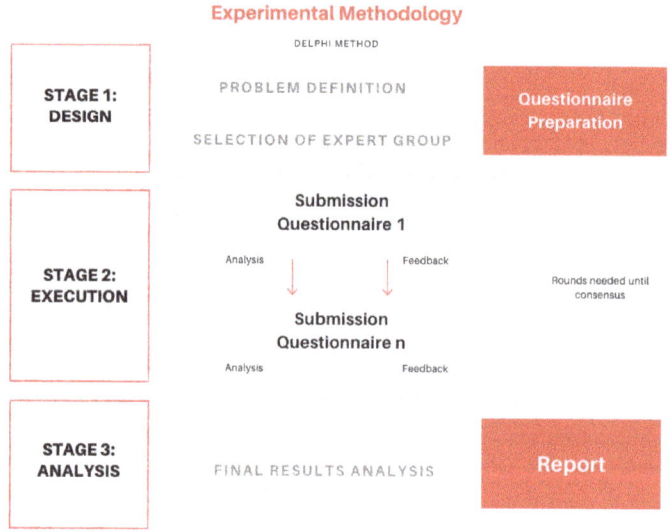

Figure 1. Scheme of the experimental methodology followed.

2.1. Stage 1: Design

The first stage of the experimental methodology begins with the definition of the problem, the selection of experts according to the problem addressed and the definition of the appropriate questionnaire that allows us to reach the solution of the problem posed.

Regarding the definition of the problem, the aim is to assess the level of adaptation of a shipbuilding company, specifically dedicated to the block assembly, to the 4.0 industry by addressing its supply chain. To this end, the first step is to define one or more of the supply chain paradigms that are most appropriate for this type of company in such a complicated sector. Specifically, the aim is to study the "LARG" paradigm formed by the combination of the Lean, Agile, Resilient, and Green paradigms.

At the same time, this stage has to establish which of the enabling technologies for Industry 4.0 facilitates and are best suited to contribute to the sustainability supply chain and subsequent implementation in the case study. There are studies that establish that there are twelve technologies that are suitable for shipbuilding [40]. These technologies are Additive Manufacturing, Big Data, Cloud Computing, Augmented Reality, Autonomous Robots, Automatic Guided Vehicles, Blockchain,

Cybersecurity, Horizontal and Vertical Integration System, Artificial Intelligence, Internet of things and Simulation.

A very important weight in the design lies on the selection of experts to participate in the surveys. It is ideal to select people who are interested in the subject matter and whose expertise includes the topic in question, as well as their impartiality. At this point, forty people take part in the survey. Twenty of them are personnel from the shipbuilding sector itself, as well as twenty from academics directly related to the sector. In this sense, one can have essential perspectives on the object of research. Finally, half of all guests agreed to participate.

It is now possible to define the questions that will constitute the questionnaire. It has seventeen sections. A first section in which an assessment is requested from nothing important to extremely important, where nothing important is weighted with 1 point and extremely important with 5 points, on the importance of each of the Supply Chain paradigms, Lean, Agile, Resilient and Green. Table 1 shows the questions in section 1 as well as the average rating values and weights of each of the paradigms posed.

Table 1. Mean rating and weightings of LARG to shipbuilding supply chain.

Paradigms	Questions	Mean Rating	Weight
Lean	How important is Lean paradigm to Shipbuilding Supply Chain	4.22	0.26
Agile	How important is Agile paradigm to Shipbuilding Supply Chain	4.15	0.26
Resilient	How important is Resilient paradigm to Shipbuilding Supply Chain	3.72	0.23
Green	How important is Green paradigm to Shipbuilding Supply Chain	3.93	0.25
	Sum	16.00	1.00

Sections two to five evaluate the implementation or non-implementation of four practices of each of the paradigms. Table 2 shows questions sections 2 to 5, indicated by P for Practice, sub-indices L, A, R, G, followed by numbering from 1 to 4, mean rating values and weightings of these practices. Finally, sections six to seventeen include the weighting of the importance of each of the twelve technologies, according to each of the supply chain paradigms. Table 3 shows the questions in sections 6 to 17 that highlight the importance of each of the industry 4.0 enabling technologies for the shipbuilding supply chain.

2.2. Stage 2: Execution

Once the questionnaire design is complete, it is ready to launch the first round of surveys to all professionals who have agreed to participate. This is sent virtually, via e-mail, where the survey is in a form linked in that e-mail.

After the first round, data is collected and analyzed. To know the level of agreement reached by the experts, the Kendall concordance coefficient is used. This coefficient indicates that the level of agreement reached in this first round is low. It is for this reason that it is necessary to carry out a new round.

In the same way, a second round is sent, in which the participants are informed of the mean rating results of the first round. In this second round, the Kendall correlation coefficient has increased in value indicating that there is greater consistency between the professionals and academics. At this point, it is decided to conclude the surveys and proceed to analyze them.

Table 2. Questions, mean rating, and weightings of LARG level of implantation to shipbuilding supply chain.

Practices	Reference	Mean Rating	Weight
P_{L1} = Just in time (in the company)	[48]	2.30	0.23
P_{L2} = Just in time (from supplier to company)	[48,49]	2.40	0.24
P_{L3} = Pull Flow	[48]	2.25	0.22
P_{L4} = Supplier relationships/long term business relationship	[50]	3.20	0.32
P_{A1} = Use of IT in design and development activities	[51,52]	2.35	0.26
P_{A2} = Capacity to change delivery times of supplier	[52]	2.35	0.26
P_{A3} = Use of IT in manufacturing activities	[52]	2.15	0.24
P_{A4} = Centralized and collaborative planning	[51]	2.10	0.23
P_{R1} = Procurement strategies that enable the change of suppliers	[53]	1.90	0.28
P_{R2} = Supply chain visibility creation	[24]	1.65	0.22
P_{R3} = Lead time reduction	[23,54]	2.10	0.26
P_{R4} = Development visibility of inventories and demand conditions	[23]	1.75	0.24
P_{G1} = Environmental collaboration with suppliers	[55]	2.15	0.22
P_{G2} = ISO 14001 certification	[28]	3.25	0.33
P_{G3} = To reduce energy consumption	[56]	2.10	0.21
P_{G4} = To reduce or recycling materials and packaging	[57]	2.40	0.24
Sum		36.40	

Table 3. Data $LARG_{SC}$ index.

Paradigm	1	2	3	4	5	6	7	8	9	10	11	12	13	14	15	16	17	18	19	20	PB_x	w_x	SC_x
Lean	3.55	2.63	3.00	4.55	3.00	2.73	1.78	2.13	2.32	3.08	2.32	1.54	2.23	2.32	2.63	2.78	3.09	2.87	1.54	1.86	2.60	0.26	0.68
Agile	2.00	2.79	2.99	1.77	2.00	1.53	1.76	2.53	3.23	1.74	2.00	2.26	2.74	1.50	3.00	2.03	2.00	2.53	2.47	2.00	2.24	0.26	0.58
Resilient	2.26	1.76	2.54	1.28	2.00	1.00	2.00	1.72	1.80	2.28	1.78	2.28	2.28	2.00	2.78	2.00	2.00	1.00	1.28	1.26	1.87	0.23	0.43
Green	2.57	3.35	4.33	2.90	3.11	1.57	1.87	2.57	2.12	2.03	1.45	2.11	2.76	2.76	3.64	2.66	2.33	2.90	2.09	2.11	2.56	0.25	0.63

2.3. Stage 3: Analysis

Data analysis has three phases. In the first one, the importance of each supply chain paradigm is analyzed separately, i.e., the importance of the Lean, Agile, Resilient, and Green paradigms for the shipbuilding supply chain. The second phase analyses how each of the previous paradigms has been implemented. After knowing this value, we obtain a general index that will allow us to know how LARG shipbuilding supply chain is.

It is in the third phase and through the twelve key enabling technologies (KETs), where we get to know the importance of each of the KETs for the supply chain of shipbuilding. In the three phases, using the same mathematical procedure, the mean rating, weightings, and consequent indices mentioned are calculated.

3. Results

3.1. How Important is Each LARG Paradigm to Shipbuilding Supply Chain

The first result found presents the importance of the Lean, Agile, Resilient, and Green paradigms. As a result of the mean rating, the weight for each paradigm is calculated according to the equation [58]:

$$w_x = \frac{M_x}{\sum_{g=1}^{n} M_g} \quad (1)$$

where w_x represents the weighting of the paradigm x, M_x represents the mean rate of that particular paradigm and $\sum_{g=1}^{n} M_g$ represents the sum of the means for each paradigm. Figure 2 shows the relative results obtained by studying the importance of LARG for Shipbuilding Supply Chain represented by its mean rating. Table 1 shows the questions in section 1, mean rating values and weightings of each paradigms asked.

As can be seen, the Lean paradigm is the most valued of all; followed closely by the Agile paradigm, even more than the Green paradigm despite the importance it must have for this sector.

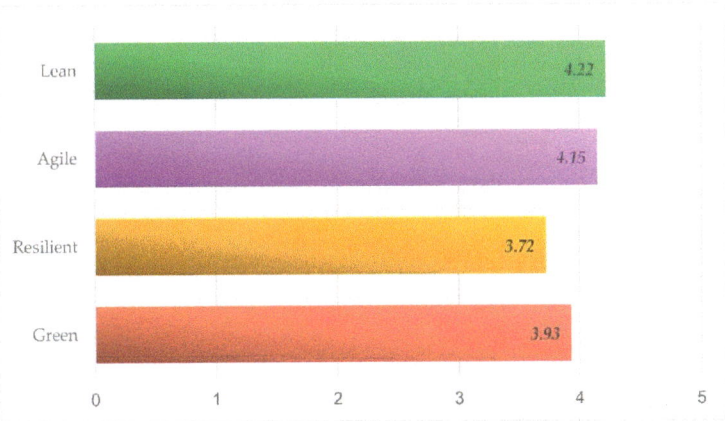

Figure 2. Importance of LARG to Shipbuilding SC.

3.2. How Implanted is Each LARG Paradigm to Shipbuilding Supply Chain

In the same way, with the values contributed by the experts with respect to the level of implantation, we are in disposition to calculate how much is each one of the studied paradigms is implemented. Figure 3 represents the level of implementation of LARG for shipbuilding supply chain represented by its mean rating. Table 2 shows the questions sections 2 to 5, indicated by P for Practice, sub-indices L, A, R, G, followed by numbering from 1 to 4, mean rating values, and weightings of these practices.

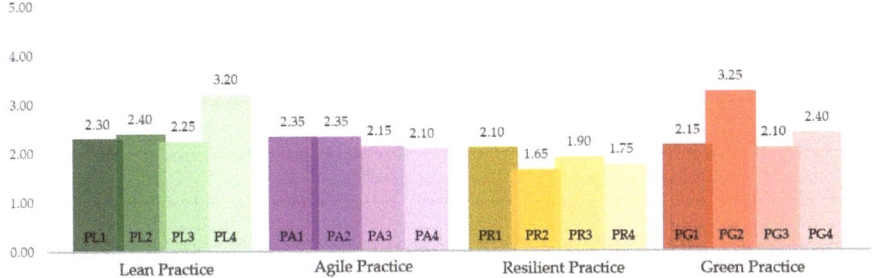

Figure 3. Level of implantation of LARG to shipbuilding SC.

It is now defined the expert behavior as:

$$EB_i = \sum_{i=1}^{n}\left(D_{ixj} \cdot w_{Pxj}\right) \quad (2)$$

where D_{ixj} is the answer of the expert i to practice j of the paradigm x. At the same time, the behavior of each paradigm is:

$$PB_x = \frac{\sum_{i=1}^{n}(EB_i)}{n} \quad (3)$$

Figure 4 shows the results obtained.

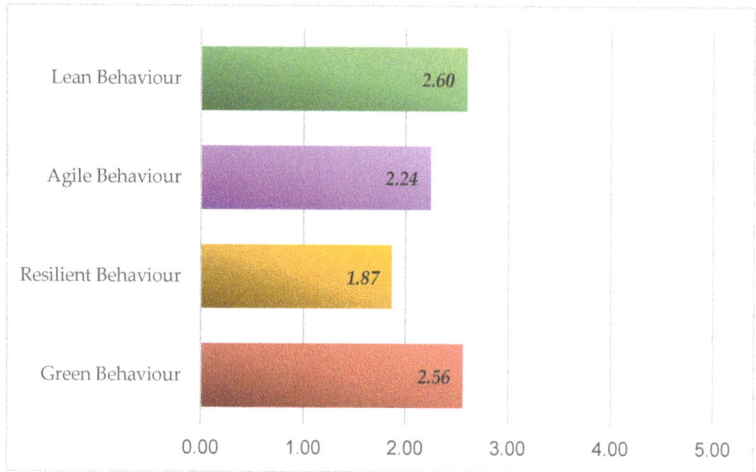

Figure 4. Paradigms behavior.

With these data, it is possible to calculate the implementation of each paradigm for the supply chain (SCx):

$$SC_x = PB_x \cdot w_x \tag{4}$$

Being LARG$_{SC}$ index:

$$LARG_{SC} = SC_L + SC_A + SC_R + SC_G \tag{5}$$

The result after the indicated calculations and according to the reflected data is LARG$_{SC}$ = 2.33. This indicates an intermediate implantation value as it is valued on a scale from 1 to 5. It is observed that despite the importance valued in the previous section, now, referring to the implementation level, Lean and Green are the most valued paradigms leaving behind the paradigms Agil and Resilient.

3.3. What Level of 4.0 is the LARG to Shipbuilding Supply Chain?

In order to know the 4.0 level of the LARG supply chain in shipbuilding, the weighting that the experts answered in this respect is used. Table 4 shows the questions in sections 6 to 17 that highlights the importance of each of the industry 4.0 enabling technologies for the shipbuilding supply chain. In the same way, calculate the mean rating, the weights using Equation (1) and the results shown in Table 5 and Figure 5 are the KETs behavior calculated in the same way as above according to the expert behavior.

Table 4. Key enabling technologies and supply chain paradigms questionnaire.

Key Enabling Technologies	Lean (L)	Agile (A)	Resilient (R)	Green (G)
K1: Additive Manufacturing	Large scale production in small batches, with a focus on the customer and the creation of more with less waste [59]	Customized products and processes [18]	Manufacturing products close to the customers geographic location reduces response time [60]	Technologies allowing techniques for the reuse and recycling of urban waste [61]
K2: Big Data	Improved information flows that enable the supply chain to operate with reduced inventory and rapid customer response [23]	Provide a single view of market trends, customer purchasing patterns and maintenance cycles, ways to reduce the costs and enable more targeted business decisions [62]	Assist in making decisions regarding pricing, optimization, reduction of operational risk, and improvement of the delivery of products and services [62]	Minimize the risk inherent in hazardous materials, associated carbon emissions, and economic cost [63]
K3: Cloud Computing	Supporting information management during the life cycle of the digital product [62]	Facilitating sharing of resources and participants collaboration throughout the supply chain lifecycle [64]	Prevent potential vulnerabilities in the cooperation of supply chain actors [65]	Evaluate the ecological performance of the supplier under economic and environmental criteria [66]
K4: Augmented Reality	Provide advance shipping instructions and times [67]	Shorten the learning curve [68]	Possibility of tracking the product [69]	Promote disintermediation [70]
K5: Autonomous Robots	Improve work cycle efficiency [71]	Increase production flexibility [72]	Improve control of interrupt detection [73]	Minimize transport time [74]
K6: Automated Vehicles	Helps achieve leaner processes [75]	Improve supply chain performance by planning and controlling movements [76]	Enable online algorithm adaptation with real-time response [77]	Guarantee the performance of environmental efficiency [78]
K7: Blockchain	Increase the efficiency, reliability and transparency of the entire supply chain, optimize input processes (ability to ensure the immutability of data) and public accessibility of data flows [79]	Provide strong process controls aligned to the interests of all operators involved [80]	Enables transactions with a unified, transparent record keeping system [81]	Allows information easily disseminated to multiple parties involved, compiling and verification of information, control of environmental quality of materials, time management for new product development projects and coordination of participants [82]
K8: Cybersecurity	Improve efficiency within the business structure [83]	Analyze the requirements of the regulations of the different countries involved [84]	Allows to take steps to build resilience [85]	Optimize overall resource, energy consumption, provider operating expenses minimizing potential loss of security in the event of a successful attack on any virtual machine [86]
K9: Horiz. & Vert. Integ. System	Improve flow and vertical integration of all departments resulting in reduced delivery time of the product or service [70]	Facilitate ability to deliver products and services on time [87]	Strengthen synergy of their networks, execute activities across the entire value chain in an intelligent manner [88]	Allow the establishment of connections with the systems of information [88]
K10: Artificial intelligence	To pursue cost reduction through the use of tools such as Just in Time [89]	Provide real-time, adaptive visibility and traceability [90]	Identifies the most resilient suppliers [91]	Select the best suppliers according to economic and environmental criteria [92]
K11: Internet of Things	Improve the efficiency and quality of production and distribution [93]	Make prompt decisions and accelerate material flows by integrating and exchanging information flows to improve effectiveness and efficiency [94]	Formulate strategies to mitigate risks [95]	Improve decision making efficiency for inventory management [96]
K12: Simulation	Facilitate a design that allows to consider pull systems [97]	Measure the efficiency of the process [98]	Increase your complexity and vulnerability to disruptions [99]	Evaluate alternative policies for long-term capacity planning [100]

Table 5. Mean rating and weight of each KET studied.

Key Enabling Technologies	Lean		Agile		Resilient		Green	
	Mean Rating	Weight	Mean Rating	Weight	Mean Rating	Weight	Mean Rating	Weight
K1	3.79	0.25	4.21	0.28	3.21	0.21	3.79	0.25
K2	4.05	0.26	3.79	0.24	3.84	0.25	3.89	0.25
K3	3.95	0.26	3.79	0.25	3.89	0.25	3.74	0.24
K4	3.58	0.25	3.47	0.24	3.79	0.26	3.47	0.24
K5	4.26	0.26	4.21	0.26	4.11	0.25	3.84	0.23
K6	3.79	0.25	4.21	0.27	3.68	0.24	3.63	0.24
K7	3.68	0.25	3.63	0.25	3.84	0.26	3.63	0.25
K8	4.16	0.26	3.84	0.24	4.16	0.26	3.79	0.24
K9	4.47	0.26	4.47	0.26	4.32	0.25	4.05	0.23
K10	4.16	0.26	4.00	0.25	3.89	0.24	3.89	0.24
K11	4.11	0.26	4.11	0.26	3.89	0.24	3.95	0.25
K12	4.05	0.25	4.26	0.26	3.95	0.25	3.84	0.24

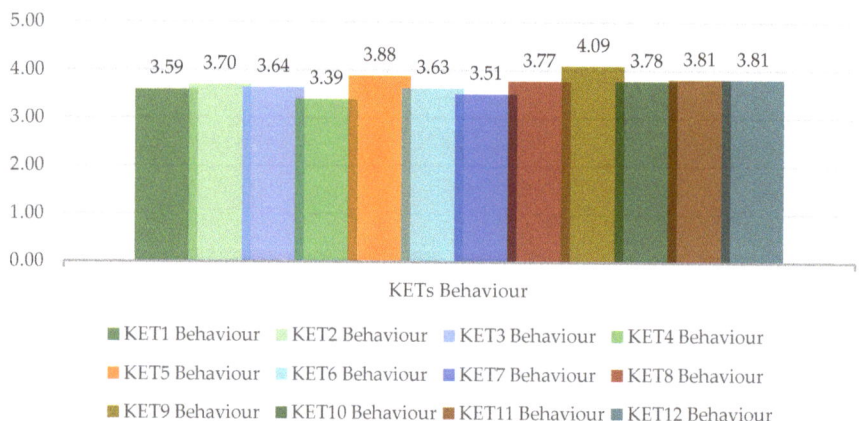

Figure 5. KET's behavior grouped for each paradigm according to the expert behavior.

With this data, it is possible to calculate the import of each KET for the SCx (KET-SCx):

$$SC_{KETi} = KETB_i \cdot w_{KETi} \quad (6)$$

Being in this case, the index $LARG_{4.0}$ is therefore defined by the following equation:

$$LARG_{4.0} = \sum_{i=1}^{n} SC_{KETi} \quad (7)$$

The result in this case of the index value that indicates the level of adaptation to industry 4.0 of the LARG supply chain has turned out to be of $LARG_{4.0} = 3.77$ indicating that this is a value on a scale of 1 to 5, which is quite interesting as we will analyze it later. Table 6 shows the data needed to calculate the $LARG_{4.0}$ index.

Table 6. LARG$_{4.0}$ index data.

KET's	1	2	3	4	5	6	7	8	9	10	11	12	13	14	15	16	17	18	19	KETB$_x$	w$_{KETi}$	SC$_{KETi}$
KET1	4.00	3.81	3.07	4.28	3.56	3.51	4.07	3.79	2.47	4.25	3.00	4.07	3.53	3.79	3.75	4.49	3.82	4.57	4.07	3.78	0.08	0.31
KET2	3.00	4.32	4.09	3.22	3.24	5.00	4.23	3.22	3.45	3.92	4.00	3.91	5.00	4.00	4.14	3.68	4.08	3.63	3.86	3.89	0.08	0.32
KET3	3.00	4.08	4.32	3.92	3.22	5.00	4.00	4.00	3.13	3.68	3.24	4.00	5.00	4.00	3.68	4.15	4.76	2.54	3.08	3.83	0.08	0.31
KET4	3.45	3.77	3.68	2.22	3.76	3.85	3.68	3.37	3.14	4.46	3.68	3.77	4.00	4.00	3.00	4.31	4.31	2.77	2.54	3.57	0.08	0.27
KET5	4.00	4.24	4.00	4.00	3.78	4.37	3.82	4.15	3.91	4.77	3.46	4.00	4.00	4.00	4.00	4.54	4.08	4.46	4.00	4.08	0.09	0.36
KET6	3.55	4.32	3.54	3.93	3.00	4.78	4.68	3.93	3.23	3.68	3.46	4.00	4.46	4.00	3.68	3.47	4.01	3.46	3.33	3.82	0.08	0.31
KET7	3.77	3.67	3.32	2.23	3.00	5.00	3.22	3.69	3.69	3.91	3.68	3.77	4.23	4.00	3.68	4.76	4.31	3.09	3.08	3.69	0.08	0.29
KET8	5.00	4.45	3.77	3.31	3.76	5.00	3.67	4.21	3.45	4.32	3.68	3.92	4.00	4.00	3.22	4.46	3.23	4.00	3.92	3.97	0.08	0.34
KET9	4.00	4.32	3.91	4.68	4.00	5.00	4.68	4.37	3.54	4.46	4.15	4.77	5.00	4.00	3.24	4.46	5.00	4.78	3.45	4.31	0.09	0.39
KET10	4.00	4.56	3.00	4.76	3.46	4.15	3.22	3.32	3.54	4.76	4.37	3.32	5.00	4.00	4.00	4.68	3.46	4.00	4.00	3.98	0.08	0.34
KET11	4.00	4.22	3.00	3.09	4.00	4.78	5.00	3.68	3.55	4.00	4.37	3.46	5.00	3.32	4.00	4.54	3.46	4.23	4.46	4.01	0.09	0.34
KET12	3.00	3.77	3.23	5.00	3.22	4.78	4.24	4.15	3.33	4.32	4.37	3.24	4.24	4.00	4.00	4.14	4.23	5.00	4.00	4.01	0.09	0.34

4. Discussion

Starting with the study of the importance for the experts of each of the paradigms consulted for the shipbuilding supply chain, the results of the survey indicate that the paradigm with the greatest weight according to the experts is Lean, closely followed by the Agile paradigm, Resilient being the one with a lower weight. Lean is the most widespread and used paradigm for the longest time, which is one of the reasons why it undoubtedly holds the first place in addition to what the paradigm itself offers [101,102]. The Agile paradigm does not seem to be entirely in line with the sector in terms of volume and the great variety of production required in this case as some authors consider [20]. It is true that Resilient is one of the paradigms that most needs SC visibility [23,24]. It is also known that this is a paradigm difficult to get into practice mainly because of the high cost associated and the possibility of bringing the practices related to it close to those of the Lean paradigm. The most impressive of these results is the lower weight of the Green paradigm, as it is considered as a key factor in shipbuilding to increase the competitiveness of the company [103] and improve the energy efficiency [104].

However, studying the implementation each paradigm has, according the supply chain behavior, it is worth mentioning that two of the practices evaluated reached higher mean rate. One of them belongs to the Lean paradigm (PL4) and the other to the Green paradigm (PG2). This shows precisely the declaration on the part of those consulted to prefer for long relationship with suppliers with the benefits that this entail, and second, to be the supporters of the certification of the corresponding regulation over the (PG3) to reduce energy consumption. In conclusion, the predominant paradigms are precisely Lean and Green, and Agile and Resilient being the least. In this case, it coincides totally with previous studies where results are the same for shipbuilding supply chain [40].

If we focus the attention on the value obtained for the LARG index that evaluates how are those paradigms implanted in SC, calculations indicate a value of 2.33. This value, on the same weight scale from 1 to 5, indicates that its implementation is on the way. It also indicates that, according to the consulted experts, the full implementation of the LARG paradigm is needed, mainly because all its benefits has not yet been achieved. This is not the case in other sectors, such as the automotive industry. Comparing the obtained values with the automotive industry ($LARG_{SC}$ = 3.75) [39], it is shown how much the automotive sector is ahead of the shipbuilding sector. The latter demonstrating to have not remained on the sidelines despite the difference between the two sectors.

Regarding the KET's, the value obtained is $LARG_{4.0}$ = 3.77. It is higher than that obtained previously ($LARG_{SC}$ = 2.33). Unlike the previous one that indicated the level of implementation of the paradigms to the supply chain, in this case it indicates the level of importance of each KET to each paradigm. Analyzing the results, it is observed that all belong to the same range. KET 9 (Horizontal and Vertical Integration System) exceeds the average value minimally. The reading in this case is that the shipbuilding is committed to adapt to Industry 4.0; however, because of the values obtained, it is not clear which technology can be more interesting to implement sooner.

Focusing on the $LARG_{4.0}$ index, as it is a new creation, it is not possible to compare the index value in itself. However, by targeting the interpretation of that value, it indicates an intermediate level that is in line with the global trend. American companies created the Industrial Internet Consortium (IIC) and some of them opted for the immediate application of the Internet of Things in the shipbuilding sector and it can be said that today shipbuilding has a great demand [105]. In Germany, one of the shipyards that stands out for technological innovation is Meyer-Werft (MW) [106], where the level of use of digital technologies throughout the production process has increased significantly. The identification of parts by radio frequency is another of the fields in which they work, increasing their control and traceability. In Korea, under the so-called "Manufacturing Innovation 3.0 Strategy" innovation is bet on the naval environment of Busan, where the shipyards of Daewoo, Samsung, and Hyundai are working toward the implementation of systems and technologies that are directly aligned with the 4.0 guidelines. Daewoo Shipbuilding and Marine Engineering (DSME) [107] develops systems according to the concept of intelligent factory and integrates technologies that allow optimizing the manufacturing processes; the key for this shipyard is to apply robotics in automating those processes that allow it. The first approaches of Hyundai Heavy Industries (HHI) [108] focused on remote monitoring, evolving into an integrated platform that contemplates fundamental aspects of the ship's life cycle [109]. The Samsung Heavy Industries (SHI) [110] shipyards quantify in 68% the automation of their productive process, motivated to a great extent by own developments of robotized solutions.

In addition to this first vision, the questionnaire aims to recognize a powerful tool for the leaders that can evaluate which are the practices that best fit in each case.

5. Conclusions

With the intention of evaluating where shipbuilding adapts the 4.0 Industry, this article develops an index that allows to know, on the one hand, how its current supply chain is and, on the other hand, how to evaluate the technologies that can allow it to adapt more quickly and efficiently to the 4.0 Industry. The development of both indices has been carried out with the collaboration of experts, both professionals from the shipbuilding sector and academics who have an important connection with this sector.

The importance that experts have given to the different paradigms studied for the supply chain has proved to be the most suitable for the sector in order to achieve a supply chain that contributes greatly to achieve the sustainability of their businesses. The value obtained for the index that indicates the level of implementation is considered satisfactory for an industrial sector in which its pace is different from that of other sectors. However, there is still room for improvement and the calculation of this index could be considered as a tool that can reflect which practices and paradigms could be more interesting to implement.

Finally, the value obtained for the index that indicates the level of adaptation of the supply chain to Industry 4.0, shows us interest as well as some disorientation. In the same way, this can be considered as an opportunity for managers to decide which technologies have priority over others in order to achieve their objectives. This also highlights the need for future research to establish different criteria to improve the index studied.

Author Contributions: M.R.-P. and M.B. conceptualized the paper. M.B. and F.J.A.F. approved the experimental procedure; M.R.-P. carried out the experiment; M.R.-P. and M.B. analyzed the data; M.R.-P. wrote the paper; M.B. and A.J.S.S. revised the paper; F.J.A.F. supervised the paper.

Acknowledgments: Navantia S.A. S.M.E. and University of Cadiz (UCA) supported this work.

Conflicts of Interest: The authors declare no conflict of interest.

References

1. Aenor Regulations. Available online: https://portal.aenormas.aenor.com/DescargasWeb/normas/Estandarizacion-para-la-industria-4_0.pdf (accessed on 26 September 2019).
2. *Factories of Future: Multi-Annual Roadmap for the Contractual PPP under Horizon 2020*; European Factories of the Future Research Association: Brussels, Belgium, 2013; p. 55.
3. Rauch, E.; Dallasega, P.; Matt, D.T. Distributed manufacturing network models of smart and agile mini-factories. *Int. J. Agil. Syst. Manag.* **2017**, *10*, 185. [CrossRef]
4. Aguayo, F.; Marcos, M.; Sánchez, M.; Lama, J. *Sistemas Avanzados de Fabricación Distribuida*; RA-MA Editorial y Publicaciones S.A.: Madrid, Spain, 2007.
5. Matt, D.T.; Rauch, E.; Dallasega, P. Trends towards distributed manufacturing systems and modern forms for their design. *Procedia CIRP* **2015**, *33*, 185–190. [CrossRef]
6. Xia, L.X.X.; Ma, B.; Lim, R. Supplier performance measurement in a supply chain. In Proceedings of the 2008 6th IEEE International Conference on Industrial Informatics, Daejeon, Korea, 13–16 July 2008; pp. 877–881.
7. Liu, J.; Xiao, B.; Bu, K.; Chen, L. Efficient distributed query processing in large RFID-enabled supply chains. In Proceedings of the IEEE INFOCOM 2014—IEEE Conference on Computer Communications, Toronto, ON, Canada, 27 April–2 May 2014; pp. 163–171.
8. Cormican, K.; Cunningham, M. Supplier performance evaluation: Lessons from a large multinational organisation. *J. Manuf. Technol. Manag.* **2007**, *18*, 352–366. [CrossRef]

9. Jones, T.C.; Riley, D.W. Using Inventory for Competitive Advantage through Supply Chain Management. *Int. J. Phys. Distrib. Mater. Manag.* **1985**, *15*, 16–26. [CrossRef]
10. Helper, S.; Sako, M. Management innovation in supply chain: Appreciating Chandler in the twenty-first century. *Ind. Corp. Chang.* **2010**, *19*, 399–429. [CrossRef]
11. Cooper, M.C.; Lambert, D.M.; Pagh, J.D. Supply Chain Management: More Than a New Name for Logistics. *Int. J. Logist. Manag.* **1997**, *8*, 1–14. [CrossRef]
12. Farmer, D. Purchasing myopia-revisited. *Eur. J. Purch. Supply Manag.* **1997**, *3*, 1–8. [CrossRef]
13. Porter, M.; Porter, M.; Kramer, M.; Kramer, M. Strategy and Society: The Link Between Competitive Advantage and Corporate Social Responsibility (HBR OnPoint Enhanced Edition). *Havard Business Rev.* **2006**, *84*, 78–92.
14. Carvalho, H.; Duarte, S.; Machado, V.C. Lean, agile, resilient and green: Divergencies and synergies. *Int. J. Lean Six Sigma* **2011**, *2*, 151–179. [CrossRef]
15. Saarni, J.; Heikkilä, K.; Kalliomäki, H.; Mäkelä, M.; Jokinen, L.; Aposto, O. *Sustainability in Shipbuilding—Observations from Project-Oriented Supply Network in Cruise Ship Construction*; Finland Futures Research Centre: Turku, Finland, 2019.
16. Ellram, L.M.; Cooper, M.C. Supply Chain Management: It 's All About the Journey. Not the Destination. *J. Supply Chain Manag.* **2014**, *50*, 8–20. [CrossRef]
17. Martínez-Jurado, P.J.; Moyano-Fuentes, J. Lean management, supply chain management and sustainability: A literature review. *J. Clean. Prod.* **2014**, *85*, 134–150. [CrossRef]
18. Christopher, M. The Agile Supply Chain Competing in Volatile Markets. *Ind. Mark. Manag.* **2000**, *29*, 37–44. [CrossRef]
19. Baramichai, M.; Zimmers, E.W.; Marangos, C.A. Agile supply chain transformation matrix: An integrated tool for creating an agile enterprise. *Supply Chain Manag.* **2007**, *12*, 334–348. [CrossRef]
20. Naylor, J.B.; Mohamed, M.; Naim, D.B. Leagility: Integrating the lean and agile manufacturing paradigms in the total supply chain. *Int. J. Prod. Econ.* **1999**, *62*, 107–118. [CrossRef]
21. Dg, D.G.L.; Pdlo, X.; Hqvd, E.; Frp, J. Supply chain improvement in LARG. In Proceedings of the 2017 6th IEEE International Conference on Advanced Logistics and Transport (ICALT), Bali, Indonesia, 24–27 July 2017.
22. Haimes, Y.Y. On the Definition of Vulnerabilities in Measuring Risks to Infrastructures. *Risk Anal.* **2006**, *26*, 293–296. [CrossRef]
23. Christopher, M.; Peck, H. Building the Resilient Supply Chain. *Int. J. Logist. Manag.* **2004**, *15*, 1–14. [CrossRef]
24. Iakovou, E.; Vlachos, D.; Xanthopoulos, A. An analytical methodological framework for the optimal design of resilient supply chains. *Int. J. Logist. Econ. Glob.* **2007**, *1*, 1–20. [CrossRef]
25. Mensah, P.; Merkuryev, Y. Developing a Resilient Supply Chain. *Procedia Soc. Behav. Sci.* **2014**, *110*, 309–319. [CrossRef]
26. Seuring, S.; Müller, M. From a literature review to a conceptual framework for sustainable supply chain management. *J Clean Prod.* **2008**, *16*, 1699–1710. [CrossRef]
27. Srivastava, S.K. Green supply-chain management: A state-of-the-art literature review. *Int. J. Manag. Rev.* **2007**, *9*, 53–80. [CrossRef]
28. Rao, P.; Holt, D. Do green supply chains lead to competitiveness and economic performance? *Int. J. Oper. Prod. Manag.* **2018**, *25*, 898–916. [CrossRef]
29. Caniëls, M.C.J.; Cleophas, E.; Semeijn, J. Implementing green supply chain practices: An empirical investigation in the shipbuilding industry. *Marit. Policy Manag.* **2016**, *43*, 1005–1020. [CrossRef]
30. Ahi, P.; Searcy, C. An analysis of metrics used to measure performance in green and sustainable supply chains. *J. Clean. Prod.* **2015**, *86*, 360–377. [CrossRef]
31. Favi, C.; Raffaeli, R.; Germani, M.; Gregori, F.; Maneri, S.; Vita, A. A Life-Cycle Model to Assess Costs and Enviromental Impacts of Different Maritime Vessel Typologies. In Proceedings of the ASME 2017 International Design Engineering Technical Conferences and Computers and Information in Engineering Conference, Cleveland, OH, USA, 6–9 August 2017; pp. 1–12.
32. Chang, S.L.; Wang, R.C.; Wang, S.Y. Applying fuzzy linguistic quantifier to select supply chain partners at different phases of product life cycle. *Int. J. Prod. Econ.* **2006**, *100*, 348–359. [CrossRef]
33. Chang, S.L.; Wang, R.C.; Wang, S.Y. Applying a direct multi-granularity linguistic and strategy-oriented aggregation approach on the assessment of supply performance. *Eur. J. Oper. Res.* **2006**, *177*, 1013–1025. [CrossRef]

34. Wang, S.Y. Applying 2-tuple multigranularity linguistic variables to determine the supply performance in dynamic environment based on product-oriented strategy. *IEEE Trans. Fuzzy Syst.* **2008**, *16*, 29–39. [CrossRef]
35. Xu, Y.; Li, C.; Wen, X. Missing values estimation and consensus building for incomplete hesitant fuzzy preference relations with multiplicative consistency. *Int. J. Comput. Intell. Syst.* **2018**, *11*, 101–119. [CrossRef]
36. Zhou, M.; Huo, J.Z. Measure product-driven supply chain performance using MEOWA. International Conference on Wireless Communications, Networking and Mobile Computing (WiCOM), Dalian, China, 12–14 October 2008.
37. Lin, C.T.; Chiu, H.; Chu, P.Y. Agility index in the supply chain. *Int. J. Prod. Econ.* **2006**, *100*, 285–299. [CrossRef]
38. Azevedo, S.G.; Govindan, K.; Carvalho, H.; Cruz-Machado, V. Ecosilient Index to assess the greenness and resilience of the upstream automotive supply chain. *J. Clean. Prod.* **2013**, *56*, 131–146. [CrossRef]
39. Azevedo, S.G.; Carvalho, H. LARG index: A benchmarking tool for improving the leanness, agilit, resilience and greenness of the automotive supply chain. *Benchmarking* **2016**, *23*, 1472–1499. [CrossRef]
40. Ramirez-Peña, M.; Sotano, A.J.S.; Pérez-Fernandez, V.; Abad, F.J.; Batista, M. Achieving a Sustainable Shipbuilding Supply Chain under I4.0 perspective. *J. Clean. Prod.* **2019**, in press.
41. Reguant-Álvarez, M.; Torrado-Fonseca, M. El método Delphi. *REIRE Rev. Innov. Recer. Educ.* **2016**, *9*. [CrossRef]
42. Webler, T.; Levine, D.; Rakel, H.; Renn, O. A novel approach to reducing uncertainty: The group Delphi. *Technol. Forecast. Soc. Chang.* **1991**, *39*, 253–263. [CrossRef]
43. Kumar, K.; Zindani, D.; Davim, J.P. Developments towards the Fourth Industrial Revolution. In *Manufacturing And Surface Engineering*; Springer Briefs in Applied Sciences and Technology; Davim, J.P., Aveiro, P., Eds.; Manufacturing and Surface Engineering: Aveiro, Portugal, 2019.
44. Anbanandam, R.; Banwet, D.K.; Shankar, R. Evaluation of supply chain collaboration: A case of apparel retail industry in India. *Int. J. Product. Perform. Manag.* **2011**, *60*, 82–98. [CrossRef]
45. Stonebraker, P.W.; Goldhar, J.; Nassos, G. Weak links in the supply chain: Measuring fragility and sustainability. *J. Manuf. Technol. Manag.* **2009**, *20*, 161–177. [CrossRef]
46. Nunlee, M.; Qualls, W.; Rosa, J.A. Antecedents of supply chain management: A performance measurement model. In Proceedings of the American Marketing Association Winter Educators' Conference, San Antonio, TX, USA, 12 February 2000; pp. 354–360.
47. Tummala, R.; Schoenherr, T. Assessing and managing risks using the Supply Chain Risk Management Process (SCRMP). *Supply Chain Manag.* **2011**, *16*, 474–483. [CrossRef]
48. Demeter, K.; Matyusz, Z. The impact of lean practices on inventory turnover. *Int. J. Prod. Econ.* **2011**, *133*, 154–163. [CrossRef]
49. Furlan, A.; Vinelli, A.; Pont, G.D. Complementarity and lean manufacturing bundles: An empirical analysis. *Int. J. Oper. Prod. Manag.* **2011**, *31*, 835–850. [CrossRef]
50. Parveen, M.; Rao, T.V.V.L.N. An integrated approach to design and analysis of lean manufacturing system: A perspective of lean supply chain. *Int. J. Serv. Oper. Manag.* **2009**, *5*, 175–208. [CrossRef]
51. Agarwal, A.; Shankar, R.; Tiwari, M.K. Modeling agility of supply chain. *Ind. Mark. Manag.* **2007**, *36*, 443–457. [CrossRef]
52. Swafford, P.M.; Ghosh, S.; Murthy, N. Achieving supply chain agility through IT integration and flexibility. *Int. J. Prod. Econ.* **2008**, *116*, 288–297. [CrossRef]
53. Rice, J.B.; Caniato, F. Building a Secure and Resilience Supply Chain.Pdf. *Supply Chain Manag. Rev.* **2003**, *5*, 22–30.
54. Taylor, P.; Boute, R.; Van Dierdonck, R.; Vereecke, A. Organising for supply chain management. *Int. J. Logist. Res. Appl.* **2012**, *5*, 37–41.
55. Hu, A.H.; Hsu, C.W. Empirical study in the critical factors of Green Supply Chain Management (GSCM) practice in the Taiwanese electrical and electronics industries. In Proceedings of the 2006 IEEE International Conference on Management of Innovation and Technology, Singapore, 21–23 June 2006; Volume 2, pp. 853–857.
56. Zhu, Q.; Sarkis, J.; Lai, K.H. Confirmation of a measurement model for green supply chain management practices implementation. *Int. J. Prod. Econ.* **2008**, *111*, 261–273. [CrossRef]
57. Holt, D.; Ghobadian, A. An empirical study of green supply chain management practices amongst UK manufacturers. *J. Manuf. Technol. Manag.* **2009**, *20*, 933–956. [CrossRef]

58. Yeung, J.F.Y.; Chan, A.P.C.; Chan, D.W.M.; Li, L.K. Development of a partnering performance index (PPI) for construction projects in Hong Kong: A Delphi study. *Constr. Manag. Econ.* **2007**, *25*, 1219–1237. [CrossRef]
59. Tuck, C.; Hague, R.; Burns, N. Rapid manufacturing: Impact on supply chain methodologies and practice. *Int. J. Serv. Oper. Manag.* **2007**, *3*, 1–22. [CrossRef]
60. Muthukumarasamy, K.; Balasubramanian, P.; Marathe, A.; Awwad, M. Additive manufacturing—A future revolution in supply chain sustainability and disaster management. In Proceedings of the 8th International Conference on Industrial Engineering and Operations Management, Bandung, Indonesia, 6–8 March 2018; Volume 5, pp. 517–523.
61. Nascimento, D.L.M.; Alencastro, V.; Quelhas, O.L.G.; Caiado, R.G.G.; Garza-Reyes, J.A.; Rocha-Lona, L.; Tortorella, G. Exploring Industry 4.0 technologies to enable circular economy practices in a manufacturing context: A business model proposal. *J. Manuf. Technol. Manag.* **2019**, *30*, 607–627. [CrossRef]
62. Wang, G.; Gunasekaran, A.; Ngai, E.W.T.; Papadopoulos, T. Big data analytics in logistics and supply chain management: Certain investigations for research and applications. *Int. J. Prod. Econ.* **2016**, *176*, 98–110. [CrossRef]
63. Zhao, R.; Liu, Y.; Zhang, N.; Huang, T. An optimization model for green supply chain management by using a big data analytic approach. *J. Clean. Prod.* **2017**, *142*, 1085–1097. [CrossRef]
64. Yan, J.; Xin, S.; Liu, Q.; Xu, W.; Yang, L.; Fan, L.; Chen, B.; Wang, Q. Intelligent supply chain integration and management based on cloud of things. *Int. J. Distrib. Sens. Netw.* **2014**, *2014*. [CrossRef]
65. Subramanian, N.; Abdulrahman, M.D. Logistics and cloud computing service providers' cooperation: A resilience perspective. *Prod. Plan. Control* **2017**, *28*, 919–928. [CrossRef]
66. Wang, K.Q.; Liu, H.C.; Liu, L.; Huang, J. Green supplier evaluation and selection using cloud model theory and the QUALIFLEX method. *Sustainability* **2017**, *9*, 688. [CrossRef]
67. Plinta, D.; Krajčovič, M. Production system designing with the use of digital factory and augmented reality technologies. *Adv. Intell. Syst. Comput.* **2015**, *350*, 187–196.
68. Mourtzis, D.; Zogopoulos, V.; Katagis, I.; Lagios, P. Augmented Reality based Visualization of CAM Instructions towards Industry 4.0 paradigm: A CNC Bending Machine case study. *Procedia CIRP* **2018**, *70*, 368–373. [CrossRef]
69. Panetto, H.; Iung, B.; Ivanov, D.; Weichhart, G.; Wang, X. Challenges for the cyber-physical manufacturing enterprises of the future. *Annu. Rev. Control* **2019**, *47*, 200–213. [CrossRef]
70. Sony, M. Industry 4.0 and lean management: A proposed integration model and research propositions. *Prod. Manuf. Res.* **2018**, *6*, 416–432. [CrossRef]
71. Nafais, S. Automated Lean Manufacturing. Ph.D. Thesis, Kingston University London, London, UK, 2017. [CrossRef]
72. FISITA. A Smart Universal Solution for Optimizing Part Delivering Routing. Available online: www.apress.com (accessed on 31 March 2019).
73. Hu, F.; Lu, Y.; Vasilakos, A.V.; Hao, Q.; Ma, R.; Patil, Y.; Zhang, T.; Lu, J.; Li, X.; Xiong, N.N. Robust Cyber-Physical Systems: Concept, models, and implementation. *Future Gener. Comput. Syst.* **2016**, *56*, 449–475. [CrossRef]
74. Yuan, Z.; Gong, Y. Improving the Speed Delivery for Robotic Warehouses. *IFAC-PapersOnLine* **2016**, *49*, 1164–1168. [CrossRef]
75. Nayak, V.H. A Smart Universal Solution for Optimizing Part Delivering Routing. In Proceedings of the FISITA World Automotive Congress 2018, Chennai, India, 2–5 October 2018.
76. Hsu, C.; Wallace, W.A. An industrial network flow information integration model for supply chain management and intelligent transportation. *Enterp. Inf. Syst.* **2007**, *1*, 327–351. [CrossRef]
77. Pendleton, S.D.; Andersen, H.; Du, X.; Shen, X.; Meghjani, M.; Eng, Y.H.; Rus, D.; Ang, M.H. Perception, planning, control, and coordination for autonomous vehicles. *Machines* **2017**, *5*, 6. [CrossRef]
78. Bechtsis, D.; Tsolakis, N.; Vlachos, D.; Iakovou, E. Sustainable supply chain management in the digitalisation era: The impact of Automated Guided Vehicles. *J. Clean. Prod.* **2017**, *142*, 3970–3984. [CrossRef]
79. Perboli, G.; Musso, S.; Rosano, M. Blockchain in Logistics and Supply Chain: A Lean Approach for Designing Real-World Use Cases. *IEEE Access* **2018**, *6*. [CrossRef]
80. Raju, K.; Ravichandran, S.; Khadri, S.P.M.S. Blockchain for on-demand small launch vehicle supply chain. In Proceedings of the 69th International Astronautical Congress Involving Everyone, Bremen, Germany, 1–5 October 2018.

81. Gao, Z.; Xu, L.; Turner, G.; Patel, B.; Diallo, N.; Chen, L.; Shi, W. Blockchain-based identity management with mobile device. In Proceedings of the 1st Workshop on Cryptocurrencies and Blockchains for Distributed Systems, Munich, Germany, 15 June 2018; pp. 66–70.
82. Kouhizadeh, M.; Sarkis, J. Blockchain practices, potentials, and perspectives in greening supply chains. *Sustainability* **2018**, *10*, 3652. [CrossRef]
83. Cavazos, C.J. Ensuring data security for drilling automation and remote drilling operations. In Proceedings of the SPE Asia Pacific Oil and Gas Conference and Exhibition, Jakarta, Indonesia, 22–24 October 2013; Volume 2, pp. 1289–1294.
84. Furfaro, A.; Gallo, T.; Garro, A.; Sacca, D.; Tundis, A. Requirements specification of a cloud service for Cyber Security compliance analysis. In Proceedings of the 2016 2nd International Conference on Cloud Computing Technologies and Applications (CloudTech), Marrakech, Morocco, 24–26 May 2016; pp. 205–212.
85. Dynes, S.; Eric Johnson, M.; Andrijcic, E.; Horowitz, B. Economic costs of firm-level information infrastructure failures: Estimates from field studies in manufacturing supply chains. *Int. J. Logist. Manag.* **2007**, *18*, 420–442. [CrossRef]
86. Maccari, M.; Polzonetti, A.; Sagratella, M. *Proceedings of the Future Technologies Conference (FTC) 2018*; Springer International Publishing: Berlin/Heidelberg, Germany, 2019; Volume 881, pp. 305–323.
87. Park, S. Development of Innovative Strategies for the Korean Manufacturing Industry by Use of the Connected Smart Factory (CSF). *Procedia Comput. Sci.* **2016**, *91*, 744–750. [CrossRef]
88. Lara, M. *Sistemas de Integración Horizontal y Vertical en Industria 4.0: Evaluación y Desarrollo*; Autonomous University of Nuevo León: San Nicolás de ñps Garza, México, 2018.
89. Zarei, M.; Fakhrzad, M.B.; Jamali Paghaleh, M. Food supply chain leanness using a developed QFD model. *J. Food Eng.* **2011**, *102*, 25–33. [CrossRef]
90. Fang, J.; Huang, G.Q.; Li, Z. Event-driven multi-agent ubiquitous manufacturing execution platform for shop floor work-in-progress management. *Int. J. Prod. Res.* **2013**, *51*, 1168–1185. [CrossRef]
91. Valipour Parkouhi, S.; Safaei Ghadikolaei, A. A resilience approach for supplier selection: Using Fuzzy Analytic Network Process and grey VIKOR techniques. *J. Clean. Prod.* **2017**, *161*, 431–451. [CrossRef]
92. Kannan, D.; Khodaverdi, R.; Olfat, L.; Jafarian, A.; Diabat, A. Integrated fuzzy multi criteria decision making method and multiobjective programming approach for supplier selection and order allocation in a green supply chain. *J. Clean. Prod.* **2013**, *47*, 355–367. [CrossRef]
93. Niranjan, S.K.; Aradhya, V.N.M. Uttar Pradesh Section, Institute of Electrical and Electronics Engineers. In Proceedings of the 2016 2nd International Conference on Contemporary Computing and Informatics (IC3I), Noida, India, 14–17 December 2016; pp. 89–94.
94. Ping, L.; Liu, Q.; Zhou, Z.; Wang, H. Agile Supply Chain Management over the Internet of Things. In Proceedings of the 2011 International Conference on Management and Service Science, Wuhan, China, 12–14 August 2011.
95. Mithun Ali, S.; Moktadir, M.A.; Kabir, G.; Chakma, J.; Rumi, M.J.U.; Islam, M.T. Framework for evaluating risks in food supply chain: Implications in food wastage reduction. *J. Clean. Prod.* **2019**, *228*, 786–800. [CrossRef]
96. Chen, R.Y. Intelligent IoT-Enabled System in Green Supply Chain using Integrated FCM Method. *Int. J. Bus. Anal.* **2015**, *2*, 47–66. [CrossRef]
97. Pool, A.; Wijngaard, J.; Van Der Zee, D.J. Lean planning in the semi-process industry, a case study. *Int. J. Prod. Econ.* **2011**, *131*, 194–203. [CrossRef]
98. Khalili-Damghani, K.; Taghavifard, M.; Olfat, L.; Feizi, K. A hybrid approach based on fuzzy dea and simulation to measure the efficiency of agility in supply chain: Real case of dairy industry. *Int. J. Manag. Sci. Eng. Manag.* **2011**, *6*, 163–172. [CrossRef]
99. Carvalho, H.; Barroso, A.P.; MacHado, V.H.; Azevedo, S.; Cruz-Machado, V. Supply chain redesign for resilience using simulation. *Comput. Ind. Eng.* **2012**, *62*, 329–341. [CrossRef]
100. Vlachos, D.; Georgiadis, P.; Iakovou, E. A system dynamics model for dynamic capacity planning of remanufacturing in closed-loop supply chains. *Comput. Oper. Res.* **2007**, *34*, 367–394. [CrossRef]
101. Womack, J.P.; Jones, D.T.; Roos, D. *La Máquina que Cambió el Mundo*; McGraw-Hill: New York, NY, USA; McGraw-Hill Management: Buenos Aires, Argentina, 1995; p. 292.
102. Womack, J.P.; Jones, D.T. From lean production to the lean enterprise. *Harvard Bus. Rev.* **1994**, *72*, 93–103.

103. Mello, M.H. Strandhagen, Supply chain management in the shipbuilding industry: Challenges and perspectives. *Proc. Inst. Mech. Eng. Part M: J. Eng. Marit. Environ.* **2011**, *225*, 261–270.
104. Xie, G.; Yue, W.; Wang, S. Energy efficiency decision and selection of main engines in a sustainable shipbuilding supply chain. *Transp. Res. Part D* **2017**, *53*, 290–305. [CrossRef]
105. Nichols, T. Shipbuilding 4.0: The Digital Thread in Shipbuilding Technology. Digital Mariner. 2017. Available online: https://blogs.plm.automation.siemens.com/t5/Digital-Transformations/Shipbuilding-4-0-the-digital-thread-in-shipbuilding-technology/ba-p/401021 (accessed on 9 september 2019).
106. Meyer Werft. Meyer Werft. Available online: https://www.meyerwerft.de/de/meyerwerft_de/index.jsp (accessed on 9 september 2019).
107. Daewoo Shipbuilding & Maritime Engineering. Available online: www.dsme.co.kr/epub/main/index.do (accessed on 10 September 2019).
108. Hyundai Heavy Industries. Available online: english.hhi.co.kr/main (accessed on 10 September 2019).
109. Cheong, Y. Connected Smart Ships. Available online: https://thedigitalship.com/conferences/presentations/2015kormarine/3.pdf (accessed on 10 September 2019).
110. Comunication Team Samsung Heavy Industries. Samsung Heavy Industries. Available online: www.samsungshi.com/Eng/default.aspx (accessed on 10 September 2019).

© 2019 by the authors. Licensee MDPI, Basel, Switzerland. This article is an open access article distributed under the terms and conditions of the Creative Commons Attribution (CC BY) license (http://creativecommons.org/licenses/by/4.0/).

Article

New Risk Methodology Based on Control Charts to Assess Occupational Risks in Manufacturing Processes

Martin Folch-Calvo [1,*], Francisco Brocal [2] and Miguel A. Sebastián [1]

1. Manufacturing and Construction Engineering Department, ETS de Ingenieros Industriales, Universidad Nacional de Educación a Distancia, Calle Juan del Rosal, 12, 28040 Madrid, Spain; msebastian@ind.uned.es
2. Department of Physics, Systems Engineering and Signal Theory, Escuela Politécnica Superior, Universidad de Alicante, Campus de Sant Vicent del Raspeig s/n, 03690 Sant Vicent del Raspeig, Alicante, Spain; francisco.brocal@ua.es
* Correspondence: mfolch15@alumno.uned.es

Received: 12 October 2019; Accepted: 7 November 2019; Published: 11 November 2019

Abstract: The accident rate in the EU-28 region of the European Union showed a value of 2 fatal accidents per 100,000 people in 2019 that mainly affect construction (24%), manufacturing (19%) and logistics (19%). To manage situations that affect occupational risk at work, a review of existing tools is first carried out taking into account three prevention, simultaneity and immediacy characteristics. As a result, a new dynamic methodology called Statistical Risk Control (SRC) based on Bayesian inference, control charts and analysis of the hidden Markov chain is presented. The objective is to detect a situation outside the limits early enough to allow corrective actions to reduce the risk before an accident occurs. A case is developed in a medium-density fiberboard (MDF) manufacturing plant, in which five inference models based on Poisson, exponential and Weibull distributions and risk parameters following gamma and normal distributions have been tested. The results show that the methodology offers all three characteristics, together with a better understanding of the evolution of the operators in the plant and the safety barriers in the scenario under study.

Keywords: Bayesian inference; control chart; dynamic methodology; hidden Markov chain; occupational accident; risk assessment; risk control; risk management

1. Introduction

The accident rate in the European Union, [1] for the EU-28 region, was 2 fatal accidents per 100,000 people employed in 2015. The most affected industrial activities were: building (24%), manufacturing (19%), transport and storage (19%), agriculture—fishing (15%), retail (9%), public administration (9%), water supply–waste management (3%) and mining (2%). In Spain, between 2014 and 2018, the most affected activities were practically the same [2].

The main causes of accidents also in Spain for the same period 2014–2018 were: inadequate movements of the human body, in actions of pushing and pulling and by inappropriate body turns, all of them under physical effort (33%); entrapment and contact with sharp areas in machine elements (22%); falls and slips (18%); loss of total or partial control of a machine (16%); breakage and sliding of a work support (6%); leaks and spillages (2%); aggression (2%); and explosions–fire (1%) [2], Figure 1.

Of the total accidents generated by these causes, (99.1%) have had minor consequences, (0.8%) have caused severe damage, and (0.1%) have been very serious with a fatal outcome. Avoiding an accident at work regardless of its severity requires specific risk management, carried out throughout the life cycle; from design, engineering and construction, commissioning, operations, logistics and final dismantling; and in which at all times it is necessary to monitor qualitatively and quantitatively the moment in which an accident risk arises.

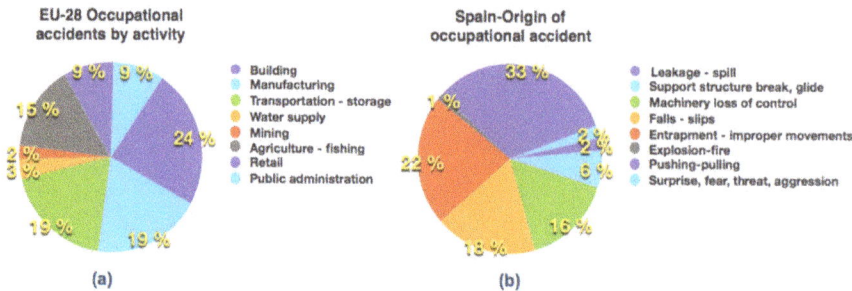

Figure 1. (**a**) Accident rate by activity in the year 2015 in the EU-28 countries of the European Union (adapted from [1]); (**b**) description of origin of occupational accidents in Spain for years 2014–2018 (adapted from [2]).

There are different points of view about the concept of risk; for example, in general a risk arises from the existence of uncertainty [3], in a more specific way a risk arises from the existence of uncertainty in the objectives [4,5], or in more practical terms, risk can be defined as the measure of lack of security [6]. The idea of security also has different meanings that can be defined simply as the lack of accidents, or from a labor point of view, how people can provide the required performance in expected and unexpected conditions [7], or quantifying safety as a numerical condition where the number of adverse outcomes is acceptably small [8] or from an analytical point of view that defines it as the study of why things go wrong [9].

Despite the definitions, the most important issue is how to manage risk in various scenarios and specifically those related to occupational safety at work. In this sense, an initial framework issued by the European Union is Directive 89/391/EEC [10] aimed at employers to establish basic principles of prevention with risk assessment and its avoidance being a preliminary and basic proposal to carry out an assessment of occupational risks.

The ISO 31000:2018 establishes the principles of risk management and the ISO/IEC 31010:2019 establishes the risk management-assessment associated techniques. It is based on the Deming cycle [11], consisting of a sequence of steps: "plan, do, check, act".

Quantitative risk assessment (QRA) is a formal and systematic risk-analysis approach to quantifying the risks associated with the industrial and human processes. The risk assessment is the general procedure that covers the risk identification process which can be performed based on historical data, through a panel of experts or using inductive cause-effect techniques; its analysis applying: qualitative methods, indicating the levels of importance of the risk and its consequences; semi-quantitative methods, indicating numerical risk rating scales and their consequences; and quantitative methods, defining the probabilities of risk-generation and its consequences [12–14], and their evaluation that implies determining the importance and prioritizing from the point of view of risk consequence or benefit–cost [15], Figure 2.

With the objective of managing occupational hazards, the ISO 45001 guideline [16] aims to provide guidelines that allow the implementation of a system of occupational health and safety (OH&S). In Spain, the main guidelines on occupational hazards come from the National Institute of Occupational Safety and Health, and are issued with the objective of providing guidelines that allow analyzing and studying health and safety conditions in the workplace, as well as their improvement [17]. In the context of manufacturing processes involving chemical agents, there are two European directives on the assessment of chemical agents at work [18] and for carcinogens and mutagens at work [19].

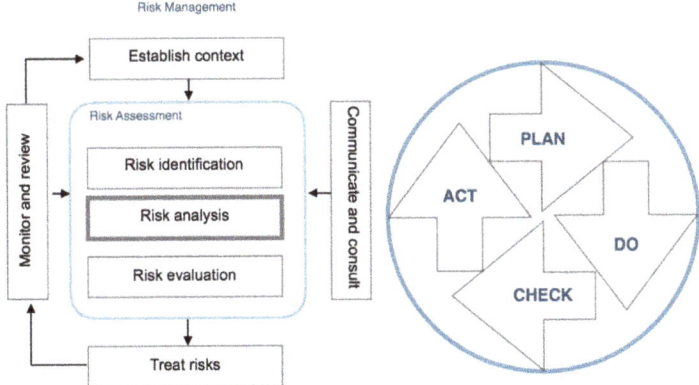

Figure 2. Risk-management process and Deming-cycle equivalence (adapted from [5]).

In this context, is especially important to prevent both occupational accidents and major accidents which can be interrelated [20], by means of adequate management and assessment methodologies. In the scientific literature, methodologies oriented to the management and assessment of occupational health and safety risks are collected with the addition of specific tools; such as a risk assessment based on fuzzy logic with application in the mining industry [21] and manufacturing plants [22]; or the two cases of risk assessment in process industries and to detect and evaluate emerging risks in industrial processes [23,24]; or the application of a multi-objective evolutionary algorithm to take into account the tasks, activities, associated risks and safety of workers with case studies associated with scaffolding falls [25,26]; the use of Bayesian networks applied to work situations on the high seas where slips, trips and falls must be avoided [27]; several cases of risk assessment applied in the mining industry [28]; in situations of construction of scaffolding falling objects and contact with moving machines and vehicles [29]; risk assessment in an aluminum process industry for workers using press extruders, forklifts, cranes, and production [30]; and the use of block diagrams to identify and evaluate fall hazards from an escalator [31]. It is applied in the construction of a natural gas pipeline using a tool based on Pythagorean fuzzy logic [32], the risk assessment produced by falls and falling objects, crane handling, and interaction with the moving parts of the machines [33–36], and risk assessment in occupational accidents related to the construction, operation and maintenance of wind farms on land [37].

However, there is no method that allows us to obtain an overview of the state of occupational risk in a manufacturing plant in the simplest way possible that is at the same time formal, which warns about the existence of a risk to avoid in advance. For this, three characteristics are considered necessary:

1. Prevention (P): be the process of avoiding or mitigating risks by reducing their probability of occurrence and their impacts on human and social; geographical and landscape; economic and infrastructure; environmental and ecosystem preservation; accident and safety (human, assets, production); perception and expectations.
2. Simultaneity (S): is the ability to update the evolution of risk according to the operation in real time.
3. Immediacy (I): is the ability to inform or infer the existence of a risk with sufficient anticipation to make the necessary corrections before the accident occurs.

To achieve this objective, it is necessary to first review what are the current tools related to risk management and evaluation and if, verifying which meet the three characteristics. Based on the most appropriate methodology, the objective of this paper is to present a new tool called Statistical Risk Control (SRC) for to manage and assess the situations of risk, focused in occupational accidents.

With this object, this work is organized as follows: in Section 2 we review the state of the existing tools; in Section 3 we examine the development of the SRC methodology applied for occupational accidents; in Section 4 we present a case study and the results; in Section 5 we discuss the results; and in Section 6 we draw conclusions.

2. Existing and Related Tools for Occupational Risk Management

2.1. Regulations and Traditional, Modern Models

The analysis covers three main groups:

1. The first group corresponds to the standards and directives whose characteristics and degree of compliance with the three characteristics (P), (S), (I) are summarized in Table 1.
2. The second group covers methodologies and models differentiating the traditional and the modern approaches [38]. The traditional approach includes the sequential and the epidemiological models, summarized in Table 2. The modern approach has five models: the systematic; cloud based; the fuzzy based, formal based and safety barrier based; summarized in Table 3.
3. The third group, which is encompassed in the modern methodologies, is specific for dynamic models and it is discussed in the next subsection.

Table 1. Existing standards and regulations.

Std's / Directives	Application	P	S	I
89/391/EEC	Occupational - basic. [10]	+	-	-
ISO 45001:2018	Implementation of a system of occupational health & safety (OH&S). [16]	+	-	-
NISHW	Spanish governmental organization of analysis and study for health and safety conditions in the workplace. [17]	+	-	-
98/24/EC, 2004/37/EC	Occupational - Chemicals and carcinogens concentration. [18,19]	+	-	-
ISO/IEC 31010:2019	Risk management process, based on a iterative cycle. Risk assessment based on identification, analysis and evaluation. General application of (QRA). [4,5,38]	+	-	-
PMBOK, PRINCE2	Documentation tailored for projects. Design, Start, Direction, Planning, Execution, Control. [39,40]	+	-	-
CCPS	Layer of Protection Analysis (LOPA) methodology. A process deviation can lead to a hazardous consequence if not interrupted by an independent protection layer (IPL). Applied in chemical process. [14,41–43]	+	-	-
NORSOK 2010	Applied in the Norwegian petroleum industry, under the idea of Operational Risk Assessment, with the aim to follow the lifecycle of a project considering planning, execution and operation. [44–46]	+	-	-
2012/18/EUCOMAH 2015	European and British Control of Major Hazards for Seveso III Directive. Emergency plan with major accident prevention policy and information mechanism to authorities and population. A 5 years safety report. [47,48]	+	-	-
CPR18E	Netherlands advisory council of dangerous substances, and the old (Commissie voor de Preventie van Rampenthat, CPR). Applied in hazardous installations and transport analyzing the loss of containment events and the modeling of the associated flammable clouds, their dispersion and toxic effects. [49]	+	-	-
EN 16991:2018	European standards for chemical, power generation and manufacturing providing guidance for the inspection and risk evaluation in operations and maintenance. [50,51]	+	-	-

Table 2. Existing traditional models.

Models	Application	P	S	I
Sequential	Are representative of the Quantitative Risk Assessment (QRA) methodology regarding accidents as outcomes of a chain of discrete events or factors that take place in a temporal order. Analyzing causes and consequences of risk.	+	+/-	-
ETA	Event Tree Analysis. Consequence analysis. General application. [14]	+	+/-	-
FTA	Fault Tree Analysis. Causes of risks for human and technical systems. Applied in occupational risk analysis in the textile industry. [14,52]	+	+/-	-
BOW-TIE	Graphic including FTA and ETA models to represent causes, safety barriers, and consequence events. [14]	+	+/-	-
THERP	(Technique for Human Error Rate Prediction) a tool based on event-tree approach for evaluating human errors alone or in connection with equipment functioning, operational procedures and practices. [53]	+	+/-	-
FMEA	Failure Mode Effect Analysis. Step-by-step approach for identifying potential failures. [54]	+	+/-	-
Check list-What if	Systematic revision to find malfunctions and compliance with a list of requirements. [54]	+	+/-	-
FMECA	Failure modes, Effects and Criticality Analysis. Upgrade of the FMEA. The criticality is determined classifying the degree of potential failures. Case application for a toxic exposure to contaminants in a drug industry. [54,55]	+	+/-	-
RA	Reliability Assessment. Quantification of the probability of failure in a system. [56]	+	+/-	-
Block Diagrams	Graphical procedure describing the function of the system and showing the logical connections of components needed to fulfill a specified system function. [57]	+	+/-	-
HAZOP/HAZID	Technique for early identification of hazards usually applied in the design, the study is carried out by an experienced multi-discipline team using a checklist of potential hazards. [58]	+	+/-	-
EBM	Energy Barrier Model defining a safety barrier management and considering that an accident occur when hazards succeed to penetrate the safety barriers deficiencies. [59,60]	+	+/-	-
MORT	Management Oversight and Risk Tree. Root cause determination. Case for an elevator incident. [61]	+	+/-	-
SCAT	Systematic Cause Analysis. Causal analysis using a poster schematic which enables the identification of relevant corrective and preventive actions. [62]	+	+/-	-
STEP	Sequential Time Events Plotting. Identification of multiple causes in occupational accidents. [63]	+	+/-	-
MTO	Man Technology and Organization. Root causes in occupational work affected by the organization; practice; management; procedures and deficiencies in work environment. [64]	+	+/-	-
SOL	Safety through Organizational Learning. Event analysis in two steps: (1) description of the actual event situation, and (2) identification of contributing factors. Application in the nuclear industry. [65]	+	+/-	-
Epidemiological	Propagation of events is analogous to a disease spreading considering their distribution and determinants. Accidents are caused by latent events under epidemic context. Application in helicopter and road accidents. [66,67]	+	+/-	-

495

Table 3. Existing modern models.

Models	Application	P	S	I
Systematic	General risk framework based on the Rasmussen's model using control theory concepts and considering that social climate is affected by government policy and budgeting, regulatory associations, organization, staff and the work operation systems for which their limitations and their interactions can allow preconditions for accidents. [68,69]	+	+/-	-
AcciMap	Cause event representation of the system interactions and how to control the hazardous processes originated into of the organizational and socio - technical system. [70]	+	+/-	-
STAMP	Systems Theoretic Accident Model. The systems are subject to external disturbances and can cause accidents due to physical, social and economic pressures and control failures in safety barriers. Human action supports part or all of the operation and actions of the system. A checklist is applied to identify control failures in safety barriers. [71–73]	+	+/-	-
CREAM	Cognitive Reliability and Error Analysis Method. Human performance is modeled to asses the consequences of the human errors. [74,75]	+	+/-	-
DREAM	Driving Reliability and Error Analysis Method. Application in driving accidents. [76]	+	+/-	-
FRAM	Functional Resonance Accident Model. As a result of the functional couplings appears resonance. The functional or basic processes in a risk scenario are identified, defining for each of them what are the inputs needed; the outputs produced; the needed resources (equipment, procedures, energy, materials and manpower); the controls to supervise, the preconditions to be fulfilled to carry the process and the time. The resonance can appear due to the variability in the dependence between processes. Application on aircraft, maritime and manufacturing. [77]	+	+/-	-
AEB	Accident Evolution and Barrier Function. Interaction between technical and human-organizational systems which may lead to an accident. The analysis needs work-team by engineers and human accidents specialists. [78]	+	+/-	-
Cloud based	Based on the FMECA. Perform a critical risk analysis based on the cloud by establishing a score based on expert knowledge. A case is presented for a gasification station. [79]	+	+/-	-
Fuzzy based	Application of fuzzy logical for define human behavior in risk situations.	+	+/-	-
HEART	Human Error and Assessment Technique. The reliability of any task can be modified by the influence of the Error Promotion Conditions (EPC). It is necessary to previously identify the tasks. For each task, with the help of a team of experts, a probability value of human error generation and the (EPC) that affect it and its relevance are defined. Fuzzy logic is applied to obtain a factor that modifies the probability of error. [53,80]	+	+/-	-
CREAM-BN	Upgrade of the systemic CREAM model. Human behavior has five components: strategic, tactical, opportunistic and scrambled. It is affected by common performance conditions (CPC) defined as: adequacy of the organization, working conditions, human-machine interface, operational support, availability of procedures, number of simultaneous objectives, available time, time of day, training and experience and quality collaboration. A Bayesian network and fuzzy logic are applied to determine the probability of human error. Scramble and opportunistic are the ones with the highest probability. Cases and examples from nuclear industry, aircraft transportation, manufacturing, retail and chemicals. [81]	+	+/-	-
Formal based	Accident causation is approached using probabilistic schemes and Bayesian networks to model the interaction between causes and effects.			
WBA	Why Because Analysis. Bayesian networks are applied considering that each component is a system is affected from the overall system environment. Application in transportation and aircraft accidents [82]	+	+/-	-

496

Table 3. Cont.

Models	Application	P	S	I
Safety Barrier				
PHPAM	Process Hazard Prevention Accident Models. Accidents are initiated by hydrocarbon release and propagation, and it is needed to establish safety barriers into five groups of prevention: release, ignition, escalation, harm and loss. Risk probabilities are evaluated before and after barriers implementation. [83]	+	+/−	−
SHIPP	System Hazard Identification Prediction and Prevention. Update of the initial probability of risk according to the actual data collected and application of Bayesian inference. [84,85]	+	+	−

2.2. The Dynamic Risk Models

This group covers a dynamic risk-analysis concept using sequential models like fault-tree analysis (FTA), event-tree analysis (ETA) and the BOW-TIE graph approaches and performing a Bayesian inference analysis to update the failure probabilities from the information collected of the named accident precursors or precursor data. This group has five models; the dynamic risk assessment (DRA) being the representative; the dynamic procedure for atypical scenarios identification (DyPASI), the dynamic risk analysis, the risk barometer methodology and the dynamic operational risk assessment, [86–88].

DRA is an extension of the risk assessment (RA). The process needs to establish a prior function for the statistical parameter that models the risk probability. The precursors, events or causes that can lead to an accident are observed and formalized through the application of Bayesian inference to obtain the posterior function of the parameter that models the risk probability, [89–94].

Figure 3 presents the main equivalences between RA and DRA models. Highlights are:

- Risk identification is similar as presented in the ISO/IEC 31010:2019. The identification of potential risks are performed by the application of HAZOP/HAZID and FMEA, FMECA techniques [95].
- Scenario consideration is similar to the answer to the "What if?" question. Scenarios reflecting the "best case", "worst case" and "expected case" may be used for quantifying the probability of potential consequences and obtain a sensitivity analysis.
- After identification of causes of risk, their paths and sequences through the safety barriers are defined using the ETA or FTA methods under a bow-tie graph and ending at the final states. Reliability data bases can be applied for human, equipment or barriers failure, or using expert judgment [96,97].
- Observation of precursor data, events and situations from the workplace or the process under analysis.
- Posterior estimation is performed using Bayesian inference through the expression:

$$f(p/Data) \propto g(Data/p) \cdot f(p) \qquad (1)$$

where p is the statistical parameter, $f(p)$ is the prior statistical distribution for the parameter p; $g(Data/p)$ is corresponding to the observed precursor data and $g(p/Data)$ is the posterior statistical distribution.

This strategy has been applied to a number of case studies in petrochemical industry including the case for a storage tank containing hazardous chemicals, a refinery, and oil-spill accidents; or additionally performing the inference using Bayesian Networks applied in offshore oil and gas accidents [98–103].

The characteristics for the four remaining models can be seen in Table 4.

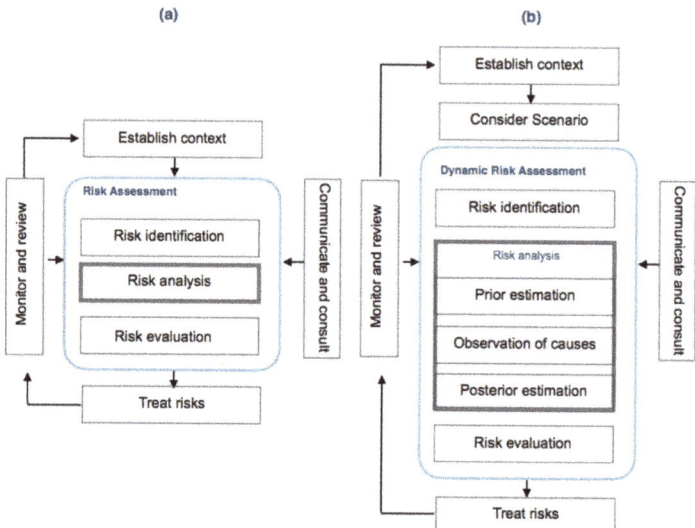

Figure 3. (a) Risk assessment; (b) dynamic risk assessment (adapted from [5]).

Table 4. Existing dynamic models.

Models	Application	P	S	I
Dynamic	Uses sequential models and the Bow-tie graph approach, performing a Bayesian inference analysis to update the failure probabilities from the information collected of the accident precursors.	+	+	-
DyPASI	Dynamic Procedure for Atypical Scenarios Identification. Identification and assessment of the potential hazards based on information from atypical accident scenarios or situations, which are not captured by conventional HAZOP/HAZID techniques. [104,105]	+	+	-
Dynamic Risk Analysis	Analysis process as a step of the Dynamic Risk Assessment methodology being a quantitative modern approach in which the frequency of accidents are updated by the application of the Bayesian theory. [106]	+	+	-
Risk Barometer	For continuously monitor the risk of failure in safety barriers based on an existing Quantitative Risk Assessment (QRA) or a Dynamic Risk Assessment (DRA) and on the Barrier and Operational Risk Analysis (BORA). The safety barriers are analyzed through influencing factors, named Risk Influencing Factors, (RIFs), that are correlated to with the estimated probabilities of failure, followed by their visualization in an equivalent barometer graph. [107]	+	+	-
Dynamic Operational Risk Assessment	Markov and MonteCarlo chain simulations applied to analyze the incidence of events and causes in each component of a system process and its behavior. The method simulates the visits in each of the four states in which they can be found: normal operation; abnormal not detected; abnormal detected and under repair. [108]	+	+	-

3. Statistical Risk Control (SRC) Methodology

3.1. General Application

There is a need to use control charts and a dynamic risk assessment model for analyze risk in general industrial situations. There is no literature in the use of the control charts for risk management but nevertheless they are applied in the earned value method in project management [109], in environmental assessment [110] or for cost control and project duration [111]. The methodology is compatible with the ISO guidelines using the Bayesian inference and Hidden Monte Carlo–Markov methods under a new concept of Statistical Risk Control (SRC) [112] (Figure 4).

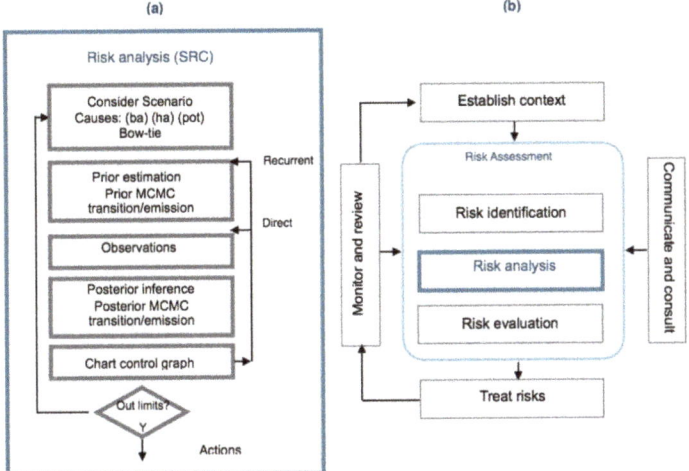

Figure 4. (**a**) Statistical Risk Control (SRC) methodology; (**b**) position in the risk-assessment scheme (adapted from [5]).

When the risk has been identified in a considered scenario, the following seven steps are applied:

1. A bow-tie analysis is performed to provide a visual representation of the causes of initiation (ic) classified as basic, human and potential that affect preventive and mitigative safety barriers and the consequences or final states when an accident occurs, Figure 5.
2. The identification and definition of the initiation causes (ic) which may be: basic events (ba) such as failures in control systems, equipment or processes; human risk factors (ha) which are human errors and the potential causes (pot), which will be defined in the following subsection. The process is iterative between step 1 and step 2 until the causes and consequences have been clearly established.
3. From the previous steps 1 and 2, the statistical parameter p that expresses the risk probability is also identified and the prior statistical distribution that reflects it can be established. Also the prior transition and emission matrices governing changes in the mitigative safety barriers can be defined.
4. The observation of the initiation causes (ic) and end states are put into effect according to a time interval.
5. From the estimated prior $f(p)$ and the observed initiation causes (ic), such as $g(data/p)$ and applying Equation (1), the posterior function for the statistical parameter (p) can be obtained and if there is not an analytical expression for it then the Metropolis–Hastings sampling method can be applied to obtain the posterior distribution and its associated parameters [113,114]. Also corresponding to the hidden Markov chain, the prior transition and emission matrices are defined for the mitigative safety barriers and the posterior transition and emission matrices are obtained using the Baum–Welch algorithm, [115,116].
6. Control chart presentation [117,118] to graph the evolution of the statistical parameter p, in a time interval.

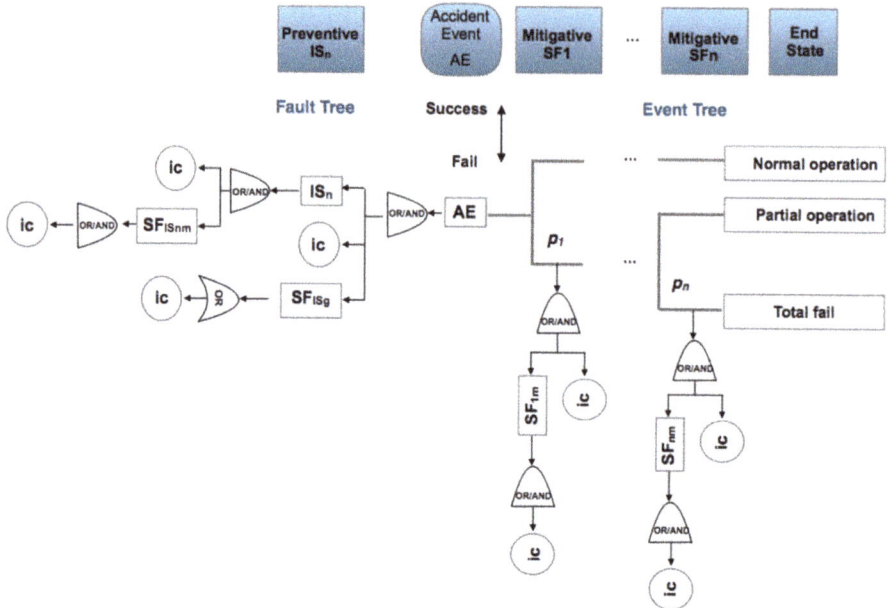

Figure 5. Safety barriers: IS_n: preventive barrier (n); SF_{ISnm}: preventive barrier (n) sub-function (m); SF_{ISg} general sub-function safety barrier; SF_n: mitigative barrier (n); SF_{nm}: mitigative barrier (n) sub-function (m); ic: initiation causes (ba, ha, pot); $p_{,s,f}$: probability of success or fail in the mitigative or reactive safety barrier.

Chart determination has two modes:

a. Direct: uses the observed data up to the analyzed interval time, but with two possibilities: the mean established in the prior function that defines the statistical parameter (p) is constant in every interval, and the standard deviation is determined using the observed data collected up to the analyzed interval, (Direct–Mean Prior) or by modifying the mean and the standard deviation also using the observed data collected up to the analyzed interval (Direct–Mean Posterior).

b. Recurrent: uses the observed data in every interval time, also with two possibilities: maintaining the mean posterior constant, and the standard deviation obtained in every interval is the new prior in the following interval (Recurrent–Mean Prior), or the mean and the standard deviation obtained in every interval are the prior values in the following interval (Recurrent–Mean Posterior).

Considering also the observed initiation causes (ic), and visualizing the bow-tie, there are two possibilities of analysis to include in the control chart:

a. For the complete bow-tie scheme, Figure 6.

    a.1. Collecting the total of the initiation causes (ic) affecting all the preventive and mitigative safety barriers and their barrier sub-functions.

    a.2. Collecting only the first level for initiation causes (ic) and fails in first level of barrier sub-functions.

b. Observing the fault tree (FT) and event tree (ET), and analyzing the response active (yes) or (no) for the preventive and mitigative safety barriers.

7. Analysis applying a hidden Monte Carlo Markov Chain for the mitigative safety barriers, also with two possibilities, Figure 7.

a. Analyzing the behavior of the mitigative safety barriers based on the end states. In this case a transition matrix is defined for the mitigative barriers and an emission matrix for the observed end states in function of the barriers' transition.
b. Analyzing the behavior of the end states based on the action of the mitigative barriers. In this case a transition matrix is defined for the end sates and an emission matrix for the observed mitigative safety barriers in function of the end states.

The Baum–Welch algorithm is applied to obtain, from the observations, the posteriors transition and emission matrices with a methodology that can also be direct or recurrent.

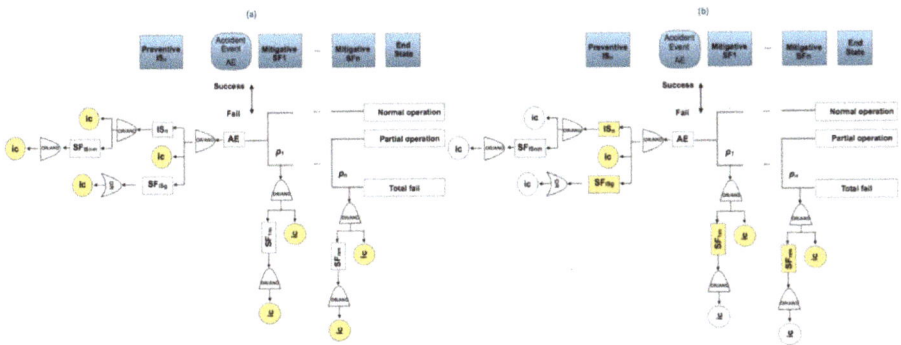

Figure 6. Analysis modes. (a) collecting total of (ic's) of preventive and mitigative barriers and their sub-functions (highlighted yellow); (b) collecting at the first level of (ic's) and barriers sub-functions (highlighted yellow).

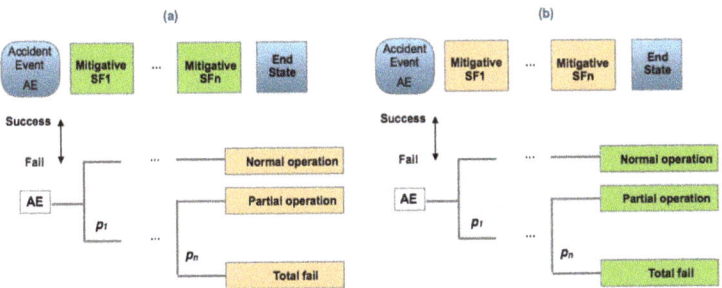

Figure 7 Analysis modes of the hidden Markov chain. (a) Defining transition probabilities for the mitigative barriers (highlighted green) and emission probabilities for the observed end sates in function of the mitigative barriers (highlighted brown); (b) defining transition probabilities for the end states (highlighted green) and emission probabilities for the observed mitigative barriers states (highlighted brown).

3.2. Potential Causes (Pot)

Some authors consider [119] that when unexpected deviations arise due to causes that are difficult to predict because of their randomness, this can be classified as a special risk [120,121]. From the point of view of the methodology (SRC), these causes that can lead to unexpected risk situations are called potential causes, which are summarized in Figure 8.

Potential Cause (pot)		Attributes
Deviation from a formal procedure	(DEF)	Changes derived from operator decisions. Procedures that are formal stablished but not well applied.
Failed test	(FTS)	Test for integrity control, performance, safety operation, production.
Technical design failures	(TFT)	Incorrect design, big repair to adapt a mechanism or process.
Immediate errors after maintenance -safety - control with repetition	(IEC)	After reparation, audit or control a new intervention is needed.
Concatenation or domino effect risk because of unplanned situations	(RCD)	In operations, maintenance. Or in projects.
Unplanned operations or application of a new technology (tested or not)	(UNT)	In operations, maintenance. Or in projects.
Lack in communication and information	(LCI)	In operations, maintenance. Or in projects. Affecting procedures, messages, safety indications, rules of actuation, policies, Or defective in planning and scheduling.
Human and social changes	(HSC)	Government changes, economical variations, migrations or immigration of people, social unrest and violence.
Investment modifications (planned or not)	(IMD)	Affecting safety, with loss of social benefits, increasing work overtime and pressure. Poor cost control.
Extreme climatic situations	(XCS)	Despite historical weather and existent climate conditions. Planned work in rough weather and climate conditions (wind, rain, sea, temperature).
Malicious acts derived from war or terrorism	(MAT)	Social upheavals, threats.

Figure 8. Potential causes (pot) and attributes defined in the Statistical Risk Control (SRC) methodology.

3.3. General Application for Occupational Accidents

In accordance with the SRC methodology, a bow-tie is defined in each particular scenario. In the case of occupational accidents before being able to work with different risk scenarios, it is necessary to analyze what human behavior is at work and what are the factors that affect it. Three views and organizations that cover the most critical have been considered. The first group compiled is general, [17,122,123]. The second group includes the factors considered critical in the 6th European Survey of Working Conditions [124], and the third group includes the emerging psychosocial risks related to occupational safety and health [125]. The aggregation and coincidence of the three groups is presented in Figure 9.

From (NISHW, 2018), (Baybutt 2013), (Kariuki and Löwe, 2007)	From (Eurofound, 2017)	From (EU-OSHA, 2009)	Attributes
Organization (ORG)	Social environment (SEN)	Emotional demands at work (EDW)	Safety policy, culture, learning policy, management supervision and quality, social behavior and support, stress
Job design (JBD)	Working time quality (WTQ) Work intensity (WKI)	Contract - job security (CJS) Poor work - life balance or ratio (PWL) Work intensification (WIT)	Irregular scheduling, shifts and overtime. Low flexibility, atypical arrangements, quantitative and emotional demands, outsourcing, part-time, temporary situations, burn-out effects.
Operator environment (OPE)	Physical environment (PHE)		Ergonomics (noise, vibration, temperature, air), biological and chemical VOC's.
Operator characteristics (OPC)	Skills and discretion (SKD) Prospects (PRO)	Ageing (AGE)	Cognitive, concentration, Expertise-aptitudes, decision level, responsibility, steadiness, participation, skills, training support.
Human system interface (HSI)			Control design interface, displays, panels, actuators. Alarms.
Information (INF)			Training procedures, labeling, communication, labeling.
Workplace design (WKD)			Layout and configuration, accesibility.
	Earnings (EAG)		Expectations

Figure 9. Factors and attributes affecting human behavior at work.

These critical factors must be taken into account as possible causes of initiation (ic) of an occupational accident and must be present in all scenarios. However, it is considered that there are four situations that occur as a consequence and symptom of the critical factors. These are: failure in the self-control of work (JSC); failure to supervise work (JSU); failure in security self-control (SSC) and failure in security supervision (SSU). In addition, these four situations that are easily observable arise in an automated process and manufacturing environments, where workers additionally perform control and supervision tasks. These four situations, pro their control and supervision function, are associated with safety barriers. The general bow-tie for occupational accidents (Figure 10), is equivalent to the general one presented above, with the differences in preventive barriers (IS_n) integrated by the four safety barriers (IS_{JSC}, IS_{JSU}; IS_{SSC}; IS_{SSU}). In this scheme, the (SF_{ISnm}) are sub-functions of the four safety barriers, which can be formed by a procedure, an automatism, an alarm indication, an actuator of a control system or the organizational culture itself. The general safety barrier (SF_{ISg}) works in parallel with the activity of the operators and may consist of automatic control systems, protections, alarms, actuators or automated operations management that guides the operator at each step of the process and does not permit the next step to be formalized if a number of conditions are not met. Sub-function barrier (SF_{nm}) covers the various functional components that integrate the mitigating safety barriers. The final states represented are bounded at one end by the absence of personal injuries, if the first mitigating safety barrier acts correctly and the other by a fatality if the failure of all mitigating safety barriers occurs. The graph must be taken as a framework and must adapt according to the processes, the occupational works and the scenarios that are being analyzed.

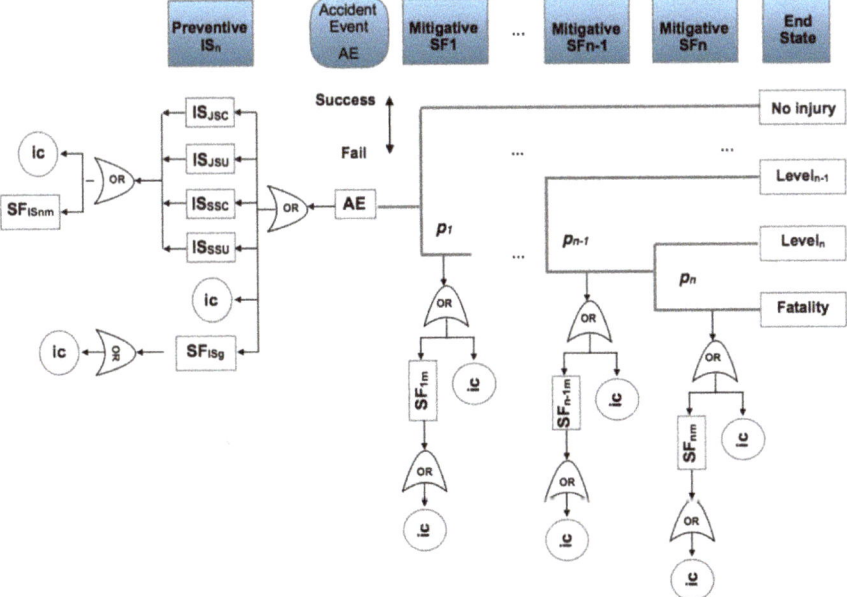

Figure 10. General bow-tie for occupational accidents. Safety barriers: IS_n: preventive barrier (n) based on job self control (JSC); job supervision (JSU); safety self control (SSC) and safety supervision (SSU). SF_{ISg} general sub-function working parallel to the human actions; SFn: mitigative barrier (n); SF_{nm}: mitigative or reactive barrier (n) sub-function (m); ic: initiation causes (ba, ha, pot); $p_{,s,f}$: probability of success or fail in the mitigative or reactive safety barrier.

4. Case Study in a Medium-Density Fiberboard (MDF) Manufacturing Process Plant

4.1. Process

The general process scheme is depicted in Figure 11. Its goal is to produce urea-melamine medium density fiber (MDF) board elements using as basic raw materials paper, wood, melamine, urea, a resin (such as a polyamide or vinyl chloroacetate) and formaldehyde. The paper is subjected to a surface printing treatment continuing with the impregnation phase performed with melamine-formaldehyde. Drying and cooling processes are executed next in a single step if only the melamine-formaldehyde polymer is added or with one additional step if the urea-formaldehyde polymer is added, and with the same impregnation, drying and subsequent cooling steps. The process continues with the cutting and winding of the paper and its stacking. At the same time, the wood is splintered by subsequently drying the material at 180 °C to reduce moisture. The dry material (8% moisture) is impregnated with the urea-formaldehyde solution and the resin. It follows a stage of forming and pressing at 200 °C. The board thus obtained is subjected to a curing process, and is completed in a union-pressing stage of the board formed with the sheet of paper.

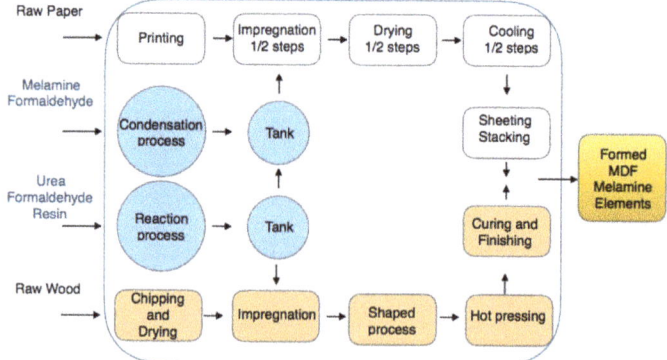

Figure 11. General production scheme of a medium-density fiberboard (MDF) urea-melamine plant.

In plant there are 42 workers distributed in two shifts. The plant is highly automated with robots for handling, feeding, palletizing and control systems in every step. The finishing area is made up of panel sectioning machines composed of vertical bar and pressure sawing machines, as well as circular saws of one or several discs, in addition to a final sanding and calibration zone. Also in the work areas and in order to maintain the correct level of particles and volatile organic compounds (VOC) emissions, there is a centralized air aspiration system with subsequent filtration an purification processes prior to its emission to the environment. Quality and safety policies are established. Workers wear personal protective equipment and there are periodic safety checks at the process plant.

4.2. Results

The analysis covers the general plant and the bow-tie is presented in Figure 12 with four final states: no injuries, minor, serious, and fatal. If an accident event (AE) is generated, the mitigating safety barriers (SF_1 SF_2 SF_3) are activated. The final states represented are bounded at one end by the absence of personal injuries, if the first mitigating safety barrier (SF1) acts correctly, in case of failure the second barrier (SF2) acts ending with a minor injury if it works correctly; in case of failure the third barrier (SF3) acts, leading to a serious problem in case of correct operation, or to a fatality in case of failure. The sub-functions (SF_{11} SF_{21} SF_{31}) correspond to automatisms, procedures, alarms and active or passive protections belonging to the main function of each of the mitigating safety barriers.

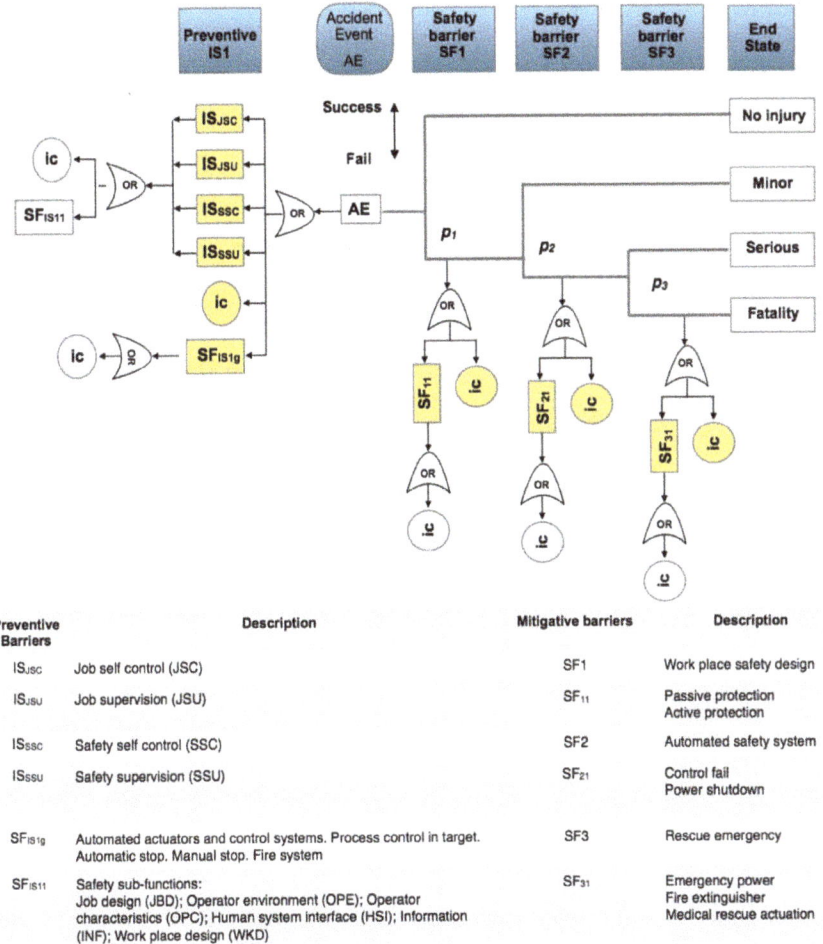

Figure 12. Bow-tie for occupational accident first level analysis in the MDF process plant.

The analysis is carried out at the first level, highlighted in yellow on the graph. The observations are made in the worst case taking the plant in general, which means that the observations are collected on the one hand for the preventive barriers, the general safety barrier (SF_{ISg}) and the initiation causes (ic) that collect all workers in one shift; and on the other hand, from each of the mitigative barriers. $SF1$–SF_{11} and associated (ic) is the group in which the general passive and active protections are located in each workplace of the plant; the $SF2$-SF_{21} and associated (ic) is the group in which the general safety controls, fire extinction, alarms and the automated or manual shutdown of every workplace are located, and the $SF3$-SF_{31} and associated (ic) is the group for the general emergency power, general fire extinction systems and internal rescue actuations. The observations are made at 10 time intervals per day, covering all shifts. Figure 13 shows three observed causes in intervals 4, 7 and 8.

Collecting observations is expected that can follow a Poisson, an Exponential or a Weibull distributions, being the $g(data/p)$ in Equation (1). And the statistical parameter p is corresponding to the rate λ or frequency of events and the prior $f(p)$ can be defined as a gamma or a normal distribution.

Interval	1	2	3	4	5	6	7	8	9	10
(ic's)	-	-	-	1 (ba)	-	-	1 IS$_{JSC}$	1(FTS)		

1 (ba) equipment failure affecting SF$_{11}$; 1 IS$_{JSC}$ job self control fail; 1 (pot) (FTS) failed test affecting SF3

Figure 13. Initiation causes (ic´s) and barrier fails collected at 10 intervals.

4.2.1. Poisson–Gamma Model

With a Poisson–gamma model the expression for $f(p/data) \propto g(data/p) \cdot f(p)$ with parameter $p = \lambda$ is;

$$f(\lambda/data) \propto g(data/\lambda) f(\lambda) \propto \lambda^{(\alpha+s)-1} e^{-\lambda(\beta+Nt)} = gam(\alpha+s, \beta+Nt) \quad (2)$$

where;

$$s = \sum data = \sum y_i \quad (3)$$

Being Nt number of interval and s the sum of initiation causes (ic's) and safety barriers incidences as data y_i in the corresponding time interval i. The values α and β are the parameters of the gamma distribution.

A recurrent method with mean prior and equal to a desired value as a target, is applied. In this case the target is for have zero accidents, then the parameters of the gamma prior are $\alpha = \beta = 0.001$. Working with $+/-1\sigma_{post}$ the posterior values are (Figure 14):

Time interval	Prior alfa	Prior beta	Posterior alfa	Posterior beta	Posterior lambda	Observed lambda	LCL	MEAN	UCL
0	0.001	0.001	-	-	0	0	-	0	-
1	0.001	0.001	0.001	1	0	0	0	0	0.032
2	0.001	1	0.001	2	0	0	0	0	0.016
3	0.001	2	0.001	3	0	0	0	0	0.011
4	0.001	3	1	4	0.25	0.25	0	0	0.25
5	1	4	1	5	0.20	0	0	0	0.20
6	1	5	1	6	0.17	0	0	0	0.17
7	1	6	2	7	0.29	0.33	0	0	0.20
8	2	7	3	8	0.38	1	0	0	0.22
9	3	8	3	9	0.33	0	0	0	0.19
10	3	9	3	10	0.30	0	0	0	0.17

Figure 14. Poisson–gamma model. Recurrent with mean prior method. Evolution of collected (ic's) and safety barrier fails in 10 intervals.

The following comments can be made for every interval.

Interval 1. With zero incidences, is, $y_i = [0]$ and posterior density for λ; $gam(\alpha + s, \beta + Nt) = gam(0.001, 1)$ with $\lambda_{post} = 0$ and $\sigma_{post} = 0.031$.

Interval 2. Also with zero incidences, is, $y_i = [0]$ and posterior density for λ; $gam(\alpha + s, \beta + Nt) = gam(0.001, 2)$ with $\lambda_{post} = 0$ and $\sigma_{post} = 0.016$.

Interval 3. With zero incidences, is, $y_i = [0]$ and posterior density for λ; $gam(\alpha + s, \beta + Nt) = gam(0.001, 3)$ with $\lambda_{post} = 0$ and $\sigma_{post} = 0.011$.

Interval 4. With one incidence in one sensor affecting the SF$_{11}$ safety barrier sub-function, is in this case, $y_i = [1]$ and posterior density for λ; $gam(\alpha + s, \beta + Nt) = gam(1, 4)$ with $\lambda_{post} = 0.25$ and $\sigma_{post} = 0.25$. Due to the no memory characteristic of the exponential, Poisson and Weibull distributions, the observed parameter value for 1 incidence in 4 time intervals is 0.25 that is coincident with the posterior. The values are in the upper control limit (UCL).

Interval 5. With zero incidences, is, $y_i = [0]$ and posterior density for λ; $gam(\alpha + s, \beta + Nt) = gam(1, 5)$ with $\lambda_{post} = 0.20$ and $\sigma_{post} = 0.20$. As a characteristic of the Bayesian inference, the posterior distribution responds softening the reduction of parameter λ from 0.25 to 0.20.

Interval 6. With zero incidences, is, $y_i = [0]$ posterior density for λ; $gam(\alpha + s, \beta + Nt) = gam(1, 6)$ with $\lambda_{post} = 0.17$ and $\sigma_{post} = 0.17$. The Bayesian inference also responds by softening the reduction of parameter λ from 0.20 to 0.17.

Interval 7. With one incidence affecting a job self control fail (JSC) in the hot pressing process, with $y_i = [1]$ and posterior density for λ; $gam(\alpha + s, \beta + Nt) = gam(2, 7)$ with $\lambda_{post} = 0.29$ and $\sigma_{post} = 0.20$. The observed parameter value for 1 incidence in $7 - 4 = 3$ time intervals is 0.33 that is practically coincident with the posterior. An out of limits is also displayed.

Interval 8. With one incidence affecting a failed test in the SF3 safety barrier, with, $y_i = [1]$ and posterior density for λ; $gam(\alpha + s, \beta + Nt) = gam(3, 8)$ with $\lambda_{post} = 0.38$ and $\sigma_{post} = 0.22$. The observed parameter value for 1 incidence in $8 - 7 = 1$ time intervals is 1. And also shows and out of limits.

Interval 9. With zero incidences, is, $y_i = [0]$ and posterior density for λ; $gam(\alpha + s, \beta + Nt) = gam(3, 9)$ with $\lambda_{post} = 0.33$ and $\sigma_{post} = 0.19$.

Interval 10. With zero incidences, is, $y_i = [0]$ and posterior density for λ; $gam(\alpha + s, \beta + Nt) = gam(3, 10)$ with $\lambda_{post} = 0.30$ and $\sigma_{post} = 0.17$.

Charts for intervals 4 and 7 for posterior and observed values are presented in Figures 15 and 16.

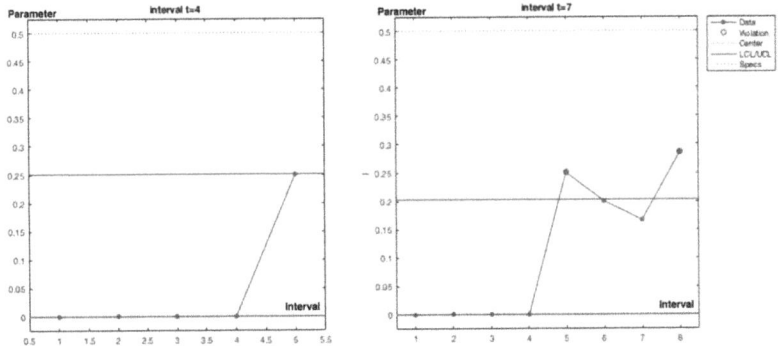

Figure 15. Poisson–gamma model. Recurrent with mean prior method. Charts based on posterior lambda evolution for intervals 4 and 7.

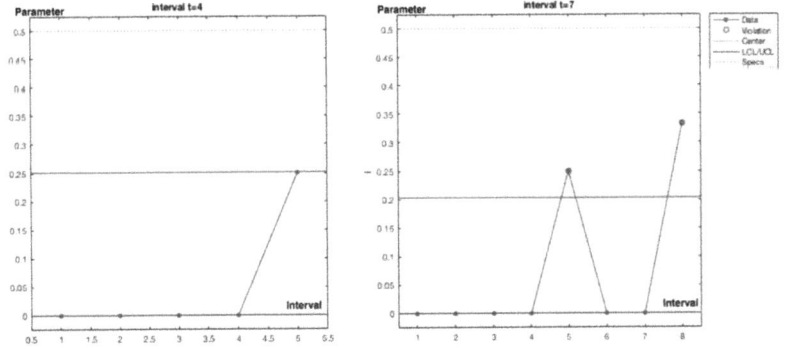

Figure 16. Poisson–gamma model. Recurrent with mean prior method. Charts based on observed lambda evolution for intervals 4 and 7.

4.2.2. Exponential–Gamma Model

With an exponential–gamma model the posterior expression $f(p/data) \propto g(data/p) \cdot f(p)$ with parameter $p = \lambda$ is:

$$f(\lambda/data)) \propto g(data/\lambda) \cdot f(\lambda) \propto \lambda^{\alpha} \, e^{-\lambda(\beta+t)} \qquad (4)$$

being α and β the parameters of the gamma distribution and t the observed time between causes.

The analysis also will be effectuated using a recurrent method with mean prior and equal to a desired value as a target, being in this example changed to a less restrictive value of 1 accident in 20 time intervals and equal to 0.05, with gamma prior parameters $\alpha = 0.5$ and $\beta = 10$. Working with $+/-1\sigma_{post}$ and using the Metropolis–Hastings (MH) sampler, Figure 17 shows the collected values and the (MH) sampling with acceptance rate (AR) 58% at interval 8 in Figure 18.

Time interval	Lambda prior	Lambda posterior	CI [5%-95%]	Observed lambda	LCL	MEAN	UCL
0	0.05	-	-	0	-	0.05	-
1	0.05	-	-	0	0.009	0.05	0.091
2	0.05	-	-	0	0.009	0.05	0.091
3	0.05	-	-	0	0.009	0.05	0.091
4	0.05	0.156	0.153-0.159	0.25	0	0.05	0.136
5	0.156	-	-	0	0	0.05	0.136
6	0.156	-	-	0	0	0.05	0.136
7	0.156	0.197	0.195-0.200	0.33	0	0.05	0.128
8	0.197	0.278	0.276-0.280	1	0	0.05	0.143
9	0.278	-	-	0	0	0.05	0.143
10	0.278	-	-	0	0	0.05	0.143

Figure 17. Exponential–gamma model. Recurrent with mean prior method. Evolution of collected (ic's) and safety barrier fails in 10 intervals. Confidence interval (CI) [5%–95%]. Out of limits framed blue.

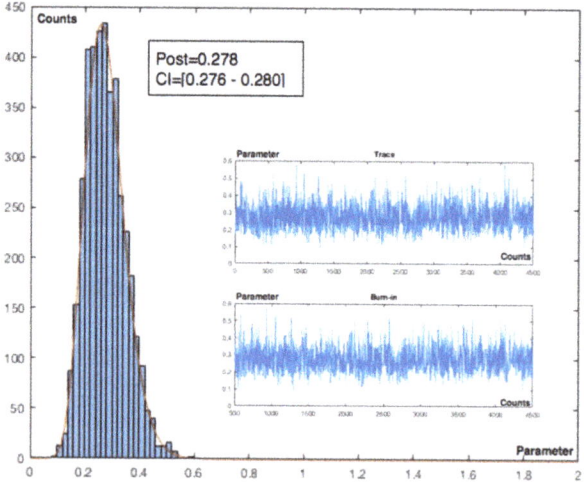

Figure 18. Metropolis–Hastings. Interval 8. Sampling n = 4500, burn = 500; 10 cycles. Acceptance rate (AR) = 58%.

4.2.3. Weibull–Gamma Model

With a Weibull–gamma model the human fatigue is considered in the analysis and the posterior expression for $f(p/data) \propto g(data/p) \cdot f(p)$ with parameter $p = \lambda$ is;

$$f(\lambda/data)) \propto g(data/\lambda) \cdot f(\lambda) \propto (\lambda)^{\alpha-1} \cdot exp(-\lambda\beta) \cdot (\lambda)^c \cdot exp(-(\lambda t)^c) \quad (5)$$

being α and β the parameters of the gamma distribution, t the observed time between causes and c is the fatigue parameter.

When c = 1 the failure rate function is constant being equivalent to an exponential–uniform model. If c > 1 the failure rate function is increasing. If 0 < c < 1 the failure rate function is decreasing. A conservative c = 1.5 value can be adopted. With a value of 1 accident in 20 time intervals equal to 0.05 and working with +/−1σ_{post} the Metropolis–Hastings sampler is applied. The Figure 19 shows the collected values and the (MH) sampling with acceptance rate (AR) 63% at interval 8 in Figure 20.

Time interval	Lambda prior	Lambda posterior	CI [5%-95%]	Observed lambda	LCL	MEAN	UCL
0	0.046	-	0.045-0.047	0	-	0.05	-
1	0.046	-	-	0	0.022	0.05	0.078
2	0.046	-	-	0	0.022	0.05	0.078
3	0.046	-	-	0	0.022	0.05	0.078
4	0.046	0,134	0.130-0.136	0.25	0	0.05	0.133
5	0.134	-	-	0	0	0.05	0.133
6	0.134	-	-	0	0	0.05	0.133
7	0.134	0,178	0.176-0.181	0.33	0	0.05	0.140
8	0.178	0,238	0.234-0.242	1	0	0.05	0.153
9	0.238	-	-	0	0	0.05	0.153
10	0.238	-	-	0	0	0.05	0.153

Figure 19. Weibull–gamma model. Recurrent with mean prior method. Evolution of collected (ic's) and safety barrier fails in 10 intervals. Confidence interval (CI) [5%–95%]. Out of limits framed blue.

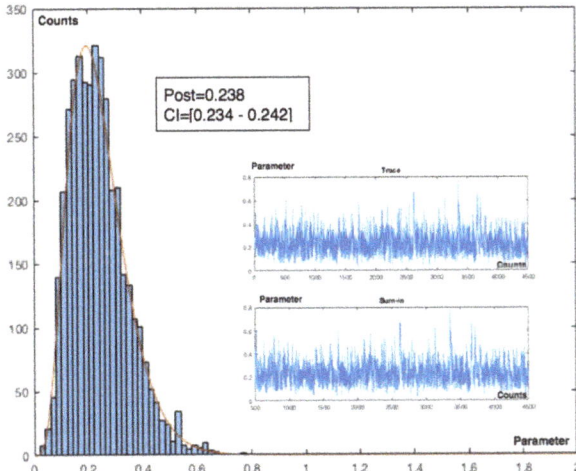

Figure 20. Metropolis–Hastings. Interval 8. Sampling n = 4500, burn = 500; 10 cycles. AR = 63%.

The out of limits are presented in the same time intervals as the previous model. It should be noted that the parameter $p = \lambda$ is modeled as a gamma distribution. The posterior distribution is also a Gamma distribution independently of the observations as Poisson, exponential or Weibull functions. Another possibility is to apply a normal distribution for the approximation to the parameter p characteristic.

4.2.4. Exponential–Normal Model

With the exponential–normal model the posterior expression $f(p/data) \propto g(data/p) \cdot f(p)$ with parameter $p = \lambda$ is;

$$f(\lambda/data)) \propto g(data/\lambda) \cdot f(\lambda) \propto \lambda exp\,(-\lambda t) \cdot 1/\sigma\, exp\,(-(x-\lambda)^2/2\sigma^2) \tag{6}$$

being t the observed time between causes and x the evolution of the parameter value.

With the same 0.05 prior value and working with $+/-1\sigma_{post}$ also in recurrent method with mean prior, the collected values are presented in Figure 21, and in Figure 22 the (MH) sampling with acceptance rate (AR) 56% at interval 4.

Time interval	Lambda prior	Lambda posterior	CI [5%-95%]	Observed lambda	LCL	MEAN	UCL
0	0.05	-	-	0	-	0.05	-
1	0.05	-	-	0	-	0.05	-
2	0.05	-	-	0	-	0.05	-
3	0.05	-	-	0	-	0.05	-
4	0.05	0.161	0.159-0.163	0.25	0	0.05	0.114
5	0.161	-	-	0	0	0.05	0.114
6	0.161	-	-	0	0	0.05	0.114
7	0.161	0.242	0.240-0.244	0.33	0	0.05	0.142
8	0.242	0.559	0.549-0.570	1	0	0.05	0.257
9	0.559	-	-	0	0	0.05	0.257
10	0.559	-	-	0	0	0.05	0.257

Figure 21. Exponential–normal model. Recurrent with mean prior method. Evolution of collected (ic's) and safety barrier fails in 10 intervals. Confidence interval (CI) [5%–95%]. Out of limits framed blue.

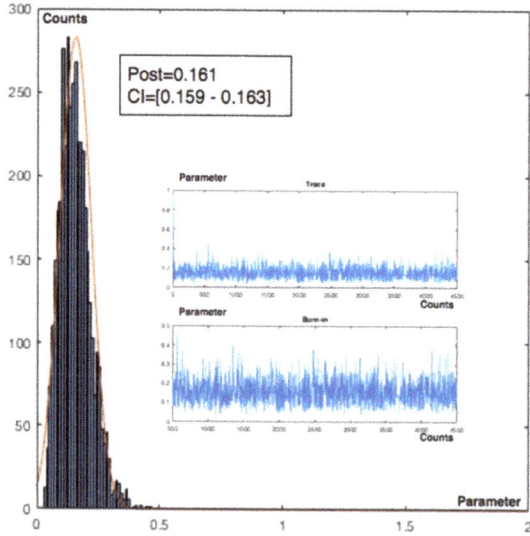

Figure 22. Metropolis–Hastings. Interval 4. Sampling n = 4500, burn = 500; 10 cycles. AR = 56%.

This model also presents out of limits at intervals 4, 7 and 8. The posterior is a normal distribution.

4.2.5. Poisson–Normal Model

In the Poisson–normal model the posterior expression $f(p/data) \propto g(data/p) \cdot f(p)$ with parameter $p = \lambda$ is;

$$f(\lambda/data)) \propto g(data/\lambda) \cdot f(\lambda) \propto (\lambda y)^n/n! \; exp\,(-\lambda y) \cdot 1/\sigma\; exp\,(-(x - \lambda)^2/2\sigma^2) \qquad (7)$$

Being y the observed data and x the evolution of the parameter value.

With the same 0.05 prior value and working with $+/-1\sigma_{post}$ in a recurrent method with mean prior, the collected values are presented in Figure 23, and in Figure 24 the (MH) sampling with acceptance rate (AR) 57% at interval 7.

Time interval	Lambda prior	Lambda posterior	CI [5%-95%]	Observed lambda	LCL	MEAN	UCL
0	0.05	-	-	0	-	0.05	-
1	0.05	-	-	0	-	0.05	-
2	0.05	-	-	0	-	0.05	-
3	0.05	-	-	0	-	0.05	-
4	0.05	0.154	0.152-0.156	0.25	0	0.05	0.127
5	0.154	-	-	0	0	0.05	0.127
6	0.154	-	-	0	0	0.05	0.127
7	0.154	0.239	0.236-0.242	0.33	0	0.05	0.155
8	0.303	0.478	0.471-0.486	1	0	0.05	0.283
9	0.478	-	-	0	0	0.05	0.283
10	0.478	-	-	0	0	0.05	0.283

Figure 23. Poisson–normal model. Recurrent with mean prior method. Evolution of collected (ic's) and safety barrier fails in 10 intervals. Confidence interval (CI) [5%–95%]. Out of limits framed blue.

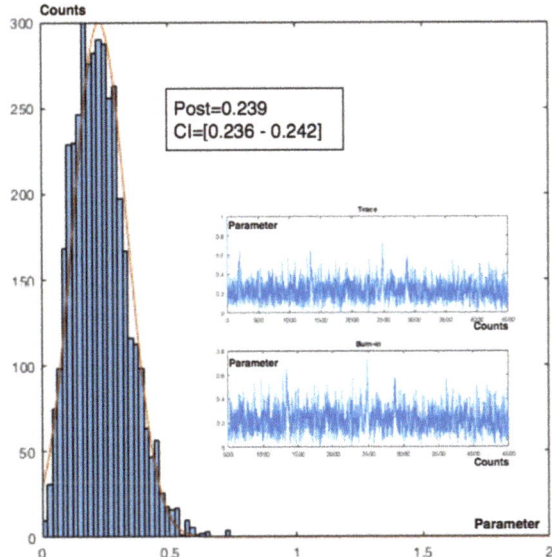

Figure 24. Interval 7. Sampling n = 4500, burn = 500; 10 cycles. AR = 57%.

This model also presents out of limits at intervals 4, 7 and 8. The posterior is a normal distribution.

4.2.6. Analysis of the Mitigative Safety Barriers Observing End States

If "0" is defined as a barrier is "correct" and active and "1" is in "fail" the possible end states are defined in Figure 25. Where "000" means that the first, second and third barriers are active and the end state is V_1 = No injury; and the same logic applies to the rest.

Barriers situation	End state	Barriers situation	End state
000	No injury (V_1)	100	Minor (V_2)
001	No injury (V_1)	101	Minor (V_2)
010	No injury (V_1)	110	Serious (V_3)
011	No injury (V_1)	111	Fatality (V_4)

Figure 25. End states in function of the mitigative safety barriers situation.

A prior transition matrix is defined for the three safety barriers. Being p_{11} the probability for the SF1 barrier to stay active in state 1; p_{12} the probability of transition from SF1 active to SF2 active because SF1 has failed; p_{13} is the probability of transition from SF1 to SF3 because SF1 has failed and also SF2 has failed, and so on. Additionally an emission matrix is defined to indicate the probabilities that a barrier S_n is active based on the observed end states (V_1 V_2 V_3 V_4), Figure 26.

$$Tran = \begin{pmatrix} p_{11} & p_{12} & p_{13} \\ p_{21} & p_{22} & p_{23} \\ p_{31} & p_{32} & p_{33} \end{pmatrix} = \begin{pmatrix} 0.8 & 0.15 & 0.05 \\ 0.55 & 0.4 & 0.05 \\ 0.2 & 0.4 & 0.4 \end{pmatrix} \quad Emiss = \begin{pmatrix} & V_1 & V_2 & V_3 & V_4 \\ S_1 & 0.8 & 0.1 & 0.05 & 0.05 \\ S_2 & 0.4 & 0.2 & 0.2 & 0.2 \\ S_3 & 0.3 & 0.4 & 0.2 & 0.1 \end{pmatrix}$$

Figure 26. Prior transition and emission matrices for mitigative safety barriers and end states.

The observations are made by creating a group of ten following a first in first out(FIFO) order. Each new one is added to the group and the oldest one disappears. The observed sequence is seqobs = [1111112311] indicating that in the six first observations times no injury has been sampled, one minor injury in the seventh, one serious in the eighth, and no injury in the next two observations. Performing a Baum–Welch algorithm, the posterior transition and emission matrices are obtained from the observed data, Figure 27.

$$Transobs = \begin{pmatrix} 0.875 & 0 & 0.1250 \\ 1 & 0 & 0 \\ 0 & 1 & 0 \end{pmatrix} \quad Emisobs = \begin{pmatrix} & V_1 & V_2 & V_3 & V_4 \\ S_1 & 1 & 0 & 0 & 0 \\ S_2 & 0 & 0 & 1 & 0 \\ S_3 & 0 & 1 & 0 & 0 \end{pmatrix}$$

Figure 27. Posterior transition and emission matrices obtained from the observed sequence.

Also, the relative occupations at steady state, after an infinite number of transitions, are obtained, being active SF1 = 80%, SF2 = 10% and SF3 = 10%. The number of visits or transitions to every state into the 10 observations are presented in Figure 28.

$$M(10) = \begin{pmatrix} p_{11} & p_{12} & p_{13} \\ p_{21} & p_{22} & p_{23} \\ p_{31} & p_{32} & p_{33} \end{pmatrix} = \begin{pmatrix} 8.04 & 0.93 & 1.03 \\ 8.24 & 0.83 & 0.93 \\ 7.44 & 1.73 & 0.83 \end{pmatrix}$$

Figure 28. Number of visits for the posterior transition matrix.

It is important also to know what will be the next passage, in observations intervals, from the SF1 and SF2 safety barriers to the most critical SF3 barrier. In this case is $m_1 = 8$ and $m_2 = 9$, meaning that, according to the observations, from SF1 a transition to SF3 can be produced in 8 intervals, and from SF2 to SF3 in 9 intervals.

The posterior transition and emission matrices in Figure 27, can be the new prior in the next sampling, in this case a minor injury (V_2) has been observed; then seqobs = [1111123112] and a new posterior for transition and emission matrices is obtained, Figure 29.

$$Transobs = \begin{pmatrix} 0.75 & 0 & 0.25 \\ 1 & 0 & 0 \\ 0 & 1 & 0 \end{pmatrix} \quad Emisobs = \begin{pmatrix} & V_1 & V_2 & V_3 & V_4 \\ S_1 & 1 & 0 & 0 & 0 \\ S_2 & 0 & 0 & 1 & 0 \\ S_3 & 0 & 1 & 0 & 0 \end{pmatrix}$$

Figure 29. Posterior transition and emission matrices obtained from a new observation.

With occupations at steady state SF1 = 67%, SF2 = 17% and SF3 = 16%, the number of visits or transitions are presented in Figure 30.

$$M(10) = \begin{pmatrix} p_{11} & p_{12} & p_{13} \\ p_{21} & p_{22} & p_{23} \\ p_{31} & p_{32} & p_{33} \end{pmatrix} = \begin{pmatrix} 6.67 & 1.58 & 1.75 \\ 7.00 & 1.42 & 1.58 \\ 6.33 & 2.25 & 1.42 \end{pmatrix}$$

Figure 30. Number of visits for the posterior transition matrix.

The next passage values for the third safety barrier are $m_1 = 4$ and $m_2 = 5$, meaning that from SF1 a transition to SF3 can be produced in 4 intervals, and from SF2 to SF3 in 5 intervals. The values are lower than the previous ones because the sampled observation has been for state V_2, a minor injury, which reduces the number of no injuries, state V_1 and makes the possible use of the third barrier more critical.

4.2.7. Analysis of the End States Observing Mitigative Safety Barriers

The prior matrices are defined as the transition matrix for the four end sates and the emission matrix indicating the probabilities to stay in an end state S_n being: S_1 = No injury; S_2 = Minor injury; S_3 = Serious injury and S_4 = Fatality; in function of the observed three active mitigative barriers (V_1 V_2 V_3), Figure 31.

$$Tran = \begin{pmatrix} 0.8 & 0.15 & 0.03 & 0.02 \\ 0.50 & 0.4 & 0.05 & 0.05 \\ 0.2 & 0.3 & 0.4 & 0.1 \\ 0.15 & 0.4 & 0.4 & 0.05 \end{pmatrix} \quad Emiss = \begin{pmatrix} & V_1 & V_2 & V_3 \\ S_1 & 0.7 & 0.2 & 0.1 \\ S_2 & 0.4 & 0.4 & 0.2 \\ S_3 & 0.3 & 0.3 & 0.4 \\ S_4 & 0.2 & 0.2 & 0.6 \end{pmatrix}$$

Figure 31. Prior transition and emission matrices for end states and mitigative safety barriers.

The sampling is obtained by sequentially observing the barriers, if the first is active, then its value is 1, if it fails and the second barrier is active, then the value is 2 and so on; a group of 10 observations is also maintained. The sampled sequence of the safety barriers is seqobs = [1111112211]. The posterior transition and emission matrices are presented on Figure 32.

$$Transobs = \begin{pmatrix} 0.875 & 0 & 0.0513 & 0.0737 \\ 1 & 0 & 0 & 0 \\ 0 & 1 & 0 & 0 \\ 0 & 1 & 0 & 0 \end{pmatrix} \quad Emisobs = \begin{pmatrix} & V_1 & V_2 & V_3 \\ S_1 & 1 & 0 & 0 \\ S_2 & 0 & 1 & 0 \\ S_3 & 0 & 1 & 0 \\ S_4 & 0 & 1 & 0 \end{pmatrix}$$

Figure 32. Posterior transition and emission matrices.

With occupations $V_1 = 80\%$, $V_2 = 10\%$, $V_3 = 4\%$ and $V_4 = 6\%$, and number of visits or transitions presented in Figure 33.

$$M(10) = \begin{pmatrix} p_{11} & p_{12} & p_{13} & p_{14} \\ p_{21} & p_{22} & p_{23} & p_{24} \\ p_{31} & p_{32} & p_{33} & p_{34} \\ p_{41} & p_{42} & p_{43} & p_{44} \end{pmatrix} = \begin{pmatrix} 8.04 & 0.93 & 0.423 & 0.607 \\ 8.24 & 0.83 & 0.3819 & 0.5481 \\ 7.44 & 1.73 & 0.3408 & 0.4892 \\ 7.44 & 1.73 & 0.3408 & 0.4892 \end{pmatrix}$$

Figure 33. Number of visits for the posterior transition matrix.

Showing a high number of visits to the end state 1 in concordance to with their occupations. The next passage values for the forth state, are $m_1 = 15$, $m_2 = 16$ and $m_3 = 17$. Making a new observation and adding it to the group is seqobs = [1111122112]. The new posterior and emission matrices are obtained if the steady state occupations are $S_1 = 0\%$, $S_2 = 100\%$, $S_3 = 0\%$ and $S_4 = 0\%$, indicating that in a steady state a high commitment is obtained in the use of the safety barrier SF2, due to the increase in the number of minor injuries corresponding to their associated final state S2.

The number of visits or transitions, are presented in Figure 34.

$$M(10) = \begin{pmatrix} p_{11} & p_{12} & p_{13} & p_{14} \\ p_{21} & p_{22} & p_{23} & p_{24} \\ p_{31} & p_{32} & p_{33} & p_{34} \\ p_{41} & p_{42} & p_{43} & p_{44} \end{pmatrix} = \begin{pmatrix} 0.7001 & 5.9137 & 1.6875 & 1.6986 \\ 0 & 10 & 0 & 0 \\ 0 & 9.3 & 0.7 & 0 \\ 0 & 7.6013 & 1.6986 & 0.7001 \end{pmatrix}$$

Figure 34. Number of visits for the posterior transition matrix.

With next passage values for the fourth state; $m_1 = 2$, $m_2 = 482$ and $m_3 = 484$. There is a correlation between the number of visits or transitions with the passage for the last barrier. In this case, in 10 intervals the transition p_{44} shows low stability in the fourth end state (Fatality) with a value $p_{44} = 0.7$, and the transitions $p_{24} = p_{34} = 0$ are indicating that there is no transition to the fourth state from the second and third and, therefore, it shows high passage intervals 482 and 484.

5. Discussion

The observations in which the risk parameter p is a frequency or ratio of risk causes requires the use of models based on Poisson, exponential and Weibull distributions that are those that have been used in this work. A recurring procedure has been used due to the characteristic of not having memory of this type of distributions, and each time interval begins as if it were new. The application of the gamma and normal distributions to characterize the risk parameter p is adequate when the variability and possible values are well collected, within the range of positive numbers, which this parameter can adopt.

The posterior distribution, because it is a reflection of the evolution of the prior, always takes the same distribution function.

The use of a control chart first requires, in every situation, previous tests to adjust the control limits. In this case the required accident rate value is very low. The control limits +/− 1sigma are restrictive in order to highlight the out of limits situation well in advance. Due to have a prior mean with a lower value near "0" and short upper limits has been possible to share the observations and the posterior mean with the same limit controls.

Despite the restrictive limits, the Poisson–gamma model showed just at the upper limit the first observation with value 0.25. As a characteristic the Bayesian inference softens the evolution of the statistical parameter based on the observations, for example, in the first observation at interval 4 with a value of 0.25, the Poisson–gamma model presents also a 0.25, showing an internal difference of a 4.5% into the group consisting of the exponential–gamma, exponential–normal and Poisson–normal but with a 60% difference with the Poisson–gamma. In the second observation at interval 7 with a value of 0.33, the inference models take values for a Poisson–gamma of 0.29, and the rest of models show an internal difference of 30% and a 135% difference with the Poisson–gamma. In the last observation with a value of 1, the Poisson–gamma shows a value of 0.38 and the exponential–normal with a value of 0.559 and Poisson-Normal with a value of 0.478 follow better this change to with the rest of inference models. At low changes the Poisson–gamma respond better and with important changes are better the Exponential-Normal and Poisson–normal.

The Metropolis–Hastings sampling has been done with a random simulation of 4500 values and removing (burning) the first 500 to guarantee the stability of the process, with a repetition cycle of 10 times, determining the average value and confidence intervals to obtain the representative posterior values. The resultant traces of the sampling processes also show a correct stochastic variability, with acceptance rates (AR) oscillating between 48% to 68% being necessary conditions to obtain a good sampling.

When initiation causes occur, at the same time information is being obtained on the status of the process and the different jobs, forcing a continuous review of its operation; for example, in the interval 4 due to a fault in a sensor that affects the hot pressing section, located in the sub-function of the SF1 mitigating safety barrier, it is a situation that requires verifying the same type of sensors. The same situation occurs when in the interval 7 there is a failure of self-control of work (JSC), then it is necessary to analyze the procedures and sub-functions that affect the design of the work, the environment of the operator, the characteristics of the operator, the interface of the system, procedures and information and design that affect workplaces. The last cause occurs in a failed rescue simulation that affects the SF3 safety barrier, which requires reviewing the performance of this procedure.

The hidden Markov chain procedure allows you to follow the situation of the mitigating barriers and obtain a map of your activity. The action of mitigating barriers is critical because when an accident occurs, they have to act with the highest probability that no failures will occur. However, it should be borne in mind that the method is an estimate of reality from the observations made allowing a quick view of which barrier is most used and likely to fail because it is showing high occupation. According to the experience gathered in the application of the Baum–Welch algorithm, the determination of the new transition and emission matrices from observations must have 10 values to obtain representative results.

If prior values for transition and emission matrices are not well known, the recurrent procedure adjusts the subsequent values according to the observations.

Collecting observations for the Bayesian inference or for the application of the hidden Markov chain can be carried out by the company's own personnel, or by automated control systems, however human failures, in general, must be collected on site by personal observation.

A control chart process could be established without the need to associate it with a Bayesian inference; however, the objective of the new method, in addition to the graphic control of the evolution of risk, allowing simultaneous and immediate corrective actions, is having, on the one hand, the objective evaluation of a statistical parameter, which has been obtained from the observations that have been made through a formal procedure, and on the other hand, being able to have a continuous map of the functioning of the barriers of safety and the causes that affect the safety and health of workers.

The new method has strengths and weaknesses. The strong point is that the objective of the new methodology is that it is easy to implement and use to quickly gain an idea of the overall risk status of the process and of the worker's situation; at the same time this is a weakness since it alone cannot be enough and must be complemented with other techniques. In addition, an important feature of the new method is that it is based on dynamic risk methodologies which, with respect to the quantitative risk analysis, allow the risk situation to be updated [87,126], and the weak points are based on the fact that by itself it is not a substitute for other methods. New methodologies are emerging as the use of the dynamic risk assessment and a decision-making trial and evaluation tool integrated into a Bayesian network for assessing the situations of leakage accidents [127], or a new assessment methodology based on the application of the FMECA including the economic valuation, [128] or the assessment of the domino effect using a graphical network of events (PETRI-NET) tool for discrete event treatment [129] or the general applications of the Bayesian networks and Petri-nets tools for risk assessment into the manufacturing and process industries [130] or the new techniques for the identification and monitoring of emerging risks over time [23,24]. These are new proposals that highlight the importance of risk assessment and the supporting role that the new SRC methodology can provide.

Future work is extensive, in a specific scenario the analysis can be carried out considering the entire industrial manufacturing process as a whole or analyzing more detailed areas. A sensitivity analysis can also be carried out determining which initiation causes (ic's) or failures in safety barriers are more common and which areas within an industrial process have more incidence of risk.

6. Conclusions

From the review of existing methodologies, the dynamic methodologies are adapted to the need to have immediacy in the information. The new methodology (SRC) applies the characteristics of this group.

The bow-tie, which is an important element of the new methodology, was first generated by establishing a framework for the general treatment of risks, followed by adaptation to occupational accidents, which is applied without major changes in the treatment of the manufacturing plant case.

The content of the initiation causes (ic's) and safety barriers have been defined and applied in the analysis of a occupational accidents in a plant for obtaining laminated paper and wood producing medium-density fiber (MDF) boards with melamine. The inference procedures applied in this case cover data observations belonging to Poisson, exponential and Weibull distributions and considering the risk statistical parameter under the evolution of gamma and normal distributions, showing that they are suitable for this type of analysis.

The application of the hidden Markov chain for the analysis of reactive safety barriers provides control for these types of barriers, since they are critical in their operation or failure in the event of an accident. The application of the hidden Markov chain for the analysis of reactive safety barriers provides control for these types of barriers, since they are critical in their operation or failure in the event of an accident. The tool, in its application to the case of the manufacturing plant, is suitable for obtaining information from the state of the different barriers or the situation of the different final states.

In conclusion, the SRC method is a formal method that allows us to meet the three characteristics of prevention, simultaneity and immediacy. It is applicable to the assessment of occupational hazards in different industrial and manufacturing scenarios, and offers an overview of the risk status in the simplest way possible and at the same time it is reliable, in accordance with the established control limits and providing warning in advance to be able to prevent risk.

This work can point towards future research actions considering its application in different types of industries and in project management, analyzing how it can complement other methodological tools.

Author Contributions: Conceptualization, M.F.-C.; Investigation, M.F.-C.; Methodology, M.F.-C.; Supervision, F.B. and M.A.S.; Validation, F.B. and M.A.S.; Writing—original draft, M.F.-C.; Writing—review and editing, M.F.-C., F.B. and M.A.S.

Funding: This research was funded by the Ministry of Economy and Competitiveness of Spain, with reference DPI2016-79824-R.

Acknowledgments: The present paper has been produced within the scope of the doctoral activities carried out by the lead author at the International Doctorate School of the Spanish National Distance-Learning University (EIDUNED_ Escuela Internacional de Doctorado de la Universidad Nacional de Educación a Distancia). The authors are grateful for the support provided by this institution.

Conflicts of Interest: The authors declare no conflict of interest.

References

1. Eurostat. Statistics. 2019. Available online: https://ec.europa.eu/eurostat/statistics-explained/index.php/ (accessed on 15 June 2019).
2. MITRAMISS. Spanish Ministery of Labour Migrations and Social Security. Statistics for Occupational Accidents 2019. Available online: http://www.mitramiss.gob.es/estadisticas/eat/welcome.htm (accessed on 19 June 2019).
3. Aven, T. Risk assessment and risk management: Review of recent advances on their foundation. *Eur. J. Oper. Res.* **2016**, *253*, 1–13. [CrossRef]
4. ISO 31000:2018. Risk Management Guidelines-International Organization for Standardization. Available online: https://www.iso.org/iso-31000-risk-management.html (accessed on 1 September 2019).
5. ISO/IEC 31010:2019. Risk Management-Risk Assessment Techniques. International Organization for Standardization. Available online: https://www.iso.org/standard/72140.html (accessed on 10 September 2019).
6. Pasman, H.; Reniers, G. Past, present and future of Quantitative Risk Assessment (QRA) and the incentive it obtained from Land-Use Planning (LUP). *J. Loss Prev. Proc. Ind.* **2014**, *28*, 2–9. [CrossRef]
7. Hollnagel, E. Is safety a subject for science? *Saf. Sci.* **2014**, *67*, 21–24. [CrossRef]
8. Ale, B.J.M.; Baksteen, H.; Bellamy, L.J.; Bloemhof, A.; Goossens, L.; Hale, A.; Mud, M.L.; Oh, J.I.H.; Papazoglou, I.A.; Post, J.; et al. Quantifying occupational risk: The development of an occupational risk model. *Saf. Sci.* **2008**, *46*, 176–185. [CrossRef]
9. Aven, T. What is safety science? *Saf. Sci.* **2014**, *67*, 15–20. [CrossRef]
10. European Agency for Safety and Health at Work. Directive 89/391/EEC of 12 June 1989 on the Introduction of Measures to Encourage Improvements in the Safety and Health of Workers at Work-"Framework Directive". Available online: https://osha.europa.eu/en/legislation/directives/the-osh-framework-directive/1 (accessed on 2 April 2019).
11. Deming, W.E. *Out of the Crisis*; MIT Press: Cambridge, MA, USA, 1986.
12. Khan, F.; Rathnayaka, S.; Ahmed, S. Methods and models in process safety and risk management: Past, present and future. *Proc. Saf. Environ. Protect.* **2015**, *98*, 116–147. [CrossRef]
13. Goerlandt, F.; Khakzad, N.; Reniers, G. Validity and validation of safety-related quantitative risk analysis: A review. *Saf. Sci.* **2017**, *99*, 127–139. [CrossRef]
14. CCPS. *Guidelines for Hazard Evaluation Procedures*; CCPS, AIChE: New York, NY, USA, 2008.
15. Proskovics, R.; Hutton, G.; Torr, R.; Niclas Scheu, M. Methodology for Risk Assessment of Substructures for Floating Wind Turbines. *Energy Procedia* **2016**, *94*, 45–52. [CrossRef]
16. ISO 45001:2018. Occupational Health and Safety Management Systems—Requirements with Guidance for Use. International Organization for Standardization. Available online: https://www.iso.org/obp/ui#iso:std:iso:45001:ed-1:v1:es (accessed on 5 May 2019).
17. NISHW. National Institute for Safety and Health at Work. Occupational Risk Assessment. Spanish Government. 2018. Available online: http://www.insht.es/portal/site/Insht/ (accessed on 21 January 2018).
18. Directive 98/24/EC of 7 April 1998 on the Risks Related to Chemical Agents at Work. European Agency for Safety and Health at Work. Available online: https://osha.europa.eu/en/legislation/directives/75 (accessed on 3 May 2019).
19. Directive 2004/37/EC of 29 April 2004 on the Carcinogens and Mutagens at Work. European Agency for Safety and Health at Work. Available online: https://osha.europa.eu/en/legislation/directives/directive-2004-37-ec-carcinogens-or-mutagens-at-work (accessed on 12 May 2019).
20. Brocal, F.; González, C.; Reniers, G.; Cozzani, V.; Sebastián, M.A. Risk Management of Hazardous Materials in Manufacturing Processes: Links and Transitional Spaces between Occupational Accidents and Major Accidents. *Materials* **2018**, *11*, 1915. [CrossRef] [PubMed]

21. Gul, M.; Fatih, A.K.M. A comparative outline for quantifying risk ratings in occupational health and safety risk assessment. *J. Clean. Prod.* **2018**, *196*, 653–664. [CrossRef]
22. Murè, S.; Demichela, M. Fuzzy Application Procedure (FAP) for the risk assessment of occupational accidents. *J. Loss Prev. Proc. Ind.* **2009**, *22*, 593–599. [CrossRef]
23. Brocal, F.; González, C.; Sebastián, M.A. Technique to identify and characterize new and emerging risks: A new tool for application in manufacturing processes. *Saf. Sci.* **2018**, *109*, 144–156. [CrossRef]
24. Brocal, F.; Sebastián, M.A.; González, C. Theoretical framework for the new and emerging occupational risk modeling and its monitoring through technology lifecycle of industrial processes. *Saf. Sci.* **2017**, *99*, 178–186. [CrossRef]
25. Papazoglou, I.A.; Aneziris, O.N.; Bellamy, L.J.; Ale, B.J.M.; Oh, J. Multi-hazard multi-person quantitative occupational risk model and risk management. *Reliab. Eng. Syst. Saf.* **2017**, *167*, 310–326. [CrossRef]
26. Papazoglou, I.A.; Aneziris, O.N.; Bellamy, L.J.; Ale, B.J.M.; Oh, J. Quantitative occupational risk model: Single hazard. *Reliab. Eng. Syst. Saf.* **2017**, *160*, 162–173. [CrossRef]
27. Song, G.; Khan, F.; Wang, H.; Leighton, S.; Yuan, Z.; Liu, H. Dynamic occupational risk model for offshore operations in harsh environments. *Reliab. Eng. Syst. Saf.* **2016**, *150*, 58–64. [CrossRef]
28. Azadeh-Fard, N.; Schuh, A.; Rashedi, E.; Camelio, J.A. Risk assessment of occupational injuries using Accident Severity Grade. *Saf. Sci.* **2015**, *76*, 160–167. [CrossRef]
29. Aneziris, O.N.; Topali, E.; Papazoglou, I.A. Occupational risk of building construction. *Reliab. Eng. Syst. Saf.* **2012**, *105*, 36–46. [CrossRef]
30. Aneziris, O.N.; Papazoglou, I.A.; Doudakmani, O. Assessment of occupational risks in an aluminium processing industry. *Int. J. Ind. Ergon.* **2010**, *40*, 321–329. [CrossRef]
31. Papazoglou, I.A.; Ale, B.J.M. A logical model for quantification of occupational risk. *Reliab. Eng. Syst. Saf.* **2007**, *92*, 785–803. [CrossRef]
32. Mete, S.; Serin, F.; Ece Oz, N.; Gul, M. A decision-support system based on Pythagorean fuzzy VIKOR for occupational risk assessment of a natural gas pipeline construction. *J. Natrl. Gas Sci. Eng.* **2019**, 71. [CrossRef]
33. Aneziris, O.N.; Papazoglou, I.A.; Baksteen, H.; Mud, M.; Ale, B.J.; Bellamy, L.J.; Hale, A.R.; Bloemhoff, A.; Post, J.; Oh, J. Quantified risk assessment for fall from height. *Saf. Sci.* **2008**, *46*, 198–220. [CrossRef]
34. Aneziris, O.N.; Papazoglou, I.A.; Mud, M.; Damen, M.; Kuiper, J.; Baksteen, H.; Ale, B.J.; Bellamy, L.J.; Hale, A.R.; Bloemhoff, A.; et al. Towards risk assessment for crane activities. *Saf. Sci.* **2008**, *46*, 872–884. [CrossRef]
35. Aneziris, O.N.; Papazoglou, I.A.; Konstandinidou, M.; Baksteen, H.; Mud, M.; Damen, M.; Bellamy, L.J.; Oh, J. Quantification of occupational risk owing to contact with moving parts of machines. *Saf. Sci.* **2013**, *51*, 382–396. [CrossRef]
36. Aneziris, O.N.; Papazoglou, I.A.; Mud, M.; Damen, M.; Manuel, H.J.; Oh, J. Occupational risk quantification owing to falling objects. *Saf. Sci.* **2014**, *69*, 57–70. [CrossRef]
37. Aneziris, O.N.; Papazoglou, I.A.; Psinias, A. Occupational risk for an onshore wind farm. *Saf. Sci.* **2016**, *88*, 188–198. [CrossRef]
38. Al-shanini, A.; Ahmad, A.; Khan, F. Accident modeling and analysis in process industries. *J. Loss Prev. Proc. Ind.* **2014**, *32*, 319–334. [CrossRef]
39. Matos, S.; Lopes, E. Prince2 or PMBOK—A question of choice. *Procedia Tech.* **2013**, *9*, 787–794. [CrossRef]
40. Aloini, D.; Dulmin, R.; Mininno, V. Risk assessment in ERP project. *Inf. Syst.* **2012**, *37*, 183–199. [CrossRef]
41. Willey, R.J. Layer of Protection Analysis. *Procedia Eng.* **2014**, *84*, 12–22. [CrossRef]
42. Jin, J.; Shuai, B.; Wang, X.; Zhu, Z. Theoretical basis of quantification for layer of protection analysis (LOPA). *Ann. Nucl. Energy* **2016**, *87*, 69–73. [CrossRef]
43. Yan, F.; Xu, K. A set pair analysis based layer of protection analysis and its application in quantitative risk assessment. *J. Loss Prev. Proc. Ind.* **2018**, *55*, 313–319. [CrossRef]
44. NORSOK. Risk and Emergency Preparedness Assessment. 2010. Available online: https://www.standard.no/en/ (accessed on 5 January 2018).
45. Yang, X.; Haugen, S. Classification of risk to support decision-making in hazardous processes. *Saf. Sci.* **2015**, *80*, 115–126. [CrossRef]
46. Yang, X.; Haugen, S.; Paltrinieri, N. Clarifying the concept of operational risk assessment in the oil and gas industry. *Saf. Sci.* **2018**, *108*, 259–268. [CrossRef]

47. Directive 2012/18/EU of 4 July 2012 on the Control of Major Accident Hazards Involving Dangerous Substances. Seveso III. Official Journal of the European Union. Available online: http://data.europa.eu/eli/dir/2012/18/oj (accessed on 9 January 2018).
48. COMAH. *The Control of Major Accident Hazard Regulations*, 3rd ed.; Health Safety Executive Books: London, UK, 2015.
49. CPR 18E. Guidelines for Quantitative Risk Assessment-Purple Book. Publication Series on Dangerous Substances. 1999. (Publicatiereeks Gevaarlijke Stoffen PGS 3). Available online: http://content.publicatiereeksgevaarlijkestoffen.nl/ (accessed on 23 March 2018).
50. EN 16991:2018. *Risk-Based Inspection Framework*; European Committee for Standardization: Geneva, Switzerland, 2018.
51. Khan, F.; Sadiq, R.; Haddara, M.M. Risk-based inspection and maintenance (RBIM). Multi-attribute Decision-making with Aggregative Risk Analysis. *Proc. Saf. Environ. Prot.* **2004**, *82*, 398–411. [CrossRef]
52. Mutlu, N.G.; Altuntas, S. Risk analysis for occupational safety and health in the textile industry: Integration of FMEA, FTA, and BIFPET methods. *Int. J. Ind. Ergon.* **2019**, *72*, 222–240. [CrossRef]
53. Castiglia, F.; Giardina, M. Analysis of operator human errors in hydrogen refuelling stations: Comparison between human rate assessment techniques. *Int. J. Hydrogen Energy* **2013**, *38*, 1166–1176. [CrossRef]
54. Rausand, M. Reliability of Safety—Critical Systems. In *Theory and Applications*; John Wiley & Sons, Inc.: Hoboken, NJ, USA, 2014. Available online: https://www.ntnu.edu/ross/books/sis (accessed on 15 December 2017).
55. Mai Le, L.M.; Reitter, D.; He, S.; Té Bonle, F.; Launois, A.; Martinez, D.; Prognon, P.; Caudron, E. Safety analysis of occupational exposure of healthcare workers to residual contaminations of cytotoxic drugs using FMECAsecurity approach. *Sci. Total Environ.* **2017**, *599*, 1939–1944. [CrossRef]
56. Zio, E. *The Monte Carlo Simulation Method for System Reliability and Risk Analysis*; Springer: London, UK, 2013.
57. Rausand, M.; Hoyland, A. System Reliability Theory. In *Models Statistical Methods and Applications*; John Wiley & Sons, Inc.: Hoboken, NJ, USA, 2004.
58. Koscielny, J.M.; Syfert, M.; Fajdek, B.; Kozak, A. The application of a graph of a process in HAZOP analysis in accident prevention system. *J. Loss Prev. Proc. Ind.* **2017**, *50*, 55–66. [CrossRef]
59. Petroleum Safety Authority. Principles for Barrier Management in the Petroleum Industry. Technical Report. 2013. Available online: http://www.ptil.no/getfile.php/PDF/Barrierenotatet%202013%20engelsk%20april.pdf (accessed on 17 January 2019).
60. Hauge, S.; Øien, K. Guidance for Barrier Management in the Petroleum Industry. 2016. SINTEF Safety Research Report A27623. Available online: https://www.sintef.no/globalassets/project/pds/reports/pds-report---guidance-for-barrier-management-in-the-petroleum-industry.pdf (accessed on 7 April 2018).
61. Ferjencik, M.; Kuracina, R. MORT WorkSheet or how to make MORT analysis easy. *J. Hazard. Mater.* **2008**, *151*, 143–154. [CrossRef] [PubMed]
62. Lees, F.P. *Loss Prevention in the Process Industries*, 4th ed.; Butterworth-Heinemann: Oxford, UK, 2012.
63. Nano, G.; Derudi, M. A Critical Analysis of Techniques for the Reconstruction of Workers Accidents. *Chem. Eng. Trans.* **2013**, *31*, 415–420. Available online: https://www.researchgate.net/publication/278081417 (accessed on 12 January 2018).
64. Sklet, S. Comparison of some selected methods for accident investigation. *J. Hazard. Mater.* **2004**, *111*, 29–37. [CrossRef] [PubMed]
65. Fahlbruch, B.; Schöbel, M. SOL—Safety through organizational learning: A method for event analysis. *Saf. Sci.* **2011**, *49*, 27–31. [CrossRef]
66. Churchwell, J.S.; Zhang, K.S.; Saleh, J.H. Epidemiology of helicopter accidents: Trends, rates, and covariates. *Reliab. Eng. Syst. Saf.* **2018**, *180*, 373–384. [CrossRef]
67. Ballester, O.C.; LLari, M.; Afquir, S.; Martin, J.L.; Bourdet, N.; Honoré, V.; Masson, C.; Arnoux, P.J. Analysis of trunk impact conditions in motorcycle road accidents based on epidemiological, accidentological data and multibody simulations. *Accid. Anal. Prev.* **2019**, *127*, 223–230. [CrossRef] [PubMed]
68. Rasmussen, J. Risk management in a dynamic society: A modeling problem. *Saf. Sci.* **1997**, *27*, 183–213. [CrossRef]
69. Waterson, P.; Jenkins, D.P.; Salmon, P.M.; Underwood, P. Remixing Rasmussen's: The Evolution of Accimaps within Systemic Accident Analysis. *Appl. Ergon.* **2017**, *59*, 483–503. [CrossRef] [PubMed]

70. Salmon, P.M.; Cornelissen, M.; Trotter, M.J. Systems-based accident analysis methods: A comparison of Accimap, HFACS, and STAMP. *Saf. Sci.* **2012**, *50*, 1158–1170. [CrossRef]
71. Leveson, N. A new accident model for engineering safer systems. *Saf. Sci.* **2004**, *42*, 237–270. [CrossRef]
72. Ouyang, M.; Hong, L.; Yu, M.H.; Fei, Q. STAMP-based analysis on the railway accident and accident spreading: Taking the China–Jiaoji railway accident for example. *Saf. Sci.* **2010**, *48*, 544–555. [CrossRef]
73. Goncalves Filho, A.P.; Jun, G.T.; Waterson, P. Four studies, two methods, one accident—An examination of the reliability and validity of Accimap and STAMP for accident analysis. *Saf. Sci.* **2019**, *113*, 310–317. [CrossRef]
74. Hollnagel, E. *Cognitive Reliability and Error Analysis Method (CREAM)*; Elsevier: Amsterdam, The Netherlands, 1998.
75. Liao, P.; Luo, X.; Wang, T.; Su, Y. The Mechanism of how Design Failures cause Unsafe Behavior: The Cognitive Reliability and Error Analysis Method (CREAM). *Procedia Eng.* **2016**, *145*, 715–722. Available online: http://creativecommons.org/licenses/by-nc-nd/4.0/ (accessed on 18 January 2019). [CrossRef]
76. Habibovic, A.; Tivesten, E.; Uchida, N.; Bärgman, J.; Aust, M.L. Driver behavior in car-to-pedestrian incidents: An application of the Driving Reliability and Error Analysis Method (DREAM). *Accid. Anal. Prev.* **2013**, *50*, 554–565. [CrossRef] [PubMed]
77. Lee, J.; Chung, H. A new methodology for accident analysis with human and systeminteraction based on FRAM: Case studies in maritime domain. *Saf. Sci.* **2018**, *109*, 57–66. [CrossRef]
78. Harms-Ringdahl, L. Analysis of safety functions and barriers in accidents. *Saf. Sci.* **2009**, *47*, 353–363. [CrossRef]
79. Yan, F.; Xu, K. Methodology and case study of quantitative preliminary hazard analysisbased on cloud model. *J. Loss Prev. Proc. Ind.* **2019**, *60*, 116–124. [CrossRef]
80. Kumar, A.M.; Rajakarunakaran, S.; Prabhu, V.A. Application of Fuzzy HEART and expert elicitation for quantifying human error probabilities in LPG refuelling station. *J. Loss Prev. Proc. Ind.* **2017**, *48*, 186–198. [CrossRef]
81. Zhou, Q.; Wong, Y.D.; Loh, H.S.; Yuen, K.F. A fuzzy and Bayesian network CREAM model for human reliability analysis—The case of tanker shipping. *Saf. Sci.* **2018**, *105*, 149–157. [CrossRef]
82. Ladkin, P.; Loer, K. *Analising Aviation Accidents Using WB-Analysis—An Application of Multimodal Reasoning*; Technical Report SS-98-04; Universität Bielefeld: Bielefeld, Germany, 1998; pp. 169–174.
83. Kujath, M.F.; Amyotte, P.; Khan, F. A conceptual offshore oil and gas process accident model. *J. Loss Prev. Proc. Ind.* **2010**, *23*, 323–330. [CrossRef]
84. Rathnayaka, S.; Khan, F.; Amyotte, P. SHIPP methodology: Predictive accident modeling approach. Part I: Methodology and model description. *Proc. Saf. Environ. Prot.* **2011**, *89*, 151–164. [CrossRef]
85. Rathnayaka, S.; Khan, F.; Amyotte, P. SHIPP methodology: Predictive accident modeling approach. Part II: Validation with case study. *Proc. Saf. Environ. Prot.* **2011**, *89*, 75–88. [CrossRef]
86. Paltrinieri, N.; Scarponi, G.E.; Khan, F.; Hauge, S. Addressing Dynamic Risk in the Petroleum Industry by Means of Innovative Analysis Solutions. *Chem. Eng. Trans.* **2014**, *36*, 451–456. [CrossRef]
87. Villa, V.; Paltrinieri, N.; Khan, F.; Cozzani, V. Towards dynamic risk analysis: A review of the risk assessment approach and its limitations in the chemical process industry. *Saf. Sci.* **2016**, *89*, 77–93. [CrossRef]
88. Misuri, A.; Khakzad, N.; Reniers, G.; Cozzani, V. A Bayesian network methodology for optimal security management of critical infrastructures. *Reliab. Eng. Syst. Saf.* **2018**, *191*, 106112. [CrossRef]
89. Bier, V.M.; Yi, W. A Bayesian method for analyzing dependencies in precursor data. *Int. J. For.* **1995**, *11*, 25–41. [CrossRef]
90. Meel, A. Dynamic Risk Assessment of Inherently Safer Chemical Processes: An Accident Precursor Approach. 2007. Available online: https://search.proquest.com/openview/b75c47f89e1984b2e6af5ea788cf26e2/1?pq-origsite=gscholar&cbl=18750&diss=y (accessed on 24 January 2018).
91. Khan, F.; Hashemi, S.J.; Paltrinieri, N.; Amyotte, P.; Cozzani, V.; Reniers, G. Dynamic risk management: A contemporary approach to process safety management. *Curr. Opin. Chem. Eng.* **2016**, *14*, 9–17. [CrossRef]
92. Khakzad, N.; Khan, F.; Amyotte, P. Safety analysis in process facilities: Comparison of fault tree and Bayesian network approaches. *Reliab. Eng. Syst. Saf.* **2011**, *96*, 925–932. [CrossRef]
93. Kanes, R.; Ramirez Marengo, M.C.; Abdel-Moati, H.; Cranefield, J.; Véchot, L. Developing a framework for dynamic risk assessment using Bayesian networks and reliability data. *J. Loss Prev. Proc. Ind.* **2017**, *50*, 142–153. [CrossRef]

94. Paltrinieri, N.; Reniers, G. Dynamic Risk Analysis for Seveso sites. *J. Loss Prev. Proc. Ind.* **2017**, *44*, 20–35. [CrossRef]
95. Pasman, H.J.; Rogers, W.J.; Sam Mannan, M. How can we improve process hazard identification? What can accidentinvestigation methods contribute and what other recent developments? Abrief historical survey and a sketch of how to advance. *J. Loss Prev. Proc. Ind.* **2018**, *55*, 80–106. [CrossRef]
96. Meel, A.; Seider, W.D. Plant—Specific dynamic failure assessment using Bayesian Theory. *Chem. Eng. Sci.* **2006**, *61*, 7036–7056. [CrossRef]
97. Kalantarnia, M.; Khan, F.; Hawboldt, K. Dynamic risk assessment using failure assessment and Bayesian theory. *J. Loss Prev. Proc. Ind.* **2009**, *22*, 600–606. [CrossRef]
98. Kalantarnia, M.; Khan, F.; Hawboldt, K. Modelling of BP Texas city refinery accident using dynamic risk assessment approach. *Proc. Saf. Environ. Prot.* **2010**, *88*, 191–199. [CrossRef]
99. Yang, M.; Khan, F.I.; Lye, L. Precursor-based hierarchical Bayesian approach for rare event frequency estimation: A case of oil spill accidents. *Proc. Saf. Environ. Prot.* **2013**, *91*, 333–342. [CrossRef]
100. Yuan, Z.; Khakzad, N.; Khan, F.; Amyotte, P. Domino effect analysis of dust explosions using Bayesian networks. *Proc. Saf. Environ. Prot.* **2016**, *100*, 108–116. [CrossRef]
101. Khakzad, N.; Khan, F.; Amyotte, P. Dynamic Safety analysis of process systems by mapping bow-tie into Bayesian network. *Proc. Saf. Environ. Prot.* **2013**, *91*, 46–53. [CrossRef]
102. Yeo, C.; Bhandari, J.; Abbasi, R.; Garaniya, V.; Chai, S.; Shomali, B. Dynamic risk analysis of offloading process in floating liquefied natural gas (FLNG) platform using Bayesian Network. *J. Loss Prev. Proc. Ind.* **2016**, *41*, 259–269. [CrossRef]
103. Barua, S.; Gao, X.; Pasman, H.; Mannan, M.S. Bayesian network based dynamic operational risk assessment. *J. Loss Prev. Proc. Ind.* **2016**, *41*, 399–410. [CrossRef]
104. Paltrinieri, N.; Tugnoli, A.; Buston, J.; Wardman, M.; Cozzani, V. Dynamic Procedure for Atypical Scenarios Identification (DyPASI): A new systematic HAZID tool. *J. Loss Prev. Proc. Ind.* **2013**, *26*, 683–695. [CrossRef]
105. Paltrinieri, N.; Tugnoli, A.; Buston, J.; Wardman, M.; Cozzani, V. DyPASI Methodology: From Information Retrieval to Integration of HAZID Process. *Chem. Eng. Trans.* **2013**, *32*, 433–438. [CrossRef]
106. Paltrinieri, N.; Khan, F.; Amyotte, P.; Cozzani, V. Dynamic approach to risk management: Application to the Hoeganaes metal dust accidents. *Proc. Saf. Environ. Prot.* **2014**, *92*, 669–679. [CrossRef]
107. Bucelli, M.; Paltrinieri, N.; Landucci, G. Integrated risk assessment for oil and gas installations in sensitive areas. *Ocean Eng.* **2018**, *150*, 377–390. [CrossRef]
108. Yang, X.; Sam Mannan, M. The development and application of dynamic operational risk assessment in oil/gas and chemical process industry. *Reliab. Eng. Syst. Saf.* **2010**, *95*, 806–815. [CrossRef]
109. Colin, J.; Vanhoucke, M. Developing a framework for statistical process control approaches in project management. *Int. J. Proj. Manag.* **2015**, *33*, 1289–1300. [CrossRef]
110. Corbett, C.J.; Pan, J. Evaluating environmental performance using statistical process control techniques. *Eur. J. Oper. Res.* **2002**, *139*, 68–83. [CrossRef]
111. Aliverdi, R.; Naeni, L.M.; Salehipour, A. Monitoring project duration and cost in construction project by applying statistical quality control charts. *Int. J. Proj. Manag.* **2013**, *31*, 411–423. [CrossRef]
112. Folch-Calvo, M.; Sebastian, M.A. Dynamic Risk Methodology through Statistical Risk Control applied to the project management in high uncertainty environments. In Proceedings of the 22nd International Congress on Project Management and Engineering—ICPME, Madrid, Spain, 11–13 July 2018. Available online: htpps://www.researchgate.net (accessed on 2 August 2018).
113. Hoff, P.D. *A First Course in Bayesian Statistical Methods*; Springer: New York, NY, USA, 2009.
114. Puza, B. *Bayesian Methods for Statistical Analysis*; Australian National University ANU eView: Canberra, Australia, 2015.
115. Rabiner, L. A tutorial on Hidden Markov Models and selected applications in speechrecognition. *Proc. IEEE* **1989**, *77*, 257–286. [CrossRef]
116. Kulkarny, V.G. *Introduction to Modeling and Analysis of Stochastic Systems*, 2nd ed.; Springer: Berlin/Heidelberg, Germany, 2011.
117. Montgomery, D.C. *Introduction to Statistical Quality Control*, 6th ed.; John Wiley & Sons: Hoboken, NJ, USA, 2009.
118. Ross, S.M. *Introduction to Probability and Statistics for Engineers and Scientists*, 4th ed.; Elsevier: Amsterdam, The Netherlands, 2009.

119. Flage, R.; Aven, T. Emerging risk—Conceptual definition and a relation to black swan type of events. *Reliab. Eng. Syst. Saf.* **2015**, *144*, 61–67. [CrossRef]
120. Hajikazemi, S.; Ekambaram, A.; Andersen, B.; Zidane, Y.J.T. The Black Swan—Knowing the unknown in projects. *Procedia Soc. Behav. Sci.* **2016**, *226*, 184–192. [CrossRef]
121. Dodson, K.; Westney, R. *Predictable Projects in a World of Black Swans*; Westney Consulting Group: Houston, TX, USA, 2009. Available online: http://www.westney.com/insights/archive (accessed on 22 June 2019).
122. Baybutt, P. The role of people and human factors in performing process hazard analysis and layers of protection analysis. *J. Loss Prev. Proc. Ind.* **2013**, *26*, 1352–1365. [CrossRef]
123. Kariuki, S.G.; Löwe, K. Integrating human factors into process hazard analysis. *Reliab. Eng. Syst. Saf.* **2007**, *92*, 1764–1773. [CrossRef]
124. Eurofound. *Sixth European Working Conditions Survey—Overview Report*; Publications Office of the European Union: Luxembourg, 2017.
125. EU-OSHA. Expert Forecast on Emerging Chemical Risks Related to Occupational Safety and Health. European Agency for Safety and Health at Work. 2009. Publications Office of the European Union. Luxembourg. Available online: https://osha.europa.eu/en/tools-andpublications/publications/reports/TE3008390ENC_chemical_risks/view (accessed on 7 May 2018).
126. Bubbico, R.; Lee, S.; Moscati, D.; Paltrinieri, N. Dynamic assessment of safety barriers preventing escalation in offshore Oil & Gas. *Saf. Sci.* **2020**, *121*, 319–330. [CrossRef]
127. Meng, X.; Chen, G.; Zhu, G.; Zhu, Y. Dynamic quantitative risk assessment of accidents induced by leakage on offshore platforms using DEMATEL-BN. *Int. J. Nav. Arch. Ocean Eng.* **2019**, *11*, 22–32. [CrossRef]
128. Di Bona, G.; Silvestri, A.; Forcina, A.; Petrillo, A. Total efficient risk priority number (TERPN): A new method for risk assessment. *J. Risk Res.* **2018**, *21*, 1384–1408. [CrossRef]
129. Zaid Kamil, M.; Taleb-Berrouane, M.; Khan, F.; Ahmed, S. Dynamic domino effect risk assessment using Petri-nets. *Proc. Saf. Environ. Prot.* **2019**, *124*, 308–316. [CrossRef]
130. Kabir, S.; Papadopoulos, Y. Applications of Bayesian networks and Petri nets in safety, reliability and risk assessments: A review. *Saf. Sci.* **2019**, *115*, 154–175. [CrossRef]

 © 2019 by the authors. Licensee MDPI, Basel, Switzerland. This article is an open access article distributed under the terms and conditions of the Creative Commons Attribution (CC BY) license (http://creativecommons.org/licenses/by/4.0/).

MDPI
St. Alban-Anlage 66
4052 Basel
Switzerland
Tel. +41 61 683 77 34
Fax +41 61 302 89 18
www.mdpi.com

Materials Editorial Office
E-mail: materials@mdpi.com
www.mdpi.com/journal/materials

www.ingramcontent.com/pod-product-compliance
Lightning Source LLC
LaVergne TN
LVHW070129100526
838202LV00016B/2250